Polymer Nanocomposites for Advanced Engineering and Military Applications

Noureddine Ramdani
Research and Development Institute of Industry and Defense Technologies, Algeria & Harbin Engineering University, China

A volume in the Advances in Chemical and Materials Engineering (ACME) Book Series

IGI Global
DISSEMINATOR OF KNOWLEDGE

Published in the United States of America by
IGI Global
Engineering Science Reference (an imprint of IGI Global)
701 E. Chocolate Avenue
Hershey PA, USA 17033
Tel: 717-533-8845
Fax: 717-533-8661
E-mail: cust@igi-global.com
Web site: http://www.igi-global.com

Library of Congress Cataloging-in-Publication Data

Names: Ramdani, Noureddine, 1986- editor.
Title: Polymer nanocomposites for advanced engineering and military applications / Noureddine Ramdani, editor.
Description: Hershey, PA : Engineering Science Reference, [2019] | Includes bibliographical references.
Identifiers: LCCN 2018036335| ISBN 9781522578383 (h/c) | ISBN 9781522578390 (eISBN)
Subjects: LCSH: Polymeric composites. | Nanocomposites (Materials)
Classification: LCC TA418.9.C6 P6475 2019 | DDC 620.1/92--dc23 LC record available at https://lccn.loc.gov/2018036335

This book is published in the IGI Global book series Advances in Chemical and Materials Engineering (ACME) (ISSN: 2327-5448; eISSN: 2327-5456)

British Cataloguing in Publication Data
A Cataloguing in Publication record for this book is available from the British Library.

Advances in Chemical and Materials Engineering (ACME) Book Series

J. Paulo Davim
University of Aveiro, Portugal

ISSN:2327-5448
EISSN:2327-5456

Mission

The cross disciplinary approach of chemical and materials engineering is rapidly growing as it applies to the study of educational, scientific and industrial research activities by solving complex chemical problems using computational techniques and statistical methods.

The **Advances in Chemical and Materials Engineering (ACME) Book Series** provides research on the recent advances throughout computational and statistical methods of analysis and modeling. This series brings together collaboration between chemists, engineers, statisticians, and computer scientists and offers a wealth of knowledge and useful tools to academics, practitioners, and professionals through high quality publications.

Coverage

- Thermo-Chemical Treatments
- Numerical Techniques
- Polymers Engineering
- Composites
- Ceramics and glasses
- Statistical methods
- Nanomaterials
- Toxicity of Materials
- Computational methods
- Heat Treatments

The Advances in Chemical and Materials Engineering (ACME) Book Series (ISSN 2327-5448) is published by IGI Global, 701 E. Chocolate Avenue, Hershey, PA 17033-1240, USA, www.igi-global.com. This series is composed of titles available for purchase individually; each title is edited to be contextually exclusive from any other title within the series. For pricing and ordering information please visit http://www.igi-global.com/book-series/advances-chemical-materials-engineering/73687. Postmaster: Send all address changes to above address.

Titles in this Series

For a list of additional titles in this series, please visit: www.igi-global.com/book-series

Process Analysis, Design, and Intensification in Microfluidics and Chemical Engineering
Harrson Silva Santana (University of Campinas, Brazil) João Lameu da Silva Jr (Federal Institute of Education, Science, and Technology of South of Minas Gerais, Brazil) and Osvaldir Pereira Taranto (University of Campinas, Brazil)
Engineering Science Reference • copyright 2019 • 367pp • H/C (ISBN: 9781522571384) • US $215.00 (our price)

Applications of Nanofluid Transportation and Heat Transfer Simulation
Mohsen Sheikholeslami (Babol Noshirvani University of Technology, Iran)
Engineering Science Reference • copyright 2019 • 692pp • H/C (ISBN: 9781522575955) • US $245.00 (our price)

Strategic Applications of Measurement Technologies and Instrumentation
Soubantika Palchoudhury (University of Tennessee at Chattanooga, USA)
Engineering Science Reference • copyright 2019 • 223pp • H/C (ISBN: 9781522554066) • US $195.00 (our price)

The Geometry of Higher-Dimensional Polytopes
Gennadiy Vladimirovich Zhizhin (Russian Academy of Sciences, Russia)
Engineering Science Reference • copyright 2019 • 286pp • H/C (ISBN: 9781522569688) • US $195.00 (our price)

Composites and Advanced Materials for Industrial Applications
K. Kumar (Birla Institute of Technology, India) and J. Paulo Davim (University of Aveiro, Portugal)
Engineering Science Reference • copyright 2018 • 402pp • H/C (ISBN: 9781522552161) • US $225.00 (our price)

Emerging Synthesis Techniques for Luminescent Materials
Ratnesh Tiwari (Bhilai Institute of Technology, India) Vikas Dubey (Bhilai Institute of Technology, India) and Sanjay J. Dhoble (Rashtrasant Tukadoji Maharaj Nagpur University, India)
Engineering Science Reference • copyright 2018 • 505pp • H/C (ISBN: 9781522551706) • US $215.00 (our price)

Handbook of Research on Ergonomics and Product Design
Juan Luis Hernández Arellano (Autonomous University of Ciudad Juárez, Mexico) Aide Aracely Maldonado Macías (Autonomous University of Ciudad Juárez, Mexico) Juan Alberto Castillo Martínez (University of Rosario, Colombia) and Porfirio Peinado Coronado (Autonomous University of Ciudad Juárez, Mexico)
Engineering Science Reference • copyright 2018 • 446pp • H/C (ISBN: 9781522552345) • US $235.00 (our price)

IGI Global
DISSEMINATOR OF KNOWLEDGE

701 East Chocolate Avenue, Hershey, PA 17033, USA
Tel: 717-533-8845 x100 • Fax: 717-533-8661
E-Mail: cust@igi-global.com • www.igi-global.com

Table of Contents

Section 2
Engineering Applications of Polymer Nanocomposites

Section 3
Defense Applications of Polymer Nanocomposites

Detailed Table of Contents

mechanical, chemical, thermal, electrical and barrier properties and their influence on fire retardancy. Hence, with the use of polymer nanocomposites, the encouraging outcomes in different sectors of automotive industry has resulted to new horizons in terms of advanced polymer nanocomposites for automobile applications. This chapter reviews advance polymer composites for automobile applications. Methods of fabricating polymer nanocomposites and several applications of polymer nanocomposites in automotive industries are discussed.

Geopolymers are amorphous aluminosilicates with some varied applications. However, the use of geopolymers in water treatment is a relatively new subject. This chapter discusses developments in synthesis, properties and applications of geopolymers and their composites for removal of heavy metals and dyes from water including reduction of hardness in water. The adsorption mechanisms and effects of various environmental conditions on adsorption efficiency are also highlighted. The chapter demonstrates that geopolymers are low cost and environmentally benign materials for wastewater treatment and offers opportunities as alternative adsorbents for sequestration of various pollutants from water subject to further investigations.

This chapter includes detailed review of the research undertaken with conducting polymer (CP) based composites with chitosan (Ch) for tissue engineering till date. The beneficial role of electrically conductive biomaterials has been discussed with the possible strategies to overcome the shortcomings of CP alone through blending with Ch due to its excellent biocompatibility, biodegradability, and bioactivity. Additionally, this embodiment deals with the optimization and characterization of electrically conductive, biocompatible and biodegradable Polyaniline: Chitosan (PAni:Ch) nanocomposites as cell culture substrates for MDA-MB-231 and NIH 3T3 fibroblast in order to examine the combined effect of nanofiber structure and surface modification on cell-biomaterial interactions. The nanocomposites were further checked as a conductive scaffold for electrical stimulation of a neuronal model PC12 cell line in order to explore the potential of the materials in neural tissue engineering.

Sky-high renewable energy demands urged the revolutionary development of environment benign, portable and miniatured, high power generating energy storage systems in the form of electrochemical capacitors commonly known as "supercapacitors" or "ultracapacitors". Supercapacitor designing requires smart electrode materials for providing higher energy and power densities, mechanical and electrochemical durability, enhanced thermal operating range with minimal production and maintenance cost. Polyaniline, as conducting polymer, particularly in nano-morphology, has been one of the pioneer electroactive materials paving the corridor for commercial development of pseudocapacitors. Considering these points, the chapter initially concentrates on basic concepts of fabrication, designing and performance evaluation procedure of supercapacitors, followed by discussion on the role of PANI as potential supercapacitor electrode material, delineating current achievements and major challenges encountered in the process of progress so as to obtain superior electrochemical energy storage systems in the near future.

Polyaniline in various forms has been widely explored as an electrode material for supercapacitors due to its high theoretical charge storage capacity, facile-cost-effective synthesis, good mechanical strength and ultrafast charge transport. However, commercialization of such pristine forms is very much restricted by low solubilities, rapid agglomeration during device design accompanied by poor electrochemical life and fast environmental decomposition. The blending with nano-carbon materials, metal oxides and other competent materials, may result in high quality materials– "nanocomposites" with superior features is ideally fit for future generation energy storage devices. The present chapter deals with detailed discussions on designing, the fabrication of such binary and ternary nanocomposites, correlating their morphology with electrochemical behavior, so as to optimize their supercapacitive performances. Such an attempt would help to outline the present status and future aspects of these materials which will be of first-hand assistance especially to the beginners to this field of research.

Corrosion is the foremost reason for the deterioration of metallic components used in harsh environments and hence research on the development of an innovative protective system with admirable performance has become a subject of leading importance. Currently, chromates free organic or the polymer coatings on the metal substrate bestow an efficient barrier amongst the metal and inhibit corrosion and save the environment. Researchers have involved in the development of polymer coating containing nanoparticles because reinforcement of nanoparticles decreases the porosity and provides a zig-zag diffusion path which in turn forms barrier layer on the metal surface to protect the metal from the harmful corrosive species. This present chapter deliberates the application of polymer-based nanocomposite coating to mitigate the corrosion of metals against harsh environment. This chapter covers the utilization of carbon-based nanoparticles, inorganic nanoparticle, conductive polymer, self-heling polymer, etc., and could shows a new insight to for anti-corrosive metal surface protective coating.

of fuel consumption and costs of fuel, thereby reducing CO2 emissions. Undoubtedly, nanocomposites applications abound in several aspects of human life and the use of nanoparticle in materials dates back to the understanding of the nature of these materials. This chapter will focus on the use of nanopolymers in the aerospace and in the military. Particular attention will be given to nano military weapons, nanocoating for military applications, nanotechnology for military drones, nanotechnology in military suits, gloves, boots and nanotechnology in armored military vehicles, aircraft, and military ships and in military medicine.

Preface

Polymer nanocomposites are produced from a polymer matrix reinforced with nano-sized inorganic fillers. They are developed by creating a very large interface between the nano-sized heterogeneities and the polymer matrix. During the processing of polymer/nanocomposites, different structure phenomena have to be considered, such as the aggregation and anisotropy orientation of nano-fillers. The properties of polymer/inorganic nanomaterials including can be generally determined according to four main factors, including component properties, nano-filler loading, nanofillers dispersion and interfacial interactions. Studies revealed that properties, like mechanical, thermal, electrical, Optical or magnetic characteristics, can change upon the decrease of particle sizes to nano-dimensions, which are, usually, of major focus in the area of nanocomposite materials. By adding inorganic nanoparticles into the polymeric matrices, this can produce high-performance materials that find applications in many advanced engineering and defense areas. In addition, the surface-treatment or chemical functionalization of nanoparticles using various surfactant, silane coupling agents, or some functionalized polymers is a promising way to tailor the properties of polymer nanocomposites. According to the type outstanding properties resulted from the combination of polymeric matrices and nanofillers, the newly produced nanomaterials meet the requirement of several potential industrial applications such as microelectronics, high-performance and anticorrosion coatings, building construction, aerospace, biomedical, defense and textiles.

ORGANIZATION OF THE BOOK

The book is organized into 11 chapters. A brief description of each of the chapters follows:

Chapter 1 gives an overview of polymer-based Nanocomposites (PMNC), focusing on the processing. Polymers such as condensation polymers, vinyl polymers, polyolefins, and specially polymers including biodegradable ones are used in production of PMNC. It is the reinforcement that is in the nano-range size in nanocomposites generally. Reinforcements used are metal powders, silica, clays, and metal oxides. The most important methods of preparing PMNC are: Intercalation of the polymer or pre-polymer from solution; In-situ intercalative polymerization; Melt intercalation; direct mixture of polymer and particulates; Template synthesis; In-situ polymerization; and Sol-gel process. The structure of polymer-based nanocomposites consists of the matrix material containing the nano-sized reinforcement components in the forms of whiskers, particles, nanotubes, fibres, etc. It is clear that polymer-based nanocomposites provide many benefits such as improved properties, minimization of solid wastes films and lower and improved manufacturing capabilities.

Chapter 2 presents structures and unique properties of commonly used nanofillers; advances, improvement potentials and characterization of nanocomposites used in energy systems. Theoretical and experimental literatures reviewed revealed that nanofillers engender improved properties in polymeric matrices, functionalization is applicable to all types of nanofillers in use today, a number of functionalized nanofillers are already commercially available; and more extensive researches are needed to achieve optimal improved results with the use of nanofillers and nanocomposites in various fields of applications.

Chapter 3 summarizes results of the studies of technology, structural features and properties of polymer/clay nanocomposites (n-PCM) prepared by melt compounding of thermoplastic polymers. Special attention is given to the analysis of the effect of nanoclays modification with surfactants on properties of nanocomposites and preparation features of nanomaterials based on polar, non-polar thermoplastics and polymer blends have been made. Effect of technological factors and special compounding regimes in the technology of n-PCM with advanced technical characteristics is also discussed. Results of the original studies of the structure and properties of the hybrid composites, filled by high modulus fibers in addition to nanoclays, are presented herein.

Chapter 4 reviews advance polymer composites for automobile applications such as the desire for light weight and low CO_2 emission from vehicles. Methods of fabricating polymer nanocomposites and several applications of polymer nanocomposites in automotive industries are discussed.

Chapter 5 concisely discusses the up to date developments in synthesis, properties and applications of geopolymers and their composites for removal of heavy metals and dyes from water including reduction of hardness in water. The adsorption mechanisms and effects of various environmental conditions on adsorption efficiency are also highlighted. The chapter demonstrates that geopolymers are low cost and environmentally benign materials for wastewater treatment and offers opportunities as alternative adsorbents for sequestration of various pollutants from water subject to further investigations.

Chapter 6 includes detailed review of the research undertaken with conducting polymer (CP) based composites with chitosan (Ch) for tissue engineering till date. The beneficial role of electrically conductive biomaterials has been discussed with the possible strategies to overcome the shortcomings of CP alone through blending with Ch due to its excellent biocompatibility, biodegradability, and bioactivity. Additionally, this embodiment deals with the optimization and characterization of electrically conductive, biocompatible and biodegradable Polyaniline: Chitosan (PAni:Ch) nanocomposites as cell culture substrates for MDA-MB-231 and NIH 3T3 fibroblast in order to examine the combined effect of nanofiber structure and surface modification on cell-biomaterial interactions. The nanocomposites were further checked as a conductive scaffold for electrical stimulation of a neuronal model PC12 cell line in order to explore the potential of the materials in neural tissue engineering.

Chapter 7 concentrates on basic concepts of fabrication, designing and performance evaluation procedure of supercapacitors, followed by discussion on the role of PANI as potential supercapacitor electrode material, delineating current achievements and major challenges encountered in the process of progress so as to obtain superior energy storage systems in the near future.

Chapter 8 deals with detailed discussions on designing, fabrication of such binary and ternary nanocomposites, correlating their morphology with electrochemical behavior, so as to optimize their supercapacitive performances. Such an attempt would help to outline the present status and future aspects of these materials which will be of first-hand assistance especially to the beginners to this research field.

Chapter 9 deliberates the application of polymer-based nanocomposite coating to mitigate the corrosion of metals against harsh environment. This chapter covers the utilization of carbon-based nanoparticles, inorganic nanoparticle, conductive polymer, self-heling polymer etc., and could shows a new insight to for anti-corrosive metal surface protective coating.

Chapter 10 outlines important aspects and progression from graphene—to—polymer/graphene nanocomposite—to—relevant defense application. Graphene is unique nanocarbon material having large surface area, high Young's modulus, thermal conductivity, electrical conductivity, and optical transmittance. Engineering thermoplastic polymers have been employed as matrices for graphene reinforcement. Various routes have been employed for graphene filled polymeric nanomaterials. Intrinsic physical properties of nanocomposite depend on graphene modification and dispersion techniques. Polymer/graphene nanocomposite may have multifunctional characteristics due to synergistic effect of polymer/graphene. The article mainly discusses nanocomposite with potential uses in soldierly applications including flame resistance, ballistic protection, electromagnetic interference shielding, electrostatic-charge dissipation, sensors, corrosion protection, fuel cell, batteries, etc. The gestalt of defense applications of polymer/graphene nanocomposite may offer future perspective to develop promising materials.

Chapter 11 focuses on the use of nano-polymers in the aerospace and in the military. Particular attention will be given to nano military weapons, nano-coating for military applications, nanotechnology for military drones, nanotechnology in military suits, gloves, boots and nanotechnology in armored military vehicles, aircraft and military ships and in military medicine.

Noureddine Ramdani
Research and Development Institute of Industry and Defense Technologies, Algeria & Harbin Engineering University, China

Acknowledgment

The editors would like to acknowledge the help of all the people involved in this project and, more specifically, to the authors and reviewers that took part in the review process. Without their support, this book would not have become a reality.

First, the editors would like to thank each one of the authors for their contributions. Our sincere gratitude goes to the chapter's authors who contributed their time and expertise to this book.

Second, the editors wish to acknowledge the valuable contributions of the reviewers regarding the improvement of quality, coherence, and content presentation of chapters. Most of the authors also served as referees; we highly appreciate their double task.

Noureddine Ramdani
Research and Development Institute of Industry and Defense Technologies, Algeria & Harbin Engineering University, China

Section 1
Processing of Polymer Nanocomposites

Chapter 1
Processing of Polymer–Based Nanocomposites in Advanced Engineering and Military Application

Francis Boluwaji Elehinafe
Covenant University, Nigeria

Augustine Omoniyi Ayeni
Covenant University, Nigeria

ABSTRACT

This chapter gives an overview of polymer-based nanocomposites (PMNC), focusing on the processing. Polymers such as condensation polymers, vinyl polymers, polyolefins, specialty polymers including biodegradable are used in production of PMNC. It is the reinforcement that is in the nanorange size in nanocomposites generally. Reinforcements used are metal powders, silica, clays, and metal oxides. The most important methods of preparing PMNC are intercalation of the polymer or pre-polymer from solution, in-situ intercalative polymerization, melt intercalation, direct mixture of polymer and particulates, template synthesis, in-situ polymerization; and sol-gel process. The structure of polymer-based nanocomposites consists of the matrix material containing the nanosized reinforcement components in the forms of whiskers, particles, nanotubes, fibers, etc. It is clear that polymer-based nanocomposites provide many benefits such as improved properties, minimization of solid wastes films, and lower and improved manufacturing capabilities.

DOI: 10.4018/978-1-5225-7838-3.ch001

INTRODUCTION

Nanocomposites are, nowadays, regarded as high-performance engineering materials. They show uncommon combinations in terms of properties and unusual design possibilities. They have high growth rate and being demanded in elastomers and engineering plastics. Their possibilities to find applications in spheres of engineering are overwhelming. Nanocomposites are special materials that have one or more of their phases in the dimension of nanometre ranging from 1 nm to 10^{-9} m (Roy et al., 1986). Nanocomposites are now emerging as dependable alternatives to overcome limitations of microcomposites and monolithics, while offering preparation challenges akin to the manipulation of elements in their compositions of phases as well as their stoichiometry. It was reported by Schmidt et al. in 2002 that the general perception of the aforementioned properties are not reached yet, though the first inference on them was reported as early as 1992 by Gleiter.

In 1991 Iijima discovered carbon nanotubes abbreviated CNTs and their subsequent use to fabricate composites exhibiting some unique CNT related properties such as thermal electrical and mechanical (Weisenberger *et al.*, 2003). The ease of incorporating CNTs into nanocomposite products made further ways for CNT-containing nanomaterials in terms of processing and subsequent applications as reported by Dalton *et al.* in 2003. Nanocomposites, nowadays, give advantages for all industrial sectors in area of manufacturing and business while the environment is compromised (Choa *et al.*, 2003). Just like microcomposites, nanocomposites are categorized into Metal Matrix Nanocomposites (MMNC), Ceramic Matrix Nanocomposites (CMNC) and Polymer Matrix Nanocomposites (PMNC) (Fischer, 2003). Since the 1990s, nanocomposites have received extensive study, the number of publications increases systematic while seminars and conferences on the subject are undertaken frequently (Fischer, 2003).

Nanocomposites that polymer-based are nanomaterials with less than ten years of advent. It pointed out by Gangopadhyay and Amitabha, (2000) that the control of the conducting nanocomposites and the optimization of their physical properties, like colloidal stability and electrical conductivity, have not been accomplished, while both their commercial availability in the closer future and quantum leap for scientific research in materials are anticipated with their suitable usage. For polymer-based nanocomposites that are biodegradable, recent developments in preparation/processing, characterization, properties, crystallization behaviour and melt rheology, of the layered and the matrix polymer-based nanocomposites were discussed by Pandey *et al.*, 2005; Ray and Bousmina, in 2005. Also, a focus on toughness and bonding (intrefacial one) between CNTs and polymer matrices is critically addressed by Andrew and Weisenberger (2004) to underscore the ability of the matrix to transfer stress and the potentials of these nanocomposites for possible CNT-polymer production on macro and nano scales. The work, on polymer-based nanocomposites and the ones with CNT reinforcements, is very limited.

An overview of polymer-based nanocomposites is given in this chapter, focusing on the processing. Also, the potential applications of polymer-based nanocomposites and the benefits they offer are also discussed.

PROCESSING OF POLYMER-BASED NANOCOMPOSITES

Raw Materials

Polymers like polyolefins, condensation polymers, vinyl polymers, speciality polymers including biodegradable ones find application in production of PMNC. The reinforcements used in nanocomposites are the components that are in the nanorange sizes generally. Natural and synthetic reinforcements have both been deployed, such as metal powders, silica, clays, and metal oxides. Although clays and stratified silicates are the most commonplace, according to Fernando & Satyanarayana (2005). These reinforcements are available with highly reduced particle sizes and intercalation chemistry (Alexandre & Dubois, 2000) and to generate better properties even when they are applied at reduced concentrations (Noh & Lee, 1999). Most of these reinforcements are prepared by methods such as mechanical (e.g. ball milling), chemical, vapour deposition, etc.

In the same vain, CNTs are produced by chemical and vapour deposition methods as shown by Alpha (2003). Analysis of CNTs conducted in 2000 by Gupta & Pangannaya, showed that about 49% of the patents produced between 1992-1999 were akin to the processing of CNTs and about 14% to their properties, models and structure. CNTs consist of graphene cylinders and there are two categories: single walled carbon nano-tubes (SWCNT) and multi walled carbon nano-tubes (MWCNT), of about 70% yield in SWCNT (Alpha, 2003). Both types exhibit physical properties of solids, either semiconducting or metallic with micro-crystalline in nature and high aspect ratios (Anya, 1999).

Modifications of reinforcements on their surfaces are usually done to give uniform distribution with less or no agglomeration, and to enhanced interfacial bonding of the matrix with the nanosized reinforcements. Use of surfactants, chemical or oxidation functionalization of surfaces are some of the techniques employed in the case of CNTs (Andrews & Weisenberger, 2004). Particularly for polymer-based naanocomposites, chemical methods may be more effective. In-situ polymerization and physical blending are used for enhancing dispersion in CNT-reinforced polymer nanocomposites, while alignment of CNTs could be achieved by methods such as ex-situ methods (force field-inducements, template and plasma-enhanced chemical vapour deposition,, filtration, etc.) (Xie *et al.*, 2005).

Processing Methods

There are many methods for the preparation of polymer-based Nanocomposites: layered materials and those incorporating CNTs (Jimenez, 1997). The best ones are: Sol-gel process; Intercalation of the polymer; In-situ intercalative polymerization; Melt intercalation as shown in Figure 1; Direct mixture of polymer and particulates; Template synthesis; andIn-situ polymerization. Various methods for the incorporation of nanoparticles into conducting polymers are also available in the work of Ogasawara et al. done in 2004. The most effective method isprobably the organic polymers incorporated in, by inorganic building blocks in. Table 1 shows the procedures employed in some of these processes, while their merits and limitations are summarized in Table 2.

Figure 1 shows intercalative procedures and processes used for the production of polymer-based nanocomposites. In this method, a number of polymer-based nanocomposites with structures from intercalated to exfoliated can be achieved. As a result, this procedure has become measure for the production of polymer-layered silicate combinations.

Figure 1. Intercalation of polymer
Source: (Ogasawara et al., 2004)

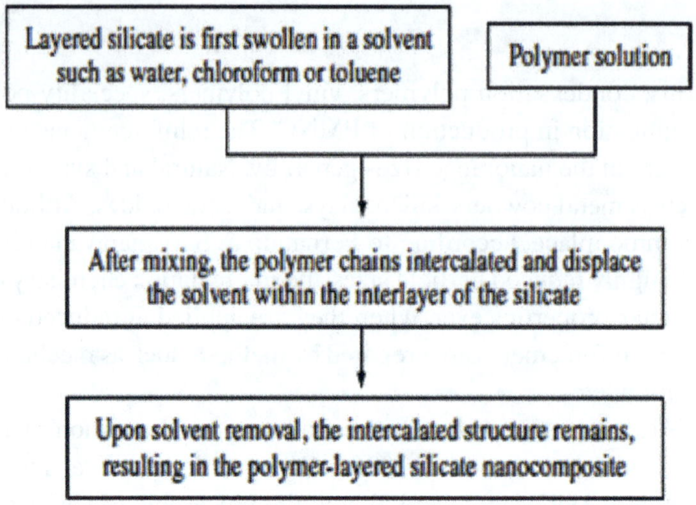

Table 1. Processing methods for polymer-based nanocomposite systems

Process	System	Procedure
Intercalation/ Prepolymer from Solution	Clay with PCL, PLA, HDPE, PEO, PVA, PVP, PVA, etc.	Employed for layered reinforcing material in which the polymer may intercalate. Mostly for layered silicates, with intercalation of the polymer or pre-polymer from solution. Use of a solvent in which the polymer or pre-polymer is soluble and the silicate layers are swellable.
In-situ Intercalative Polymerization	Montmorillonate with N6/ PCL/PMMA /PU/Epoxy	Encasing of the layered silicate within the liquid monomer or a monomer solution → formation of polymer between the intercalated sheets. Polymerization by heat or radiation, by diffusion of a suitable initiator or by a catalyst fixed through cation exchange inside the interlayer, before the swelling step.
Melt Intercalation	Montmorillonate with PS/ PEO/PP/ PVP, Clay-PVPH	Annealing of a mixture of the polymer and the layered host above the softening point of the polymer, statically or under shear. Diffusion of polymer chains from the bulk polymer melt into the galleries between the host layers during annealing (Figure 1).
Template Synthesis	Hectorite with PVPR, HPMC, PAN, PDDA, PANI	In situ formation of the layered structure of the inorganic material in an aqueous solution containing the polymer. The water soluble polymer acts as a template for the formation of layers. Widely used for the synthesis of LDH nanocomposites, but less developed for layered silicates.
(a) Mixing (b) In situ polymerization	PVA)/Ag; PMMA/Pd Polyester/TiO2 PET/CaCO3, Epoxy vinyl ester/Fe3O4 ; Epoxy vinyl ester/γ-Fe2O3; Poly (acrylic acid)(PAA)/Ag, PAA/Ni and PAA/Cu AgNO3, NiSO4 and CuSO4;	(a) Mixing of either polymer or monomer with reinforcing materials; (b1) Dispersion of inorganic particles into a precursor of the polymeric matrix (monomer); (b2) Polymerization of the mixture by addition of an appropriate catalyst; (b3) Processing of this material by conventional moulding technologies. Use of ultrasonics for dispersion in epoxy systems. Exposition of AG systems to 60 Co γ-ray to promote simultaneous polymerization and metal nanoparticle formation.
Sol-Gel Process	Polyimide/SiO2; 2-hydroxyethyl acrylate (HEA)/SiO2, polyimide/ silica. PMMA/ SiO2, polyethylacrylate/ SiO2, polycarbonate / SiO2 and poly (amide-imide)/TiO2	Embedding of organic molecules and monomers on sol-gel matrices; introduction of organic groups by formation of chemical bonds → In-situ formation of sol-gel matrix within the polymer and/or simultaneous generation of inorganic /organic networks.

Source: Ogasawara et al., 2004

The production of CNT-reinforced polymer nanocomposites is carried out by various methods, including direct mixing, melt-mixing, in-situ polymerization and solution mixing. These, as applicable to various systems (Roslaniec *et al.*, 2003), are listed in Table 3. Similarly, different processing methods, mostly chemical and electrochemical methods, are deployed for the production of polymer nanocomposites that have conductivity (Gangopadhyay & Amitabha, 2000). Table 4 gives the summary these techniques, while a germane classification of these polymer-based nanocomposites is summarized in Table 5.

Table 2. Advantages and limitations of polymer-based nanocomposite processing methods

Process	Advantages	Limitations
Intercalation / Prepolymer from Solution	Synthesis of intercalated nanocomposites based on polymers with low or even no polarity. Preparation of homogeneous dispersions of the filler.	Industrial use of large amounts of solvents.
In-situ Intercalative Polymerization	Easy procedure, based on the dispersion of the filler in the polymer precursors.	Difficult control of intragallery polymerization. Limited applications.
Melt Intercalation	Environmentally benign; use of polymers not suited for other processes; compatible with industrial polymer processes.	Limited applications to polyolefins, who represent the majority of used polymers.
Template Synthesis	Large scale production; easy procedure.	Limited applications; based mainly in water soluble polymers, contaminated by side products.
Sol-Gel Process	Simple, low processing temperature; versatile; high chemical homogeneity; rigorous stoichiometry control; high purity products; formation of three dimensional polymers containing metal-oxygen bonds. Single or multiple matrices. Applicable specifically for the production of composite materials with liquids or with viscous fluids that are derived from alkoxides.	Greater shrinkage and lower amount of voids, compared to the mixing method.

Source: Ogasawara et al., 2004

Table 3. Processing methods for polymer-CNT nanocomposite systems

Process	System	Procedure	Remarks
Direct Mixing	Thermoset Resins	Dispersion of CNTs; Cure.	
Solution Mixing	Thermoplastic Resins (PS/Epoxy)	Dispersion of 0.2-1% CNTs, (100 nm dia, 10 μm long); Removal of solvent or precipitation of polymer; Cure.	Modification of polymer behaviour; synergistic effect; shape memory nanocomposites.
Melt Mixing	Polymers, N6	Mechanical mixing of CNTs with prepolymer melt followed by extrusion, injection or compression moulding.	Use of 0.2-2.0% MWCNT, twin screw mixer.
In-situ Polymerization	Polyaniline-CNT, MMA-CNT, Epoxy-CNT, Poly(ether-ester)	Use of ultrasonics for dispersion in monomer/matrix; Cure.	Preparation of the polymer with CNT, good chemical bonding.
Others	PP-CNT, PVK-SWCNT, iPP-SWCNT, PANi-SWCNT	Solid-state mechanochemical pulverization; blending + sonication; melt blending; VDP.	0-10 wt. (%) CNTs

Source: Roslaniec et al., 2003

Table 4. Summary of processing methods for conducting polymer nanocomposites

Polymer of Interest (Shell)	Inorganic Particle (Core)	Significant Characterization / Applications
PPy and PAn	SiO_2 (1 μm, 35 nm, 20 nm), SnO_2-Sb(10 nm), Stringy SiO_2 (40-300 nm long)	Stable colloidal form, 'raspberry morphology' and inorganic stable rich surface.
PPy and PAn	CeO_2 (0.52 μm), CuO (1.6 μm), α-Fe_2O_3 (Sph, Polyhedral and spindle shaped), NiO (3.8 μm), SiO_2 (0.46 μm)	Colloidally stable nanocomposite with low dc conductivity and formed without a polymerisation initiator.
PPy and PAn	$BaSO_4$(20 nm), Colloidal gold (7-9 nm), Al_2O_3 membrane	In situ formation of colloidal nanocomposite within the microemulsion or inside the Al_2O_3 membrane.
Ppy, Pan, NVC and PPV	ZrO_2 (20-30 nm), Fe_2O_3 (25-50 nm), SiO_2, n-TiO_2 (~10 nm), Al_2O_3 (35-40 nm), MgO (2-4 μm), CB	Nanocomposites in macroscopic precipitate form or with limited colloidal stability but improved thermal and electrical properties and novel transport properties.
PPy and PAOABSA	MS (15-30 nm), FexOy (14 nm), Fe_2O_3 (~15-50 nm), γ-Fe_2O_3 (85 nm),	Nanocomposites with significant magnetic susceptibility.
PPy and PAn	BT (~1 μm), $LiMnO_2$, $LiMnO_4$, V_2O_5, β-MnO_2, PMo12, $H_3PMo_{12}O_{40}$, CB, Fe_2O_3 (4 nm, 40 nm)	Nanocomposites with important charge storage and dielectric properties, suitable for cathode applications.
Ppy, Pan, PTh and PEDOT	Pt (~4 nm), PtO_2, Pt, Cu, Pd, SiO_2 (20 nm), & bimetallic couples	Nanocomposites with catalytically important metals; catalytic applications.
PPy and PAn	SiO_2 (20 nm)	Grafted surface nanocomposites - important for immunodiagnostic assays.
PPy	SiO_2, PB, MnO_2, Ta_2O_5, TiO_2	Electrochemically synthesised composite films with improved charge storage properties.
PPy and PAn	WO_3	Nanocomposite films with important ECD application and optical activity.

Source: (Gangopadhyay & Amitabha, 2000)

Table 5. Categorization of processing methods for conducting polymer nanocomposites

Chemical Preparation	Electrochemical Preparation
Nanocomposites with colloidal stability (SiO_2, SnO_2, $BaSO_4$, etc., as core)	Nanocomposites with charge storage, optical and electrochromic activities (incorporation of MnO_2, SnO_2, CB, PB, WO_3, SiO_2, etc.)
Nanocomposites with improved physical and mechanical properties (Fe_2O_3, ZrO_2, TiO_2, etc., as incorporated materials)	Nanocomposites with catalytic activities (incorporation of catalytically active Pt, Pd, Cu, etc. microparticles and some bimetallic couples like Pd/Cu, etc.)
Nanocomposites with magnetic susceptibility (using Fe_2O_3, γ-Fe_2O_3, etc., magnetic particles)	Nanocomposites with magnetic susceptibility (γ-Fe_2O_3 magnetic macroanion)
Nanocomposites with dielectric, energy storage, piezoresistive and catalytic activities (with BT, POM, PtO_2, TiO_2, Pd, Pt, etc., incorporation)	-
Nanocomposites with grafted surface (NH_2/COOH functional groups on surface and colloidal silica as core)	-

Source: (Gangopadhyay & Amitabha, 2000)

For polymer-based nanocomposites which contain layered reinforcements, depending on the components (polymer matrix, layered silicate, and organic cation), the method of processing and the strength of interfacial interactions of the polymer matrix with the layered silicate either modified or not. As illustrated in Figure 2 (Alexandre & Dubois, 2000) three different types of nanocomposites may be obtained. When the polymer is not able to intercalate between the sheets of the silicate, a phase-separated nonacomposite is obtained (Figure 2), whose characteristics remain as that of conventional microcomposites. In another vain, in intercalated nanocomposites, the insertion of a polymer matrix into the layered silicate structure occurs in a crystallographically regular manner, without to polymer-clay ratio. An orderly multilayer morphology built up with alternating inorganic and polymeric layers is generated. Usually, only a few molecular layers of polymer can be dispersed by intercalation in these engineering materials. The in-situ techniques can be used with vegetal/mineral fibres, with a view to attaching the polymer to the grafted surface through coupling agents, which consequently optimize the interface bonding and the mechanical properties (Fernando & Satyanarayana, 2005)

Despite the successful use of these different techniques for the preparation of polymer-based nanocomposites, information on different factors is still not forthcoming, such as: the use of a suitable techniques for a specific matrix-reinforcement combination and the maximum levels of reinforcements to arrive at optimum property combinations and reduce the cost of the processes, etc.

Figure 2. Different types of composites arising from the interaction of layered silicates, polymers: a) Phase-separated microcomposite; b) Intercalated nanocomposite, and c) Exfoliated nanocomposite
Source: (Alexandre & Dubois, 2000)

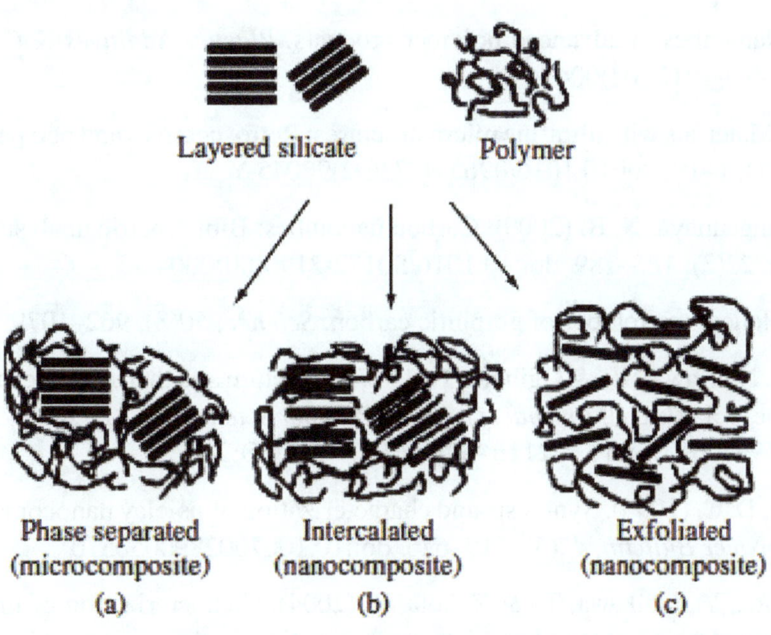

REFERENCES

Alexandre, M., & Dubois, P. (2000). Polymer-layered silicate nanocomposites: Preparation, properties and uses of a new class of materials. *Materials Science and Engineering*, *28*(1-2), 1–63. doi:10.1016/S0927-796X(00)00012-7

Andrews, R., & Weisenberger, M. C. (2004). Carbon nanotube polymer composites. *Current Opinion in Solid State and Materials Science*, *8*(1), 31–37. doi:10.1016/j.cossms.2003.10.006

Anya, C. C. (1999). Microstructural nature of strengthening and toughening in Al2O3-SiC(P) nanocomposites. *Journal of Materials Science*, *34*(22), 5557–5567. doi:10.1023/A:1004729015686

Choa, Y. H., Yang, J. K., Kim, B. H., Jeong, Y. K., Lee, J. S., Nakayama, T., ... Niihara, K. (2003). Preparation and characterization of metal: Ceramic nanoporous nanocomposite powders. *Journal of Magnetism and Magnetic Materials*, *266*(1-2), 12–19. doi:10.1016/S0304-8853(03)00450-5

Fernando, W., & Satyanarayana, K. G. (2005). Functionalization of single layers and nanofibers: A new strategy to produce polymer nanaocomposites with optimized properties. *Journal of Colloid and Interface Science*, *285*(1), 532–543. PMID:15837469

Fischer, H. (2003). Polymer nanocomposites: From fundamental research to specific applications. *Materials Science and Engineering C*, *23*(6-8), 763–772. doi:10.1016/j.msec.2003.09.148

Gangopadhyay, R., & Amitabha, D. (2000). Conducting polymer nanocomposites. A brief overview. *Chemistry of Materials*, *12*(7), 608–622. doi:10.1021/cm990537f

Gary, A. (2003). Nanotubes of advanced polymer products. *Plastics Additives & Compounding*, *5*(4), 12. doi:10.1016/S1464-391X(03)00414-8

Gleiter, H. (1992). Materials with ultrafine microstructures: Retrospectives and perspectives. *Nanostructured Materials*, *1*(1), 1–19. doi:10.1016/0965-9773(92)90045-Y

Gupta, V. K., & Pangannaya, N. B. (2000). Carbon nanotubes: Bibliometric analysis of patents. *World Patent Information*, *22*(3), 185–189. doi:10.1016/S0172-2190(00)00040-5

Iijima, S. (1991). Helical microtubes of graphitic carbon. *Science*, *50*(8), 962–1079.

Jimenez, G., Ogata, N., Kawai, H., & Ogihara, T. (1997). Structure and thermal/mechanical properties of poly (ε-caprolactone) - clay blend. *Journal of Applied Polymer Science*, *64*(11), 2211–2220. doi:10.1002/(SICI)1097-4628(19970613)64:11<2211::AID-APP17>3.0.CO;2-6

Noh, M. W., & Lee, D. C. (1999). Synthesis and characterization of ps-clay nanocomposite by emulsion polymerization. *Polymer Bulletin*, *42*(5), 619–626. doi:10.1007002890050510

Ogasawara, T., Ishida, Y., Ishikawa, T., & Yokota, R. (2004). Characterization of multi-walled carbon nanotube/phenylethynyl terminated polyimide composites. *Composites part A - Applied Science*, *35*(1), 67–74. doi:10.1016/j.compositesa.2003.09.003

Pandey, J. K., Kumar, A. P., Misra, M., Mohanty, A. K., Drzal, L. T., & Singh, R. P. (2005). Recent advances in biodegradable nanocomposites. *Journal of Nanoscience and Nanotechnology*, *5*(4), 497–526. doi:10.1166/jnn.2005.111 PMID:16004113

Ray, S. S., & Bousmina, M. (2005). *Biodegradable polymers and their layered silicate nanocomposites: in greening the 21st century materials world.* Progress in Matererials.

Roslaniec, Z., Broza, G., & Schulte, K. (2003). Nanocomposites based on multiblock polyester elastomers (PEE) and carbon nanotubes (CNT). *Composite Interfaces*, *10*(1), 95–102. doi:10.1163/156855403763586819

Roy, R., Roy, R. A., & Roy, D. M. (1986). Alternative perspectives on "quasi-crystallinity": Non- uniformity and nanocomposites. *Materials Letters*, *4*(8-9), 323–328. doi:10.1016/0167-577X(86)90063-7

Schmidt, D., Shah, D., & Giannelis, E. P. (2002). New advances in polymer/layered silicate nanocomposites. *Current Opinion in Solid State and Materials Science*, *6*(3), 205–212. doi:10.1016/S1359-0286(02)00049-9

Weisenberger, M. C., Grulke, E. A., Jacques, D., Ramtell, T., & Andrews, R. (2003). Enhanced mechanical properties of polyacrylonitrile: Multiwall carbon nanotube composite fibers. *Journal of Nanoscience and Nanotechnology*, *3*(6), 535–539. doi:10.1166/jnn.2003.239 PMID:15002136

Xie, X. L., Mai, Y. W., & Zhou, X. P. (2005). Dispersion and alignment of carbon nanotubes in polymer matrix: A review. *Materials Science and Engineering R Reports*, *49*(4), 89–112. doi:10.1016/j.mser.2005.04.002

Chapter 2
Relevance of Chemically Functionalized Nano-Fillers and Modified Nanocomposite in Energy Systems

Damilola Elizabeth Babatunde
Covenant University, Nigeria

Iheanacho Henry Denwigwe
University of Lagos, Nigeria

Olubayo Moses Babatunde
University of Lagos, Nigeria

Oluranti Agboola
Covenant University, Nigeria

Gbemisola Deborah Akinsipe
Covenant University, Nigeria

ABSTRACT

Reliable energy systems and advances in nanotechnology together will play key role in channeling future cutting edge inventions and developments in all spheres. In this review article, the pertinence of functionalizing nanofillers and modifying nanocomposites for improved performance in various energy applications such as energy conversion, energy efficiency, energy storage, alternative energy and energy saving are expounded. This article also presents structures and unique properties of commonly used nanofillers; advances, improvement potentials and characterization of nanocomposites used in energy systems. Theoretical and experimental literature reviewed revealed that nanofillers engender improved properties in polymeric matrices. Functionalization is applicable to all types of nanofillers in use today, a number of functionalized nanofillers are already commercially available; and more extensive research is needed to achieve optimal improved results with the use of nanofillers and nanocomposites in various fields of applications.

DOI: 10.4018/978-1-5225-7838-3.ch002

INTRODUCTION

Global development, industrial revolution and population increase have led to massive consumption of goods and energy. By 2050 and 2100, the global energy demands are estimated to be 30 and 46 Terawatts respectively (Sahaym & Norton, 2008). The issue of efficiency and sustainability in energy systems is therefore gaining more relevance due to environmental pressures, the ever - growing demands for energy and the finite nature of fossil-based energy sources. Adequate supply of energy is a major challenge of this century because of the pivotal role it plays in continued development and economic growth; alleviation of poverty; balanced ecosystem; meeting security demands and quality health delivery. The quest to abate environmental pollution; find diverse sustainable alternatives to the unrenewable sources of energy; improve reliability and efficiency of energy systems and use of existing sources energy; and effectively combat social, political and security risks has necessitated concerted efforts by researchers, industrialist and several stakeholders to proffer suitable solutions.

Nanotechnology provides an array of resources to resolve numerous energy-related problems. The technology possesses highly important improvement potentials for the development of both renewable and non – renewable energy sources. The ability to manipulate and fabricate materials with unique properties at nanoscale, among other things, has made it possible to explore more efficient, cost effective, and environmentally benign technologies. Components of appliances smaller than 100 nm which can harness, store and transfer energy in a new way are being developed (Abdin et al., 2013). For over ten years, studies on nanocomposites materials have been extensive, chiefly because of the notable enhancement of properties they possess relative to natural and synthetic polymeric materials, even at low loading of nanoparticles. Nanocomposites with desired and improved physical, chemical and mechanical properties can be made for diverse fields of applications from an extensive range of polymers. More than ordinary synthetic and natural polymers, polymeric nanocomposites with appropriate filler, filler size, dispersion of filler and enhanced filler/matrix interaction have greater prospect for applications in various energy fields. Such polymeric composites are attractive and potential materials for fuel cells, solar cells and effective energy storage which is of particular and great importance in mobile electronic devices, hybrid electric transport systems, immobile power systems and pulsed power systems for several commercial and military applications (Barber et al., 2009). This paper presents information on functionalization of nanofillers, modification of nanocomposite in relation to their application in energy systems.

CHEMICAL FUNCTIONALIZATIONS OF NANOFILLERS

Chemical Functionalization of nanofillers has to do with the modification of the surface of nanoparticles to enhance application in different areas. The importance of carrying out this surface modification has been researched severally and found critical in many operations.

Functionalization of Nanofillers

The characteristics of nanocomposites is dependent on the dispersal of the nanofillers within them and the interaction between these fillers and the matrix. Both of these determinants are subject to the application of chemical treatment at the surface of the filler. By functionalization, suitable organic links are introduced between the matrix and the filler leading to improved interposition of polymer chains between the layers. Prior to the use of nanofillers in the formation of nanocomposites, it is required that they are pre-treated to improve the interactions at the interface between the matrix and the filler (Marquis et al., 2011). Functionalization is applicable to all types of nanofillers in use today. These nanofillers include nanoclays, nano-oxides, carbon nanotubes, and metallic nanoparticles among others. The structure and unique properties of these particles will be discussed later in the chapter. A brief description of production techniques of some of these nanofillers and their functionalization are discussed below.

Clays are traditionally known to be inorganic fraction of soil with less than 2 mm particle size. Clay mineral, also referred to as phyllosilicate usually consists of hydrous, magnesium or aluminum silicates. Generally, the nanolayered silicates are known as nanoclays which may be from natural clay or synthetic clays of transition metals (Kurahatti et al., 2010). Techniques used for extracting and processing of nanoclays involves energetic stirring with subsequent centrifugation and freeze-drying; centrifugation and cross-flow filtration; and ultracentrifugation (Floody et al. 2009). The physical mixture of a polymer and layered silicate may not necessarily form nanocomposites because of the characteristic hydrophilic surface of the silicates. Failure to achieve intercalation between fillers and polymers results in formation of an aggregated composites with almost similar properties as those of conventional microcomposites. Unlike aggregated composites, fillers are totally and homogenously dispersed in the polymer matrix of exfoliated nanocomposites and this ensures good quality of composites (Khan et al., 2017). Nanoclays are functionalized in order to gain improved interaction between the nanofillers and the matrix at the nanometric scale so that an intercalated structure is achieved. For this desired incorporation to occur, it is important to replace the substitutable cations or alkaline earth in the nanoclay with organophilic cations. For example, Montmorillonite (MMT) after extraction from bentonite and activation by sodium ions (Na^+) to enhance its swelling is functionalized (Kornmann et al., 1998). This functionalization is achieved by the insertion of organic cations, such as silanes, ammonium bromides and ammonium chlorides (Marquis et al., 2011). MMT has been functionalized with ammonium bromides such as cetyltrimethylammonium bromide (CTAB), (11-acryloyloxyundecyl)dimethyl-(2-hydroxyethyl)ammonium bromide, and (11-acryloyloxyundecyl)-dimethylethylammonium bromide (Samakande, 2005; Wang et al., 2008; Kenawy, et al., 2018). Fu and Qutubuddin functionalized MMT using ammonium chlorides: vinylbenzyldimethyldodecylammonium chloride (VDAC) and vinylbenzyloctadecyldimethyl ammonium chloride (VOAC) (Fu & Qutubuddin, 2001; Fu & Qutubuddin, 2004). Nanoclays have also been functionalized with ionic liquids (Ha & Xanthos, 2009) such as 1-ethyl-3-methylimidazolium tetrafluoroborate (an ambient temperature ionic liquid prepared by reacting 1-ethyl-3-methylimidazolium bromide with tetrafluoroboric acid) (Jarosovaa & Swain, 2015; Holbrey & Seddon, 1999; Wang et al., 2008). These ions enhance ingress inbetween the layers of monomers and polymers having varying polarities by reducing the surface energy of clay. This surface energy reduction makes the nanoclays more organophilic. Pre-modified MMTs are commercially available and often used. Examples include Closite®, Garamite® (Marquis et al., 2011). Figure 1 shows dispersion types in a polymer matrix.

Like other nano materials, nano – oxides are of much interest for the various distinctive properties they possess and also because of their usefulness as fillers which enhances the properties of polymers. Chemical precipitation, sol-gel, vapour deposition, solvo thermal and solid-state reaction have been mentioned as methods adopted for the synthesis of these oxides (Bhattacharya et al., 2013). Nano-oxides require surface functionalization in order to maximize exfoliation and obtain full dispersion into the matrix (Marquis et al., 2011). The functionalization of nano-oxides has been studied and applied severally. Guo et al. functionalized zinc oxide (ZnO) and iron oxide (Fe_2O_3) nanoparticles with methacryloxypropyltrimethoxysilane (MPS) and then employed them to create a vinyl-ester resin polymeric nanocomposite.

Figure 1. Types of dispersion of clay layers in polymer matrix
(Source: Khan et al., 2017)

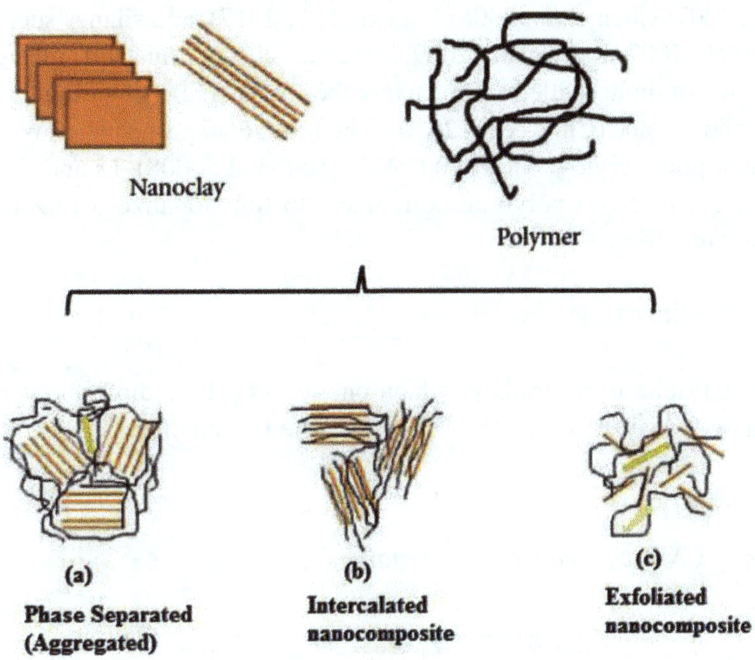

They obtained enhanced properties in the resulting nanocomposite (Guo et al., 2007; Guo et al., 2008). Brassard and coworkers functionalized hydrophilic silica (SiO_2) and Zinc Oxide (ZnO) nanoparticles using fluoroalkylsilane molecules and stearic acid molecules respectively (Brassard et al., 2012). Alumina nanoparticles were treated with fluorosilane and found to engender unprecedented wear resistance in polytetrafluoroethylene (PTFE) (Burris et al., 2009).

Carbon nanotubes are major discoveries in the field of nanotechnology because of their versatility. Several methods are available for their production but electric arc-discharge, laser ablation and chemical vapor deposition (CVD) are mostly adopted. Electric arc-discharge technique involves the creation of vapour by an arc discharge between two carbon electrodes which occurs with or without the presence of catalysts. Laser ablation is a relatively newer technique for nanotubes production and it is comparable to the arc-discharge. The technique involves vaporization of the carbon and subsequent deposition onto a substrate. The CVD basically involves thermal cracking of gaseous hydrocarbon feed into carbon and hydrogen. The temperature required for the thermal dehydrogenation reaction is lowered by the presence of transition metal catalyst like iron, nickel, or cobalt. Detailed description of these techniques is found in literature (Mubarak et al., 2014) and summarized using Figure 2, Figure 3 and Figure 4. In terms of carbon source, reaction temperature, nanotube selectivity, nanotubes yield, purification of nanotubes, process parameter control, energy requirement, reactor design, per unit design cost and ease of reproducibility, CVD method appears to be the most promising for producing carbon nanotubes (Mubarak et al., 2014). Carbon nanotubes are functionalized to maximize their dispersion into the matrix and improve interfacial interactions. This can be achieved using three different techniques: chemical functionalization, physical functionalization and surface deposition (Marquis et al., 2011). Different researches have been conducted to examine the chemical functionalization of carbon nanotubes. Sham and Kim functionalized a carbon nanotube (CNT) surface via ultraviolet (UV)/Ozone and triethylenetetramine (TETA) treatment (Sham & Kim, 2006). CNTs have also been functionalized using amines (Fiedler et al., 2006; Shen et al., 2007; Chen et al., 2008; Maleki et al., 2017), and silanes such as 3-aminopropyl-triethoxysilane (Ma et al., 2007; Kathi et al., 2009, Lee et al., 2011). Chua and colleagues carried out the surface functionalization of multi-walled carbon nanotubes (MWNTs) using diphenyl-carbinol followed by 3-aminopropyltriethoxisilane (Chua et al., 2010). The functional groups on MWNTs were magnified by inserting a bridge of poly(acryloyl chloride) (PAC) (Zou et al., 2008). Li and Coleman reported the use of diamines, dianhydrides and polyimide oligomers to functionalize oxidized carbon nanofibers (OCNFs) (Li & Coleman, 2008).

Structures of Nanofillers

Nanocomposites are multiphase materials in which one, two or three dimensions of less than 100nm exist in one of its phases (Ajayan et al., 2003). They are the results of the incorporation of nanofillers

Figure 2. Production of CNTs by electric arc-discharge
(Source: Mubarak et al., 2014)

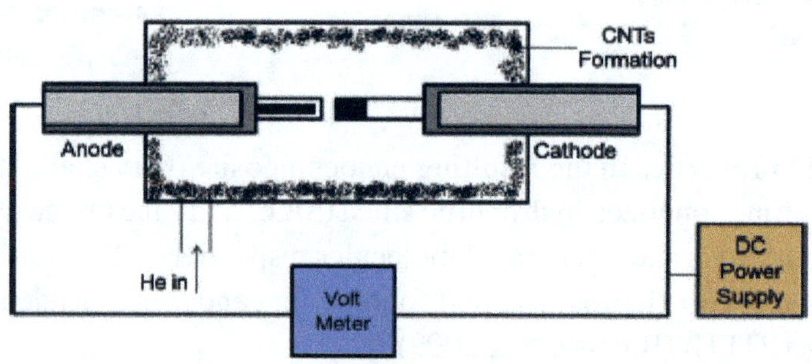

Figure 3. Production of CNTs by CVD laser ablation
(Source: Mubarak et al., 2014)

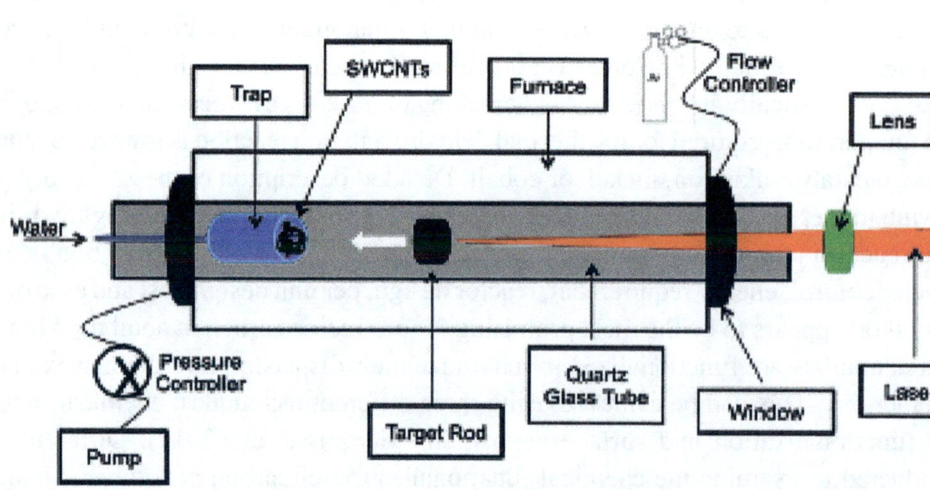

Figure 4. Production of CNTs by CVD
(Source: Mubarak et al., 2014)

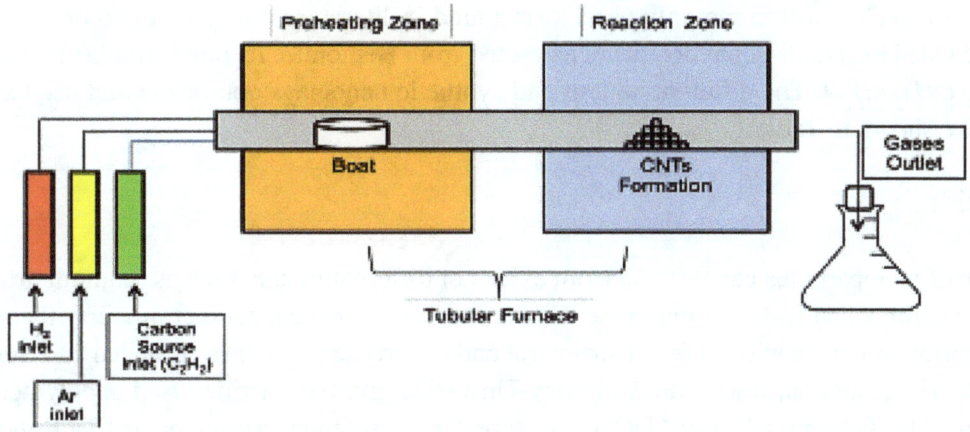

into polymer chains. These nanocomposites can be classified based on the structure of nanofillers they contain as given in reference (ISO/TS80004-2, 2015):

- **Unidimensional Nanofillers:** They exist form of plates, laminas and/or shells;
- **2-Dimensional Nanofillers:** Nanotubes and nanofibres with diameter lesser than 0.1 μm;
- **3-Dimensional Nanofillers:** Isodimensional nanoparticles like nanometric silica beads belong to this category. Figure 5 show dimensional classification of nanofillers.

Some of the diverse nanofillers in use today are described in the subsequent paragraphs.

Nanoclays

These are unidimensional nanofillers which take the form of plates. They possess crystalline structures and can be natural or synthetic clays, as well as phosphates of transition metals. Nanoclays are commonly used because they are naturally abundant and possess very high form factor. Phyllosilicates (smectites) which are the most widely used, have a shell-shaped crystalline structure with nanometric thickness. Clays are categorized in relation to their crystalline structures as well as by the quantity and position

Figure 5. Classification of nanofillers based on dimensions
(Source: Marquis et al., 2011)

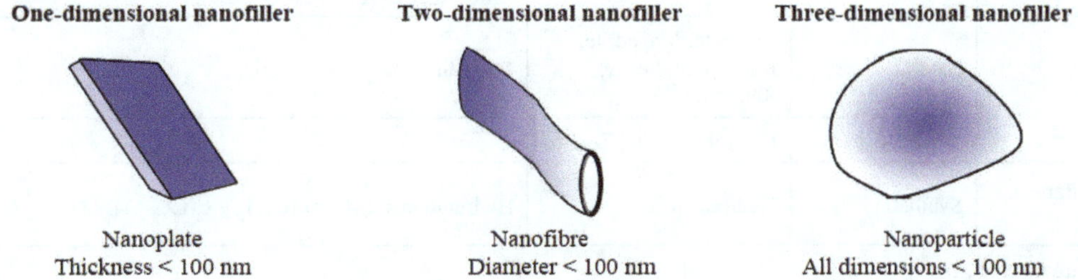

of the ions within the elementary mesh. This elementary mesh refers to the simplest atomic geometric pattern, which is sufficient for duplicating the crystalline network, by repeating itself indefinitely in the three directions. The most commonly used form found in literature concerns organomodified Montmorillonite (MMT), a natural phyllosilicate extracted from Bentonite. Its raw formula is $(Na, Ca)_{0,3}(Al, Mg)_2Si_4O_{10}(OH)_2,nH_2O$. The different natural and synthetic nanoclays obtainable and used as fillers in polymers are shown in Table 1.

Nano-Oxides

Fabrication of nanoparticles can be done from oxides of different metals such as titanium, iron, manganese, magnesium, silica and aluminium etc. Elements of metal have a distinct property that will result to the formation of oxides with different structural and electronic properties (Kaushal and Singh 2017). Titanium dioxide, alumina, silica, and Antimony-Tin-Oxide give nano-oxides used in polymer matrixes. The nanoparticle of titanium dioxide (TiO_2) is anatase. It is a tetragonal crystal form of TiO_2 and is usually used as spherical particles with diameter of about 20 *nm.* It has photocatalytic properties. Mogilevsky and colleagues reported that they constructed titania nanotubes by hydrothermal synthesis of delaminated anatase; therefore, validating the fact that anatase can be converted into titanium nanotubes (Mogilevsky et al., 2008). These nanotubes have an outer diameter of 10 – 20 *nm*, an inner diameter of 5 – 8 *nm* and a length of 1 *μm.* Rutile, another tetragonal crystal form of TiO_2 is also usually used in polymers. It is employed as a white pigment with its particles ranging from 200 – 300 *nm* (Marquis et al., 2011). Alumina is among the usually used mineral particles in composite industry. Alumina nanoparticles are usually used in polymer matrix because the spherical shape of alumina nanoparticle will produce a simple composite system which is devoid of orientation effect (Mirjalili et al., 2014). Hence, nanoalumina are

Table 1. Nanoclays identification

Family		Group	Formula
Phyllosilicates	TO (1:1)	Kaolinite	The reference plate is consist of a tetrahedral plate T and an octahedral plate O. The layer is about 0.7 *nm* thick. Kaolinite $Al_4Si_4O_{10}.(OH)_8$
	TOT(2:1)	Smectite (Talc, Mica, Mommorillonite, Sepiolite)	The reference plate is formed by two tetrahedral plates T in both sides of an octahedral plate O. The layer is about 1 nm thick. The group includes several minerals that are key constituents of clays.
	TOT:O(2:1:1)	Chlorite, Bentonite, Saponite	The reference plate is composed of three plates TOT and another isolated O plate. The layer is about 1.4 *nm* thick. Chlorite di-tri $Al_2Mg_3Si_4O_{10}(OH)_8$
Polysilicate	Natural	Kenyaite, Magadiite, Kanaemite, Ilerite, Silhydrite, Zeolite	Magadiite $(Na_2Si_{14}O_{29}H_2O)$
	Synthetic	FluoroHectorite, Zeolite	
Double lamellar hydroxide	Synthetic	Hydrotalcite	Hydrotalcites: $(Mg_6Al_2(OH)_{16})(CO_3^{2-})4H_2O$

(Source: Marquis et al., 2011)

alumina nanoparticles composed of spherical crystal particles of Al_2O_3 and are often used as inert fillers in polymers. They are usually employed in a variety of sizes ranging from 20 *nm* to micrometric sizes (Marquis et al., 2011). Antimony-tin dioxide has appealed substantial attention due to its prospective applications as gas sensor, humidity sensor, solar battery, transparent electrode, electricity conducting coatings (Li et al., 2006). Nano-antimony-Tin Oxide (ATO) nanoparticle is a tetragonal crystal particle of about 15 *nm* diameter with raw formula given as Sb_2O_5/SnO_2. It is commonly used in polymers as flame retardant (Marquis et al., 2011). Nanosilica represents a wide range of nanoparticles from diverse origins. There are both natural and synthetic ones. The most frequently used occurs naturally and is called diatomite. Diatomite originates from the skeleton of a unicellular algae known as the diatomea, developing into sedimentary layers. It is constituted from ultrafine particles of 750 *nm*. Two groups of synthetic nanosilicas generally used are pyrogenic silica and silica fume, forming particles of about 5 to 100*nm* and 100*nm* respectively (Marquis et al., 2011).

Carbon Nanotubes

These are formed by a catalytic chemical vapour decomposition process at medium temperatures (600-1000°*C*) or by laser ablation at 1200 – 3000 °*C* or by electric discharge (arc) process under helium at high temperature (3000 to 4000°*C*). Each process produces a mix of the different kinds of carbon nanotubes with surface defaults (e.g. some pentagonal cycles in place of aromatic rings), and present important catalytic residues. Generally, three kinds of carbon nanotubes are encountered: Single-wall carbon nanotubes (SWCNT) which have diameter between 1 and 2 *nm*; Double-wall carbon nanotubes (DWCNT) with diameter between 2 and 4 *nm*; and Multi-wall carbon nanotubes (MWCNT) with diameter between 4 and 150 *nm*. Carbon nanotubes possess a theoretical range of incredible properties such as Young modulus up to 1 *TPa*, heat conductivity of 3000 $W.m^{-1}.K^{-1}$, and electric conductivity of 107 $S.m^{-1}$ (Marquis et al., 2011). Carbon nanotube is theoretically unique as a cylinder synthesized of rolled up grapheme sheet. Carbon can bond in diverse ways in order to build structures with totally different properties. The sp^2 hybridization of carbon makes a layered construction with weak out of plane bonding of the van der Waals form and strong in-plane bounds (Eatemadi et al., 2014). Furthermore, elasticity is another incredible property of carbon nanotubes. With the influence of high force and press sitting when exposed to great axial compressive forces, it can bend, twist, kink, and finally buckle without damaging the nanotube. Then, the nanotube will return to its initial structure; however, an elasticity of nanotubes does have a restriction, and with application of a very physically powerful forces presses, there is possibility of temporary deformation of the shape of a nanotube.

Nanotube's strength could be weakened by certain defects in the structure of the nanotube, example of such is the defects in atomic vacancies or a reorganization of the carbon bonds (Eatemadi et al., 2014). Hence, it is important to use appropriate tools for characterizing the properties of nanostructures. Different methods such as Transmission Electron Microscopy (TEM), Atomic Force Microscopy (AFM) and scanning electron microscopy (SEM) can be used for characterizing and measuring of carbon nanotubes basic parameters (nanotube diameter, nanotube length, surface roughness, surface morphology, and elemental composition). Study have shown that TEM is the best method for detailed analysis of carbon nanotubes (Safarova et al., 2007). As a result of high resolution of transmission electron microscope, it is possible to examine the sample even in atomic resolution. The use of TEM is based on its capability to measure nanotube diameter in the bundle. From the TEM image (Figure 6) the diameter of a nanotube and bundle diameter can be determined directly. As a result of this information, number of nanotubes in the bundle can be found out. The main information SEM gives is about surface morphology of the sample (Figure 7) and about chemical composition of the sample (Figure 8) (Safarova et al., 2007).

Other Nanofillers

Apart from the types of nanofillers mentioned above, some others have been found in literature, though less prominent. These include:

- **Metallic Nanoparticles:** Nanosilver, nanozinc and nanogold fillers fall under this category. They exhibit catalytic behavior, electrical and magnetic properties which make them beneficial for nanocomposite creation. The term metal nanoparticle is used to characterized nanosized metals with dimensions (length, width or thickness) within the size range 1-100 nm. The characteristics of metallic nanoparticles are large surface area to volume ratio as compared to the bulk equivalents and large surface energies. The metallic nanoparticles having large surface energy coalesce

Figure 6. TEM image of the single-walled carbon nanotubes, a) marker 20nm, b) marker 10 nm
(Source: Safarova et al., 2007)

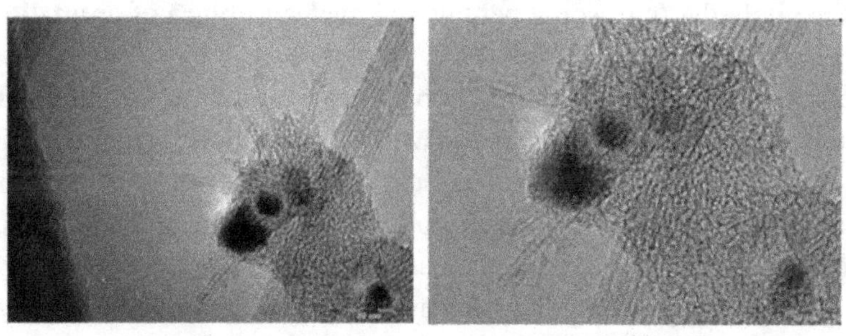

Figure 7. a) SEM images of SWCNTs with metallic catalyst particles acquired in secondary electron mode, b) SEM images of SWCNTs with metallic catalyst particles using backscattered electron mode
(Source: Safarova et al., 2007)

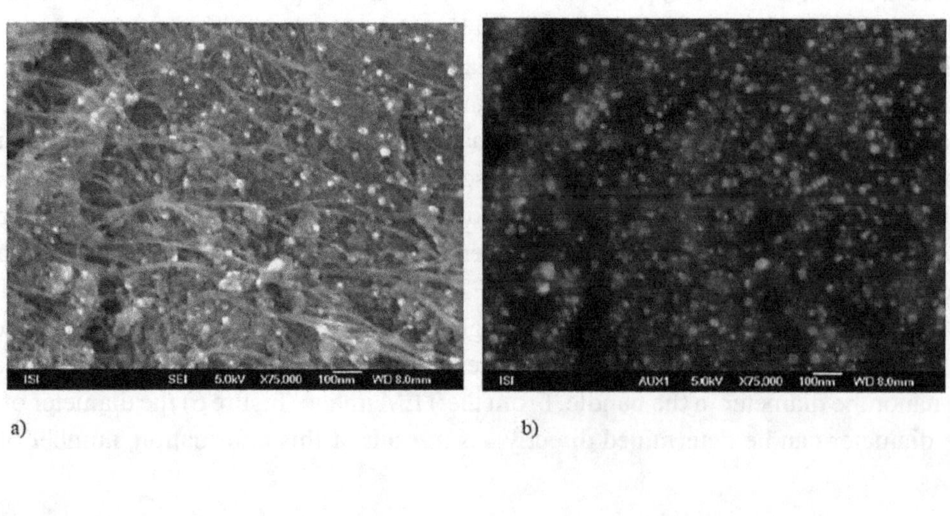

Figure 8. EDX spectrum of the sample, Ni and Y belonging to metallic catalyst particles
(Source: Safarova et al., 2007)

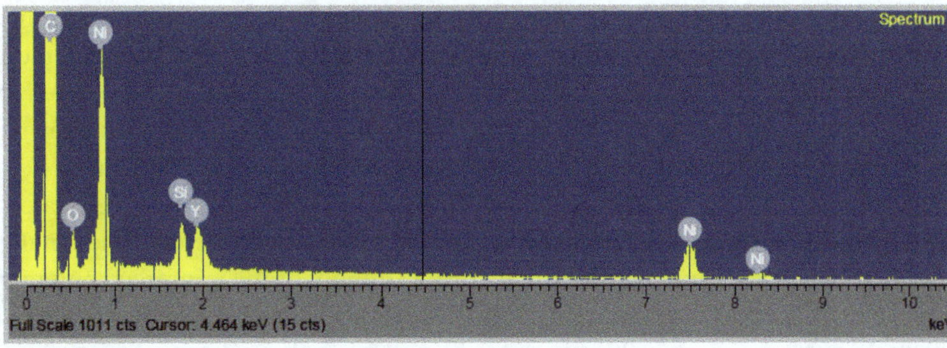

in order to provide thermodynamically favored bulk particle. In the absence of any counteractive repulsive forces the van der Waals forces between two metal nanoparticles would lead towards coagulation. Hence for spatial confinement of the particles in nano range, it is important to stabilize the particles. Thus, this stabilization can be attained either by steric exclusion or electrostatic stabilization by applying a capping agent such as surfactant, polymer, solid support or ligand with suitable functional groups (Harish et al., 2018).

- **Silsesquioxanes:** Silsesquioxanes are big synthetic cage macromolecules with raw formula found to be $R - SiO_{3/2}$ where R typically represents alkyl or organo-functional groups. They possess high three-dimensional symmetry, controlled orientation in 3-D space and a diversity of functional groups. These properties allow for highly tailored nanometer-by-nanometer construction in all three dimensions. Commonly found examples include polyhedral oligosilsesquioxanes (POSS), octasilsesquioxanes of the $[ROSiO_{1.5}]_8$ type with its polymeric derivatives, and $[R - phenylSiO_{1.5}]_8$ type with its polymeric derivatives. They find application in making structured $3- D$ nanocomposites for thin films, monoliths, and fibre reinforced composites. They are presently limited in usage due to their expensive price limits (Marquis et al., 2011).
- **Boehmite:** Boehmite is a mineral aluminium hydroxide orthorhombic dipyramidal crystal that can be obtained naturally or synthesized chemically from saturated solutions. It is an aluminum oxide hydroxide particle containing extra hydroxyl groups on its surface with a raw formula of, $\gamma - AlO(OH)$ and is used as a nanofiller in nanocomposites (Marquis et al., 2011; Vatanpour et al., 2012). The use of boehmite nanoparticles as novel nanofiller for preparation of antifouling mixed matrix membrane has been investigated. Its influences as innovative nanofiller for fabrication of polyethersulfone (PES) blended membranes with respect to hydrophilicity, permeation performance, antifouling property and membrane morphology show that hydrophilicity and pure water flux of the membranes were enhanced by integration of boehmite nanoparticles; PES membrane embedded with nanoboehmite exhibited superior antifouling properties compared to gamma – Al_2O_3/PES membranes and also possess similar typical assymetric structure as bare PES membrane (Vatanpour et al., 2012).

Adhesion Between Filler and Matrix

Filler – polymer matrix adhesion is of great concern in most engineering and military applications. For instance, the use of polymers as multilayered structure dielectrics for electronic purposes requires strong interfacial adhesion because the interfaces are expected to be able to withstand accelerated thermal and humidity cycling as indicators of their reliability. Particle/matrix adhesion has been shown to strongly affect composite strength, toughness and some other mechanical properties. The impact of interfacial adhesion on the mechanical properties of polypropylene/glass composites was investigated in a study which involves treatment of glass particles with two silanes possessing different functional groups, hydrocarbons and fluorocarbons. Findings show that tensile strength of the reinforced polymer increased as the interface adhesion increased (Thio et al., 2004). Nanosilica particles have been treated by various monomers to obtain different interfacial interactions in a study on the influence of interfacial adhesion on the strength of PP/silica nanocomposites (Rong et al., 2004; Wu et al., 2005). The introduction of amino-silane as coupling agent to treat particles have been reported to increase composite tensile strength of nylon 6,6/kaolin as the interface adhesion improved (Buggy et al., 2005). These outcomes are expected because strength is determined by effective stress transfer between filler and matrix; and effective stress transfer is hinged on strong interfacial bonding between particles and polymer matrix.

Basic mechanisms at polymer surface responsible for adhesion at molecular level, approaches for evaluating adhesion between different materials and critical discussion on measurement of work of adhesion are found in literatures (Creton et al., 2001; Fu et al., 2008; Packham, 1996). Interface molecular interaction has been mentioned as the cause of adhesion and cannot be directly investigated because of the thermodynamic irreversibility accompanying the process of interface formation (Creton et al., 2001). Adhesion experiments could give an important parameter, the critical energy release rate, Gc which is the energy necessary to grow a crack by a unit. The actual failure of the interface, however requires a molecular event such as chain pullout or chain scission within, or adjacent to, the plastic zone. This molecular event could be considered as being triggered by a critical force on the connector. In obtaining optimum interface reinforcement, the highest possible areal density of molecular connectors with a large critical failure force may be provided so that a large stress on the interface can develop in, or adjacent to, the plastic zone, which would permit the plastic zone to grow as large as possible before the interface fails. Understanding what the maximum number of connectors could be at the interface per unit area, and how effective these connectors are at transmitting the stress is therefore paramount. Lack of entanglement across the interface is considered responsible for very low adhesion, and increasing the adhesion would normally require replacement of the entanglement network strands which can be achieved by chemical reaction at the interface (Creton et al., 2001). Method of processing also play crucial role in determining the strength of nanocomposite, for instance, in a study cited by Fu et al. (2008) involving preparation of polyurethane/nanosilica composites, at the same filler weight percentage, tensile strength was higher in the composite obtained through *in situ polymerization* than what was obtained by blending. This was because the *in situ polymerization* method resulted in more polyester chains which are chemically bonded to nanosilica and are more homogeneously dispersed in the polymer matrix. Detailed review of empirical relationships on particle/matrix adhesion, and particle–matrix interfacial bonding can be found in Fu et al., 2008.

Types of Nanofillers

Nanofillers are incorporated into polymer matrices to reinforce its chains at the molecular scale. Therefore, they can greatly improve or modify various properties of the materials into which they are inserted. The nanocomposites formed from this method usually have enhanced optical, electrical, mechanical or thermal properties and are consequently employed in varied applications, such as energy storage, efficiency and conversion (Marquis et al., 2011).

Nanofillers in Energy Conversion

A promising energy source in the search of sustainable, environment-friendly and clean energy sources to meet the energy challenge of the 21st century is solar energy. The sun supplies 1.2×10^5 Terawatt of light to the earth and just a small fraction of that would be sufficient to solve the global energy crisis if efficiently used for direct electricity generation through solar cell panels and/or clean fuel (e.g., hydrogen production through water splitting) (Pan & Gupta, 2012). Notable advancement has been made towards the development of efficient and economic solar energy conversion systems. Solar cells have been developed as low-cost, stable and highly efficient alternatives to crystalline silicon in capturing solar energy for the generation of electricity. The use of these solar cells is however inhibited by amorphous silicon used for the solar cell panels because they have limited efficiency and energy-intensive production techniques. Organic photovoltaics (OPVs) have also been developed as alternatives but these are limited by poor-charge transport mobility and the conflicting need for a thick donor layer for maximum light absorption and the restriction brought about by short exciton diffusion distance (Forrest, 2005; Pan & Gupta, 2012). New solar energy conversion devices that can aid efficient light absorption and hole carrier transport are highly desired (Pan & Gupta, 2012). Direct water spotting through photoelectrochemical systems has the potential to become a practical method of obtaining high purity hydrogen and consequently energy. It is also auspicious as a method for industrial processes and storage.

A proposed technique for amplifying the conversion efficiency of photovoltaics and photocatalysts is the incorporation of metallic nanoparticles, which have been shown to improve the efficiency of solar energy systems due to their surface plasmonic effects. Standridge and colleagues (2009) fabricated TiO_2 based dye-sensitized solar cells with incorporated corrosion protected silver nanoparticles. They obtained clear improvement of the dye extinction in the presence of plasmonic particles over the range of TiO_2 coating thickness examined (2 to 8 nm). The enhancements increased as the TiO_2 decreased. The optical improvements observed, translated into photocurrent enhancements, that is, an expansion in the cells' solar energy response wavelength range and light absorption cross-section. This amplified the overall efficiency of the dye-sensitized solar cell (Standridge et al., 2009). Yakimov and Forrest obtained high power conversion efficiencies with high open circuit voltage in the presence of silver nanoparticles (Yakimov & Forrest, 2002).

The energy conversion efficiency and short-circuit current density in amorphous silicon solar cells were improved via the deposition of gold nanoparticles above the amorphous silicon film. Forward scattering by surface plasmon polariton modes in the gold nanoparticles was said to have caused the improvement. Approximately 8% increase was obtained for both short-circuit current density and energy conversion efficiency with gold nanoparticles density of $\sim 3.7 \times 10^8 \text{ cm}^{-2}$. It was stipulated that even larger improvements would be attained for higher nanoparticle densities (Derkacs et al., 2006).

Chan et al. (2015) investigated the effects of nanofillers in the performance of dye-sensitized solar cells by modifying polymer electrolytes with nanosized titanium dioxide, iron (III) oxide and halloysite fillers. The results showed that the incorporation of fillers resulted in increased ionic conductivity of the cells and improved dissociation of iodide ions (Chan et al., 2015). However, it has been noted that the metallic nanoparticles should be encapsulated on their surfaces to prevent formation of charge traps or charge recombination sites which may lower the efficiency of the solar cells (Thomann et al., 2011; Pan & Gupta, 2012; Chan et al., 2015). Zhao and coworkers' research on increased charge transfer of poly(vinylidene fluoride-co-hexafluoro propylene) (PVDF-HFP) based electrolyte by addition of graphite nanofiber (GNF) and its usefulness in dye sensitized solar cells (DSSCs) produced results that corroborated the points made above. They claimed to have specifically added GNF to the polymer electrolyte in order to increase the ionic conductivity and charge transfer. They obtained the highest energy conversion efficiency from the polymer electrolyte containing GNF (Zhao et al., 2013).

Electrostrictive polymers, such as polyurethane (PU), have exhibited an ability to convert mechanical energy into electrical and are more recently being investigated as materials for energy harvesting. Guyomar and coworkers developed a method for performing energy conversion which consisted of employing PU and polymers loaded with nanofillers (nanosized SiC and carbon nanopowder). Such materials are known to convert surrounding ambient vibration energy into electrical energy. They obtained that the C nanofillers and SiC fillers led to dielectric constant in the composites that was greater than that of pure PU. It was also observed that the incorporation of nanofillers increased the obtained current. It was concluded that the composites loaded with nanofillers had improved results of current harvesting and energy conversion (Guyomar et al., 2009).

Nanofillers for Energy Efficiency

Energy efficiency refers to the use of less energy to provide the same function. It is a key tool for the reduction of greenhouse gas emissions. Nanomaterials are promising in the field of energy efficiency. They have been investigated for abilities to improve the efficiency of fuel production through better catalysis, to enhance fuel consumption efficiency in vehicles, to facilitate more efficient lighting systems among others. They have already found application in the development of more efficient, lighter weight batteries and in the design of more efficient prototype solar panels.

Incorporating carbon black, carbon nanofibers (CNF), nanoalumina, graphene silica and organosilane to rubber composition at nanometric dimensions has been found to significantly improve tyre properties. These nanoparticles possess coarser surfaces and higher surface energy than traditional tyre components and therefore reduce inner friction. This reduced inner friction gives rise to better tyre rolling properties and consequently higher fuel efficiency in vehicles. However, surface modification of the particles is important as it affects their interaction with rubber matrix and its final properties. Tyres with better traction on wet and icy roads can be created using nanoparticles. This results in the reduction of car stopping distance by 15-20% and fuel consumption by 5% (Mohseni et al., 2012; Zhou et al., 2010; Das et al., 2008).

Fluids containing nanosized materials like nanofibers, nanotubes, nanowires, nanorods and nanosheets, popularly known as nanofluids, possess unique characteristics and have aided the design of highly efficient cooling systems for cars. Nanofluids make it possible for smaller sized compact cooling system for cars' radiator to be designed. The improvement of the thermal conductivity of ethylene glycol-based fluids using nanomaterials has drawn much attention as an engine coolant. Nanofluids based coolants have higher boiling points than conventional ones and can therefore reject more heat through the coolant system. Nanofluids could lead to fuel savings of about 6% (Mohseni et al., 2012; Yu & Xie, 2011; Kole & Dey, 2010; Kakac & Pramuanjaroenkij, 2009; Wang & Mujumdar, 2008). Several existing chemical processes waste a lot in terms of energy, solvents and precursors. Catalysts offer significant assistance in this as they can speed up a reaction, reduce the temperature needed to convert raw materials into fuel among others. Improving catalysts holds the promise of providing more efficient processes. Nanomaterials contribute to energy efficiency by increasing the effectiveness of catalysts. It has been found that catalysts made from nanoparticles have greater surface area to interact with reacting chemicals than catalysts made from larger particles. This enhanced interaction makes the catalyst more effective and can therefore make the overall process more energy efficient. Wong and colleagues reported that laying of nanometer clusters of tungsten oxide on zirconium oxide produces a highly efficient catalyst that turns straight-line molecules of n-pentane, one of the numerous hydrocarbons in gasoline, into better-burning branched n-pentane. This process makes the manufacturing of gasoline more efficient and environment-friendly. It leads to the production of higher-octane gasoline and saves money for the industry. Tungsten oxide has been famous for its catalytic capabilities for an extended duration of time, however, it has been discovered that nanoparticles are required to maximize those capabilities (Boysen, n.d.; Understanding Nano, 2010; Wong et al., 2010).

Effective waste heat recovery from various sources can improve energy efficiency and reduce fossil fuel consumption by increasing the market share of renewable energy systems. Nanofillers offer enhancements to the properties of polymers considered for waste heat recovery and are therefore important in this field also. It has been presented that electroactive polymers are suitable for waste heat recovery for low-grade sources (sources with temperature range from 40 - 250°C). However, the performance of these polymers is based on the modification methods and modifier character. Polyaniline (PANI) is considered to be one of the most fitting of such polymers (Kolasińska & Kolasiński, 2016; Hostler et al., 2006; Kaiser & Park, 2005; Rogers & Kaiser, 2004). Properly incorporating nanoparticles into the polymer matrix (as a form of modification) improves the thermoelectric performance as well as the mechanical and thermal resistance. PANI composites with nanofillers display higher electrical conductivity compared to microfillers because the fine fillers interact strongly with the polymer. Metallic nanoparticles when added to PANI build a conducting network in the matrix that amplifies the inherent properties on the atomic (quantum) scale (Kolasińska & Kolasiński, 2016).

Hostler and colleagues (2006) investigated the effects of the inclusion of nanofillers on the thermal, electrical and thermoelectric characteristics of polymer nanocomposites. They embedded bismuth nanoparticles into thin films of camphorsulfonic acid protonated polyaniline and measured the resulting thermopower, thermal and electrical conductivities with different volume fractions of the nanofillers. The results obtained portrayed that the properties of polyaniline are significantly influenced by the incorporation of nanoparticles. Increasing the nanofiller load (up to 10 vol %) positively affected the electric conductivity of the composite and enhanced the arrangement of polyaniline particles leading to improved charge transport. However, it was observed that increase of Bi load beyond the 10 vol% mark negatively influenced the electrical conductivity. A probable cause given for this occurrence was the scattering effect of the nanoparticles (Hostler et al., 2006; Kolasińska & Kolasiński, 2016).

In addition to these, nanofillers improve the efficiency of several operations such as solar cells conversion and heat management in batteries which are also discussed in this chapter.

Nanofillers in Energy Storage

While discovering new and renewable energy sources is of paramount importance, another viable option to tackle the growing global energy demand is to develop energy storage devices. The storage of energy in forms that can easily be converted to required energy forms, is a current challenge to scientists. Energy storage not only reduces the disparity between supply and demand but also improves the performance and reliability of energy systems. It also plays a crucial role in conserving energy. By storing energy, premium fuels are saved and systems become more cost effective due to reduction in wastage of energy and capital cost. A power generation plant for example, would have improved performance as a result of load leveling, higher efficiency and lower generation cost brought about by energy storage. One of promising techniques of storing thermal energy is the application of phase change materials (PCMs). Unfortunately, prior to the large-scale practical application of this technology, it is required that several problems are resolved at the research and development stage (Lingayat & Suple, 2013). Use of nanofillers have been established as a feasible way to solve one of such problems.

A field in which energy storage via PCMs has become increasingly important is in the thermal management of battery packs for portable electronics to improve their efficiency. These batteries, typically Lithium—ion (Li-ion) batteries, suffer from temperature rises beyond the normal operating range and sometimes lead to operational damages. In some cases, uncontrollable temperature inside these batteries can result in fire or even explosion (Shirazi, Mohebbi, Kakavand, He, & Rabczuk, 2016). Latent heat storage system using PCMs is an effective way of storing thermal energy due to its high storage density and isothermal storage process. PCMs therefore, appear to be a possible solution to the overheating of battery packs. PCMs can absorb high amounts of heat with barely any change in their temperature. An example is paraffin wax which is cheap and has moderate thermal energy storage density (Sharma & Sagara, 2005; Farid et al., 2004; Renteria et al., 2014). However, a major drawback of this approach is low thermal conductivity. The common PCMs, including paraffin waxes, have very low thermal conductivity, K, with typical values in the range of 0.17–0.35 W/mK at room temperature and hence, require large surface area (Sharma & Sagara, 2005; Farid et al., 2004). The properties of PCMs can be varied by changing their chemical composition or by adding components such as nanofillers. Fabricating nanocomposite structures by adding nanofillers to the PCMs improves the thermal conductivity. Carbon based nanostructures such as graphene and carbon nanotubes offer unusually high thermal conductivities and can therefore be considered as the best candidates to enhance heat conduction for PCMs (Shirazi et al., 2016; Ghosh et al., 2010; Marquis et al., 2011; Pop et al., 2012).

High latent heat of phase change materials (PCMs) has made them as one of the utmost significant materials for thermal management purposes. Nonetheless, PCMs' low thermal diffusivities could restrict their use in applications which need fast thermal response Samimi et al. (2016). Goli and coworkers (2014) enhanced the performance of lithium-ion (Li-ion) batteries using paraffin PCMs with graphene fillers. They demonstrated that the use of graphene fillers in organic phase change material increases its thermal conductivity by more than two orders of magnitude while preserving its latent heat storage ability. This enhancement is achieved via easy binding of graphene flakes to paraffinic hydrocarbons resulting in good thermal coupling. The exceptionally high thermal conductivity of graphene improves the heat conduction ability of paraffins and consequently decreases temperature rise inside a typical Li-ion battery pack (See Figure 9). Their described combined heat storage – heat conduction approach may lead to a transformative change in thermal management of batteries (Goli et al., 2014). Samimi et al. (2016) studied the thermal management analysis of a Li-ion battery cell using phase change material loaded with carbon fibers by simulating thermal performance of a lithium ion battery cell in the presence of carbon fiber-PCM composites. They found that the presence of carbon fibers increases the effective thermal conductivity of PCM, thus, impacts temperature distribution within the cell. PCM composites comprising higher percentages of carbon fibers present a more uniform temperature distribution. Shirazi et al. (2016) also established that an appropriate solution to control the temperature changes during the operation of Li-ion batteries is to embed the batteries inside a paraffin matrix to absorb and dissipate heat. They carried out a computational investigation to ascertain the possibility of making paraffin nanocomposites for better heat management of a Li-ion battery pack. Using Newman's well established electrochemical pseudo-2D model, they simulated the heat generation during a battery's charging/discharging cycles and then coupled the model to a 3D heat transfer model to predict the temperature evolution during the battery operation. They then considered different paraffin nanocomposites structures made by the addition of graphene, carbon nanotubes, and fullerene to enhance thermal conductivity, using the 3D model. Their results assessed the degree of enhancement in heat dissipation of Li-ion batteries through the use of paraffin nanocomposites. They obtained that as the filler's volume fraction inside the paraffin increased, the amount of heat dissipation to the ambient environment increased, due to the increase in the thermal conductivity of PCMs. They found that the fabrication of paraffin nanocomposites can enhance the heat management of batteries as seen in the case of the carbon nanotubes (Shirazi et al., 2016). Alipanah and Li (2016) also used numerical method to investigates thermal management systems (TMS) of lithium-ion battery made from pure octadecane, pure gallium and octadecane–Al foam composite materials. The numerical simulation was based on SIMPLE (Semi-Implicit Method for Pressure Linked Equations) algorithm, staggered grid, and temperature transforming model. Three different heat fluxes of 400, 600, and 800 W/m^2 at the left and right boundaries of the computational domain are considered to simulate the heat released from the battery. They also studied different TMS's thicknesses. The time variations of battery surface temperatures are compared with different phase change materials (PCMs) to compare the effectivity of the TMS. Their results showed that the discharge time before the average battery surface temperature reaches above 60°C increases with an increasing thickness (between 7.5 and 15 mm) of the TMS (See Figure 10). The result with pure octadecane indicates that the discharge time was increased by 87 percent when the thickness of the TMS was increased from 7.5 mm to 15 mm. The surface temperature of the battery is more uniform and the discharge time is 4.7 times longer when gallium is used as the PCM, compared with those with octadecane for all thicknesses at the heat flux of 600 W/m^2 and thickness of 12.5 mm.

Another prospective technique for storing energy is the use of dielectric materials. Over the past decade, dielectric materials have attracted significant attention in terms of their energy storage applications (Bi et al., 2017; Wang et al., 2017; Dang et al., 2012). Wang et al. (2017) stated that among various physical properties, high dielectric constant, low dielectric loss and high electric breakdown strength have stood out as three essential parameters for dielectric materials. Dielectric ceramic materials can exhibit these qualities but their inherent brittleness, which is common to inorganic materials, limits their application in soft/flexible devices. Flexibility is crucial to the development of the electric charge storage capacitors with miniature, lightweight and ultrathin features required in the electronics industry. Therefore, the making of composites by combining high permittivity inorganic or conductive nanofillers with flexible polymers has been investigated as a solution (Wang et al., 2017). Generally, the energy storage density of a dielectric material U_e is related to the applied electric field E_b and the electric displacement D, given by equation 1. Electric displacement, D is dependent on electric polarization and

Figure 9. Thermal properties of hybrid graphene-PCM with respect to the Thermal conductivity of the graphene-paraffin composites with different graphene loading as the function of temperature
(Source: Goli et al., 2014)

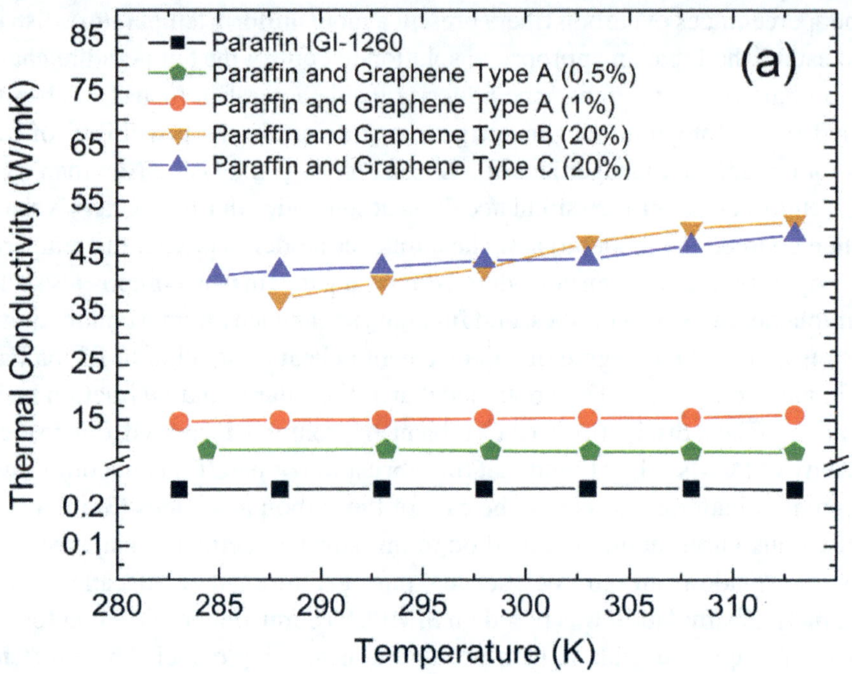

Figure 10. Surface temperature history for (a) different thicknesses (q = 600 W/m²) and (b) different heat fluxes (L = 12.5 mm)
(Source: Alipanah and Li, 2016)

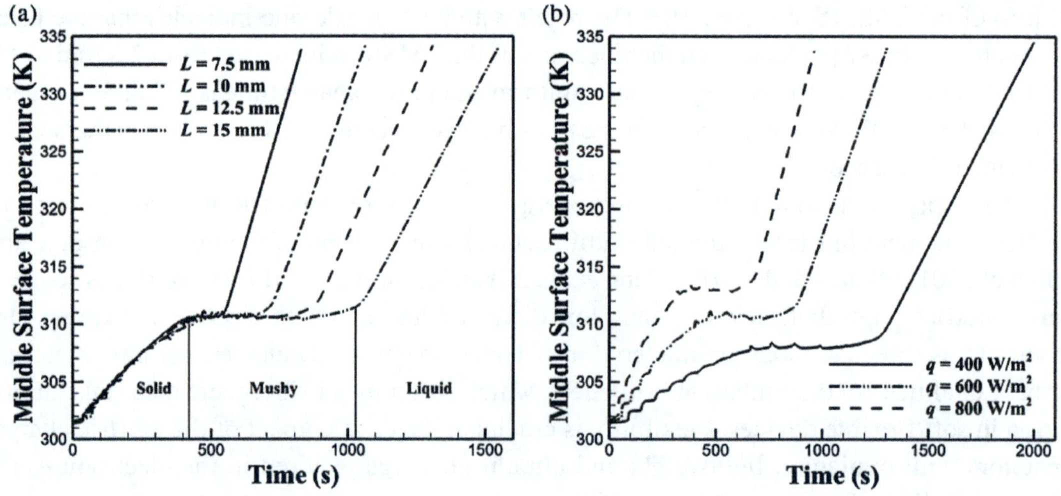

relative permittivity. Therefore, it is necessary to combine high breakdown strength and high dielectric permittivity to obtain desirable energy storage density (Zhang et al., 2017; Bi et al., 2017).

$$U_e = \int_{D_{max}}^{0} E_b D \tag{1}$$

Typical ceramic materials usually possess high permittivity but limited breakdown strength. Polymer dielectrics, in contrast, have high breakdown strength, good flexibility but low permittivity values. In order to solve these issues, nanocomposites which combine the high dielectric permittivity of ceramic materials and the potentially high dielectric strength/electric field of the polymer matrix are created. Ceramic nanoparticles are introduced into the polymer matrix to improve the electric displacement and enhance the energy density of the composites (Bi et al., 2017). The nanoscale particles are dispersed into flexible polymer matrices so that both the physical/chemical characteristics of the nanoparticles and the interaction between the nanoparticles and the polymers have crucial effects on the microstructures and final properties (Dang et al., 2013). The resulting composites have distinct electric properties such as high dielectric constant, low dielectric loss, outstanding breakdown strength, high machinability and flexibility. By careful design and preparation, these composites possess enhanced energy storage capacity and can act as energy-storage dielectrics for pulsed power systems that are used in the electrical industry and renewable energy field (Hao et al., 2015; Bi et al., 2017). Nanodielectrics is therefore a promising field with emerging applications in capacitors, gate dielectrics, energy storage, alternatives to Li-ion batteries, and frequency modulation in communications devices (Huang et al., 2010).

Calcium barium zirconate titanate ($Ba_{0.95}Ca_{0.05}Zr_{0.15}Ti_{0.85}O_3$, BCZT) ceramic particles were prepared, modified by dopamine and incorporated into PVDF to obtain composite flexible films. SEM results showed that the ceramic particles were dispersed homogeneously in the matrix. Luo and colleagues observed that the dielectric constant increased with the increase of BCZT contents, while the loss tangent remained constant in the frequency range of 10^3 to 10^5 Hz. They obtained that the breakdown strength decreased then stayed over 60 kV mm^{-1}, thus proving that the flexible film possessed high dielectric constant and low dielectric loss (Luo et al., 2014). Hao and colleagues studied these nanocomposites as energy-storage dielectrics in high-performance capacitors. They presented a gradated multilayer $BaTiO_3$/poly(vinylidene fluoride) (PVDF) thin film structure as a means to achieve both a higher breakdown strength and a superior energy-storage capability. The film was composed of a high-volume fraction of 6–10 nm $BaTiO_3$ nanocrystals and PVDF, with a significantly low volume fraction of $BaTiO_3$, at the central and outer layers respectively. The film obtained was mechanically flexible, and could be removed from the substrate, with total thicknesses in the range of 1.2–1.5 μm. Parallel plate capacitance devices displayed highly improved dielectric performances with low-frequency permittivity values of 20–25, a maximal discharge energy density of 19.37 J cm^{-3} and dielectric (breakdown) strengths of up to 495 kV mm^{-1} (Hao et al., 2015). Hao and colleagues (2017) explained that the commonly used high permittivity particle fillers in flexible polymer matrix actually cause inevitable dielectric strength deterioration and impede the energy density and reliability of the nanocomposite. They depicted that the solution to this is to employ ultimate sized ferroelectric nanofillers. They introduced ultimate sized ferroelectric nanofillers, 6.9 nm $BaTiO_3$ into a poly(vinylidene fluoride-co-hexafluoro propylene) (PVDF-HFP) polymer matrix. The introduction of smaller dispersed nanocrystals assisted in the formation of denser composite structures and finer defects in the nanocomposite films. PVDF-HFP, which helps as the matrix phase in this method, is a ferroelectric polymer that possesses a dielectric constant around 10, high dielectric strength, but large dissipation (~20%) at high frequency, as shown in Figure 11. They found that the breakdown strengths (≥330 kV/mm) in the nanocomposite films containing 10-40% $BaTiO_3$ nanofillers were significantly enhanced when compared to polymer-ceramic composites with coarse particle fillers. A maximal discharged energy density of 9.7 J/cm^3 was obtained. This confirmed that ultimate sized nanocrystals can perform as superior high permittivity fillers in nanocomposites for energy storage applications (Hao et al., 2017).

The effect of nanofillers particle size on the energy storage performance of polymer nanocomposites was further studied by Hao and coworkers (Hao et al., 2017). BaTiO3/polyvinylidene fluoride (BT/PVDF) nanocomposites with 92.3 nm, 17.8 nm and 5.9 nm BT particle fillers were fabricated and compared. It was observed that the smaller BT fillers promoted electric field uniformity and dielectric homogeneity in the nanocomposite, and thereby improved the dielectric breakdown strength and energy storage density of the nanocomposite films better than the coarse BT fillers. The maximum discharged energy density, 4.12 J cm^{-3} was derived from the BT/PVDF nanocomposite film with 5.9 nm BT fillers at a volume fraction of 30%. The study provided experimental evidence for the application of ultrafine nanofillers in nanocomposites for upcoming energy storage systems (Bi et al., 2017).

Further improvements of the energy storage capabilities of these ceramic/polymer composites can be obtained via surface modification of the nanoparticles. Phosphonic acids can serve as robust surface modifiers. Kim et al. (2007) propounded that on application of phosphonic acids, homogeneous, high-volume-fraction composites of $BaTiO_3$ nanoparticles in polymeric hosts were formed by simple solution processing. They obtained that pentafluorobenzyl phosphonic acid-modified BT nanocomposite films in poly(vinylidenefluoride-co-hexafluoropropylene) (PVDF-HFP) show large relative permittivities and unusually high dielectric breakdown strengths (Kim et al., 2007). Polymer nanocomposites were prepared by Wang and coworkers using surface-functionalized $BaTiO_3$ nanoparticles and ferroelectric polymers. The resulting nanocomposite containing 30 vol % $BaTiO_3$ at 150 MV/m achieved an energy density of 7 J/cm^3 which signified about 120% improvement in comparison with that in the neat polymer (Li et al., 2008). Wang et al. (2017) investigated novel polymer-based nanocomposites with both homogenously dispersed conductive nanofillers and ion-conductive nanodomains within the polymer

Figure 11. (a) Dielectric constants and (b) losses of the BT/PVDF-HFP nanocomposite films with various BT nanocrystal volume fractions
(Source: Hao et al., 2017)

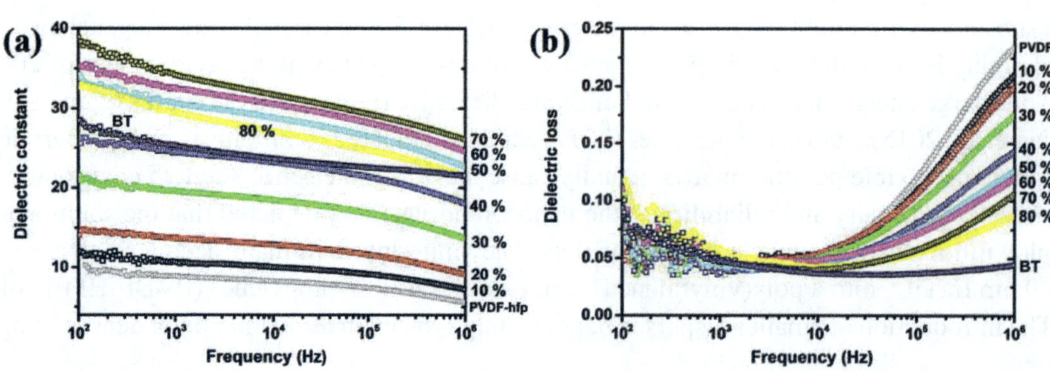

matrix. They reported that multi-walled carbon nanotubes (CNTs) were prepared and coated with ionic liquid (ILs), 1-vinyl-3-ethylimidazolium tetrafluoroborate ([VEIM][BF_4]) by mechanical grinding. The ILs coated CNTs were then well dispersed in PVDF matrix. By electron beam irradiation (EBI), the ILs on the surface of the CNTs were afterwards grafted onto the PVDF chains and these later aggregated into ionic nanodomains. The resulting PVDF nanocomposites displayed high dielectric constant, low dielectric loss and good flexibility. Wang and coworkers, explained that both carbon nanotubes and ionic nanodomains contributed to the enhancement of the dielectric constant of PVDF significantly, and also decreased current leakage which led to low dielectric loss (Wang et al., 2017).

Polyethylene nanocomposites are being considered for their energy storage capabilities. The use of nanofillers in them stops partial discharge degradation, increases voltage endurance and improves space-charge formation. Tomer et al. (2011) explored polyethylene/montmorillonite nanocomposites (PE/oMMT) and investigated how to obtain desirable dielectric behavior from them. They concluded that tailoring nanofillers orientation together with the PE/oMMT interface nature and mechanical properties can enhance high-performance for energy storage (Tomer et al., 2011).

Nanofillers as Alternative Energy

A critical issue of the 21st century is the generation of energy to meet the growing universal energy demand while preserving the environment. The International Energy Agency reports that global energy needs would expand by 30% between now and 2040. This is analogical to adding another China and India to today's global demand (International Energy Agency, 2017). This challenge necessitates the exploration of alternative energy sources that would be consequential in meeting the global energy needs. Alternatives such as wind power, solar energy, fuel cells are in continuous development.

Nanofillers have found increased use in this field of research, for example in Direct Alcohol Fuel Cells (DAFCs). A fuel cell directly converts the energy of fuel into electricity direct current in an electrochemical reaction without employing combustion (Demirbas, 2008). Power generation using Fuel cells has greater energy conversion efficiency and is eco-friendlier than other systems. Therefore, they form a power source that has the potential to replace the current battery. Fuel cells (FCs) are likely to meet the necessities for future sustainable power sources (Zakaria et al., 2016). FCs due to their characteristics are about to create significant change in the field of electricity. They may be classified based on different parameters which include the kind of electrolyte used in fuel cell, the sort of ion exchanged through the electrolyte, the type of reactants (e.g. primary fuels and oxidants), and operating temperature and pressure. Fuel cells are commonly grouped and nominated based on the qualities of the electrolyte used in them (Peighambardoust et al., 2010). Two types of fuel cells are proton exchange membrane fuel cells (PEMFCs) and alkaline fuel cells (AFCs). Due to their low working temperature, non-corrosiveness and flexibility, FCs are suitable for application in portable device. Hydrogen (H_2) is the prototypical fuel for both types of FCs because it is the most electrochemically active fuel regarding its rate to ionize at low voltage loss during reactions. However, hydrogen is not a naturally existing gaseous fuel and therefore requires an external reforming process. Furthermore, it is difficult to store and entails bulkier systems with increased production costs. These disadvantages pose serious challenges. Due to a number of attractive properties, alcohols represent another fuel that has been widely applied for FCs in ambient conditions. Alcohols because they are liquids at ambient conditions are easy to handle, store and transport. They also have higher volumetric energy density than H_2 at room temperature and pressure (Zakaria et al., 2016; Zhou et al., 2003). Alcohol fuel cells are typically direct methanol fuel cell (DMFC) and ethanol fuel cell

(DEFC) (Saisirirat & Joommanee, 2017). Wide range of research has been conducted to determine the viability of nanofillers' introduction into both types of alcohol fuel cells and the results are promising.

Poly(vinyl alcohol) (PVA) has good film-forming property and after doping with potassium hydroxide (KOH) solution facilitates good stability and efficient electrolytes. This PVA membrane separator has been well proven for various alkaline batteries and fuel cells. After being functionalized with ammonium groups, quaternized PVA (QPVA) membranes exhibited high alkaline fuel cell performance and stable energy output. The quaternized PVA (QPVA) membrane exhibited improved hydrophilicity, water permselectivity, higher ionic conductivity and reduced methanol permeability compared to a pristine PVA membrane (as cited by Kumar et al., 2017; Kumar et al., 2016; Liao et al., 2015). Incorporating organic/inorganic nanofillers into a PVA-based matrix has been shown to improve chemical stability, physicochemical, and electrochemical properties for Direct Alkaline Alcohol Fuel Cells (DAAFC) applications (as cited by Kumar et al., 2017).

A unique polymer composite electrolyte comprising of quaternized polyvinyl alcohol (Q-PVA) and quaternized chitosan nanoparticles (Q-chitosan) was synthesized for use in a direct alcohol alkaline fuel cell (DAAFC). Incorporation of 5% Q-chitosan nanoparticles into the Q-PVA matrix resulted in an increased ion-exchange capacity, decreased polymer crystallinity, higher free volume hole density and consequently, enhanced ion conduction. The resulting maximum power densities were higher than most literature data. The alkaline fuel cell demonstrated stable long-term power output for at least 230 h. Q-PVA/Q-chitosan electrolyte exhibits great prospect for imminent commercialization. (Liao et al., 2015). Yang and Lin prepared a novel poly(vinyl alcohol)/titanium dioxide (PVA/TiO_2) composite by a sol-gel method and investigated its characteristic properties. They discovered that the addition of the TiO_2 fillers into the PVA polymer matrix enhanced their ionic conductivities, thermal and mechanical properties, and dimensional stability. The alkaline DMFC with the PVA/TiO_2 composite polymer membrane exhibited superb electrochemical performance at ambient temperature and atmospheric pressure. Consequently, the maximum power density of alkaline DMFC in a mixture solution of 8M KOH + 2M CH_3OH is 18.42mW cm^{-2} at 60°C and ambient air (Yang & Lin, 2008). Kumar et al. (2016) prepared quaternized poly(vinyl alcohol)/fumed silica (QPVA/FS) based membrane and investigated its physico-chemical properties. The QPVA/FS membrane exhibited high ionic conductivity which depended significantly on the concentration of fumed silica in the QPVA matrix. They obtained a maximum conductivity of 3.50×10^{-2} S/cm for QPVA/5%FS at 60°C when it was doped with 6 M KOH. As fumed silica content increased, the permeabilities of methanol and ethanol reduced. A maximum power density of 96.8 mW/cm^2 was attained with QPVA/5% FS electrolyte using 2 M methanol + 6 M KOH as fuel at 80°C. A peak power density of 79 mW/cm^2 was obtained using the QPVA/5% FS electrolyte with 3 M ethanol + 5 M KOH as fuel. The resulting peak power densities were higher than the majority of published reports. Their results confirmed that QPVA/FS shows prospect as a future polymeric electrolyte for use in direct alkaline alcoholic fuel cells (Kumar et al., 2016).

Higher improvements were recorded with functionalization of the nanofillers. Multi-walled carbon nanotubes (CNTs) were functionalized with pendant Fe_3O_4 nanoparticles, incorporated into a PVA solution and used to prepare a cast film. The films were then doped with a potassium hydroxide (KOH) solution to prepare hydroxide-conducting solid electrolytes and applied in methanol alkaline fuel cells. It was noted that the ionic conductivity of the KOH-doped PVA–functionalized CNTs (FeCNT) composite membrane was enhanced by the FeCNT addition. This enhancement was attributed to the increased polymeric amorphous region with the ionic conducting micro-channels provided by the FeCNT. An open-circuit potential and a peak power density of 0.87 V and 87.8 mW cm^{-2}, respectively, were obtained using a 2

M methanol fuel in 6 M KOH at 60 °C with the PVA-based electrolyte containing 0.15% FeCNT. This peak power density was more than double the value observed without the FeCNT incorporation (Lo et al., 2013). Kumar at al. (2017) in a more recent study incorporated Perovskite $LaFeO_3$ nanofillers (0.1%) into a QPVA matrix for use as hydroxide-conducting membranes in Direct Alkaline Methanol Fuel Cells (DAMFCs). Two forms of the nanofillers were employed: the amorphous, as-synthesized $LaFeO_3$ nanofillers functionalized with cetyltrimethylammonium bromide (CTAB) surfactant and the crystalline, annealed $LaFeO_3$ nanofillers without CTAB. They found that the QPVA/CTAB-coated $LaFeO_3$ composite membrane showed a defect-free structure while the QPVA/annealed $LaFeO_3$ film had voids at the interfaces between the soft polymer and rigid nanofillers. The QPVA/CTAB-coated $LaFeO_3$ composite had lower methanol permeability and higher ionic conductivity than the pure QPVA and QPVA/annealed $LaFeO_3$ films. They suggested that the functionalized $LaFeO_3$ particles increased polymer free volume, contributed to the ammonium group and served as a plasticizer to enhance interfacial compatibility. They then concluded that the composite containing CTAB-coated $LaFeO_3$ resulted in superior cell performance. A maximum power density of 272 mW cm^{-2} is achieved, which is among the highest power outputs reported for DAMFCs in the literature (Kumar et al., 2017).

Another source of energy that is growing rapidly is wind energy. It accounted for the largest share of renewable growth in 2017 (Richard, 2018). The market and need for wind energy has increased in the past few decades and continuous effort is being made to maximize it. The wind turbine blade is a crucial part of the wind turbine and accounts for more than 20% of the wind turbine manufacturing. It has become necessary to develop large size and super large size (with length greater than 70 meter) wind turbine blades for the enhancement of wind power (Chinese Academy of Science, 2014). The incorporation of nanofillers into polymer matrices to improve wind turbine blades holds promise. Analysis on the feasibility and advantages of carbon nanotubes/polymer nanocomposites for wind turbine blade manufacturing shows that adding even small amounts of CNTs to wind turbine blade materials –fiber reinforced composite (FRP) component (fiber or matrix) is capable of enhancing the mechanical properties (modulus, strength, toughness, and fatigue resistance) of the matrix. It is also able to improve the interfacial interaction between the fibers and the FRP matrix. Furthermore, the researchers postulated that CNTs would amplify the lightning protection of wind turbine blades, as well as the anti-moisture and anti-heat properties and aid internal structure defects detection (Ma & Zhang, 2014; Chinese Academy of Science, 2014). Dai and Mishnaevsky Jr. (2015) explored the ability of advanced carbon/glass hybrid reinforced composites with secondary carbon nanotube reinforcement for wind energy applications using computational experiments. The results depicted that composites with the secondary carbon nanotube reinforcements (especially, aligned tubes) have higher fatigue resistances than those without reinforcements and are therefore viable for wind energy applications (Dai & Mishnaevsky Jr., 2015). Yavari and coworkers reported the incorporation of different mass fractions of graphene platelets into fiberglass/epoxy resin. They obtained that ~0.2% (with regard to the epoxy resin weight and ~0.02% with regard to the whole laminate weight) of graphene additives improved the fatigue life of the composite in the flexural bending mode up to 1200-fold. They explained that the graphene network toughened the fiberglass/epoxy-matrix interface and prevented the delamination/buckling of the glass microfibers under compressive stress. Such materials which are resistant to fatigue will be instrumental in enhancing the safety, reliability, and cost effectiveness of fiber-reinforced composites which are increasingly employed in wind energy industries (Yavari et al., 2010).

MODIFICATION OF NANOCOMPOSITE IN ENERGY SYSTEM

Chemical Modification of Nanocomposite

The use of nanocomposites cuts across several fields such as cosmetics, food packaging, agriculture, medicine, optics, catalysis, textiles, aerospace, semiconductor devices, construction and optoelectronic devices. Chemical modification of nanocomposites is useful in ensuring that nanomaterials have anti-corrosion properties, low toxicity, high level of performance, thermal stability, increase in macro hardness, resistance to scratch, adhesion resistance, flexibility and functional compatibility. Chemical modification of nanocomposites involves the chemical reaction of nanoparticles and chemical reagents to form a new class of materials which produces an improved and effective performance when compared to particles of microscale. This means that the properties of nanoparticles are fine-tuned to suit different nanotechnology applications. There exists an improvement in the electrical, rheological, thermal, optical, fire retardancy, mechanical and catalytic properties of the formed composite. These properties are dependent on the type, size, shape and concentration of nanoparticles used for the chemical reaction, and the way they interact with chemical reagents. Chemical modifications of nanocomposites are also capable of advancing engineering applications. Chemical modification of nanocomposites is capable of reducing agglomeration through bonding which occurs on the surface of the nanoparticle, bonding reduces the force of hydroxyl groups which occurs on the surface of the nanoparticle, thereby eliminating the presence of hydrogen bonds occurring between particles and preventing the formation of oxygen bridge bonds during the drying process of nanoparticles. The major aim of the chemical modification process is to make nanocomposites anti-corrosive, hydrophobic, hydrophilic and conductive (Pleşa et al., 2016). Oliveira et al. (2013) explained by experiments how the chemical modification (through acid-base reactions) of cobalt ferrite nanoparticles could increase the lipophilic nature of the nanoparticles. The acid-base reactions insert hydrocarbon chains in the surfaces of the nanoparticles which therefore increases lipophilicity in the cobalt ferrite nanoparticles without bringing about a change in the nanometric size of the nanoparticles. Lipophilicity has applications in medicine through the production of drugs used for the treatment of skin cancer. The increased lipophilic nature also makes the nanoparticles to be potentially able to act as flocculants for use in refining oil.

The different methods for chemical modification according to Habibi (2014) include:

1. **Non-Covalent Surface Modification:** Involves the adsorption of poly-electrolytes and surfactants. The adsorption process is driven by van der Waal forces or hydrogen bonds, electrostatic attractions and hydrophilic affinity. This could be applied in the production of dyes where the microbead gels are preserved by the adsorption process.
2. **Sulfonation:** A reaction involving the replacing atoms of hydrogen with sulfuric acid to form esters which have stable colloidal suspension properties making them useful as solvents in paints, lacquers and varnishes. The formation of esters is subject to factors such as temperature, acid concentration and time; making it very difficult to accurately control the number of sulfate groups which are formed. Sulfonation can be used as a pre-treatment method for promoting the mechanical disintegration of cellulose fibers into nanofibrils.
3. **TEMPO (2, 2, 6, 6-Tetramethylpiperidine-1-oxyl) - Mediated Oxidation:** Involves a simple oxidation reaction for the conversion of hydroxymethyl groups to carboxylic form. This method is very cost-effective for chemically treating nanofibers which ensures the breakdown of nanofibers

into individual particles. This is done by breaking down the hydrogen bond network of the nanofiber and reducing the hardness of the nanofiber's rigid structure. This method supports the hydration and swelling of nanofibers thereby increasing their flexibility and making their crystalline region more accessible.

4. **Esterification:** Involves the formation of ester sulfonates or acetylation which is the reaction between dry mixtures acetic anhydride and acetic acid using sulfuric acid as a catalyst.
5. **Etherification:** Involves the activation of nanofibers using sodium hydroxide or any aqueous alkali hydroxide which allows the conversion of hydroxyl groups into carboxylmethyl groups

Nanofillers can also be modified to enhance the properties of polymer. Some of the benefits of nanofillers in polymers include strong thermal conductivity, strong fluorescence spectroscopy, fire retardancy, strong electrical and mechanical properties. Jo et al. (2008) illustrated the modification of nanoclays using treated montmorillonite for the improvement of polymers (Jo et al., 2008). The treated montmorillonite causes an increase in the modulus and strength of concrete compression. Treated nanoclays involves increasing the adhesion of fillers and brings about a reduction in shrinkage which can occur in polyester resins (Schubel et al., 2006). Surface treatment of nano-oxides increases the elastic modulus, abrasion resistance, and the impact, flexural and tensile strength of polyester resin (Zang et al., 2004; Yinghong et al. 2003). Surface treatment of carbon nanotubes increases toughness in polymers (Seyhan et al., 2007). In summary, the improvement of polymers by nanofillers depends on the surface treatment of nanofillers which increases the matrix adhesion of the fillers and the dispersion of nanofillers which comes with a high amount of energy.

Nanotechnology offers excellent tools for compounding different types of materials to obtain unique properties and chemical modification of micro- and nanoparticles has been a useful strategy to prevent phase segregation during incorporation of these particles into polymer matrices. In a study conducted to investigate the effect of chemical modification on banana fiber/polyester composite, maximum conductivity was observed with methacryloxy silane treated fiber and the dielectric constant values of the treated fiber composites were lower compared to those of the untreated composites. The improved mechanical properties which were observed were linked to enhanced interaction between the fiber and the matrix. The enhanced interaction was discovered to depend on the surface polarity of the fibers modified chemically (Pothan et al., 2007). Research interests on graphene, attracted by its extraordinary structural, mechanical and electronic properties, and its applicability where high conductivities are required are being further expanded through applications of modified and functionalized forms of graphene. Structural modification of graphene through chemical functionalization result in graphene derivatives with defined structures, opening up several new possibilities of graphene application, ranging from solar cells, hydrogen storage and bio – imaging (Lonkar, 2014).

Functionalization Techniques

This involves methods of adding functional groups to the properties of nanocomposites. These techniques allow modification of nanocomposites for applications in diverse fields of medicine and technology. The chemical functionalization of nanomaterials can be achieved using different forms of organic functional groups which includes sulfonic acid, amines, thiols, alcohol and carboxylic acids and amines. Chemical modification of nanocomposites is based on (i) Functionalization of nanoclays, nanotubes, types of lamellar structure – based compounds etc using organic methods and so on (ii) copolymerization involving macromonomers, alkoxides of metal and functional organosilanes (iii) Functionalization of organic components obtained from oxides of metal.

Bhartacharya summarized the different kinds of functionalization techniques used for the chemical modification of nanomaterials (which allows nanofillers and polymers to interact). According to Bhartacharya, a polymer matrix where individual nanoparticles coated by the polymer are dispersed will bring about uniform stress distribution as a result of proper load transfer caused by dispersion on the nanofiller network (Bhartacharya, 2016). The dispersion can be realized through the addition of functional groups (Surfactants) such as carboxylic acid, peptides, poly-vinyl alcohol, proteins, starch, hydroxyl etc. to carbon nanotubes (Vaisman et al., 2006; Wang, 2009; Tkalya et al., 2012; Georgakilas et al., 2012). A nanocomposite having a maximized interfacial area between the matrix and nanofiller will reduce the mismatch between polymer and nanoparticles which increases the strength of the nanocomposites. Alignment is a functionalization technique which increases the strength and modulus of the polymer matrix of the nanocomposite but makes the nanocomposite have different physical properties in different directions (Lu et al., 2007). Hapsin – Tsai, Kerner, Mori- Tanaka and Nielsen are mathematical models that were developed for carrying out a qualitative analysis of Young's modulus of the geometry of the nanofiller, interfacial and dispersion properties in a polymer matrix of nanocomposites (Ward & Sweeney, 2004). The results of these models show that Young's modulus is directly proportional to the properties of the nanocomposites (Bigg, 1987). The interaction between polymers and fillers can be increased by using polymers as carbon nanotubes, this can be done either by grafting (where polymer chains are prepared for reaction with the nanotube surface) or by reactions of condensation, esterification, cycloaddition, sonochemical reactions and nucleophilic addition (Spitalsky et al., 2010; Bahr & Tour, 2002; Hirsch & Vostrowsky, 2005).

Several processing methods for nanocomposites include

1. **Melt Blending:** Involves heating the mixture of the nanoparticle and the polymer together to form a melt structure. The mixture between the polymer and nanoparticle assists in breaking and uniformly dispersing the agglomerates (obtained from the nanofillers) in the polymer matrix. An organic solvent is not required for processing in this method. Melt blending is very environmental-friendly and economical and is preferred for use in many industries (Villmow et al., 2008; Wu et al., 2009).
2. **Casting of Solvent or Solution Mixing:** Involves the suspension or dissolution of nanoparticles in a solution containing polymer and all components dissolved in solvents which are volatile e.g. water or ethanol. This method has applications in medical and pharmaceutical environments especially in the manufacture of films which are capable of preventing strains which could lead to damage in medical equipment (Pazourková et al., 2015).
3. **In-Situ Polymerization:** Involves the dispersion of nanotubes in monomers for the production of polymers. Nanotubes have microwave absorption properties which increase temperature and bring about polymerization (polymerization is induced by an increase in temperature). This method is very important in the production of thermostats and thermoplastic materials (Bhartacharya, 2016).
4. **Electrospinning:** Carbon nanotubes are aligned along the axis of the nanofiber using an electrode with a high-power supply which contains a polymeric solution. Factors such as ambient variables (humidity, air velocity and temperature), solution properties (conductivity, surface tension, viscosity and elasticity) and system properties (applied voltage and hydrostatic pressure affect nanofiber formation in the process of electrospinning (Pazourková et al., 2015).

5. **Layer by Layer (LBL) Assembly:** Multi-material and multi-layer thin film are assembled through the immersion of positively or negatively charged substrate in a polyelectrolyte which has a charge opposite to the charge on the substrate. The charge on polyelectrolyte is therefore accumulated on the substrate and charge equilibrium is reached. The equilibrium of charges brings about the removal, rinsing and drying of the substrate in the polyelectrolyte solution. The continuous repetition of this process brings about a desired thickness in the film which depends on the ionic strength, pH level, the polymer's molecular strength, temperature during assembly of particles and polymer's concentration in the solution. The method of LBL assembly is very versatile in the combination of different materials like clays, proteins, Carbon Nanotubes, nanoparticles, polymers etc. for the production of nanocomposites. LBL assembly also has the potential of determining the type of material structure that exists at nanoscale level to ensure that the structures are well-arranged and compact when compared to structures formed from other solution-based methods. Thin films produced from the method of LBL assembly find applications in the production of super-capacitors and Field Effect Transistors (FETs) (Bhartacharya, 2016).

Other techniques for functionalization include

- **Direct Covalent Sidewall Functionalization:** Involves the reaction of high chemical reactivity molecules such as fluorine which brings about a change in hybridization and a loss of pi-conjugation occurring simultaneously with the change in hybridization (Ma et al., 2010).
- **Defective Functionalization:** Involves the stabilization of defects (such as holes or open ends on the sidewalls of the nanotube, irregularities in the framework of the nanomaterial) occurring on carbon nanotubes. The stabilization is done by bonding using hydroxyl and carboxylic acid groups (Chen et al., 2012). The functional groups provide the modified nanocomposites with high mechanical and chemical properties by producing interfacial bonds for reaction with polymers, the modified nanocomposites can be used to initiate additional chemical reactions like esterification, grafting, alkylation, arylation, silanation and thiolation which leads to the production of more stable compounds (Liu, 2005; Mc Carthy et al., 2001). This method is very easy to use.
- **Polymer Wrapping:** Involves the formation of supramolecular compounds of carbon nanotubes through polymers wrapped around carbon nanotubes which occurs when carbon nanotubes are suspended in polymers. The interaction between van der Waal forces and the pi-pi stacking occurring between carbon nanotubes and polymer chains having aromatic rings is responsible for the wrapping of polymers around the carbon nanotubes (Zhi et al,. 2005; Yi et al., 2010). This method assists in tuning the interfacial properties of nanotubes.
- **Endohedral Method:** Involves the use of capillary effect for the storage of guest molecules in the inner cavity of carbon nanotubes. An example of endohedral method of functionalization is the insertion of inorganic molecules (such as silver, platinum, fullerene, gold etc.) into carbon nanotubes and the use of adsorption for the entrapment of biomolecules like protein in the nanotube's hollow channel. This method is useful for excellent performance of hybrid materials in nanotechnology, catalysis, molecular scale devices and energy storage (Ma and Kim, 2011).
- **Latex Technology:** Involves the separate dispersion of polymer particles in a colloid solution. The carbon nanotubes are first dispersed in a solution of aqueous surfactants, after which they (the dispersed carbon nanotubes treated with surfactants) are mixed with polymer matrix. The mixture then produces a nanocomposite (made up of carbon nanotubes in a polymer matrix) after under-

going freeze-drying and melt-processing. Because water is the solvent used in this method, latex technology is deemed to be environmentally-friendly, cheap and safe. Other advantages of this method include reproducibility, versatility and reliability. Due to the method basically consisting of a simple mixture of two components, it is very easy to use and allows separate carbon nanotubes to be integrated into a polymer matrix that is highly viscous (Grossiord and Hermant, 2016).

CHARACTERIZATIONS OF NANOCOMPOSITES USED IN ENERGY SYSTEMS

Different types of instruments with different levels of sophistication are used in the characterization of nanocomposites to obtain their shape, bonding environment, surface structure, chemical composition, absorption, properties of diffraction and scattering, electron band gap, actual size, light emission, and valency. They include Scanning Electron Microscopy (SEM), Transmission Electron Microscopy (TEM), ultraviolet and visible spectroscopy (UV-Vis), transmission electron microscopy (TEM), Field Emission Transmission Electron Microscopy (FE-SEM), X-Ray Diffraction (XRD), Fourier Transform Infrared Spectroscopy (FTIR).

Scanning Electron Microscopy (SEM)

Scanning electron microscopy involves the use of an electron microscope for the scanning of specimen samples with a beam of electrons focused on the samples to produce a high-quality resolution image of the sample which has a three-dimensional appearance. SEMs make use of electron beams which is focused on the examined specimen rather than a beam of light. The electron microscope makes use of an electron gun to expose beams which are highly concentrated in electrons. A vacuum chamber is usually required for an SEM to function as the vacuum chamber prevents the obstruction of an electron beam when the beam moves through the body of a microscope. Within a vacuum chamber, the electron microscope has a series of lenses which is used for the maximization of efficiency by directing the electrons towards the samples of the specimen. SEM uses primary backscatter electrons and secondary electrons which are X-rays emitted when a beam of electrons hits a sample specimen.

A major challenge always encountered since the innovation of scanning electron microscopes (SEM) is the contamination of the microscope by the electron beam-induced specimen which can be caused by outgassing of internal component parts of the SEM, preparation and handling of the sample, and the pumping system. Postek outlined three ways to ensure the reduction of contamination in a scanning electron microscopes (SEM) (Postek, 2006). The approach includes:

1. The use of a dry nitrogen purge system near the mechanical pump of the SEM to maintain the foreline of the microscope at the point of measurement in order to prevent the backflow of oil molecule which causes backstreaming (a type of contamination). The dry nitrogen purge system designed by Postek is very straightforward and easy to manufacture, the system also has digital thermocouple gauges for gas flow monitoring and provides a guideline for pressure adjustment during the procedure of nitrogen purging. The system is easy and cheap to design.
2. Replacing the liquid nitrogen cryotap with a standard liquid nitrogen cold trap which allows the maintenance of the liquid nitrogen for a minimum of 24 hours on one liquid nitrogen filling. The

gold-plated interior surfaces of the liquid nitrogen cold trap which faces the vacuum space reduce heat transfer by radiation.

3. The use of a nitrogen purge system which can reverse an SEM contaminated by backstreaming through the removal of oil backstreams using the scrubbing or nitrogen purge procedure. Under conditions of purging, the viscous gas flow desorbs hydrocarbon contamination from the surfaces of the specimen. This ensures that the vacuum pump line and all parts of the SEM are properly cleaned.

The thickness of the layer which causes contamination in SEMs can be recorded using the signal of backscattered or secondary electrons. The signals of Secondary electrons can also measure the effect of electron beam energy, specimen cooling and current density on the rate of contamination (Hirsch et al., 1994). Hirsch et al. proposed four methods for treating specimens to reduce contaminations in SEMs (Hirsch et al., 1994). They include:

- Heating of specimens to about 200°C
- Cooling an anti-contamination blade around the specimen with liquid nitrogen
- Cooling of specimen and anti-contamination blade to a low temperature which is the same for both entities
- Washing the specimen overnight with methanol or ethanol to drastically reduce deposited organic molecules from the atmosphere

Nanotips are known to produce high-resolution images with effective signal-to-noise ratio (SNR) in SEMs because of the presence of sharp emitter electron tips in their structural makeup. Vladár et al. summarized the possibilities presented by the use of nanotip electron emitters in Scanning Electron Microscope (SEM) (Vladár et al., 2006). The use of nanotips as an electron source improves the size and brightness of the electron beam which therefore improves the performance of the SEM. According to Vladár et al., nanotips are potentially able to:

- Enhance image resolutions and ensure the improvement of the SNR.
- Cause changes in the electron optics of the SEM column when it replaces the original field emission tip.
- Adjust the excitation of the condenser lens in order to make the crossover points of the machine identical to bring about increased optical brightness in the SEM which positively affects the probe size.
- Decrease the size of the final probe of the SEM at an accelerating voltage of 1 kV and operating voltage of between 150 and 500 eV. A small probe size positively affects the image resolution.
- Delivery of a primary electron beam current which is six times higher than that of the conventional S6000 gun into a reduced probe size which would bring about an improvement in the SNR by at least a factor of two.

The possibility of having the imaging of secondary electron in a variable pressure scanning electron microscope is very low as a result of the shortness of the mean free path of the gas secondaries which gives the secondary electrons access to reach the detector. Mohan et al. (2006) reviewed an alternative method (which involves the modification of the electron microscope by adding a biased electrode at a level above the sample in order to obtain a resulting specimen current signal) of image production which contains substantial amounts of secondary electrons alternative strategy for producing an image containing significant amounts of secondary electron (Mohan et al., 2006). The method involves the use of a detector system which detects ion signals to produce images. The detector system uses a specimen current amplifier (SCA) (placed in the line that connects the sample to ground) to collect signals. The system also has an annular ring (which has a positive bias potential in the range of 200–400 V) that produces a field above the specimen. The biased specimen-current detector is capable of producing images which are similar to those obtained by conventional detectors in SEMs. The detector system does not interfere with any of the other modes of operation involved with the scanning electron microscope.

Transmission Electron Microscopy (TEM)

Transmission electron microscopy (TEM) is a technique of microscopy which involves the transmission of a beam of electrons through an ultra-thin specimen to bring about an interaction with the specimen as the beam passes through. The interaction of the electron beam with the specimen produces an image which is magnified and focused on a device for imaging. TEMs have the ability to produce images with resolutions that are substantially higher than those produced by light microscopes, which provides the user of the instrument with the opportunity to examine an important detail which can be as little as the single column of atoms. Advanced TEMs have the capacity to produce high spatial resolution of 0.05 nm and high energy-resolution of 7 meV.

There are three types of techniques involved in the use of TEM (Su, 2017), namely:

- **Techniques of Imaging:** This includes scanning transmission electron microscopy (STEM) imaging, 3D electron tomography technique, and high-resolution transmission electron microscopy (HRTEM).
- **Techniques of Spectroscopy:** This includes Electron energy-loss spectroscopy (EELS), TEM-cathodoluminescence microscopy (TEM- CL) and X-ray energy dispersive spectroscopy (EDS).
- **Techniques of Diffraction:** This includes nano-beam electron diffraction (NBED), convergent beam electron diffraction (CBED) and their combinations, and selected area electron diffraction (SAED).

It is publicly known that Size distribution in nanomaterials varies inversely with the portions occupied by the surface atoms. X-Ray diffraction can be used to calculate the average particle size distribution of nanomaterials using line-broadening fitting with Scherrer equation, and averaging the results over the entire sample (Su, 2017). The localized information produced by TEM/STEM images can also be used to calculate the size distribution by counting the number of particles having different sizes respectively (Liu et al., 2012).

X-ray energy dispersive spectroscopy (EDS) or electron energy-loss spectroscopy (EELS) can be used to evaluate the compositional distribution in nanomaterials. STEM-EELS or STEM-EDS properly characterizes the composition of materials in nanostructures. This could be done using EELS or EDS detectors which collect signals for scanning electron probes in the nanocatalyst using 2D mapping or 1D line profile modes to obtain the elemental distribution. The method of STEM-EELS or STEM-EDS is used in industries to investigate the electrode materials which are present in batteries where valence change occurring 3d metals brings about redox reaction during charge/discharge (Keast et al., 2001).

Phase in the structural point of view means a lattice with uniform sequence and uniform parameters. Phase identification in crystalline structures having a relative large structure with grain size greater than 5 nm can be carried out using selected area electron diffraction (SAED) which is very powerful for local phase identification in crystals (Ding et al., 2010), while crystalline structures having small grain size and poor crystallinity will have their phase identification carried out by a combination of High Resolution- scanning transmission electron microscopy (HR-STEM) imaging and STEM-EELS mapping (He et al., 2015).

The shape and surface aspects of a structure represent the term "morphology" in nanomaterials. Nanoparticles (0D), nanowire / nanotube / nanorod (1D), and nano-sheet / nanoplates (2D) represent different kinds of morphology in nanomaterial structure, which allows nanocrystal activities to be controlled and therefore ensures that the catalysts are stabilized (Xia et al., 2008). The 3D morphology of nanoparticles can be decoded using the following techniques.

- Reconstruction of a 3D structure from a series of high-resolution transmission electron microscopy (HRTEM) images having a different focus (Van Aert et al., 2011)
- Reconstruction done directly from modelled (HAADF)-STEM (high-angle annular dark-field imaging - scanning transmission electron microscopy) atomic image resolution (Van Aert et al., 2011).
- Titling the sample or specimen using titling the specimen (Jinschek et al., 2008).

The interaction of matter (either electrons or nuclei) and an electron beam of TEM with high voltage brings about particular structural changes in the TEM specimen (Egerton, 2013). These changes are known as Electron-beam induced damage and can either be temporal or permanent occurring in bulk or on the surface (Jiang, 2013). Samples composed of elements or molecules (like organic materials, alkali, water etc.) and some techniques (like 3D electron tomography and STEM-EELS mapping or STEM-EDS) requiring a lengthy time of collection, suffer severe form electron-beam induced damage. The total rate of damage is dependent on the time of irradiation, acceleration voltage, the sample geometry/ composition and the dose of the electron beam (decided by the intensity of the electron beam) (Su et al., 2013). The damage on the electron beam can be reduced by minimizing the intensity of the electron beam and lessening the time of illustration (which represents the dose and rate of dose respectively). An electron beam has the potential of promoting the transitions of phase in nanoparticles in some cases, the change in TEM/STEM images or EELS can be used to monitor these transitions which lead to damages. The electron-beam induced damage is fundamentally a limitation to the production of resolutions and even the possibility of most techniques based on TEM (Jiang et al., 2012).

The natural components of an interface are very essential in determining the properties of epitaxial thin films which are used for purposes of harvesting energy. Interfaces can also determine the behaviors of alloys, coreshells, electrode materials and composited catalysts in batteries. Vibrational spectroscopy, X-ray absorption spectroscopy, and atomic force microscopy are known comprehensively analyze structures of interfaces but TEM provides novel information as regards atoms and molecules (Hÿtch et al., 1998). TEM images and SAED patterns give solid identification of interfaces in nanomaterials. SAED patterns and high-resolution images can identify strains (which are biaxial and large) which affect the performance of batteries (Johnson et al., 2008).

The principle of in-situ electron microscopy involves a technique of investigation in which a sample's reaction to stimulus in real time is checked using an electron microscopy. Information on the response of a material's structural dynamics at a high spatial resolution can be supplied by In-situ TEM (Jakubovics, 2008). With In situ TEM, great opportunities exist in form of characterization of mechanical properties under the application of stress, heat, optical excitation, and magnetic or electric fields, switching and behavior of ferroelectric and ferromagnetic, and characterization of solid-state batteries (Zheng et al., 2015). The reactions of gas-solid interfaces can be studied using direct information on catalyst mechanisms which are provided by in situ TEM would give the direct mechanistic information on how catalysts work (Sharma, 2001), however, In situ TEM encounters challenges with the introduction of gas into the specimen area of the microscope as it experiences difficulties in maintaining the high vacuum existing at the TEM column as a high vacuum is needed by the high energy electron beam to prevent an unwanted scattering of gas molecules. This challenge can be solved using an environmental TEM which is specially equipped pumped vacuum systems (which ensures that a high vacuum is maintained at the electron source while the gas is pumped into the specimen area) and sealed gas cell which is integrated with a special TEM holder (Wagner et al., 2012).

Field Emission Transmission Electron Microscopy (FE-SEM)

Field emission scanning electron microscopy (FE-SEM) with magnifying abilities of imaging and observation between 10x to 300,000x provides information on elements and topography, having a virtual depth of field which is unlimited. To obtain an FE-SEM, a field-emission cathode is placed in the electron gun of a scanning electron microscope for the provision of probing beams which are narrower at low and high electron energy which produces spatial resolutions which are improved and minimizes sample damage and charging. FE-SEM produces images which are clearer and less electrostatically distorted (as low as 1.5 nanometers making 3 to 6 times better than SEM) as compared to the scanning electron microscopy (SEM). FE-SEM has the capacity to carry out the examination of contamination spots (occurring in small areas) at accelerating voltages of an electron which is consistent with energy dispersive spectroscopy (EDS). FE-SEM ensures that the penetration of electron probes with low kinetic energy which is closer to the next material surface occurs at a reduced rate. FE-SEM has the ability to produce images having high quality and low voltage (which ranges from 0.5 kV to 30 kV) with electrical charging of samples which is negligible. FE-SEM has the ability to provide protection on insulating materials by ensuring there is no need for using conducting coatings. FE-SEM can be applied in 1) Analysis of cross-sections for film thicknesses, construction details, gate-widths and gate oxides in semiconductors; 2) Determination of structural uniformity and provision of coating having advanced thickness; 3) Measurement of elemental composition and provision of feature geometry for small contamination; 4) Evaluation of laser and weld resistance; 5) Analysis of Printed and Integrated Circuits; 6) Characterization of fracture in polymers and very small components; 7) Study of microstructure.

Bandurin et al. (2013) used the principle of FE-SEM to study the field emission properties of nanographite films of carbon nanotubes. Voltage maps which were measured at a constant current of 10nA at a varying distance for the same film area were used to determine the emission sites density for nanographite films. The maps are made up of emission sites which are round in shape having a central part which corresponds to voltages which are low in amount and peripherals which correspond to voltages which are high in amount. The average size of the emission sites is directly proportional to the distance of the anode-sample. The value of the field enhancement factor and its statistical dispersion is directly proportional to the thickness of the film. The screening effect and strong overlapping of nanotubes present in the films cause the field enhancement to vary (Bandurin et al., 2013).

FE-SEM is also very valuable as an approach used for the visualization of biological structures at nanoscale such as DNA material and cell organelles. Yao and Kimura (Yao & Kimura, 2007) analyzed the characterization of constructed 3D superlattices using FE-SEM. Observations from FE-SEM were used to analyze and characterize the morphology and packing structure of the superlattices. The observations of Morphology obtained from FE-SEM highlighted a variety of well-defined, plate-like and complicated shapes existing in the 3D nanoparticle (superlattices). Symmetric fivefold structures like decahedron, pentagonal rod and icosahedron were most likely developed by multiple twinning. According to observations from FE-SEM images, the symmetric structures presented nanoparticles whose arrangements were excellent and properly-ordered. This FE-SEM therefore provides a new development in the field of nanotechnology by making new research findings in the very wide and complicated field of fivefold twinned structures. FE-SEM measurements are very important for analysis as they are very suitable for use when observing nanomaterials which are based on carbon in situations which do not require a high beam of energy as electron with a small amount of energy easily pass through materials which are light and are based on atoms.

X-Ray Diffraction (XRD)

X-Ray Diffraction as a method of characterization has a traditional history (which is still very relevant at present) of identifying crystalline forms or phases which are present in particles and materials. XRD remains the most convenient means of measuring crystallinity which is a fundamental property in materials such as glass ceramics and polymer. XRD refines crystal structure parameters (which is considered a basis for relationships of structure and property e.g. formation of solid solutions) which provide a comprehensive and comprehensive information about the crystalline nature of the material. Detailed and comprehensive knowledge of material structures which are present at the atomic level is required for the proper determination of the properties and characteristics of a material. Conventional X-Ray diffraction methods for analysis in ordinary materials can be used to provide information on the crystalline nature of materials. These methods basically involve the measurement of distances of material atoms with reference to the wavelength of X-Rays. Since the inception of XRD, there has been development and application of several XRD techniques to different materials ranging from simple solids to complex proteins. XRD has in many cases used for the successful exploration of nanomaterial structure. XRD ensures the use of materials showing periodic atomic properties at long range for grating purposes. The structural coherence in length of several nanometers by Nanomaterials shows that nanomaterials fall between crystals and non-crystals. The structural coherence in nanomaterials which is limited in the degree of structural coherence is linked to the proper ordering of atoms in few nanometers distance, this therefore leads to grating in nanomaterials causing the provision of Bragg peaks and diffuse components which are brought about by the production of XRD patterns.

The number of sharp Bragg peaks are low in XRD patterns produced by nanomaterials as compared to XRD patterns produced by regular crystals. However, XRD patterns produced by nanomaterials provide diffused components which are very strong and possess similarities with those of non-crystal materials. The measurement of the positions and intensities of the Bragg peaks can be used to determining the spatial characteristics of grating in nanomaterials which therefore allows the arrangement of atoms in three-dimensional structures to be studied and analyzed. The techniques involved in XRD are not only versatile but equally non-destructive and allows the provision of comprehensive and detailed information on materials as regards crystallography, types of synthesis, microstructures and chemical composition.

The method of XRD has shown itself to be a very important tool for research in the study and analysis of nanostructures and has therefore attracted significant interests in research when compared to other techniques of microscopy due to its accuracy it provides as regards information relating to the location and positioning of atoms in space.

Mocuta et al. (2008) used the principle of XRD with the combination of focused X-Ray beams for the localization of nanostructure in order to carry out analysis on their composition and strain. Their method involved the combination of local resolution with composition sensitivity and strain which can be used for both buried structures and free surfaces as the method needs no preparation for sample, is sensitive to depth and is non-destructive. With this method, the combination of XRD in complement with microscopy techniques (such as SEM, near-field microscopies, atomic-force microscopy and micro-photo luminescence) on the exact same nano or microstructure is a huge possibility. Their method in summary properly resolved and compared the differences between strain, chemical composition and morphology of individual objects which other techniques of microscopy find difficult to achieve. Lynch et al. reviewed the properties of the high-spatial resolution produced by a microdiffraction laboratory system which uses spot capabilities (which are small in size) of an X-ray source (with a focus at micro-scale) having a gain in flux obtained from the optic of a single bounce capillary (Lynch et al., 2007). The continuum radiation from the X-ray source was used to obtain polychromatic data. The polychromatic data in combination with the high-spatial resolution enabled measurements of diffraction to be carried out by the system which was used to study the strain and orientation of the individual grain in a polycrystalline magnesium alloy using the indexing software package named X-Ray Micro-Diffraction Analysis Software (XMAS).

Fourier Transform Infrared Spectroscopy (FTIR)

FTIR is a method of infrared spectroscopy which has wide chemical applications ranging from polymers to organic compounds. The process of FTIR involves the passing of infrared radiation through a sample (this means the sample is directly placed into an infrared beam) in which the sample absorbs some radiation and transmits some radiation. The signal produced by the radiation at the FTIR detector is a spectrum which represents the sample's molecular fingerprint. The different spectral fingerprints produced by the spectroscopy is very useful in chemical structures and molecules. The Fourier Transform aspect of FTIR basically converts the output of the detector to a spectrum which can be interpreted and understood. The FTIR is therefore known for generating spectra patterns which provide a structural and compositional understanding of materials (Duygu et al., 2009). The FTIR uses the process of interferometry to keep record of information on any material positioned in the infrared beam. With the results obtained from Fourier Transform, analysts are provided with the opportunity of identifying or quantifying the material. The decoding of interferograms in spectra which can be recognized results in the production of FTIR (Rocha et al., 2017). The spectra produce patterns which assist in the identification of the sample as specific infrared fingerprints are exhibited by molecules. FTIR has four major techniques for sampling which includes

- **Transmission:** This technique involves the generation of a spectrum resulting from the measurement of energy which is transmitted when the infrared beam passes through the sample. The rate of quality which can be obtained in samples using the technique of transmission is in most cases excellent. The technique of transmission can be used separately or in combination with gas or liquid cells and microscopes for the analysis of thin films of polymer, thermoplastic powders, soluble polymers, prepared polymers with regular and irregular shapes, viscous or free flowing liquids, organic powders, gases and dark films made from polymer. The cells and mounts used in transmission are generally cheap making the technique very economical. The technique has a long history of use in sample measurement making it well-established. The spectral information it produces in most or all cases are excellent making it a very convenient technique for carrying out quality measurements. Many procedures which are considered as standard are based on the technique of transmission making it a superior option for works which are quantitative (Yandao et al., 1998).

- **Attenuated Total Reflection (ATR):** An ATR involves the measurement of changes that take place in an infrared beam (which is internally reflected) that has contact with a sample. In ATR, there is a direction of the infrared beam to a crystal having an optically dense nature and a refractive index which is high at a particular angle. An evanescent wave is then produced by the internal reflectance stretches its length towards the sample which is in contact with the crystal. There exists an attenuation of the evanescent wave in infrared spectrum regions where there is absorption of energy by the sample. After this action, the beam which was attenuated leaves the crystal and gets a direction to the detector located in the infrared spectrometer. The infrared beam which was attenuated is then recorded as an interferogram signal by the detector. It is this interferogram signal is used for the generation of an infrared spectrum. ATR works very conveniently for samples which are thick or have strong rates of absorption, the peaks produced by these samples are usually very sharp and strong when they are measured by the technique of transmission. The exponential decay of intensity which occur with distance from ATR crystal surfaces in evanescent waves produced by samples makes the technique of ATR insensitive to thickness. This therefore makes ATR very convenient for thick samples. ATR is also convenient for solids such as plastics, laminates, rubbers, natural powders, diamonds, paints and coatings. ATR can also be used for analysis in biological materials, viscous liquids and free-flowing aqueous solutions as analysis of liquids requires the placing of just a drop of liquid on the crystal. The preparation of samples in ATR is very easy as it involves the placement of the sample on the crystal and collection of data. At the end of the process, cleanup is easy and fast as it involves no stress other than the removal of the sample and cleaning the crystal surface. The samples are analyzed in their natural states which involves no heating and grinding in order to obtain spectrum. The method of ATR is the best for samples with strong rate of absorption (Grdadolnik, 2002).
- **Specular Reflection:** The technique of specular reflection uses the theory of reflective efficiencies for operation. This theory of reflective efficiencies states that there is a variation which exists between the refractive index and the frequency of light (to which a sample is exposed to) in any sample. Specular reflection evaluates the refractive index of a sample or the energy reflected off a sample surface rather than investigate the energy which goes through the sample. Assumptions concerning the sample's nature of absorption can be made by investigating the bands of frequency using qualitative data which are produced by specular reflection. Specular reflectance is very convenient and frequently used for the analysis of both organic and inorganic samples with metallic surfaces, silicon wafers, materials laminated on metals and thin films existing on reflective substrates. Specular reflection has the ability to detect thick coatings existing on metal substrates making it sensitive to monolayer samples. The analysis has no destructive tendencies as no contact is made which therefore prevents damage to the sample. Combination with other microscopy techniques is also possible but all depends on the sample size and surface layer thickness (Rytter & Einarsrud, 1988).

- **Diffuse Reflectance:** The focus of an infrared beam on a fine particle matter material causes interaction with the particle either by reflection or radiation off the particle's top surface without the particle being penetrated or multiple reflections off surface particles by light without the particle being penetrated. The penetration of sample particles by radiation incident the surface which causes the sample matrix to be scattered subsequently results to diffuse reflectance. A Diffusion reflectance microscopy operates by giving direction to the infrared energy into a sample. An interaction between the radiation and particles which causes a reflection of the infrared radiation off the particle surfaces, which results in a diffusion of light throughout the sample. The diffused light is then directed to the spectrometer detector which records the attenuated infrared beam as an interferogram signal which generates a spectrum. Proper preparation of sample with this method ensures an excellent collection of quantitative and qualitative data. Due to pathlength which is longer in techniques of ATR and transmission, ATR and transmission techniques are better than diffuse reflectance in obtaining quantitative data. Diffuse reflectance can be used to analyze organic and inorganic samples such as tablets, rigid polymers, powders, varnish surfaces and paint. Diffuse reflectance involves very simple preparation, easy cleanup and is a relatively easy process (Leyden & Murthy, 1988).

The application of FTIR ranges from being used as a single purpose tool to a research machine which is highly flexible. With FTIR configurations, unknown samples can be identified, quantitative information on additives or contaminants can be obtained, information on growth or decay existing in infrared absorptions can be obtained and complex information involving attachment with other devices can be obtained. FTIR can therefore be summarized as cost-effective information machine (Maria, 2012).

APPLICATION OF NANOCOMPOSITES IN ENERGY SYSTEMS

The distinctive physical and chemical properties of nanocomposites which have independent capabilities and are deficient in individual components have brought about research interests as a result of their potentials for wide-ranging applications in energy systems related to electromagnetic absorption, energy storage, information, environmental remediation, novel catalysts, transportation and safety, sensing and actuation, industry and defense systems. Polymer nanocomposites can be used in energy saving (for the capture of carbon dioxide and formation of electrochromic devices), anticorrosion (for application in conductive and non-conductive polymer nanocomposite with anticorrosive coatings) and energy storage (for application in electrochemical capacitors and batteries) (Yang et al., 2015).

Nanotechnologies in Solar Cell

The use of Nanotechnology is very valuable in the design and manufacture of thin-film photovoltaics cells. The use of nanomaterials like quantum dots, radial junctions and nanowires are beneficial in solar cell technologies as they increase the efficiency of photovoltaics (Tsakalakos, 2010). With the use of Nanostructures, more conventional and cheaper materials like titanium dioxide and silicon can be used to produce more efficient solar cells (Jadhav et al., 2011). Cheaper raw materials reduce the cost of solar cells in the market thereby reducing any cost limitation involved in the development of techniques used in the mass-production of nano-enhanced PV cells. Nanomaterials are used for the production of flexible rolls which replaces crystalline panels (which are inflexible in nature). These flexible rolls reduce the costs of installation as they are light in weight (Kim et al., 2015). In addition, the application of ultraviolet light in conventional solar cells causes the production of damaging heat rather than electricity through the absorption of the ultraviolet light by the silicon in the solar cell. The application of nanoparticles with solar cells leads to a 60% improvement of the cell's performance in power within the range of the ultraviolet spectrum (Sethi et al., 2011).

The use of quantum dots reduces the wavelength of the ultraviolet light (which increases the efficiency as lights with short wavelength pass through the cell more effectively than lights of longer wavelength) which therefore increases the solar cell efficiency through the adjustment of band gaps. Carbon nanotubes equip the solar cells with a high capacity of current density and a superior property of speedily transporting electrons along the axis of the solar cell surface without any serious form of loss. The conversion of longer photovoltaic wavelengths to shorter ones is higher in efficiency when carbon nanotubes are properly aligned with polymer composites. The contact area for conversion of energy and better transfer of charges is increased with the use of polymer nanocomposites. The use of group 3 and group 4 chemical element semiconductors and other combinations of materials like Silicon/Berylium, Telurium/Selenium and Silicon/Germanium can be applied in quantum dot Solar cells (Raffaelle et al., 2002). The advantages involved in the use of quantum dot solar cells include (Romero & Al-Jassim, 2003):

- Production of silicon solar cells which have high quality and are compatible for long-term use.
- Absorption of a higher amount of light in the region containing infrared spectra.
- Improved hardness in radiation which is better than conventional solar cells.
- The photo currents are increased at higher temperatures.

In summary, the real-life application of nanotechnologies in solar cells include

- The preservation of the environment using cheap solar cells made up of nanomaterials.
- Reduction of pollution and reliance on electric grids powered by fossil fuel through the coating of roofing materials with cheap photovoltaic cells made up of nanomaterials.
- Solar cells with flexible rolls used for producing a clean, green and efficient source of energy which reduces carbon gas emissions.
- The use of nanostructure flexible rolls which are light in weight for producing light weight solar military equipment especially batteries which are used by soldiers in special operations which ensures speed and increased mobility of the soldier.

Zhang et al. (2009) reviewed the application of Zinc Oxide (ZnO) nanostructures in Dye-sensitized solar cell (DSC). DSC is a system based on light, electricity and chemicals which combines oxide films with molecules of dye for photosensitive electron reactions. Dye-sensitized solar cells consisting of ZnO and dyes (either organic or complex dyes) are capable of reducing costs of production as a result of their high efficiency in solar to energy conversion. The advantages of DSCs as photovoltaic devices range from being practically high efficiency and cost-effective devices. DSCs can form interconnected networks (which are porous in nature) for an increase in the surface area of films thereby increasing efficiency in conversion (Zhang et al., 2009).

Nanocomposites for Energy Storage Applications

Innovations in the field of nanotechnology now largely contribute to improvements in energy conversion, transmission and storage. Currently, various forms of research and development in nanotechnology are addressing the development and improvement initiatives in rechargeable energy storage devices (Serrano et al., 2009). The applications of nanotechnology in area of energy storage include (Kalaiselvam & Parameshwaran, 2014):

- The use of nanostructured separators and electrode materials for mass storage in batteries and supercapacitors. For example, the application of the nanostructuring process for capacity optimization in lithium-ion batteries can ensure mass storage in automotive drives. Also, electrolytic systems (Separators) and electrode materials made from nanocomposites can be used to develop high-efficiency batteries such as lithium-sulphur and lithium-air batteries. Electrode materials made from Carbon nanotubes can produce capacities having a high amount of charges in batteries and supercapacitors.
- Nanoporous metal-organic structures can be used for mobile gas storages such as fuel tanks.

Electrolytes are very important for the storage capabilities of lithium-ion batteries as they enable the movement of lithium ions between the regions of the anode and cathode. The use of polymer nanocomposites for the production of electrolytes can ensure quality properties in the lithium-ion batteries (Cho et al., 2017). Some of these properties include anti-corrosion in batteries, the absence of leakage in batteries, full construction of batteries in solid state, the high energy density in batteries, batteries can have different varieties of shapes and sizes, and there is a very thin possibility for the occurrence of internal short-circuits which means greater safety in the battery. The use of inorganic nanoscale fillers in electrolytes which are free from solvents increases conductivity in electrolytes thereby improving the transport qualities in electrolytes (Prabaharan & Michael, 2013). Polymer nanocomposites can be used in lithium-ion batteries which are applied to flywheel energy storage, pumped hydro storage, compressed air energy storage and thermal energy storage (Daberkow et al., 2013). Intrinsically conductive polymers contain organic functional groups which are electroactive and have oxidation-reduction reactions which are faster in kinetics than the inorganic electrode materials found in conventional lithium-ion batteries, the polymer chain structure of the intrinsically conductive polymers is flexible and more stable than hard crystal particles during the fast process involving charging and discharging, these intrinsically conductive polymers which form polymer nanocomposites provide the lithium-ion batteries with high capacity, long cycling life, high working voltage and low toxicity (Shi et al., 2015). The presence of poly (3,4-ethylenedioxythiophene) also known as PEDOT in polymer nanocomposites is potentially able to reduce the process of electrode polarization and can elongate the discharge plateau above 3.0volts when applied to lithium-ion batteries thereby bringing about high voltage capacity and energy density in the battery (Xing et al., 1997). Intrinsically conductive polymers nanocomposites can form a very complex network having a conductive structure which is porous and unique, this network is very profitable as it improves the cell performance by ensuring a high rate of performance and long-term stability in cycling (Shi et al., 2015).

Supercapacitors are highly important in providing system support in terms of voltage in all cases of increased loads which occur in portable equipment and electric-powered vehicles (Salunkhe et al., 2014). The design and manufacture of batteries and supercapacitors are very similar as they both have two electrodes, electrolyte and separator. Supercapacitors are designed having long life cycle (which 100 times greater than the life of a battery) and high level of power, these two qualities prevent supercapacitors from possessing the quality of high energy density and cyclic stability (Xu et al., 2017). The use of nanoparticles such as nanopowders and nanostructured materials (such as nanotemplates, aerogels and carbon nanotubes) can ensure energy density in supercapacitors. The use of nanomaterials which are pseudo-capacitive, having high surface area electrodes and high-level electrochemical performance replace the negative electrodes in supercapacitors to bring about high energy density and cyclic stability in supercapacitors (Goikolea & Mysyk, 2017).

Fuel cells are very efficient for energy conversion and storage applications in portable power sources and transportation (Ahluwalia & Wang, 2008). However, fuel cells have limitations such as the need for expensive platinum materials for use as catalysts in the electrodes of fuel cells, the specific need for hydrogen (N.B: any fuel cell other than the one made from hydrogen can cause electrode damage) which has expensive cost of production and storage, and lifespan reduction as a result of component corrosion due to operation of the fuel cell at high temperature (Burchardt et al., 2002). The use of nanostructures such as carbon nanotube for supporting the platinum on a porous surface, makes the platinum surfaces more accessible thereby reducing the amount of high-cost metal required to make the catalytic electrode effective. Also, carbon nanotubes with modified properties (through doping with nitrogen or using polymers so they can function effectively as catalysts) can act as a replacement for platinum thereby reducing the cost of the fuel cell by 25% which is the percentage in cost which platinum accounts for in fuel cells. Nanoparticles made from barium oxide can be used to modify electrode surfaces which would have the property of eliminating deposits of carbon which are formed from carbon soot and carbon monoxide (which can damage the electrodes) obtained from cheap fuel cells using natural gas or methanol. The coating property of the nanoparticle forms regions which allows the occurrence of reactions while enabling reactions of oxidation to eliminate deposits formed on the surface. The use of nanoscale materials for producing very thin electrolyte films in fuel cells favours the weight and final size of the fuel cell thereby increasing the electrical resistance of the fuel cell at high temperatures which are needed for the efficiency in fuel cell operation (Sahaym & Norton, 2008; Will Soutter, 2012).

Nanocomposites for Energy Saving Applications

Polymer nanocomposites also have a broad range of applications in energy-saving activities. When intrinsically conductive polymers experience changes in their pi-electronic character which follows a reversible insertion and extraction of ions which occurs through electrochemical reduction and oxidation (redox) reactions in polymer films, the polymer nanocomposite is said to be "electrochromic". The property of electrochromism is very important as it has energy-saving applications through the reduction of lighting expenses, consumption of energy and cooling/heating loads (Soliman et al., 2010). Electrochromic nanocomposites are applied in light-reflective devices which can be used for optical storage and information. It also finds applications as light-reflective or light-transmissive components in protective glasses for military use, anti-glare car rearview mirrors, smart windows for cars and buildings, controllable canopies for aircrafts, and glare-reduction systems used in offices (Quintanilha et al., 2014). Polymer nanocomposites also have the ability to capture carbon in order to reduce carbon emissions. Removal of CO_2 or carbon capture can be achieved through functionalization techniques such as membrane separation, solution absorption (using amine groups), solid sorbents and cryogenic processing methods (Ashley et al., 2012).

In transportation, fuel economy and low – emissions are global expectations. Nanocomposites have been classified as distinctive polymeric materials displaying superior mechanical, thermal, and processing properties, fit to replace metals in automobiles. The use of nanocomposites in automotive parts and systems is likely to reduce weight, enhance speed, improve environmental and thermal stability. Inasmuch as the larger percentage of total energy consumed by an automobile in its life cycle is associated with fuel consumption, weight reduction and improved speed may significantly result in energy savings for the automotive industry and vehicle users (Garces et al., 2000). Furthermore, the application of coatings from nanostructures to components of a vehicle engine is potentially able to control the transfer of heat thereby increasing the horsepower of the engine. Nordin (1980) reported that the application of the coating to the vehicle's air conditioning system ensures a reduction in the consumption of fuel.

The use of insulation materials in some systems such as the solar energy systems are known to lower energy consumption. Nanomaterials have the potentials of functioning as insulation materials which can be used as thin-films on surfaces used for coating windows and solar cells. These nano-based insulators lower the amount of energy consumed by ensuring that the heat in windows and solar energy systems is filtered (Baneyx, 2011). Experimental results have revealed that nano-fillers enhance insulation breakdown properties in nanocomposites and nano- and micro- filler mixed composites insulation breakdown properties are greater to those of traditional insulating materials (Imai et al., 2008)

In petroleum and chemical industries, intensification of processes to save energy is considered important for economic viability. The cost of goods consumed and petroleum products have direct impact on the economy and is dependent on the extent of optimized reactions achieved in these industries. Increasing the efficiency of processes taking place in these industries adds to the economy and an approach to this is effective use of catalysts. Catalysts made from nanomaterials lowers the amount of energy needed to produce chemicals owing to the property of large surface areas possessed by the nanoparticles which provide a large area needed for the occurrence of catalytic reactions (Fayette & Robinson, 2014). Efforts are continuously being geared towards developing efficient, stable and cost-effective catalysts for industrial purposes. For instance, Li et al. (2016) have reported preparing Ni–Mn_3O_4 nanocomposites on Ni foam as an electrocatalysts for hydrogen production in alkaline environments for improved energy efficiency for water splitting. The Ni–Mn_3O_4 nanocomposites was found to exhibit an excellent hydrogen evolution reaction catalytic activity.

Lighting using conventional bulbs and devices are characterized by excessive heat generation resulting in loss of limited energy supply. The use of nanomaterials (such as nanofiber) and nanocomposites have been found to alleviate the issue of excessive loss of heat. Nanofibers have nanoscale diameters and macro-scale lengths. Polymers of nanofibers are capable of becoming mass producible materials with exceptional optical properties when formed into non-woven fabrics, and can be used for various applications. A distinctive nanocomposite that offers numerous benefits to energy efficient lighting devices can be created by combining nanofibers with luminescent nanoparticles to form photoluminescent nanofiber substrates. Nanofibers and nanocomposites manufactured from nanofibers are evolving as a major enabling technology for the future generation of high efficiency lighting devices. The emergence of these materials in lighting applications is hinged on the point that nanofiber substrates have exceptional optical characteristics (Davis et al., 2010). In addition to the use of nanofibre based materials as lighting devices, quantum dots-based lighting system have been reported to produce light having 100% efficiency which prevents wastes in the form of heat (Ko et al., 2011).

CONCLUSION

For more than ten years, researches on use of nanofillers to enhance the properties of clean polymers have been ongoing with nanoclays, nano-oxides and carbon nanotubes being the most frequently studied nanofillers in the development of polymeric nanocomposites. Well-developed techniques, each with its uniqueness and varied level of sophistication, for characterization of nanocomposite may have facilitated and is still facilitating researches in this area. In the process of developing polymeric nanocomposites, pretreatment; functional group on nanofiller; functionalization techniques; modification of nanocomposites; in addition to the type of fillers and quantity of filler used, are some of the factors that determine the properties of resulting nanocomposites.

Functionalization have been used to improve intercalated structure of nanoclays, maximize exfoliation and obtain full dispersion of nano-oxides into polymer matrix and improve interfacial interactions between nanotubes and the matrix in which they are dispersed. Chemical modification of nanocomposites, among other things, is acclaimed to improve wear resistance, fire retardancy, anti-corrosive, hydrophobic, hydrophilic and conductive properties of the polymeric composites. Among several methods of processing nanocomposites that have been put forward, melt blending is considered very environmental-friendly and economical while layer by layer (LBL) assembly find wide applications in the production of supercapacitors and Field Effect Transistors (FETs).

A number of pre - modified nanofillers are commercially available and several nanocomposites have been successfully developed for a range of engineering and military applications. Inferences drawn from literature search indicate that extensive researches are still needed to further unlock more incredible potentials of nanocomposites and to achieve optimal improved results with their use in various fields of applications.

REFERENCES

Abdin, Z., Alim, M. A., Saidur, R., Islam, M. R., Rashmi, W., Mekhilef, S., & Wadi, A. (2013). Solar energy harvesting with the application of nanotechnology. *Renewable & Sustainable Energy Reviews*, *26*, 837–852. doi:10.1016/j.rser.2013.06.023

Ahluwalia, R. K., & Wang, X. (2008). Fuel cell systems for transportation: Status and trends. *Journal of Power Sources*, *177*(1), 167–176. doi:10.1016/j.jpowsour.2007.10.026

Ajayan, P. M., Schadler, L. S., & Braun, P. V. (2003). Nanocomposite Science and Technology. Wiley. doi:10.1002/3527602127

Alipanah, M., & Li, X. (2016). Numerical studies of lithium-ion battery thermal management systems using phase change materials and metal foams. *International Journal of Heat and Mass Transfer*, *102*, 1159–1168. doi:10.1016/j.ijheatmasstransfer.2016.07.010

Ashley, M., Magiera, C., Ramidi, P., Blackburn, G., Scott, T. G., Gupta, R., ... Biswas, A. (2012). Nanomaterials and processes for carbon capture and conversion into useful by-products for a sustainable energy future. *Greenhouse Gases: Science and Technology, 2*(6), 419–444. doi:10.1002/ghg.1317

Bahr, J. L., & Tour, J. M. (2002). Covalent chemistry of single-wall carbon nanotubes. *Journal of Materials Chemistry, 12*(7), 1952–1958. doi:10.1039/b201013p

Bandurin, D. A., Kleshch, V. I., Smolnikova, E. A., Obronov, I. V., Nasibulin, A. G., Kauppinen, E. I., & Obraztsov, A. (2013). Scanning anode field emission microscopy of nanocarbons. *Journal of Nanoelectronics and Optoelectronics, 8*(1), 114–118. doi:10.1166/jno.2013.1431

Baneyx, F. (2011). Review of biomimetic and bioinspired nanomaterials. *Journal of the American Chemical Society, 133*(13), 5161–5161. doi:10.1021/ja2018338

Barber, P., Balasubramanian, S., Anguchamy, Y., Gong, S., Wibowo, A., Gao, H., ... zur Loye, H.-C. (2009). Polymer composite and nanocomposite dielectric materials for pulse power energy storage. *Materials (Basel), 2*(4), 1697–1733. doi:10.3390/ma2041697

Bhattacharya, M. (2016). Polymer nanocomposites-A comparison between carbon nanotubes, graphene, and clay as nanofillers. *Materials (Basel), 9*(4), 262. doi:10.3390/ma9040262 PMID:28773388

Bhattacharya, S., Saha, I., Mukhopadhyay, A., Chattopadhyay, D., Ghosh, U.C. & Chatterjee, D. (2013). Role of nanotechnology in water treatment and purification: Potential applications and implications. *International Journal of Chemical Science and Technology.*

Bi, M., Hao, Y., Zhang, J., Lei, M., & Bi, K. (2017). Particle size effect of $BaTiO_3$ nanofillers on the energy storage performance of polymer nanocomposites. *Nanoscale, 9*(42), 16386–16395. doi:10.1039/C7NR05212J PMID:29053167

Bigg, D. M. (1987). Mechanical properties of particulate filled polymers. *Polymer Composites, 8*(2), 115–122. doi:10.1002/pc.750080208

Boysen, E. (n.d.). Fuel and nanotechnology. *Understanding Nano*. Retrieved from http://www.understandingnano.com/fuel.html

Brassard, J.-D., Sarkar, D.K. & Perron, J. (2012). Surface modification and functionalization of oxide nanoparticles for superhydrophobic applications. *Advanced Materials Research, 409*, 469-473. doi:10.4028/www.scientific.net/AMR.409.469

Buggy, M., Bradley, G., & Sullivan, A. (2005). Polymer–filler interactions in kaolin/nylon 6,6 composites containing a silane coupling agent. *Composite Part A, 36*(4), 437–442. doi:10.1016/j.compositesa.2004.10.002

Burchardt, T., Gouérec, P., Sanchez-Cortezon, E., Karichev, Z., & Miners, J. H. (2002). Alkaline fuel cells: Contemporary advancement and limitations. *Fuel, 81*(17), 2151–2155. doi:10.1016/S0016-2361(02)00163-1

Burris, D., Zhao, S., Duncan, R., Lowitz, J., Perry, S., Schadler, L., & Sawyer, W. (2009). A route to wear resistant PTFE via trace loadings of functionalized nanofillers. *Wear, 267*(1-4), 653–660. doi:10.1016/j.wear.2008.12.116

Carbajal, G., Martínez-Villafañe, A., Gonzalez-Rodriguez, J. G., & Castaño, V. M. (2001). Corrosion-resistant coatings: A nanotechnology approach. *Anti-Corrosion Methods and Materials, 48*(4), 241–244. doi:10.1108/00035590110398899

Chan, S., Bantang, J. P., & Camacho, D. (2015). Influence of nanomaterial fillers in biopolymer electrolyte system for squaraine-based dye-sensitized solar cells. *International Journal of Electrochemical Science, 10*, 7696–7706.

Chen, D., Feng, H., & Li, J. (2012). Graphene oxide: Preparation, functionalization, and electrochemical applications. *Chemical Reviews, 112*(11), 6027–6053. doi:10.1021/cr300115g PMID:22889102

Chen, X., Wang, J., Lin, M., Zhong, W., Feng, T., Chen, X., ... Xue, F. (2008). Mechanical and thermal properties of epoxy nanocomposites reinforced with amino-functionalized multi-walled carbon nanotubes. *Materials Science and Engineering A, 492*(1-2), 236–242. doi:10.1016/j.msea.2008.04.044

Chinese Academy of Science. (2014, July 10). Wind turbine blade carbon nanotubes / polymer nano composites developed. JEC Group. Retrieved from http://www.jeccomposites.com/knowledge/international-composites-news/wind-turbine-blade-carbon-nanotubes-polymer-nano-composites

Cho, S.-J., Uddin, M.-J., & Alaboina, P. (2017). Review of nanotechnology for cathode materials in batteries. In Emerging nanotechnologies in rechargeable energy storage systems (pp. 83–129).

Chua, T., Mariatti, M., Azizan, A., & Rashid, A. (2010). Effects of surface-functionalized multi-walled carbon nanotubes on the properties of poly(dimethyl siloxane) nanocomposites. *Composites Science and Technology, 70*(4), 671–677. doi:10.1016/j.compscitech.2009.12.023

Creton, C., Kramer, E. J., Brown, H. R., & Hui, C. Y. (2001). Adhesion and fracture of interfaces between immiscible polymers: From the molecular to the continuum scale. *Advances in Polymer Science, 156*, 53–136. doi:10.1007/3-540-45141-2_2

Daberkow, A., Ehlert, M., & Kaise, D. (2013). Electric car operation and flywheel energy storage. In M. Lienkamp (Ed.), *Conference on Future Automotive Technology* (pp. 19–30). 10.1007/978-3-658-01141-3_2

Dai, G., & Mishnaevsky, L. M. Jr. (2015). Carbon nanotube reinforced hybrid composites: Computational modeling of environmental fatigue and usability for wind blades. *Composites. Part B, Engineering, 78*, 349–360. doi:10.1016/j.compositesb.2015.03.073

Dang, Z. M., Yuan, J. K., Yao, S. H., & Liao, R. J. (2013). Flexible nanodielectric materials with high permittivity for power energy storage. *Advanced Materials, 25*(44), 6334–6365. doi:10.1002/adma.201301752 PMID:24038139

Dang, Z. M., Yuan, J. K., Zha, J. W., Zhou, T., Li, S. T., & Hu, G. H. (2012). Fundamentals, processes and applications of high-permittivity polymer–matrix composites. *Progress in Materials Science, 57*(4), 660–723. doi:10.1016/j.pmatsci.2011.08.001

Das, A., Stockelhuber, K. W., Jurk, R., & Heinrich, G. (2008). Modified and unmodified multiwalled carbon nanotubes in high performance solution-styrene-butadiene and butadiene rubber blends. *Polymer, 49*(24), 5276–5283. doi:10.1016/j.polymer.2008.09.031

Davis, J. L., Guzan, K., Mills, K., & Han, L. (2010). *Advanced Nanocomposite Materials Tailored to Solid-State Lighting Applications.*

Demirbas, A. (2008). Direct use of methanol in fuel cells. *Energy Sources. Part A, Recovery, Utilization, and Environmental Effects*, *30*(6), 529–535. doi:10.1080/15567030600817159

Derkacs, D., Lim, S. H., Matheu, P., Mar, W., & Yu, E. T. (2006). Improved performance of amorphous silicon solar cells via scattering from surface plasmon polaritons in nearby metallic nanoparticles. *Applied Physics Letters*, *89*(9), 0931031–0931033. doi:10.1063/1.2336629

Ding, Y., Fan, F., Tian, Z., & Wang, Z. L. (2010). Atomic structure of Au-Pd bimetallic alloyed nanoparticles. *Journal of the American Chemical Society*, *132*(35), 12480–12486. doi:10.1021/ja105614q PMID:20712315

Duygu, D., Baykal, T., Açikgöz, I., & Yildiz, K. (2009). Fourier transform infrared (FT-IR) spectroscopy for biological studies. *G.U. Journal of Science*, *22*(3), 117–121.

Eatemadi, A., Daraee, H., Karimkhanloo, H., Kouhi, M., Zarghami, N., Akbarzadeh, A., ... Joo, S. W. (2014). Carbon nanotubes: Properties, synthesis, purification, and medical applications. *Nanoscale Research Letters*, *9*(393), 1–13. PMID:25170330

Egerton, R. F. (2013). Control of radiation damage in the TEM. *Ultramicroscopy*, *127*, 100–108. doi:10.1016/j.ultramic.2012.07.006 PMID:22910614

Farid, M. M., Khudhair, A. M., Razack, S. A., & Al-Hallaj, S. (2004). A review on phase change energy storage: Materials and applications. *Energy Conversion and Management*, *45*(9-10), 1597–1615. doi:10.1016/j.enconman.2003.09.015

Fayette, M., & Robinson, R. D. (2014). Chemical transformations of nanomaterials for energy applications. *Journal of Materials Chemistry. A, Materials for Energy and Sustainability*, *2*(17), 5965–5978. doi:10.1039/C3TA13982D

Fiedler, B., Gojny, F. H., Wichmann, M. H., Nolte, M. C., & Schulte, K. (2006). Fundamental aspects of nano-reinforced composites. *Composites Science and Technology*, *66*(16), 3115–3125. doi:10.1016/j.compscitech.2005.01.014

Floody, M. C., Theng, B. K. G., Reyes, P., & Mora, M. L. (2009). Natural nanoclays: Applications and future trends - a Chilean perspective. *Clay Minerals*, *44*(02), 161–176. doi:10.1180/claymin.2009.044.2.161

Forrest, S. R. (2005). The Limits to organic photovoltaic cell efficiency. *MRS Bulletin*, *30*(1), 28–32. doi:10.1557/mrs2005.5

Fu, S.-Y., Feng, X.-Q., Lauke, B., & Mai, Y.-W. (2008). Effects of particle size, particle/matrix interface adhesion and particle loading on mechanical properties of particulate–polymer composites. *Composites. Part B, Engineering*, *39*(6), 933–961. doi:10.1016/j.compositesb.2008.01.002

Fu, X. A., & Qutubuddin, S. (2001). Polymer–clay nanocomposites: Exfoliation of organophilic montmorillonite nanolayers in polystyrene. *Polymer*, *42*(2), 807–813. doi:10.1016/S0032-3861(00)00385-2

Fu, X. A., & Qutubuddin, S. (2004). Synthesis of unsaturated polyester-clay nanocomposites using reactive organoclays. *Polymer Engineering and Science*, *44*(2), 345–351. doi:10.1002/pen.20031

Garces, J. M., Moll, D. J., Bicerano, J., Fibiger, R., & McLeod, D. G. (2000). Polymeric nanocomposites for automotive applications. *Advanced Materials*, *12*(23), 1835–1839. doi:10.1002/1521-4095(200012)12:23<1835::AID-ADMA1835>3.0.CO;2-T

Georgakilas, V., Otyepka, M., Bourlinos, A. B., Chandra, V., Kim, N., Kemp, K. C., ... Kim, K. S. (2012). Functionalization of graphene: Covalent and non-covalent approaches, derivatives and applications. *Chemical Reviews*, *112*(11), 6156–6214. doi:10.1021/cr3000412 PMID:23009634

Ghosh, S., Bao, W., Nika, D. L., Subrina, S., Pokatilov, E. P., Lau, C. N., & Balandin, A. A. (2010). Dimensional crossover of thermal transport in few-layer graphene. *Nature Materials*, *9*(7), 555–558. doi:10.1038/nmat2753 PMID:20453845

Goikolea, E., & Mysyk, R. (2017). *Nanotechnology in Electrochemical Capacitors*. In *Emerging Nanotechnologies in Rechargeable Energy Storage Systems* (pp. 131–169).

Goli, P., Legedza, S., Dhar, A., Salgado, R., Renteria, J., & Balandin, A. A. (2014). Graphene-enhanced hybrid phase change materials for thermal management of li-ion batteries. *Journal of Power Sources*, *248*, 37–43. doi:10.1016/j.jpowsour.2013.08.135

Grdadolnik, J. (2002). ATR-FTIR spectroscopy: Its advantages and limitations. *Acta Chimica Slovenica*, *49*(3), 631–642.

Grossiord, N., & Hermant, M. C. (2016). *Latex technology for conductive polymer nanocomposites*. John Wiley & Sons, Inc. doi:10.1002/0471440264.pst506.pub2

Guo, Z., Lei, K., Li, Y., Ng, H. W., Prikhodko, S., & Hahn, H. T. (2008). Fabrication and characterization of iron oxide nanoparticles reinforced vinyl-ester resin nanocomposites. *Composites Science and Technology*, *68*(6), 1513–1520. doi:10.1016/j.compscitech.2007.10.018

Guo, Z., Wei, S., Shedd, B., Scaffaro, R., Pereira, T., & Hahn, H. T. (2007). Particle surface engineering effect on the mechanical, optical and photoluminescent properties of ZnO/vinyl-ester resin nanocomposites. *Journal of Materials Chemistry*, *17*(8), 806–813. doi:10.1039/B613286C

Guyomar, D., Lebrun, L., Putson, C., Cottinet, P. J., Guiffard, B., & Muensit, S. (2009). Electrostrictive energy conversion in polyurethane nanocomposites. *Journal of Applied Physics*, *106*(1), 014910–014919. doi:10.1063/1.3159900

Ha, J. U., & Xanthos, M. (2009). Functionalization of nanoclays with ionic liquids for polypropylene composites. *Polymer Composites*, *30*(5), 534–542. doi:10.1002/pc.20583

Habibi, Y. (2014). Key advances in the chemical modification of nanocelluloses. *Chemical Society Reviews*, *43*(5), 1519–1542. Retrieved from http://xlink.rsc.org/?DOI=C3CS60204D doi:10.1039/C3CS60204D PMID:24316693

Hao, Y., Wang, X., Bi, K., Zhang, J., Huang, Y., Wu, L., ... Li, L. (2017). Significantly enhanced energy storage performance promoted by ultimate sized ferroelectric BaTiO3 fillers in nanocomposite films. *Nano Energy*, *31*, 49–56. doi:10.1016/j.nanoen.2016.11.008

Hao, Y. N., Wang, X. H., O'Brien, S., Lombardi, J., & Li, L. T. (2015). Flexible BaTiO3/PVDF gradated multilayer nanocomposite film with enhanced dielectric strength and high energy density. *Journal of Materials Chemistry. C, Materials for Optical and Electronic Devices*, 37.

Harish, K. K., Venkatesh, N., Bhowmik, H., & Kuila, A. (2018). Metallic Nanoparticle: A Review. *Biomedical Journal of Scientific and Technical Research*, *4*(2), 1–11. PMID:30370423

He, K., Lin, F., Zhu, Y., Yu, X., Li, J., Lin, R., ... Su, D. (2015). Sodiation kinetics of metal oxide conversion electrodes: A comparative study with lithiation. *Nano Letters*, *15*(9), 5755–5763. doi:10.1021/acs.nanolett.5b01709 PMID:26288360

Hirsch, A., & Vostrowsky, O. (2005). Functionalization of carbon nanotubes. *Topics in Current Chemistry*, *245*, 193–237. doi:10.1007/b98169

Hirsch, P., Kässens, M., Püttmann, M., & Reimer, L. (1994). Contamination in a scanning electron microscope and the influence of specimen cooling. *Scanning*, *16*(2), 101–110. doi:10.1002ca.4950160207

Holbrey, J. D., & Seddon, K. R. (1999). The phase behaviour of 1-alkyl-3-methylimidazolium tetrafluoroborates; ionic liquids and ionic liquid crystals. *Journal of the Chemical Society, Dalton Transactions: Inorganic Chemistry*, (13): 2133–2140. doi:10.1039/a902818h

Hostler, S., Kaul, P., Qu, V., Cullen, C., & Abramson, A. (2006). Thermal and electrical characterization of nanocomposites for thermoelectrics. In *Proceedings of The Tenth Intersociety Conference on Thermal and Thermomechanical Phenomena in Electronics Systems, ITHERM'06*, San Diego, CA (pp. 1400–1405). 10.1109/ITHERM.2006.1645508

Huang, L., Jia, Z., Kymissis, I., & O'Brien, S. (2010). High K capacitors and OFET gate dielectrics from self-assembled BaTiO3 and (Ba, Sr) TiO3 nanocrystals in the superparaelectric limit. *Advanced Functional Materials*, *20*(4), 554–560.

Hÿtch, M. J., Snoeck, E., & Kilaas, R. (1998). Quantitative measurement of displacement and strain fields from HREM micrographs. *Ultramicroscopy*, *74*(3), 131–146. doi:10.1016/S0304-3991(98)00035-7

Imai, T., Komiya, G., Murayama, K., Ozaki, T., Sawa, F., Shimizu, T., ... Tanaka, T. (2008) Improving Epoxy-based Insulating Materials with Nano-fillers toward Practical Application. In *IEEE International Symposium on Electrical Insulation.* 10.1109/ELINSL.2008.4570310

International Energy Agency. (2017). *World Energy Outlook 2017*. Retrieved from https://www.iea.org/weo2017/

ISO. (2015). ISO/TS80004-2 Nanotechnologies—Vocabulary—Part 2: Nano-objects.

Jadhav, M. V., Todkar, A. S., Gambhire, V. R., & Sawant, S. Y. (2011). Nanotechnology for powerful solar energy. *International Journal of Advanced Biotechnology and Research*, *2*(1), 208–212.

Jakubovics, J. P. (2008). Lorentz Microscopy. In *Handbook of Microscopy Set* (pp. 505–514). Wiley.

Jarosovaa, R., & Swain, G. M. (2015). Rapid preparation of room temperature ionic liquids with low water content as characterized with a ta-C:N Electrode. *Journal of the Electrochemical Society*, *162*(8), H507–H511. doi:10.1149/2.0191508jes

Jiang, N. (2013). Damage mechanisms in electron microscopy of insulating materials. *Journal of Physics. D, Applied Physics*, *46*(30), 305502. doi:10.1088/0022-3727/46/30/305502

Jiang, N., Zhou, S., Su, D., & Qiu, J. (2012). Do Eu dopants prefer the precipitated LaF_3 nanocrystals in glass ceramics? *Physica Status Solidi. Rapid Research Letters*, *6*(12), 487–489. doi:10.1002/pssr.201206476

Jinschek, J. R., Batenburg, K. J., Calderon, H. A., Kilaas, R., Radmilovic, V., & Kisielowski, C. (2008). 3-D reconstruction of the atomic positions in a simulated gold nanocrystal based on discrete tomography: Prospects of atomic resolution electron tomography. *Ultramicroscopy*, *108*(6), 589–604. doi:10.1016/j.ultramic.2007.10.002 PMID:18082327

Jo, B., Park, S., & Kim, D. (2008). Mechanical properties of nano-MMT reinforced polymer composite and polymer concrete. *Construction & Building Materials*, *22*(1), 14–20. doi:10.1016/j.conbuildmat.2007.02.009

Johnson, C. L., Snoeck, E., Ezcurdia, M., Rodríguez-González, B., Pastoriza-Santos, I., Liz-Marzán, L. M., & Hÿtch, M. J. (2008). Effects of elastic anisotropy on strain distributions in decahedral gold nanoparticles. *Nature Materials*, *7*(2), 120–124. doi:10.1038/nmat2083 PMID:18084296

Kaiser, A., & Park, Y. (2005). Current-voltage characteristics of conducting polymers and carbon nanotubes. *Synthetic Metals*, *152*(1-3), 181–184. doi:10.1016/j.synthmet.2005.07.245

Kakac, S., & Pramuanjaroenkij, A. (2009). Review of convective heat transfer enhancement with nanofluids. *International Journal of Heat and Mass Transfer*, *52*(13-14), 3187–3196. doi:10.1016/j.ijheatmasstransfer.2009.02.006

Kalaiselvam, S., & Parameshwaran, R. (2014). Nanotechnology in thermal energy storage. In Thermal Energy Storage Technologies for Sustainability (pp. 163-202). doi:10.1016/B978-0-12-417291-3.00008-6

Kathi, J., Rhee, K. Y., & Lee, J. H. (2009). Effect of chemical functionalization of multi-walled carbon nanotubes with 3-aminopropyltriethoxysilane on mechanical and morphological properties of epoxy nanocomposites. *Composites. Part A, Applied Science and Manufacturing*, *40*(6-7), 800–809. doi:10.1016/j.compositesa.2009.04.001

Kaushal, A., & Singh, S. K. (2017). Removal of heavy metals by nanoadsorbents. A review. *Journal of Environmental Biotechnology Research*, *6*(1), 96–104.

Keast, V. J., Scott, A. J., Brydson, R., Williams, D. B., & Bruley, J. (2001). Electron energy-loss near-edge structure - A tool for the investigation of electronic structure on the nanometre scale. *Journal of Microscopy*, *203*(2), 135–175. doi:10.1046/j.1365-2818.2001.00898.x PMID:11489072

Kenawy, E., Ghfar, A., Wabaidur, S., Khan, M., Siddiqui, M., Alothman, Z., ... Hamid, M. (2018). Cetyltrimethylammonium bromide intercalated and branched polyhydroxystyrene functionalized montmorillonite clay to sequester cationic dyes. *Journal of Environmental Management*, *219*, 285–293. doi:10.1016/j.jenvman.2018.04.121 PMID:29751259

Khan, W. S., Hamadneh, N. N., & Khan, W. A. (2016). Polymer nanocomposites–synthesis techniques, classification and properties. One Central Press.

Kim, P., Jones, S., Hotchkiss, P., Haddock, J., Kippelen, B., Marder, S., & Perry, J. (2007). Phosphonic acid-modified barium titanate polymer nanocomposites with high permittivity and dielectric strength. *Advanced Materials, 19*(7), 1001–1005. doi:10.1002/adma.200602422

Kim, T., Kim, J. H., Kang, T. E., Lee, C., Kang, H., Shin, M., ... Kim, B. J. (2015). Flexible, highly efficient all-polymer solar cells. *Nature Communications*, 6. PMID:26449658

Ko, Y. H., Kim, J.-H., Jin, L.-H., Ko, S.-M., Kwon, B.-J., Kim, J., ... Cho, Y.-H. (2011). Electrically driven quantum dot/wire/well hybrid light-emitting diodes. *Advanced Materials, 23*(45), 5364–5369. doi:10.1002/adma.201102534 PMID:22002261

Kolasińska, E., & Kolasiński, P. (2016). A review on electroactive polymers for waste heat recovery. *Materials (Basel), 9*(6), 485. doi:10.3390/ma9060485 PMID:28773605

Kole, M., & Dey, T. K. (2010). Thermal conductivity and viscosity of Al_2O_3 nanofluid based on car engine coolant. *Journal of Physics D, 43*(31), 315501. doi:10.1088/0022-3727/43/31/315501

Kornmann, X., Berglund, L. A., Sterte, J., & Giannelis, E. P. (1998). Nanocomposites based on montmorillonite and unsaturated polyester. *Polymer Engineering and Science, 38*(8), 1351–1358. doi:10.1002/pen.10305

Kumar, S. R., Juan, C.-H., Liao, G.-M., Lin, J.-S., Yang, C.-C., Ma, W.-T., ... Lue, S. J. (2016). Fumed silica nanoparticles incorporated in quaternized poly(vinyl alcohol) nanocomposite membrane for enhanced power densities in direct alcohol alkaline fuel cells. *Energies, 9*(15), 1–19. doi:10.3390/en9010015

Kumar, S. R., Ma, W.-T., Lu, H.-C., Teng, L.-W., Hsu, H.-C., Shih, C.-M., ... Lue, S. J. (2017). Surfactant-assisted perovskite nanofillers incorporated in quaternized poly (vinyl alcohol) composite membrane as an effective hydroxide-conducting electrolyte. *Energies, 10*(615), 1–22. doi:10.3390/en10050615

Kurahatti, R. V., Surendranathan, A. O., Kori, S. A., Singh, N., Kumar, A. V. R., & Srivastava, S. (2010). Defence applications of polymer nanocomposites. *Defence Science Journal, 60*(5), 551–563. doi:10.14429/dsj.60.578

Lee, J.-H., Rhee, K. Y., & Park, S. J. (2011). Silane modification of carbon nanotubes and its effects on the material properties of carbon/CNT/epoxy three-phase composites. *Composites. Part A, Applied Science and Manufacturing, 42*(5), 478–483. doi:10.1016/j.compositesa.2011.01.004

Leyden, D. E., & Murthy, R. S. S. (1988). Diffuse reflectance fourier transform IR spectroscopy. *Trends in Analytical Chemistry, 7*(5), 164–169. doi:10.1016/0165-9936(88)85044-1

Li, J., Claude, J., Norena-Franco, L. E., Seok, S. I., & Wang, Q. (2008). Electrical energy storage in ferroelectric polymer nanocomposites containing surface-functionalized $BaTiO_3$ nanoparticles. *Chemistry of Materials, 20*(20), 6304–6306. doi:10.1021/cm8021648

Li, L., Liu, J., Su, Y., Li, G., Chen, X., Qiu, X., & Yan, T. (2009). Surface doping for photocatalytic purposes: Relations between particle size, surface modifications, and photoactivity of SnO_2:Zn^{2+}nanocrystals. *Nanotechnology, 20*(15), 155706. doi:10.1088/0957-4484/20/15/155706

Li, L. L., Li, M., & Chen, D. X. (2006). Solvothermal synthesis and characterization of Sb-doped SnO2 nanoparticles used as transparent conductive films. *Materials Research Bulletin, 41*(3), 541–546. doi:10.1016/j.materresbull.2005.09.011

Li, X., & Coleman, M. R. (2008). Functionalization of carbon nanofibers with diamine and polyimide oligmer. *Carbon, 46*(8), 1115–1125. doi:10.1016/j.carbon.2008.01.035

Li, X., Liu, P. F., Zhang, L., Zu, M. Y., Yang, Y. X., & Yang, H. G. (2016). Enhancing alkaline hydrogen evolution reaction activity through Ni–Mn_3O_4 nanocomposites. *Chemical Communications.* PMID:27500290

Liao, G.-M., Yang, C.-C., Hu, C.-C., Pai, Y.-L., & Lue, S. J. (2015). Novel quaternized polyvinyl alcohol/quaternized chitosan nano-composite as an effective hydroxide-conducting electrolyte. *Journal of Membrane Science, 485*, 17–29. doi:10.1016/j.memsci.2015.02.043

Lingayat, A. B., & Suple, Y. R. (2013). Review on phase change material as thermal energy storage medium: Materials, application. *International Journal of Engineering Research and Applications, 3*(4), 916–921.

Liu, C., Sungchul Lee, S., Su, D., Zhang, Z., Pfefferle, L., & Haller, G. L. (2012). Synthesis and characterization of nanocomposites with strong interfacial interaction: Sulfated ZrO_2 nanoparticles supported on multiwalled carbon nanotubes. *The Journal of Physical Chemistry C, 116*(41), 21742–21752. doi:10.1021/jp302377p

Liu, P. (2005). Modifications of carbon nanotubes with polymers. *European Polymer Journal, 41*(11), 2693–2703. doi:10.1016/j.eurpolymj.2005.05.017

Lo, C.-F., Wu, J.-F., Li, H.-Y., Hung, W.-S., Shih, C.-M., Hu, C.-C., ... Lue, S. J. (2013). Novel polyvinyl alcohol nanocomposites containing carbon nano-tubes with Fe_3O_4 pendants for alkaline fuel cell applications. *Journal of Membrane Science, 444*, 41–49. doi:10.1016/j.memsci.2013.05.001

Lonkar, S. P., Deshmukh, Y. S., & Abdala, A. A. (2014). Recent Advances in chemical modifications of graphene. *Nano Research.*

Lu, A. H., Salabas, E. L., & Schüth, F. (2007). Magnetic nanoparticles: Synthesis, protection, functionalization, and application. *Angewandte Chemie International Edition, 46*(8), 1222–1244. doi:10.1002/anie.200602866 PMID:17278160

Luo, B., Wang, X., Wang, Y., & Li, L. (2014). Fabrication, characterization, properties and theoretical analysis of ceramic/PVDF composite flexible films with high dielectric constant and low dielectric loss. *Journal of Materials Chemistry. A, Materials for Energy and Sustainability, 2*(2), 510–519. doi:10.1039/C3TA14107A

Lynch, P. A., Stevenson, A. W., Liang, D., Parry, D., Wilkins, S., & Tamura, N. (2007). A laboratory based system for Laue micro x-ray diffraction. *The Review of Scientific Instruments, 78*(2), 023904. doi:10.1063/1.2437777 PMID:17578120

Ma, P.-C., & Kim, J.-K. (2011). *Carbon nanotubes for polymer reinforcement.* Taylor and Francis Group, LLC.

Ma, P. C., Kim, J.-K., & Tang, B. Z. (2007). Effects of silane functionalization on the properties of carbon nanotube/epoxy nanocomposites. *Composites Science and Technology*, *67*(14), 2965–2972. doi:10.1016/j.compscitech.2007.05.006

Ma, P. C., Siddiqui, N. A., Marom, G., & Kim, J.-K. (2010). Dispersion and functionalization of carbon nanotubes for polymer-based nanocomposites: A review. *Composites. Part A, Applied Science and Manufacturing*, *41*(10), 1345–1367. doi:10.1016/j.compositesa.2010.07.003

Ma, P.-C., & Zhang, Y. (2014). Perspectives of carbon nanotubes/polymer nanocomposites for wind blade materials. *Renewable & Sustainable Energy Reviews*, *30*, 651–660. doi:10.1016/j.rser.2013.11.008

Maleki, A., Hamesadeghi, U., Daraei, H., Hayati, B., Najafi, F., McKay, G., & Rezaee, R. (2017). Amine functionalized multi-walled carbon nanotubes: Single and binary systems for high capacity dye removal. *Chemical Engineering Journal*, *313*, 826–835. doi:10.1016/j.cej.2016.10.058

Maria, C. (2012). Application of FTIR spectroscopy in environmental studies. In Advanced Aspects of Spectroscopy. doi:10.5772/48331

Marquis, D. M., Guillaume, E., & Chivas-Joly, C. (2011). Properties of nanofillers in polymer. In J. Cuppoletti (Ed.), *Nanocomposites and Polymers with Analytical Methods*. IntechOpen; doi:10.5772/21694

Mc Carthy, B., Coleman, J. N., Czerw, R., Dalton, A. B., Carroll, D. L., & Blau, W. J. (2001). Microscopy studies of nanotube-conjugated polymer interactions. *Synthetic Metals*, *121*(1-3), 1225–1226. doi:10.1016/S0379-6779(00)00906-1

Mirjalili, F., Chuah, L., & Salahi, E. (2014). Mechanical and Morphological Properties of Polypropylene/Nano α-Al$2_0$3 Composites. T*he Scientific World Journal, 2014,* 1–12. doi:10.1155/2014/718765 PMID:24688421

Mocuta, C., Stangl, J., Mundboth, K., Metzger, T. H., Bauer, G., Vartanyants, I. A., ... Boeck, T. (2008). Beyond the ensemble average: X-ray microdiffraction analysis of single SiGe islands. *Physical Review B: Condensed Matter and Materials Physics*, *77*(24), 245425. doi:10.1103/PhysRevB.77.245425

Mogilevsky, G., Chen, Q., Kleinhammes, A., & Wu, Y. (2008). The structure of multilayered titania nanotubes based on delaminated anatase. *Chemical Physics Letters*, *460*(4-6), 517–520. doi:10.1016/j.cplett.2008.06.063

Mohan, A., Khanna, N., Hwu, J., & Joy, D. C. (2006). Secondary electron imaging in the variable pressure scanning electron microscope. *Scanning*, *20*(6), 436–441. doi:10.1002ca.1998.4950200603

Mohseni, M., Ramezanzadeh, B., Yari, H., & Gudarzi, M. M. (2012). The role of nanotechnology in automotive industries. In J. Carmo (Ed.), *New Advances in Vehicular Technology and Automotive Engineering*. IntechOpen. doi:10.5772/49939

Mubarak, N. M., Abdullah, E. C., Jayakumar, N. S., & Sahu, J. N. (2014). An overview on methods for the production of carbon nanotubes. *Journal of Industrial and Engineering Chemistry*, *20*(4), 1186–1197. doi:10.1016/j.jiec.2013.09.001

Nordin, T. (1980). Development of energy-saving vehicles. *ASEA Journal*, *53*(4-5).

Oliveira, G. E., Clarindo, J. E. S., Santo, K. S. E., & Souza, F. G. Jr. (2013). Chemical modification of cobalt ferrite nanoparticles with possible application as asphaltene flocculant agent. *Materials Research*, *16*(3), 668–671. Retrieved from http://www.scielo.br/scielo.php?script=sci_arttext&pid=S1516-14392013000300020&nrm=iso doi:10.1590/S1516-14392013005000048

Packham, D. E. (1996). Work of adhesion: Contact angles and contact mechanics. *Int. J. Adhes*, *16*(2), 121–128. doi:10.1016/0143-7496(95)00034-8

Pan, S., & Gupta, A. (2012). Surface-enhanced solar energy conversion systems using gold and silver nanoparticles. *Material Matters*, *7*(4), 64–66.

Pazourková, L., Martynková, G.S., Plachá, D. (2015). Preparation and mechanical properties of polymeric nanocomposites with hydroxyapatite and hydroxyapatite/clay mineral fillers – review. *HSOA Journal of Nanotechnology: Nanomedicine & Nanobiotechnology*.

Peighambardoust, S., Rowshanzamir, S., & Amjadi, M. (2010). Review of the proton exchange membranes for fuel cell applications. *International Journal of Hydrogen Energy*, *35*(17), 9349–9384. doi:10.1016/j.ijhydene.2010.05.017

Pleşa, I., Noţingher, P. V., Schlögl, S., Sumereder, C., & Muhr, M. (2016). Properties of polymer composites used in high-voltage applications. *Polymers*.

Pop, E., Varshney, V., & Roy, A. K. (2012). Thermal properties of graphene: Fundamentals and applications. *MRS Bulletin*, *37*(12), 1273–1281. doi:10.1557/mrs.2012.203

Postek, M. T. (2006). An approach to the reduction of hydrocarbon contamination in the scanning electron microscope. *Scanning*, *18*(4), 269–274. doi:10.1002ca.1996.4950180402

Pothan, L. A., George, C. N., Jacob, M., & Thomas, S. (2007). Effect of chemical modification on the mechanical and electrical properties of banana fiber polyester composites. *Journal of Composite Materials*, *41*(19), 2371–2386. doi:10.1177/0021998307075456

Prabaharan, S. R. S., & Michael, M. S. (Eds.). (2013). *Nanotechnology in advanced electrochemical power sources*. Taylor & Francis Group, LLC.

Quintanilha, R. C., Rocha, I., Vichessi, R. B., Lucht, E., Naidek, K., Winnischofer, H., & Vidotti, M. (2014). Electrochromism: Basis and application of nanomaterials in development of high performance electrodes. *Quimica Nova*, *37*(4), 677–688. doi:10.5935/0100-4042.20140112

Raffaelle, R. P., Castro, S. L., Hepp, A. F., & Bailey, S. G. (2002). Quantum dot solar cells. *Progress in Photovoltaics: Research and Applications*, *10*(6), 433–439. doi:10.1002/pip.452

Renteria, J. D., Nika, D. L., & Balandin, A. A. (2014). Graphene thermal properties: Applications in Thermal management and energy storage. *Applied Sciences*, *4*(4), 525–547. doi:10.3390/app4040525

Richard, C. (2018, March 22). Renewables fastest growing energy source in 2017. *Wind Power Monthly*. Retrieved from https://www.windpowermonthly.com/article/1460160/renewables-fastest-growing-energy-source-2017

Rocha, F. S., Mota, H. O., & Vasconcelos, F. H. (2017). The Monte Carlo method to uncertainty calculation of the displacement measurement interferometry in FTIR spectrometry systems. In *I2MTC 2017 - 2017 IEEE International Instrumentation and Measurement Technology Conference, Proceedings*. 10.1109/I2MTC.2017.7969829

Rogers, S. A., & Kaiser, A. B. (2004). Thermopower and resistivity of carbon nanotube networks and organic conducting polymers. *Current Applied Physics*, *4*(2-4), 407–410. doi:10.1016/j.cap.2003.11.060

Romero, M. J., & Al-Jassim, M. M. (2003). Advantages of using piezoelectric quantum structures for photovoltaics. *Journal of Applied Physics*, *93*(1), 626–631. doi:10.1063/1.1524705

Rong, M. Z., Zhang, M. Q., Pan, S. L., Lehmann, B., & Friedrich, K. (2004). Analysis of the interfacial interactions in polypropylene/silica nanocomposites. *Polymer International*, *53*(2), 176–183. doi:10.1002/pi.1307

Rongzhi, X., Yandao, G., Xiufang, Z., & Nanming, Z. (1998). Fibrinogen adsorption on biomaterials as characterized by transmission FTIR spectroscopy. *Tsinghua Science and Technology*, *3*(3), 1115–1120.

Rytter, E., & Einarsrud, M. A. (1988). Band distortions in emission and specular reflection FTIR spectroscopy. *Mikrochimica Acta*, *95*(1-6), 307–309. doi:10.1007/BF01349775

Safarova, K., Dvorak, A., Kubinek, R., Vujtek, M., & Rek, A. (2007). Usage of AFM, SEM and TEM for the research of carbon nanotubes. *Modern Research and Educational Topics in Microscopy*, *2*, 513–519.

Sahaym, U., & Norton, M. (2008). Advances in the application of nanotechnology in enabling a 'hydrogen economy.' *Journal of Materials Science*, *43*(16), 5395–5429. doi:10.100710853-008-2749-0

Saisirirat, P., & Joommanee, B. (2017). Study on the performance of the micro direct ethanol fuel cell (Micro-DEFC) for applying with the portable electronic devices. *Energy Procedia*, *138*, 187–192. doi:10.1016/j.egypro.2017.10.148

Salunkhe, R. R., Lee, Y. H., Chang, K. H., Li, J. M., Simon, P., Tang, J., ... Yamauchi, Y. (2014). Nanoarchitectured graphene-based supercapacitors for next-generation energy-storage applications. *Chemistry (Weinheim an der Bergstrasse, Germany)*, *20*(43), 13838–13852. doi:10.1002/chem.201403649 PMID:25251360

Samakande, A. (2005). Synthesis and Characterization of Surfmers for the Synthesis of Polystrene-Clay Nanocomposites [Masters' Thesis]. University of Stellenbosch.

Samimi, F., Babapoor, A., Azizi, M., & Karimi, G. (2016). Thermal management analysis of a Li-ion battery cell using phase change material loaded with carbon fibers. *Energy*, *96*, 355–371. doi:10.1016/j.energy.2015.12.064

Schubel, P. J., Johnson, M. S., Warrior, N. A., & Rudd, C. D. (2006). Characterisation of thermoset laminates for cosmetic automotive applications: Part III - shrinkage control via nanoscale reinforcement. *Composites. Part A, Applied Science and Manufacturing*, *37*(10), 1757–1772. doi:10.1016/j.compositesa.2005.09.014

Serrano, E., Rus, G., & García-Martínez, J. (2009). Nanotechnology for sustainable energy. *Renewable & Sustainable Energy Reviews*, *13*(9), 2373–2384. doi:10.1016/j.rser.2009.06.003

Sethi, V. K., Pandey, M., & Shukla, P. (2011). Use of nanotechnology in solar PV cell. *International Journal of Chemical Engineering and Applications*, *2*(2).

Seyhan, A., Gojny, F., Tanoglu, M., & Schulte, K. (2007b). Rheological and dynamic-mechanical behavior of carbon nanotube/vinyl ester-polyester suspensions and their nanocomposites. *European Polymer Journal*, *43*(7), 2836–2847. doi:10.1016/j.eurpolymj.2007.04.022

Sham, M. L., & Kim, J. K. (2006). Surface functionalities of multi-wall carbon nanotubes after UV/Ozone and TETA treatments. *Carbon*, *44*(4), 768–777. doi:10.1016/j.carbon.2005.09.013

Sharma, R. (2001). Design and applications of environmental cell transmission electron microscope for in situ observations of gas-solid reactions. *Microscopy and Microanalysis*, *7*(6), 494–506. PMID:12597794

Sharma, S., & Sagara, K. (2005). Latent heat storage materials and systems: A review. *International Journal of Green Energy*, *2*(1), 1–56. doi:10.1081/GE-200051299

Shen, J., Huang, W., Wu, L., Hu, Y., & Ye, M. (2007). The reinforcement role of different amino-functionalized multi-walled carbon nanotubes in epoxy nanocomposites. *Composites Science and Technology*, *67*(15-16), 3041–3050. doi:10.1016/j.compscitech.2007.04.025

Shi, Y., Peng, L., Ding, Y., Zhao, Y., & Yu, G. (2015). Nanostructured conductive polymers for advanced energy storage. *Chem. Soc. Rev. Chem. Soc. Rev*, *6684*(44), 6684–6696. doi:10.1039/C5CS00362H PMID:26119242

Shirazi, A., Mohebbi, F., Kakavand, M. A., He, B., & Rabczuk, T. (2016). Paraffin nanocomposites for heat management of lithium-ion batteries: A computational investigation. *Journal of Nanomaterials*, *2016*, 1–12. doi:10.1155/2016/2131946

Soliman, H. M. A., Kashyout, A. B., El Nouby, M. S., & Abosehly, A. M. (2010). Preparation and characterizations of tungsten oxide electrochromic nanomaterials. *Journal of Materials Science Materials in Electronics*, *21*(12), 1313–1321. doi:10.100710854-010-0068-0

Soultanidis, N., Zhou, W., Psarras, A. C., Gonzalez, A. J., Iliopoulou, E. F., Kiely, C. J., ... Wong, M. S. (2010). Relating n-pentane isomerization activity to the tungsten surface density of WOx/ZrO_2. *Journal of the American Chemical Society*, *132*(38), 13462–13471. doi:10.1021/ja105519y PMID:20815386

Spitalsky, Z., Tasis, D., Papagelis, K., & Galiotis, C. (2010). Carbon nanotube-polymer composites: Chemistry, processing, mechanical and electrical properties. *Progress in Polymer Science*, *35*(3), 357–401. doi:10.1016/j.progpolymsci.2009.09.003

Standridge, S. D., Schatz, G. C., & Hupp, J. T. (2009). Distance dependence of plasmon-enhanced photocurrent in dye-sensitized solar cells. *Journal of the American Chemical Society*, *131*(24), 8407–8409. doi:10.1021/ja9022072 PMID:19473006

Su, D. (2017). Advanced electron microscopy characterization of nanomaterials for catalysis. *Green Energy & Environment*, *2*(2), 70–83. doi:10.1016/j.gee.2017.02.001

Su, D., Wang, F., Ma, C., & Jiang, N. (2013). Engineering nano-composite $Li_4Ti_5O_{12}$ anodes via scanning electron-probe fabrication. *Nano Energy*, *2*(3), 343–350. doi:10.1016/j.nanoen.2012.10.009

Thio, Y. S., Argon, A. S., & Cohen, R. E. (2004). Role of interfacial adhesion strength on toughening polypropylene with rigid particles. *Polymer*, *45*(10), 3139–3147. doi:10.1016/j.polymer.2004.02.064

Thomann, I., Pinaud, B. A., Chen, Z., Clemens, B. M., Jaramillo, T. F., & Brongersma, M. L. (2011). Plasmon enhanced solar-to-fuel energy conversion. *Nano Letters*, *11*(8), 3440–3446. doi:10.1021/nl201908s PMID:21749077

Tkalya, E. E., Ghislandi, M., de With, G., & Koning, C. E. (2012). The use of surfactants for dispersing carbon nanotubes and graphene to make conductive nanocomposites. *Current Opinion in Colloid & Interface Science*, *17*(4), 225–232. doi:10.1016/j.cocis.2012.03.001

Tomer, V., Polizos, G., Randall, C. A., & Manias, E. (2011). Polyethylene nanocomposite dielectrics: Implications of nanofiller orientation on high field properties and energy storage. *Journal of Applied Physics*, *109*(7), 074113. doi:10.1063/1.3569696

Tsakalakos, L. (2010). *Nanotechnology for photovoltaics*. CRC Press. doi:10.1201/9781420076752

Understanding Nano. (2010, September). *Rice U. formula could make fuel manufacturing better, greener*. Retrieved from http://www.understandingnano.com/nanocatalyst-tunsten-oxide.html

Vaisman, L., Wagner, H. D., & Marom, G. (2006). The role of surfactants in dispersion of carbon nanotubes. *Advances in Colloid and Interface Science*, *128-130*, 37–46. doi:10.1016/j.cis.2006.11.007 PMID:17222381

Van Aert, S., Batenburg, K. J., Rossell, M. D., Erni, R., & Van Tendeloo, G. (2011). Three-dimensional atomic imaging of crystalline nanoparticles. *Nature*, *470*(7334), 374–377. doi:10.1038/nature09741 PMID:21289625

Vatanpour, V., Madaeni, S. S., Rajabi, L., Zinadini, S., & Derakhshan, A. A. (2012). Boehmite nanoparticles as a new nanofiller for preparation of antifouling mixed matrix membranes. *Journal of Membrane Science*, *401*, 132–143. doi:10.1016/j.memsci.2012.01.040

Villmow, T., Pötschke, P., Pegel, S., Häussler, L., & Kretzschmar, B. (2008). Influence of twin-screw extrusion conditions on the dispersion of multi-walled carbon nanotubes in a poly(lactic acid) matrix. *Polymer*, *49*(16), 3500–3509. doi:10.1016/j.polymer.2008.06.010

Vladár, A. E., Radi, Z., Postek, M. T., & Joy, D. C. (2006). Nanotip electron gun for the scanning electron microscope. *Scanning*, *28*(3), 133–141. doi:10.1002ca.4950280301 PMID:16878784

Wagner, J. B., Cavalca, F., Damsgaard, C. D., Duchstein, L. D. L., & Hansen, T. W. (2012). Exploring the environmental transmission electron microscope. *Micron (Oxford, England)*, *43*(11), 1169–1175. doi:10.1016/j.micron.2012.02.008 PMID:22459250

Wang, H. (2009). Dispersing carbon nanotubes using surfactants. *Current Opinion in Colloid & Interface Science*, *14*(5), 364–371. doi:10.1016/j.cocis.2009.06.004

Wang, X., Du, Y., Yang, J., Tang, Y., & Luo, J. (2008). Preparation, characterization, and antimicrobial activity of quaternized chitosan/organic montmorillonite nanocomposites. *Journal of Biomedical Materials Research*, *84A*(2), 384–390. doi:10.1002/jbm.a.31326 PMID:17618481

Wang, X. Q., & Mujumdar, A. S. (2008). A review on nanofluids—part I: Theoretical and numerical investigations. *Brazilian Journal of Chemical Engineering, 25*(4), 613–630. doi:10.1590/S0104-66322008000400001

Wang, Y., Xing, C., Guan, J., & Li, Y. (2017). Towards flexible dielectric materials with high dielectric constant and low loss: PVDF nanocomposites with both homogenously dispersed CNTs and ionic liquids nanodomains. *Polymers, 9*(562). doi:10.3390/polym9110562

Wang, Y.-P., Gao, X.-H., Wang, R.-M., Liu, H.-G., Yang, C., & Xiong, Y.-B. (2008). Effect of functionalized montmorillonite addition on the thermal properties and ionic conductivity of PVDF–PEG polymer electrolyte. *Reactive & Functional Polymers, 68*(7), 1170–1177. doi:10.1016/j.reactfunctpolym.2008.04.002

Ward, I. M., & Sweeney, J. (2004). *An introduction to the mechanical properties of solid polymers* (2nd ed.). Wiley.

Will Soutter. (2012). Nanotechnology in fuel cells. Retrieved from https://www.azonano.com/article.aspx?ArticleID=3032

Wu, C. L., Zhang, M. Q., Rong, M. Z., & Friedrich, K. (2005). Silica nanoparticles filled polypropylene: Effects of particle surface treatment, matrix ductility and particle species on mechanical performance of the composites. *Composites Science and Technology, 65*(3-4), 635–645. doi:10.1016/j.compscitech.2004.09.004

Wu, D., Sun, Y., & Zhang, M. (2009). Kinetics study on melt compounding of carbon nanotube/polypropylene nanocomposites. *Journal of Polymer Science. Part B, Polymer Physics, 47*(6), 608–618. doi:10.1002/polb.21668

Xia, Y., Xiong, Y., Lim, B., & Skrabalak, S. E. (2008). Shape-controlled synthesis of metal nanocrystals: Simple chemistry meets complex physics? *Angewandte Chemie International Edition, 47*, 2–46. PMID:19053095

Xing, W., Wilson, A. M., Eguchi, K., Zank, G., & Dahn, J. R. (1997). Pyrolyzed polysiloxanes for use as anode materials in lithium-ion batteries. *Journal of the Electrochemical Society, 144*(7), 2410–2416. doi:10.1149/1.1837828

Xu, G., Nie, P., Dou, H., Ding, B., Li, L., & Zhang, X. (2017). Exploring metal organic frameworks for energy storage in batteries and supercapacitors. *Materials Today, 20*(4), 191–209. doi:10.1016/j.mattod.2016.10.003

Yakimov, A. & Forrest, S.R. (2002). High photovoltage multiple-heterojunction organic solar cells incorporating interfacial metallic nanoclusters. *Applied Physics Letters, 80*(1667). doi:10.1063/1.1457531

Yang, C., Wei, H., Guan, L., Guo, J., Wang, Y., Yan, X., Zhang, X., Wei, S. & Guo, Z. (2015). Polymer nanocomposite for energy storage, energy saving, and anticorrosion. *Journal of material Science Chemistry A, 3*(29), 14929-14941.

Yang, C.-C., & Lin, C.-T. (2008). Preparation and characterization of the PVA/TiO_2 composite polymer membrane and its application on alkaline DMFC. *ECS Transactions, 13*(23). doi:10.1149/1.3010874

Yao, H., & Kimura, K. (2007). Field emission scanning electron microscopy for structural characterization of 3D gold nanoparticle superlattices. In Modern Research and Educational Topics in Microscopy (pp. 568-575).

Yavari, F., Rafiee, M. A., Rafiee, J., Yu, Z. Z., & Koratkar, N. (2010). Dramatic increase in fatigue life in hierarchical graphene composites. *ACS Applied Materials & Interfaces*, *2*(10), 2738–2743. doi:10.1021/am100728r PMID:20863061

Yi, W., Malkovskiy, A., Xu, Y., Wang, X.-Q., Sokolov, A. P., Lebron-Colon, M., ... Pang, Y. (2010). Polymer conformation-assisted wrapping of single-walled carbon nanotube: The impact of cis-vinylene linkage. *Polymer*, *51*(2), 475–481. doi:10.1016/j.polymer.2009.11.052

Yinghong, X., Xin, W., Xujie, Y., & Lude, L. (2003). Nanometre-sized TiO_2 as applied to the modification of unsatured polyester resin. *Materials Chemistry and Physics*, *77*(2), 609–611. doi:10.1016/S0254-0584(02)00114-1

Yu, W., & Xie, H. (2011). A review on nanofluids: Preparation, stability mechanisms, and applications. *Journal of Nanomaterials*. doi:10.1155/2012/435873

Zakaria, Z., Kamarudin, S., & Timmiati, S. (2016). Membranes for direct ethanol fuel cells: An overview. *Applied Energy*, *163*, 334–342. doi:10.1016/j.apenergy.2015.10.124

Zang, Z., Breidt, C., Chang, L., Haupert, F., & Friedrich, K. (2004). *Enhancement of the wear resistance of epoxy: short carbon fibre, graphite, ptfe and nano-TiO_2. In Composites Part A Applied Science and Manufacturing*. Elsevier. doi:10.1016/j.compositesa.2004.05.005

Zhang, Q., Dandeneau, C. S., Zhou, X., & Cao, G. (2009). ZnO nanostructures for dye-sensitized solar cells. *Advanced Materials*, *21*(41), 4087–4108. doi:10.1002/adma.200803827

Zhang, Y., Chi, Q., Liu, L., Zhang, C., Chen, C., Wang, X., & Lei, Q. (2017). Enhanced electric polarization and breakdown strength in the all-organic sandwich-structured poly(vinylidene fluoride)-based dielectric film for high energy density capacitor. *APL Materials*, *5*(7), 076109. doi:10.1063/1.4995653

Zhao, X. G., Jin, E. M., & Gu, H.-B. (2013). Increased charge transfer of PVDF-HFP based electrolyte by addition of graphite nanofiber and its application in dye-sensitized solar cells. *Applied Surface Science*, *287*, 8–12. doi:10.1016/j.apsusc.2013.09.017

Zheng, H., Meng, Y. S., & Zhu, Y. (2015). Frontiers of in situ electron microscopy. *MRS Bulletin*, *40*(1), 12–18. doi:10.1557/mrs.2014.305

Zhi, C., Bando, Y., Tang, C., Xie, R., Sekiguchi, T., & Golberg, D. (2005). Perfectly dissolved boron nitride nanotubes due to polymer wrapping. *Journal of the American Chemical Society*, *127*(46), 15996–15997. doi:10.1021/ja053917c PMID:16287265

Zhou, W., Zhou, Z., Song, S., Li, W., Sun, G., Tsiakaras, P., & Xin, Q. (2003). Pt based anode catalysts for direct ethanol fuel cells. *Applied Catalysis B: Environmental*, *46*(2), 273–285. doi:10.1016/S0926-3373(03)00218-2

Zhou, X., Zhu, Y., Liang, J., & Yu, S. (2010). New fabrication and mechanical properties of styrene-butadiene rubber/carbon nanotubes nanocomposite. *Journal of Materials Science and Technology*, *26*(12), 1127–1132. doi:10.1016/S1005-0302(11)60012-1

Zou, W., Du, Z., Liu, Y., Yang, X., Li, H., & Zhang, C. (2008). Functionalization of MWNTs using polyacryloyl chloride and the properties of CNT–epoxy matrix nanocomposites. *Composites Science and Technology*, *68*(15-16), 3259–3264. doi:10.1016/j.compscitech.2008.08.011

Chapter 3
Polymer/Clay Nanocomposites Produced by Dispersing Layered Silicates in Thermoplastic Melts

S. S. Pesetskii
V. A. Belyi Metal Polymer Research Institute of National Academy of Sciences of Belarus, Belarus

S. P. Bogdanovich
V. A. Belyi Metal Polymer Research Institute of National Academy of Sciences of Belarus, Belarus

V. N. Aderikha
V. A. Belyi Metal Polymer Research Institute of National Academy of Sciences of Belarus, Belarus

ABSTRACT

Results of the studies of technology, structural features and properties of polymer/clay nanocomposites (n-PCM) prepared by melt compounding of thermoplastic polymers are systematized. Special attention is given to the analysis of the effect of nanoclays modification with surfactants on properties of nanocomposites and preparation features of nanomaterials based on polar, non-polar thermoplastics and polymer blends. Effect of technological factors and special compounding regimes in the technology of n-PCM with advanced technical characteristics is considered. Results of the original studies of the structure and properties of the hybrid composites, filled by high modulus fibers in addition to nanoclays, are presented.

INTRODUCTION

Polymer/clay nanocomposites (n-PCM) contain a small amount of a layered clay silicate (LCS) arbitrarily dispersed in a polymer matrix. In the molded state, mechanical, thermal, barrier and other properties of n-PCM are significantly improved compared to traditional composites. Some of the first results in the field published in (Okada & Kawasumi, 1987) were amazing and, at least, initially considered as surprising, due to the "nano-effect." They became widely known since the publication of several reviews on technology, structure and properties of n-PCM (Okada & Usuki, 2006; Pavlidou & Papaspyrides, 2008; Paul & Robeson, 2008).

DOI: 10.4018/978-1-5225-7838-3.ch003

A significant interest in these materials is stimulated, primarily, by the desire to prepare materials, featuring a low cost combined with a superior set of performance characteristics compared to traditional composites (Gerasin et al., 2013; Pesetskii, Bogdanovich & Myshkin, 2013).

Homogeneous distribution of LCS nanoparticles in a polymer matrix produces a vast interphase boundary and intense interphase interactions can be expected. For the nanoparticles to maintain such a state it is important that direct interparticle contact is avoided. To estimate the upper concentration of a nanofiller Okada and Uzuki assumed that the shape of the particles is tetrahedrical measuring 1×100×100 nm. In that case, full interparticle contact along all the planes is achieved in nanocomposites at ≈7.5 wt.%, while in case when the particles contact with all the vertices of the tetrahedron the upper concentration limit is achieved at 1.5 vol.% (3.8 wt.%). It led them to a conclusion that the upper limit of a clay content in n-PCM should not exceed several percent.

Methods of preparing n-PCM are divided into three main categories: in-situ polymerization (polycondensation) and intercalation in a polymer solution or in the melt. Melt intercalation possesses evident advantages. It was first implemented by Vaia et al. (Vaia, Ishii & Giannelis, 1993). Melt intercalation is not applicable for thermosetting matrices because of accompaning crosslinking processes, so n-PCM based on thermosets are most often prepared using the clays pre-swelled in resin solvents or using water suspensions of clays in a resin solvent (Domenech, Peuvrel-Disdier & Vergnes, 2011). Elastomer-based n-PCMs may be produced via similar routes, although the more promising techniques in this case are latex intercalation and melt compounding (Finnigan, 2004; Fischer, 2003). It should be marked that due to the specific technological equipment used in preparation of rubber compounds and the molecular structure of rubber (low shear effects on the melt, high molecular weight and melt viscosity of rubber), it is difficult to produce an exfoliated LCS in the final rubber (Finnigan, 2004).

It should be noted that despite a large number of original publications reviewing some aspects of technology, structure and properties of n-PCM prepared by melt compounding, the up-to-date overview information in this area is rather limited and focuses mainly on particular types of n-PCM (Frontini & Pouzara, 2015; Ray, 2015; Lu, Wang & Chua, 2008), which makes it difficult to analyze the overall situation and prospects for the development of this field. The present review attempts to analyze and generalize the published results of research and original results of the authors, and consider the most important directions of the development of the technology of n-PCM prepared by compounding LCS in thermoplastic melts.

BRIEF DESCRIPTION OF NANOCLAYS AND TECHNIQUES OF N-PCM ANALYSIS

Layered silicates routinely used for preparing nanocomposites include montmorillonite (MMT), hectorite and saponite, which belong to the series of 2:1 clay minerals or phyllosilicates. They possess two main properties, i.e. the ability to disperse to individual layers in the polymer bulk (layer thickness ≈1 nm, lateral size from 30 nm to several microns), the ability to adjust the chemical activity through reactions of ionic exchange with organic and inorganic cations. The gap between the layers or the gallery may be varied by ionic exchange.

The aspect or the characteristic ratio (P) in a commercial organoclay (surfactant-treated clay) of montmorillonite type amounts to about 280 ($p = d/h \approx 280$, where d – diameter, h – thickness of an individual clay platelet). An average platelet of a smectic clay is about 1 nm thick, 150 nm wide and 500 nm long.

The aspect ratio in commercial clays may reach 6000. Natural clays may contain three main types of impurities: platelets with crystallographic defects, sand (approximately from 0.3 to 2 mass %, mainly of quartz, with the diameter greater than 300 nm) and amorphous silicates (Jlassi, Chehimi & Thomas, 2017). Natural MMT is quite abundant, and it is relatively cheap, however, it may contain significant quantities of impurities and its composition is variable. This necessitates thorough purification of the natural MMT intended for preparation of n-PCM.

Peculiarities of LCS chemical structure are well described in publications (Jlassi, Chehimi & Thomas, 2017). Montmorillonite clay represents a crystalline solid composed by platelets having an inner octahedral layer of aluminum oxide (alumina) fused to two tetrahedrally coordinated layers of silicate (silicon dioxide, silica) via edge-shared silicon atoms. The octahedral layer may be regarded as a layer of alumina in which several atoms of aluminum are substituted by magnesium. The difference in valence between Al and Mg produces a negative charge distributed within the layer counterbalanced by cations (often by Na^+ ions) located between the platelets (in the galleries). In a natural state, this clay exists as piles of numerous platelets. Hydration of sodium ions causes expansion of galleries and swelling of clay; moreover, these platelets can be completely dispersed (exfoliated) in water. Ions of Na^+ may be exchanged for organic cations, e.g. of ammonium, bonded with hydrocarbon radicals in the structure of ammonium surfactants producing a so-called organomodified clay or organoclay (Yariv & Cross, 2002).

The value of the negative surface charge is known as the cation exchange capacity, CEC. The distance between the platelets determined by X-ray analysis (d_{001}) in the completely dried sodium montmorillonite amounts to 0.96 nm and the platelet thickness to 0.94 nm.

It is important to note that the chemical processes occurring at the surface of clay particles can be controlled through ion exchange reactions with various inorganic and organic cations, which allows to control the interfacial interaction with macromolecules of different types.

When clay is introduced into the polymer matrix, materials with three different types of structures can be produced: microcomposites with separated phases, intercalated and exfoliated nanocomposites (Kotal & Bhowmick, 2015).

When macromolecules do not penetrate the clay galleries, the clay remains in the form of a densely packed structure and the resulting material represents a microcomposite with separated phases. The properties of such composites are little different from those of traditional composites filled with fine mineral fillers. When macromolecules penetrate the spaces between the clay layers then compounding produces an intercalated nanocomposite, in which the clay preserves an ordered multilayered structure. However, the presence of polymer in the interlayer space causes a certain increase in the interlayer distance of the clay. If penetration of the macromolecular chains into the clay galleries is more complete, the clay tactoids may be separated into single layers about 1 nm thick producing an exfoliated nanocomposite. It should be marked that, in practice, the nanocomposites based on polymer and clay often display a mixed morphology.

The nature of clay distribution in a polymer matrix is most often analyzed by wide-angle X-ray diffraction and transmission electron microscopy (TEM). X-ray diffraction allows to determine the interlayer distance in the original clay and in the produced composites. The angle of diffraction (θ) relates to interlayer distance (d) by Bragg equation: $\lambda = 2d \cdot \sin\theta$, where λ denotes the wave length of the radiation used. The width of the diffraction peak is a measure of the degree of ordering of the layers, and the intensity of the peaks gives information on the number of layers in the multilayer structure (content of non-exfoliated clay). The layered structure can be analyzed using the X-ray diffraction method if the layers are arranged in parallel and approximately at the same distance from each other. In exfoliated nanocomposites, the initial structure of the filler is completely destroyed (exfoliated) and LCS particles are randomly distributed in the polymer matrix. Therefore, when analyzing such materials by X-ray diffraction, the reflexes of the original crystal structure are not detected in X-ray diffraction patterns. As for intercalated nanocomposites, the distance between individual layers of the clay increases as the result of the penetration (intercalation) of the polymer macromolecules into the gallery. If clay particles are weakly bonded to the polymer matrix, the diffraction peak is observed at approximately the same values of θ as in the clay samples in absence of the polymer (Figure 1, B).

In the x-ray diffraction analysis of the degree of nanoclay exfoliation, two variants of changes in the diffraction curves can be observed (Figure 1): the displacement of the angular position of the 001 reflex to the smaller angles (Figure 1, A), which is a consequence of the penetration of macrochains into the galleries, and a decrease in the intensity of the diffraction peak, due to mechanical delamination of clay particles under the action of shear forces (Figure 1, B).

The X-ray technique allows to analyze the interlayer distance in the filler, both in clays and in n-PCM. However, the information obtained by this method does not allow to draw any conclusions on the spatial distribution of the filler layers in the polymer matrix. Therefore, the results obtained by X-ray diffraction are usually supplemented with TEM data.

When interpreting the data obtained, several features should be taken into account, which may cast doubt on the correctness of the conclusions. As shown in Figure 1, the organoclay has a characteristic maximum, indicating remoteness of the plates from each other by a distance of d_{001}. The presence of the same maximum in n-PCM is an irrefutable confirmation that the nanocomposite contains organoclay tactoids. However, the absence of a maximum is not a full proof of the existence of a completely exfoliated structure (Morgan & Gilman, 2003). Many factors need to be taken into account when interpreting these data. If the sensitivity of the instrument is low, the existing maximum may not be observed. When the tactoids are internally disordered or not sufficiently aligned with each other, the intensity of the maximum will be low, or it may be absent altogether (Lee, Fasulo, Rodgers & Paul, 2005).

Figure 1. Two mechanisms of intercalation of nanoclay in the polymer bulk; the maxima on the curves are the angular position of the reflection 001. (a) - the increase in the interlayer distance due to the penetration of macrochains into the gallery: (b) - clay delamination due to shear mixing in the melt

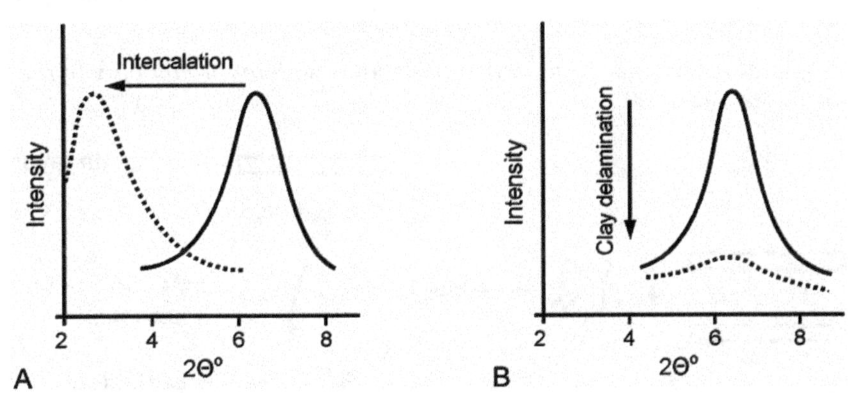

In some cases, wide-angle X-ray diffraction can reveal a reverse shift (a peak shift to larger angles) can also occur, and this is usually associated with a loss of the unbound (not chemisorbed on the particle surface) surfactant from the gallery or with its thermal decomposition (Yoon, Hunter & Paul, 2003; Shah & Paul, 2006). All these processes can occur in aggregate, causing uncertainty in intercalation estimates.

It should be noted that small-angle X-ray scattering can be more informative than the wide-angle diffraction, as stressed by many authors (Lincoln, Vaia & Krishnamoorti, 2004; Vaia, Liu & Koerner, 2003). However, this technique is not widely used, except for individual laboratories, probably because most research centers lack the equipment or experience in interpreting the results. Other methods, such as nuclear magnetic resonance (NMR) in the solid state and neutron scattering, have also been used to a limited extent to investigate the clay dispersion (VanderHart, Asano & Gilman, 2001; Schmidt, Nakatani, Butler & Han, 2002; Corté & Leibler, 2005). NMR results correspond with TEM images, however, it should be noted that the advantage of NMR is that it yields the data typical for the entire specimen and ensures its accurate quantitative interpretation.

The use of TEM reveals morphology in a very small area, however, this disadvantage can be overcome by imaging at different magnifications, different locations and orientations. When the exfoliation of clay is complete, the TEM images display clay platelets ≈1 nm thick as dark lines (the microtome cut perpendicular to the platelets). The image analysis can be used to quantify the distribution of platelet lengths, but significant statistical data require the analysis of several hundred particles. However, it should be kept in mind that the observed dimensions reflect an arbitrary cut of an irregular platelet, and the maximum size is rarely observed. The platelet size distribution observed by TEM therefore gives smaller values than the actual ones.

FEATURES OF EXTRUSION COMPOUNDING OF N-PCM IN POLYMER MELTS: ADVANTAGES OF TECHNOLOGY

The preparation of n-PCM by melt intercalation involves the step of mixing the polymer binder and the LCS at a temperature above the temperature of melting (viscous flow) of the polymer. During mixing, the polymer chains diffuse into the galleries between the silicate layers, which is shown schematically in Figure 2.

Figure 2. Scheme of interaction of polymer macromolecules with an organoclay in the melt (Pesetskii, Bogdanovich & Myshkin, 2013)

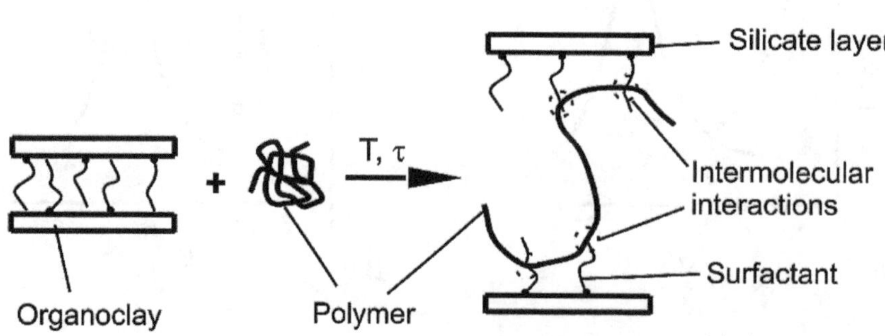

Melt compounding can be much more economical and easier to implement than other methods. At the same time, capital costs are reduced to a minimum, taking into account the possibility of implementing the method using serial compounding and mixing equipment. Producers of n-PCM have more opportunities to create a wide range of end products by varying the nature of the polymer, organoclay, its concentration and modifying additives. This technology can be implemented in medium and small enterprises in contrast to the synthesis of n-PCM in situ, which can be carried out only at specialized enterprises for the synthesis of polymers. Melt compounding is environmentally friendly, since the use of solvents is excluded (Pesetskii, Bogdanovich & Myshkin, 2013; Liu & Wu, 2002; Fornes, Yoon, Keskkula & Paul, 2001; McNally, Raymond Murphy, Lew, Turner & Brennan, 2003).

It should be noted that technology of melt intercalation is used in manufacture of n-PCM with most thermoplastic matrices, including polyamides (PA), polyethylene terephthalate (PET) (Davis et al., 2002), recycled PET (Pegoretti, Kolarik, Peroni & Migliaresi, 2004), ethylene vinyl acetate copolymer (EVA) (Srivastava, Pramanik & Acharya, 2005; Tang et al., 2003), thermoplastic polyurethane (TPU) (Finnigan, 2004), polyolefins (PO) (Gopakumar, Lee, Kontopoulou & Parent, 2002), polylactides (PLA) (Sinha Ray, Maiti, Okamoto, Yamada & Ueda, 2002; Pluta, 2004), polycaprolactone (PCL) (Di, Iannace, Di Maio & Nicolais, 2003), blends of thermoplastic polymers (Chow & Mohd Ishak, 2015), elastomers, including tire rubbers (Galimberti, 2011). Let us consider the most important factors determining the mechanism and degree of LCS intercalation in thermoplastic melts, and hence the properties of n-PCM.

THERMODYNAMIC ASPECTS OF N-PCM FORMATION

From the viewpoint of thermodynamics, clay exfoliation should proceed at a negative value of free energy in the polymer/clay mixtures: $\Delta G = \Delta H - T\Delta S < 0$. In the process of nanocomposite formation surfactant molecules acquire greater conformational freedom as the volume of interlayer space increases, whereas macromolecules from the melt penetrate a narrow gap between the nanoparticles and their mobility is reduced. The total entropy change (ΔS) in the system of the clay-polymer-surfactant is negative. It is, therefore, necessary that interaction energy in the system compensated the entropy barrier to yield a thermodynamically stable nanocomposite. In this regard, the ability of the polymer to penetrate the interlayer space and the conformational behavior of macromolecules in it, is primarily determined by the polarity of the macromolecules, the type of clays and surfactants used for their organomodification, and also the compounding technology, above all the degree of thermal and force factors, uniformity of the distribution of nanoparticles in the polymer matrix bulk (Fischer, 2003).

It should be noted that even when the alkyl or other surfactant radicals located in the gallery can mix with the macromolecules of the polymer matrix, intensive interactions at the polymer-filler surface are required to overcome the entropic barrier and ensure complete separation of the clay platelets. If these interactions are not sufficient, good dispersion of the particles can be produced by using strong shear forces during compounding of n-PCM, but the system remains thermodynamically unstable (Fischer, 2003).

EFFECT OF SURFACTANTS

The initial LCSs usually contain hydrated Na^+ and K^+ ions. Therefore, in the original (not modified) state, layered silicates mix only with hydrophilic polymers. To improve the wettability of melts of other (including weakly polar) polymer matrices, the surface of the layered mineral is rendered organophilic by exchange reactions with cationic surfactants, which usually include primary, secondary, tertiary or quaternary ammonium or alkylphosphonium cations (Pavlidou & Papaspyrides, 2008).

Factors determining the effectiveness of surfactants: NH_4^+-R are usually most effective; the length of the alkyl chain in the ammonium surfactant should be more than eight carbon atoms ($> C_8$);, the thermal stability of a surfactant is of great importance in melt compounding (when the surfactant degrades in the gallery, the interplanar distance decreases); an optimum density (concentration) of the chemisorbed surfactant on the clay particle should be ensured. Other types of LCS modifiers: polyethylene oxide (PEO) - nonionic surfactant, polymeric (oligomeric) ammonium salts, partially fluorinated surfactants, silane coupling agents, organotitanates, stabilized surfactants (Pavlidou & Papaspyrides, 2008; Gerasin et al., 2013).

Scheme of chemical interaction (ion-exchange reaction) of an ammonium surfactant with clay particles is shown in Figure 3. Inorganic, relatively small cations, for example, Na^+, are exchanged in the interlayer space (gallery) with bulkier organic cations (Figure 3A). The ion-exchange reaction leads to two results: first, the gap between the layers increases, enabling penetration of polymer chains, and secondly, the surface properties of each layer change from hydrophilic to hydrophobic.

Possible structures formed by surfactants in the interlayer space of nanoclays are shown in Figure 3B (Vaia, Teukolsky & Giannelis, 1994): horizontal monolayer (1), horizontal bilayer (2), horizontal pseudo-triple layer (3), monolayer of "paraffin" type ("pseudo-double") (4), bilayer of "paraffin" type (5).

Figure 3. Scheme of interaction of a cationic surfactant with clay particles (explanations in the text) (Vaia, Teukolsky & Giannelis, 1994; Fischer, 2003)

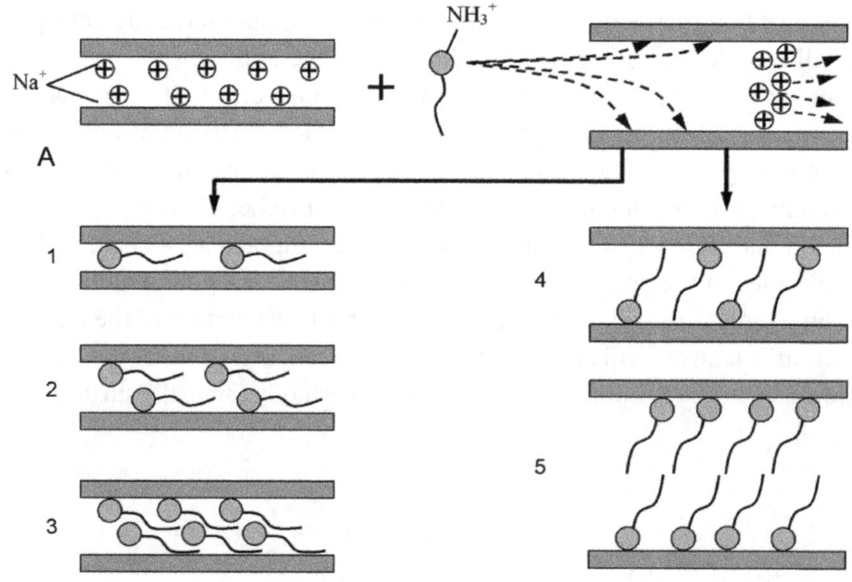

It should be noted that in order to completely replace alkali metal ions, the surfactant content in the clay should be about 35-45 wt.% based on the clay CEC (Yu, Yang, Zhang, Zhao & Mai, 2003). In commercial organoclays, though, the surfactant concentration very often exceeds this value and reaches ≈ 60 wt.% (Sodyleva, Kuznetsova, Bogdanovich, Okhlopkova & Pesetskii, 2016). In general, the longer the length of the surfactant chain and the higher the charge density of the clay, the potentially greater is the distance the clay monolayers can be separated to.

For example, Wang et al. (Wang, Choi, Koo, Choi & Chung, 2001) studied n-PCM based on polyethylene and organoclays with different lengths of the alkylammonium chain, as well as Cloisite 20A organoclay, which has two long alkyl chains (each containing two methyl and octadecyl radicals). It was found that the interlayer distance d_{001} increases with the length of the alkylamine chain and reaches 1.36, 1.79, 1.85 and 2.47 nm for the surfactants with 12, 16 and 18 carbon atoms in the alkylammonium chain and Cloisite 20A, respectively. It also follows from this work that the maximum values of d_{001} are reached when hydrocarbon radicals of short length (C_1) are present along with the long-chain ones (C_{18}) in the surfactant structure.

Therefore, the most important types of industrial surfactants, as a rule, include radicals of different length and polarity. These are mainly quaternary ammonium salts of hydrochloric acid, in which the nitrogen cation is bonded with methyl, hydroxyethyl and longer saturated and unsaturated hydrocarbon radicals produced from tallow, rapeseed or coconut oils.

Industrial clays contain sorbed water in the galleries despite the hydrophobic nature of most surfactants. Its concentration may reach several percent, due to the hydrophilicity of the clay itself, and complete removal of the sorbed water is very problematic. In addition, the process of sorption-desorption of water by clay has a dynamic character. Dry clay relatively quickly absorbs water in the air. This should be taken into account in practice, especially in production of n-PCM based on hydrolysis-sensitive polymers. Removal of water does not significantly affect the volume of the gallery (Sodyleva, Kuznetsova, Bogdanovich, Okhlopkova & Pesetskii, 2016).

Selection of surfactants should be given special attention in the production of n-PCM based on both polar and nonpolar polymers.

Surfactants play a very important role in polyester/clay nanocomposites owing to high processing temperatures (the melting point of PET is ≈255 °C), and to the sensitivity of ester bonds in the composition of macromolecules to thermal, mechanical and hydrolytic degradation (Pesetskii, Shevchenko & Dubrovsky, 2015). These circumstances complicate the correct choice of surfactants. In contrast to aliphatic polyamides, the use of alkyl ammonium compounds as surfactants for PBTs and especially PET is less effective because of the possible aminolysis of the ester bond resulting in excessive degradation of the binder. Therefore, 1,2-dimethyl-3-N-alkylimidazole (Davis et al., 2002), pentaerythritol and maleic anhydride (Sanchez-Solis, Garcia-Rejon & Manero, 2003), cetylpyridinium chloride (Xiao et al., 2005) have been tested for treating clay mineral particles to control their interaction with PET.

To ensure good dispersion of clay in the PP melt, particles of layered silicates are successfully treated by ammonium surfactants with a long hydrocarbon radical (stearylammonium) (Kawasumi, Hasegawa, Kato, Usuki & Okada, 1997). Exfoliation of the mineral is facilitated and bonding between the polymer matrix and nanoparticles is enhanced when PP is modified with maleic anhydride (MA)-grafted PP. Grafting of polar fragments to PO macromolecules intensifies their adhesion interaction with nanoparticles and improves the materials strength. In (Gerasin et al., 2007) attention is drawn to the concentration of surfactants introduced into the gallery during the production of organoclays. It is determined in fractions of CEC. For effective intercalation of nonpolar polymers, it is necessary that aliphatic chains of surfactants are oriented in clay galleries at an angle close to 90°, which ensures the maximum distance between the layered silicate platelets. The concentration of the surfactant should be approximately 1 CEC, which allows to produce adsorption layers of surfactants with large free volume.

It is important to note that in addition to the chemical activity of the clay, its quantity introduced into the polymer matrix also plays a decisive role. It is reported (Fischer, 2003) that low clay concentrations contribute to exfoliation, whereas an increased amount of clay (typically higher than 10 wt%) allows only intercalation thereof. The exfoliation of clay platelets is inhibited, apparently due to percolation phenomena.

EFFECT OF THE POLYMER MATRIX

Parameters of the polymer matrix, such as molecular weight, polarity, melt viscosity, thermal stability can determine the process of mixing LCS in the polymer melt and the properties of the n-PCM produced.

In (Fornes, Yoon, Keskkula & Paul, 2001), n-PCM based on three grades of PA6 with different molecular weights (low, medium and high, LMW, MMW, HMW) were studied. The results of wide-angle X-ray diffraction indicate that the higher degrees of delamination of clay particles for nanocomposites are attained in the order of HMW> MMW> LMW. The composites based on HMW polymer demonstrate the highest tensile strength.

The most important condition for the formation of intercalated and especially exfoliated clays in melt compounding is the realization of the polar type interactions (stronger than the van der Waals ones). In this context, polar polymers having groups capable of associated type interactions, such as Lewis interactions or hydrogen bonding, facilitate the intercalation of macromolecules into silicate galleries (Pavlidou & Papaspyrides, 2008).

The importance of polar interactions is well illustrated by the example of a copolymer of ethylene with vinyl acetate (EVA) (Chaudhary, Prasad, Gupta & Bhattacharya, 2005). It was found that the presence of polar groups (ester groups of vinyl acetate fragments) improves its ability to intercalate organo-modified montmorillonites (the higher the content of vinyl acetate (VA), the greater was the increase in the interlayer distance d_{001}). The increase in VA content in EVA from 9.18 to 28 mol.% reduces the thermodynamic barrier of clay-polymer interaction and allows a relatively larger number of polymer chains migrate into the gallery to form partially and/or completely exfoliated clay particles.

CHALLENGES IN CLAY INTERCALATION WITH NONPOLAR POLYMERS

At present, the most promising strategy of compounding involves additional introduction of a small amount of a polyolefin (PO) with a grafted polar monomer (more often MA), which is well compatible with the base PO. Polar MA provides affinity with clay materials, so that functionalized PO (PO-g-MA) can perform the functions of a compatibilizer between the matrix and the filler (Lu, Wang & Chua, 2008; Hu, Lu & Song, 2008). Functionalized PO are effective additives not only for PO polymers and copolymers but also for other polymers, in particular, PS (Wang & Wilkie, 2003).

In addition to PO-g-MA, other types of functionalized polyolefins (FPO), as well as additions of polar polymers (oligomers) and a variety of compounding techniques are used to improve exfoliation of clays.

According to (García-López, 2003), polar oligomers were added into PP matrix in order to prepare PP/clay nanocomposites with an improved dispersion of clay. A significant improvement in the mechanical properties of PP/clay nanocomposites was achieved by using PP-g-MA with grafted (aminopropyl) triethoxy silane (APTS) as a compatibilizer (Liaw, Huang, Chen, Lo & Chang, 2008). PPs with grafted diethyl maleate (García-López, 2003), itaconic acid (Moncada, Quijada, Lieberwirth & Yazdani-Pedram, 2006), functionalized side chains (containing hydroxyl, chloride and sulfochloride groups) (Kotek, Kelnar, Studenovský & Baldrian, 2005), as well as terminal functional groups (Wang, Nakajima, Manias & Chung, 2003).

BLENDS OF THERMOPLASTIC POLYMERS

Studies of the structure and properties of polymer blends containing LCS nanoparticles draw increased attention, which is confirmed by a large number of original works and individual surveys in this field (Pesetskii, Bogdanovich & Myshkin, 2013; Chow & Mohd Ishak, 2015; Pesetskii, Shevchenko & Dubrovsky, 2015). In general case, addition of nanoclay into a binary blend of incompatible polymers may form three types of structure depending upon the nature of preferential distribution of clay between the components: (1) in one of the phases, (2) in both polymer phases and (3) at the interphase boundary. Concurrently, at least three different types of changes in the morphology of nanocomposites can occur. Within the first type of morphology change, the particle size of the dispersed phase is reduced, which indicates the compatibilizing effect of LCS. The second type of morphology change represents the transition from fibrillated to co-continuous morphology. The third type is phase inversion, which is characterized by a transition from fibrillated to co-continuous morphology and return to fibrillated morphology at high shear rates in the melt.

It is believed that the thermodynamic stabilization of a nano-filled polymer blend is enhanced due to the large specific surface area of nanoclays and the anchoring of macromolecules of polymer components on the surface of solid nanoparticles due to physicochemical interaction (Nuzzo, Bilotti, Peijs, Acierno & Filippone, 2014).

It has been experimentally confirmed that nanoclays can significantly reduce the size of the dispersed phase in the polymer blend (Yoo, Tiwari, Yoo & Paul, 2010). This change can occur due to the barrier effect, which limits the contact of the polymer phases. The effect of compatibilization and the resulting properties are determined by the location of the clay and the degree of dispersion strongly influenced by the clay-polymer affinity. The addition of small amounts of nanoparticles can also affect the microstructure of incompatible polymer blends by stimulating the formation of co-continuous morphologies.

The sequence of mixing steps often plays a very important role in determining the phase morphology. In principle, it is possible to use two compounding technologies: one-stage, based on direct mixing of all components in the melt, and two-stage, based on the use of clay masterbatch based on one or both polymer components of the blend (Chow & Mohd Ishak, 2015).

The chemical properties of the nanofiller surface (the nature of the surfactant used) have a significant effect on the localization of nanoparticles in the polymer blend, hence, on their possible compatibilizing role and final properties of n-PCM. In most cases, the localization of the nanofiller at the interphase boundary is not clearly fixed, and the size and chemical activity of the surface are not well controlled parameters; therefore, nanoclays with the desired surface properties, controlled localization and dispersion in blends need to be developed and improved.

EFFECT OF HYBRID FILLING WITH ORGANOCLAY AND SHORT FIBERS

Synergistic strengthening effects were observed on concurrent filling of thermoplastics with organoclays and short glass fibers (GF) (Cho & Paul, 2001; Pedrazzoli & Pegoretti, 2013). According to (Cho & Paul, 2001) addition of 5 wt.% of clay increases the elasticity modulus by ≈38% compared to pure PA6 (for the PA6/GF composite the respective increase equals 22%), while for the triple PA6/clay/GF system, the modulus increases by 81%, which significantly exceeds the expected value based on the principle of additivity (Cho & Paul, 2001). The joint addition of nanoclay and GF in thermoplastics is manifested not only in non-additive changes in the mechanical strength of the composites but also in the shrinkage phenomena in injection molding (Garcia, Netto & Pontes, 2017).

The effects of synergistic reinforcement of thermoplastic n-PCM were recently observed in composites filled with basalt fiber (BF) and organoclay (Eslami-Farsani, Reza Khalili, Hedayatnasab & Soleimani, 2014; Pesetskii, Bogdanovich & Sodyleva, 2017), and in materials filled with carbon fiber (CF) and organoclay (Pesetskii et al., 2016).

In particular, in Pesetskii et al. (Pesetskii, Bogdanovich & Sodyleva, 2017) investigated the effect of short BF and organoclay on structure and mechanical properties of PA6. The organoclay used, Cloisite 30B (Cl30B) by Southern Clay Products Inc. USA, represents MMT modified with $(HE)_2M_1T_1$ where HE – hydroxyethyl $-CH_2CH_2OH$, M – methyl, T – tallil (65% C_{18}, 30% C_{16}, 5%C_{14}); The fibrous filler represented BF roving from basalt complex threads grade NRB 13-25 20-KB-11, produced by Kamenny Vek LLC, Russia (diameter of monofilaments 13 microns, linear density 2530 tex.). The organoclay content was varied from 0.5 to 6 wt.%, and that of BF was 15 wt.% and 35 wt.%. The composites were produced by melt compounding in a co-rotating twin screw extruder with the screw diameter D=35 mm and L/D=40.

It has been established that the addition of organoclay promotes the growth of tensile and flexural strength characteristics, and the reinforcing effect on hybrid filling is more pronounced at a lower BF content in composites. For example, for n-PCM with 35% and 15% BF, the maximum increase in bending strength (σ_b) upon the addition of Cl30B reaches 114% and 133%, and that of elasticity modulus 126% and 138% respectively.

It is also noted that organoclay additions cause shrinkage reduction during injection molding. Moreover, the strongest decline in this indicator by 1.6-1.7 times is observed with a minimum (0.5 wt.%) concentration of Cl30B in the composites. It is obvious that the presence of organoclay affects the structure of the composites, despite the presence of a significant amount of short BF, which can nucleate the crystallization of thermoplastics. According to (Pesetskii, Bogdanovich & Sodyleva, 2017), BF additives contribute to a slight increase in the degree of intercalation of the organoclay. The wide-angle X-ray diffraction data also indicate pronounced nucleating activity of the organoclay on crystallization of PA6 in the surface layers of the molded blocks, compared to the crystallization of PA6 in PA6/BF composites although the crystallinity of the underlying layers remains unchanged. Thus, the increase in the strength characteristics of PA6 on hybrid filling is not explained by the increase in the crystallinity of the binder.

From Figure 4A it is clear, that an increase in the concentration of organoclay in the composite is accompanied by an increase in the glass transition temperature (T_g) of PA6 amorphous phase, regardless of BF concentration in the material. This fact is an unquestionable consequence of the restriction of segmental mobility in the amorphous phase of PA6 induced by addition of the organoclay. The heat deflection temperature of the hybrid composites increases in comparison with PA6/BF materials due to intensive interphase interaction of macromolecules with the surface of nanoparticles, revealed by an increased level of the dynamic shear modulus of PA6/BF/C130B composites at the temperature of 160 °C, which substantially exceeds T_g of the neat PA6 (Figure 4B).

The noticeable effect of organoclay addition on interfacial adhesion is evidenced by the data of (Pesetskii et al., 2016), in which hybrid PA6 composites with short CF, organoclay and small addition of an isocyanate chain extender (0.4 wt.% of PA6) were studied. It is shown that melt compounding of n-PCM based on PA6 with CF and C130B organoclay produces composites with the synergistic improvement of static and impact strength. This effect is especially pronounced at a low (15 wt.%) CF content in the composites. The maximum increase in flexural strength upon addition of 6 wt.% of C130B into composites with 40 wt.% and 15 wt.% of CF reaches 106% and 132%, and that of tensile modulus 109% and 120%, respectively.

For high-filled (40 wt.%) CF-composites, the Charpy impact strength values somewhat decrease as the C130B concentration increases from 0.5 wt.% to 6 wt.%, and in low-filled (15 wt.%) CF-composites, it increases up to 1.7 times compared to the clay-free composite.

It has been found that, in spite of the expectation, addition of organoclay to the CF-filled PA6 in presence of an isocyanate-containing chain extender not only not increased, but even slightly reduced the polymer crystallinity according to DSC and XRD data, which is attributed to the steric hindrance to crystallization of PA6 macromolecules. In presence of the clay, the average CF length increases (more significantly, by a factor of 1.7 for composites with 40 wt.% of CF), and, according to DMA, the seg-

Figure 4. The effect of C130B organoclay concentration on the glass transition temperature (a) and the dynamic shear modulus (b, c) of n-PCM; a – BF concentration is 15 wt.% (1) and 35 wt.% (2); b, c – the concentration of BF is 15 wt.% (1.1') and 35wt.% (2.2'), test temperatures are (b) – 30 °C, (c) – 160 °C (Pesetskii, Bogdanovich & Sodyleva, 2017)

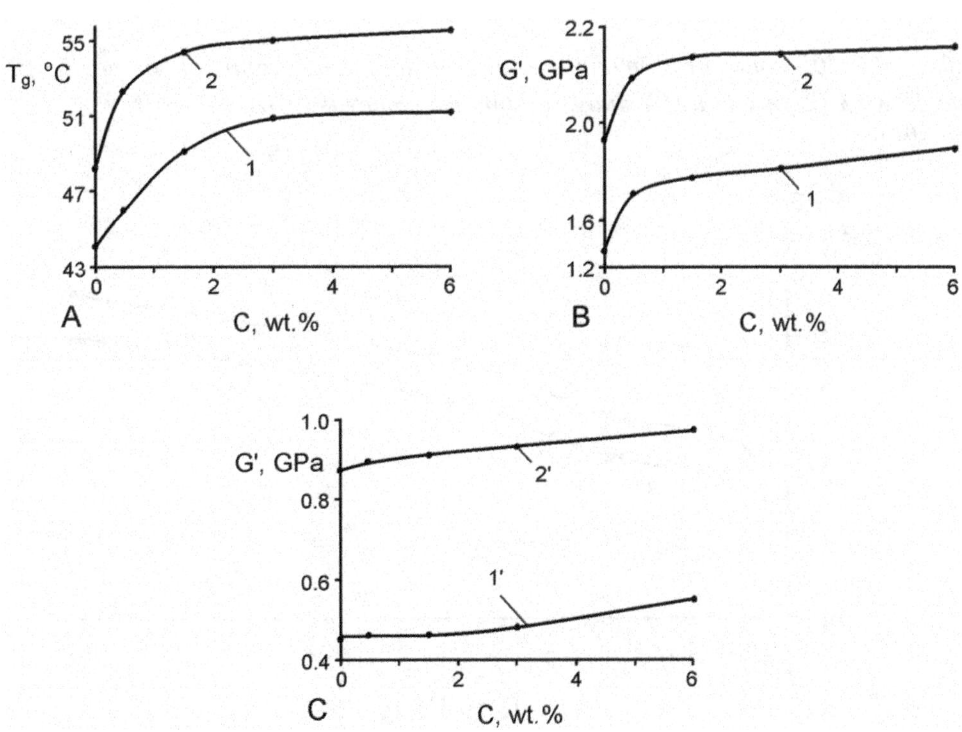

mental mobility of PA6 macromolecules is restricted and the interfacial adhesion in the polymer/hybrid filler system is increased.

It follows from Figure 5, that an increase in the organoclay concentration in the composite is accompanied by an increase in the glass transition temperature of the PA6 amorphous phase, and the nature of the change depends on the CF concentration in the material.

This fact is a consequence of the restriction of segmental mobility in the amorphous phase of PA6 with the addition of organoclay. At CF content of 40 wt.%, a larger increase (by 5-7 °C) in the T_g value is observed upon addition of Cl30B.

Interfacial interaction of macromolecules with the nanoparticles surface should increase the heat deflection temperature (HDT) of the hybrid composites as compared to PA6/CF materials which is revealed by increased values of the dynamic shear modulus G' of PA6/CF/Cl30B composites not only at temperatures below (30 °C) or close (90 °C) to T_g but also at T = 160 °C which significantly exceeds T_g of PA6 (Figure 6).

Interphase interaction of macromolecules with nanoparticles surface can be evaluated in a wide temperature range irrespective of orientation and geometrical parameters of the particles based on the data of relaxation spectrometry (DMA) by calculating the parameter A (Pesetskii et al., 2016; Kubát, Rigdahl & Welander, 1990):

$$A = \frac{\varphi_i}{1-\varphi_f} \cdot \frac{tg\delta_i}{tg\delta_m} = \frac{1}{1-\varphi_f} \cdot \frac{tg\delta_c}{tg\delta_m} \tag{1}$$

where ϕ_f, ϕ_i – volume fraction of the filler and the interphase layer respectively; $tg\delta_c, \cdot tg\delta_m, \cdot tg\delta_i$ – tangent of the angle of mechanical losses of the composite, matrix and the interphase layer.

Figure 5. Effect of Cl30B concentration on the glass transition temperature of composites filled with 40 wt.% (1) and 15 wt.% (2) of CF. DMA study, loading frequency is 1 Hz.
(Pesetskii et al., 2016)

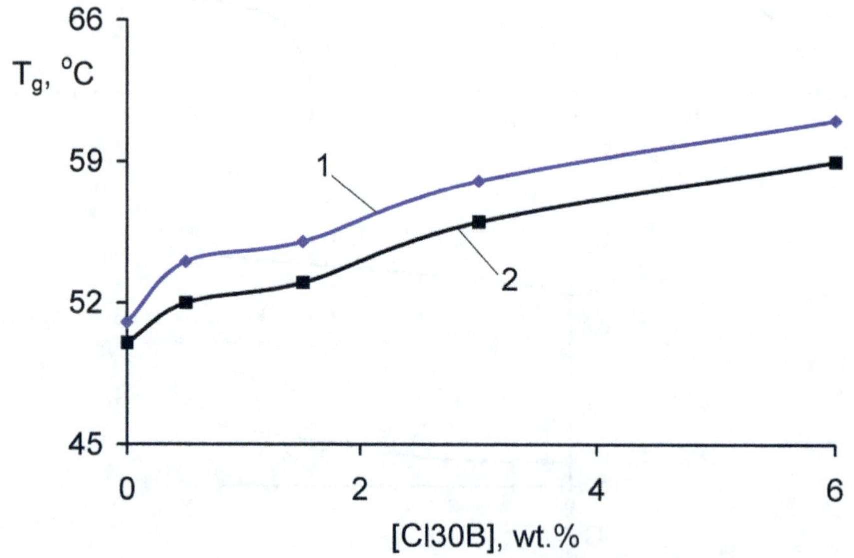

Figure 6. Effect of Cl30B concentration on the dynamic shear modulus (G') of PA6/CF/Cl30B composites at the temperature of 30 (a), 90 (b) and 160 °C (c). DMA study, loading frequency is 1 Hz
(Pesetskii et al., 2016)

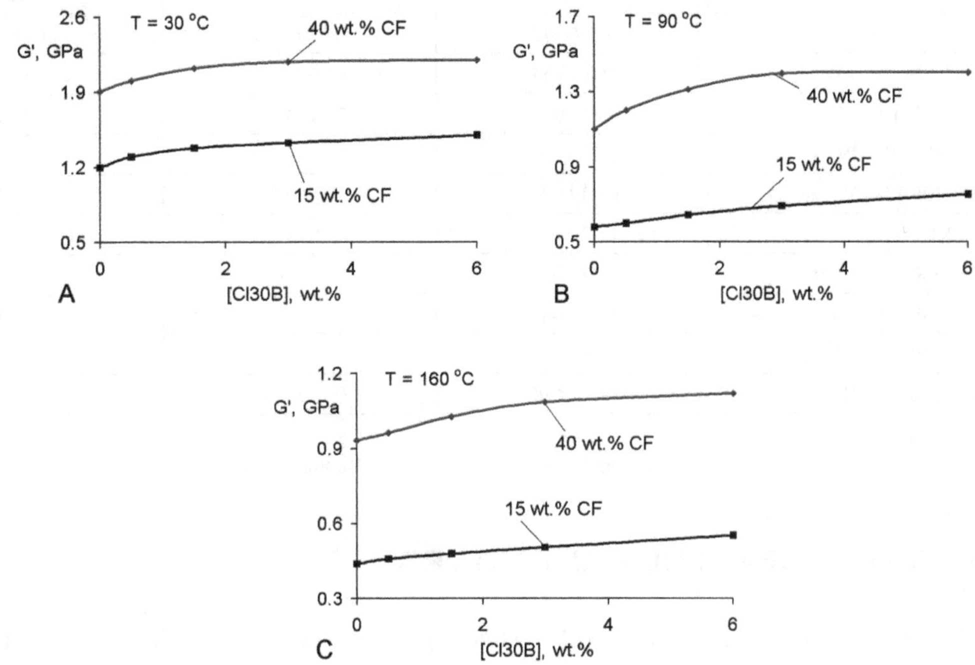

Parameter A in equation 1 characterizes the interphase adhesion between the polymer and the filler. Strong interactions between the matrix and the filler reduce the molecular mobility at the interface. It leads to reduction in the value of tgδ hence of parameter A. Low values of A point to the high level of the interphase adhesion between the polymer and CF surface and visa versa, increased values of A indicate reduced adhesive strength (Kubát, Rigdahl & Welander, 1990). Results of the experimental evaluation of the parameter A at different temperatures are presented in Table 1.

Analysis of the concentration dependence of parameter A allows to conclude that the interfacial adhesion in the system PA6/GF depends on the test temperature and increases with addition of organoclay at T = 30 °C and 90 °C. However, at T = 160 °C the addition of organoclay is not accompanied by a decrease of parameter A, and thus by an increase in interfacial adhesion. Besides, from Table 1 it follows that the minimum values of this parameter depend on the concentration of fillers. At $T < T_g$ the parameter A is much greater for composites with 15 wt.% CF, which indicates a weaker interfacial adhesion. The minimum value of A is found at a temperature of 90 °C, that is close to the T_g of the composites. Interestingly, at T = 160 °C the values of parameter A for materials with 40% CF (Table 1, compounds I-V) exceed those for low-filled CF materials containing organoclay. Consequently, the interfacial interactions in highly filled composites are weakened with increasing temperature to a greater extent than in low-filled ones.

Table 1. Values of parameter A characterizing the interfacial adhesion in composites

Composition of a Composite and Its Symbol	A, rel. un., at T, °C		
	30	90	160
PA6/CF-40% - I	1.1 / 0.77	0.91 / 0.84	1.06 / 0.81
PA6/CF -40%/Cl30B-0.5% - II	0.9 / 0.54	0.89 / 0.77	1.12 / 0.84
PA6/CF -40%/Cl30B-1.5% - III	0.84 / 0.51	0.87 / 0.76	1.13 / 0.87
PA6/CF -40%/Cl30B-3% - IV	0.97 / 0.57	0.91 / 0.79	1.17 / 0.88
PA6/CF -40%/Cl30B-6% - V	0.93 / 0.59	0.9 / 0.78	1.23 / 0.98
PA6/CF -15% - VI	1.40 / 1.16	0.79 / 0.75	0.79 / 0.60
PA6/CF -15%/Cl30B-0.5% - VII	1.27 / 0.94	0.77 / 0.75	0.75 / 0.59
PA6/CF -15%/Cl30B-1.5% - VIII	1.21 / 0.81	0.76 / 0.72	0.78 / 0.66
PA6/CF -15%/Cl30B-3% - IX	1.18 / 0.79	0.73 / 0.68	0.85 / 0.66
PA6/CF -15%/Cl30B-6% - X	1.09 / 0.72	0.71 / 0.65	0.87 / 0.69

(Pesetskii et al., 2016)

Note. The values of A in the numerator refer to a loading frequency of 1 Hz, in the denominator to that at 10 Hz

THE EFFECT OF TECHNOLOGICAL FACTORS

A variety of devices providing a certain level of shearing effects on the melt, for example, extruders, mixers, ultrasound sonicators, etc., are used for the preparation of polymer/clay nanocomposites, among which the twin screw extruders proved to be the most effective for dispersion and delamination of LCS (Cho & Paul, 2001).

Two different mixing schemes are possible in a twin-screw extruder: with co- and counter rotation. In the first one, the two screws rotate in one direction, the material moving from one screw to the other, following the trajectory above and under the screws. This results in a large area of polymer contact with the body of the material cylinder of the extruder, which improves the heating efficiency. The melt flow path also provides for a relatively constant level of shear in most part of the polymer melt, as it moves between the screws and the wall of the cylinder. The co-rotating screws are more fully self-cleaning compared to counter-rotating ones, therefore, it is less likely that the material may end up in stagnant zones. This makes the co-rotating extruders more efficient than the counter-rotating ones.

As a rule, increasing the dwell time of the melt in the material cylinder of the extruder enhances the organoclay delamination. However, when the optimum level of the shear stress on the melt is exceeded, dispersion and delamination of the filler deteriorate (Dennis et al., 2001).

For example, the study of nanocomposites of PP and maleated PP with organomodified montmorillonite (clay Dellite 67G) prepared with a co-rotating twin screw extruder at a mass ratio of components of 85/10/5 shows that the degree of intercalation increases with increasing the values of both the screws' rotation speed and the feed rate up to a certain critical value, after which it remains practically constant (Domenech, Peuvrel-Disdier & Vergnes, 2011).

It is shown in (Modesti, Lorenzetti, Bon & Besco, 2005) that the shear stresses arising in PP bulk during the melt compounding of the components (PP/PP-g-MA/clay) exert a more pronounced effect on the degree of the clay exfoliation than the residence time of the material in the material cylinder.

The shear rate is another important factor on which to dwell. The study of the influence of the shear rate on the characteristics of PP-based n-PCM shows that an increase in the shear rate (screw rotation speed) improves the quality of the clay distribution in the polymer matrix (Lertwimolnun & Vergnes, 2005).

Wang et al. (Wang, Chen & Wu, 2004) investigated the influence of the location of the components feed zone in the material cylinder, as well as the order of the components input into the extruder. Two sequences of the components' input are realized: 1) feed through the hopper, in which the matrix PP is fed to the material cylinder through the main hopper while the pre-mixed clay and PP-g-MA compatibilizer are fed into PP melt through a side feeder and 2) in which the mixture of PP-g-MA/organoclay is fed through the main hopper while the matrix PP is introduced through the side feeder.

Subsequent TEM analysis revealed that the n-PCM prepared according to scheme 2 featured a greater homogeneity of the clay distribution in the polymer. It was assumed, that in this case mixing of clay and PP-g-MA proceeded over a longer period of time which ensured a greater area of contact between the compatibilizer and the organoclay, which facilitated its dispersion.

SPECIAL TECHNOLOGICAL COMPOUNDING TECHNIQUES

An increased efficiency of LCS dispersion is affected using ultrasound-assisted extrusion technology (Zhao, Li, Guo & Du, 2006), in which case the extruder head is equipped with an ultrasonic sonicator, generating ultrasonic vibrations in the melt. Sonication of the melt creates shear stresses facilitating clay delamination and distribution in the polymer bulk.

The method of direct compounding is described in (Kato, Matsushita & Fukumori, 2004), which can preferably be used in the preparation of nanocomposites based on hydrophobic and hydrolytically stable polymers, for example, PP. Its implementation does not require preliminary organomodification of the clay. PP, PP-g-MA, untreated clay and alkylammonium salt are fed into the extruder's first zone and pre-mixed. In the second zone, water is supplied to the material cylinder, which allows to transfer the clay into a suspension and distribute it relatively homogeneously in the PP bulk. Next, water is evaporated in the third and fourth zones of the extruder. The resulting materials have approximately the same mechanical properties as nanocomposites based on PP and organoclays produced by the conventional method.

To improve the technical characteristics of n-PCM, the already mentioned two-stage compounding process based on nanoclay masterbatches can be used (Bogdanovich, Kuznetsova, Usova & Pesetskii, 2018). It includes preparation of a highly concentrated nanocomposite based on the matrix polymer and clay during the first stage, which is subsequently diluted with the neat polymer in the second compounding step. The two-stage mixing process yields a higher degree of clay exfoliation in n-PCM.

Preparation of nanocomposites based on polyolefins and clay may be implemented by several original methods based on reactive compounding. In particular, the authors of (Liu & Wu, 2001) describe a technology of preparing a PP-based n-PCM, which combines melt compounding with grafting of a polar monomer to the macromolecules. The process of intercalation is carried out by introducing a surfactant (conventional alkylammonium), an unsaturated monomer (epoxypropyl methacrylate) and a free radical initiator into the reaction mixture. According to XRD and TEM, the interlayer distance in the clay increases so does the interaction of the components as a result of polar monomer grafting to PP macromolecules, which significantly improves the clay distribution in PP matrix.

The authors of (Chen, Wang, Toh & He, 2005) propose another method of preparing nanocomposites based on PP and clay using reactive compounding with the formation of an intermediate superconcentrate. The superconcentrate is prepared using the method of “compounding in suspension”. In this case, epoxy oligomer is used as the carrier polymer (base of the concentrate). Next, PP and superconcentrate are compounded in the melt in the presence of a compatibilizer (PP-g-MA). It is assumed that MA fragments grafted to PP chain can react with epoxy groups of the oligomer, producing a block copolymer of PP and epoxy resin during melt compounding. The copolymer produced, on one hand, acts as a compatibilizer, improving the compatibility of PP and epoxy resin, which allows to increase the degree and quality of the filler distribution in the polymer bulk; on the other hand, the copolymerization process minimizes the negative consequences of the low molecular weight of the components, such as PP-g-MA and epoxy oligomer.

Practical implementation of the method of melt compounding of n-PCM, especially those based on hydrolytically and thermally unstable thermoplastics, processed at high temperatures, may be hampered by excessive degradation of macromolecules activated by LCSs. Such polymers include, in particular, polyethylene terephthalate. Therefore, it is very important to use compounding technology that minimizes the negative consequences of degradation processes.

In (Pesetskii, Bogdanovich, Aderikha & Koval, 2016), in particular, the authors analyzed the influence of the temperature regime of the operation of a twin screw compounding extruder on structural features and mechanical properties of PET/clay nanocomposites. In a series of experiments, three fundamentally different temperature regimes of operation of the material cylinder of the extruder were used (Figure 7). Original Na+-montmorillonite and Cloisite 30B organoclay were used as LCS fillers. Compounding of materials was carried out with a temperature profile ranging from the temperature above the melting point of PET (extrusion in the melt) to temperatures $\approx 50°$ and $\approx 100°C$ below the melting point (low-temperature extrusion) in the main mixing zones. It is shown that the transition from the regime typical for extrusion in the melt to low-temperature extrusion improves the strength characteristics of PET/clay nanocomposites (Figure 8). According to XRD analysis, low-temperature extrusion regime increases the degree of intercalation of nanoclays in the PET matrix.

The use of the low-temperature extrusion regimes allows not only to reduce the negative consequences of the degradation of PET macromolecules and, as a result, increase the mechanical properties of the materials produced, but also to improve dispersion and delamination of nanoparticle aggregates in the polymer bulk by using nanoclays not subjected to surfactant treatment.

It follows from the above analysis that a number of factors influence the final morphology of the nanostructures, as well as the thermodynamic and mechanical properties of n-PCM, the most important of which are the chemical nature of the polymer, the method of preparation of the organoclay, the level of the values of thermal and force (shear) factors acting on the melt in the process of compounding in an extrusion mixing reactor, the structural features of the latter. The above data generally answer the key question of how to best select an organoclay and optimal technological conditions for producing specific systems. By applying the described special compounding techniques, it is possible to control the degree of LCS intercalation in a thermoplastic bulk and to expediently influence the most important properties of n-PCM.

Figure 7. Position of the individual sections of the material cylinder and the temperature profile along the length of the screws; I – III - designations of temperature regimes of compounding
(Pesetskii, Bogdanovich, Aderikha & Koval, 2016)

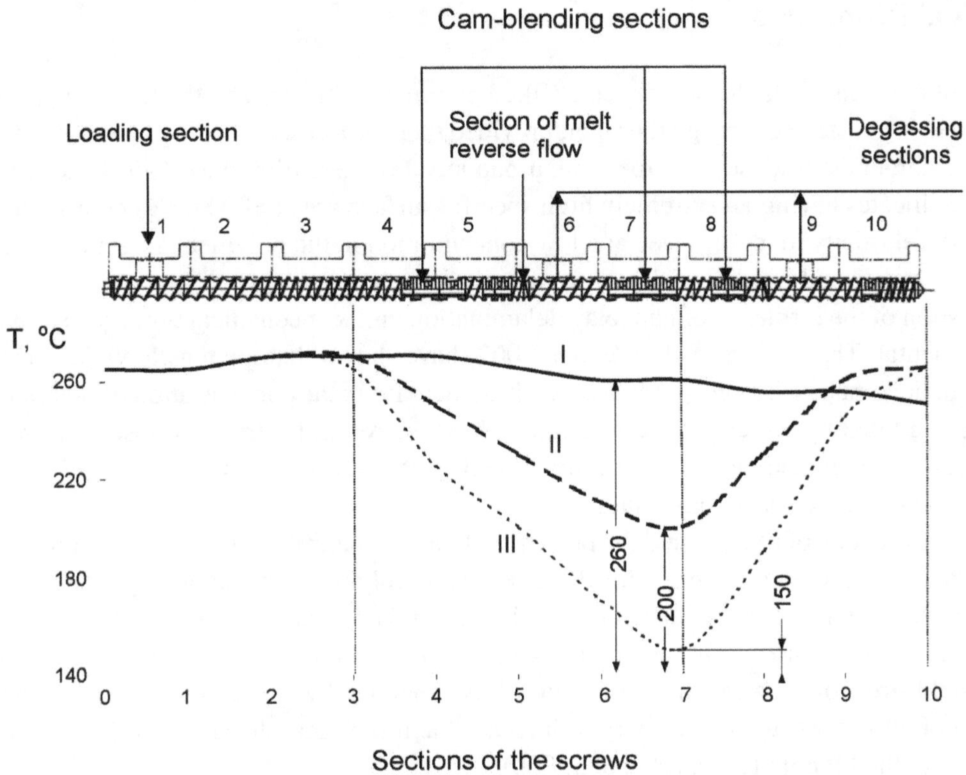

Figure 8. The stress-strain curves of PET/Cl30B – 1 wt.% composites at tension; the designations I – III of the curves are the temperature regimes of compounding, shown in Figure 7
(Pesetskii, Bogdanovich, Aderikha & Koval, 2016)

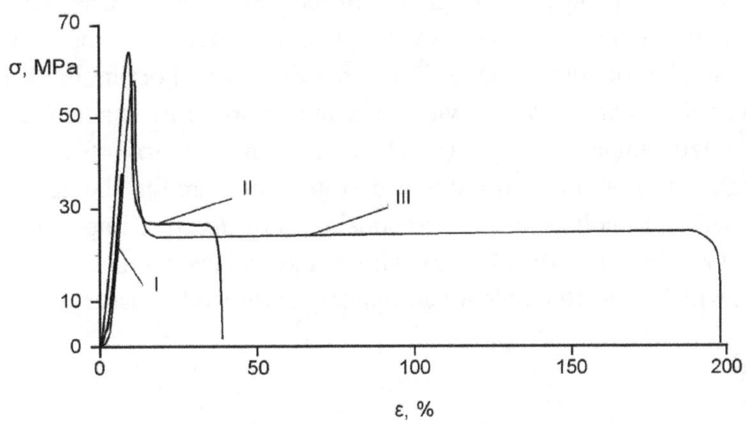

PROPERTIES AND APPLICATION OF POLYMER/CLAY NANOCOMPOSITES PREPARED BY MELT COMPOUNDING

Mechanical Properties

Improvement in mechanical characteristics of filled polymers is ensured by the fact that part of the mechanical stress is transferred through the filler, provided there is a strong bonding between the matrix and the filler. The larger the filler surface, the more pronounced is the reinforcing effect. This partly explains why layered silicates having an extremely high specific surface area (≈800 m^2/g) contribute to a sharp increase in the elasticity modulus, even at a low concentration in the polymer. As a rule, nanoclays also increase the strength and rigidity of the thermoplastic polymers.

The question of the influence of nanoclay delamination on the mechanical properties of nanomaterials is fundamental. The authors of (Liu & Wu, 2002) have shown that the tensile yield stress of PA66/organoclay nanocomposites rapidly increases with an increase in the concentration of o-MMT, however, the growth is limited by the clay concentration of 5 wt.%. A similar trend is observed for the elastic modulus. Consequently, the factors determining the degree of the clay dispersion greatly influence the strength characteristics of nanocomposites.

Organomodification of the clay and the presence of the compatibilizer in the polymer matrix can play a decisive role in improving the mechanical properties of polymer/clay nanocomposites. For example, the data in Figure 9 indicate that with an increase in the number of carbon atoms in the chain of substituents from 8 to 12 in the surfactant molecule, there is a sharp increase in Young's elastic modulus and the upper yield stress of PP/clay nanocomposites. It is explained by the achievement of a certain length of the chain of alkyl substituent necessary to increase the interplanar distance and high degree of LCS dispersion of in the PP bulk (Reichert et al., 2000).

Figure 9 also shows the determining role of the compatibilizer, particularly PP-g-MA. In its absence, the modulus of elasticity and the upper yield stress, vary insignificantly in response to the type of surfactant used. Addition of the compatibilizer intensifies interaction of the nonpolar PP with the surface of the organoclay. As a result, the material is a significantly reinforced.

Addition of clay typically reduces the toughness of thermoplastic n-PCM. The reason for this may be a too tiny size of nanoparticles, to increase the toughness by increasing the curvature of the fracture path. Nevertheless, a number of authors have shown that the optimal organomodification of nanoclays provides for an increased toughness, along with other indicators of mechanical properties. According to (Liu & Wu, 2002), Izod impact strength (notched) rises from 98 to 146 J/m for PA66/organoclay composite containing 5 wt.% of clay and remains elevated compared to the neat PA66 even at higher clay concentrations. The morphology of nanocomposites is of decisive importance in the mechanism of failure under the impact loading. Morphology with an intercalated nanoclay can provide an improvement of this property, which is unattainable for completely exfoliated structures (Zerda & Lesser, 2001).

Heat Resistance

When nanoclay is introduced into a semicrystalline thermoplastic polymer it has practically no effect on the melting point. However, the fact of a significant increase in heat deformation resistance for nanocomposites based on both polar and nonpolar thermoplastics has been repeatedly confirmed (Okada & Usuki, 2006; Pavlidou & Papaspyrides, 2008). According to the data of, for example, (Okada & Usuki,

Figure 9. Effect of the length of alkyl substituents in the ammonium surfactant on the tensile elasticity modulus (a) and the tensile upper yield stress (b) of n-PCM PP/o-MM –10 wt.%; C_n is the number of carbon atoms in the alkyl radical; H is the compatibilizer (PP-g-MA), branded Hostaprime HC5; the concentration of H is 20 wt.%
(Reichert et al., 2000)

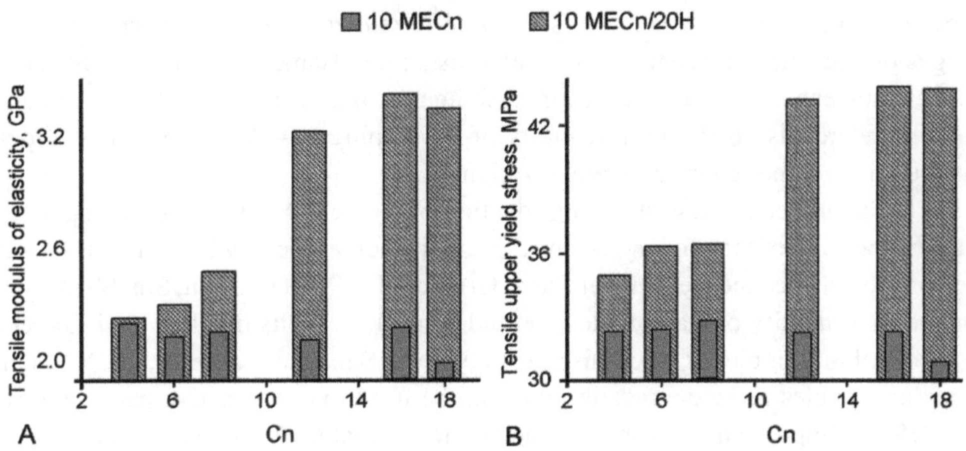

2006), the introduction of 4 wt.% of organoclay in PA6 (MMT treated with tetraphenyl phosphonium) increases its HDT from 62 °C to 112 °C. The effect of such a strong increase in HDT of semicrystalline thermoplastics on addition of relatively small amounts of nanoclay is extraordinary and is certainly caused by the formation of a developed interphase boundary in the nanofiller-matrix contact zones.

Barrier Properties

The distribution of the platelets of exfoliated clay in the polymer bulk leads, as a rule, to a decrease in the material permeability to gases and water vapor (Pavlidou & Papaspyrides, 2008). According to (Ke & Yongping, 2005), small additions of o-MMT to PET effectively reduce the oxygen permeability of a packaging film. With an o-MMT concentration of 3 wt.%, the decrease reaches 50% compared to pure PET. Such effects are explained by the significantly increased trajectory along which the gas (vapor) molecules are forced to diffuse into the matrix polymer in presence of exfoliated nanoclay, as their penetration directly through the clay platelets is excluded. The outstanding barrier properties of n-PCM predetermine wide prospects for their use in film or sheet materials for a variety of packaging, elastomers, and the need for new technical developments in this field (Vannini, Marchese, Celli, Marega & Lorenzetti, 2018).

Flame Retardancy

Nanoclay additions to thermoplastics, as a rule, improve their flame retardancy. Nanoclay accelerates the formation of coke (carbonation) on the surface of a burning specimen, which prevents the inflow of combustible products to the combustion zone (Beyer, 2002). The carbonized layer formed on the surface during combustion is enriched with silicates and can be regarded as a sort of a carbon-silicate nanomaterial. It is this layer that provides barrier protection for the burning specimen, reducing heat and mass transfer in the combustion zone. The coke insulates the underlying layer of polymer from heat fluxes and also slows down the diffusion of oxygen to the polymer and that of combustible gaseous products of the polymer degradation into the combustion zone.

It should be noted that the addition of nanoclays alone cannot provide for the highest category of polymers flame retardancy (V-0 by UL-94). However, with the combined use of nanoclays and traditional microdispersible flame retardants, it is possible to realize synergistic effects of enhancing the extinguishing action and to produce materials with outstanding flame retardancy (Zanetti et al., 2002). Effects of synergistic enhancement of fire extinguishing action on combined addition of nanoclays and traditional flame retardants create the possibility of developing a wide range of new types of flame retarding formulations suitable for commercialization.

In addition to the properties described, introduction of nanoclays during melt compounding allows to control the crystallinity of the semicrystalline thermoplastics, especially in combination with conventional nucleators, to enhance the mechanical characteristics of n-PCM (Xu, Yu, Shi, Shi & Nie, 2018); to promote the thermal stability of macromolecules and control the melts rheology (Silvano et al., 2013); to improve compatibility in blends of immiscible polymers (Babaei & Arefazar, 2015); to increase the ability of macromolecules to biodegradation and can be used to create active packaging (Ray, 2015; Saha et al., 2018); to improve tribological characteristics of materials, etc. (Pesetskii, Bogdanovich & Myshkin, 2013; Azam & Samad, 2018).

Application of Nanocomposites

The aforementioned and many other properties of polymer/clay nanocomposites, the possibility of achieving technical advantages over analogues, low cost and economic benefits of the technology of melt compounding open up a number of promising areas of their application. These are, first of all, automotive parts, as well as products for special equipment with high strength and heat resistance. Barrier characteristics of n-PCM with respect to gases with a high penetrating power (oxygen, water vapor, carbon dioxide, etc.), enhanced with good exfoliation of organoclays and their proper preferential orientation, predetermine the possibility of their use in thin-film materials for food packaging. Increased fire resistance and environmental safety of self-extinguishing n-PCM open up wide opportunities for the production of parts for household electrical appliances, as well as electrical insulation products for electric power and electric transport. Great prospects for practical application are associated with biodegradable n-PCM, since inclusion of the optimal type of organoclay in a biodegradable polymer matrix not only improves its mechanical, barrier properties and thermal stability during processing, but also significantly increases the degree of decomposition during composting (Domenech, Peuvrel-Disdier & Vergnes, 2011).

Additions of organoclays to thermoplastics improve their technological properties and formability of materials during melt processing, increase the spatial stability of geometric dimensions due to the stabilization of thermal expansion coefficient values, which is especially important for microelectronics products. Increased thermal stability, improved tribotechnical characteristics, the possibility of creating smart materials for optical and biomedical applications, all this contributes to the constant expansion of applications of n-PCM (Domenech, Peuvrel-Disdier & Vergnes, 2011).

CONCLUSION

The method of compounding based on the reactive blending in the polymer melt is the main one in the technology of thermoplastic n-PCM and has found wide practical applications. The technology can be successfully implemented at small and medium-sized enterprises, and not only at large firms for the synthesis of polymers. The most important advantages of this technology are environmental safety, lack of need to create special technological equipment, almost unlimited possibilities for development of a wide range of technically valuable n-PCM.

To date, the technical information has been accumulated and systematized, which allows to expediently select the organoclays for the production of n-PCM, to assign the level of the values in the formulation and that of technological factors which determine the degree of LCS exfoliation in the polymer matrix, the character and stability of the morphological structures, distribution of nanoparticles in individual phases of incompatible polymer blends and consumer properties of n-PCM.

Melt compounding allows to produce n-PCM with a set of technical characteristics, not inferior to those prepared by other methods, e.g. by in situ polymerization. The most important advantages of n-PCM produced by melt compounding include increased strength, modulus of elasticity, heat deflection temperature, flame retardancy, biodegradability, high barrier effect, improved gloss of block and film products, the ability to compatibilize blends of incompatible polymers and improve tribological behavior.

Melt compounding of n-PCM should be preferably implemented with the co-rotating twin-screw extruders. The use of special compounding techniques implemented in extrusion equipment, such as, for example, ultrasonic sonication-assisted dispersion, water-assisted extrusion, low temperature extrusion, use of supercritical CO_2 help to produce n-PCM with improved characteristics. The application of these approaches, as well as, apparently, new technological solutions still being developed will make it possible to successfully solve the problem of aggregation of nanoparticles in a polymer matrix. The prospects for further improvement of commercial polymer/clay n-PCM technologies will depend on the results of studies aimed at deepening the understanding of the nanoclays dispersion in polymer melts and the interaction of macromolecules with the surface of clay particles. These questions are of great interest for n-PCM based on polymers blends, including incompatible or partially compatible ones.

Of great interest is the deepening of research in the field of technology, structure and properties of n-PCM containing hybrid fillers - organoclays and fibrous fillers, including materials based on polymer blends. The key problem here is to find out the reasons behind the manifestation of synergism in the action of fillers, which is necessary for the implementation of n-PCM technology with record mechanical and other characteristics.

NOMENCLATURE

APTS: (aminopropyl) triethoxy silane,
VA: vinyl acetate,
CEC: cation exchange capacity,
LCS: layered clay silicate;
MA: maleic anhydride,
MMT: montmorillonite,
o-MMT: organomodified (surfactant-treated) montmorillonite,

n-PCM: nanocomposite,
PA: polyamide,
PA6, 66, 11 and 12: polyamides 6, 66, 11 and 12,
PBT: polybutylene terephthalate,
PI: polyimide,
PC: polycarbonate,
PO: polyolefin,
PP: polypropylene.
PS: polystyrene,
HDPE: high density polyethylene,
PEI: polyetherimide,
PEO: polyethylene oxide,
PET: polyethylene terephthalate,
EVA: ethylene vinyl acetate copolymer,
FPO: functionalized polyolefin (PO with grafted polar monomer),
FPP: functionalized polypropylene,
TEM: transmission electron microscopy,
D, L: screw diameter and lengths,
d and h: diameter and thickness of an individual clay platelet,
p = d/h: aspect or characteristic ratio in nanoclay,
d_{001}: distance between individual clay platelets.

REFERENCES

Azam, M., & Samad, M. (2018). A novel organoclay reinforced UHMWPE nanocomposite coating for tribological applications. *Progress in Organic Coatings*, *118*, 97–107. doi:10.1016/j.porgcoat.2018.01.028

Babaei, A., & Arefazar, A. (2015). Structural, rheological, and mechanical properties of PA6/SAN/SEBS ternary blend/organoclay nanocomposites. *Journal of Applied Polymer Science*, *132*(20). doi:10.1002/app.41969

Beyer, G. (2002). Nanocomposites: A new class of flame retardants for polymers. *Plastics Additives & Compounding*, *4*(10), 22–28. doi:10.1016/S1464-391X(02)80151-9

Bogdanovich, S., Kuznetsova, U., Usova, V., & Pesetskii, S. (2018). Influence of technology of melt compounding of PA6 / organoclay nanocomposites on their structure, mechanical and rheological properties. *Polymer Materials and Technologies*, *4*(2), 50–60. doi:10.32864/polymmattech-2018-4-2-50-60

Chaudhary, D., Prasad, R., Gupta, R., & Bhattacharya, S. (2005). Clay intercalation and influence on crystallinity of EVA-based clay nanocomposites. *Thermochimica Acta*, *433*(1-2), 187–195. doi:10.1016/j.tca.2005.02.031

Chen, L., Wang, K., Toh, M., & He, C. (2005). Poly(propylene)/clay nanocomposites prepared by reactive compounding with an epoxy based masterbatch. *Macromolecular Materials and Engineering*, *290*(10), 1029–1036. doi:10.1002/mame.200500176

Cho, J., & Paul, D. (2001). Nylon 6 nanocomposites by melt compounding. *Polymer*, *42*(3), 1083–1094. doi:10.1016/S0032-3861(00)00380-3

Chow, W., & Mohd Ishak, Z. (2015). Polyamide blend-based nanocomposites: A review. *Express Polymer Letters*, *9*(3), 211–232. doi:10.3144/expresspolymlett.2015.22

Corté, L., & Leibler, L. (2005). Analysis of polymer blend morphologies from transmission electron micrographs. *Polymer*, *46*(17), 6360–6368. doi:10.1016/j.polymer.2005.05.109

Davis, C., Mathias, L., Gilman, J., Schiraldi, D., Shields, J., Trulove, P., ... Delong, H. C. (2002). Effects of melt-processing conditions on the quality of poly(ethylene terephthalate) montmorillonite clay nanocomposites. *Journal of Polymer Science. Part B, Polymer Physics*, *40*(23), 2661–2666. doi:10.1002/polb.10331

Dennis, H., Hunter, D., Chang, D., Kim, S., White, J., Cho, J., & Paul, D. (2001). Effect of melt processing conditions on the extent of exfoliation in organoclay-based nanocomposites. *Polymer*, *42*(23), 9513–9522. doi:10.1016/S0032-3861(01)00473-6

Di, Y., Iannace, S., Di Maio, E., & Nicolais, L. (2003). Nanocomposites by melt intercalation based on polycaprolactone and organoclay. *Journal of Polymer Science. Part B, Polymer Physics*, *41*(7), 670–678. doi:10.1002/polb.10420

Domenech, T., Peuvrel-Disdier, E., & Vergnes, B. (2011). Influence of twin-screw processing conditions on structure and properties of polypropylene-organoclay nanocomposites. In *27th World Congress of the Polymer Processing Society* (pp. 1-10). Marrakech: Polymer Processing Society.

Eslami-Farsani, R., Reza Khalili, S., Hedayatnasab, Z., & Soleimani, N. (2014). Influence of thermal conditions on the tensile properties of basalt fiber reinforced polypropylene–clay nanocomposites. *Materials & Design*, *53*, 540–549. doi:10.1016/j.matdes.2013.07.012

Finnigan, B., Martin, D., Halley, P., Truss, R., & Campbell, K. (2004). Morphology and properties of thermoplastic polyurethane nanocomposites incorporating hydrophilic layered silicates. *Polymer*, *45*(7), 2249–2260. doi:10.1016/j.polymer.2004.01.049

Fischer, H. (2003). Polymer nanocomposites: From fundamental research to specific applications. *Materials Science and Engineering C*, *23*(6-8), 763–772. doi:10.1016/j.msec.2003.09.148

Fornes, T., Yoon, P., Keskkula, H., & Paul, D. (2001). Nylon 6 nanocomposites: The effect of matrix molecular weight. *Polymer*, *42*(25), 09929–09940. doi:10.1016/S0032-3861(01)00552-3

Frontini, P., & Pouzara, A. (2015). Trends in the multifunctional performance of polyolefin/clay nanocomposite injection moldings. In K. Friedrich & U. Breuer (Eds.), *Multifunctionality of polymer composites. Challenges and new solutions* (pp. 213–245). Oxford: Elsevier. doi:10.1016/B978-0-323-26434-1.00006-4

Galimberti, M. (Ed.). (2011). *Rubber-Clay Nanocomposites: Science, Technology, Applications*. New Jersey: Wiley & Sons. doi:10.1002/9781118092866

Garcia, M., Netto, A., & Pontes, A. (2017). Experimental study of shrinkage and ejection forces of reinforced polypropylene based on nanoclays and short glass fibers. *Polymer Engineering and Science*, *58*(1), 55–62. doi:10.1002/pen.24530

García-López, D. (2003). Polypropylene–clay nanocomposites: Effect of compatibilizing agents on clay dispersion. *European Polymer Journal*, *39*(5), 945–950. doi:10.1016/S0014-3057(02)00333-6

Gerasin, V., Antipov, E., Karbushev, V., Kulichikhin, V., Karpacheva, G., Talroze, R., & Kudryavtsev, Y. (2013). New approaches to the development of hybrid nanocomposites: From structural materials to high-tech applications. *Russian Chemical Reviews*, *82*(4), 303–332. doi:10.1070/RC2013v082n04ABEH004322

Gerasin, V., Zubova, T., Bakhov, F., Barannikov, A., Merekalova, N., Korolev, Y., & Antipov, E. (2007). Structure of polymer/Na+-montmorillonite nanocomposites, produced by blending in melt. *Nanotechnologies in Russia*, *2*(1-2), 90–105.

Gopakumar, T., Lee, J., Kontopoulou, M., & Parent, J. (2002). Influence of clay exfoliation on the physical properties of montmorillonite/polyethylene composites. *Polymer*, *43*(20), 5483–5491. doi:10.1016/S0032-3861(02)00403-2

Hu, Y., Lu, H., & Song, L. (2008). Polyethylene/layered silicate nanocomposites: preparation and properties. In D. Nwabunma & T. Kyu (Eds.), *Polyolefin Composites* (pp. 228–250). Hoboken, N.J.: Wiley-Interscience.

Jlassi, K., Chehimi, M., & Thomas, S. (Eds.). (2017). *Clay-polymer nanocomposites*. Philadelphia, PA: Elsevier Science.

Kato, M., Matsushita, M., & Fukumori, K. (2004). Development of a new production method for a polypropylene-clay nanocomposite. *Polymer Engineering and Science*, *44*(7), 1205–1211. doi:10.1002/pen.20115

Kawasumi, M., Hasegawa, N., Kato, M., Usuki, A., & Okada, A. (1997). Preparation and mechanical properties of polypropylene-clay hybrids. *Macromolecules*, *30*(20), 6333–6338. doi:10.1021/ma961786h

Ke, Z., & Yongping, B. (2005). Improve the gas barrier property of PET film with montmorillonite by in situ interlayer polymerization. *Materials Letters*, *59*(27), 3348–3351. doi:10.1016/j.matlet.2005.05.070

Kotal, M., & Bhowmick, A. (2015). Polymer nanocomposites from modified clays: Recent advances and challenges. *Progress in Polymer Science*, *51*, 127–187. doi:10.1016/j.progpolymsci.2015.10.001

Kotek, J., Kelnar, I., Studenovský, M., & Baldrian, J. (2005). Chlorosulfonated polypropylene: Preparation and its application as a coupling agent in polypropylene–clay nanocomposites. *Polymer*, *46*(13), 4876–4881. doi:10.1016/j.polymer.2005.02.119

Kubát, J., Rigdahl, M., & Welander, M. (1990). Characterization of interfacial interactions in high density polyethylene filled with glass spheres using dynamic-mechanical analysis. *Journal of Applied Polymer Science*, *39*(7), 1527–1539. doi:10.1002/app.1990.070390711

Lee, H., Fasulo, P., Rodgers, W., & Paul, D. (2005). TPO based nanocomposites. Part 1. Morphology and mechanical properties. *Polymer*, *46*(25), 11673–11689. doi:10.1016/j.polymer.2005.09.068

Lertwimolnun, W., & Vergnes, B. (2005). Influence of compatibilizer and processing conditions on the dispersion of nanoclay in a polypropylene matrix. *Polymer*, *46*(10), 3462–3471. doi:10.1016/j.polymer.2005.02.018

Liaw, W., Huang, P., Chen, C., Lo, C., & Chang, J. (2008). PPgMA/APTS compound coupling compatabilizer in PP/clay hybrid nanocomposite. *Journal of Applied Polymer Science*, *109*(3), 1871–1880. doi:10.1002/app.28294

Lincoln, D., Vaia, R., & Krishnamoorti, R. (2004). Isothermal Crystallization of Nylon-6/Montmorillonite Nanocomposites. *Macromolecules*, *37*(12), 4554–4561. doi:10.1021/ma049768k

Liu, X., & Wu, Q. (2001). PP/clay nanocomposites prepared by grafting-melt intercalation. *Polymer*, *42*(25), 10013–10019. doi:10.1016/S0032-3861(01)00561-4

Liu, X., & Wu, Q. (2002). Polyamide 66/Clay Nanocomposites via Melt Intercalation. *Macromolecular Materials and Engineering*, *287*(3), 180. doi:10.1002/1439-2054(20020301)287:3<180::AID-MAME180>3.0.CO;2-T

Lu, X., Wang, K., & Chua, Y. (2008). Polypropylene/Clay Nanocomposites. In D. Nwabunma & T. Kyu (Eds.), *Polyolefin composites* (pp. 251–282). Hoboken, NJ: Wiley-Interscience.

McNally, T., Raymond Murphy, W., Lew, C., Turner, R., & Brennan, G. (2003). Polyamide-12 layered silicate nanocomposites by melt blending. *Polymer*, *44*(9), 2761–2772. doi:10.1016/S0032-3861(03)00170-8

Modesti, M., Lorenzetti, A., Bon, D., & Besco, S. (2005). Effect of processing conditions on morphology and mechanical properties of compatibilized polypropylene nanocomposites. *Polymer*, *46*(23), 10237–10245. doi:10.1016/j.polymer.2005.08.035

Moncada, E., Quijada, R., Lieberwirth, I., & Yazdani-Pedram, M. (2006). Use of PP grafted with itaconic acid as a new compatibilizer for PP/Clay nanocomposites. *Macromolecular Chemistry and Physics*, *207*(15), 1376–1386. doi:10.1002/macp.200600150

Morgan, A., & Gilman, J. (2003). Characterization of polymer-layered silicate (clay) nanocomposites by transmission electron microscopy and X-ray diffraction: A comparative study. *Journal of Applied Polymer Science*, *87*(8), 1329–1338. doi:10.1002/app.11884

Nuzzo, A., Bilotti, E., Peijs, T., Acierno, D., & Filippone, G. (2014). Nanoparticle-induced co-continuity in immiscible polymer blends – A comparative study on bio-based PLA-PA11 blends filled with organoclay, sepiolite, and carbon nanotubes. *Polymer*, *55*(19), 4908–4919. doi:10.1016/j.polymer.2014.07.036

Okada, A., & Kawasumi, M. (1987). Synthesis and characterization of a nylon 6-clay hybrid. *Polymeric Preprints*, *28*(2), 447–448.

Okada, A., & Usuki, A. (2006). Twenty years of polymer-clay nanocomposites. *Macromolecular Materials and Engineering*, *291*(12), 1449–1476. doi:10.1002/mame.200600260

Paul, D., & Robeson, L. (2008). Polymer nanotechnology: Nanocomposites. *Polymer*, *49*(15), 3187–3204. doi:10.1016/j.polymer.2008.04.017

Pavlidou, S., & Papaspyrides, C. (2008). A review on polymer–layered silicate nanocomposites. *Progress in Polymer Science*, *33*(12), 1119–1198. doi:10.1016/j.progpolymsci.2008.07.008

Pedrazzoli, D., & Pegoretti, A. (2013). Silica nanoparticles as coupling agents for polypropylene/glass composites. *Composites Science and Technology*, *76*, 77–83. doi:10.1016/j.compscitech.2012.12.016

Pegoretti, A., Kolarik, J., Peroni, C., & Migliaresi, C. (2004). Recycled poly(ethylene terephthalate)/layered silicate nanocomposites: Morphology and tensile mechanical properties. *Polymer*, *45*(8), 2751–2759. doi:10.1016/j.polymer.2004.02.015

Pesetskii, S., Bogdanovich, S., Aderikha, V., & Koval, V. (2016). Low-temperature extrusion in technology of polyethylene terephtalate/clay nanocomposites. *Doklady of the National Academy of Sciences of Belarus*, *60*(1), 102–108.

Pesetskii, S., Bogdanovich, S., Dubrovskii, V., Sodyleva, T., Aderikha, V., & Usova, V. (2016). Morphology and properties of PA6 hybrid composites filled with short carbon fibers and organoclay. *Polymer Materials and Technologies*, *2*(3), 45–57. doi:10.32864/polymmattech-2016-2-3-47-57

Pesetskii, S., Bogdanovich, S., & Myshkin, N. (2013). Tribological behavior of polymer nanocomposites produced by dispersion of nanofillers in molten thermoplastics. In K. Friedrich & A. Schlarb (Eds.), *Tribology of polymeric nanocomposites* (2nd ed., pp. 119–162). Oxford: Elsevier. doi:10.1016/B978-0-444-59455-6.00005-2

Pesetskii, S., Bogdanovich, S., & Sodyleva, T. (2017). Polyamide 6 reinforcement by hybrid short basalt fiber and organoclay filling. *Doklady of the National Academy of Sciences of Belarus*, *61*(2), 74–83.

Pesetskii, S., Shevchenko, V., & Dubrovsky, V. (2015). Reactive compatibilization in technology of poly(alkylene terephthalate)-based composites: polyester blends, short fiber-filled materials, and nanomaterials. In K. Friedrich & U. Breuer (Eds.), *Multifunctionality of Polymer Composites. Challenges and New Solutions* (1st ed., pp. 302–337). Oxford: Elsevier. doi:10.1016/B978-0-323-26434-1.00009-X

Pluta, M. (2004). Morphology and properties of polylactide modified by thermal treatment, filling with layered silicates and plasticization. *Polymer*, *45*(24), 8239–8251. doi:10.1016/j.polymer.2004.09.057

Ray, S. (2015). Multifunctional nanobiocomposites of biodegradable polylactide and nanoclay. In K. Friedrich & U. Breuer (Eds.), *Multifunctionality of polymer composites. Challenges and new solutions* (pp. 144–212). Oxford: Elsevier. doi:10.1016/B978-0-323-26434-1.00005-2

Reichert, P., Nitz, H., Klinke, S., Brandsch, R., Thomann, R., & Mülhaupt, R. (2000). Poly(propylene)/organoclay nanocomposite formation: Influence of compatibilizer functionality and organoclay modification. *Macromolecular Materials and Engineering*, *275*(1), 8–17. doi:10.1002/(SICI)1439-2054(20000201)275:1<8::AID-MAME8>3.0.CO;2-6

Saha, N., Roy, I., Sarkar, G., Bhattacharyya, A., Das, R., Rana, D., ... Chattopadhyay, D. (2018). Development of active packaging material based on cellulose acetate butyrate/polyethylene glycol/aryl ammonium cation modified clay. *Carbohydrate Polymers*, *187*, 8–18. doi:10.1016/j.carbpol.2018.01.065 PMID:29486847

Sanchez-Solis, A., Garcia-Rejon, A., & Manero, O. (2003). Production of nanocomposites of PET-montmorillonite clay by an extrusion process. *Macromolecular Symposia*, *192*(1), 281–292. doi:10.1002/masy.200390038

Schmidt, G., Nakatani, A., Butler, P., & Han, C. (2002). Small-angle neutron scattering from viscoelastic polymer–clay solutions. *Macromolecules*, *35*(12), 4725–4732. doi:10.1021/ma0115141

Shah, R., & Paul, D. (2006). Organoclay degradation in melt processed polyethylene nanocomposites. *Polymer*, *47*(11), 4075–4084. doi:10.1016/j.polymer.2006.02.031

Silvano, J., Rodrigues, S., Marini, J., Bretas, R., Canevarolo, S., Carvalho, B., & Pinheiro, L. (2013). Effect of reprocessing and clay concentration on the degradation of polypropylene/montmorillonite nanocomposites during twin screw extrusion. *Polymer Degradation & Stability*, *98*(3), 801–808. doi:10.1016/j.polymdegradstab.2012.12.009

Sinha Ray, S., Maiti, P., Okamoto, M., Yamada, K., & Ueda, K. (2002). New polylactide/layered silicate nanocomposites. 1. Preparation, characterization, and properties. *Macromolecules*, *35*(8), 3104–3110. doi:10.1021/ma011613e

Sodyleva, T., Kuznetsova, Y., Bogdanovich, S., Okhlopkova, A., & Pesetskii, S. (2016). On effect of excess surfactant in the organoclay on properties of nanocomposites based on PA6, obtained by reactive extrusion. *Polymer Materials and Technologies*, *2*(1), 36–47. doi:10.32864/polymmattech-2016-2-1-36-47

Srivastava, S., Pramanik, M., & Acharya, H. (2005). Ethylene/vinyl acetate copolymer/clay nanocomposites. *Journal of Polymer Science. Part B, Polymer Physics*, *44*(3), 471–480. doi:10.1002/polb.20702

Vaia, R., Ishii, H., & Giannelis, E. (1993). Synthesis and properties of two-dimensional nanostructures by direct intercalation of polymer melts in layered silicates. *Chemistry of Materials*, *5*(12), 1694–1696. doi:10.1021/cm00036a004

Vaia, R., Liu, W., & Koerner, H. (2003). Analysis of small-angle scattering of suspensions of organically modified montmorillonite: Implications to phase behavior of polymer nanocomposites. *Journal of Polymer Science. Part B, Polymer Physics*, *41*(24), 3214–3236. doi:10.1002/polb.10698

Vaia, R., Teukolsky, R., & Giannelis, E. (1994). Interlayer structure and molecular environment of alkylammonium layered silicates. *Chemistry of Materials*, *6*(7), 1017–1022. doi:10.1021/cm00043a025

VanderHart, D., Asano, A., & Gilman, J. (2001). Solid-State NMR Investigation of Paramagnetic Nylon-6 Clay Nanocomposites. 1. Crystallinity, Morphology, and the Direct Influence of Fe3+on Nuclear Spins. *Chemistry of Materials*, *13*(10), 3781–3795. doi:10.1021/cm0110775

Vannini, M., Marchese, P., Celli, A., Marega, C., & Lorenzetti, C. (2018). Strategy to improve PA6 performances by melt compounding. *Polymer Testing*, *67*, 84–91. doi:10.1016/j.polymertesting.2018.02.021

Wang, D., & Wilkie, C. (2003). In-situ reactive blending to prepare polystyrene–clay and polypropylene–clay nanocomposites. *Polymer Degradation & Stability*, *80*(1), 171–182. doi:10.1016/S0141-3910(02)00399-3

Wang, K., Choi, M., Koo, C., Choi, Y., & Chung, I. (2001). Synthesis and characterization of maleated polyethylene/clay nanocomposites. *Polymer*, *42*(24), 9819–9826. doi:10.1016/S0032-3861(01)00509-2

Wang, Y., Chen, F., & Wu, K. (2004). Twin-screw extrusion compounding of polypropylene/organoclay nanocomposites modified by maleated polypropylenes. *Journal of Applied Polymer Science*, *93*(1), 100–112. doi:10.1002/app.20407

Wang, Z., Nakajima, H., Manias, E., & Chung, T. (2003). Exfoliated PP/clay nanocomposites using ammonium-terminated PP as the organic modification for montmorillonite. *Macromolecules*, *36*(24), 8919–8922. doi:10.1021/ma0352911

Xiao, J., Hu, Y., Wang, Z., Tang, Y., Chen, Z., & Fan, W. (2005). Preparation and characterization of poly(butylene terephthalate) nanocomposites from thermally stable organic-modified montmorillonite. *European Polymer Journal*, *41*(5), 1030–1035. doi:10.1016/j.eurpolymj.2004.11.025

Xu, W., Yu, M., Shi, S., Shi, H., & Nie, M. (2018). Simultaneous durability and strength enhancement of β-polypropylene through montmorillonite and melt-soluble β-nucleating agent addition. *Polymer Testing*, *65*, 150–155. doi:10.1016/j.polymertesting.2017.11.024

Yariv, S., & Cross, H. (Eds.). (2002). Organo-clay complexes and interactions. New York: Marcel Dekker.

Yoo, Y., Tiwari, R., Yoo, Y., & Paul, D. (2010). Effect of organoclay structure and mixing protocol on the toughening of amorphous polyamide/elastomer blends. *Polymer*, *51*(21), 4907–4915. doi:10.1016/j.polymer.2010.08.036

Yoon, P., Hunter, D., & Paul, D. (2003). Polycarbonate nanocomposites: Part 2. Degradation and color formation. *Polymer*, *44*(18), 5341–5354. doi:10.1016/S0032-3861(03)00523-8

Yu, Z., Yang, M., Zhang, Q., Zhao, C., & Mai, Y. (2003). Dispersion and distribution of organically modified montmorillonite in nylon-66 matrix. *Journal of Polymer Science. Part B, Polymer Physics*, *41*(11), 1234–1243. doi:10.1002/polb.10480

Zanetti, M., Camino, G., Canavese, D., Morgan, A., Lamelas, F., & Wilkie, C. (2002). Fire retardant halogen−antimony−clay synergism in polypropylene layered silicate nanocomposites. *Chemistry of Materials*, *14*(1), 189–193. doi:10.1021/cm011124t

Zerda, A., & Lesser, A. (2001). Intercalated clay nanocomposites: Morphology, mechanics, and fracture behavior. *Journal of Polymer Science. Part B, Polymer Physics*, *39*(11), 1137–1146. doi:10.1002/polb.1090

Zhao, L., Li, J., Guo, S., & Du, Q. (2006). Ultrasonic oscillations induced morphology and property development of polypropylene/montmorillonite nanocomposites. *Polymer*, *47*(7), 2460–2469. doi:10.1016/j.polymer.2006.02.011

Section 2

Engineering Applications of Polymer Nanocomposites

Chapter 4
Polymer Nanocomposites for Advanced Automobile Applications

Oluranti Agboola
Covenant University, Nigeria & Tshwane University of Technology, South Africa

Patricia Popoola
Tshwane University of Technology, South Africa

Rotimi Sadiku
Tshwane University of Technology, South Africa

Samuel Eshorame Sanni
Covenant University, Nigeria

Damilola E. Babatunde
Covenant University, Nigeria

Olubunmi G. Abatan
Covenant University, Nigeria

Sunday Ojo Fayomi
Tshwane University of Technology, South Africa & Covenant University, Nigeria

Oluwaseun V. Fasiku
University of Kwazulu Natal, South Africa

ABSTRACT

The automotive industry is one major sector that consume material such as plastic at the advanced technological level. Hence, automotive plastics are now gaining attention due to the desire for light weight and low CO_2 emission from vehicles. It is therefore anticipated that polymer nanocomposites will significantly enhance the performance of current technologies for car industries due to their excellent mechanical, chemical, thermal, electrical and barrier properties and their influence on fire retardancy. Hence, with the use of polymer nanocomposites, the encouraging outcomes in different sectors of automotive industry has resulted to new horizons in terms of advanced polymer nanocomposites for automobile applications. This chapter reviews advance polymer composites for automobile applications. Methods of fabricating polymer nanocomposites and several applications of polymer nanocomposites in automotive industries are discussed.

DOI: 10.4018/978-1-5225-7838-3.ch004

INTRODUCTION

A nanocomposite is a matrix to which nano-sized particles have been added in order to enhance a specific property of a material. Nanocomposites are usually found in nature, as a multiphase solid material in which one of the phases has one, two or three dimensions of less than 100 nm, or structures having nano-scale repeat distances between the different phases that make up the material (Majumder, Majumder, & Karan, 2013). The properties of nanocomposites depend on a range of variables, mainly the matrix material, which can exhibit nanoscale dimensions, size, shape, loading, degree of dispersion and orientation of the nanoscale second phase together with interactions between the matrix and the second phase (Ashby, Ferreira, & Schodek, 2009). Hence, the properties of composites are a function of the properties of the constituent phases, their relative amounts, and the geometry of the dispersed phase. Nanocomposite matrix is classified into metal matrix, ceramic matrix and polymer matrix. Polymer-based nanocomposites are high performance materials that exhibit unusual property combinations and a distinctive design possibility (Camargo, Satyanarayana, & Wypych, 2009). Their potential is so promising that they are useful in several applications ranging from automobile, to aerospace and biomedical. Hence, various nanocomposite materials have been developed for use in aerospace industries in the past five years. Nonetheless, the raw materials are expensive and the manufacturing process needs long production cycle times. These considerations have made nanocomposites exorbitant in automotive industries, where the components need to be mass-produced at much lower cost (see Figure 1) (Naskar, Keum, & Boeman, 2016).

Suspended nanoparticles in polymer melt or solution usually yield nanocomposites comprising of random mixtures of structural building blocks of particles and polymers with slight improvement in mechanical toughness. The solvent-based gel of the building blocks is usually cast into a film or tape, with the solvent being taken out via evaporation or coagulation. Melt-based nanocomposites, when extruded

Figure 1. Relative performance and cost profiles in nanocomposite manufacturing
(Naskar, Keum, & Boeman, 2016)

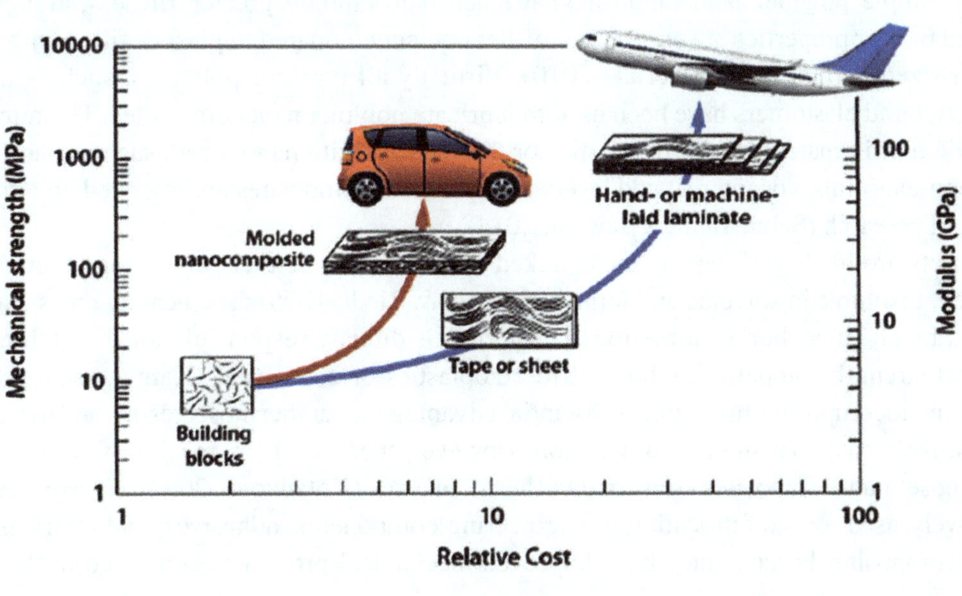

in sheet form, achieve positioning and ordered structure via shear-induced alignment of components in the material (Naskar, Keum, & Boeman, 2016).

Polymer-Based Nanocomposite

Polymer-based nanocomposites are new key materials for different advanced applications. These nanocomposites offer the possibility of substantial improvements in material properties such as shear and bulk modulus, yield greater strength, toughness, film scratch resistance, optical properties, gas and solvent transport, among many with only very small amounts of nanoscale material dispersed in the polymer matrix (Jancar et al., 2010). Studies have shown that at the nanoscale (below about 100 nm), a material's property can change dramatically. With only a reduction in size and no change in the substance itself, these materials can manifest additional new properties such as insulating behaviour, elasticity, electrical conductivity, different colour, and greater reactivity-characteristics that the very same substances do not exhibit at the microscale or macroscale. Furthermore, as dimensions reach the nanometer level, interactions at interfaces of phases become largely improved, and this is very significant in enhancing materials properties. Hence, the surface area/volume ratio of reinforcement materials employed in the preparation of nanocomposites is crucial in understanding their structural property relationships (Anandhan, & Bandyopadhyay, 2011). Their benefit over traditional composite materials originates from the extremely high surface area of the fillers, which have proportionally more surface atoms than their microscale counterparts, hence permitting intimate interphase interactions and conferring extraordinary properties to the polymer (Coiai et al., 2015). Composites have been going through several transitions such as the transition from micro-scale reinforcement fillers to nano-scale reinforcement fillers, resulting in the nanocomposite. The effectiveness of the nano-sized fillers in composites can be described by one of their exceptional geometric properties: the length-to-thickness aspect ratio. Hence, nano-sized fillers have outstanding higher reinforcing efficiency than the conventional, large fillers (Pilla, & Lu, 2015). The efficiency of the nano-sized fillers in composites is also as a result of their large surface area and surface energy. It has been known for about ten decades that the addition of nanofillers such as carbon black to rubbery polymers has a strong impact on the properties of these polymer materials (Mullins, 1969). For example, polymer nanocomposites can reach a substantially greater stiffness, strength, thermal stability and barrier properties at very low nanofiller content compared to plastics filled with traditional micrometer-sized particles (Jancar et al., 2010). Virtually all types of polymers, such as thermosets, thermoplastics and elastomers have been used to fabricate polymer nanocomposites. The reinforcement of polymeric matrix materials (thermoplastics or thermosets) with nano-sized, such as nano-sized particles, carbon nano-tubes or intercalated layers to forms nanocomposites are regarded an attractive and active area of research (Saba, Tahir, & Jawaid, 2014).

Thermosets are hard, stiff heavily cross-linked materials that neither soften nor become moldable when heated, insoluble in solvents and infusible (Peng, & Riedl, 1995). The heavily cross-linked structure formed by chemical bonds in thermoset materials is directly responsible for the high mechanical and physical strength compared to those of thermoplastics or elastomers (Kargarzadeh et al., 2017). Thermosets nanocomposites have some substantial advantages over thermoset resins, and these materials may be relatively easy to bring into production. However, there has been much less commercial interest in thermoset nanocomposites compared to thermoplastics (Kotsilkova, 2007). Thermoset polymers are extensively used for the fabrication of engineering components, adhesives, and matrices for fiber-reinforced composites because they have improved mechanical properties compared to thermoplastic polymers, they are however, usually brittle and vulnerable to cracking (Kargarzadeh et al., 2017). The incorporation of elastomers (Yang et al., 2015), inorganic nanoparticles (Haupert, & Wetzel, 2005), and natural or synthetic fibers (Divya & Suresha, 2016; Deepak, Reddy, & Naidu, 2015) into thermoset matrixes enhances fracture toughness. Furthermore, a range of nanofiber, nanoparticle and nano-platelet reinforcements with different shapes have been used in fabricating polymer nanocomposites (Anandhan, &. Bandyopadhyay, 2011).

Nanofiber Reinforced Composites

The most advanced and adaptable engineering materials are reinforced composite materials. The perfect integration of reinforcing natural fibers in a plastic polymeric matrix produces composites, possessing the finest properties of the material component. The term "nanofiber reinforced composite" usually refers to fibers with diameter in the nanometer range in any sort of polymeric matrix (thermoset or thermoplastic; synthetic or natural). Nevertheless, these composites found to be eco-friendly to a greater degree (Saba, Tahir, & Jawaid, 2014). These materials provide many of the same and equivalent advantages with respect to strength and toughness as conventional composites together with their own exceptional advantages including lower density (Alcock et al., 2006), better matrix–fiber compatibility, and recyclability (Abraham, & George, 2009; Matabola et al., 2009). Nanofiber-reinforced composites have proven to have good mechanical properties under static and dynamic loading conditions for an extensive range of matrix materials and for a wide range of engineering applications (Pike, & Caglar-Oskay, 2014). One of the main advantages of these composites is that they can be tailored to optimally perform under variety of loading and environmental conditions. Furthermore, fiber reinforcements offer additional exceptional properties such as self-control of cracks, self-sensing and electromagnetic field shielding, making them attractive for multifunctional applications (Reza et al., 2004).

Most investigations are aimed at obtaining basic building blocks in order to be able to fabricate complex devices and systems. The growing interest in the electrospinning process is as a result of the possibility of aiming new materials applications in nanotechnology, hence, nanofibers and nanostructures of various polymers can be obtained by this process (Neves et al., 2014). The choice of the polymer solutions, co-processing of polymer mixtures, chemical cross-linking of the fabricated nanofibers, etc., can offer a variety of pathways for controlling the chemical composition of electrospun nanofibers with a wide range of properties (such as strength, weight, elasticity, porosity, charged surface area, etc.) (Chronakis, 2005). Recently, the interest of electrospun nanofibers has grown as reinforcements for composite materials owing to their small diameter (20 nm - 1 mm) and high aspect ratio (Yao et al., 2015). Such nanocomposites have potentially great influence properties due to the high interfacial area, leading to high fiber debonding and pull-out energies during composite fracture (Stachewicz et al., 2012). Fiber aspect ratio, fiber orientation and the interfacial bonding between fiber and matrix have the capability of greatly influencing mechanical properties of fiber reinforced composites. The aspect ratios of electrospun fibers are extremely high as they are generally continuous, and much greater than the aspect ratios of other nano-reinforcements such as graphene, carbon nanotubes and nanoclays, making them excellent candidates for composite reinforcement (Yao et al., 2015). In addition, the electrospun polymer nanofibers can directly be collected as uniaxially aligned arrays by properly designing the conductive collector (Li et al., 2004).

Nanoparticle Reinforced Composites

Recently, the novel developments in polymers have fabricated nanocomposites, which are composites reinforced by nanoparticles. The nanoparticles in these materials, have shown great enhancements in thermal properties, mechanical properties and electrical properties. The addition of nanoparticles as reinforcement to conventional composites have shown great promise, not only to enhance the existing properties of these composites, but to also add many other properties that would maximize their applications in engineering and technology (Dante Rene Guerra, 2009). The reinforcing material can be made of particles such as minerals; sheets such as exfoliated clay stacks; and fibres such as carbon nanotubes or electrospun fibres. The production of polymeric nanocomposites by the use of nano scale reinforcement fillers has evolved with high level of interest. Nano composites are particle-filled polymers with at least one dimension of the dispersed particle in nano-meter scale (Kashiwagi, 2005). Different range of nanoparticles such as titanium, silica, zinc, alumina, micro and nano-sized silicon carbide, carbon nanotube, carbon black nanoparticles, calcium carbonate etc. are usually used as fillers to reinforce polymer nanocomposites in order to enhanced material properties. The polymer nanocomposites technology is a relatively new branch of material science which has an inherent adaptability in terms of new prospects (Mishra et al., 2017).

Nanoparticles take advantage of their huge surface area to volume ratio in order for their optical properties to become a function of the particle diameter. When integrated into a bulk material, they can strongly influence the mechanical properties such as stiffness or elasticity. Traditional polymers, for example, can be reinforced by nanoparticles leading to novel materials to be used as lightweight replacements for metals. Such improved materials will enable a weight reduction together with an increase in durability and enhanced functionality (Mohseni et al., 2012). Nanoparticle-reinforced polymer composites (NRPs) are getting special attention, particularly for bonding applications in automotive, aerospace, marine and oil and gas industries. Apart from improving mechanical, electrical, thermal properties, NRPs are known to improve the permeability properties and diffusion barrier attributes of their host polymers. The validation of mechanical properties gained by the inclusion of nanoparticles comprises of enhanced strength and stiffness to weight and cost ratios, enhanced fatigue and corrosion resistance, more controllable damage mechanism, and enhancement of the energy-absorption capacity of their host polymer (Soltannia et al., 2016). Hence, these ascribed properties make them effective candidates for reinforcing polymers (Choudalakis, & Gotsis, 2009; Loos, & Schulte Lu et al., 2013; Shadlou, 2014). Nanoparticle-reinforced polymer composites are progressively being used in various engineering applications, specifically in the form of adhesives. Adhesives are now becoming increasingly more popular for joining structural components because of the recent advancements in techniques used for toughening them (Soltannia et al., 2016).

The extent of the interface area between the matrix and reinforcement phase(s) is naturally an order of magnitude, which is greater than conventional composite materials. The properties of the matrix material are significantly influenced within the vicinity of the reinforcement. Hence, this huge amount of reinforcement surface area implies that a relatively small amount of nanoscale reinforcement can have a noticeable effect on the macroscale properties of the composite. The percentage by weight (known as mass fraction) of the nanoparticulate introduced, can remain very low (on the order of 0.5% to 5%) as a result of the low filler percolation threshold, specifically for the most commonly used non-spherical, high aspect ratio fillers; such as nano-meter-thin platelets, or nano-meter-diameter cylinders. Therefore, the orientation and arrangement of asymmetric nanoparticles, interface density per unit volume of nanocomposite, thermal property mismatch at the interface, and polydispersity of nanoparticles significantly affect the effective thermal conductivity of nanocomposites (Tian et al., 2013).

Functional Polymer and Nanocomposites

Functional polymers are macromolecules that bears specific chemical group with unique properties such as advanced optic and/or electronic properties, and hence, have unique uses. A functional polymer has physical properties of the polymer support associated with the chemical reactivities of the attached functional group. The study of the chemical reactions of polymers has appeared to be one of the most active fields in polymer science because of its exceptional ability to produce special polymers with desirable chemical and physical properties (Kumar, 1992). Hence, the application of a functional polymer depends on the physical properties and the chemical composition of the polymers. The use of functional group transformations done in order to introduce required functional or reactive groups is to alter polymer surfaces, provide graft and unusual side chain substituents (Kumar, 1992). This will aid in the structural characterization of polymers that will represent significant portion in the application of functionalized polymer in science and engineering. The different types of functionalized polymer and nanocomposites have grown both in the nature of polymer-composite backbone and in the arrangement of functional group.

Graphene is a molecular sheet of graphite that has potential applications in different fields of material science and engineering for fabrication of nanocomposites, sensors, supercapacitors, hydrogen storage and optoelectronic device (Ramanathan et al., 2008). The function of graphene in different applications is based on its one or more fundamental properties such as excellent mechanical (Papageorgiou, Kinloch, & Young, 2017), chemical (Zhu et al., 2010), thermal conductivity (Balandin et al., 2008), electrical (Wang et al., 2017), thermal and thermoelectric transport (Xu, Li, & Duan, 2014). Due to these intriguing properties, graphene research has generated a great deal of interest. These remarkable properties of graphene offer infinite potentials for several applications in areas of science and engineering such as electronics, energy storage and conversion, biotechnology and especially improvement in composite fiber materials (Ji et al., 2016). In order to exploit these important properties, the inclusion of graphene sheets within a polymer matrix with homogeneous dispersion and a fine interface control is utmost important (Layek, & Nandi, 2013). In graphite, it is difficult to get exfoliated graphene sheets because the graphene layers have strong cohesive force. In order to reduce the cohesive force between the graphene sheets and to also impart specific interaction with the host, polymer matrix functionalization of graphene is very important (Layek, & Nandi, 2013).

The structure and properties of graphite oxide depend on a particular method of production and the degree of oxidation. It usually conserves the layer structure of the parent graphite, nonetheless the layers are buckled and the interlayer spacing is about two times larger (~0.7 nm) than that of graphite (Kumar, Woltornist, & Adamson, 2015). On oxidation, graphite produces graphite oxide (GO) having hydrophilic functional groups (-OH, epoxide, -COOH) which promotes the intercalation of water molecules into the gallery and the graphene sheets can be easily separated from each other via sonication, hence producing highly dispersible GO sheets in aqueous medium (Szabo et al., 2006; Hernandez et al., 2008; Kudin et al., 2008). Hence, graphene oxide is made of a monolayer of graphite oxide which is usually achieved due to the presence of oxygen functionalities by oxidative treatments of graphite (Zhang, 2013). The properties of graphene can be altered by the functionalization of graphene oxide. The chemically-changed graphene obtained by this method can possibly be used in more than a few applications. The graphene oxide can be functionalized in a number of ways, depending on the intended application (Azo-Nano, 2015).

These exfoliated GO sheets are typically used for diverse applications and/or are further functionalized to be used for targeted applications (Rama, & Nandi, 2013). The utmost attractive property of GO is that it can be (partly) reduced to graphene-like sheets by removing the oxygen-containing groups with the recovery of a conjugated structure (Pei, & Cheng, 2012). The GO is electrically insulating but becomes conducting when it is reduced to produce reduced graphene oxide (rGO) by sodium borohydride and/or hydrazine hydrate (Rama, & Nandi, 2013). The reduced GO (rGO) sheets are usually considered as one kind of chemically derived graphene. Some other names have also been given to rGO, such as functionalized chemically modified graphene (Pei, & Cheng, 2012). The GO sheets in organic solvent and rGO sheets in both aqueous and organic solvent suffer aggregation owing to the high cohesive interaction, which makes them difficult to disperse. This lack of homogenous dispersion restricts its exploitation in many potential commercial applications as a result of the fragile interfacial interaction between the GO or rGO and the host polymer matrix cast from sodium borohydride and/or hydrazine hydrate (Li et al., 2008). In order to attain optimum physical and mechanical properties of graphene polymer nanocomposites, the solubility of graphene sheets should be maximized in a common solvent with that of the matrix polymer. Furthermore, the stress transfer between the sheet and polymer matrix should also be maximized through the interface due to the specific interaction between the sheet and polymer matrix (Rama, & Nandi, 2013).

Carbon nanotubes (CNTs) are allotropes of carbon with cylindrical nanostructure. The integration of a minute amount of these inorganic materials (CNTs) into the polymer matrix significantly enhance the performance of the polymer materials as a result of their exceptional properties such as high mechanical strength, electrical and thermal conductivity. Thus, polymer nanocomposites can be used for variety of applications depending on the inorganic materials present in the polymers (Yeum et al., 2012). Good dispersion and interfacial adhesion between CNTs and the polymer matrix is important in order to improve the mechanical properties of CNT polymer nanocomposites (Gupta, & Kumar, 2018). Due to the nature of CNTs, CNTs are inherently insoluble in most organic, aqueous solvents and are poorly chemically compatible with the polymer matrix. Hence, surface modification of CNTs is required in order to enhance the compatibility and solubility (Le et al., 2013).

The functionalization of carbon nanotubes has gained substantial interest in the fields of physics, chemistry, engineering, material science and biology. The functionalized CNTs exhibit enhanced properties that allows facile construction of novel nanomaterials and nanodevices. At the moment, most of the functionalization methods developed can be grouped into the covalent attachment of functional groups and the non-covalent adsorption of various functional molecules onto the surface of CNTs (Meng, Fu, & Lu, 2009). Covalent functionalization depends on the formation of a covalent connection between functional entities and the carbon skeleton of nanotubes. It can be divided into direct covalent sidewall functionalization and indirect covalent functionalization with carboxylic groups on the surface of CNTs (Meng, Fu, & Lu, 2009). Direct covalent sidewall functionalization is connected with a change of hybridization from sp2 to sp3 and a concurrent loss of p-conjugation system on graphene layer. This process can be achieved by reacting the graphene with some molecules of a high chemical reactivity (Jeon, et al., 2011). Indirect covalent functionalization uses chemical transformations of carboxylic groups at the open ends and holes in the sidewalls. These carboxylic groups could have occurred on the as-grown CNTs, furthermore, the carboxylic groups can be generated during oxidative purification. The carboxylic acid groups usually need to be converted into acid chloride and then undergo an esterification or amidation reaction in order to increase the reactivity of CNTs (Kim et al., 2008). The disadvantage of covalent functionalization is that the perfect structure of CNTs has to be destroyed, resulting in substantial changes in their physical properties (Meng, Fu, & Lu, 2009). Non-covalent functionalization is primarily based on supramolecular complexation using various adsorption forces, such as hydrogen bonds, electrostatic force. Van der Waals force and p-stacking interactions (Mu et al., 2008; Yu et al., 2008; Georgakilas et al., 2016). Compared to the covalent functionalization, non-covalent functionalization has the advantages to be functioned under moderately mild reaction conditions and the perfect graphitic structure of CNTs can be maintained (Meng, Fu, & Lu, 2009). In any case, non-covalent or covalent functionalization of CNTs can enhance their compatibility and solubility in polymer (Hong et al., 2006).

Fabrication of Polymer Nanocomposites

Polymer nanocomposites can be fabricated either by chemical or mechanical process. The major problem encountered in polymer nanocomposite fabrication is homogeneous and even dispersion of nanoparticles in the polymer matrix. The Nanofillers have a tendency to aggregate and form micron size filler cluster, which limit the dispersion of nanoparticles in the polymer matrix thereby limiting the performance of nanocomposite properties (Khan, Hamadneh, & Khan, 2017). The most common methods used in the fabrication of nanocomposites can be grouped into the following approaches. (1) fabrication of nanocomposites by induced intercalation method, (2) fabrication of nanocomposites via in situ polymerization (3) fabrication of nanocomposites by mechanical/melting process (Tanahashi, 2010).

Induced Intercalation Method

Usually, intercalation method is a typical top-down approach based on the downsizing of fillers to nano dimensions involving the dispersion of nanoplatelets types of nanomaterials into the polymer matrix. In intercalation of polymer from solution approach of nanocomposite blending, the organically modified silicate is dispersed in a solvent in which the polymer is also soluble (Mittal, 2012). Hence, intercalation method requires surface modification of nanoplatelets for homogeneous dispersion of plate-like nanofillers in the polymer matrix (Khan, Hamadneh, & Khan, 2017). The polymer then adsorbs onto the delaminated sheets, then the evaporation of the solvent takes place. When the solvent is evaporated, the sheets reassemble. This process traps the polymer chains in between, hence, an ordered multilayer structure is generally formed using this method (Mittal, 2012). Intercalated morphology takes place when polymer chains diffuse into the gallery spacing of structured layers. The polymer chains loose entropy in the course of intercalation, which is compensated by the increase in the entropy of the solvent molecules as a result of their desorption from the filler interlayers (Mittal, 2012). The nanoplatelets filler interlayers can be homogeneously dispersed either by chemical or mechanical technique (Alexandre, & Dubois, 2000).

Chemical - In-Situ Polymerization Technique

Chemical technique involves the in-situ polymerization method in which nanoparticles are dispersed in liquid monomer by employing polymerization reaction. This technique occurs via a chemical reaction resulting in the development of very fine and thermodynamically stable reinforcing phase within a matrix. In this technique, nanoplatelets are dispersed into polymer followed by additional polymerization process. Polymerization process starts either using heat, radiation, initiator diffusion, or by organic initiator or catalyst fixed through cationic exchange (Alexandre, & Dubois, 2000). This method was reported by a group of Toyota researchers for the synthesis of polyamide nanocomposites that led to the exponential growth in the nanocomposites research (Mittal, 2012). In this method, nanocomposites can be made by the in-situ intercalation polymerization method, where the monomer is initially intercalated in the clay, and afterward polymerized in situ. Hence, nanocomposites should be net-shape moldable and may be extruded, permitting consolidation of parts and minimizing the assembly steps. Figure 2 shows the main steps of the conceptual process needed to fabricate nanocomposites for automotive application (Garcés et al. 2000). For the generation of polymer nanocomposites by this technique, the layered silicate mineral is swollen in monomer. The polymerization of the monomer is initiated after swelling. The generated structure is exfoliated or significantly intercalated with the presence of monomer in and out of the filler interlayers (Mittal, 2012). The nanoplatelets are swollen in monomer solution and the polymer formation occurs between the interlayers forming intercalated or exfoliated nanocomposites by polymerization technique (Khan, Hamadneh, & Khan, 2017). The advantage of this technique lies in the better exfoliation achieved.

Mechanical/Melt Technique

Mechanical technique involves direct intercalation of polymer with nanoplatelets taking place through solution mixing in a suitable solvent. The polymer is dissolved in a co-solvent and nanoplatelets sheets are swollen in the solvent, then these two solutions are mixed together and the polymer chains in the solution intercalate into the nanoplatelets layers and displace the solvent (Khan, Hamadneh, & Khan, 2017). This technique can also be achieved by a melt intercalation method through melt-compounding of the polymer with the nanoplatelets in the absence of a solvent under a high shear condition (Ray, & Okamoto, 2003). Melt intercalation is known to be most commonly used techniques for the synthesis of polymer nanocomposites (Hausner et al., 2015; Xing, Deng, & Liu, 2015; Zhang et al., 2017). The high molecular weight polymer is melted at high temperature and the filler is then blended with the melted polymer at high temperature under shear. Therefore, this mode of composite generation does not need any chemical synthesis or solvent. Nonetheless, there is still a challenge with the intercalation of high molecular weight polymer chains in the filler interlayers because thermodynamic and kinetic factors

Figure 2. Nanocomposites process diagram for automotive applications
(Garcés et al. 2000)

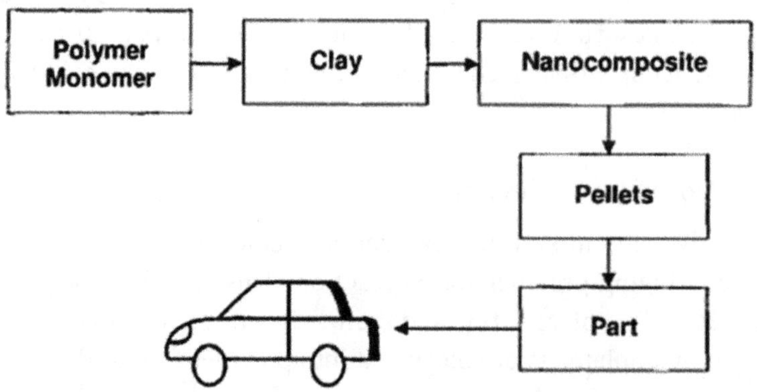

have impact on the intercalation. Therefore, it is essential to modify the filler in such a way that it would be able to exfoliate in the polymer matrix by the action of shear (Mittal, 2012). However, polymeric systems not appropriate for in situ polymerization can be used for the fabrication of nanocomposites using this technique (Tanahashi, 2010).

Applications of Polymer Nanocomposites in Automotive Industries

With their high specific strength and stiffness, nanocomposites have the capability to significantly lower vehicle weight, which can have a dramatic effect on improving fuel efficiency and reducing greenhouse gas emissions (Pilla, & Lu, 2015). For example, nanoparticles can be used to reinforce traditional polymers that result to novel materials that can be used as lightweight replacements for metals. Such improved materials will allow a weight reduction, increase in durability and enhanced functionality of automobile parts. The basic trends that polymer matrix nanocomposite allows for the automobile are: car structure (front and rear structure: front seat frames and rear seat lower frame), coating in car engine applications, improved energy efficiency, lighter but stronger materials (for tyres, better fuel consumption and increased safety), improved and miniaturized electronic systems (Mohseni et al., 2012). The first company to commercialise polymer/layered silicate nanocomposites was Toyota (Okada et al., 1998), which used nanocomposite parts in the production of their novel car models (Kotsilkova, 2007).

The commercialization of nanocomposites production really started over the last decade. A one-piece compression moulded rear floor assembly was made in 2009 by General Motor (GM) for their Pontiac Solace using nano-enhanced Sheet Moulding Compounds (SMCs) developed by Molded Fiber Glass Companies (MFG), Ohio (nano werk, 2012). Furthermore, this technology is used on GM's Chevrolet Corvette Coupe and Corvette ZO6. The nano-filled SMCs significantly evinced lower density than conventional SMCs leading to an enhanced fuel efficiency (Fukushima & Inagaki, 1987). The automotive industry can benefit from polymer nanocomposites in numerous applications such as combustion engines and powertrain, structural components (frames and body parts), tires, interior and exterior panels, suspension and breaking systems, exhaust systems and catalytic converters, paints and coatings, lubrication, and electric and electronic equipment. Figure 3 shows the usage of polymer nanocomposites in vehicle parts.

Polymer Nanocomposite for Automobile in Combustion Engine

The components of an engine are very complex and need numerous design iterations and testing before they can be concluded. A timing belt, a part of an internal combustion engine, has the function of controlling the timing of the engine's valves (Galimberti, Cipolletti, & Coomb, 2013). Nanocomposites for timing belt can be produced by integrating nanometer-size clay particles in polymers such as polypropylene (PP), polyethylene (PE), polyesters and epoxies (Giannelis, 1996). Nanoclays are the most used commercial nanomaterials. It is as strong as 7% glass filled polyamide, with a density of 1.14 g/cm; it is an excellent light weight substitute for polyamide added to 30% glass beads. Hence, nanoclay takes over the traditional fillers at a 3:1 ratio (Müller et al., 2017). The clay materials can be dispersed and exfoliated into polymers through the following techniques: conventional melt compounding, solution methods, by the in-situ intercalation polymerization method (Garcés et al., 2000). This method was pioneered by Toyota Motor Company to create nylon 6-clay hybrid, used for manufacturing a timing-belt cover, the first practical commercial example of polymeric nanocomposites for automotive applications (Kurauchi et al., 1991). In 2004, the Chevrolet had a body side trim with nanoclays, and with a new design part,

Figure 3. Illustrations of the application of polymer nanocomposites by automotive industry
(Adapted from nano werk, 2012)

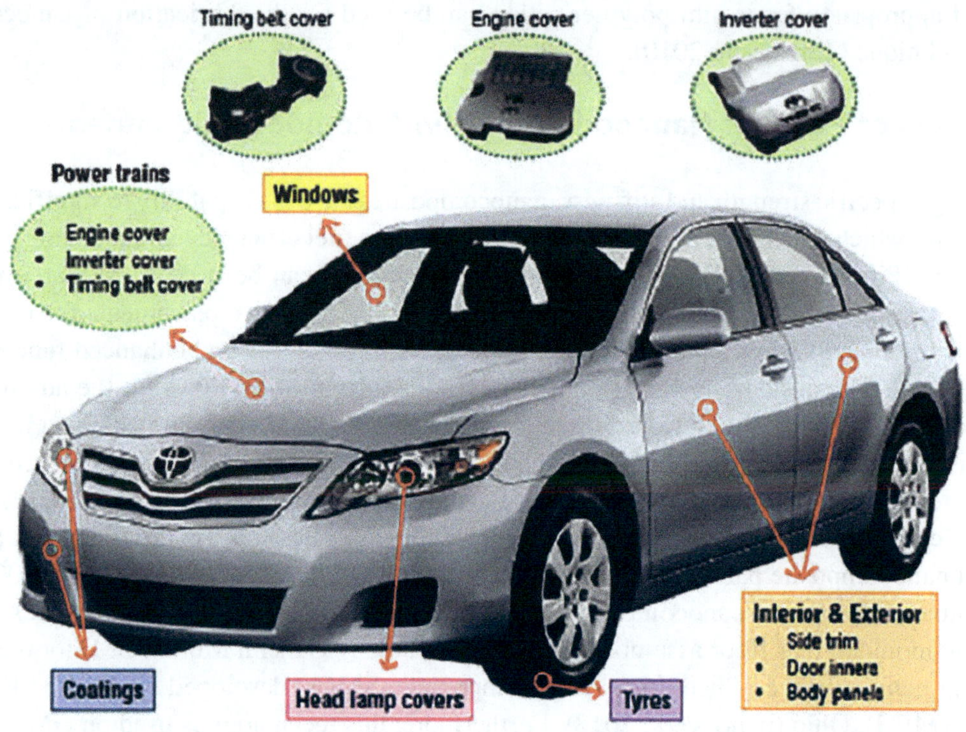

a weight saving of 25% was attained. Another example with the same nanofiller was shown again for general motors in 2005 to diminish the weight (Müller et al., 2017). Maserati engine bay covers were made with a nanocomposite of nylon 6 and nanoclays, reducing the weight, and increasing mechanical properties (Müller et al., 2017).

There is presently an increasing interest to reduce the weight of the components in a car in order to reduce the fuel consumption. Hence, with time, plastics will become more incorporated in the design of engines. The reason for a higher incorporation of plastics in this automotive industry is due to the fact that they are very adaptable to the required needs of the engine. Furthermore, plastics can be manipulated chemically in order to suit the strength or toughness required by a component. As a result of the low cost and enhanced physico-mechanical properties, polyolefins are now attractive thermoplastic in nanocomposites, greater than nylon 6/clay nanocomposites. Furthermore, lubricants from polymer-derived nanocomposite reduce friction and wear in applications involving elevated temperatures, such as within an internal combustion engine. Hence, polymer-derived nanocomposite lubricants advantageously extend the life span of an engine by lubricating surfaces of the engine. For example, polymer derived nanocomposite lubricants could reduce the friction between a piston and cylinder or lubricate engine components, such as cam shafts and valve stems (Raj, & Shah, 2012). A US patent reported a silicon polymer-based nanocomposite precursor material consisting either of organopolysilaZane, polysilaZane, polycarbosilane, polysilane, polysiloxane, and combinations thereof, in liquid form, was mixed with engine oil that was placed into an internal combustion engine, such as a four cycles engine. The engine was operated to heat the motor oil to a temperature sufficient enough to improve lubricity of motor oil and enhances wear resistance properties (Raj, & Shah, 2012).

Polymer Matrix Nanocomposite for Automotive Structural Components

In the ongoing mission to squeeze more miles out of a gallon of fuel, auto manufacturers are progressively turning to strong, lightweight plastics and plastic composites in an effort to reduce the weight of the vehicles they produce. According to the U.S. Department of Energy, reducing the weight of a vehicle can improve its fuel efficiency by as much as 6-8% for every 10 percent reduction in poundage. Hence, reducing the weight of a vehicle can have significant long-term benefits for consumers and the environment (Plastics, 2015). In today's automobiles, plastics can be found in headlamps and tail lamps; fuel lines and fuel tanks; windows; housings for electronics and braking systems; body panels and bumpers; and foam fill material for hollow structures such as areas between windows, door panels, bumpers and frames (Plastics, 2015).

The automotive industry has expended significant effort in order to develop lightweight parts from new easy-to-process polymeric nanocomposites. These polymeric nanocomposites have been principally attractive due to their increase in fuel efficiency and reduction of greenhouse gas emissions (Naskar, Keum, & Boeman, 2016). Polymer matrix composites are made up of thermoset or thermoplastic matrix resins reinforced by fibers that are much stronger and stiffer than the matrix. As a result of their characteristics such as lighter nature, stronger, and stiffer than the unreinforced polymers or conventional metals, polymer matrix composites are attractive. Additional advantage to their properties is that polymer matrix composites can be tailored to meet the needs of a specific application (Park, & Seo, 2011). Hence, polymer matrix nanocomposite promises breakthroughs in areas such as materials, structure and manufacturing. Currently, the automotive industry focuses on replacing of traditionally used metal alloys with polymer composites, temperature resistant and high-strength thermoplastics. It is well known that the use of polymer composites, temperature resistant and high-strength thermoplastics in the automotive market is not only a trend, but also a necessity (Koniuszewska, & Kaczmar, 2016). Furthermore, polymer matrix composites are characterised by good corrosion resistance, extremely good formability and damping of mechanical vibrations (Paduchowicz et al., 2012). In the automotive industry, the relative high specific strength of polymer matrix composite makes possible the reduction of weight of final elements of car structures (Koniuszewska, & Kaczmar, 2016).

The use of reinforced composite materials has been gaining attention by auto manufacturers. Hence, automotive industries have been using reinforced composite materials for the construction of important elements of a vehicle. For example, carbon fiber-reinforced plastics used in race cars are currently making their way into use for the production of vehicles because of their strength and light weight (Plastics, 2015). These auto manufacturing companies combine exceptional design with innovative technologies while saving energy and weight. Companies such as Smart and BASF, in Germany, have jointly developed and used composite materials as a concept that combines exceptional design with innovative technologies in the areas of energy efficiency, lightweight construction and temperature management for vehicles (Smart, & BASF, 2018). Smart and BASF used epoxy systems reinforced with carbon fibers to produce components such as chassis and external framework. The addition of the carbon fibres strengthens the high-performance polymers with comparable properties to that of the light metal alloys. The carbon fibres are impregnated and joined together with the epoxy resin for further processing (Koniuszewska, & Kaczmar, 2016). Mass saving of up to 150 kg per chassis was attainable with the utilization of composite materials and at the same time, the strength equivalent to that of metal alloy was maintained (Zindani, & Kumar, 2018). Polymeric wheel fabricated from another high performing material named Ultramid® Structure is another innovative technology that is made by using polyamide matrix with long glass fiber as reinforcing material. The polyamide reinforcing material has excellent chemical and thermal stability, dynamic strength, good toughness and sustaining endurance. This gave a weight saving of 12 kg per car (Koniuszewska, & Kaczmar, 2016; Zindani, & Kumar, 2018).

Polymer Composites (Fiber Reinforced Composites) in Exterior and Interior Panels

Reinforced composites of natural fiber and nanocomposites are more ecologically pleasant, they are thus often used in military and transportation (automobiles, railway coaches, aerospace) applications (Kozłowski, & Władyka-przybylak, 2008). Various researchers have developed hybrid natural fiber composite by joining natural fibers/natural fiber and natural fibers/synthetic fibers with phenolic, poly vinyl ester, epoxy, polyester and poly urethane resins (Sathishkumar, Naveen, & Satheeshkumar, 2014; Jawaid, & Abdul Khalil). These natural fiber/nanofiller-based hybrid composites can be implemented in advanced automobile applications; they could also be used to produce acoustic insulator and extremely thermally stable materials (Saba et al., 2014). Table 1 shows the list of some natural fiber, their origin and their applications in automotive industries.

Glass-fiber-reinforced plastics have proven to meet the demands of structural and durability of automobile interior and exterior parts. Good mechanical properties and a well-developed, installed manufacturing base have assisted in the incorporation of glass-fiber-reinforced plastics within the automotive industry. The chemical composition of the glass fiber has a diverse stage of strength. The chemical compositions of glass-fiber in wt% are shown in Table 2. However, glass-reinforced plastics have shown some shortcomings such as their relatively high fiber density; with approximately 40% higher than natural fibers, difficulty to machine, poor recycling properties and the potential health

Table 1. List of some natural fiber, their origin and their applications in automotive industries

Natural Fibers	Origin	Applications in Automotive Industries
Sisal Fiber	Sisal plant	Sisal fibers are used in automotive industry with fiberglass in composite materials for carpets.
Kenaf Fiber	Deccan plant	Kenaf fiber composites are used to improve the fuel consumption and emission in automotive industries, as a result of their light weight.
Hemp Fiber	Cannabaceae plant	Hemp fibres are utilized by the automotive industry as reinforcement in moulded composites through thermoplastic and thermoset matrices. They are also used in the automotive industries for reinforcement of passenger rear decks, body panels, pillars, seats and boot linings.
Palm fibers	Palm tree	Palm fibers are used for insulation purposes in automotive sectors (Faris, AL-Oqla, & Sapuan, 2014).
Wool fibers	Sheep	Wool fibers are utilized by the automotive industries for upholstery.
Cotton fiber	Seeds of the cotton plants of the genus *Gossypium* in the mallow family *Malvaceae*.	Cotton fiber are utilized by the automotive industries for sound insulator and absorber.
Milkweed fiber	Seeds of any of several milkweed plants of the genus Asclepias	The potentiality of milkweed fiber as reinforcement for sound insulator and absorber automotive applications has been explored.

hazards from glass-fiber particulates (Holbery, & Houston, 2006). The utilization of lightweight, low-cost natural fibers provides a very good prospect of replacing a large segment of the glass in several automotive interior and exterior parts. Natural fibers such as sisal, hemp, kenaf, flax and jute provide unique benefits such as reductions in weight, cost, and CO_2, less reliance on foreign oil sources, and recyclability. Though natural-fiber-reinforced plastic parts has many unique benefits than fiber glass, numerous major technical considerations must however, be addressed before the engineering, scientific, and commercial communities gain the confidence to allow wide-scale acceptance (Holbery, & Houston, 2006). Hence, it is usually essential to compatibilized or couple the blend due to the interfacial adhesion between the natural fiber and polymer matrix which determines the composite physical properties (Zafeiropoulos et al., 2002). The composition of the natural fibers listed in Table 1 are given in Table 3. The composition of palm fiber and wool fiber are quite different from the listed natural fibers and they are given in Table 4a and Table 4b.

Table 2. Chemical compositions of glass fibers in wt%.

Types	SiO_2	Al_2O_3	TiO_2	B_2O_3	CaO	MgO	Na_2O	K_2O	Fe_2O_3
E-glass	55.2	14.8	0.2	7.3	18.7	3.3	0.3	0.2	-
C-glass	64.6	4.1	-	5.0	13.4	3.3	9.6	0.5	-
S-glass	65.0	25.0	-	-	-	10.0	-	-	-
A-glass	67.5	3.5	-	1.5	6.5	4.5	13.5	3.0	-
D-glass	74.0	-	-	22.5	-	-	1.5	2.0	-
R-glass	60.0	24.0	-	-	9.0	6.0	0.5	0.1	-
EGR-glass	58.4	11.0	2.1	0.09	22.0	2.2	0.9	0.5	0.26
Basalt	52.0	17.2	1.0	-	8.6	5.2	5.0	1.0	5.0

(Imtiaz, 2014)

Table 3. The composition of the natural fibers used in automotive industries

Fiber	Cellulose (wt%)	Hemicellulose (wt%)	Lignin (wt%)	Pectin (wt%)	Moisture (wt%)	Waxes (wt%)
Sisal fiber	66-78	10-14	10-14	10	10-22	2
Kenaf fiber	45-47	21.5	8-13	3-5	-	-
Hemp fiber	70-74	17.9-20.4	3.7-5.7	0.9	6.2-12	0.8
Cotton fiber	85-90	5.7	-	0.1	7.85-8.5	0.6
Milkweed fiber	55	24	18	3	-	-

Table 4a. The composition of the palm fiber used in automotive industries

Fiber	Lignin (wt%)	Hemicellulose (wt%)	Lipid (wt%)	Protein (wt%)	Ash (wt%)	Others (wt%)
Palm fiber	28.5	20.8	1.9	3.6	5.6	-

(Koba & Ishizaki, 2014)

Table 4b. The composition of the wool fiber used in automotive industries

Fiber	Keratin (wt%)	Dust (wt%)	Suint (wt%)	Fat (wt%)	Mineral Matter (wt%)
Wool fiber	33	26	28	12	1

(Hossain, 2014)

Car manufacturing companies in Europe also embraced the use of natural-fiber composites with thermoplastic and thermoset matrices in order to enhance their applications in automotive industry, especially in car interior such as seat backs, parcel shelves, boot linens, front and rear door linens, truck linens, and door-trim panels (Davoodi et al., 2011). Furthermore, apart from the use for car interior parts in automobile industry, natural fiber integrated in polymers has been used for high requirement applications for exterior auto body components, such as the middle section between the headlights above the fender of a passenger bus (Graupner, Herrmann, & Mussig, 2009). Most of the car manufacturing companies in the world have done a lot of investigation in order to incorporate the natural-fiber polymer composites in their products (Mohammed et al., 2015). Germany is well known to be a leader in the use of natural fiber composites. The German auto-manufacturers; such as Mercedes, Volkswagen, BMW and Audi stepped forward to introduce natural fiber composites for interior and exterior applications (Sanjay et al., 2016). For example, the first commercial application of natural fiber composite is the inner door panel of the 1999 S-Class MercedesBenz, made in Germany. The composite was made up of 35% Baypreg F semi-rigid (PUR) elastomer from Bayer and 65% of a blend of flax, hemp and sisal. Furthermore, it should be highlighted that luxury automotive manufacturers are now on board, this can be seen as an evidence that natural fiber composites are being used for environmental needs and not to lower costs (Puglia, Biagiotti, & Kenny, 2004). In 1996, Mercedes-Benz used an epoxy matrix with the addition of jute in the door panels in its E-class vehicles back. In 2000, another example of natural fiber composites' application appeared commercially, when Audi company uses flax/sisal material reinforced polyurethane composite with a mix to make door trim panels. Toyota also developed an eco-plastic made from sugar cane to line the interiors of the cars (Koronis, Silva, & Fontul, 2013).

Automotive industry is under increasing pressure to fulfil environmental, performance demands and higher fuel efficiency at competitive costs. Hence, the expansion of the development of high-performance materials produced from natural resources worldwide (Fogorasi & Barbu, 2017). The development of high performance natural fibre composites resulted in the collaboration of automotive industry in Europe and research institutions in European Commission FP5 GROWTH Projects. The 'ECOFINA' Project particularly addresses the substitution of mineral fillers and fibres presently used in automotive parts made with organic matrices, with polymeric matrix composites based on annually renewable natural fibres. This substitution will result in the production of vehicle components with potentially complete recyclability. Furthermore, the lower weight attainable due to the low density of the natural fibres will allow a lowering of gas emission in vehicles, enhancing quality of life (Puglia, Biagiotti, & Kenny, 2004). In addition, the use of long fibres in composites imparts enhanced mechanical properties and capability to adhesion over the matrix to the fiber (Jeyanthi, & Rani, 2012). Hence, the orientation and the length of natural fibre is an important feature for their selection in composites for automotive application (Laine, & Rozite, 2010).

Polymer Nanocomposites in the Automotive Tyre Industry

Rubber nanocomposites is the integration of naturally derived nanofillers such as cellulose and starch, in a rubber matrix. The development of rubber/bio-nanofillers is relatively still in its early stage as a result of the difficulty in the extraction of nanofillers from naturally occurring fibers or from other sources. There are some challenges such as the complete exfoliation and uniform dispersion of nanofillers in rubber, in the area of rubber nanocomposites. Hence, effective surfactants have to be designed for an exceptional dispersion. The orientation of nanoplatelets in rubber matrix by special extrusion is also one of the major challenges. Hence, the in-situ monitoring of the flow of rubber nanocomposites during processing requires a lot of attention (Stephen, & Thomas, 2010). However, tyre industries consume more than 60% of the rubber product; hence, the prime factors of consideration are safety and tread life (Pal, & Pal, 2009). In addition, among all the polymer-based automotive components, tire is the most vital one because a tire must deliver high traction on dry and wet roads, commonly called as wet and dry grip. The high traction force between tread and road is required for avoiding slippage while driving on the road (Chandra, & Kumar, 2017). Furthermore, 30% of the mechanical work available from the fuel is dissipated in the tyres and a reduction of 15% in tyre rolling resistance may improve fuel consumption by 4.5% (Pal, & Pal, 2009).

According to the statistical survey of 2006; the consumption of rubber worldwide, rose to 21.5 million tonnes. This statistical report discovered the inevitability of rubber goods in the day to day activities (Stephen, & Thomas, 2010). Potential industrial applications of rubber nanocomposites include a large number of latex and dry rubber goods such as, foams, tubes, tire inner tubes, automobile components, electrical devices, applications where electrical heating is required, electrical shielding, sensing devices and related products (Alex, 2010). The automotive industry has the leading market for rubber with respect to the manufacturing of tires and inner tubes (Stephen, & Thomas, 2010). Nanocomposite technology can infiltrate the automotive industries especially pneumatic tire, as barrier improvement is vital for air retention in tires. The outstanding air retention properties of butyl/halo butyl and chemically modified rubbers are known in the tire industries because these rubbers are extensively used in the inner tube and inner liner in automotive tires (Nah, & Abdul Kader, 2010). Elastomers such as, epichlorohydrin rubber, polysulfide rubber and polyurethane rubber; have very good and even better air retention properties but are not used in the tire industries. Furthermore, elastomers such as nitrile rubber and ethylene/acrylate copolymers provide moderate air retention properties. These polymers can be used for automotive components by integrating nanofillers. When nanofillers are used in rubbers having moderate air retention properties, the final product can provide a substantial increase in air retention properties (Nah, & Abdul Kader, 2010).

Furthermore, the addition of nanofillers increases electrical, thermal, mechanical, optical, air permeability and flame retardancy exponentially, especially, at a lower filler dosage. Hence, the possibility of producing lighter weight tyre with enhanced rolling resistance and extended life, at the same time. The reduction of filler will in turn reduce the specific gravity and heat generation, which can lead to the performance enhancement of the of different tyre compounds and ultimately result in better fuel-efficient tyre. Carbon or silica nanoparticles have been used as fillers for tyre rubber for many years, assisting in improving the performance and durability of tyre. Silica, which does not naturally mix well with rubber, is now often used in combination with organosilanes, which can bind well to both components, creating a more stable material (Soutter, 2012). Hence the reason why polymer / layered silicate nano-composites is considered one of the most important types of nano fillers. As such, polymer / layered silicate nanocomposites have gained strong momentum in achieving material properties desired for tyre applications. The application of nano-composite / nano-materials like nano zinc oxide, carbon nanotube will help tyre industry to a great extent in the future (Chandra, 2017). Hence, the use of elastomer nanocomposites in automotive tires is anticipated to contribute to market growth as they consist of polymeric chains that display a high degree of flexibility, mobility and rubber-like elasticity if the chains are linked together in a network structure.

Polymer Nanocomposites in the Automotive Coating

The purpose of automotive coatings has been twofold: decoration and conservation. Coating for decoration provide a coloured and smooth surface to the car body. Furthermore, it accentuates the shape of the car body via a viewing angle dependent brightness and colour (Hintze-Bruening, & Leroux, 2012). Coatings in shielding a metallic car body are primarily based on the protection of the substrate against corrosion (electrochemical degradation) either actively or by providing a barrier layer. Hence, a best possible limitation of the coatings defects produced in the wake of dramatic mechanical perturbations such as impacting gravel stones. In addition, mounted plastic parts or body panels have to be protected against chemical degradation activated by exposure to ultraviolet light, physical erosion by swelling solvents such as fuel or water depending on the nature of the polymers and the morphology of the bulk. Furthermore, chemical degradation can be activated by catastrophic failure of the part upon mechanical impacts such as crash behaviour (Hintze-Bruening, & Leroux, 2012). Coating degradation depends on three main factors: coating formula, the intensities of the environment or situation, and their duration (Akafuah et al., 2016). Furthermore, Paints can alter the heat-reflecting properties depending on the intensity of the incident sunlight. This assists in regulating the temperature of the vehicle, hence, making the role of the air conditioning system easier and thus saving fuel. Nanotechnology research has been working towards coatings and paints which are highly scratch resistant and dirt repellant. This technology would allow automotive paint on vehicle to last long without aging and require cleaning much less often (Soutter, 2012).

In today's car manufacturing process, automotive coatings have been optimized in an evolutionary process over the past decades leading to a typical stack of highly specialized individual coating layers. This is made up of an initially cathodic electrodeposited material that offers adhesion and active corrosion inhibition (ED coat) which is then covered by a spray coated layer (known as primer) for the smoothening and protection of the ED coat towards ultraviolet light. Apart from the smoothening and the protection of the ED coating layer against ultraviolet light, the stone chip resistance of an automotive coating system is usually attributed to the primer layer (Hintze-Bruening & Leroux, 2012). However, the failure mode of any multilayer coating upon such a mechanical impact is based on the characteristics of all individual layers (Rutherford et al., 1997). The typical composition and the morphology of an automotive primer layer are in a way appropriate to mitigate the action of propagating cracks, the dislocation and crushing of the material.

Recently, nanocomposite coatings have shown increasing attention in many strategic industries such as automotive industry. Nanocomposite coatings are tailored to offer applications that have attractive and cost-effective functional surface coatings with excellent properties for anticorrosion, antimicrobial, antifogging, and adhesive. The unique features of nanocomposite coatings include increased heat, enhanced mechanical strength, weight reduction, improved barrier properties, wear, and scratch resistance for lifelong performance (Abu-Thabit, & Makhlouf, 2015). Polymer nanocomposites and nanoparticles can be used as nanocoatings by depositing nanoscale layer on selected substrates to reach specific surface behaviour (Schmid et al., 2014). Advanced nanotechnology now uses hard, water-repelling polymer nanocomposites or quartz nanoparticles which could enable coatings on vehicle to remain clean and protect the vehicle against scratches and chips; hence, reduce corrosion, without altering the appearance of the paint underneath. Furthermore, polymer nanocomposites can rectify any minor damage which a vehicle sustain. This is attained by inserting ceramic nanoparticles in a flexible polymer, which can flow over itself to rectify cracks and scratches whilst remaining robust. Hence, polymer nanocomposites coatings for vehicles can self-clean as shown in Figure 4. The surface is made superhydrophobic, causing water to form beads on the surface and run easily off, collecting and removing dirt in the process. Cleaning is therefore much easier and required less frequently (Soutter, 2012). When preparing superhydrophobic coatings for windshields in automobiles, high transparency is needed. In order to achieve high transparency, the coating must be composed of low-light absorbing materials with refractive indices spanning the refractive indices of air to that of the substrates (Müller et al., 2017).

Polymer Nanocomposites for Weather Resistance Automotive Coating

The method using nanoparticles as dirt-resistant paints for cars in order to create a hydrophobic (water-resistant) surface has also been applied to glass for vehicle windows. The effect of beading for water on the surface implies that water runs off much more easily and does not diminish visibility during heavy rain or spray. In addition, it reduces wear on windscreen wipers, which are barely required at all, as most road speeds the air flow past the car would be sufficient to clear the beaded water droplets from the glass. Nanocomposite coatings on the inside of the glass can apply a similar method to prevent water vapour from condensing on the glass in humid conditions (Soutter, 2012).

The insertion of nanoparticles into polymeric coatings is one of the most promising approaches to enhance both their weathering and mechanical properties (Mirabedini et al., 2008). Some of the nanoparticles used in coatings are SiO_2 (Vega, Olaya, & Ruiz, 2018), TiO_2, (Pavlovska et al., 2014; Mirabedini et al., 2018), ZrO_2 (El-Lateef & Khalaf, 2015), ZnO (Stieberova et al., 2017), and Al_2O_3 (Dhoke, Sinha, & Khanna, 2009). However, the choice of nanoparticles depends on the intrinsic properties they possess (Asmatulu et al., 2016). Some nanomaterial additives that can be incorporated to coatings for ultraviolet protection are ZnO (Mohseni et al., 2012) and TiO_2 (Abbas et al., 2018). Though they enhance the anti-weathering properties of the coatings, anatase Titania nanomaterials produce free radicals as a result of

Figure 4. Illustration of self-clean nanocomposites coatings for vehicles. The surface is made superhydrophobic, causing water to form beads on the surface and run easily off, collecting and removing dirt in the process
(Adapted from Soutter, 2012).

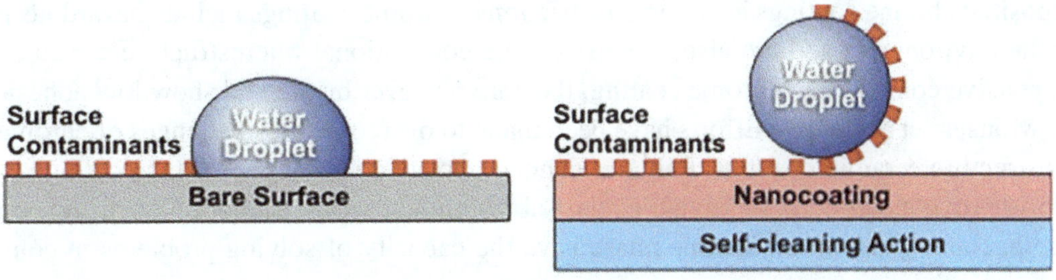

their photocatalytic properties (Asmatulu et al., 2016). Furthermore, carbon nanofillers such as carbon black (Naik, Jagtap & Ratna, 2015), graphene (Chang et al., 2012) and multi-wall carbon nanotube (Naik, Jagtap & Ratna, 2015) incorporated in polymer coatings further increase the protection ability. These fillers can perform multifunctional components in polymer coatings (Vijayan et al., 2017). Hence, these fillers are effective in improving the corrosion resistance, adhesion of the polymer coating, barrier and weathering properties.

Among these nanoparticles and fillers, carbon black and ZnO and have been successfully used to block the coating from ultraviolet irradiation (Ghasemi-Kahrizsangi et al., 2015; Rashvand, Ranjbar, & Rastegar, 2012). These nanofillers act as UV absorbers and prevent degradation of epoxy resin (Vijayan et al., 2017). Several kinds of approaches can be attempted to protect coatings against weathering conditions and radiation failures (Lee & Kim, 2006; Merlatti et al., 2008; Mirabedini et al., 2008). In order to reduce ultraviolet degradation, ultraviolet screeners are embedded into the bulk polymer (Ammala et al., 2002). Normally, these additives absorb ultraviolet light by themselves and reduce the quantity of ultraviolet light absorbed on the polymer surface. Ultraviolet screeners are high in cost and degrade as a result of the heavy ultraviolet exposure (Lee & Kim, 2006; Merlatti et al., 2008).

Nuraje et al. (2013) evaluated the addition of graphene nanoflakes in different weight percentages to polyurethane top coatings. The coatings were relatively exposed to QUV accelerated weathering cabinet and a corrosion test was carried out in a salt spray chamber. Their results showed that the addition of graphene enhanced the resistance of the coatings against ultraviolet degradation and corrosion. They concluded that the process will enhance the properties of the polyurethane top coating used in many industries against environmental factors. Hence, Nanocomposites can be used to enhance the properties of the coating as a result of their outstandingly high surface area-to-volume ratio which gives rise to exceptional properties that will enhance the resistance of the coatings against ultraviolet degradation and corrosion.

Polymer Nanocomposites Coating in Car Engine Application

Several researchers have studied coating for engine applications (Dahotre & Nayak, 2005; Louda, 2007). Different kinds of materials are available for car engine coating; however, the selection of material is based on the compatibility with the base material and the techniques used for coating. There are advantages and disadvantages of conventional materials systems and techniques such as plasma sprayed metal matrix composite coating and chemical-vapor-deposited diamond-like carbon coating (Coelho et al., 2012). The most used kinds of protective coatings used for engine parts are electrodeposited hard chrome and microstructure ceramic coatings. The ceramic coatings are often used on metal parts by thermal spray. In plasma spraying, the coating powder reinforced with ceramic particles is injected into a plasma stream, subsequently heating and accelerating takes place toward the metal substrate. The ceramic quickly cools and yield a coating layer over the substrate. Nonetheless, the use of microstructure ceramic and electrodeposited chrome coatings have some restrictions. Chrome coatings include hazardous materials affecting the environment and are also expensive. The conventional microstructure ceramic coatings are not expensive compared to chrome coating, they are however brittle and show low adhesion to the substrate (Mohseni et al., 2012). Efforts have been made to overcome the challenges of chrome coating and microstructure ceramic coatings. Polymer nanocomposite containing ceramic coatings have been utilized to improve metal parts of engine against abrasion and wear (Rathod, Kumar, & Jain, 2017). Furthermore, Nano-grains of crystalline phase have the capacity of solving problems in conventional coatings. These coatings offer high toughness, high hardness, low residual stress and display a very good tribological properties (Coelho et al., 2012; Paiva et al., 2018).

The roles of nanocomposites coatings in engine application are: (1) reduction of frictional loss, (2) Higher operating temperature, (3) surface finish and affinity/oil and (4) replacement of cast iron block/liner. Engine of a car is known to operate at higher temperatures by reducing external heat removal. Lightweight materials in engine are used to reduce load and heat losses. Frictional loss is another method used in improving fuel efficiency. An important factor used in improving fuel efficiency is the reduction of engine weight. Substituting cast iron (with density of 7.8 g/cm^3) used in engine blocks with low-cast aluminum-silicone (with density of 2.79 g/cm^3) is one good prospect for engine weight reduction. Nonetheless, aluminum alloys do not have adequate wear resistance and high seizure loads to be used in the cylinder bores. Hence, cylinder bores are made of cast iron liners which have good wear resistance (Mohseni et al., 2012). Attempts have therefore been made to improve aluminum bars properties using new composites and/or monolithic coatings (Rajan, Pillai, & Pai, 1998; Dahotre & Nayak, 2005; Paramsothy, Gupta, & Srikanth, 2008). Nanomaterials can be employed to achieve extraordinary properties for aluminum bars (Mohseni et al., 2012).

Nanocomposites Coat in High Scratch Resistance

Many coating systems requires coating that protects the underlying layers or substrate from chemical and ultraviolet degradation and provides protection from mechanical damage that can result in surface scratches. Presently, different types of plastic material are available for use as alternatives for mineral glass, due to their excellent properties, such as lightness, ease of mass production and lower price. Poly(methylmethacrylate) and polycarbonate are the two plastic material that are widely used with excellent optical properties (Tanglumlert et al., 2006). Nonetheless, one of the main shortcomings of the organic glasses is their weakness to scratch. Hence, enhancing surface resistance against mechanically induced scratches, without any trade off in transparency and optical properties, still remains a challenge in organic coating (Mohamadpour, Pourabbas, & Fabbri, 2011). Furthermore, inorganic coatings done via gas-phase or vacuum deposition are extensively used to improve scratch resistance; but, the shortcomings of these techniques are the poor adhesion of the final coat to the substrate and high cost production process (Messori et al., 2003). However, consumers want a permanent, scratch-free finish on all parts of their vehicles, which is demonstrated by scratch performance becoming the highest rated customer concern for automotive paint systems (Courter, 1997).

Recently, polymeric nanocomposite coatings have drawn substantial attention, owing to enhancements in scratch resistance. Bautista et al. (2011) studied the relationship between the wear resistance and the scratch behaviour of nanocomposites with different organic matrices and different inorganic nanofillers. They found that the elastic and plastic behaviour of the nanocomposite coatings have a strong influence in the scratch and wear resistance. However, the main shortcomings in developing hard coatings based on nanocomposites are related to the effects of filler distribution on the optical properties of the pristine substrate. Aggregates of filler seriously affect both optical and mechanical properties, thus demanding their homogeneous distribution and dispersion. Because of polymeric nanocomposites hydrophilic character and strong tendency to agglomeration, dried nanoparticles powders are hardly dispersed in organic matrix (Tana et al., 2017). A non-aqueous sol-gel method is one captivating technique that can be used to overcome these shortcomings. This technique permits obtaining ready-to-use crystalline nanoparticles, often made of pure crystal phases dispersed in organic media without requiring further functionalization procedures or calcination steps (Tana et al., 2017). Hence, hybrid materials can be gotten by direct insertion of inorganic fillers into coat or by in-situ production of inorganic domain in a sol-gel processing method. Amerio et al. (2008) obtained organic-inorganic hybrid coatings by a dual-curing process combining the sol-gel reaction with the ultraviolet-induced polymerization technique. The hybrid coating started from bisphenol A ethoxylate (15 EO/phenol) dimethacrylate, as a former organic network, methacryloyloxypropyltrimethoxysilane, as coupling agent and tetraethoxysilane, as inorganic silica network precursor. For comparison, Tetraethoxysilane was also substituted with preformed silica nanoparticles. Scratch test was carried out in order to study the scratch resistance of that silica reinforced acrylic resins. Excellent scratch resistant coatings were obtained by ultraviolet and sol-gel dual curing process. Tana et al. (2017) recently designed Hybrid transparent coatings based on zirconia nanoparticle fillers and epoxy resin to increase scratch resistance of commodity polymers without impairing their optical properties. The obtained coatings showed a homogeneous adhesion and a negligible effect on the transparency of these polymers. An increased scratch-resistance was obtained by increasing the ZrO_2 nanoparticles content.

Polymer Nanocomposites for Energy Efficiency in Vehicles

Markets for vehicle safety, performance, and fuel efficiency offer a demand for new and advanced materials in automotive applications (Lloyd & Lave, 2003). Hence, the automotive industries are continuously aspiring to develop safe, clean, more energy efficient and affordable automobile. In the past five years, numerous vehicles have considerably improved their fuel efficiency; however, the average fleet fuel efficiency has increased from 24.2 to 25.2. In addition, the average number of advanced technology safety attributes has simultaneously increased from 19.5% to over 64%. Today's fleet of cars comprises more fuel efficient vehicles and considerably higher percentages of advanced safety attributes. The gradually increasing fuel efficiency requirements from 2012 to 2018 did not have any undesirable effect on the safety of today's vehicles (Gillis, 2018). New vehicles offer great evidence of manufacturers' ability to enhance fuel efficiency as these vehicles have been designed and built with the new fiber composites. Figure 5 presents a model that shows how much vehicles have improved in both safety and fuel efficiency for vehicles of 2011 and 2018 version (Gillis, 2018). 2018 version was significantly more than the average of 7.4 safety features in their 2011 versions. These same vehicles increased their fuel efficiency by 15 percent from an average of 21.8, to an average of 25.1 mile per gallon (MPG).

Electric vehicles have the capacity to drastically increase the energy efficiency of the transport industry. Electric vehicles are much more efficient at converting stored energy into motion than the normal use of internal combustion engines (Soutter, 2013). The batteries that are currently used in electric vehicles are based on nickel metal hydride or lithium ion technology. Modern examples virtually use lithium ion batteries, exclusively; which provide numerous benefits over nickel metal hydride. Nonetheless, there is currently no battery technology that is actually good enough to make electric vehicles feasible for the replacement for internal combustion. The range and recharging time of electric cars restricts their use to short trips where recharging points are always nearby (Soutter, 2013).

In addition, power sources such as batteries experience some serious challenge during operation such as shock, power turn off and voltage drop. Thus, in energy back up applications, supercapacitors are linked parallel with the main power source in order to deliver energy to the system when the main power source stops. A large percentage of supercapacitors act as back up devices for electronic industries (Cherusseri, Sharma, & Kar, 2015). Other application of supercapacitors as a main power source

Figure 5. A model safety and fuel efficiency 2011 vs. 2018 vehicles
(Gillis, 2018)

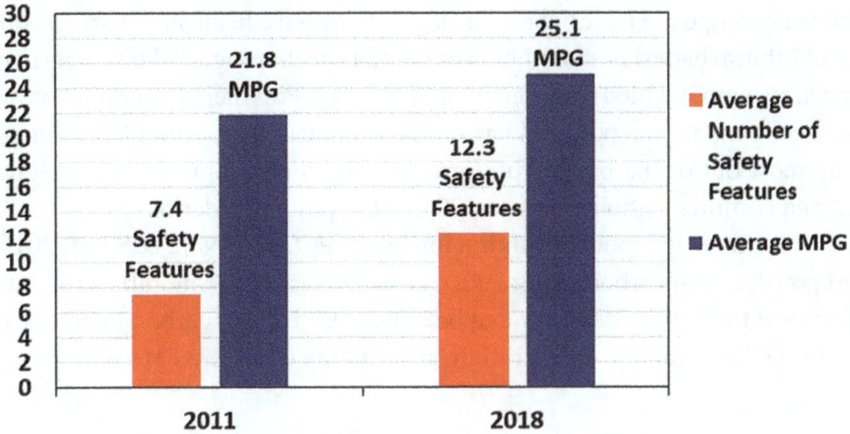

are found in automobile industries. For example, supercapacitors having an energy storage capacity of 75-150 Wh when used in mild hybrid electric vehicles (Burke, 2007). Supercapacitors have the capacity to be used in automobiles as they have advantage of comparable energy efficiency, high power density, absorb excess power from stack and the regenerative braking energy which in turns provide a fraction of transient power for vehicle acceleration (Zhao & Burke, 2010). Furthermore, for the consideration of the need for long range between refueling and the capability to handle rapid changes in the demand of power and frequent capture of regenerative braking energy, the combination of fuel cells and supercapacitors/batteries could take advantage of the strengths of each of the technologies and attain performance in automotive applications better than that of internal combustion engines (Zhao & Burke, 2010). The electrochemical performance of supercapacitors is determined by several factors such as specific surface area, pore size distribution and geometry, electrical conductivity, wettability and doping of electrodes (Zhang & Samori, 2017).

Numerous materials such as carbon-based materials, metal oxides, and conducting polymers, etc. have been used as electrode materials for supercapacitors. Carbon based materials have the storage mechanism of electric double-layer capacitors (EDLCs). These materials normally have high power densities, good cycling performance, however, they have unsatisfactory energy densities. Metal oxides and conducting polymers are the typical materials showing a pseudo-capacitive behavior. They have high specific capacitance yet limited cycling stability (Zhang & Samori, 2017). Graphene is attractive for supercapacitors owing to its excellent electrical conductivity, superior mechanical properties, large specific surface area and good electrochemical performance (Li et al., 2016). Graphene/polymer hybrid nanocomposites have attracted extraordinary interest for applications in transparent conductive coatings, sensors, catalysts and memory devices, especially in energy conversion and storage (Wang et al., 2014). Combining graphene and polymers as hybrid systems have shown obvious benefits. The nanocomposite electrodes have the capacity to combine both EDLCs from graphene and pseudo-capacitance from polymers, hence increase the overall electrochemical performance of the nanocomposites by taking full advantage of both energy storage mechanisms (Zhang & Samori, 2017).

Hence, polymer nanotechnology (polymer nanocomposites) could hold the key to making electric vehicles to be extensively usable by increasing the performance of battery. Enhanced electrolytes using nanoparticles and nanocomposite materials such as graphene have been shown to considerably improve specific qualities of lithium batteries and other novel battery technologies. The performance of a vehicle can also be enhanced by nano-enabled components and systems. Polymer nanocomposites, nanostructured metals, nano-enhanced sensors and power electronics can contribute to a number of important measures in order to get the most out of the battery performance, such as weight reduction, energy efficiency, improved control and communications (Soutter, 2013). Graphite powder has been used as an intercalation material on the negative electrode for lithium ion batteries (Jeon, Noh, & Song, 2017). By replacing micrometer sized powder with carbon nanomaterials such as carbon nanotubes, the rate of removal or insertion of lithium and the battery capacity can be enhanced. Hence, carbon nanotubes can bind much higher concentrations of lithium due to their high surface area (Soutter, 2013).

CONCLUSION

Polymer nanocomposites technology have a great prospect and have attracted considerable attention worldwide as a result of their recent development in the automotive industry. Awareness of polymer nanocomposites has resulted in different innovative methods of fabricating new materials with excellent properties. The interest in light weight parts for automobile has given advanced polymer nanocomposites an advantage in automotive industry as a result of their light weight with fuel saving and high strength properties. Hence, polymer nanocomposites application has empowered the automobile parts in lightweight vehicles, reduction in CO_2 emissions and enhanced engine efficiency leading to excellent performance.

REFERENCES

Abbas, M., Iftikhar, H., Malik, M. H., & Nazir, A. (2018). Surface coatings of TiO_2 nanoparticles onto the designed fabrics for enhanced self-cleaning properties. *Coatings*, *8*(35), 1–9.

Abd El-Lateef, H. M., & Khalaf, M. M. (2015). Corrosion resistance of ZrO_2-TiO_2 nanocomposite multilayer thin films coated on carbon steel in hydrochloric acid solution. *Materials Characterization*, *108*, 29–41. doi:10.1016/j.matchar.2015.08.010

Abraham, T. N., & George, K. E. (2009). Studies on recyclable nylon-reinforced PP composites: Effect of fiber diameter. *Journal of Thermoplastic Composite Materials*, *22*(1), 5–20. doi:10.1177/0892705708091603

Abu-Thabit, N. Y., & Makhlouf, A. S. H. (2015). Recent advances in nanocomposite coatings for corrosion protection application. In *Handbook of nanoceramics and nanocomposite coatings and materials* (pp. 515–549). Amsterdam: Elsevier. doi:10.1016/B978-0-12-799947-0.00024-9

Akafuah, N. K., Poozesh, S., Salaimeh, A., Patrick, C., Lawler, K., & Saito, K. (2016). Evolution of the Automotive Body Coating Process-A Review. *Coatings*, *6*(24), 1–22.

AL-Oqla, F. M., & Sapuan, S. M. (2014). Natural fiber reinforced polymer composites in industrial applications: Feasibility of date palm fibers for sustainable automotive industry. *Journal of Cleaner Production*, *66*, 347–354. doi:10.1016/j.jclepro.2013.10.050

Alcock, B., Cabrera, N. O., Barkoula, N.-M., Loos, J., & Peijs, T. (2006). The mechanical properties of unidirectional all-polypropylene composites. *Composites. Part A, Applied Science and Manufacturing*, *37*(5), 716–726. doi:10.1016/j.compositesa.2005.07.002

Alex, R. (2010). Nanofilters in rubber and rubber blends. In Rubber nanocomposites-Preparation, properties, and applications (pp. 231-233). Singapore: John Wiley & Sons (Asia) Pte Ltd.

Alexandre, M., & Dubois, P. (2000). Polymer-layered silicate nanocomposites: Preparation, properties and uses of a new class of materials. *Materials Science and Engineering*, *28*(1-2), 1–63. doi:10.1016/S0927-796X(00)00012-7

Amerio, E., Fabbri, P., Malucelli, G., Messori, M., Sangermano, M., & Taurino, R. (2008). Scratch resistance of nano-silica reinforced acrylic coatings. *Progress in Organic Coatings*, *62*(2), 129–133. doi:10.1016/j.porgcoat.2007.10.003

Ammala, A., Hill, A. J., Meakin, P., Pas, S. A., & Turney, T. W. (2002). Degradation studies of polyolefins incorporating transparent nanoparticulate zinc oxide UV stabilizers. *Journal of Nanoparticle Research*, *4*(1-2), 167–174. doi:10.1023/A:1020121700825

Anandhan, S., & Bandyopadhyay, S. (2011). Polymer Nanocomposites: From Synthesis to Applications, Nanocomposites and Polymers with Analytical Methods. InTech.

Ashby, M. F., Ferreira, P. J., & Schodek, D. L. (2009). *Nanomaterials, Nanotechnologies, and design: An introduction for engineers and architects*. Elsevier.

Asmatulu, R., Diouf, D., Moniruddin, M., & Nuraje, N. (2016). Enhanced anti-weathering of nanocomposite coatings with silanized graphene nanomaterials. *International Journal of Engineering Research and Application*, *6*(6), 23–36.

Azo-Nano. (2015). Properties and Applications of Graphene Oxide and Functionalized Graphene. Retrieved from https://www.azonano.com/article.aspx?ArticleID=4044

Balandin, A. A., Ghosh, S., Bao, W., Calizo, I., Teweldebrhan, D., Miao, F., & Lau, C. N. (2008). Superior thermal conductivity of single-layer graphene. *Nano Letters*, *8*(3), 902–907. doi:10.1021/nl0731872 PMID:18284217

Bautista, Y., Gonzalez, J., Gilabert, J., Ibanez, M. J., & Sanz, V. (2011). Correlation between the wear resistance, and the scratch resistance, for nanocomposite coatings. *Progress in Organic Coatings*, *70*(4), 178–185. doi:10.1016/j.porgcoat.2010.09.022

Burke, A. F. (2007). Batteries and ultracapacitors for electric, hybrid, and fuel cell vehicles. *Proceedings of the IEEE*, *95*(4), 806–820. doi:10.1109/JPROC.2007.892490

Camargo, P. H. C., Satyanarayana, K. G., & Wypych, F. (2009). Nanocomposites: Synthesis, structure, properties and new application opportunities. *Materials Research*, *12*(1), 1–39. doi:10.1590/S1516-14392009000100002

Chandra, A. K. (2017). Nanomaterial adding to efficiency. *Rubber Asia*. Retrieved from https://www.rubberasia.com/2017/02/13/nano-materials-add-efficiency/

Chandra, A. K., & Kumar, N. R. (2017). Polymer nanocomposites for automobile engineering applications. In *Properties and applications of polymer nanocomposites*. Germany: Springer. doi:10.1007/978-3-662-53517-2_7

Chang, C.-H., Huang, T.-C., Peng, C.-W., Yeh, T.-C., Lu, H.-I., Hung, W.-I., ... Yeh, J.-M. (2012). Novel anticorrosion coatings prepared from polyaniline/graphene composites. *Carbon*, *50*(14), 5044–5051. doi:10.1016/j.carbon.2012.06.043

Cherusseri, J., Sharma, R., & Kar, K. K. (2015). Nanotechnology advancement on carbon nanotube/polypyrrole composite electrodes for supercapacitors. In *Handbook of polymer nanocomposites. Processing, performance and application* (p. 483). London: Springer. doi:10.1007/978-3-642-45229-1_22

Choudalakis, G., & Gotsis, A. D. (2009). Permeability of polymer/clay nanocomposites: A review. *European Polymer Journal*, *45*(4), 967–984. doi:10.1016/j.eurpolymj.2009.01.027

Chronakis, I. S. (2005). Novel nanocomposites and nanoceramics based on polymer nanofibers using electrospinning process-A review. *Journal of Materials Processing Technology*, *167*(2-3), 283–293. doi:10.1016/j.jmatprotec.2005.06.053

Coelho, M. C., Torrão, G., Emami, N., & Grácio, J. (2012). Nanotechnology in automotive Industry: Research strategy and trends for the future-Small Objects, big impacts. *Journal of Nanoscience and Nanotechnology*, *12*(8), 1–10. doi:10.1166/jnn.2012.4573 PMID:22962798

Coiai, S., Passaglia, E., Pucci, A., & Ruggeri, G. (2015). Nanocomposites based on thermoplastic polymers and functional nanofiller for sensor applications. *Materials (Basel)*, *8*(6), 3377–3427. doi:10.3390/ma8063377

Courter, J. L. (1997). Mar resistance of automotive clearcoats: I. Relationship to coating mechanical properties. *Journal of Coatings Technology*, *69*, 57–63.

Dahotre, N. B., & Nayak, S. (2005). Nanocoatings for engine application. *Surface and Coatings Technology*, *194*(1), 58–67. doi:10.1016/j.surfcoat.2004.05.006

Dante Rene Guerra, B. S. (2009). Influence of nanoparticles on the physical properties of fiber reinforced polymer composites [Master Thesis]. The Graduate School of The Ohio State University.

Davoodi, M. M., Sapuan, S. M., Ahmad, D., Aidy, A., Khalina, A., & Jonoobi, M. (2011). Concept selection of car bumper beam with developed hybrid bio-composite material. *Materials & Design*, *32*(10), 4857–4865. doi:10.1016/j.matdes.2011.06.011

Deepak, K., Reddy, N. S., & Naidu, T. V. S. (2015). Thermosetting polymer and nano clay based natural fiber bio-composites. *Procedia Materials Science*, *10*, 626–631. doi:10.1016/j.mspro.2015.06.095

Dhoke, S. K., Sinha, T. J. M., & Khanna, A. S. (2009). Effect of nano-Al_2O_3 particles on the corrosion behavior of alkyd based waterborne coatings. *Journal of Coatings Technology and Research*, *6*(3), 353–368. doi:10.100711998-008-9127-3

Divya G.S., & Suresha, B. (2016). Recent developments of natural fiber reinforced thermoset polymer composites and their mechanical properties. *Indian Journal of Advances in Chemical Science, S1*, 267-274.

Fogorasi, M. S., & Barbu, I. (2017). The potential of natural fibres for automotive sector – Review. IOP Conf. Series. *Materials Science and Engineering, 252*(012044), 1–11.

Fukushima, Y., & Inagaki, S. (1987). Synthesis of an intercalated compound of montmorillonite and 6-Polyamide. *Journal of Inclusion Phenomena, 5*(4), 473–482. doi:10.1007/BF00664105

Galimberti, M., Cipolletti, V. R., & Coomb, M. (2013). *Applications of clay-polymer nanocomposites. In Handbook of clay science.* (*Vol. 5B*, p. 557). The Netherlands: Elsevier. doi:10.1016/B978-0-08-098259-5.00020-2

Garcés, J. M., Moll, D. J., Bicerano, J., Fibiger, R., & McLeod, D. G. (2000). Polymeric Nanocomposites for Automotive Applications. *Advanced Materials, 12*(23), 1835–1839. doi:10.1002/1521-4095(200012)12:23<1835::AID-ADMA1835>3.0.CO;2-T

Georgakilas, V., Tiwari, J. N., Kemp, K. C., Perman, J. A., Bourlinos, A. B., Kim, K. S., & Zboril, R. (2016). Noncovalent Functionalization of Graphene and Graphene Oxide for Energy Materials, Biosensing, Catalytic, and Biomedical Applications. *Chemical Reviews, 116*(9), 5464–5519. doi:10.1021/acs.chemrev.5b00620 PMID:27033639

Ghasemi-Kahrizsangi, A., Shariatpanahi, H., Neshati, J., & Akbarinezhad, E. (2015). Degradation of modified carbon black/epoxy nanocomposite coatings under ultraviolet exposure. *Applied Surface Science, 353*, 530–539. doi:10.1016/j.apsusc.2015.06.029

Giannelis, E. P. (1996). Polymer layered silicate nanocomposites. *Advanced Materials, 8*(1), 29–35. doi:10.1002/adma.19960080104

Gillis, J. (2018). Fuel economy standards: there is no tradeoff with safety, cost, and fleet turnover. Consumer Federation of America. Retrieved from https://consumerfed.org/wp-content/uploads/2018/05/fuel-efficiency-vs-safety-cost-and-fleet-turnover-1.pdf

Graupner, N., Herrmann, A. S., & Mussig, J. (2009). Natural and man-made cellulose fibre-reinforced poly (lactic acid) (PLA) composites: An overview about mechanical characteristics and application areas. *Composites. Part A, Applied Science and Manufacturing, 40*(6), 810–821. doi:10.1016/j.compositesa.2009.04.003

Gupta, T. K., & Kumar, S. (2018). 4 - Fabrication of Carbon Nanotube/Polymer Nanocomposites. In *Carbon nanotubes-reinforced polymers* (pp. 61–81). doi:10.1016/B978-0-323-48221-9.00004-2

Haupert, F., & Wetzel, B. (2005). Reinforcement of thermosetting polymers by the incorporation of micro- and nanoparticles. In *Polymer composites: from nano- to macroscale* (pp. 45–50). USA: Springer. doi:10.1007/0-387-26213-X_3

Hausner, J., Ziadeh, M., Fischer, B., Kalo, H., Schmid, J., Kunz, R., ... Breu, J. (2015). Transfer batch blending, an innovative solvent/solid assisted method for melt compounding to achieve good dispersion quality for polymerclay-nanocomposites. *Composites Science and Technology, 114*, 34–41. doi:10.1016/j.compscitech.2015.03.019

Hernandez, Y., Nicolosi, V., Lotya, M., Blighe, F. M., Sun, Z., McGovern, I. T., ... Byrne, M. (2008). High-yield production of graphene by liquid-phase exfoliation of graphite. *Nature Nanotechnology*, *3*(9), 563–568. doi:10.1038/nnano.2008.215 PMID:18772919

Hintze-Bruening, H., & Leroux, F. (2012). Nanocomposite based multifunctional coatings. In *New advances in vehicular technology and automotive engineering* (pp. 56–70). Croatia: Intech Publisher. doi:10.5772/48567

Holbery, J., & Houston, D. (2006). Natural fiber reinforced polymer composites in automotive applications. *JOM*, *58*(11), 80–86. doi:10.100711837-006-0234-2

Hong, C. Y., You, Y. Z., & Pan, C. Y. (2006). A new approach to functionalize multi walled carbon nanotubes by the use of functional polymers. *Polymer*, *47*(2), 4300–4309. doi:10.1016/j.polymer.2006.04.006

Hossain, M. Y. (2014). Natural Textile fibers (Cotton, Jute, Silk, Wool). Retrieved from https://www.slideshare.net/YousufTE/natural-textile-fibers-cotton-jute-silk-wool

Imtiaz, S. (2014). Glass Fibre. Retrieved from https://www.academia.edu/7463236/GLASS_Fibres_manufacturing_properties_and_applications

Jancar, J., Douglas, J. F., Starr, F. W., Kumar, S. K., Cassagnau, P., Lesser, A. J., ... Buehler, M. J. (2010). Current issues in research on structure-property relationships in polymer nanocomposites. *Polymer*, *51*(15), 3321–3343. doi:10.1016/j.polymer.2010.04.074

Jawaid, M., & Abdul Khalil, H. P. S. (2011). Cellulosic/synthetic fibre reinforced polymer hybrid composites: A review. *Carbohydrate Polymers*, *86*(1), 1–18. doi:10.1016/j.carbpol.2011.04.043

Jeon, I.-Y., Chang, D. W., Kumar, A. S., & Beak, J.-B. (2011). *Functionalization of Carbon Nanotubes* (p. 96). London: Intech.

Jeon, Y., Noh, H. K., & Song, H.-K. (2017). A lithium-ion battery using partially lithiated graphite anode and amphi-redox $LiMn_2O_4$ cathode. *Scientific Reports*, *7*(1), 1–9. doi:10.103841598-017-14741-x PMID:29093574

Jeyanthi, S., & Rani, J. J. (2012). Influence of natural long fiber in mechanical, thermal and recycling properties of thermoplastic composites in automotive components. *International Journal of Physical Sciences*, *7*(43), 5765–5771.

Ji, X., Xu, Y., Zhang, W., Cui, L., & Liu, J. (2016). Review of functionalization, structure and properties of graphene/ polymer composite fibers. *Composite: Part A*, *87*, 29–45. doi:10.1016/j.compositesa.2016.04.011

Kargarzadeh, H., Mariano, M., Huang, J., Lin, N., Ahmad, I., Dufresne, A., & Thomas, S. (2017). Recent developments on nanocellulose reinforced polymer nanocomposites: A review. *Polymer*, *132*, 368–393. doi:10.1016/j.polymer.2017.09.043

Kashiwagi, T. (2005). Flammability of nanocomposites: Effect of the shape of nanoparticles. In Fire retardancy polymer: New applications of mineral fillers (p. 81). The royal society of chemistry, UK

Khan, W. S., Hamadneh, N. N., & Khan, W. A. (2017). Polymer nanocomposites - synthesis techniques, classification and properties. In *Science and applications of tailored nanostructures* (p. 63). UK: One Central Press.

Kim, W., Nair, N., Lee, C. Y., & Strano, M. S. (2008). Covalent functionalization of single-walled carbon nanotubes alters their densities allowing electronic and other types of separation. *The Journal of Physical Chemistry C*, *112*(19), 7326–7331. doi:10.1021/jp710919b

Koba, Y., & Ishizaki, A. (2014). Chemical Composition of Palm Fiber and Its Feasibility as Cellulosic Raw Material for Sugar Production. *Agricultural and Biological Chemistry*, *54*(5), 1183–1187.

Koniuszewska, A. G., & Kaczmar, J. W. (2016). Application of polymer based composite materials in transportation. *Progress in Rubber, Plastics and Recycling Technology*, *32*(1), 1–24. doi:10.1177/147776061603200101

Koronis, G., Silva, A., & Fontul, M. (2013). Green composites: A review of adequate materials for automotive applications. *Composites. Part B, Engineering*, *44*(1), 120–127. doi:10.1016/j.compositesb.2012.07.004

Kotsilkova, R. (2007). *Thermoset nanocomposites for engineering applications*. UK: Smithers Rapra Technology Limited Publisher.

Kozłowski, R., & Władyka-przybylak, M. (2008). Review flammability and fire resistance of composites reinforced by natural fibers. *Polymers for Advanced Technologies*, *19*(6), 446–453. doi:10.1002/pat.1135

Kudin, K. N., Ozbas, B., Schniepp, H. C., Prud'homme, R. K., Aksay, I. A., & Car, R. (2008). Raman spectra of graphite oxide and functionalized graphene sheets. *Nano Letters*, *8*(1), 36–41. doi:10.1021/nl071822y PMID:18154315

Kumar, G. S. (1992). *Azo functional polymers: Functional group approach in macromolecular design*. USA: Technomic Publishing.

Kumar, H. V., Woltornist, S. J., & Adamson, D. H. (2015). Fractionation and characterization of graphene oxide by oxidation extent through emulsion stabilization. *Carbon*, *98*, 491–495. doi:10.1016/j.carbon.2015.10.083

Kurauchi, T., Okada, A., Nomura, T., Nishio, T., Saegusa, S., & Deguchi, R. (1991). Nylon 6-clay hybrid-Synthesis, properties and application to automotive timing belt cover. *SAE*.

Laine, L., & Rozite, L. (2010). State of the art, eco-efficient composite materials. Retrieved from https://people.arcada.fi/~rene/biocomposites/lectures/STATE%20OF%20THE%20ART_material_examples.pdf

Layek, R. K., & Nandi, A. K. (2013). A review on synthesis and properties of polymer functionalized graphene. *Polymer*, *54*(19), 5087–5103. doi:10.1016/j.polymer.2013.06.027

Le, V. T., Ngo, C. L., Le, Q. T., Ngo, T. T., Nguyen, D. N., & Vu, M. T. (2013). Surface modification and functionalization of carbon nanotube with some organic compounds. *Advances in Natural Sciences: Nanoscience and Nanotechnology*, *4*, 1–5.

Lee, B. H., & Kim, H. J. (2006). Influence of isocyanate type of acrylated urethane oligomer and of additives on weathering of UV-cured films. *Polymer Degradation & Stability*, *91*(5), 1025–1035. doi:10.1016/j.polymdegradstab.2005.08.002

Li, D., Muller, M. B., Gilje, S., Kaner, R. B., & Wallace, G. G. (2008). Processable aqueous dispersions of graphene nanosheets. *Nature Nanotechnology*, *3*(2), 101–105. doi:10.1038/nnano.2007.451 PMID:18654470

Li, D., Wang, Y. L., & Xia, Y. N. (2004). Electrospinning nanofibers as uniaxially aligned arrays and layer-by-layer stacked films. *Advanced Materials*, *16*(4), 361–366. doi:10.1002/adma.200306226

Li, J., Haung, X., Cui, L., Chen, N., & Qu, L. (2016). Preparation and supercapacitor performance of assembled graphene fiber and foam [Review]. *Progress in Natural Science: Materials International*, *26*(3), 212–220. doi:10.1016/j.pnsc.2016.05.006

Lloyd, S. M., & Lave, L. B. (2003). Life cycle economic and environmental implications of using nanocomposites in automobiles. *Environmental Science & Technology*, *37*(15), 3458–3466. doi:10.1021/es026023q PMID:12966996

Loos, M. R., & Schulte, K. (2011). Is it worth the effort to reinforce polymers with carbon nanotubes? *Macromolecular Theory and Simulations*, *20*(5), 350–362. doi:10.1002/mats.201100007

Louda, P. (2007). Application of thin coating in automotive industry. *Journal of Achievements in Materials and Manufacturing Engineering*, *24*(1), 50–56.

Lu, P., Leong, Y. W., Pallathadka, P. K., & He, C. B. (2013). Effective moduli of nanoparticle reinforced composites considering interphase effect by extended double-inclusion model – Theory and explicit expressions. *International Journal of Engineering Science*, *73*, 33–55. doi:10.1016/j.ijengsci.2013.08.003

Majumder, D. D., & Karan, S. (2013). Magnetic properties of ceramic nanocomposites. In *Ceramic Nanocomposites* (pp. 51–91). Elsevier.

Matabola, K. P., Vries, A. R., Moolman, F. S., & Luyt, A. S. (2009). Single polymer composites: A review. *Journal of Materials Science*, *44*(23), 6213–6222. doi:10.100710853-009-3792-1

Meng, L., Fu, C., & Lu, Q. (2009). Advanced technology for functionalization of carbon nanotubes: A Review. *Progress in Natural Science*, *19*(7), 801–810. doi:10.1016/j.pnsc.2008.08.011

Merlatti, C., Perrin, F. X., Aragon, E., & Margaillan, A. (2008). Natural and artificial weathering characteristics of stabilized acrylicurethane paints. *Polymer Degradation & Stability*, *93*(5), 896–903. doi:10.1016/j.polymdegradstab.2008.02.008

Messori, M., Toselli, M., Pilati, F., Fabbri, E., Fabbri, P., & Busoli, S. (2003). Poly (caprolactone)/silica organic–inorganic hybrids as protective coatings for poly (methyl methacrylate) substrates. *Surface Coatings International Part B-Coatings Transactions*, *86*(3), 181–186. doi:10.1007/BF02699651

Mirabedini, S. M., Mohseni, M., PazokiFard, S., & Esfandeh, M. (2008). Effect of TiO_2 on the mechanical and adhesion properties of RTV silicone elastomer coatings. *Colloids and Surfaces. A, Physicochemical and Engineering Aspects*, *317*(1-3), 80–86. doi:10.1016/j.colsurfa.2007.09.044

Mishra, R., Jamshaid, H., & Militky, K. (2017). Basalt nanoparticle reinforced hybrid woven composites: Mechanical and thermo-mechanical performance. *Fibers and Polymers*, *18*(12), 2433–2442. doi:10.100712221-017-7545-5

Mittal, V. (2012). In-situ synthesis of polymer nanocomposites (1st ed.). Wiley-VCH Verlag GmbH & Co.

Mohamadpour, S., Pourabbas, B., & Fabbri, P. (2011). Anti-scratch and adhesion properties of photo-curable polymer/clay nanocomposite coatings based on methacrylate monomers. *Scientia Iranica F*, *18*(3), 765–771. doi:10.1016/j.scient.2011.06.001

Mohammed, L., Ansari, M. N. M., Pau, G., Jawaid, M., & Saiful Islam, M. (2015). A Review on natural fiber reinforced polymer composite and Its applications. *International Journal of Polymer Science*.

Mohseni, M., Ramezanzadeh, B., Yari, H., & Gudarzi, M. M. (2012). The role of nanotechnology in automotive industries. In *New advances in vehicular technology and automotive engineering* (p. 4). London: Intech. doi:10.5772/49939

Mu, Q. X., Liu, W., Xing, Y. H., Zhou, H., Li, Z., Zang, Y., ... Yan, B. (2008). Protein binding by functionalized multiwalled carbon nanotubes is governed by the surface chemistry of both parties and the nanotube diameter. *The Journal of Physical Chemistry C*, *112*(9), 3300–3307. doi:10.1021/jp710541j

Müller, K., Bugnicourt, E., Latorre, M., Jorda, M., Sanz, Y. E., Lagaron, J. M., ... Schmid, M. (2017). Review on the processing and properties of polymer nanocomposites and nanocoatings and their applications in the packaging, automotive and solar energy fields [Review]. *Nanomaterials (Basel, Switzerland)*, *7*(74), 1–47. PMID:28362331

Mullins, L. (1969). Softening of rubber by deformation. *Rubber Chemistry and Technology*, *42*(1), 339–362. doi:10.5254/1.3539210

Nah, C., & Abdul Kader, M. (2010). Barrier properties of rubber nanocomposites. In Rubber nanocomposites-Preparation, properties, and applications (pp. 500-521). Singapore: John Wiley & Sons (Asia) Pte Ltd.

Naik, R. B., Jagtap, S. B., & Ratna, D. (2015). Effect of carbon nano - fillers on anticorrosive and physico-mechanical properties of hyperbranched urethane alkyd coatings. *Progress in Organic Coatings*, *87*, 28–35. doi:10.1016/j.porgcoat.2015.05.001

nano werk. (2012). Polymer nanocomposites drive opportunities in the automotive sector. Retrieved from https://www.nanowerk.com/spotlight/spotid=23934.php

Naskar, A. K., Keum, J. K., & Boeman, R. G. (2016). Polymer matrix nanocomposites for automotive structural components. *Nature Nanotechnology*, *11*(12), 1026–1030. doi:10.1038/nnano.2016.262 PMID:27920443

Neves, N. M., Campos, R., Pedro, A., Cunha, J., Macedo, F., & Reis, R. L. (2007). Patterning of polymer nanofiber meshes by electrospinning for biomedical applications. *International Journal of Nanomedicine*, *2*(3), 433–448. PMID:18019842

Nuraje, N., Khan, S. I., Misak, H., & Asmatulu, R. (2013). The addition of graphene to polymer coatings for improved weathering. *ISRN Polymer Science*.

Okada, A., Fukushima, Y., Kawasumi, M., Inagaki, S., Usuki, A., Sugiyama, S., . . . Kamigaito, O. (1988). U.S. Patent No. 4,739,007. Washington, DC: U.S. Patent and Trademark Office.

Paduchowicz, M., Górski, A., Czmochowski, J., & Rusiński, E. (2012). Numerical and experimental identification of vibration convection chamber of fluid power bowler. *Journal of Vibroengineering*, *14*, 151–157.

Paiva, J. M., Fox-Rabinovich, G., Locks Junior, E., Stolf, P., Ahmed, Y. S., Martins, M. M., ... Veldhuis, S. (2018). Tribological and wear performance of nanocomposite PVD hard coatings deposited on aluminum die casting tool. *Materials (Basel)*, *11*(358), 1–15. PMID:29495620

Pal, S. K., & Pal, K. (2009). Status of abrasion resistant tyres for mining industries: A review. In Tyres in mining and allied sectors: Status and outlook. Allied Publishers, Pvt Ltd.

Papageorgiou, D. G., Kinloch, I. A., & Young, R. J. (2017). Mechanical properties of graphene and graphene-based nanocomposites. *Progress in Materials Science*, *90*, 75–127. doi:10.1016/j.pmatsci.2017.07.004

Paramsothy, M., Gupta, M. M., & Srikanth, N. (2008). Processing, microstructure, and properties of a Mg/Al bimetal macrocomposite. *Journal of Composite Materials*, *42*(24), 2567–2584. doi:10.1177/0021998308098369

Park, S.-J., & Seo, M.-K. (2011). Types of composites. *Interface Science and Technology*, *18*, 501–629. doi:10.1016/B978-0-12-375049-5.00007-4

Pavlovska, I., Malnieks, K., Mezinskis, G., Bidermanis, L., & Karpe, M. (2014). Hard TiO_2-SiO_2 sol-gel coatings for enamel against chemical corrosion. *Surface and Coatings Technology*, *258*, 206–210. doi:10.1016/j.surfcoat.2014.09.027

Pei, S., & Cheng, H.-M. (2012). The reduction in graphene oxide. *Carbon*, *50*(9), 3210–3228. doi:10.1016/j.carbon.2011.11.010

Peng, W., & Riedl, B. (1995). Thermosetting resins. *Journal of Chemical Education*, *72*(7), 587–592. doi:10.1021/ed072p587

Pike, M. G., & Oskay, C. (2014). Modeling random short nanofiber- and microfiber-reinforced composites using the extended finite-element method. *Journal of Nanomechanics & Micromechanics*, *5*(1), 1–11.

Pilla, S., & Lu, Y. C. (2015). The use of nano composites in automotive applications. SAE International Publisher. doi:10.4271/PT-165

Plastics Make It Possible. (2015). Make it possible. Automakers turn to plastics for lighter vehicles, increased fuel efficiency. Retrieved from https://www.plasticsmakeitpossible.com/whats-new-cool/automotive/fuel-efficiency/automakers-turn-to-plastics-for-lighter-vehicles-increased-fuel-efficiency/

Plastics Make It Possible. (2016). Make it possible. Automotive plastics play key role as automakers seek to build fuel-efficient vehicles. Retrieved from https://www.plasticsmakeitpossible.com/whats-new-cool/automotive/plastics-play-key-role-automakers-seek-build-fuel-efficient-vehicles/

Puglia, D., Biagiotti, J., & Kenny, J. M. (2004). A review on natural fibre-based Composites-Part II: Application of natural reinforcements in composite materials for automotive industry. *Journal of Natural Fibers, 1*(3), 23–64. doi:10.1300/J395v01n03_03

Raj, R., & Shah, S. R. (2012). U.S. Patent No. 8,232,234. Washington, DC: U.S. Patent and Trademark Office.

Rajan, T. P. D., Pillai, R. M., & Pai, B. C. (1998). Reinforcement coatings and interfaces in aluminium metal matrix composites [Review]. *Journal of Materials Science, 33*(14), 3491–3503. doi:10.1023/A:1004674822751

Rama, K. L., & Nandi, A. K. (2013). A review on synthesis and properties of polymer functionalized graphene. *Polymer, 54*(19), 5087–5103. doi:10.1016/j.polymer.2013.06.027

Ramanathan, T., Abdala, A. A., Stankovich, S. D., Dikin, A., Herrera-Alonso, M., Piner, R. D., ... Aksay, I. A. (2008). Functionalized graphene sheets for polymer nanocomposites. *Nature Nanotechnology, 3*(6), 327–331. doi:10.1038/nnano.2008.96 PMID:18654541

Rashvand, M., Ranjbar, Z., & Rastegar, S. (2012). Preserving anticorrosion properties of epoxy based coatings simultaneously exposed to humidity and UV-radiation using nano zinc oxide. *Journal of the Electrochemical Society, 159*(3), 129–132. doi:10.1149/2.093203jes

Rathod, V. T., Kumar, J. S., & Jain, A. (2017). Polymer and ceramic nanocomposites for aerospace applications. *Applied Nanoscience, 7*(8), 519–548. doi:10.100713204-017-0592-9

Ray, S. S., & Okamoto, M. (2003). Polymer/layered silicate nanocomposites: A review from preparation to processing. *Progress in Polymer Science, 28*(11), 1539–1641. doi:10.1016/j.progpolymsci.2003.08.002

Reza, F., Yamamuro, J. A., & Batson, G. B. (2004). Electrical resistance change in compact tension specimens of carbon fiber cement composites. *Cement and Concrete Composites, 26*(7), 873–881. doi:10.1016/j.cemconcomp.2003.06.002

Rutherford, K. L., Trezona, R. I., Ramamurthy, A. C., & Hutchings, I. M. (1997). The abrasive and erosive wear of polymeric paint films. *Wear, 203-204*, 325–334. doi:10.1016/S0043-1648(96)07369-3

Saba, N., Tahir, P. M., & Jawaid, M. (2014). A Review on Potentiality of Nano Filler/Natural Fiber Filled Polymer Hybrid Composites. *Polymers, 6*(8), 2247–2273. doi:10.3390/polym6082247

Sanjay, M. R., Arpitha, G. R., Naik, L. L., Gopalakrishna, K., & Yogesha, B. (2016). Applications of natural fibers and its composites: An overview. *Natural Resources, 7*(03), 108–114. doi:10.4236/nr.2016.73011

Sathishkumar, T., Naveen, J., & Satheeshkumar, S. (2014). Hybrid fiber reinforced polymer composites-A review. *Journal of Reinforced Plastics and Composites, 33*(5), 454–471. doi:10.1177/0731684413516393

Schmid, M., Sängerlaub, S., Miesbauer, O., Jost, V., Werthan, J., Stinga, C., ... Müller, K. (2014). Water repellence and oxygen and water vapor barrier of pvoh-coated substrates before and after surface esterification. *Polymers, 6*(11), 2764–2783. doi:10.3390/polym6112764

Shadlou, S., Ahmadi-Moghadam, B., & Taheri, F. (2014). Nanoenhanced adhesives: A critical review. *Review of Adhesion and Adhesives, 2*(3), 371–412. doi:10.7569/RAA.2014.097307

Smart and BASF. (2018). Retrieved from http://www.smartforvision.basf.com/#polymer_composite_frame

Soltannia, B., Haji Gholami, I., Masajedian, S., Mertiny, P., Sameoto, D., & Taheri, F. (2016). Parametric study of strain rate effects on nanoparticle-reinforced polymer composites. *Journal of Nanomaterials, 2016*, 1–9. doi:10.1155/2016/9841972

Soutter, W. (2012). Nanotechnology in the automotive industry. *AZO NANO*. Retrieved from https://www.azonano.com/article.aspx?ArticleID=3031

Soutter, W. (2013). Nanotechnology in electric vehicle batteries. *AZO NANO*. Retrieved from https://www.azonano.com/article.aspx?ArticleID=3157

Stachewicz, U., Modaresifar, F., Bailey, R. J., Peijs, T., & Barber, A. H. (2012). Manufacture of void-free electrospun polymer nanofiber composites with optimized mechanical properties. *ACS Applied Materials & Interfaces, 4*(5), 2577–2582. doi:10.1021/am300235r PMID:22524442

Stephen, R., & Thomas, S. (2010). Nanocomposites: State of the art, new challenges and opportunities. In Rubber nanocomposites-Preparation, properties, and applications (pp 8-13). Singapore: John Wiley & Sons (Asia) Pte Ltd.

Stieberova, B., Zillka, M., Ticha, M., Freiberg, F., Caramazana-Gonzalez, P., Mckechnie, J., & Lester, E. (2017). Application of ZnO Nanoparticles in a self-cleaning coating on a metal panel: An assessment of environmental benefits. *ACS Sustainable Chemistry & Engineering, 5*(3), 2493–2500. doi:10.1021/acssuschemeng.6b02848

Szabo, T., Berkesi, O., Forgo, P., Josepovits, K., Sanakis, Y., Petridis, D., & Dekany, I. (2006). Evolution of Surface Functional Groups in a Series of Progressively Oxidized Graphite Oxides. *Chemistry of Materials, 18*(11), 2740–2749. doi:10.1021/cm060258+

Tana, F., Messori, M., Contini, D., Cigada, A., Valente, T., Variola, F., ... Bondioli, F. (2017). Synthesis and characterization of scratch-resistant hybrid coatings based on non-hydrolytic sol-gel ZrO_2 nanoparticles. *Progress in Organic Coatings, 103*, 60–68. doi:10.1016/j.porgcoat.2016.11.022

Tanahashi, M. (2010). Development of fabrication methods of filler/polymer nanocomposites: With focus on simple melt-compounding based approach without surface modification of nanofillers. *Materials (Basel), 3*(3), 1593–1619. doi:10.3390/ma3031593

Tanglumlert, W., Prasassarakich, P., Supaphol, P., & Wongkasemjit, S. (2006). Hard-coating materials for poly (methyl methacrylate) from glycidoxypropyltrimethoxysilane-modified silatrane via a sol-gel process. *Surface and Coatings Technology, 200*(8), 2784–2790. doi:10.1016/j.surfcoat.2004.11.018

Tian, Z., Hu, H., & Sun, Y. (2013). A molecular dynamics study of effective thermal conductivity in nanocomposites. *International Journal of Heat and Mass Transfer, 61*, 577–582. doi:10.1016/j.ijheatmasstransfer.2013.02.023

Vega, F. L. A., Olaya, J. J., & Ruiz, J. B. (2018). Synthesis and corrosion resistance of SiO_2-TiO_2-ZrO_2-Bi_2O_3 coatings spin-coated on Ti6Al4V alloy. *Ceramics International, 44*(2), 2123–2131. doi:10.1016/j.ceramint.2017.10.161

Vijayan, P. P., El-Gawady, Y. M. H., & Al-Maadeed, M. A. S. A. (2017). A comparative study on long term stability of self-healing epoxy coating with different inorganic nanotubes as healing agent reservoirs. *Express Polymer Letters*, *11*(11), 863–872. doi:10.3144/expresspolymlett.2017.83

Wang, J., Ma, F., Liang, W., & Sun, M. (2017). Electrical properties and applications of graphene, hexagonal boron nitride (h-BN), and graphene/h-BN heterostructures. *Materials Today Physics*, *2*, 6–34. doi:10.1016/j.mtphys.2017.07.001

Wang, L., Lu, X., Lei, S., & Song, Y. (2014). Graphene-based polyaniline nanocomposites: Preparation, properties and applications. *Journal of Materials Chemistry. A, Materials for Energy and Sustainability*, *2*(13), 4491–4509. doi:10.1039/C3TA13462H

Xing, J., Deng, B., & Liu, Q. (2015). Effect of benzimidazolium salt on dispersion and properties of polyphenylene sulfide/organic clay nanocomposites via melt intercalation. *Fibers and Polymers*, *16*(6), 1220–1229. doi:10.100712221-015-1220-5

Xu, Y., Li, Z., & Duan, W. (2014). Thermal and thermoelectric properties of graphene: A review. *Small*, *10*(11), 2182–2199. doi:10.1002mll.201303701 PMID:24610791

Yang, J., Lee, S., Choi, W. J., Seo, H., Kim, P., Kim, G.-J., ... Shin, J. (2015). Thermoset elastomers derived from Carvomenthide. *Biomacromolecules*, *16*(1), 246–256. doi:10.1021/bm501450c PMID:25514368

Yao, J., Guang, L., Bastiaansen, C. W. M., & Peijs, T. (2015). High performance co-polyimide nanofiber reinforced composites. *Polymer*, *76*, 46–51. doi:10.1016/j.polymer.2015.08.054

Yeum, J. H., Park, S. M., Kwon, I. I. J., Kim, J. W., Kim, H. W., Rabbani, M. M., … Oh, W. (2012). Carbon Nanotube Embedded Multi-Functional Polymer Nanocomposites. In Nanocomposites. London: Intech.

Yu, J. G., Huang, K. L., Liu, S. Q., & Tang, J. C. (2008). Preparation and characterization of polycarbonate modified multiple-walled carbon nanotubes. *Chinese Journal of Chemistry*, *26*(3), 560–563. doi:10.1002/cjoc.200890105

Zafeiropoulos, N. E., Williams, D. R., Baillie, C. A., & Matthews, F. L. (2002). Engineering and characterisation of the interface in flax fibre/polypropylene composite materials. Part I. Development and investigation of surface treatments. *Composites. Part A, Applied Science and Manufacturing*, *33*(8), 1083–1093. doi:10.1016/S1359-835X(02)00082-9

Zhang, G., Wu, T., Lin, W., Tan, Y., Chen, R., Huang, Z., ... Qu, J. (2017). Preparation of polymer/clay nanocomposites via melt intercalation under continuous elongation flow. *Composites Science and Technology*, *145*, 157–164. doi:10.1016/j.compscitech.2017.04.005

Zhang, K., Heo, N., Shi, X., & Park, J. H. (2013). Chemically modified graphene oxide-wrapped quasi-micro Ag decorated silver trimolybdate nanowires for photocatalytic applications. *The Journal of Physical Chemistry C*, *117*(45), 24023–24032. doi:10.1021/jp407038c

Zhang, X., & Samori, P. (2017). Graphene/Polymer nanocomposites for supercapacitors. *ChemNanoMat*, *3*(6), 362–372. doi:10.1002/cnma.201700055

Zhao, H., & Burke, A. (2010). *Fuel cell powered vehicles using supercapacitors: Device characteristics, control strategies, and simulation results. Research Report – UCD-ITS-RR-10-01*. Davis: Institute of Transportation Studies, University of California. Retrieved from www.its.ucdavis.edu

Zhu, Y., Murali, S., Cai, W., Li, X., Suk, J. W., Potts, J. R., & Ruoff, R. S. (2010). Graphene and graphene oxide: Synthesis, properties, and applications. *Advanced Materials*, *22*(35), 3906–3924. doi:10.1002/adma.201001068 PMID:20706983

Zindani, D., & Kumar, K. (2018). Material application of polymer composite material. In *Composites and advanced materials for industrial applications* (p. 8). Hershey, PA: IGI Global. doi:10.4018/978-1-5225-5216-1.ch001

Chapter 5
Application of Geopolymer Composites in Wastewater Treatment:
Trends, Opportunities, and Challenges

Victor Odhiambo Shikuku
Kaimosi Friends University College, Kenya

Tome Sylvain
University of Douala, Cameroon

ABSTRACT

Geopolymers are amorphous aluminosilicates with some varied applications. However, the use of geopolymers in water treatment is a relatively new subject. This chapter discusses developments in synthesis, properties and applications of geopolymers and their composites for removal of heavy metals and dyes from water including reduction of hardness in water. The adsorption mechanisms and effects of various environmental conditions on adsorption efficiency are also highlighted. The chapter demonstrates that geopolymers are low cost and environmentally benign materials for wastewater treatment and offers opportunities as alternative adsorbents for sequestration of various pollutants from water subject to further investigations.

INTRODUCTION

Geopolymer Chemistry

Geopolymer is a class of largely X-ray amorphous aluminosilicate materials, generally synthesised at ambient or slightly elevated temperature by reaction of a solid aluminosilicate powder with a concentrated alkali metal silicate or hydroxide solution (Provis et al., 2005). This inorganic material was created by the French scientist Joseph Davidovits in 1970s. The aim of its creation was to develop a fire-resistant alternative to organic polymers in the aftermath of various catastrophic fires in France between 1970

DOI: 10.4018/978-1-5225-7838-3.ch005

and 1972.Since the period following the work carried out by Wastiels et al. (1994), the application of geopolymer shifted to uses in construction as green alternative material to ordinary Portland cement. Nowadays, its plethora applications include: fire resistant materials, decorative stone artefacts, thermal insulation, low-tech building materials, low energy ceramic tiles, refractory items, thermal shock refractories, bio-technologies (materials for medicinal applications),foundry industry, cements and concretes, composites for infrastructures repair and strengthening, high-tech composites for aircraft interior and automobile, high-tech resin systems, radioactive and toxic waste containment, arts and decoration, cultural heritage, archaeology, history of sciences and adsorbent material for water treatment. The wide variety of potential application gives rise to referred name. In literature, geopolymer is described by different terminologies such as mineral polymer, inorganic polymer, inorganic polymer glasses, alkali ash material, soil cement etc.

Mechanism of Formation of Geopolymer

The formation of geopolymer involves the mixing of solid precursors (aluminosilicates sources) such as metakaolin, fly ash, volcanic ash, slag etc with alkaline activator solution (alkali hydroxide and metasilicate). After the mixing, multi steps of reactions occur namely:

1. Dissolution of glassy aluminosilicates;
2. Reorganization and diffusion of dissolved ions with formation of small coagulated structures;
3. Polycondensation to form aluminosilicate gel phases; and
4. Solid state transformation and hardening to form hard solid.

The aforementioned geopolymerization reaction steps are summarized in Figure1 (van Deventer et al., 2007).

Structure of Geopolymer

The structure of geopolymer materials are similar to zeolites, but the microstructure of a geopolymeric material is a three-dimensional silico-aluminate amorphous or semi-amorphous structure instead of crystalline (Provis et al., 2005; Xu, 2002). The geopolymer network consists of SiO_4 and AlO_4 tetrahedra linked alternately by sharing all the oxygens. Positive ions (Na^+, K^+, Li^+, Ca^{2+}, Ba^{2+}, NH_4^+, H_3O^+) must be present in the framework cavities to balance the negative charge of Al^{3+} in IV-fold coordination. The poly(sialate) is designation of geopolymers based on silico-aluminates where sialate is an abbreviation for silicon-oxo-aluminate. The empirical formula of poly(sialates) is $Mn\{-(SiO_2)z\text{-}AlO_2\}n$, wH_2O which M is a cation such as potassium, sodium or calcium, and «n» is a degree of polycondensation; «z» is an integer 1, 2, 3 (Davidovits, 2017). Geopolymers comprise the following molecular units:siloxo (-Si-O-Si-O-), sialate (-Si-O-Al-O-), sialate-siloxo (-Si-O-Al-O-Si-O-), sialate-disiloxo (-Si-O-Al-O-Si-O-Si-O-),ferro-sialate(-Fe-O-Si-O-Al-O-Si-O-). Figure 2 is a representation of the structure of geopolymer as proposed by Davidovict in 1994 (Davidovits, 1994).

Figure 1. Schematic outline of the reaction processes involved in geopolymerisation

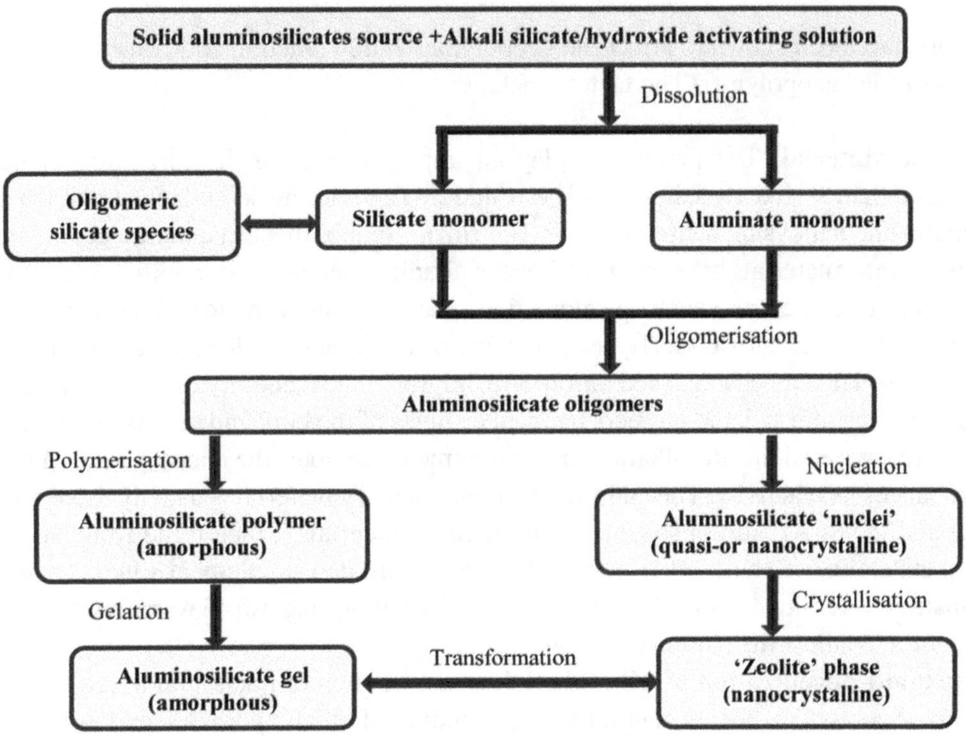

Figure 2. Structure of geopolymer proposed by Davidovits

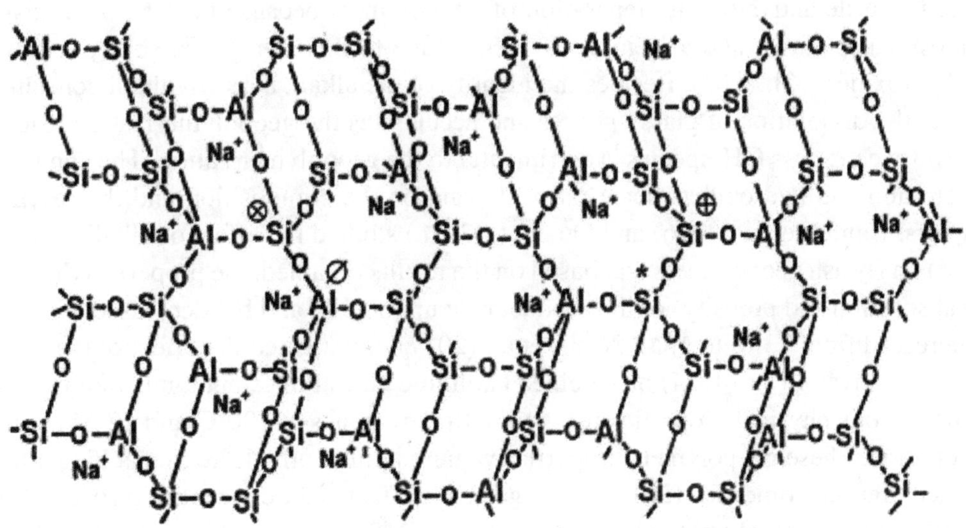

Factors Affecting the Properties of Geopolymer

Several factors have been shown to affect the geopolymerization reaction and consequently the properties of the so-formed geopolymer. The factors include:

1. **Feedstock Material:** The precursors play an important role in the formation of geopolymer. Materials rich in Si (like fly ash, volcanic ash, and slag and rice husk) and materials rich in Al (clays like kaolin and halloysite, bentonites) are the primary requirement to undergo geopolymerization. However, some materials have low reactivity in alkaline medium and require pre-treatment before using at room temperature such as calcination (clay, volcanic ash etc), alkaline fusion (volcanic ash) and mechanical activation (fly ash, volcanic ash). Kumar and Kumar (2011) and Djobo et al. (2016) showed that mechanical activation cure, improves the reactivity of coal fly ash and volcanic ash, respectively through the physicochemical changes of the bulk and reactive surface. Tchakoute et al. (2013) revealed that the alkaline fusion treatment promotes the glassy phases in volcanic ash, thus enhances its reactivity. The addition of supplementary material is usually used to improve the setting and the Si/Al ratio in some aluminosilicates materials. Silica fume from smelting operations, metakaolin, rice husk ashes, waste glass etc. are used to supplement silicon content. Sodium aluminates (Bayer liquor), metakaolin, aluminium smelting slag, waste water treatment sludge etc. can be used to adjust the alumina content.
2. **Nature and Concentration of Alkaline Solutions:** Sodium or potassium hydroxide, water glass (sodium or potassium silicate solution), and a mixture of alkali hydroxide and water glass are the commonly used activator liquids in the synthesis of geopolymers. The alkaline liquid containing the silicate species, i.e. water glass, increases the rate of the reaction compared to the alkaline solutions containing sodium or potassium hydroxide only. Sodium containing solutions are generally the favourite activators in preparation of geopolymers because they promote dissolution of aluminosilicates glassy and also invite low costs due to their wide availability relative to other metal hydroxides (Xu, 2002). Besides the identity of the alkali, a rise in alkali concentration also increases the dissolution of glassy phases and accelerates the geopolymerization reaction rate as well. However, excess OH^- species is detrimental to the geopolymerization. The appropriate alkali concentration and hydroxide/silicate ratios depend on the composition and the reactivity of the precursors. Hamidi et al. (2016) and Ou et al. (2018) studied the effect of alkali concentration (4 to 18 M) on fly ash geopolymers and based on the results obtained, the properties (higher strength, water absorption and porosity) of the product rose up to 12 M and but decreased as the concentration increased from 13 M to 18M. Nmiri et al. (2017) investigated the effect of the concentration of alkaline activator (5 to 18M) on calcined kaolinitic clay at different temperatures (550 to 1100 °C).The various physical properties reached optimum values at 700°C and 13 M alkali activator concentration. These geopolymers properties values significantly decreased at higher temperature and concentration. Tome et al. (2018) investigated the effect of the concentration of alkaline solution (6 to 10M) on municipal waste incineration fly ash (MSWI-FA): volcanic ash based geopolymer at room temperature and the results revealed that the 6M concentration promoted the best textural properties. Low alkali concentrations are also beneficial in terms of chemical safety.
3. **Solid/Activator Ratios:** The fraction of solid to liquid has also been documented to have significant effect on the formation and properties of geopolymers. According to Yao et al. (2009), the high solid/activator (S/A) ratio had a low viscosity, while geopolymer paste with a low S/A ratio resulted

in high viscosity. Besides that, geopolymer with a low S/A ratio could accelerate the dissolution rate, but it was not applicable to the polycondensation process when high concentrations of NaOH were used (Zuhua et al., 2009). In addition, the low S/A ratio resulted in low strength geopolymers due to insufficient formation of binders and increase in water in system (Provis et al., 2009). The S/A ratio value vary from one solid to another and also depend on the physical properties of the solid (specific surface area, particle shape). The S/A ratio by mass in order to achieve the optimum setting time and compressive strength in fly ash geopolymers is recommended to be maintained in the range of 0.30 to 0.45 (Xu, 2012; Rahmiatti et al., 2014; Duxson et al., 2007). Metakaolin (calcined clay) has higher liquid demand than fly ash due to the difference in particle shape and a lower S/A ratio to provide the near optimum strength and good workability is 0.80 (Kong et al., 2007). The surplus of activator promotes the free alkali ions and water in the system, which affect the strength and microstructure of geopolymer.

4. **Curing Temperature:** Temperature plays a vital role in geopolymer formation. Generally, elevating curing temperature (27°C ˂ T < 100°C) accelerates the dissolution, polymerization and precipitation processes of the geopolymerization reaction (Mo et al., 2014). However, the hydrothermal synthesis promotes the coexistence of three dimensional amorphous aluminosilicate (geopolymer) and the nanocrystalline structure such zeolite. Noe et al. (2016) studied the reactivity of volcanic ash at 27-80 °C range. The authors reported that the degree of crystallinity and the extent of geopolymerization increase with curing temperature. Mustafa Al Bakriand co-workers (2011) reported that the maximum compressive strength (67.04 MPa) was achieved at a temperature of 60 °C in fly ash geopolymer and the strength decreases at temperatures above 60 °C.

Preparation of Geopolymers

The porous geopolymers is one of the relevant geopolymer materials, due to its low-cost and niche applications such as catalyst supports or membranes, filtration of liquid or gases, adsorption and insulation. The aluminosilicates material and alkaline solution are used as in traditional geopolymer. The difference between traditional geopolymer and porous geopolymer lies in the adjuvant and curing temperature used for pore generation. In literature, the processing methods used for the fabrication of porous geopolymer can be generally divided into five approaches, namely: (a) Direct foaming, (b) Replica method, (c) Sacrificial filler method, (d) Additive manufacturing, and (e) Other methods.

1. **Direct Foaming:** According to the literature data, the main raw materials used for direct foaming process method are metakaolin, fly ash and slag. The chemical blowing agents that have been used the most are metal powders, such as silicon (Medri et al., 2013), powders or Si-containing compounds such as silica fume (SF), SiC, FeSi alloy, (Prud et al., 2010; Prud et al., 2015 Gualtieri et al., 2016) or hydrogen peroxide (H_2O_2) (Bai and Colombo, 2016), NaOCl (Boke et al., 2015), sodium perborate (Abdollahnejad et al., 2015) and AlN and $FeSO_3$(Bumanis et al., 2017). Schematic diagram of foaming approach is shown in Figure 3.

The stabilizing agents (surfactant, fibres etc) are used to stabilize the wet foam in order to avoid large pores (hundreds of microns) and as a result a wide range of pore sizes are typically present in the final porous geopolymer body. The curing temperature range used, depending on the bulk composition, is 22-100 °C.

Figure 3. The direct foaming fabrication route of porous geopolymers

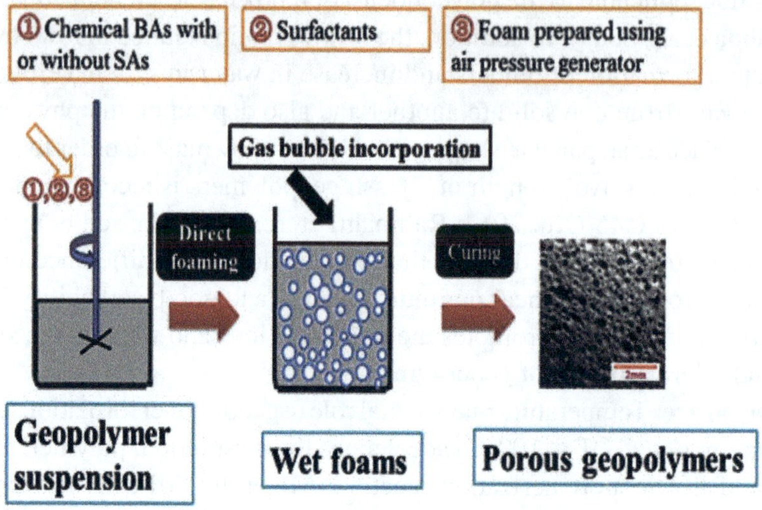

2. **Replica Method:** The product of replica method is not considered as geopolymer material due to the high temperature (1100-1300 °C) used in the last step of its synthesis. However, the intermediate steps involve geopolymerization. Figure 4 depicts the schematic diagram of its process as described by Kovarik et al., 2017.
3. **Sacrificial Filler Method:** Here, the porosity is generated by extracting the fillers, either by dissolution, melting or thermal decomposition, from a dense biphasic composite, comprising a continuous geopolymer matrix and a dispersed sacrificial phase, that could be interconnected or not (Bai and Colombo, 2018). Figure 5 depicts the process method for the production of porous geopolymer using sacrificial filler method.

Figure 4. The schematic diagram of the fabrication process of porous geopolymers by templating (with final conversion to ceramic foams)

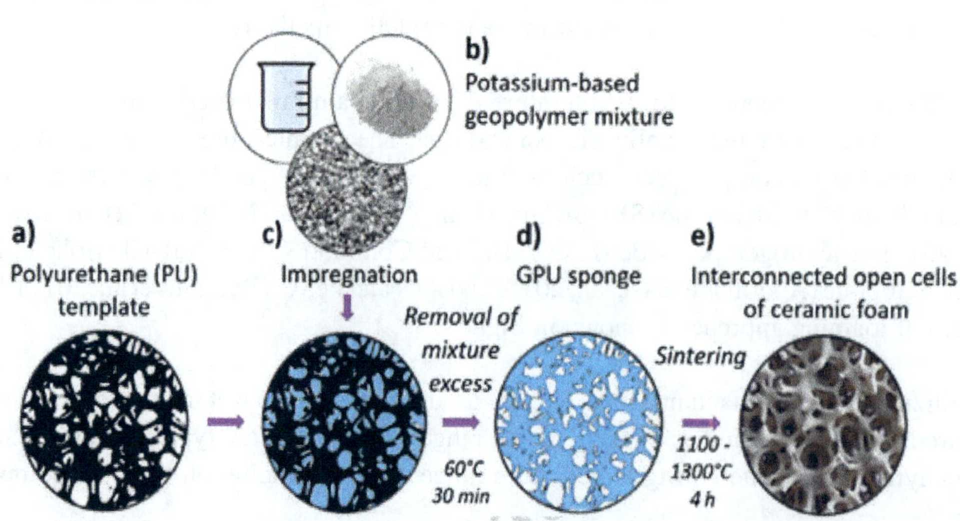

Figure 5. Process method for the production of porous geopolymer using sacrificial filler method

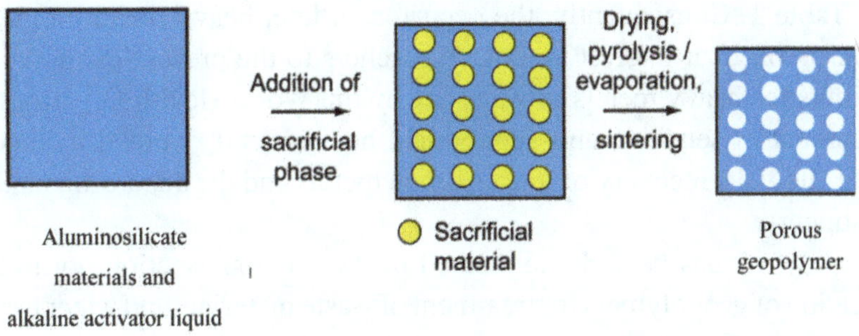

4. **Additive Manufacturing:** The additive manufacturing process involves the use of 3D printing process. It takes advantage of the fact that additive manufacturing technologies can produce structures which cannot be obtained by traditional processing technologies, with complex, non-stochastic porosity and precisely controlled pore dimension, shape, and amount (Zocca et al., 2015).
5. **Other Methods:** Other processes include the use of the reaction of saponification to form the surfactant molecule and H_2O_2 as the blowing agent (Bai and Combo, 2018; Strozi et al., 2014; Cilla and Colombo, 2017). Figure 6 shows the multiscale pore generation mechanisms for the saponification/peroxide combined route.

GEOPOLYMERS FOR WATER TREATMENT

Water is one of the most important chemical compounds necessary for life. Therefore, drinking water quality is as important for public health as raw water is to aquatic life. As such, systematic water quality protection programs and efficient drinking and wastewater treatment strategies are crucial for both humans and aquatic life. Most heavy metals are generally toxic and are potentially carcinogenic under

Figure 6. The mechanism for the generation of hierarchical porosity by direct foaming/reactive emulsion templating combined route

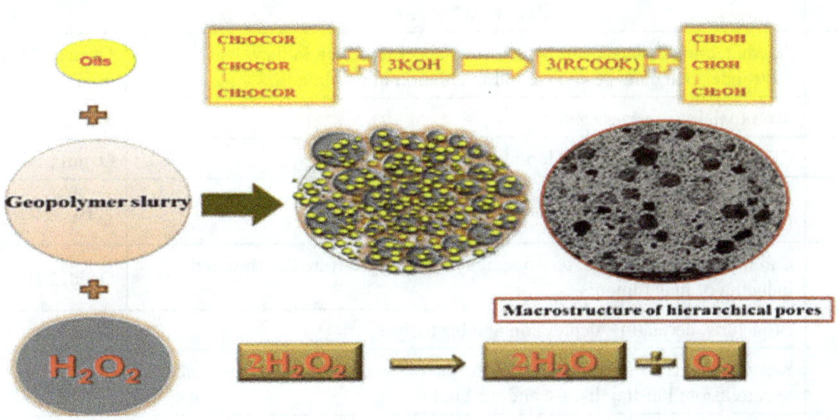

prolonged exposure. Some of the toxicity effects reported for some heavy metals discussed in this chapter are summarized in Table 1. Consequently, the occurrence, fate, heavy metal monitoring and removal from water are subjects of great interest among researchers to the present (Shikuku et al., 2017). The considerably low limits for heavy metals in water set by the World Health Organization (WHO, 2008) necessitates development of sensitive and economical heavy metal removal methods that reach such molecular limits and allow the recovery of the adsorbed metals and the adsorbing materials to maximize on the capital investment.

Synthesis of geopolymer has been demonstrated in the previous section not to be a new concept. However, the application of geopolymers for treatment of waste materials and water treatment is relatively recent. The intrinsic properties of geopolymers awaken much interest amongst researchers in various disciplines as a subject of study. Of particular interest, the capacity of geopolymers to immobilize most heavy metal ions contained within the geopolymeric structure makes them suitable candidates for water treatment. The main rationale for application of geopolymers as adsorbents in water treatment systems is based on their cation exchange capacities. Relative to conventional powdered adsorbents, such as activated carbon, porous monoliths (such as honeycombs), geopolymers exhibit superior performance for they reduce pressure drop, increase contact time and enhance thermal/mass transfer, and they are easily recoverable hence can be recycled (Twigg and Richardson, 2007).

In this section, the use of geopolymers and their composites for sequestration of various contaminants from water are discussed.

Geopolymers for Heavy Metal Removal

Scientific literature presents a significant number of studies on application of geopolymeric materials for removal of heavy metals from aqueous solution. Al-Zboon et al. (2001) reported the synthesis of fly-ash based geopolymer for removal of Pb from water. In the study, the synthesized geopolymer had a higher adsorption capacity for Pb than the raw coal fly ash. From the various experimental conditions, the removal efficiency increased with increase in contact time, geopolymer dosage and temperature. Thermodynamically, the increase in adsorption efficiency with rise in temperature depicts the process to be endothermic in nature and favourable at high temperatures. On the contrary, the removal efficiency

Table 1. Toxicity effects of selected heavy metals

Heavy Metal	Reported Toxicity	Reference
Lead	Anemia, brain damage, anorexia, loss of appetite, Liver, kidney, gastrointestinal damage and mental retardation in children	Chen and Hao, 1998; Godt et al., 2007
Copper	neurotoxicity, dizziness and diarrhea	Sharma, 1995
Nickel	Chronic bronchitis, reduced lung function and lung cancer,	Ozturk, 2007
Mercury	Damage to nervous system, corrosive to skin, eyes, muscles, dermatitis and kidney damage	WHO, 2001
Chromium	Carcinogenic, mutagenic, teratogenicity, vomiting, severe diarrhea and inducing of lung tumours	Gadd, 2010
Arsenic	Bronchitis, dermatitis, depression and haemolysis	Robertson, 1989
Cadmium	Kidney damage, bronchitis, Gastrointestinal disorder, bone marrow cancer, hypertension, Itai-Itai disease and weight loss	WHO, 2001

decreased with increase in Pb^{2+} initial concentration, from 88.8% at 100 mg/L to 81%, at 140 mg/L, due to occupancy of the available pores. This is also a testament that less energetically favoured sites take part in the adsorption process with increased concentration. Ultimately, the adsorption process was describable by the Langmuir isotherm model implying monolayer adsorption with a maximum adsorption capacity of 40.76 mg/g and 45.52 mg/g at 25°C and 45°C, respectively. Similar findings have been reported for adsorption of Cu(II) onto a fly-ash derived amorphous geopolymer (Al-Harahsheh et al., 2015). The highly amorphous structure observed was attributed to dissolution of fly ash glass phase. Noteworthy, the geopolymer exhibited a higher adsorption capacity for Cu(II) removal compared to the raw fly ash with Langmuir maximum adsorption capacity of 152 mg/g reported at pH 6. Table 2 is a summary of comparison of the adsorption capacities of different adsorbents for Cu(II) removal from water. Geopolymers are demonstrably comparable and excellent adsorbents.

In another study, porous geopolymeric spheres, as shown by SEM and BJH pore-size distribution studies, were synthesized by suspension and solidification method and applied for Cu(II) ions removal from water (Ge et al., 2015). The synthesized geopolymer exhibited higher adsorption higher adsorption capacity for Cu(II) than commercial spherical molecular sieves. In terms of reaction kinetics and adsorption mechanism, the adsorption process was described by the pseudo-second order kinetics model and the Langmuir isotherm, respectively. Leaching tests depicted higher leaching ratios pH<4 with highest leaching of 3.01% at pH 2. High leaching at low pH is attributed to structural damages of the spheres due to possible leaching of fractions of aluminium (Cheng et al., 2012). Overall, the geopolymeric spheres had acceptable stability in acidic conditions suitable for removal of heavy metals from acidic industrial effluents which tend to be acidic. The porous geopolymer spheres were demonstrated to be low-cost, convenient, environmentally benign and efficient for heavy metals removal from water. Liu and co-workers (2016) reported a comparative study on the adsorption of Pb onto fly ash, geopolymer and faujasite block transformed from geopolymer. The adsorption equilibrium data of fly ash, geopolymer and faujasite well fitted to the Langmuir model. The maximum adsorption capacities of fly ash, geopolymer and faujasite block were 49.8, 118.6 and 143.3 mg/g, respectively. The Gibbs free energy values of fly ash were higher than -15 kJ/mol, suggesting that adsorption sites and metal ion involved physical interactions. On the other hand, the Gibbs free energy values of geopolymer and faujasite are between -15 kJ/mol and -30 kJ/mol, suggesting that the adsorption rate is predominantly controlled by chemical interactions (both ionic and covalent). The reported negative ΔG values imply the adsorption is

Table 2. Adsorption capacities of various adsorbents for copper removal from water

Adsorbent	Adsorption Capacity (mg/g)	Reference
Activated carbon	54	Patnukao et al., 2008
Peanut pellets	12.1	Johnson et al., 2002
Crosslinked chitosan polymer	248.3	Lee et al., 2001
Chitosan powder	45.1	Huang et al., 1994
Kaolinite clay	10.8	Yavuz et al., 2003
Zeolite	20	Doula et al., 2002
Geopolymer	98.4	Wang et al., 2007
Geopolymer	152.3	Al-Harahsheh et al., 2015

thermodynamically spontaneous. The calculated enthalpy change (ΔH) was 30.43 kJ/mol; a magnitude which corroborates the aforementioned claim that the adsorption of Pb^{2+}onto the geopolymer is a chemisorption process. Noteworthy, the study proved that geopolymer had the same adsorption mechanisms as faujasite or similar zeolitic materials. The study also demonstrated that geopolymer technology is an efficient energy-saving process for development of adsorbents for water treatment. Novais et al. (2016) developed a novel porous fly ash-containing porous geopolymer monoliths using H_2O_2 as blowing agent and demonstrated its stupendous adsorbing properties for lead removal.

Composite materials of hierarchically porous geopolymer and amorphous hydrous ferric oxide (HFO) were synthesized by Medpelli and co-researchers (2015) as adsorbents for arsenic (As) removal from water using batch reactor experiments and laboratory scale continuous flow experiments. It has been shown in previous studies that the pore characteristics hierarchically porous geopolymers could be controlled by adjustment of the synthetic parameters (Medpelli et al., 2014). The BET surface area increased after impregnation without exception. As for the adsorption characteristic, all synthesized materials yielded Freundlich isotherm intensity (1/n) parameters ranging between 1.05 and 1.3, suggesting that the materials may be suitable for treatment of waters exhibiting higher than typically reported arsenic concentrations in natural waters. However, the adsorption capacities did not correlate directly with BET surface area and iron content. This is a clear demonstration that synthesis of materials with high specific surface area and high metal (hydr)oxide content does not necessarily result in highest adsorption capacity because much of the metal (hydr)oxide may not be accessible for sorption of arsenic as a result of pore clogging during impregnation or in material with limited meso/macroporosity. Noteworthy, the study provides opportunities for inclusion of active species into the hierarchically network of geopolymers thus imbuing them with increased functionalities. Another metal ion of health concern in water is nickel. Ge et al. (2015) investigated the adsorption of Ni^{2+}onto a self-supporting metakaolin-geopolymer membrane, fabricated without a blowing agent, with a total porosity of 62 vol%.

Kara and co-workers (2018) fabricated a geopolymer using metakaolin as precursor material and evaluated its performance for Mn(II) and Co(II) removal in batch and continuous mode. The experimental data conformed to the Langmuir isotherm model with maximum adsorption capacity of 72.35 and 69.23 mg/g for Mn(II) and Co(II), respectively. Adsorption yield varied slightly with temperature and ionic strength and no pH adjustment was needed to achieve high adsorption yield. Here, ion exchange was postulated to be the effective adsorption mechanism for the metal ions adsorption onto the geopolymer. The study effectively demonstrated metakaolin derived geopolymer is an effective adsorbent for Mn(II) and Co(II) ions polluted waters.

The effect of occurrence of other mineralogical phases on the textural and adsorption properties of geopolymers has also been reported in literature. For instance, the effects of natural zeolite, clinoptilolite, on the microstructural properties of metakaolin based geopolymer and its adsorption characteristics for Pb^{2+}, Cd^{2+}, Zn^{2+}, Cu^{2+} and $Cr^{3+}$was reported by Andrejkovicová et al. (2016). Here, clinoptilolite was used to replace metakaolin in varied proportions (0, 25. 50 and 75%). It was noted that addition of zeolite resulted in decrease in the water adsorption and BET surface areas. In terms of adsorption capacity, the maximum adsorption capacity for Cu^{2+} and Cr^{3+} increased consistently with increase in metakaolin (decrease in clinoptilolite) content in the structure. This observation contradicts and is a reverse of those reported by El-Eswed et al. (2012) for zeolite-based studies. This shows that the effect of zeolite content on the adsorption properties of geopolymers must be obtained empirically and may not be determined *a priori*. Furthermore, the amount of Cr^{3+} adsorbed by all samples showed marginal difference and therefore is arguably independent of the adsorbent composition and microstructure. This is possibly due to complexation of Cr^{3+} in water which consequently acts as a limiting step in the adsorption reaction. As for Pb^{2+}, Cd^{2+} and Zn^{2+}, maximum adsorption was realized in samples with 25% zeolite. The adsorption of the heavy metals was in the order $Cr^{3+}>Cu^{2+}> Zn^{2+}>Cd^{2+}>Pb^{2+}$in agreement to the reverse order of ionic potential (Cheng et al., 2012). Table 3 highlights a comparison of the adsorption capacities for various organic polymer/layered compound hybrids for chromium removal from water relative to geopolymers material. Maingi and co-workers (2018) reported the adsorptive removal of Cd(II) ions onto clay and rice husks based geopolymers. They reported an increase in initial adsorption rates, obtained from the pseudo-second order kinetic model, with increase in Si/Al ratios of the geopolymers with an approximately 90% removal efficiency within 80 min. The processes were thermodynamically spontaneous and endothermic in nature. All these findings are a testament on the suitability of geopolymers for heavy metals removal from water.

Geopolymers for Dye Removal

Many industrial processes, such as textile and printing, generate highly coloured effluents causing environmental problems. Various methods for discoloration of effluents have been proposed. These include; aerobic and anaerobic microbial degradation, coagulation and chemical oxidation, adsorption, membrane separation, filtration, electrochemical degradation, photocatalytic degradation and reverse osmosis. Inevitably, each method has its inherent limitations. There is a handful of publications on geopolymers for dye removal relative to heavy metals and fewer studies are highlighted here. Li et al. (2006) reported the application of fly ash derived geopolymeric materials for removal of methylene blue and crystal violet dyes from aqueous media. The adsorption capacity of the prepared adsorbents was demonstrated to be dependent on the preparation conditions such as fusion temperature and fly ash: NaOH ratio. The geopolymers exhibited higher adsorption capacities than raw fly ash and natural zeolites under similar conditions. The adsorption isotherm data was best fitted to the two-site Langmuir isotherm following nonlinear regression method. Conformity to the two-site Langmuir isotherm suggests that some heterogeneous sites, such as metal oxides and unburnt carbon, in the adsorbent play a role in the adsorption process. Elsewhere, the photocatalytic properties of fly ash based geopolymer for degradation of methyl blue dye was reported (Zhang and Liu, 2013). A degradation efficiency of 92.79% was realized under UV irradiation attributed to the synergistic effects of adsorption and photocatalysis. The photocatalytic

Table 3. Adsorption capacities of organic polymer/layered compound hybrids for chromium removal from water

Adsorbent	Adsorption Capacity (mg/g)	Reference
Chitosan/cloisite 10 A	357.14	Pandey and Mishra, 2011
Chitosan/perlite	153.8	Hasan and Krishnaiah, 2003
Chitosan/alumina	9.71	Gandhi et al., 2010
Chitosan/alumina	153.85	Boddu et al., 2003
Chitosan/montmorillonite	40.65	Fan et al., 2006
Polyethylenimine800/magnetite-montmorillonite	8.77	Larraza et al., 2012
Geopolymer	21.84	Andrejkovicová et al., 2016

degradation process obeyed the third-order reaction kinetics. The above studies underscore the fact that geopolymers present facile materials both as adsorbents and photocatalysts for dye removal from water. Khan et al. (2015) fabricated a metakaolin based phosphoric acid activated geopolymer and evaluated its adsorptive features for removal of methylene blue (MB) dye from wastewater. The adsorbent posted an adsorption capacity of 2.84 mg/g and upon regeneration by thermal treatment the adsorption capacity increased from 2.84 to 4.92 mg/g. Studies by Maingi et al. (2017) on adsorption of methylene blue dye onto geopolymers of different Si/Al ratios prepared from clay and rice husk ash revealed relatively high adsorption capacities. The batch equilibrium data was described by the Sips isotherm model while the adsorption kinetics followed the pseudo-second order kinetics model. Dye removal efficiencies in the range 85-99% was reported.

Geopolymers as Water Softeners

While there is sizeable bulk information on a geopolymer chemistry and application for heavy metals and dye removal, there is a paucity of information on the use of geopolymeric materials for water softening. As far as the knowledge of the present authors is concerned, to date, we could only find one article with systematic application of geopolymeric materials for water softening in model and real environmental samples. Naghsh and Shams (2017) synthesized a kaolin-based geopolymer using a two-step method consisting of a fusion step and a hydration and dealkalization step. The adsorption efficiency of the as-synthesized geopolymer for Mg^{2+} and Ca^{2+}, the ions responsible for water hardness, was examined in synthetic model water and in ground water. The findings showed that NaOH: kaolin ratio, fusion time and temperature have significant impact on the efficacy of the geopolymer in reduction of water hardness. Response Surface Methodology (RSM) revealed that NaOH: kaolin ratio above 2.2, the fusion time beyond 14 h and the fusion temperature range of 500–700 °C provide the optimal synthesis conditions. From the Langmuir isotherm model, the maximum adsorption density was 76.34 and 39.68 mg/g for Ca^{2+} and Mg^{2+} at 25°C, respectively. Furthermore, the exhausted geopolymer was regenerated by use of sodium chloride solution suggesting that ion exchange was the dominant mechanism of water softening process by the geopolymer. As far water softening is concerned, efficiency of other geopolymeric materials for reduction of water hardness is a subject of further research.

FUTURE AND CHALLENGES OF GEOPOLYMER ADSORBENTS

The studies so far demonstrate that geopolymers possess stupendous adsorption capacities for removal of heavy metals and dyes. However, real wastewaters are loaded with more contaminants besides heavy metals and dyes. Therefore, a broad spectrum of pollutants such as pharmaceutical ingredients, phenolic compounds and the emerging contaminants must be tested on geopolymers. Furthermore, geopolymers possess a negative charge resulting from charge efficiency of Al imbuing them with high affinity for cationic substances. However, wastewaters contain many anionic contaminants such as chlorides, nitrates and phosphates. Tunability of geopolymers for removal of anionic adsorbates should be investigated. There is scarcity of data on adsorption of anionic adsorbates onto geopolymers relative to cationic ones. Geopolymers also suffer the inherent limitation of conventional powdered adsorbents. That is, it is difficult to recover the spent adsorbent in a continuous flowing system. Adsorbent recovery steps, such as impregnation with magnetically responsive metal oxides, should be considered. Since the properties of geopolymers depend on the feedstock used, different aluminosilicate source materials should also be explored for the development of geopolymers in order to promote their formation and enhance their adsorption capabilities. Noteworthy, all the works reported so far are based on adsorption in single solute adsorbate systems in batch mode. Future work should focus also on geopolymer adsorbents in fixed bed, continuous and multi-component solute systems. Furthermore, besides the proof of concept in laboratory scale synthetic wastewaters, future research should validate the findings on real wastewaters. A crude-to-grave assessment of aluminosilicate-based waste as geopolymer materials should also be investigated for effective reuse, reduce and disposal of such materials.

CONCLUSION

The need for water treatment technologies that are cost-effective, environmentally friendly and that completely eliminates or reduces pollutant loads to molecular levels safe for human and other life forms has seen the emergence of geopolymeric materials as alternative adsorbents, photocatalysts and water softeners. Geopolymers possess the inherent advantages of immobilization of heavy metals, utilization of otherwise waste materials as precursor materials besides possessing comparable adsorption capacity as other competing adsorbents. In terms of synthesis routes, presently, direct forming seems to be the simplest and widely used method for porous geopolymer development. Novel processing methods of geopolymers is a subject of ongoing research and the trend is expected to change in coming years. The factors affecting the structural properties including additives exhibit complex interplay and in certain cases may not be determined *a priori*. Nevertheless, further research may fill in the literature gaps on the synergistic and antagonistic effects of various parameters affecting the structural features of geopolymers. Literature presents paucity of information on the application of geopolymers for water softening. This subject requires further investigation. The chapter demonstrate that geopolymeric materials possess untapped potential for application in wastewater treatment and provide opportunity for improvement of traditional water purification strategies with concomitant benefit in waste management.

REFERENCES

Abdollahnejad, Z., Pacheco-torgal, F., Félix, T., Tahri, W., & Aguiar, J. B. (2015). Mix design, properties and cost analysis of fly ash-based geopolymer foam. *Construction & Building Materials*, *80*, 18–30. doi:10.1016/j.conbuildmat.2015.01.063

Al-Harahsheh, M., Al-Zboon, K., Al-Makhadmeh, L., Muhannad, H., & Mehaysen, M. (2015). Fly ash based geopolymer for heavy metal removal: A case study on copper removal. *Journal of Environmental Chemical Engineering*, *3*(3), 1669–1677. doi:10.1016/j.jece.2015.06.005

Al-Zboon, K., Al-Harahsheh, M. S., & Falah, B. H. (2011). Fly ash-based geopolymer for Pb removal from aqueous solution. *Journal of Hazardous Materials*, *188*(1-3), 414–421. doi:10.1016/j.jhazmat.2011.01.133 PMID:21349635

Andrejkovicová, S., Sudagar, A., Rocha, J., Patinha, C., Hajjaji, W., Ferreira da Silva, E., ... Rocha, F. (2016). The effect of natural zeolite on microstructure, mechanical and heavy metals adsorption properties of metakaolin based geopolymers. *Applied Clay Science*, *126*, 141–152. doi:10.1016/j.clay.2016.03.009

Bai, C., & Colombo, P. (2016). High-porosity geopolymer membrane supports by peroxide route with the addition of egg white as surfactant. *Ceramics International*, 1–7. doi:10.1016/j.ceramint.2016.10.205

Bai, C., & Colombo, P. (2018). Processing, properties and applications of highly porous geopolymers: A review. *Ceramics International*, *44*(14), 16103–16118. doi:10.1016/j.ceramint.2018.05.219

Boddu, V. M., Abburi, K., Talbott, J. L., & Smith, E. D. (2003). Removal of hexavalent chromium from wastewater using a new composite chitosan biosorbent. *Environmental Science & Technology*, *37*(19), 4449–4456. doi:10.1021/es021013a PMID:14572099

Böke, N., Birch, G. D., Nyale, S. M., & Petrik, L. F. (2015). New synthesis method for the production of coal fly ash-based foamed geopolymers. *Construction & Building Materials*, *75*, 189–199. doi:10.1016/j.conbuildmat.2014.07.041

Bumanis, G., Vitola, L., Bajare, D., Dembovska, L., & Pundiene, I. (2017). Impact of reactive SiO2/Al2O3 ratio in precursor on durability of porous alkali activated materials. *Ceramics International*, *43*(7), 5471–5477.

Chen, J. M., & Hao, O. J. (1998). Microbial chromium (VI) reduction. *Critical Reviews in Environmental Science and Technology*, *28*(3), 219–251. doi:10.1080/10643389891254214

Cheng, T. W., Lee, M. L., Ko, M. S., Ueng, T. H., & Yang, S. F. (2012). The heavy metal adsorption characteristics on metakaolin-based geopolymer. *Applied Clay Science*, *56*, 90–96. doi:10.1016/j.clay.2011.11.027

Cilla, M. S., de Mello Innocentini, M. D., Morelli, M. R., & Colombo, P. (2017). Geopolymer foams obtained by the saponification/peroxide/gelcasting combined route using different soap foam precursors. *Journal of the American Ceramic Society*, *100*(8), 3440–3450.

Davidovits, J. (1994, October). Properties of geopolymer cements. In *First international conference on alkaline cements and concretes* (Vol. 1, pp. 131-149). Kiev State Technical University, Ukraine: Scientific Research Institute on Binders and Materials.

Davidovits, J. (2017). Geopolymers: Ceramic-like inorganic polymers. *J. Ceram. Sci. Technol.*, *8*, 335–350. doi:10.4416/JCST2017-00038

Djobo, J. N. Y., Elimbi, A., Tchakouté, H. K., & Kumar, S. (2016). Reactivity of volcanic ash in alkaline medium, microstructural and strength characteristics of resulting geopolymers under different synthesis conditions. *Journal of Materials Science*, *51*(22), 10301–10317. doi:10.100710853-016-0257-1

Djobo, J. N. Y., Elimbi, A., Tchakouté, H. K., & Kumar, S. (2016). Reactivity of volcanic ash in alkaline medium, microstructural and strength characteristics of resulting geopolymers under different synthesis conditions. *Journal of Materials Science*, *51*(22), 10301–10317. doi:10.1088/1757-899X/343/1/012013

Doula, M., Ioannou, A., & Dimirkou, A. (2002). Copper adsorption and Si, Al, Ca, Mg, and Na release from clinoptilolite. *Journal of Colloid and Interface Science*, *245*(2), 237–250. doi:10.1006/jcis.2001.7961 PMID:16290357

Duxson, P., Fernández-Jiménez, A., Provis, J. L., Lukey, G. C., Palomo, A., & Van Deventer, J. S. J. (2007). Geopolymer technology: The current state of the art. *Journal of Materials Science*, *42*(9), 2917–2933. doi:10.100710853-006-0637-z

El-Eswed, B., Alshaaer, M., Yousef, R. I., Hamadneh, I., & Khalili, F. (2012). Adsorption of Cu(II), Ni(II), Zn(II), Cd(II) and Pb(II) onto kaolin/zeolite based-geopolymers. *Adv. Mater. Phys. Chem.*, *2*(04), 119–125. doi:10.4236/ampc.2012.24B032

Fan, D., Zhu, X., Xu, M., & Yan, J. (2006). Adsorption properties of chromium (VI) by chitosan coated montmorillonite. *The Journal of Biological Sciences*, *6*(5), 941–945. doi:10.3923/jbs.2006.941.945

Gadd, G. M. (2010). Metals, minerals and microbes: Geomicrobiology and bioremediation. *Microbiology*, *156*(3), 609–643. doi:10.1099/mic.0.037143-0 PMID:20019082

Gandhi, M. R., Viswanathanb, N., & Meenakshi, S. (2010). Preparation and application of alumina/chitosan biocomposite. *International Journal of Biological Macromolecules*, *47*(2), 146–154. doi:10.1016/j.ijbiomac.2010.05.008 PMID:20471415

Ge, Y., Yuan, Y., Wang, K., He, Y., & Cui, X. (2015). Preparation of geopolymer-based inorganic membrane for removing Ni2+ from wastewater. *Journal of Hazardous Materials*, *299*, 711–718.

Godt, J., Scheidig, F., Grosse-Siestrup, C., Esche, V., Brandenburg, P., Reich, A., ... Stournaras, C. J. (2007). Copper and cadmium adsorption on pellets made from fired coal fly ash. *Journal of Hazardous Materials*, *148*(3), 538–547. doi:10.1016/j.jhazmat.2007.03.020 PMID:17416461

Gualtieri, M. L., Cavallini, A., & Romagnoli, M. (2016). Interactive powder mixture concept for the preparation of geopolymers with fine porosity. *Journal of the European Ceramic Society*, *2*, 2–7. doi:10.1016/j.jeurceramsoc.2016.03.030

Hamidi, R. M., Man, Z., & Azizli, K. A. (2016). Concentration of NaOH and the Effect on the Properties of Fly Ash Based Geopolymer. *Procedia Engineering*, *148*, 189–193. doi:10.1016/j.proeng.2016.06.568

Hasan, S., Krishnaiah, A., Ghosh, T. K., Viswanath, D. S., Boddu, V. M., & Smith, E. D. (2003). Adsorption of chromium (VI) on chitosan-coated perlite. *Separation Science and Technology*, *38*(15), 3775–3793. doi:10.1081/SS-120024229

Huang, C., Liou, M. R., & Liu, C. B. (1994). In C. P. Huang (Ed.), *Hazardous Industrial Wastes* (p. 275). Lancaster: Technomic Publishing Company.

Johnson, P. D., Watson, M. A., Brown, J., & Jefcoat, I. A. (2002). Peanut hull pellets as a single use sorbent for the capture of Cu(II) from wastewater. *Waste Management (New York, N.Y.)*, *22*(5), 471–480. doi:10.1016/S0956-053X(01)00036-8 PMID:12092755

Kara, I., Yilmazer, D., & Akar, S. T. (2017). Metakaolin based geopolymer as an effective adsorbent for adsorption of zinc(II) and nickel(II) ions from aqueous solutions. *Applied Clay Science*, *139*, 54–63. doi:10.1016/j.clay.2017.01.008

Khan, M. I., Min, T. K., Azizli, K., Sufian, S., Ullah, H., & Man, Z. (2015). Effective removal of methylene blue from water using phosphoric acid based geopolymers: Synthesis, characterizations and adsorption studies. *RSC Advances*, *5*(75), 61410–61420. doi:10.1039/C5RA08255B

Kong, D. L. Y., Sanjayan, J. G., & Sagoe-Crentsil, K. (2007). Comparative performance of geopolymers made with metakaolin and fly ash after exposure to elevated temperatures. *Cement and Concrete Research*, *37*(12), 1583–1589. doi:10.1016/j.cemconres.2007.08.021

Kovářík, T., Křenek, T., Rieger, D., Pola, M., Říha, J., Svoboda, M., ... Kadlec, J. (2017). Synthesis of open-cell ceramic foam derived from geopolymer precursor via replica technique. *Materials Letters*, *209*, 497–500. doi:10.1016/j.matlet.2017.08.081

Kumar, S., & Kumar, R. (2011). Mechanical activation of fly ash: Effect on reaction, structure and properties of resulting geopolymer. *Ceramics International*, *37*(2), 533–541. doi:10.1016/j.ceramint.2010.09.038

Larraza, I., López-Gónzalez, M., Corrales, T., & Marcelo, G. (2012). Hybrid materials: Magnetite-polyethylenimine-montmorillonite, as magnetic adsorbents for Cr(VI) water treatment. *Journal of Colloid and Interface Science*, *385*(1), 24–33. doi:10.1016/j.jcis.2012.06.050 PMID:22841705

Lee, S. T., Mi, F.-L., Shen, Y.-J., & Shyu, S.-S. (2001). Equilibrium and kinetic studies of copper(II) ion uptake by chitosan-tripolyphosphate chelating resin. *Polymer*, *42*(5), 1879–1892. doi:10.1016/S0032-3861(00)00402-X

Li, L., Wang, S., & Zhu, Z. (2006). Geopolymeric adsorbents from fly ash for dye removal from aqueous solution. *Journal of Colloid and Interface Science*, *300*(1), 52–59. doi:10.1016/j.jcis.2006.03.062 PMID:16626729

Maingi, F. M., Mbuvi, H. M., Ng'ang'a, M. M., & Mwangi, H. (2017). Adsorption Kinetics and Isotherms of Methylene Blue by Geopolymers Derived from Common Clay and Rice Husk Ash. *Phys. Chem.*, *7*(4), 87–97. doi:10.5923/j.pc.20170704.02

Maingi, F. M., Mbuvi, H. M., Ng'ang'a, M. M., & Mwangi, H. (2018). Adsorption of Cadmium ions on Geopolymers derived from ordinary clay and rice husk. *Int. J. Mater. Chem.*, *8*(1), 1–9. doi:10.5923/j.ijmc.20180801.01

Medpelli, D., Sandoval, R., Sherrill, L., Hristovski, K., & Seo, D.-K. (2015). Iron oxide-modified nanoporous geopolymers for arsenic removal from ground water. *Resour. Efficient Technol.*, *1*(1), 19–27. doi:10.1016/j.reffit.2015.06.007

Medpelli, D., Seo, J. M., & Seo, D. K. (2014). Geopolymer with hierarchically meso-/microporous structures from reactive emulsion templating. *Journal of the American Ceramic Society*, *97*(1), 70–73. doi:10.1111/jace.12724

Medri, V., Papa, E., Dedecek, J., Jirglova, H., Benito, P., Vaccari, A., & Landi, E. (2013). Effect of metallic Si addition on polymerization degree of in situ foamed alkali-aluminosilicates. *Ceramics International*, *39*(7), 7657–7668. doi:10.1016/j.ceramint.2013.02.104

Mo, B. H., Zhu, H., Cui, X. M., He, Y., & Gong, S. Y. (2014). Effect of curing temperature on geopolymerization of metakaolin-based geopolymers. *Applied Clay Science*, *99*, 144–148. doi:10.1016/j.clay.2014.06.024

Mustafa Al Bakria, A. M., Kamarudin, H., Bin Hussain, M., Khairul Nizar, I., Zarina, Y., & Rafiza, A. R. (2011). The effect of curing temperature on physical and chemical properties of geopolymers. *Physics Procedia*, *22*, 286–291. doi:10.1016/j.phpro.2011.11.045

Naghsh, M., & Shams, K. (2017). Synthesis of a kaolin-based geopolymer using a novel fusion method and its application in effective water softening. *Applied Clay Science*, *146*, 238–245. doi:10.1016/j.clay.2017.06.008

Nmiri, A., Hamdi, N., Yazoghli-Marzouk, O., Duc, M., & Srasra, E. (2017). Synthesis and characterization of kaolinite-based geopolymer: Alkaline activation effect on calcined kaolinitic clay at different temperatures. *J. Mater. Environ. Sci.*, *8*, 276–290.

Novais, R. M., Buruberri, L. H., Seabra, M. P., & Labrincha, J. A. (2016). Novel porous fly-ash containing geopolymer monoliths for lead adsorption from wastewaters. *Journal of Hazardous Materials*, *318*, 631–640.

Öztürk, A. (2007). Removal of nickel from aqueous solution by the bacterium Bacillus thuringiensis. *Journal of Hazardous Materials*, *147*(1-2), 518–523.

Pandey, S., & Mishra, S. B. (2011). Organic-inorganic hybrid of chitosan/organoclaybionanocomposites for hexavalent chromium uptake. *Journal of Colloid and Interface Science*, *361*(2), 509–520. doi:10.1016/j.jcis.2011.05.031 PMID:21679960

Patnukao, P., Kongsuwan, A., & Pavasant, P. (2008). Batch studies of adsorption of copper and lead on activated carbon from Eucalyptus camaldulensisDehn.bark. *Journal of Environmental Sciences (China)*, *20*(9), 1028–1034. doi:10.1016/S1001-0742(08)62145-2 PMID:19143307

Provis, J. L., Lukey, G. C., & van Deventer, J. S. J. (2005). Do Geopolymers Actually Contain Nanocrystalline Zeolites? A Reexamination of Existing Results. *Chemistry of Materials*, *17*(12), 3075–3085. doi:10.1021/cm050230i

Provis, J. L., Yong, C. Z., Duxson, P., & van Deventer, J. S. J. (2009). Correlating mechanical and thermal properties of sodium silicate-fly ash geopolymers. *Colloids Surfaces A Physicochem. Eng. Asp.*, *336*(1-3), 57–63. doi:10.1016/j.colsurfa.2008.11.019

Prud, E., Michaud, P., Joussein, E., Peyratout, C., Smith, A., Arrii-clacens, S., ... Rossignol, S. (2010). Silica fume as porogent agent in geo-materials at low temperature. *Journal of the European Ceramic Society*, *30*(7), 1641–1648. doi:10.1016/j.jeurceramsoc.2010.01.014

Prud'homme, E., Joussein, E., & Rossignol, S. (2015). Use of silicon carbide sludge to form porous alkali-activated materials for insulating application. *The European Physical Journal. Special Topics*, *224*(9), 1725–1735.

Rahmiati, T., Azizli, K. A., Man, Z., Ismail, L., & Nuruddin, M. F. (2014, July). Effect of Solid/liquid ratio during curing time fly ash based geopolymer on mechanical property. In *Materials Science Forum* (Vol. 803).

Robertson, F. N. (1989). Arsenic in ground water under oxidizing conditions, southwest United States. *Environmental Geochemistry and Health, 11*(3-4), 171–176. doi:10.1007/BF01758668 PMID:24202428

Sharma, Y. C. (1995). Economic treatment of Cadmium (II)-rich hazardous waste by indigenous materials. *Journal of Colloid and Interface Science, 173*(1), 66–70. doi:10.1006/jcis.1995.1297

Shikuku, V. O., Achieng', G. O., Ng'eno, E., Okowa, G. M., Masitsa, G. A., & Owuor, J. J. (2017). Distribution and Removal Efficiency of Heavy Metals by a Conventional Activated Sludge at a Municipal Wastewater Treatment Plant in Kisumu City-Kenya. *Res. J. Chem. Sci., 7*(8), 19–25.

Strozi, M., Raymundo, M., & Colombo, P. (2014). Open cell geopolymer foams by a novel saponification / peroxide / gelcasting combined route. *Journal of the European Ceramic Society, 1–5*. doi:10.1016/j.jeurceramsoc.2014.04.001

Tchakoute, H. K., Elimbi, A., Diffo Kenne, B. B., Mbey, J. A., & Njopwouo, D. (2013). Synthesis of geopolymers from volcanic ash via the alkaline fusion method: Effect of Al_2O_3/Na_2O molar ratio of soda–volcanic ash. *Ceramics International, 39*(1), 269–276. doi:10.1016/j.ceramint.2012.06.021

Tome, S., Etoh, M., Etame, J., & Sanjay, K. (2018). Characterization and leachability behaviour of geopolymer cement synthesized from municipal solid waste incinerator fly ash and volcanic ash blends. *Recycling, 3*(4), 50. doi:10.3390/recycling3040050

Twigg, M. V., & Richardson, J. T. (2007). Fundamentals and applications of structured ceramic foam-catalysts. *Industrial & Engineering Chemistry Research, 46*(12), 4166–4177. doi:10.1021/ie061122o

van Deventer, J. S. J., Provis, J. L., Duxson, P., & Lukey, G. C. (2007). Reaction mechanisms in the geopolymeric conversion of inorganic waste to useful products. *Journal of Hazardous Materials, 139*(3), 506–513. doi:10.1016/j.jhazmat.2006.02.044 PMID:16600483

Wang, S., Li, L., & Zhu, Z. H. (2007). Solid-state conversion of fly ash to effective adsorbents for Cu removal from wastewater. *Journal of Hazardous Materials, 139*(2), 254–259. doi:10.1016/j.jhazmat.2006.06.018 PMID:16839666

Wastiels, J., Wu, X., Faignet, S., & Patfoort, G. (1994). Mineral polymer based on fly ash. *J. Resour. Manag. Technol., 22*, 135–141.

WHO. (2001). Environmental Health Criteria 221: Zinc. Geneva. *Health, 11*, 171–176.

WHO. (2008). *Guidelines for Drinking-Water Quality*. Geneva, Switzerland: World Health Organization.

Xu, H. (2002). *Geopolymerisation of Aluminosilicate Minerals* [PhD Thesis]. The University of Melbourne.

Yankwa Djobo, J. N., Elimbi, A., Tchakouté, H. K., & Kumar, S. (2016). Mechanical activation of volcanic ash for geopolymer synthesis: Effect on reaction kinetics, gel characteristics, physical and mechanical properties. *RSC Advances, 6*(45), 39106–39117. doi:10.1039/C6RA03667H

Yao, X., Zhang, Z., Zhu, H., & Chen, Y. (2009). Geopolymerization process of alkali-metakaolinite characterized by isothermal calorimetry. *Thermochimica Acta*, *493*(1-2), 49–54. doi:10.1016/j.tca.2009.04.002

Yavuz, Ö., Altunkaynak, Y., & Güzel, F. (2003). Removal of copper, nickel, cobalt and Manganese from aqueous solution by kaolinite. *Water Research*, *37*(4), 948–952. doi:10.1016/S0043-1354(02)00409-8 PMID:12531278

Zhang, Y., & Liu, L. (2013). Fly ash-based geopolymer as a novel photocatalyst for degradation of dye from wastewater. *Particuology*, *11*(3), 353–358.

Zocca, A., Colombo, P., Gomes, C. M., & Jens, G. (2015). Additive Manufacturing of Ceramics. *Issues, Potentialities, and Opportunities*, *2001*. doi:10.1111/jace.13700

Zuhua, Z., Xiao, Y., Huajun, Z., & Yue, C. (2009). Role of water in the synthesis of calcined kaolin-based geopolymer. *Applied Clay Science*, *43*(2), 218–223. doi:10.1016/j.clay.2008.09.003

Chapter 6
Enhanced Cellular Activity on Conducting Polymer:
Chitosan Nanocomposite Scaffolds for Tissue Engineering Application

Rajiv Borah
Tezpur University, India

Ashok Kumar
Tezpur University, India

ABSTRACT

This chapter includes detailed review of the research undertaken with conducting polymer (CP) based composites with chitosan (Ch) for tissue engineering till date. The beneficial role of electrically conductive biomaterials has been discussed with the possible strategies to overcome the shortcomings of CP alone through blending with Ch due to its excellent biocompatibility, biodegradability, and bioactivity. Additionally, this embodiment deals with the optimization and characterization of electrically conductive, biocompatible and biodegradable Polyaniline: Chitosan (PAni:Ch) nanocomposites as cell culture substrates for MDA-MB-231 and NIH 3T3 fibroblast in order to examine the combined effect of nanofiber structure and surface modification on cell-biomaterial interactions. The nanocomposites were further checked as a conductive scaffold for electrical stimulation of a neuronal model PC12 cell line in order to explore the potential of the materials in neural tissue engineering.

INTRODUCTION

Tissue engineering is an interdisciplinary field that applies the principles and innovations from engineering and life-sciences toward the development of biological substitutes that restore, maintain or improve tissue function (Vacanti & Langer, 1999; Olson, Atala, & Yoo, 2011; Howard, Buttery, & Shakesheff, 2008). In recent years, the multidisciplinary knowledge obtained from micro and nanoscience and technology are being also involved in tissue engineering (Ratner, 2004). Typically, three approaches are investigated

DOI: 10.4018/978-1-5225-7838-3.ch006

singularly or in combination: cells that create tissue, biomaterial scaffolds that give structural support to cells, and growth factors and cell-matrix (scaffold) interactions to create an environment that promotes the regeneration of functional tissues and organs. Therefore, one of the basic and sensitive strategies of the tissue engineering is the selection and construction of a biomaterial scaffold with the desired features in the design and fabrication of neotissues/organs. International Union of Pure and Applied Chemistry (IUPAC) defined biomaterial as a material exploited in contact with living tissues, organisms or microorganisms (Vert, Doi, & Hellwich, 2012). Biomaterials facilitate exciting new opportunities for repairing and reconstruction of damaged tissues by fabrication of biomimetic, biocomposite scaffolds upon which new cells can regenerate in vivo. Therefore, the ideal biomaterial should be biocompatible, biodegradable, highly porous with a large surface area to volume ratio, mechanically strong and capable of being formed into desired shapes. The biomaterial should possess appropriate surface properties to favor cellular attachment, proliferation, and differentiation (Yang, Leong, & Du, 2001). Moreover, the biomaterial scaffold should degrade at the rate of tissue formation as the cells synthesize natural matrix structure around themselves, the scaffold shall provide structural integrity within the body and eventually break down leaving the neo-tissue.

Since the first use of biomaterial based medical devices on human in the late 1940s and early 1950, the biomaterials field has acquired widespread scientific and technological applications such as cardiovascular prostheses, intraocular lenses, joint replacements, dental implants, scaffolds for in vivo and in vitro cell growth, skin substitutes, sutures, blood bags, bone cement, etc. (Ratner, Hoffman, & Schoen, 2004). With the advancement in the last few decades in areas such as medicine, cell and molecular biology, chemistry, materials science and engineering, biomaterials research has significantly evolved. The stages of development of biomaterials research can be classified in three generations, where each generation differs from the previous one by the specific purpose it tries to accomplish, thus enlarging the field of applications, and temporal overlapping can occur. Till now, the field is continuously updating with the design of novel biomaterials and emerging technologies.

The 1st generation of biomaterials during 1950-1960, comprised of widely available industrial materials which, although not being designed for a specific medical application. These materials induced minimal foreign body response and indicated potential in tissue replacement due to their suitable chemical and physical properties (Narayan, 2010). For example, silicone rubber used in medical components and assemblies or pyrolytic carbon for mechanical heart valves are examples of 1st generation biomaterials. During 1980-1990, the 2nd generation of biomaterials was designed with an aim to achieve bioactivity for favorable cell-biomaterial interactions (Hench, & Thompson, 2010). Hence, biomaterials are conceptualized as systems for controlled drug release and gene therapy, delivering pharmacologic agents such as drugs, active proteins, growth factors, and other macromolecules of interest to localized areas. For example, bioactive glasses and ceramics are widely used in orthopedic and dental procedures (Kaur, Pandey, & Singh, 2014). Drug-eluting stent coatings designed to prevent endovascular restenosis (blood vessel closure) or biodegradable sutures are also found among this group (Daemen, & Serruys, 2007). From 2000 onwards, the 3rd generation biomaterials research is focused on developing novel biomaterials with innovative strategies to support functional tissue regeneration as opposed to replacement (1st generation), or either resorbability or bioactivity (2nd generation) (Hench, & Polak, 2002). Herein, the designing of the biomaterial scaffold involves the reassemble the complexity of natural tissue assemblages to mimic the native extracellular matrix (ECM), and stimulate specific cellular responses at the biointerfaces to promote cell adhesion, proliferation, migration and differentiation processes (Moroni, De Wijn, & Van Blitterswijk, 2008). Several new strategies are adapted to design as a smart interface or proactive cell-instructive material including the use a series of chemical, biochemical and biophysical signals that actively guide the whole tissue repairing process (Chai & Leong, 2007; Place, Evans, & Stevens, 2009; Wu, He, & Ren, 2014).

Various materials such as metals (titanium, stainless steel, platinum, platinum-iridium), ceramics (calcium phosphates), glasses, carbons, polymers (polyester, poly(lactic acid), polyurethane, silicone, polypropylene, Teflon®, poly(methyl methacrylate), nylon) and biopolymers have been utilized as biomaterials throughout these three generations (Chen, Ushida, & Tateishi, 2002). Out of which polymeric biomaterials are idiosyncratic due to ease of the flexibility in formulation into a wide range of shapes, viz., coatings, fibers, films, membranes, foams, nanoparticles or hydrogels. Although the polymeric biomaterials may be either synthetic or natural, the synthetic polymers offer some extraordinary advantages such as the potential of being customized within a wide range of physicochemical properties, being fabricated into various shapes with desired morphologic features and uniformity.

From 2000 onwards, the polymeric materials have the edge over all other materials in biomaterials research and the list has become extensive by now. Regarding tissue engineering applications, most of them have been utilized in a passive way, just as support for the cells and tissues. However, the active behaviour or unique properties intrinsic to the material can be taken into account for some specific applications as an advantage to the conventional polymeric biomaterials. This idea initiated the need for the development of smart biomaterials.

In this regard, with the similar electrical and optical properties of metals and inorganic semiconductors and flexibility in ease of synthesis and processing of conventional polymers, conducting polymers (CPs) have been proved to be innovative bioactive materials or functional biointerfaces. CPs are the recent generation of organic polymers, which combine the electrical, electronic, magnetic, and optical properties of metals and inorganic semiconductors with the mechanical properties, processability, etc. of traditional polymers (MacDiarmid, 2001). The structure of this special class of polymers, considered as fourth generation of polymers, is entirely different from traditional polymers or blend of insulating polymer with a conducting material such as a metal or carbon powder. The alternating single and double bonds along the highly conjugated backbone in CPs aid electron mobility and charge transport within and between the polymer chains, leading to high electrical conductivity (Shirakawa, Louis, & MacDiarmid, 1977; Xia, Wei, & Wan, 2010). The conductivities of the CPs are in the range 1-10^3 S/cm, while the traditional insulating polymers have much fewer conductivities in the range of 10^{-20} to 10^{-6} S/cm (Le, Kim, & Yoon, 2017). It has been verified that CPs have all the desired features of intelligent materials, because (Wallace, Teasdale, & Spinks, 2002):

- CPs can be engineered at the molecular level to recognize specific stimuli.
- They are conductive; they facilitate transport of electrical information.
- They are capable of localized processing as well as actuation of response mechanisms.

Due to this unique combination of beneficial properties, CPs have found a diverse range of applications in the microelectronics industry (Molapo, Ndangili, & Ajayi, 2012; Bao, Dodabalapur, & Lovinger, 1996), battery technology (Frackowiak, Khomenko, & Jurewicz, 2006; Snook, Kao, & Best, 2011), photovoltaic devices (Wang, Zhi, & Müllen, 2008; Huynh, Dittmer, & Alivisatos, 2002), light emitting diodes (Gustafsson, Cao, & Treacy, 1992), and electrochromic displays (Mortimer, Dyer, & Reynolds, 2006; Pei, Zuccarello, & Ahlskog, 1994), and more recently in the biomedical field (Wang, Roberge, & Wan, 2003; Hardy, Mouser, & Arroyo-Currás, 2014; Hardy, Lee, & Schmidt, 2013). The evidence of compatibility of CPs with many biological molecules in 1980s initiated the research on biomedical applications of CPs. Since then, CPs have been investigated extensively for evaluation of their biocompatibility due to their intrinsic the ability to entrap and controllably release biological molecules through reversible doping, ability to charge transfer from a biochemical reaction, and the flexibility to easily alter the electrical, chemical, physical, optical and other physicochemical properties of the CPs to better fit the nature of the desired application. After the assessment of the biocompatibility of CPs, their utilization in biomedical applications has been remarkably increased. Thus, the combination of their outstanding properties (good stability, intrinsic high conductivities and ease of synthesis) with additional features such as biocompatibility, high sensitivity/selectivity towards specific analyte, reproducibility of the electrode response and redox stability, turned them into potential candidates in biosensors (Contractor, Sureshkumar, & Narayanan, 1994; Gerard, Chaubey, & Malhotra, 2002), tissue engineering (Hardy, Mouser, & Arroyo-Currás, 2014; Hardy, Lee, & Schmidt, 2013; Ghasemi-Mobarakeh, Prabhakaran, & Morshed, 2011), neural probes (Green, Lovell, & Wallace, 2008; Cui, Lee, & Raphael, 2001), drug delivery (Hardy, Mouser, & Arroyo-Currás, 2014; Svirskis, Travas-Sejdic, & Rodgers, 2010), bioactuator devices (Shin, Spinks, & Shin, 2009). Thereby, during past 20 years, CPs such as polyaniline (PAni), polypyrrole (PPy), polythiophene (PT) and poly(3,4-ethylenedioxythiophene) (PEDOT) have gained significant importance in fabrication of biomaterials for biomedical applications. Particularly for tissue engineering applications, CP based biomaterials are important in that they can induce electrical, electrochemical and electromechanical stimulation of cellular activities (Hardy, Lee, & Schmidt, 2013; Prabhakaran, Ghasemi-Mobarakeh, & Ramakrishna, 2011; Ghasemi-Mobarakeh, Prabhakaran, & Morshed, 2009).

However, there exist practical problems in the preparation of clinically relevant CP based tissue scaffolds with biomimetic chemical, mechanical, and topological properties, which make the design and fabrication of CP based scaffolds for tissue engineering purpose is a challenging task. The characteristics

Figure 1. Various parameters and factors in fabrication of electrically conductive CP based biomaterials

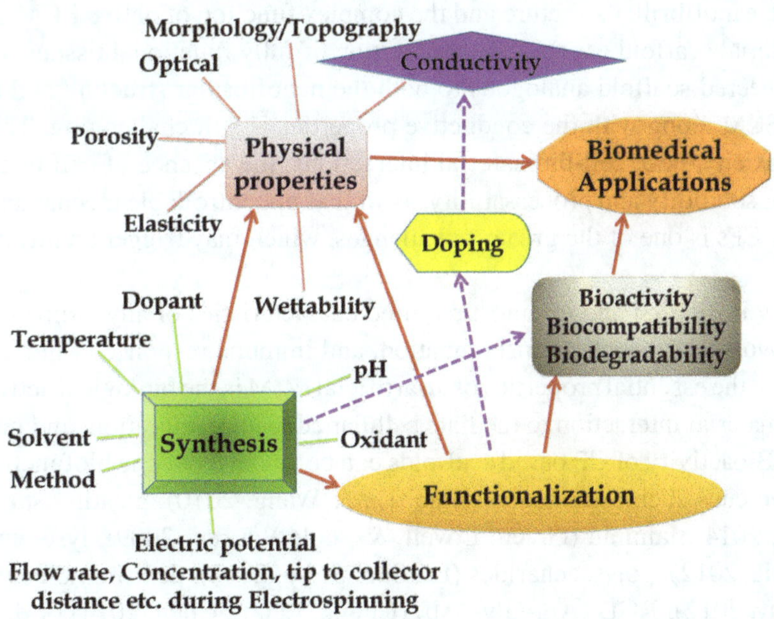

of the biomaterial scaffolds are governed by the fabrication technique. Therefore, during design, it is the utmost importance to incorporate all well-defined features to the biomaterial scaffold that mimic the native ECM. While synthesis of CP based biomaterials is highly sensitive to the choice, it is important to bear in mind that due to the vast versatility of CPs, a number of influencing parameters such as purity of the dopant, solvent, the oxidant, the relative concentration of the reagents, reaction time, temperature, stirring rate, etc. are interconnected in the fabrication process (Figure 1) (Balint, Cassidy, & Cartmell, 2014). The challenge is to balance all these parameters to fabricate a functional CP based biomaterial. The selection of dopant determines the properties of a CP for a specific application. Using electrical stimulation, the dopant can be expelled and incorporated again into the polymer, allowing the control of these preset physical properties. If the chosen biomolecule cannot be used as the dopant, it can still be incorporated using an intermediating doping molecule. If doping is not the right way to make the conductive polymer better suited for an application, physical adsorption, entrapment and covalent bonding offer alternative routes. Synthesizing chemically or electrochemically, perhaps as composites, electrospun fibers or hydrogels, can also be used to improve the usefulness of the end product (Balint, Cassidy, & Cartmell, 2014). Many synthetic strategies (both physical and chemical) have been adopted for the preparation of CP based biomaterials viz. electrospinning, hard physical template-guided synthesis and soft chemical template synthesis (e.g., interfacial polymerization, template-free method, dilute polymerization, reverse emulsion polymerization, etc.), and a variety of lithography techniques. The ideal tissue engineered scaffold as a temporary support should be able to control the interplay between cells, materials, and delivery of growth factors to create environments that promote the regeneration of functional tissues and organs (Balint, Cassidy, & Cartmell, 2014; Guarino, Zuppolini, & Borriello, 2016). The native ECM in the body is a complex and dynamic arrangement of proteins and polysaccharides such as collagen, glycosaminoglycans, hyaluronic acid, proteoglycans, elastin along with a complex mixture of nano features of pores, ridges and fibers of nano-scaled sizes, from several to more than 100 nm and thereby, provides substrates with specific ligands for cell adhesion and migration, regulates cellular proliferation and function by providing various growth factors (Freed, Marquis, & Nohria, 1993; Zhang, & Zhang, 2002; Murugan, & Ramakrishna, 2007; Martino, D'Angelo, & Armentano, 2012). Therefore, mimicking both the nanofibrillar structure and the complex function of native ECM are required to create an ideal biomaterial scaffold capable of regeneration of fully functional tissues.

The tissue engineered scaffold analogous to both the nanofibrillar structure and the complex function of the native ECM along with the conductive properties is still challenging. The main drawbacks with the CPs are toxicity, poor cell-biomaterial interactions, the absence of cell interaction sites, poor hydrophilicity, poor solubility and processability, as well as uncontrollable mechanical properties. Non-biodegradability of CPs is one of the greatest challenges, which may trigger an inflammatory response inside the body.

Biocompatibility is the first and foremost required characteristics of any biomaterial scaffolds such that it does not provoke any rejection, inflammation, and immune responses when in contact with the living system. One of the essential properties of an artificial ECM is the biological activity to facilitate the necessary cell-biomaterial interaction to mediate cellular adhesion, migration, and proliferation (Elbert & Hubbell, 1996). Bioactivity of CP based scaffolds can be achieved by the biofunctionalization of CPs using bio-regulative cues such as collagen (Xiao, Li, & Wang, 2010), gelatin (Stavrinidou, Winther-Jensen, & Shekibi, 2014), laminin (Green, Lovell, & Poole-Warren, 2009), lysozyme (Teixeira-Dias, del Valle, & Aradilla, 2012), polysaccharides (Lv, Sun, & Yu, 2012), dextrin molecules (Teixeira-Dias, del Valle, & Estrany, 2012), RGD (Arg-Gly-Asp) (Chung, Lai, & Chen, 2014) or doping with heparin

(Shamaeli, & Alizadeh, 2014), hyaluronic acid (HA) and chondroitin sulphate A (Gilmore, Kita, & Han, 2009) to promote cellular adhesion, while nerve growth factors (NGF) have been used to improve compatibility with neural cells (Kim, Richardson-Burns, & Hendricks, 2007). It should provide a 3D template in nanofibriliar feature with the surface to volume ratio for maximum accommodation of cells and cell-biomaterial interaction. The biomaterial scaffold should possess a high level of porosity (with pore size for efficient influx of anabolic nutrients and outflow of catabolic waste (approx. 10-1000 nm). Furthermore, the scaffold should be biodegradable such that the degradation rate of the scaffolds matches the rate of tissue regeneration to maintain the tissue functionality. It is worth pointing out the necessary mechanical stability and elasticity of the scaffold to withstand in vivo biological forces, and to support the cells to synthesize specific proteins including other biochemical and biological processes needed for a healthy tissue function and growth. In addition, it should be sterilizable as well to avoid toxic contaminations without compromising any physicochemical properties. Specifically, to utilize the conductive property of CP based scaffold as a smart biomaterial, it should be designed in such a manner that charge transfer takes place efficiently between the plasma membrane of cells and the biomaterial to stimulate the normal cellular activities just like longer neurite outgrowth of neurons under electrical stimulation. Moreover, the stability of the CP based biomaterials in physiological condition required without any effect on its conductivities. Last but not least, the production process CP based biomaterial scaffold incorporating all the above essential features must be performed in a reproducible, economical, and scalable manner.

In the present thesis, the synthesis and modification of one dimensional (1D) CP nanostructures have been emphasized to use as potential biomaterial scaffold and accordingly, the following sections, subsections of this chapter deals with the brief discussion of the advantages of 1D CP nanostructures over other nanostructures and its bulk counterpart followed by the synthesis and modification strategies of 1D CP nanostructured biomaterials.

As discussed above, although a range of biocompatible and biodegradable polymers have been explored with CPs for tissue engineering applications, biopolymers are of particular interest due to the biological and chemical similarities to natural tissues (Krajewska, 2005). In this regard, chitosan (Ch) has some of the superior properties to be blended with CPs in fabrication of electrically conductive biocompatible, bioactive, and biodegradable tissue engineered scaffolds (Martino, Sittinger, & Risbud, 2005; Kim, Seo, & Moon, 2008; Madihally & Matthew, 1999; Suh & Matthew, 2000). Ch is a partially de-acetylated derivative of chitin, which is semicrystalline in nature. Chitin, the second most abundant organic materials produced annually by biosynthesis after cellulose, is an important constituent of the exoskeleton in animals, especially in crustacean, molluscs and insects. In Ch, D-glucosamine and randomly located N-acetyl-glucosamine groups are linked through β (1-4) glycosidic bonds, which makes some of its characteristics similar to glycosaminoglycans (GAG) and hyaluronic acid present in living tissue (Chandy & Sharma, 1990). In fact, the N-acetylglucosamine moiety in chitosan is a structural feature also found in the GAGs. Moreover, the cationic nature of Ch enables it to interact electrostatically interactions with anionic GAG, proteoglycans and other negatively charged molecules. Since, GAG is linked to a large number of cytokines/growth factors, use of Ch with GAG is always preferred as it helps to retain and concentrate growth factors secreted by colonizing cells (Madihally & Matthew, 1999; Suh & Matthew, 2000). Apart from advantageous bioactivity, Ch has beneficial a stimulatory effect on macrophages and is well known as chemo attractant for neutrophils both *in vitro* and *in vivo* (Usami, Okamoto, & Minami, 1994; Usami, Okamoto, & Minami, 1994). Thus, Ch possesses unique biological properties including biocompatibility, biodegradability to non-toxic products, nontoxicity, physiological inertness, remarkable affinity to proteins, antibacterial, haemostatic, fungistatic, antitumoral and anticholesteremic properties (Nishimura, Nishimura, & Nishi, 1984; Tanigawa, 1992; Mori, Okumura, & Matsuura, 1997; Singla & Chawla, 2001; Shi, Zhu, & Ran, 2006). Due to such versatile properties, Ch and its derivatives have been explored extensively in almost all sectors of tissue engineering such as skin (Mizuno, Yamamura, & Yano, 2003), bone (Martino, Sittinger, & Risbud, 2005), cartilage (Mattioli-Belmonte, Gigante, & Muzzarelli, 1999), liver (Li, Pan, & Zhang, 2003), nerve (Haipeng, Yinghui, & Jianchun, 2000) and blood vessel (Madihally, Flake, & Matthew, 1999). Additionally, the easy processability due to pH dependent solubility, flexibility in processing in desired forms and good mechanical properties, makes Ch a suitable component in fabrication of CP based tissue engineered scaffolds.

In the present chapter, every kind of biological cell related works including the advanced tissue engineering applications accomplished till date using CP based composites with Ch have been reviewed in details along with a gist of the authors' own investigations on different variations of Polyaniline:Chitosan (PAni:Ch) nanocomposites for tissue engineering applications. The rationale behind using CP based electroconductive biomaterials in tissue engineering applications has been summarized in the "Background" section. Thereafter, composites of Ch with three widely investigated CPs, viz. PPy, PEDOT and PAni for tissue engineering related research has been discussed followed by the authors own research in the concerned topic.

Briefly, the proposed chapter also embodies the strategies adopted by our research group for development of CP based nanocomposites for tissue engineering applications to overcome the challenges associated with CP based biomaterials. It also tries to gain deeper insight into the synthesis and optimization of ECM analogue nanostructured CP based biomaterials with considerable focus on the improvement of cell-biomaterial interactions. After incorporating adequate optimization, the present thesis work further emphasizes on the potential of CPs in neuronal stimulation under external electrical field for nerve repairing applications. Herein, PAni nanofibers synthesized by dilute polymerization method have been blended with Ch, a natural biopolymer, at a concentration of 4% and 6% (w/v) to confer bioactivity and biodegradability for improved function as biomaterial scaffold. The tissue engineering aspect of PAni:Ch nancomposites has not been investigated till now. Whilst the potential biomedical applications for CPs are known ways in which to improve their physicochemical and biological properties using novel nanocomposite and surface functionalization routes have yet to be studied extensively. Within this study, an electrically conductive PAni:Ch nanocomposite based biodegradable material has been surface functionalized with glutaraldehyde and glycine N-hydroxysuccinimide (NHS) ester, separately, in order to investigate the hypothesis that surface functionalization of these conductive materials will enhance bioactivity and improve cell-biomaterial interactions. The effect of surface functionalization of PAni:Ch nanocomposites on their biological properties has been investigated in terms of blood compatibility by Hemolysis assay, cytotoxicity by MTS proliferation assay with MDA-MB-231 cells along with cell adhesion and spreading by acridine orange/ethidium bromide (AO/EtBr) staining and SEM. Similarly, the effect of altered surface properties of glycine NHS ester functionalized PAni:Ch (EFPAni:Ch) nanocomposites has been investigated to modulate the blood compatibility by Hemolysis assay, cytotoxicity by MTS proliferation assay with 3T3 fibroblasts and rat neuronal model PC12 cells, cell adhesion, spreading and proliferation by live/dead assay and SEM. Electrically stimulation of PC12 cells through electrically conductive collagen I coated PAni:Ch nanocomposites has been performed to explore the potential of these materials in nerve repairing.

BACKGROUND

Donor dependent organ transplantation, the major current clinical treatments for damaged tissues or organs, is subject to a major organ shortage crisis worldwide. According to reports published in 2008, the incidence of renal failure 380-430 million/y, whereas the supply of organs from deceased donors (DCD) is 38-53 million/y (Bendorf, Kelly, & Kerridge, 2013); Hernandez-Alejandro, Caumartin, & Marotta, 2010). In India, 4 lacs people across the nation die each year, while waiting for an organ transplant (Goklany, 2017). The depressing statistics in India as published in 2016 are as: 1,00,000 people die of liver diseases, while 1,000 get a liver transplant; 2,20,000 awaits kidney transplants, while 15,000 get a transplant; while people awaiting corneal transplant 10 lacs, heart transplants 50,000, lung transplants 20,000 (Nisha, 2017; Ahuja, 2017). There is a wide gap between the number of transplants awaited and the organs available. While host rejection is one of the main problems, other clinical treatments available such as autografting and allografting have also several disadvantages including loss of function at the donor site and the need for multiple surgeries. Also, when the damaged area is large, these techniques are not suitable. Tissue engineering, as a solution to address the above limitations, aims to create living, functional tissues which can be used in the replacement or repairing of tissues lost due to disease, age, congenital defects, or physical damage by combining scaffolds, cells and bioactive molecules. Biomaterials provide exciting new opportunities for tissue repair and reconstruction by fabrication of biomimetic, biocomposite scaffolds upon which new cells can regenerate in vivo.

The global regenerative medicine market with a worth \$18.9 billion globally at present, is growing at rapid pace and is expected to grow tremendously at the CAGR (Compound Annual Growth Rate) of 23.6% during 2016-2021 (GlobalData, France, 2017). But, most of the tissue engineered products concentrated in developed regions such as Europe and North America, India is yet to witness widespread penetration of tissue repair biomaterial products.

Thus, future development of regenerative medicine is focused to address the worldwide crisis of organs and the consequent rise of the risk of organ trafficking along with the shortcomings of the current clinical treatments for damaged tissue or organ. Although there are considerable evidences of efficient working of tissue engineered products such as biomaterials with biological tissues, the key challenges are to ensure its safety to use in vivo and to meet the Food and Drug Administration (FDA) approved quality standards. Efforts to resolve these issues are in progress, and if successful, tissue engineering will have the major impact on future healthcare practices.

Additionally, it is well known that electricity plays a vital role in normal functioning of our body. In fact, without electricity our brain cannot work, muscles cannot contract, we cannot visualize things with our eyes, we cannot sense, the rhythm of our heartbeat and the blood transportation through our circulatory system could not be controlled. All the biological processes in the body are controlled by charge transport phenomena such as transport of ions across the plasma membranes and of electrons along biomolecules. Electric potentials exist inside and outside cells and continuous electrical signals through neurons are responsible for the internal body function. Therefore, due to the ability to regulate cell growth and function, there is a growing interest of utilizing the beneficial effect of electrical stimulation for wound healing, stem cell differentiation and nerve repairing purposes. Electrical stimulation (ES) is one of the potential methods to improve the rate of nerve regeneration and the restoration of function (Al-Majed, Neumann, & Brushart, 2000; Schmidt & Leach, 2003). The concept of using ES for nerve regeneration is based on the fact that the bioelectricity plays an integral role in maintaining normal biological functions via cranial, spinal and peripheral neural networks which signal for ex-

Figure 2. Statistics on the literature published on (a) CPs in tissue engineering applications and (b) CP based biomaterials for electrical stimulation during the period 1993-2017(search made using the Web of Science using as keywords: (a) Conducting polymers + Tissue engineering or (b) Electrical stimulation + Conducting polymers).

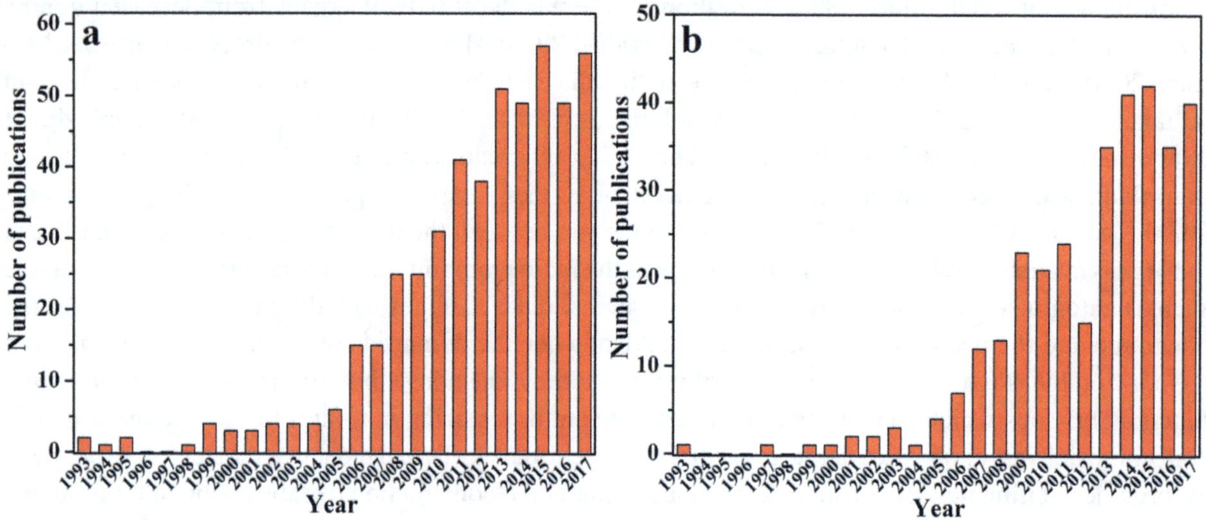

ample muscle contraction and wound healing in bone, cartilage, skin, connective tissue (Shi, Zhang, & Rouabhia, 2008). ES activates neurons through membrane depolarization which ultimately promotes the speed and accuracy of motor and sensory axon regeneration both *in vitro and in vivo*. Specifically, the electrical signal in the peripheral nervous system accelerates axonal regeneration and elongation and enhances expression of neurotrophic factors and the biological activity of Schwann cells (Haastert-Talini, & Grothe, 2013). In the context of cellular activities, the beneficial effects of electrical stimulation have been demonstrated in more than 50,000 articles in the form of scientific journals, patents, books etc [Source: Web of Science]. Consequently, during past 25 years, especially from 2006 onwards, research on CP in tissue engineering (Figure 2(a)) and CP based biomaterials for electrical stimulation has been increasing significantly (Figure 2 (b)).

Unlike natural biodegradable polymers, electroactive biomaterials are a part of a new generation of ''smart'' biomaterials that delivers the electrical, electrochemical and electromechanical stimulation to cells directly. There are other electroactive biomaterials such as electrets (materials with a quasi-permanent surface charge provided by trapped charge carriers) and piezoelectric materials (materials which generate transient electrical charges under mechanical deformation) that can deliver electrical stimulation without the need for an external power source but with limited control over the stimulus (Balint, Cassidy, & Cartmell, 2014).

However, CPs exhibit many advantages over these materials in terms of excellent control of the electrical stimulus over level and duration, good electrical and optical properties, a high conductivity/weight ratio, ability to entrap and controllably release biological molecules via reversible doping, ability to transfer charge from a biochemical reaction, and the potential to easily alter its electrical, chemical, physical, and other properties needed for desired specific application. Furthermore, CPs can be made biocompatible, biodegradable and porous and additionally, their properties can further be altered and controlled through stimulation (e.g. electricity, light, pH) even after synthesis (Balint, Cassidy, & Cartmell, 2014). Thus, there is a tremendous potential of development of conductive tissue engineered scaffold for modifying the regeneration, differentiation, or function of cells both in vivo and in vitro at a faster rate than conventional non-conductive scaffolds. Research in this direction using CPs has been started in the past decade enabling to exercise control over the level and duration of the electrical stimulation for improved nerve regeneration, bone and cartilage repairing. Intended for nerve regeneration, CPs have a higher charge injection limit with improved charge-discharge characteristics leading to enhanced charge transportation to cells for membrane depolarisation. This, in turn, can improve the adhesion and proliferation of nerve cells including the promotion of axonal growth (Samba, Herrmann, & Zeck, 2015; Yan, Zhao, & Liu, 2016). In combination with the properties of novel scaffold and options for electrical stimulation, CPs lend themselves as one of the most promising biomaterials in the future development of regenerative medicine to address the worldwide crisis of organs and the consequent rise of the risk of organ trafficking along with the shortcomings of the current clinical treatments for damaged tissue or organ.

CONDUCTING POLYMER: CHITOSAN COMPOSITES FOR TISSUE ENGINERING APPLICATIONS

Polypyrrole: Chitosan (PPy:Ch) Composites

Polypyrrole (PPy) is a well-known conductive polymer owing to its attractive properties such as high conductivity, excellent redox activity, easy processability, good thermal and electrochemical stability, low toxicity to biological environment etc. (Ateh, Navsaria, & Vadgama, 2006). Although, PPy is non-thermoplastic, mechanically rigid, brittle and sometimes less biocompatible, when conjugated with biocompatible biopolymers such as collagen, fibronectin, cellulose, gelatin, chitosan, etc., its biocompatibility is improved without significantly affecting its conductivity (Ateh, Navsaria, & Vadgama, 2006; Liu, Yue, & Higgins, 2011). Moreover, these shortcomings of the PPy have been overcome when blended with synthetic polymers such as polycaprolactone (PCL), polyethylene terephthalate (PET), poly(Lactide-co-Glycolide) (PLGA), poly-DL-lactic acid (PDLLA) etc. (Balint, Cassidy, & Cartmell, 2014). Thus, it is a very promising candidate as an electrically conductive smart biomaterial for various tissue engineering applications such as neural, cardiac, bone and cartilage tissue engineering.

Chitin, the second most abundant natural polysaccharide with N-acetyl-D-glucosamine monomeric units, is a major structural component in the exoskeleton of arthropods and cell walls of fungi and yeast. Due to its non-solubility to common solvents, a deacetylated derivative chitosan (Ch) is more commonly used and due to its solubility in acidic, neutral, and alkaline solutions, Ch is preferred over chitin for a wide range of applications (Muzzarelli, 2009). Ch promotes cell adhesion, proliferation and differentiation and is also biodegradable, biocompatible, nonantigenic and nontoxic (Chen, Mo, & Qing, 2007). Due to these properties, Ch is used in a range of fields related to biomedicine and is used as a composite to form hybrid materials for tissue engineered scaffolding, drug delivery, wound dressings, separation membranes and antibacterial coatings, stent coatings and sensors (Chen, Mo, & Qing, 2007; Muzzarelli, 2009; Muzzarelli, 2011). Recently a number of PPy:Ch composites have been reported by different groups for supercapacitors, sensors and dye removal applications (Ji, Xiong, & Zhu, 2018; Sall, Diaw, &

Gningue-Sall, 2018; Xiao, Feng, & Liu, 2018). However, there are limited studies of PPy in conjugation with chitosan for tissue engineering applications, which have been reviewed below.

Mihic et al. synthesized PPy:Ch hydrogel having semiconducting and good mechanical propertiesl lacking in Ch alone (Mihic, Cui, & Jun, 2015). They showed enhanced Ca^{2+} signal conduction in neonatal rat cardiomyocytes cultured on PPy:Ch. They demonstrated that PPy:Ch enhanced biological *conduction in vitro* and *in vivo*. The injected PPy:Ch around the scar significantly improved border zone conduction velocity and better maintained heart function after myocardial infarction than Ch alone. Very recently, the same group demonstrated enhanced synchronous contraction between physically separated beating neonatal rat cardiomyocytes clusters *in vitro* on PPy:Ch (Cui, Ni, & Wu, 2018). They further showed that the injected PPy:Ch following cardiac injury improved electrical impulse propagation of scar tissue *in vivo*. In both the studies, PPy was achieved through oxidative polymerization of pyrrole in Ch solution and hydrogel was achieved through glutaraldehyde cross-linking. Zhang et. al. polymerized PPy on genipin cross-linked Ch scaffold to obtain electrically conductive PPy:Ch scaffold for co-culture of adipose-derived mesenchymal stem cells (ADMSCs) and umbilical vein endothelial cells (HUVECs) under electrical stimulation (Zhang, Neoh, & Kang, 2018). They demonstrated remarkable increase in calcium deposition and CD31 expression in the co-culture of ADMSCs and HUVECs upon application of electrical signal (ES) of 200 μA for 4 h/day through the electrically conductive PPy:Ch scaffold. Thereby, they were able to show that the synergistic effect of ES and co-culture of ADMSCs and HUVECs enhanced the autocrine signaling, resulting in further promotion of cellular functions, which could be a promising therapeutic solution for bone defect regeneration. Kumar *et. al.* synthesized PPy:Ch bioactive composites through *in-situ* electrochemical polymerization in oxalic acid medium and studied the growth of MG-63 human osteoblast cells (Kumar, Suresh, & Das, 2017). Owing to electroactivity, PPy:Ch was also checked for its potential in nerve regeneration. In a recent study, a mixture of sodium aligante (SA) and carboxymethyl chitosan (CMCS) was doped with chemically polymerized PPy to obtain an electrically conductive SA/CMCS/PPy hydrogel for potential application as a artificial nerve graft (Bu, Xu, & Li, 2018). It was demonstrated that the physico-chemical properties such as gelation time, swelling ratio, porosity, mechanical and mechanical properties of the hydrogel was dictated by PPy content. Upon electrical signal, the SA/CMCS/PPy hydrogel supported excellent cell adhesion and proliferation of PC12, RSC96 and BMMSC cells *in vitro*. The *in vivo* biocompatibility of the hydrogel was demonstrated in adult sprague dawley rats with the help of subcutaneous inflammatory reaction indicating the potential of this conductive hydrogel as a filling material in the nerve conduit to provide strong support for peripheral nerve regeneration. In another study, Qi *et. al.* demonstrated electrical regulation of olfactory ensheathing cells using PPy:Ch composites for its potential in the treatment of lengthy nerve defects (Qi, Wang, & Ma, 2013). They synthesized PPy nanoparticles by microemulsion technique followed by blending with Ch solution for culturing olfactory ensheathing cells (OECs), which enhance axonal regeneration and to produce myelin after transplantation. They reported remarkable enhancement of the expression and secretion of brain-derived neurotrophic factor (BDNF), nerve growth factor (NGF), neural cell adhesion molecule (N-CAM) and vascular endothelial growth factor (VEGF) on PPy:Ch scaffold upon application of external ES indicating its potential in nerve repairing due to increased neurotrophin secretion.

POLY(3,4-ETHYLENEDIOXYTHIOPHENE): CHITOSAN COMPOSITES

Poly(3,4-ethylenedioxythiophene) (PEDOT) is a polythiophene derivative conducting polymer (CP), which has highest electrical conductivity and stability when compared to the other CPs. Recently, the physicochemical and biological functionality of various PEDOT materials were shown to be enhanced using various biological molecules such as amino acids, proteins, polysaccharide and neurotrophic substances and organic dopants such as small molecule salt ions, macromolecule. In order to eliminate the drawbacks such as poor mechanical properties, poor processability, hydrophobicity and non-degradability, PEDOT has been belnded with various biocompatible polymers such as poly lactide (PLA), chitosan, gelatin, collagen, hyaluronic acid (HA), etc. Xiao et al. showed PEDOT:Collagen matrix for improved PC12 adhesion (Xiao, Li, & Wang, 2010). Similarly, PEDOT was doped using a negatively charged ECM component such as HA, which has vital role in wound healing, tissue regeneration and angiogenesis for the fabrication of synthetic matrices (Wang, Guan, & Wang, 2017). There are extensive studies that were undertaken on PEDOT:Ch composites for various energy related, sensing and biomedical applications. For instances, PEDOT along with carboxymethyl Ch was reported for the $LiFePO_4$ (LFP) cathode in Li-ion batteries (Zhong, He, & Lu, 2016). Khan et al. reported highly electroactive PEDOT:Ch composites for antibacterial application (Khan & Narula, 2016). Kuralay et al. showed PEDOT coated Ch for the investigation of DNA sensing and DNA-anticancer drug interaction (Kuralaya, Demirci, & Kiristi, 2014). Similarly, an electrochemical biosensor based on PEDOT:Ch was demonstrated for H_2O_2 detection (Zhang, Xu, & Fan, 2018). However, there are extremely few reports that can be found on PEDOT:Ch composites for tissue engineering applications, which are reviewed below.

Rad et al. demonstrated the beneficial role chemically synthesized tissue-engineered scaffolds of chitosan, gelatin, hyaluronic acid, collagen and PEDOT to provide the best condition for attachment and proliferation of the liver cells (Tahmasbi Rad, Ali, & Kotturi, 2014). They worked on the hypothesis that electrical signal applied through an electrically conductive scaffold controls the electrical communication among the liver cells and showed that the excellent cell attachment and proliferation of hepatocyte cells on the blend of gelatin-chitosan-PEDOT-collagen. They stated that the external electrical field regulated the transfer of ions and their movement across the plasma membrane, which subsequently influence the cellular behaviors, such as cell attachment, cell proliferation and proteins expression. In another study, electrospun PEDOT:Ch microfibers coated with PLGA were shown to be biocompatible with neuroblastoma cells in addition to release of an antibiotic drug ciprofloxacin hydrochloride (Esrafilzadeh, Jalili, & Liu, 2016). Lari et al. fabricated a 3D, highly porous and interconnected scaffolds by introducing PEDOT onto the nanohydroxyapatite/chitosan (nHA/CS) composite scaffolds through a freezing and lyophilization technique (Lari, Sun, & Sultana, 2016). Apart from the excellent physico-chemical properties, they showed that the mouse fibroblast (L929) cells adhered, spread and proliferated well on the conductive scaffold. Among the other CPs, PEDOT has been found to be considerably explored more for neural tissue engineering applications owing to its excellent biocompatibility and electrical conductivity. In a study, Wang *et. al.* reported the PEDOT-HA/poly(L-lactic acid) (PLLA) composite for its potential use in nerve repairing (Wang, Guan, & Liu, 2017). They have synthesized PEDOT nanoparticles by chemical oxidative polymerization using HA as dopant and blended with PLLA. The resultant electrically conductive scaffold showed that neuron-like pheochromocytoma (PC12) cells adhered and spread well on it. Additionally, they demonstrated more elongated cell morphology after modifying the scaffold with laminin. They applied electrical signal through this scaffold and showed enhanced neurite outgrowth along with considerable increase in mRNA expression of growth-associated protein (GAP43) and synaptophysin (SYP) in electrically stimulated PC12 cells suggesting their potential in nerve regeneration. In a similar study they replaced the PLA by Ch and gelatin (Gel) and performed similar experiments (Wang, Guan, & Zhu, 2017). The study demonstrated enhanced neurite outgrowth and higher expression of GAP43 and SYP upon application of external electric field. This group further explored the proliferation and differentiation of neural stem cell on the PEDOT-HA/Cs/Gel scaffold for possible neural tissue engineering applications (Wang, Guan, & Xu, 2017). They experimentally showed that the promoted NSCs proliferation with up-regulated expression of Ki67 and differentiation into neurons and astrocytes with up-regulated expression of β tubulin-III and GFAP, respectively. In a recently published literature, the same group prepared semi-interpenetrating network hydrogels of PEDOT with carboxymethyl Ch, which is biodegradable and electroconductive (Xu, Guan, & Wang, 2018). They have evaluated the physicochemical and electrochemical properties of hydrogels including chemical structure, compressive modulus, electrical conductivity, surface morphology, water absorption ability and biodegradation followed by cell culture study with PC12 cells. They successfully showed the improved spreading and proliferation of PC12 cells on the as-synthesized scaffold.

POLYANILINE: CHITOSAN COMPOSITES

Polyaniline (PAni) is one of the most adaptive CPs due to its diverse structural forms, ease of synthesis, high environmental stability and excellent charge transport property by the doping/dedoping process (Huang, Humphrey, & MacDiarmid, 1986; Ansari & Keivani, 2006; Bober, Humpolíček, & Pacherník, 2015). One of the major advantages of PAni is its solubility in a few selected organic solvents, which makes possible it to process in free standing film form with higher degree of stability on account of insolubility in most of the solvents including water (Chiou & Epstein, 2005). The in vitro biocompatibility of pure PAni or it derivatives has been investigated with various cell types including H9c2 cardiac myoblasts (Bidez, Li, & MacDiarmid, 2006), C2C12 myoblasts (Jeong, Jun, & Choi, 2008), L929 murine fibroblasts (Qazi, Rai, & Boccaccini, 2014; Xia, Lu, & Zhu, 2013), human mesenchymal stromal cells (Thrivikraman, Madras, & Basu, 2014), rat nerve stem cells (Qazi, Rai, & Boccaccini, 2014), PC12 cells (Bhang, Jeong, & Lee, 2012), etc. However, issues such as biocompatibility, biodegradability including any unexpected side effects related to bioactivity are important when considering biomedical applications for these materials. Recently, to overcome the poor biodegradability and biocompatibility, PAni has been blended with natural and synthetic biopolymers for various biomedical applications including tissue engineering, biosensing and drug delivery. If processed all these properties, PAni based materials have the potential to fulfill the technological expectations for biomedical applications.

Like PPy and PEDOT, there have been extensive studies of PAni:Ch composites for a range of applications from supercapacitor, catalytic activity, sensing, antibacterial activity and tissue engineering applications. For example, Ayad et al. synthesized PAni by oxidative *in-situ* polymerization in Ch solution followed by Pd incorporation (Ayad, Amer, & Kotp, 2017). They showed the reduction of the toxic 4-nitrophenol to a safer form, 4-amino phenol using the above nanocomposite. In another similar study, PAni:Ch composite was decorated with silver for the reduction of 4-nitrophenol (Ayad, Amer, & Kotp, 2017). PAni:Ch composite was also shown to successfully remove Congo Red, Coomassie Brilliant Blue, Remazol Brilliant Blue R, and Methylene Blue from aqueous solution through adsorption (Janaki, Oh, & Shanthi, 2012). Chemically synthesized ternary composite of PAni with Ch and ZnO was also demonstrated to remove reactive orange 16 dye (Kannusamy & Sivalingam, 2013). Recently, a honeycomb nanocomposite of PAni:Ch:Nickel oxide synthesized through chemical route was shown to exhibit specific capacitance of 554 F/g and energy density of 76.9 Wh/kg at 1 A/g current density, and high power density of 2.00 kW/kg at 4 A/g, indicating the possible use of the nanocomposite in supercapacitor (Vellakkat & Hundekal, 2017). PAni in conjunction with Ch has been studied in biosensing as well. For instances, Srivastava et al. reported chemically synthesized PAni:Au:Ch for cholesterol oxidase immobilization for the voltammetric detection of cholesterol (Srivastava, Srivastava, & Nirala, 2014).

There are also reports of use of PAni with Ch in antibacterial and wound healing applications. Recently, Moutsatsou et al. demonstrated electrospun PAni:Ch nanofibers for enhanced cell attachment, morphology and proliferation of osteoblasts and fibroblasts (Moutsatsou, Coopman, & Georgiadou, 2017). They stated that the presence of conducting PAni stimulated electrical signalling communication, and subsequently helped to attach and proliferate rapidly, which is useful for possible wound healing applications. In an interesting study, an injectable self-healing hydrogel dressing with multifunctional properties including anti-infection, antioxidative and conductivity was prepared by mixing quaternized Ch grafted PAni with benzaldehyde group functionalized poly(ethylene glycol)-co-poly(glycerol sebacate) (Zhao, Wu, & Guo, 2017). They demonstrated that use of the above composite offered excellent *in vivo* blood clotting capacity and wound healing process by upregulating the gene expression of growth factors including vascular endothelial growth factor (VEGF), epidermal growth factor (EGF) and transforming growth factor-β (TGF-β) and then promoting granulation tissue thickness and collagen deposition. The same group developed quaternized Ch grafted PAni with oxidized dextran as crosslinker and demonstrated its enhanced biocompatibility and cell adhesion with adipose derived mesenchymal stem cells and C2C12 myoblasts (Zhao, Li, & Guo, 2015). Ulutürk et al. showed *in-situ* polymerization of PAni in photocrosslinked network Ch grafted glycidyl methacrylate and poly(ethylene glycol)diacrylate, which was cytocompatible with L929 fibroblast cells (Ulutürk & Alemdar, 2018). In another recent investigation, Gu et al. demonstrated the positive contribution electrospun aligned PAni:Chitin nanofibers for the directional guide of human dermal fibroblasts (Gu, Park, & Kim, 2018). A 3D Ch-Gel:nHA-PAni was synthesized by freeze-drying method with improved swelling ratio, density, biodegradation, biomineralization, and mechanical behavior (Farshi Azhar, Olad, & Salehi, 2014). The same scaffold was shown to be biocompatible with bovine leukemia virus fetal lamb kidney cell line, while dental pulp stem cells were attached and proliferated well on it. A few studies have been found reporting PAni in conjunction with Ch for neural tissue engineering applications. Baniasadi et al. demonstrated electrically conductive PAni:Graphene:Ch:Gel composite for potential neural tissue engineering application (Baniasadi, SA, & Mashayekhan, 2015). They showed *in-situ* polymerization of PAni graphene dispersion followed by subsequent blending with Ch and Gel. It has been proved that the addition of PAni and graphene into Ch:Gel matrix increased the electrical and mechanical properties with slower biodegradability. The adhesion of Schwann cells was greatly enhanced on this scaffold suggesting its potential nerve repairing. Hu *et. al.* performed condensation polymerization in an acetic acid/DMF/DMSO solution in which aniline pentamer was cross-linked with Ch (Hu, Huang & Zhuang, 2008). They showed improved neuronal differentiation of PC12 on this scaffold.

While most of the studies reporting CP composites with Ch were in bulk, there have been a few studies were reported on their nanocomposites for tissue engineering applications till date. Therefore, the author and their group have undertaken the synthesis and surface functionalization of PAni:Ch nanocomposites for tissue engineering applications including electrical stimulation of neuronal cells and those have been discussed in the subsequent sections.

ENHANCED CELLULAR ACTIVITIES ON SURFACE FUNCTIONALIZED POLYANILINE: CHITOSAN NANOCOMPOSITES

Polyaniline (PAni) nanofibers were synthesized using the method described by Chiou et al. (Chiou & Epstein, 2005; Chiou, 2006) using HCl as dopant and ammonium peroxydisulfate (APS) as oxidant. Polyaniline:Chitosan (PAni:Ch) nanocomposites were prepared by mixing 4% and 6% (w/v) of PAni in 2% (v/v) chitosan solution in acetic acid. One set of the as-synthesized PAni:Ch nanocomposites was surface functionalized using 1% glutaraldehyde in phosphate buffer saline (PBS, pH = 7.4), while the another set was surface functionalized using BOC glycine N-hydroxysuccinimide (NHS) ester. The schematic illustrations of surface functionalization of PAni:Ch nanocomposites using glutaraldehyde and glycine NHS ester were shown in Figure 3 and Figure 4. It is to be noted that PAni:Ch nanocomposites with 4 wt% and 6 wt% PAni content, have been named as PAni:Ch-4 wt% and PAni:Ch-6 wt%, respectively. Consequently, the glutaraldehyde functionalized PAni:Ch nanocomposites with 4 wt% and 6 wt% PAni content, have been named as GFPAni:Ch-4 wt% and GFPAni:Ch-6 wt%, respectively. Similarly, the glycine NHS ester functionalized PAni:Ch nanocomposites with 4 wt% and 6 wt% PAni content, have been named as EFPAni:Ch-4 wt% and EFPAni:Ch-6 wt%, respectively, throughout the discussion.

SEM revealed interconnected networks of PAni nanofibers soaked in Ch matrix with rough and porous surface when compared to smooth pure Ch as shown in Figure 5. Figure 6 displayed the room temperature (300 K) non-linear I-V characteristics of pristine PAni:Ch nanocomposites films along with glutaraldehyde functionalized and Glycine NHS ester functionalized nanocomposites films i.e., GFPAni:Ch and EFPAni:Ch nanocomposites with variation in PAni content. The I-V characteristics of all the materials are almost symmetric upon reversal of the applied voltage. The I-V characteristics the nanocomposites with higher PAni content exhibit higher value of current at a particular applied voltage, while after surface functionalization both GFPAni:Ch and EFPAni:Ch nanocomposites with different PAni

Figure 3. Scheme of surface functionalization of PAni:Ch nanocomposites using glutaraldehyde to introduce aldehyde functionality on the surface
(Borah, 2018)

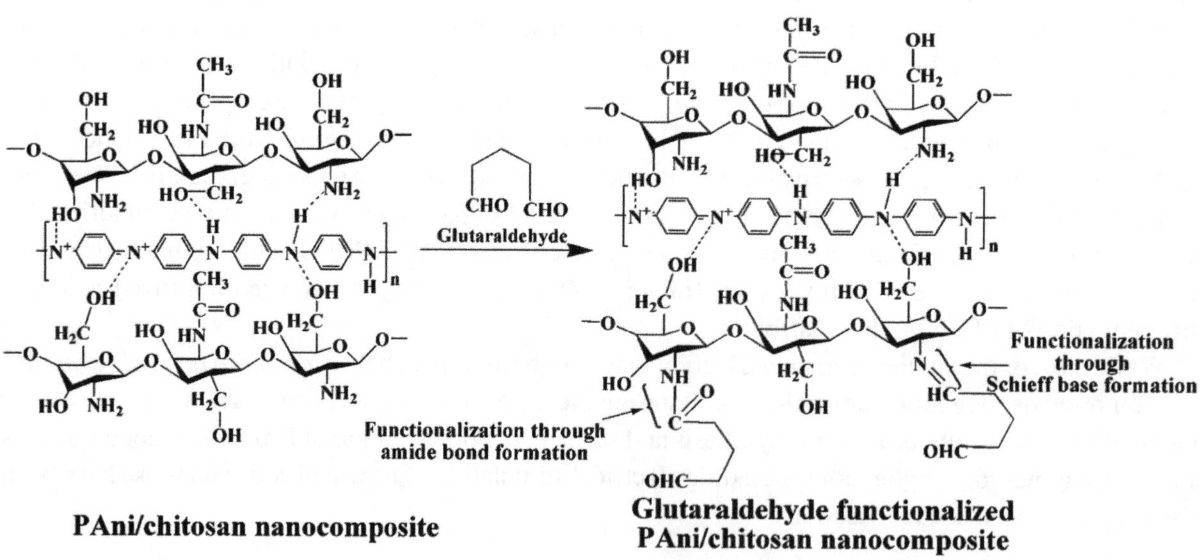

Figure 4. Scheme of surface functionalization of PAni:Ch nanocomposites using glycine NHS ester to introduce amine and carboxyl functionalities on the surface
(Borah, 2018)

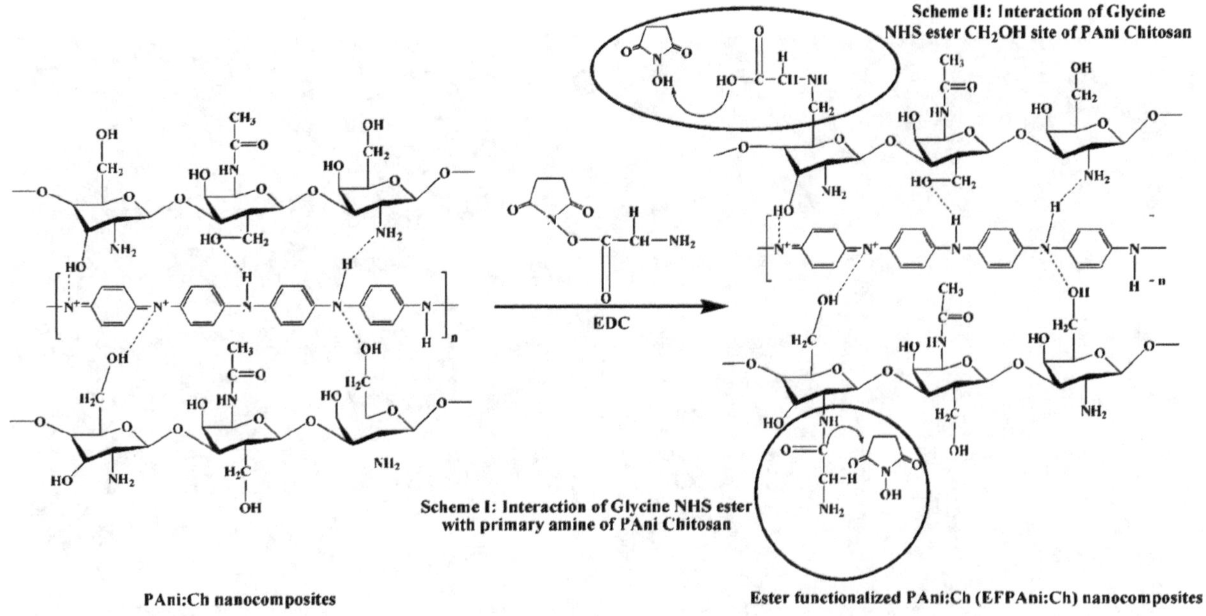

content slight decrease in the current at a particular applied voltage, which may be due to loss in $\pi-\pi$ conjugation in PAni chains during surface functionalization process. The results indicated that surface functionalization was performed without much affecting the conductive properties of the nanocomposites.

GLUTARALDEHYDE FUNCTIONALIZED POLYANILINE: CHITOSAN NANOCOMPOSITES

Hemocompatibility, also known as blood compatibility, is an important property of biomaterials, which strongly depends on the materials chosen. Therefore, materials intended for biomedical applications involving direct contact with living tissues, are to be checked for its blood compatibility. Hemolysis is the rupturing of RBCs or erythrocytes and thereby, releases the cytoplasmic materials, mainly hemoglobin from within the red blood cells into the blood plasma. Thus, the materials, which cause damage of host cytoplasmic membrane of RBCs and cell lysis and death, are known as hemolysins. The hemocompatibility of the non-functionalized and functionalized PAni:Ch nanocomposites has been tested in terms of their hemolysis activity with mammalian red bold cells (RBCs) and the results are presented in Figure 7. The glutaraldehyde functionalized nanocomposites, i.e., GFPAni:Ch-4 wt% and GFPAni:Ch-6 wt% demonstrate very less hemolysis activity of 0.86% ± 1.02 and 0.92% ± 1.52, which is nearly equal to the hemolysis activity exhibited by the negative control (0.41% ± 0.07), indicating non-hemolytic nature of the nanocomposites. On the other hand, the non-functionalized nanocomposites, i.e., PAni:Ch-4 wt% and PAni:Ch-6 wt% demonstrate hemolysis activity more than 5% suggesting hemolytic nature of the nanocomposites. To evaluate the biocompatibility of the non-functionalized and glutaraldehyde functionalized

Figure 5. Scanning electron micrographs of (a) pure chitosan (Ch), (b) powdered PAni nanofibers and PAni:Ch nanocomposites with (c) 4 wt% and (d) 6wt% of PAni content
(Kumar & Borah, 2016).
Reproduced from Kumar et al. by permission VBRI press

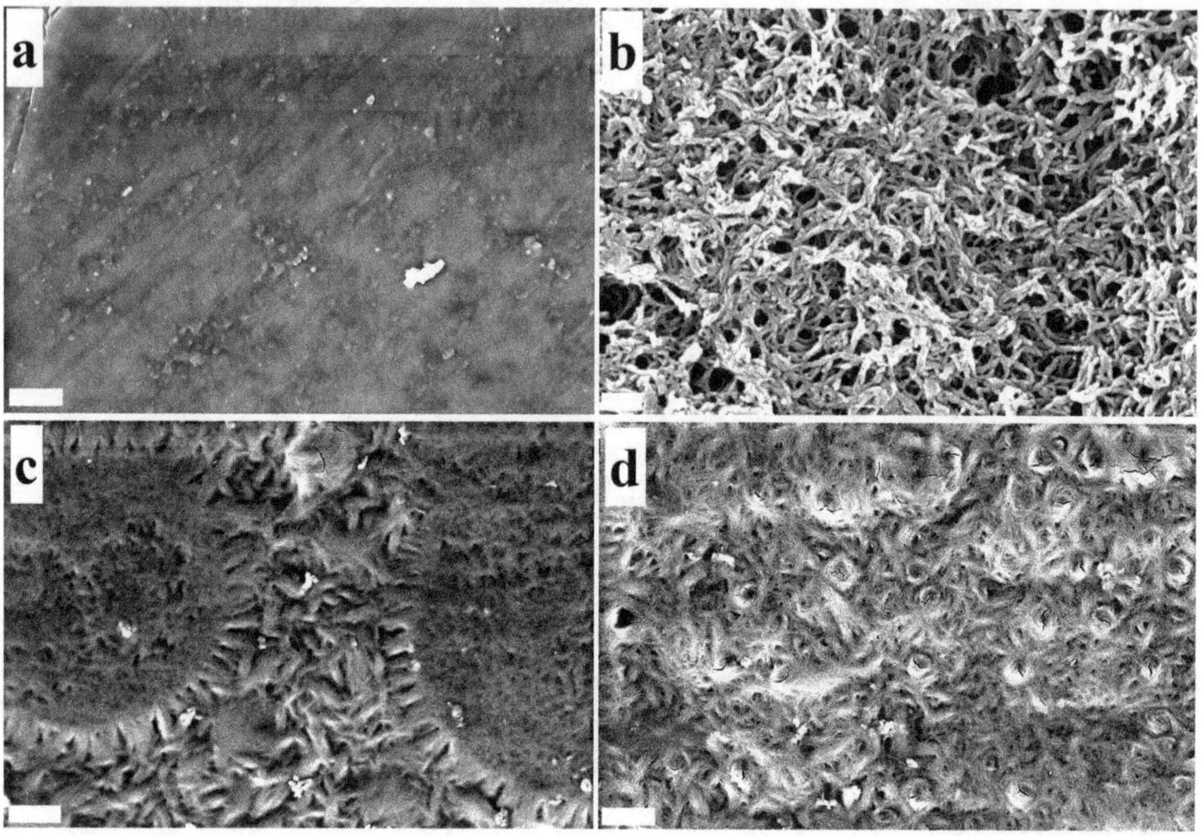

nanocomposites, MDA-MB-231 cell viability was measured in direct contact with the different nanocomposites films after 24 h of culture. This has been performed by MTS cell proliferation assay, where the metabolically active cells reduce the tetrazolium compound (MTS) into a water-soluble formazan product and the non-viable cells rapidly lose their ability to do that. The colored formazan product gives absorption maximum at 490 nm and the production of the colored formazan product is proportional to the number of viable cells (Barltrop, Owen, & Cory, 1991). The negative control is believed to induce maximum cell viability, whereas the positive control (*tert* butyl maleate) causes maximum cell death. The percentage of MDA-MB-231 cell viability on different nanocomposites films are shown in Figure 8 along with raw PAni nanofibers and Ch. The MTS results with MDA-MB-231 cells indicates significantly better cell viability and proliferation on glutaraldehyde functionalized nanocomposites than that on the non-functionalized nanocomposites and raw PAni nanofibers, at $p<0.05$ ($p=0.0046$). However, non-functionalized nanocomposites and pure Ch have been also found to be biocompatible as they support approximately 80% cell viability (Bae, Che, & Seo, 2012). It was also noted that nanocomposites with lower PAni content, i.e., 4 wt% nanocomposites (both non-functionalized & functionalized) show slightly better cell viability than that of nanocomposites with 6 wt% PAni content (both non-functionalized &

Figure 6. Room temperature (300 K) I-V characteristics of (a) GFPAni:Ch-4 wt% (red), (b) GFPAni:Ch-6 wt% (red), (c) EFPAni:Ch-4 wt% (red) and (d) EFPAni:Ch-6 wt% (red) with comparison to their non-functionalized counterparts (black)
(Borah, 2018)

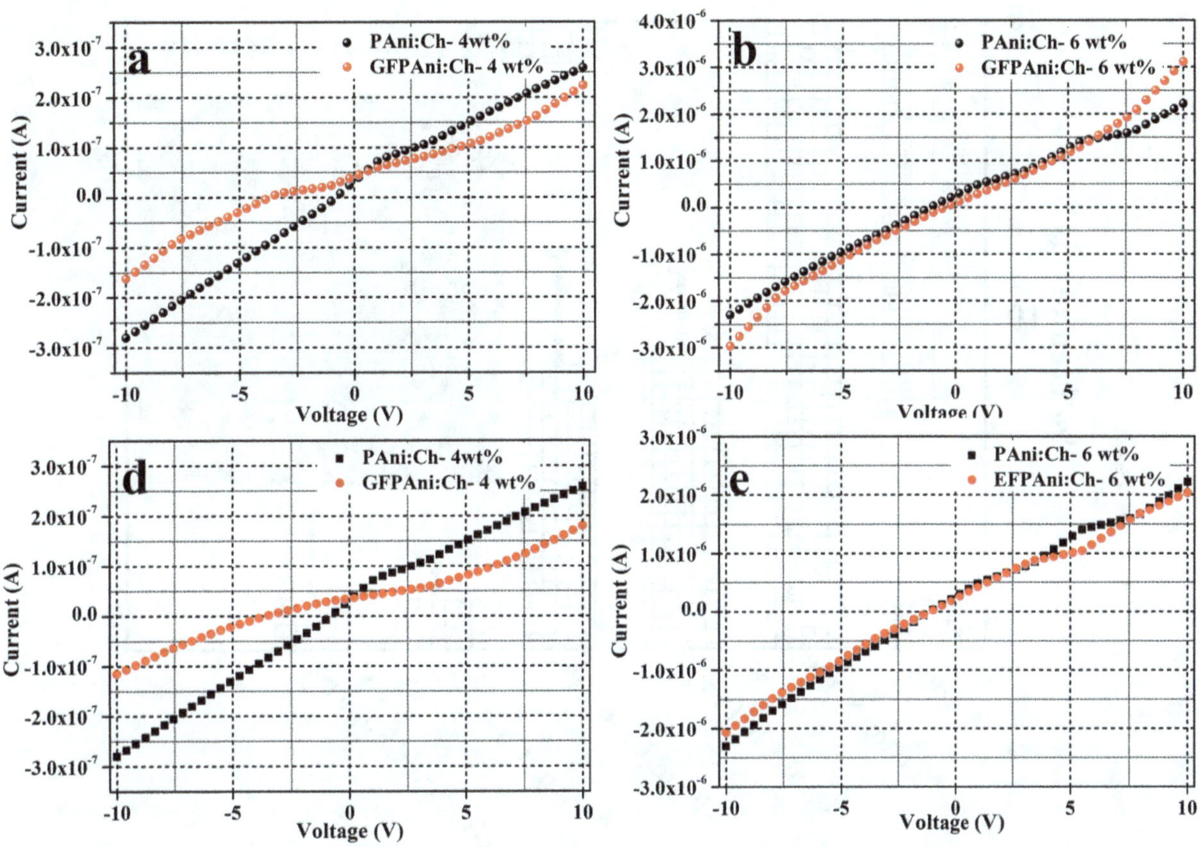

functionalized). Overall, the significant improvement in cell viability on glutaraldehyde functionalized nanocomposites supports the hypothesis that enhanced surface hydrophilicity conferred onto the surface of the nanocomposites significantly increases adhesive, metabolising cell fraction.

To visualize the MDA-MB-231 cells cultured on different PAni:Ch nanocomposites, the cells were stained with acridine orange (AO) and ethidium bromide (EtBr) for nucleic acid staining after 48 h of culture. The fluorescence micrographs of AO/EtBr stained MDA-MB-231 cells on the non-functionalized and glutaraldehyde functionalized nanocomposites films after 48 h of culture are shown in Figure 9. The staining can be visualized throughout the whole body of the MDA-MB-231 cells. It has been observed that the cells cultured on glutaraldehyde functionalized nanocomposites, i.e., GFPAni:Ch-4 wt% and GFPAni:Ch-6 wt% [Figure 9 (c) & (d)] possess elongated or polygonal morphology with higher cell counts, indicating healthier morphology than that of cultured on the non-functionalized nanocomposites, i.e., PAni:Ch-4 wt% and PAni:Ch-6 wt%. The cells cultured on the non-functionalized nanocomposites possess round morphology with irregular shapes [Figure 9 (a) & (b)]. To confirm the cell attachment, SEM of the MDA-MB-231 cells cultured on the different PAni:Ch nanocomposites after 48 h, were performed and the results are shown in Figure 5.23. The MDA-MB-231 cells adhere, spread well and

Figure 7. Percentage hemolysis (Mean ± S.D, n=4) shown by the non-functionalized and functionalized PAni:Ch nanocomposites as compared to negative control (Phosphate buffer saline, pH= 7.4) and positive control (Triton X 100). Inset shows photographs of RBC treatment with different materials as labeled (Borah, 2018)

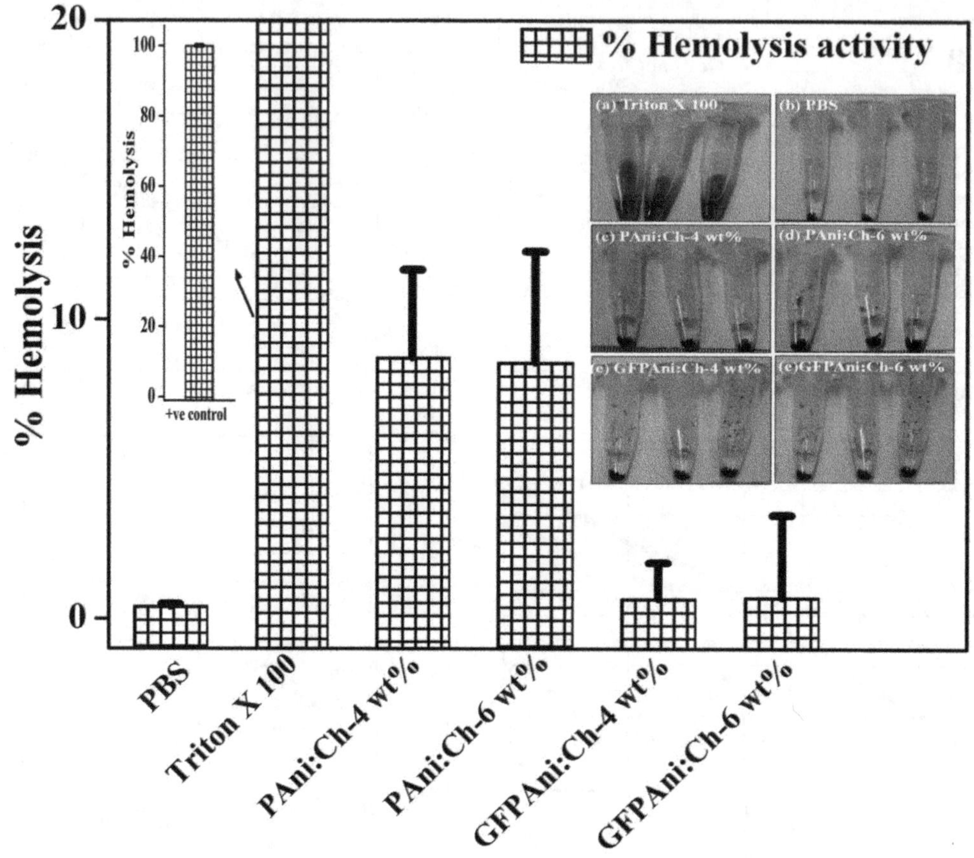

adopt elongated or polygonal morphology on the glutaraldehyde functionalized PAni:Ch nanocomposites [Figure 10 (c) & (d)] as compared to the round and irregular shapes of cells on the non-functionalized nanocomposites [Figure 10 (a) & (b)]. Thereby, SEM confirms the observation obtained by AO/EtBr staining. The enhanced cellular activities on the glutaraldehyde functionalized nanocomposites indicate the potential of these materials in tissue engineering applications.

GLYCINE NHS ESTER FUNCTIONALIZED POLYANILINE: CHITOSAN NANOCOMPOSITES

The hemolysis activity of the glycine NHS ester functionalized PAni:Ch nanocomposites were comparable to non-hemolytic control PBS (0.41% ± 0.07) as depicted in Figure 11. The enhanced surface hydrophilicity due to incorporation of polar functionalities such as carboxyl and amino groups and softer nature of the nanocomposites after surface functionalization (lower stiffness constant) contribute towards the lower hemolysis activity of the glycine NHS ester functionalized PAni:Ch nanocomposites, indicat-

Figure 8. Percentage of MDA-MB-231 cell viability on the non-functionalized and glutaraldehyde functionalized PAni:Ch nanocomposites in direct contact after 24 h of culture as compared to tissue culture plastic (TCP) as a negative control and tert butyl maleate as a positive control. Data were Mean ± S.D, n=3.≤0.05*
(Borah, 2018)

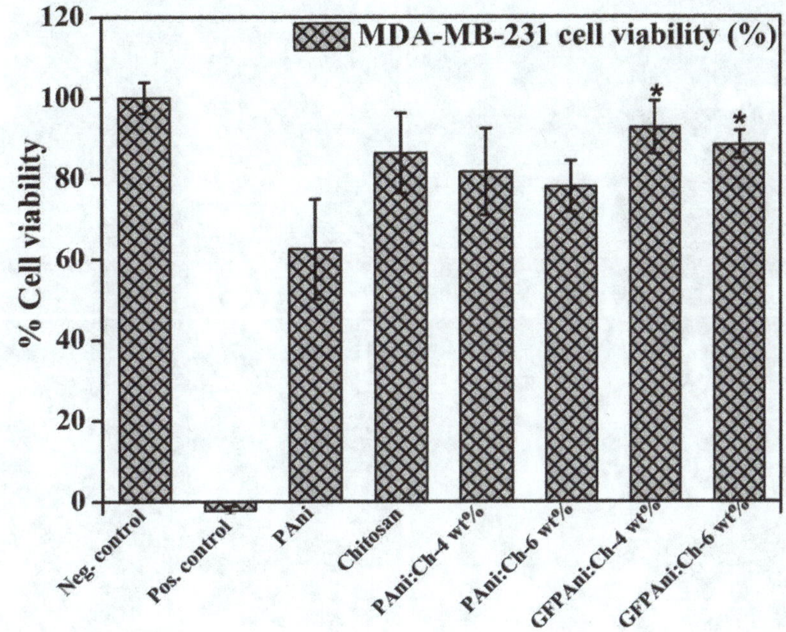

Figure 9. Fluorescence micrographs of AO/EtBr stained MDA-MB-231 cells on (a) PAni:Ch-4 wt%, (b) PAni:Ch-6 wt%, (c) GFPAni:Ch-4 wt% and (d) GFPAni:Ch-6 wt% after 48 h of culture
(Borah, 2018)

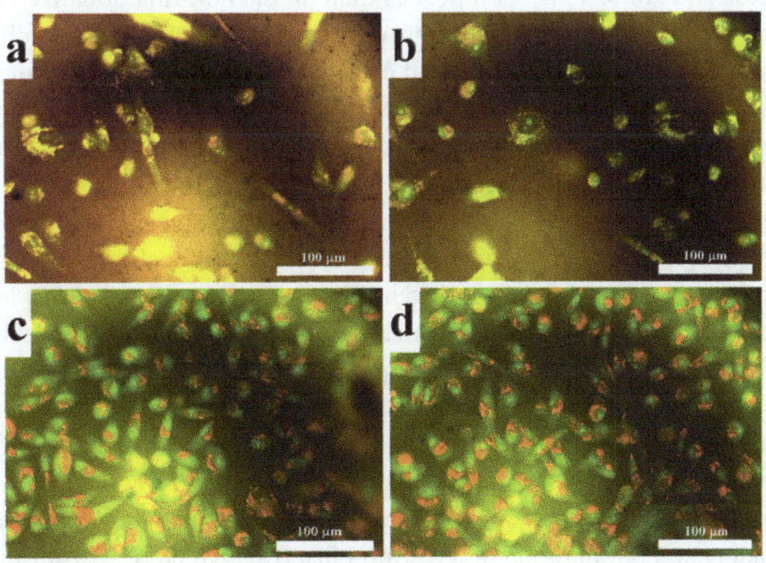

Figure 10. Scanning electron micrograph of MDA-MB-231 cells cultured on (a) PAni:Ch-4 wt% (Scale bar = 50 μm), (b) PAni:Ch-6 wt% (Scale bar = 20 μm), (c) GFPAni:Ch-4 wt% (Scale bar = 50 μm) and (d) GFPAni:Ch-6 wt% (Scale bar = 20 μm) after 48 h
(Borah, 2018)

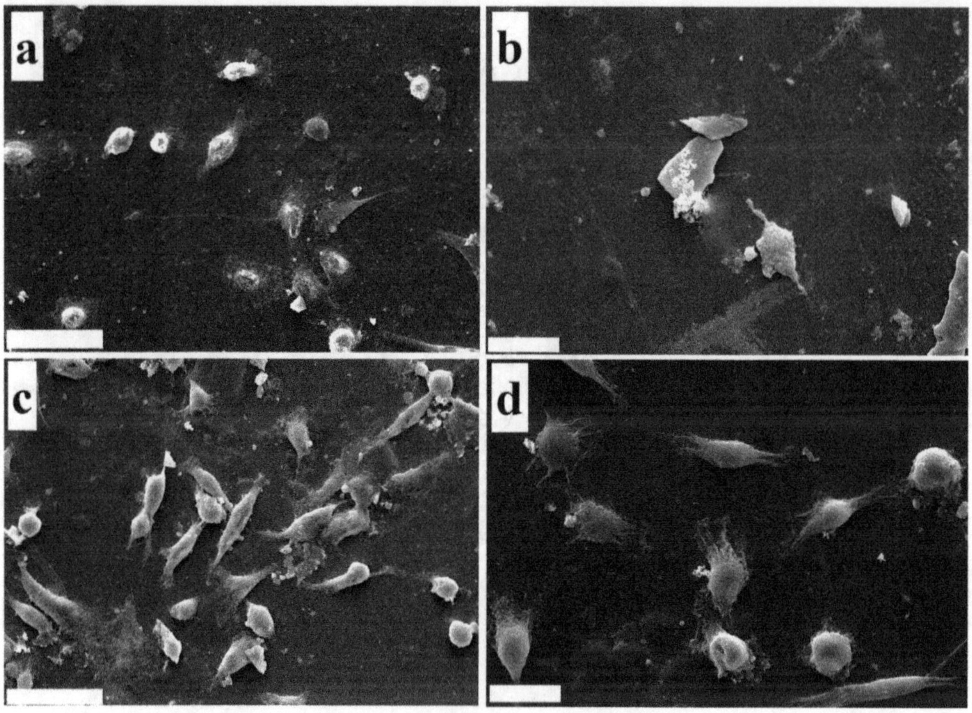

ing their improved blood compatibility (Upadhyay, Kumar, & Gupta, 2015). The MTS results with 3T3 fibroblasts and PC12 cells confirm that significantly better cell viability and proliferation on glycine NHS ester functionalized PAni:Ch nanocomposites than on the non-functionalized nanocomposites, at $p<0.05$ [Figure 12]. Overall, the significant improvement in cell viability on glycine NHS ester functionalized supports the hypothesis that amine and carboxyl functionality incorporated onto the surface of the nanocomposites significantly increases adhesive, metabolising cell fraction.

To better confirm about the cell adhesion, spreading and proliferation, 3T3 fibroblasts were cultured at lower concentration than that of the live/dead assay on the scaffolds as discussed above and stained with calcein AM after 24 h and 48 h of culture to monitor the changes in cell morphology with time. Quantitative analysis of cell morphology was carried out using confocal images of calcein AM stained cells seeded on different substrates for 24 h and 48 h [Figure 13 (a)], with relatively lower cell density so that analysis was restricted to only those cells, which were not in contact with more than one cell. Figure 13 (b) & (c) indicated that projected cell area and percent cell spreading were significantly higher for cells on glycine NHS ester functionalized nanocomposites compared to cells on the non-functionalized nanocomposites, at $p \leq 0.05$. Projected cell area for cells on functionalized nanocomposites is also significantly higher than the cells on TCP and Ch, at $p \leq 0.05$; whereas percent cell spreading for cells on modified nanocomposites, TCP and Ch are not significantly different. These observations demonstrate that adhesion, spreading and proliferation of 3T3 fibroblasts on glycine NHS ester functionalized nanocomposites occur more favorably compared to pristine nanocomposites. Quantitative analysis of cell

Figure 11. Percentage hemolysis (n = 4, mean ± S.D) shown by the non-functionalized and glycine NHS ester functionalized PAni:Ch nanocomposites as compared to negative control (Phosphate buffer saline, pH = 7.4) and positive control (Triton X 100). Inset shows photographs of RBC treatment with different materials as labeled
(Borah, 2018).

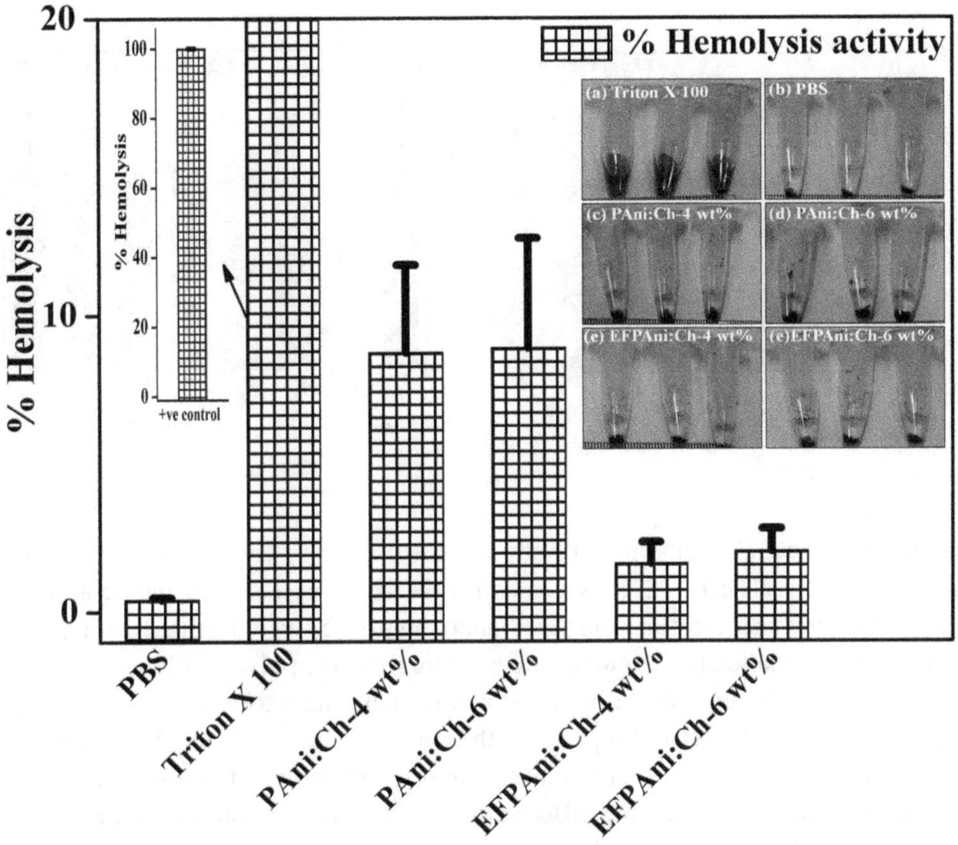

density confirms better cell proliferation on glycine NHS ester functionalized nanocomposites owing to improved cell adhesion and spreading, than that on the non-functionalized nanocomposites. Cell densities, after cultured for 24 h and 48 h, on EFPAni:Ch-4 wt% (79 ± 22 and 156 ± 45) and EFPAni:Ch-6 wt% (76 ± 28 and 137 ± 41) statistically differ from that of PAni:Ch-4 wt% (43±17 and 71±19) and PAni:Ch-6 wt% (48 ± 19 and 77 ± 18) at $p \leq 0.05$.

The PC12 cells were cultured for 7 days in differentiating medium on different PAni:Ch nanocomposites films with and without collagen I coating to investigate the suitability of these materials for neural tissue engineering. This cell line has been widely used as a model neuronal system. PC12 cells cultured with NGF (Nerve growth factor) develop long neurite outgrowth, become electrically excitable and take on many of the biochemical traits of sympathetic noradrenergic neurons (Cheever & Koshland, 1992). Since, PC12 cells readily adhere to collagen, all the electrospun meshes were coated with collagen I. Collagen I is a fibril forming collagen present in the ECM of the peripheral nervous system (PNS) and plays an important role in the development of the peripheral nervous system as well as in the maintenance of normal peripheral nerve function during adulthood (Koopmans, Hasse, & Sinis, 2009).

*Figure 12. Percentage cell viability after 24 h of culture on glycine NHS ester functionalized nanocomposites compared to non-functionalized nanocomposites, raw materials (n = 4, mean ± SD) with (a) 3T3 fibroblasts and (b) PC12 cells. *Significant differences between negative control and test materials at p<0.05; #Significant differences between unmodified nanocomposites and modified nanocomposites at p<0.05*
(Borah, 2018).

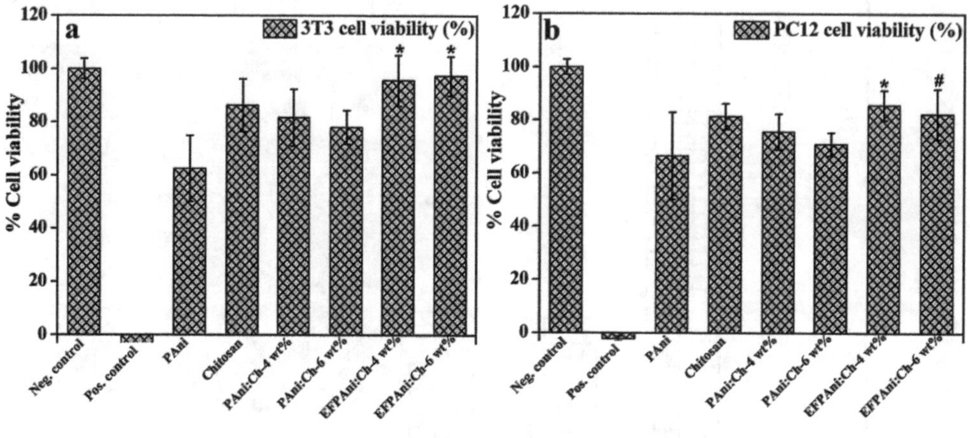

Immunolabelling with beta (III) tubulin antibody was performed to confirm PC12 cell differentiation into sympathetic neurons. The PC12 cells grown on the collagen I coated nanocomposites, glycine NHS ester functionalized and non-functionalized nanocomposites were labelled with beta (III) tubulin antibody to visualize cytoskeletal microtubules, which are dynamic polymer filaments of alpha and beta tubulin subunits that drive neurite outgrowth and control neuronal morphology (Gladwin, Whitby, & Mikhalovsky, 2013). Figure 14 (a1-b3) represents the confocal images of beta III tubulin stained PC12 cells cultured on different substrates after seven days to study the effect of surface functionalization and collagen coating of substrates on neuronal differentiation. This observation reveals that collagen coating affects adhesion of PC12 cells significantly and the surface functionalization by glycine NHS ester has no effect on PC12 cell adhesion. The results demonstrate that the average neurite lengths per cell on the collagen coated PAni:Ch nanocomposites films are statistically different from those of all uncoated nanocomposites films, at $p \leq 0.05$.

Analysis of neurite formation shows that percent of neurite bearing cells on collagen coated PAni:Ch-4 wt% (34% ± 15), PAni:Ch-6 wt% (35% ± 9) are significantly greater that of uncoated PAni:Ch-4 wt% (12% ± 5), PAni:Ch-6 wt% (10% ± 4), EFPAni:Ch-4 wt% (18% ± 6) and EFPAni:Ch (15% ± 10) at $p \leq 0.05$ [Figure 14 (c)]. The average neurite length per cell as presented in Figure 14 (d) were as: 57 ± 4 µm (Collagen coated PAni:Ch-4 wt%), 53 ± 8 µm (Collagen coated PAni:Ch-6 wt%), 22 ± 7µm (uncoated PAni:Ch-4 wt%), 19 ± 5 µm (uncoated PAni:Ch-6 wt%), 32 ± 5 µm (uncoated EFPAni:Ch-4 wt%) and 28 ± 6 µm (uncoated EFPAni:Ch-6 wt%). However, there was no significant difference on neurite formation and neurite outgrowth on the collagen coated nanocomposites films and collagen coated glass. It shows that PC12 cells are differentiated well only on collagen coated PAni:Ch nanocomposites indicating potential of these materials for electrical stimulation of PC12 cells.

The electrical stimulation of differentiated PC12 cells was carried out though collagen coated PAni:Ch nanocomposites using bespoke electrical stimulation set up as shown in Figure 15 (a-c). A constant

*Figure 13. (a) Representative confocal images of calcein AM stained viable cells seeded on different scaffolds (labeled in the photographs) for 24 h and 48 h, used for analysis of cell morphology(Scale bar = 75 µm). (b) Histogram showing area covered by single cell on different scaffolds after 24 h and 48 h. c. Histogram of percent of cell spreading on different scaffolds after 24 h and 48 h. *indicates significant differences between the non-functionalized and functionalized nanocomposites at p≤0.05*
(Borah, 2018).

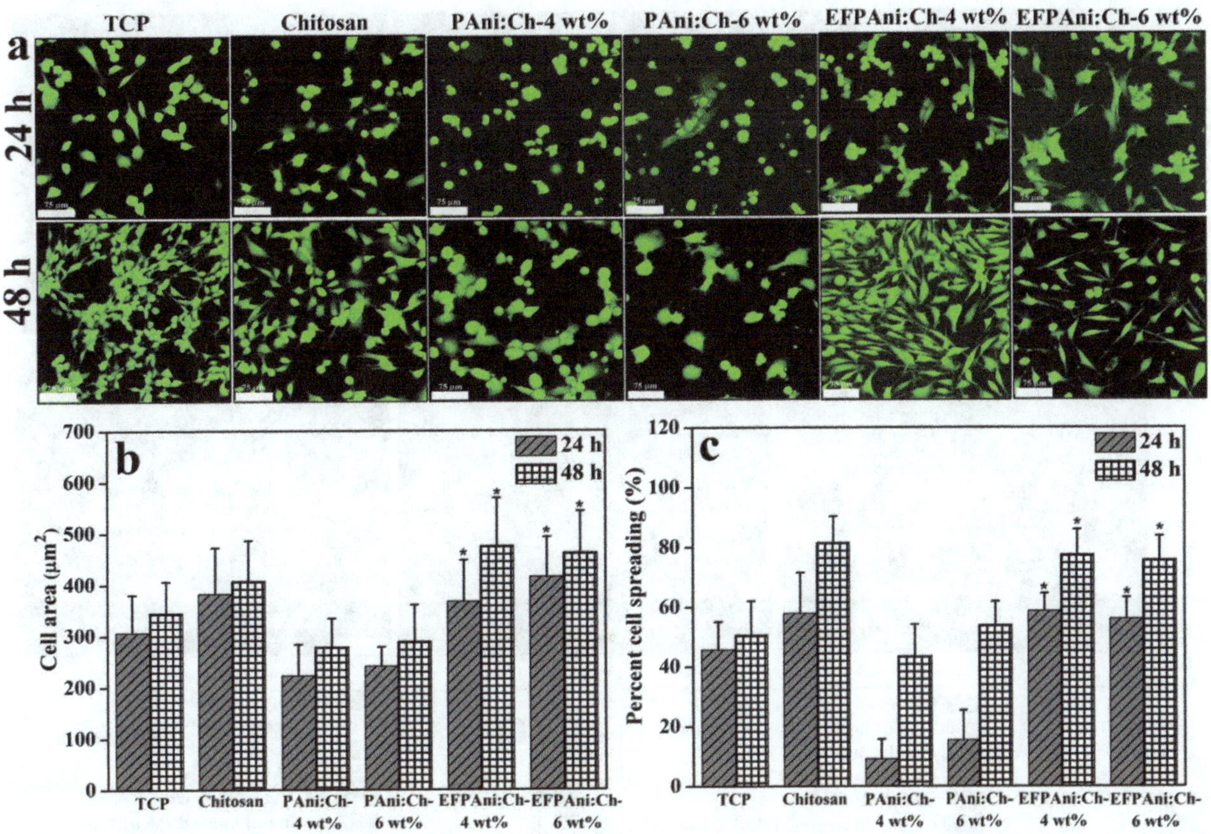

electrical potential of 500 mV/cm was applied across the electrodes for 2 h/day for 3 consecutive days using a portable bipotentiostat (EmStat Blue, PalmSens BV, Netherlands) [Figure 15 (a)]. The electrical stimulation was carried out by a double pulsed potential chronoamperometric technique in an incubator [Figure 15 (c)]. For comparison, PC12 cells on coated PAni:Ch nanocomposites without electrical stimulation are also cultured under the same condition for 7 days treated as control. Representative confocal images of the electrically stimulated and unstimulated PC12 cells on collagen coated various PAni:Ch nanocomposites are shown in Figure 15 (a1-b2). It appears from the confocal images that more neurite formation on electrically stimulated PC12 cells on collagen coated PAni:Ch nanocomposites as compared to the unstimulated PC12 cells. Therefore, the effect of electrical stimulation of PC12 cells through the conductive nanocomposites has been assessed in terms of quantitative analysis of percentage of neurite bearing cells, neurite per cell, neurite length per cell and median neurite lengths from the confocal images of beta tubulin stained PC12 cells on electrospun meshes and the results are shown in [Figure 15 (c-f)].

Figure 14. Beta (III) tubulin immunostaining of PC12 cells cultured for 7 days on non-functionalized PAni:Ch-4 wt% (a1) and PAni:Ch-6 wt% (b1), EFPAni:Ch-4 wt% (a2), EFPAni:Ch-6 wt% (b2) and collagen I coated PAni:Ch-4 wt% (a3) and PAni:Ch-6 wt% (b3) [Scale bar = 75 μm]. Inset of a1 shows the immnulabelling of PC12 cells cultured on collagen coated glass cover slip. Quantitative analysis of differentiated PC12 cells: (c) % neurite bearing cells and (d) average neurite length per cell
(Borah, 2018).

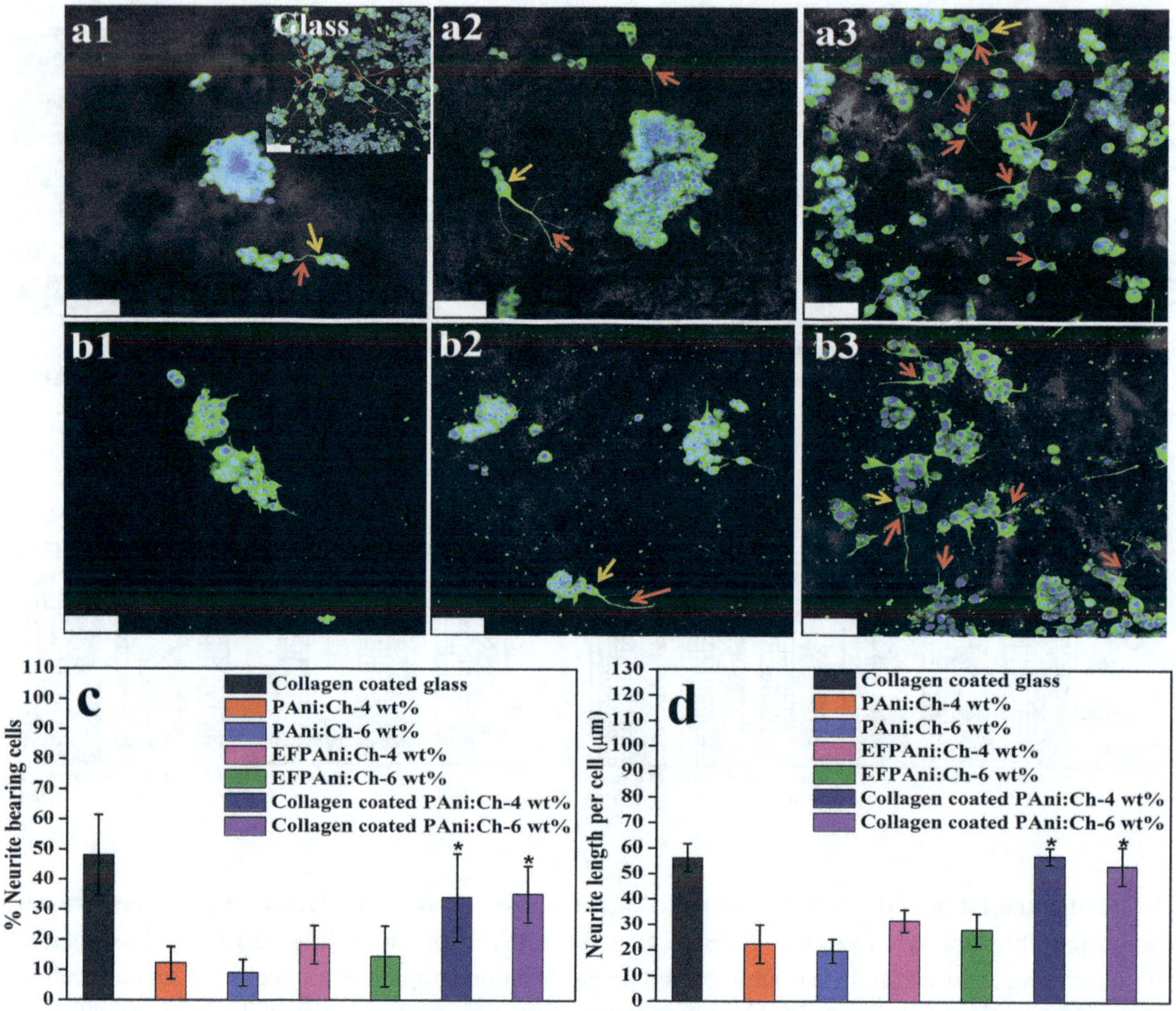

Quantitative analysis indicates that electrical stimulation for 2 h/day for 3 consecutive days through conductive PAni:Ch nanocomposites scaffolds induced enhancement in the neurite formation and neurite outgrowth. The electrically stimulated PC12 cells have significantly greater number of neurite bearing cells on coated PAni:Ch-4 wt% (62% ± 13, N=388, m=318) and PAni:Ch-6 wt% (66% ± 17, N=413, m=356) when compared to the unstimulated cells on coated PAni:Ch-4 wt% (41% ± 10, N=364, m=209) and PAni:Ch-6 wt% (39% ± 8, N=359, m=194), at $p \leq 0.05$, where N and m denote number of cells and number of neurites analyzed [Figure 16 (c)]. Neurite lengths per cell and median neurite lengths of

Figure 15. (a) Schematic illustration of the electrical stimulation experiment using a custom made electrical stimulation set up ; (b) photograph of self made cell culture plate with PAni:Ch nanocomposites with 4 wt% and 6 wt% PAni content (dark blue colur) fixed on it for electrical stimulation experiment; (c) photograph of the electrical stimulation experiment in situ
(Borah, 2018).

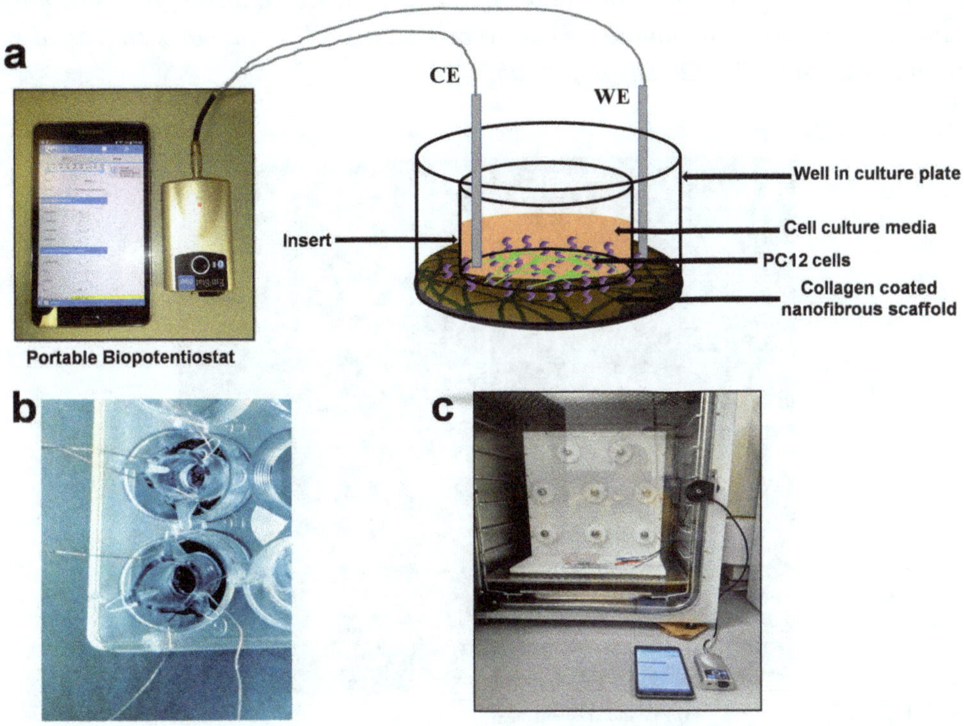

electrically stimulated cells on coated PAni:Ch-4 wt% (117 ± 23 and 70 ± 19 μm) and PAni:Ch-6 wt% (138 ± 36 and 73 ± 14 μm) were statistically different ($p \leq 0.05$) from unstimulated cells on coated PAni:Ch-4 wt% (61 ± 14 and 42 ± 15 μm), and PAni:Ch-6 wt% (55 ± 12 and 37 ± 12 μm) [Figure 16 (d) & (e)]. Analysis of neurites per cell reveals that each electrically stimulated PC12 cells has more number of neurites on coated PAni:Ch-4 wt% (1.85) and PAni:Ch-6 wt% (1.97) than those of unstimulated cells on coated PAni:Ch-4 wt% (1.16) and PAni:Ch-6 wt% (1.09) [Figure 16 (f)]. The results confirm that PC12 cells under electrical stimulation had more neurites per cell than those grown under no electrical stimulation and the results statistically different at $p \leq 0.05$. Ultimately, it can be concluded that that electrical stimulation of PC12 cells through these conductive nanocomposites scaffolds has a beneficial effect on neurite formation and neurite outgrowth. Moreover, it has been noticed that PAni:Ch-6 wt%, having lower sheet resistance as compared to that of PAni:Ch-4 wt%, provides improved differentiation of PC12 cells with more neurite bearing cells, neurite length per cell and median neurite length than those of PAni:Ch-4 wt%. Interestingly, this difference is statistically significant at $p \leq 0.01$. This significant change in neurite formation and neurite outgrowth on coated PAni:Ch-6 wt% under electrical stimulation is referred to the improved conductive behaviour of the nanocomposites due to higher conductive PAni content. It is also evident from the current signals during stimulation, which shows that more current flow through coated PAni:Ch-6 wt% (~30 μA) than PAni:Ch-4 wt% (~3 μA). Thus, the neurite

*Figure 16. Confocal images with phase contrast overlay of beta (III) tubulin immunostained PC12 cells cultured for 7 days under no electrical stimulation (a1-collagen coated PAni:Ch-4 wt%, b1-collagen coated PAni:Ch-6 wt%) and under electrical stimulation of 500 mV/cm for 2h/day for 3 cosecutive days (a2-collagen coated PAni:Ch-4 wt%, b2-collagen coated PAni:Ch-6 wt%) [Scale bar = 75 μm]; (c) Percentage of neurite bearing cells, (d) Neurite length per cell, (e) Median neurite length and (f) Neurite per cell of differentiated PC12 cells on collagen coated PAni:Ch nanocomposites without electrical stimulation and with electrical stimulation. Data were Mean ± S.D. *indicates statistical significance difference from unstimulated PC12 cells at p≤0.05*
(Borah, 2018).

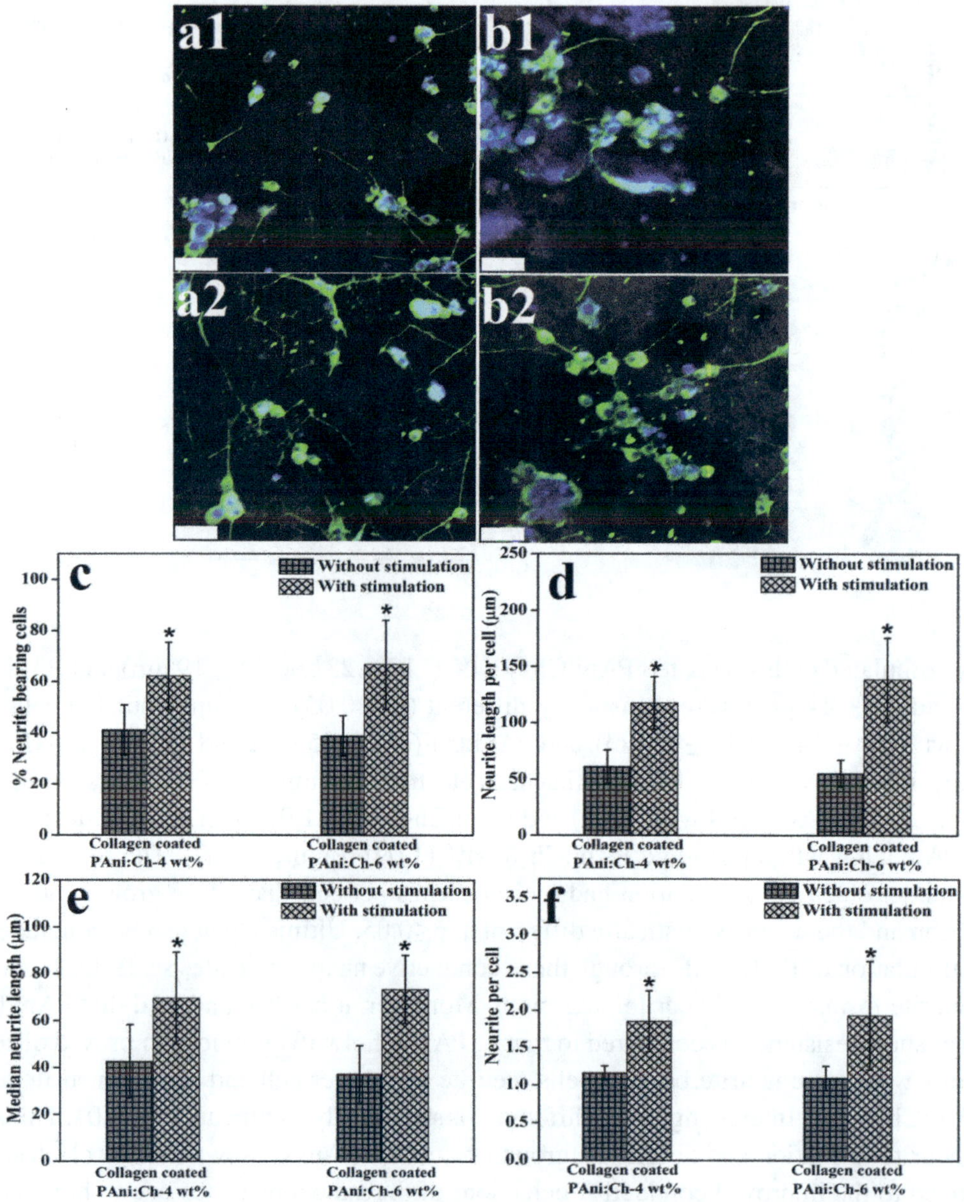

formation and outgrowth increases with an increase in the current signal. Under electrical stimulation of 500 mV/cm, the availability of free charge carriers in PAni:Ch nanocomposites (more in PAni:Ch-6 wt%) offers more charge-transport between the scaffold and the cell membrane, resulting in the change in the resting membrane potential of differentiated PC12 cells. Consequently, the cell membrane undergoes an intensity-dependent depolarization resulting action potential, which is responsible for axonal growth. This change in cell membrane potential is also believed to activate growth-controlling transport processes across the plasma membrane and cause redistribution of cytoplasmic materials (Lee, Bashur, & Goldstein, 2009; Haastert-Talini & Grothe, 2013; Rajnicek, Robinson, & McCaig, 1998; McCaig & Zhao, 1997). According to Patel et al., ES causes electrophoretic accumulation of surface molecules on the working electrode (scaffold). All these are likely to be responsible for longer neurite growth or cell–substratum adhesion.

CONCLUSION

The detailed literature review on tissue engineering applications of CP based composites with Ch demonstrates that composites of PAni, PPy and PEDOT with Ch are biocompatible enough to mimic the natural ECM both *in vitro*. However, there are only a few studies *in vivo*. Taking the initial indication, there are scopes to carry similar research further for practical realization of the application. The author and his group have developed an electrically conductive, biocompatible, porous, nanofibrous, bioactive and mechanically strong biomaterial scaffolds of PAni:Ch nanocomposites with detailed physicochemical and biological characterization of these materials. It has been shown for the first time that surface functionalization as an attractive technique to improve cell-biomaterial interaction using simplest wet chemical method through incorporation of polar functionalities such as carboxyl (-COOH), amine (-NH_2), aldehyde (-CHO), hydroxyl (-OH) groups. It has the potential to replace the need of coating the biomaterial scaffold with costly biomolecules such as collagen, laminin, fibronectin etc. The combined effect of nanofiber feature, surface amination and electrical stimulation along with good electrical conductivity, mechanical stability, porosity makes it a potential alternative to the conventional technique for nerve regeneration. However, there are further scopes in research using these materials regarding the following concerns:

- There is still need of some more research in the effect of surface modification on neuronal growth.
- Optimization of electrical stimulation with different form of electrical signal is still undone using these materials.
- Biodegradability of these materials is still unexplored.
- May become suitable material for nerve tissue engineering.

ACKNOWLEDGMENT

The authors are very grateful Mr. Oukil Khaled Ibn Elwalid for his grateful help. In addition, we are grateful to all the publishers, including Advanced Materials Proceedings and Tezpur University to reuse in our review illustrations issued from their publications.

REFERENCES

Ahuja, A. (2017, November 26). Lack of organ donation in India: Here is why half a million people die annually in India due to unavailability of organs. *NDTV*. Retrieved from http://sites.ndtv.com/moreto-give/lack-organ-donation-india-half-million-people-die-annually-india-due-unavailability-organs-2107/

Al-Majed, A. A., Neumann, C. M., Brushart, T. M., & Gordon, T. (2000). Brief electrical stimulation promotes the speed and accuracy of motor axonal regeneration. *The Journal of Neuroscience*, *20*(7), 2602–2608. doi:10.1523/JNEUROSCI.20-07-02602.2000 PMID:10729340

Ansari, R., & Keivani, M. B. (2006). Polyaniline conducting electroactive polymers thermal and environmental stability studies. *Journal of Chemistry*, *3*(4), 202–217.

Ateh, D. D., Navsaria, H. A., & Vadgama, P. (2006). Polypyrrole-based conducting polymers and interactions with biological tissues. *Journal of the Royal Society, Interface*, *3*(11), 741–752. doi:10.1098/rsif.2006.0141 PMID:17015302

Ayad, M. M., Amer, W. A., & Kotp, M. G. (2017). Magnetic polyaniline-chitosan nanocomposite decorated with palladium nanoparticles for enhanced catalytic reduction of 4-nitrophenol. *Molecular Catalysis*, *439*, 72–80. doi:10.1016/j.mcat.2017.06.023

Ayad, M. M., Amer, W. A., Kotp, M. G., Minisy, I. M., Rehab, A. F., Kopecký, D., & Fitl, P. (2017). Synthesis of silver-anchored polyaniline–chitosan magnetic nanocomposite: A smart system for catalysis. *RSC Advances*, *7*(30), 18553–18560. doi:10.1039/C7RA02575K

Bae, S. H., Che, J. H., Seo, J. M., Jeong, J., Kim, E. T., Lee, S. W., ... Chung, H. (2012). *In vitro* biocompatibility of various polymer-based microelectrode arrays for retinal prosthesis. *Investigative Ophthalmology & Visual Science*, *53*(6), 2653–2657. doi:10.1167/iovs.11-9341 PMID:22427592

Balint, R., Cassidy, N. J., & Cartmell, S. H. (2014). Conductive polymers: Towards a smart biomaterial for tissue engineering. *Acta Biomaterialia*, *10*(6), 2341–2353. doi:10.1016/j.actbio.2014.02.015 PMID:24556448

Baniasadi, H., SA, A. R. & Mashayekhan, S. (2015). Fabrication and characterization of conductive chitosan/gelatin-based scaffolds for nerve tissue engineering. *International Journal of Biological Macromolecules*, *74*, 360–366. doi:10.1016/j.ijbiomac.2014.12.014 PMID:25553968

Bao, Z., Dodabalapur, A., & Lovinger, A. J. (1996). Soluble and processable regioregular poly (3-hexylthiophene) for thin film field-effect transistor applications with high mobility. *Applied Physics Letters*, *69*(26), 4108–4110. doi:10.1063/1.117834

Barltrop, J. A., Owen, T. C., Cory, A. H., & Cory, J. G. (1991). 5-(3-carboxymethoxyphenyl)-2-(4, 5-dimethylthiazolyl)-3-(4-sulfophenyl) tetrazolium, inner salt (MTS) and related analogs of 3-(4, 5-dimethylthiazolyl)-2, 5-diphenyltetrazolium bromide (MTT) reducing to purple water-soluble formazans as cell-viability indicators. *Bioorganic & Medicinal Chemistry Letters*, *1*(11), 611–614. doi:10.1016/S0960-894X(01)81162-8

Bendorf, A., Kelly, P. J., Kerridge, I. H., McCaughan, G. W., Myerson, B., Stewart, C., & Pussell, B. A. (2013). An international comparison of the effect of policy shifts to organ donation following cardiocirculatory death (DCD) on donation rates after brain death (DBD) and transplantation rates. *PLoS One*, *8*(5), e62010. doi:10.1371/journal.pone.0062010 PMID:23667452

Bhang, S. H., Jeong, S. I., Lee, T. J., Jun, I., Lee, Y. B., Kim, B. S., & Shin, H. (2012). Electroactive Electrospun Polyaniline/Poly [(L-lactide)-co-(ε-caprolactone)] Fibers for Control of Neural Cell Function. *Macromolecular Bioscience*, *12*(3), 402–411. doi:10.1002/mabi.201100333 PMID:22213547

Bidez, P. R., Li, S., MacDiarmid, A. G., Venancio, E. C., Wei, Y., & Lelkes, P. I. (2006). Polyaniline, an electroactive polymer, supports adhesion and proliferation of cardiac myoblasts. *Journal of Biomaterials Science. Polymer Edition*, *17*(1-2), 199–212. doi:10.1163/156856206774879180 PMID:16411609

Bober, P., Humpolíček, P., Pacherník, J., Stejskal, J., & Lindfors, T. (2015). Conducting polyaniline based cell culture substrate for embryonic stem cells and embryoid bodies. *RSC Advances*, *5*(62), 50328–50335. doi:10.1039/C5RA07504A

Borah, R. (2018). Synthesis and surface functionalization of conducting polymer nanostructures for biomedical applications [Ph.D. Thesis]. Tezpur University.

Bu, Y., Xu, H. X., Li, X., Xu, W. J., Yin, Y. X., Dai, H. L., ... Xu, P. H. (2018). A conductive sodium alginate and carboxymethyl chitosan hydrogel doped with polypyrrole for peripheral nerve regeneration. *RSC Advances*, *8*(20), 10806–10817. doi:10.1039/C8RA01059E

Chai, C., & Leong, K. W. (2007). Biomaterials approach to expand and direct differentiation of stem cells. *Molecular Therapy*, *15*(3), 467–480. doi:10.1038j.mt.6300084 PMID:17264853

Chandy, T., & Sharma, C. P. (1990). Chitosan-as a biomaterial. *Biomaterials, Artificial Cells, and Artificial Organs*, *18*(1), 1–24. doi:10.3109/10731199009117286 PMID:2185854

Cheever, L., & Koshland, D. E. (1992). Retention of habituation in PC12 cells. *Proceedings of the National Academy of Sciences of the United States of America*, *89*(21), 10084–10088. doi:10.1073/pnas.89.21.10084 PMID:1438197

Chen, G., Ushida, T., & Tateishi, T. (2002). Scaffold design for tissue engineering. *Macromolecular Bioscience*, *2*(2), 67–77. doi:10.1002/1616-5195(20020201)2:2<67::AID-MABI67>3.0.CO;2-F

Chen, Z., Mo, X., & Qing, F. (2007). Electrospinning of collagen–chitosan complex. *Materials Letters*, *61*(16), 3490–3494. doi:10.1016/j.matlet.2006.11.104

Chiou, N. R. (2006). Aligned and oriented polyaniline nanofibers: fabrication and applications [PhD thesis]. The Ohio State University.

Chiou, N. R., & Epstein, A. J. (2005). Polyaniline nanofibers prepared by dilute polymerization. *Advanced Materials*, *17*(13), 1679–1683. doi:10.1002/adma.200401000

Chung, T. W., Lai, D. M., Chen, S. D., & Lin, Y. I. (2014). Poly (ε-caprolactone) scaffolds functionalized by grafting NGF and GRGD promote growth and differentiation of PC12 cells. *Journal of Biomedical Materials Research. Part A*, *102*(2), 315–323. doi:10.1002/jbm.a.34693 PMID:23468336

Contractor, A. Q., Sureshkumar, T. N., Narayanan, R., Sukeerthi, S., Lal, R., & Srinivasa, R. S. (1994). Conducting polymer-based biosensors. *Electrochimica Acta*, *39*(8-9), 1321–1324. doi:10.1016/0013-4686(94)E0054-4

CountryFocus. (2017). *Healthcare*. GlobalData, France: Regulatory and Reimbursement Landscape-France.

Cui, X., Lee, V. A., Raphael, Y., Wiler, J. A., Hetke, J. F., Anderson, D. J., & Martin, D. C. (2001). Surface modification of neural recording electrodes with conducting polymer/biomolecule blends. *Journal of Biomedical Materials Research. Part A*, *56*(2), 261–272. doi:10.1002/1097-4636(200108)56:2<261::AID-JBM1094>3.0.CO;2-I PMID:11340598

Cui, Z., Ni, N. C., Wu, J., Du, G. Q., He, S., Yau, T. M., ... Li, R. K. (2018). Polypyrrole-chitosan conductive biomaterial synchronizes cardiomyocyte contraction and improves myocardial electrical impulse propagation. *Theranostics*, *8*(10), 2752–2764. doi:10.7150/thno.22599 PMID:29774073

Daemen, J., & Serruys, P. W. (2007). Drug-eluting stent update 2007: part II: unsettled issues. *Circulation*, *116*(8), 961–968. doi:10.1161/CIRCULATIONAHA.107.691451 PMID:17709651

Di Martino, A., Sittinger, M., & Risbud, M. V. (2005). Chitosan: A versatile biopolymer for orthopaedic tissue-engineering. *Biomaterials*, *26*(30), 5983–5990. doi:10.1016/j.biomaterials.2005.03.016 PMID:15894370

Elbert, D. L., & Hubbell, J. A. (1996). Surface treatments of polymers for biocompatibility. *Annual Review of Materials Science*, *26*(1), 365–394. doi:10.1146/annurev.ms.26.080196.002053

Esrafilzadeh, D., Jalili, R., Liu, X., Gilmore, K. J., Razal, J. M., Moulton, S. E., & Wallace, G. G. (2016). A novel and facile approach to fabricate a conductive and biomimetic fibrous platform with sub-micron and micron features. *Journal of Materials Chemistry. B, Materials for Biology and Medicine*, *4*(6), 1056–1063. doi:10.1039/C5TB02237A

Farshi Azhar, F., Olad, A., & Salehi, R. (2014). Fabrication and characterization of chitosan–gelatin/nanohydroxyapatite–polyaniline composite with potential application in tissue engineering scaffolds. *Designed Monomers and Polymers*, *17*(7), 654–667. doi:10.1080/15685551.2014.907621

Frackowiak, E., Khomenko, V., Jurewicz, K., Lota, K., & Béguin, F. (2006). Supercapacitors based on conducting polymers/nanotubes composites. *Journal of Power Sources*, *153*(2), 413–418. doi:10.1016/j.jpowsour.2005.05.030

Freed, L. E., Marquis, J. C., Nohria, A., Emmanual, J., Mikos, A. G., & Langer, R. (1993). Neocartilage formation in vitro and in vivo using cells cultured on synthetic biodegradable polymers. *Journal of Biomedical Materials Research. Part A*, *27*(1), 11–23. doi:10.1002/jbm.820270104 PMID:8380593

Gerard, M., Chaubey, A., & Malhotra, B. D. (2002). Application of conducting polymers to biosensors. *Biosensors & Bioelectronics*, *17*(5), 345–359. doi:10.1016/S0956-5663(01)00312-8 PMID:11888724

Ghasemi-Mobarakeh, L., Prabhakaran, M. P., Morshed, M., Nasr-Esfahani, M. H., Baharvand, H., Kiani, S., ... Ramakrishna, S. (2011). Application of conductive polymers, scaffolds and electrical stimulation for nerve tissue engineering. *Journal of Tissue Engineering and Regenerative Medicine*, *5*(4), e17–e35. doi:10.1002/term.383 PMID:21413155

Ghasemi-Mobarakeh, L., Prabhakaran, M. P., Morshed, M., Nasr-Esfahani, M. H., & Ramakrishna, S. (2009). Electrical stimulation of nerve cells using conductive nanofibrous scaffolds for nerve tissue engineering. *Tissue Engineering. Part A*, *15*(11), 3605–3619. doi:10.1089/ten.tea.2008.0689 PMID:19496678

Gilmore, K. J., Kita, M., Han, Y., Gelmi, A., Higgins, M. J., Moulton, S. E., ... Wallace, G. G. (2009). Skeletal muscle cell proliferation and differentiation on polypyrrole substrates doped with extracellular matrix components. *Biomaterials*, *30*(29), 5292–5304. doi:10.1016/j.biomaterials.2009.06.059 PMID:19643473

Gladwin, K. M., Whitby, R. L., Mikhalovsky, S. V., Tomlins, P., & Adu, J. (2013). *In vitro* biocompatibility of multiwalled carbon nanotubes with sensory neurons. *Advanced Healthcare Materials*, *2*(5), 728–735. doi:10.1002/adhm.201200233 PMID:23184463

Goklany, T. (2017, August 22). In Numbers: The Status of Organ Donation in India. NDTV. from http://sites.ndtv.com/moretogive/numbers-status-organ-donation-india-221/

Green, R. A., Lovell, N. H., & Poole-Warren, L. A. (2009). Cell attachment functionality of bioactive conducting polymers for neural interfaces. *Biomaterials*, *30*(22), 3637–3644. doi:10.1016/j.biomaterials.2009.03.043 PMID:19375160

Green, R. A., Lovell, N. H., Wallace, G. G., & Poole-Warren, L. A. (2008). Conducting polymers for neural interfaces: Challenges in developing an effective long-term implant. *Biomaterials*, *29*(24-25), 3393–3399. doi:10.1016/j.biomaterials.2008.04.047 PMID:18501423

Gu, B. K., Park, S. J., & Kim, C. H. (2018). Beneficial effect of aligned nanofiber scaffolds with electrical conductivity for the directional guide of cells. *Journal of Biomaterials Science. Polymer Edition*, *29*(7-9), 1053–1065. doi:10.1080/09205063.2017.1364097 PMID:28771114

Guarino, V., Zuppolini, S., Borriello, A., & Ambrosio, L. (2016). Electro-active polymers (EAPs): A promising route to design bio-organic/bioinspired platforms with on demand functionalities. *Polymers*, *8*(5), 185. doi:10.3390/polym8050185

Gustafsson, G., Cao, Y., Treacy, G. M., Klavetter, F., Colaneri, N., & Heeger, A. J. (1992). Flexible light-emitting diodes made from soluble conducting polymers. *Nature*, *357*(6378), 477–479. doi:10.1038/357477a0

Haastert-Talini, K., & Grothe, C. (2013). Electrical stimulation for promoting peripheral nerve regeneration. *International Review of Neurobiology*, *109*, 111–124. doi:10.1016/B978-0-12-420045-6.00005-5 PMID:24093609

Haipeng, G., Yinghui, Z., Jianchun, L., Yandao, G., Nanming, Z., & Xiufang, Z. (2000). Studies on nerve cell affinity of chitosan-derived materials. *Journal of Biomedical Materials Research*, *52*(2), 285–295. doi:10.1002/1097-4636(200011)52:2<285::AID-JBM7>3.0.CO;2-G PMID:10951367

Hardy, J. G., Lee, J. Y., & Schmidt, C. E. (2013). Biomimetic conducting polymer-based tissue scaffolds. *Current Opinion in Biotechnology*, *24*(5), 847–854. doi:10.1016/j.copbio.2013.03.011 PMID:23578463

Hardy, J. G., Mouser, D. J., Arroyo-Currás, N., Geissler, S., Chow, J. K., Nguy, L., ... Schmidt, C. E. (2014). Biodegradable electroactive polymers for electrochemically-triggered drug delivery. *Journal of Materials Chemistry. B, Materials for Biology and Medicine*, *2*(39), 6809–6822. doi:10.1039/C4TB00355A

Hench, L. L., & Polak, J. M. (2002). Third-generation biomedical materials. *Science*, *295*(5557), 1014–1017. doi:10.1126cience.1067404 PMID:11834817

Hench, L. L., & Thompson, I. (2010). Twenty-first century challenges for biomaterials. *Journal of the Royal Society, Interface*, *7*(Suppl. 4), S379–S391. doi:10.1098/rsif.2010.0151.focus PMID:20484227

Hernandez-Alejandro, R., Caumartin, Y., Marotta, P. J., Ghent, C., Levstik, M. A., Quan, D., ... Luke, P. P. (2010). Kidney and liver transplants from donors after cardiac death: Initial experience at the London Health Sciences Centre. *Canadian Journal of Surgery*, *53*(2), 93. PMID:20334741

Howard, D., Buttery, L. D., Shakesheff, K. M., & Roberts, S. J. (2008). Tissue engineering: Strategies, stem cells and scaffolds. *Journal of Anatomy*, *213*(1), 66–72. doi:10.1111/j.1469-7580.2008.00878.x PMID:18422523

Hu, J., Huang, L., Zhuang, X., Zhang, P., Lang, L., Chen, X., ... Jing, X. (2008). Electroactive aniline pentamer cross-linking chitosan for stimulation growth of electrically sensitive cells. *Biomacromolecules*, *9*(10), 2637–2644. doi:10.1021/bm800705t PMID:18698845

Huang, W. S., Humphrey, B. D., & MacDiarmid, A. G. (1986). Polyaniline, a novel conducting polymer. Morphology and chemistry of its oxidation and reduction in aqueous electrolytes. *Journal of the Chemical Society, Faraday Transactions 1. Physical Chemistry in Condensed Phases*, *82*(8), 2385–2400.

Huynh, W. U., Dittmer, J. J., & Alivisatos, A. P. (2002). Hybrid nanorod-polymer solar cells. *Science*, *295*(5564), 2425–2427. doi:10.1126cience.1069156 PMID:11923531

Janaki, V., Oh, B. T., Shanthi, K., Lee, K. J., Ramasamy, A. K., & Kamala-Kannan, S. (2012). Polyaniline/chitosan composite: An eco-friendly polymer for enhanced removal of dyes from aqueous solution. *Synthetic Metals*, *162*(11-12), 974–980. doi:10.1016/j.synthmet.2012.04.015

Jeong, S. I., Jun, I. D., Choi, M. J., Nho, Y. C., Lee, Y. M., & Shin, H. (2008). Development of Electroactive and Elastic Nanofibers that contain Polyaniline and Poly (L-lactide-co-ε-caprolactone) for the Control of Cell Adhesion. *Macromolecular Bioscience*, *8*(7), 627–637. doi:10.1002/mabi.200800005 PMID:18401867

Ji, J., Xiong, H., Zhu, Z., Li, L., Huang, Y., & Yu, X. (2018). Fabrication of Polypyrrole/Chitosan Nanocomposite Aerogel Monolith for Removal of Cr (VI). *Journal of Polymers and the Environment*, *26*(5), 1979–1985. doi:10.100710924-017-1095-1

Kannusamy, P., & Sivalingam, T. (2013). Synthesis of porous chitosan–polyaniline/ZnO hybrid composite and application for removal of reactive orange 16 dye. *Colloids and Surfaces. B, Biointerfaces*, *108*, 229–238. doi:10.1016/j.colsurfb.2013.03.015 PMID:23563288

Kaur, G., Pandey, O. P., Singh, K., Homa, D., Scott, B., & Pickrell, G. (2014). A review of bioactive glasses: Their structure, properties, fabrication and apatite formation. *Journal of Biomedical Materials Research. Part A*, *102*(1), 254–274. doi:10.1002/jbm.a.34690 PMID:23468256

Khan, S., & Narula, A. K. (2016). Bio-hybrid blended transparent and conductive films PEDOT:PSS:Chitosan exhibiting electro-active and antibacterial properties. *European Polymer Journal*, *81*, 161–172. doi:10.1016/j.eurpolymj.2016.06.005

Kim, D. H., Richardson-Burns, S. M., Hendricks, J. L., Sequera, C., & Martin, D. C. (2007). Effect of immobilized nerve growth factor on conductive polymers: Electrical properties and cellular response. *Advanced Functional Materials*, *17*(1), 79–86. doi:10.1002/adfm.200500594

Kim, I. Y., Seo, S. J., Moon, H. S., Yoo, M. K., Park, I. Y., Kim, B. C., & Cho, C. S. (2008). Chitosan and its derivatives for tissue engineering applications. *Biotechnology Advances*, *26*(1), 1–21. doi:10.1016/j.biotechadv.2007.07.009 PMID:17884325

Koopmans, G., Hasse, B., & Sinis, N. (2009). The role of collagen in peripheral nerve repair. *International Review of Neurobiology*, *87*, 363–379. doi:10.1016/S0074-7742(09)87019-0 PMID:19682648

Krajewska, B. (2005). Membrane-based processes performed with use of chitin/chitosan materials. *Separation and Purification Technology*, *41*(3), 305–312. doi:10.1016/j.seppur.2004.03.019

Kumar, A., & Borah, R. (2016). Effect of plasma irradiation on biocompatibility and cell adhesion of polyaniline/chitosan nanocomposites towards Hep G2 and PBMC cells. *Advanced Materials & Processes*, *1*(2), 146–155. doi:10.5185/amp.2016/206

Kumar, A. M., Suresh, B., Das, S., Obot, I. B., Adesina, A. Y., & Ramakrishna, S. (2017). Promising bio-composites of polypyrrole and chitosan: Surface protective and *in vitro* biocompatibility performance on 316L SS implants. *Carbohydrate Polymers*, *173*, 121–130. doi:10.1016/j.carbpol.2017.05.083 PMID:28732850

Kuralay, F., Demirci, S., Kiristi, M., Oksuz, L., & Oksuz, A. U. (2014). Poly(3,4-ethylenedioxythiophene) coated chitosan modified disposable electrodes for DNA and DNA-drug interaction sensing. *Colloids and Surfaces. B, Biointerfaces*, *123*, 825–830. doi:10.1016/j.colsurfb.2014.10.021 PMID:25454660

Lari, A., Sun, T., & Sultana, N. (2016). PEDOT: PSS-Containing Nanohydroxyapatite/Chitosan Conductive Bionanocomposite Scaffold: Fabrication and Evaluation. *Journal of Nanomaterials*.

Le, T. H., Kim, Y., & Yoon, H. (2017). Electrical and electrochemical properties of conducting polymers. *Polymers*, *9*(4), 150. doi:10.3390/polym9040150

Lee, J. Y., Bashur, C. A., Goldstein, A. S., & Schmidt, C. E. (2009). Polypyrrole-coated electrospun PLGA nanofibers for neural tissue applications. *Biomaterials*, *30*(26), 4325–4335. doi:10.1016/j.biomaterials.2009.04.042 PMID:19501901

Li, J., Pan, J., Zhang, L., Guo, X., & Yu, Y. (2003). Culture of primary rat hepatocytes within porous chitosan scaffolds. *Journal of Biomedical Materials Research. Part A*, *67*(3), 938–943. doi:10.1002/jbm.a.10076 PMID:14613242

Liu, X., Yue, Z., Higgins, M. J., & Wallace, G. G. (2011). Conducting polymers with immobilised fibrillar collagen for enhanced neural interfacing. *Biomaterials*, *32*(30), 7309–7317. doi:10.1016/j.biomaterials.2011.06.047 PMID:21745688

Lv, R., Sun, Y., Yu, F., & Zhang, H. (2012). Fabrication of poly (3, 4-ethylenedioxythiophene)-polysaccharide composites. *Journal of Applied Polymer Science*, *124*(1), 855–863. doi:10.1002/app.35117

MacDiarmid, A. G. (2001). Nobel Lecture: "Synthetic metals": A novel role for organic polymers. *Reviews of Modern Physics*, *73*(3), 701–712. doi:10.1103/RevModPhys.73.701

Madihally, S. V., Flake, A. W., & Matthew, H. W. (1999). Maintenance of CD34 expression during proliferation of CD34+ cord blood cells on glycosaminoglycan surfaces. *Stem Cells (Dayton, Ohio)*, *17*(5), 295–305. doi:10.1002tem.170295 PMID:10527464

Madihally, S. V., & Matthew, H. W. (1999). Porous chitosan scaffolds for tissue engineering. *Biomaterials*, *20*(12), 1133–1142. doi:10.1016/S0142-9612(99)00011-3 PMID:10382829

Martino, S., D'Angelo, F., Armentano, I., Kenny, J. M., & Orlacchio, A. (2012). Stem cell-biomaterial interactions for regenerative medicine. *Biotechnology Advances*, *30*(1), 338–351. doi:10.1016/j.biotechadv.2011.06.015 PMID:21740963

Mattioli-Belmonte, M., Gigante, A., Muzzarelli, R. A. A., Politano, R., De Benedittis, A., Specchia, N., ... Greco, F. (1999). N, N-dicarboxymethyl chitosan as delivery agent for bone morphogenetic protein in the repair of articular cartilage. *Medical & Biological Engineering & Computing*, *37*(1), 130–134. doi:10.1007/BF02513279 PMID:10396855

McCaig, C. D., & Zhao, M. (1997). Physiological electrical fields modify cell behaviour. *BioEssays*, *19*(9), 819–826. doi:10.1002/bies.950190912 PMID:9297973

Mihic, A., Cui, Z., Wu, J., Vlacic, G., Miyagi, Y., Li, S. H., ... Li, R. K. (2015). A conductive polymer hydrogel supports cell electrical signaling and improves cardiac function after implantation into myocardial infarct. *Circulation*, *132*(8), 772–784. doi:10.1161/CIRCULATIONAHA.114.014937 PMID:26304669

Mizuno, K., Yamamura, K., Yano, K., Osada, T., Saeki, S., Takimoto, N., ... Nimura, Y. (2003). Effect of chitosan film containing basic fibroblast growth factor on wound healing in genetically diabetic mice. *Journal of Biomedical Materials Research. Part A*, *64*(1), 177–181. doi:10.1002/jbm.a.10396 PMID:12483711

Molapo, K. M., Ndangili, P. M., Ajayi, R. F., Mbambisa, G., Mailu, S. M., Njomo, N., ... Iwuoha, E. I. (2012). Electronics of conjugated polymers (I): Polyaniline. *International Journal of Electrochemical Science*, *7*(12), 11859–11875.

Mori, T., Okumura, M., Matsuura, M., Ueno, K., Tokura, S., Okamoto, Y., ... Fujinaga, T. (1997). Effects of chitin and its derivatives on the proliferation and cytokine production of fibroblasts in vitro. *Biomaterials*, *18*(13), 947–951. doi:10.1016/S0142-9612(97)00017-3 PMID:9199765

Moroni, L., De Wijn, J. R., & Van Blitterswijk, C. A. (2008). Integrating novel technologies to fabricate smart scaffolds. *Journal of Biomaterials Science. Polymer Edition*, *19*(5), 543–572. doi:10.1163/156856208784089571 PMID:18419938

Mortimer, R. J., Dyer, A. L., & Reynolds, J. R. (2006). Electrochromic organic and polymeric materials for display applications. *Displays*, *27*(1), 2–18. doi:10.1016/j.displa.2005.03.003

Moutsatsou, P., Coopman, K., & Georgiadou, S. (2017). Biocompatibility assessment of conducting PANI/Chitosan nanofibers for wound healing applications. *Polymers*, *9*(12), 687. doi:10.3390/polym9120687

Murugan, R., & Ramakrishna, S. (2007). Design strategies of tissue engineering scaffolds with controlled fiber orientation. *Tissue Engineering*, *13*(8), 1845–1866. doi:10.1089/ten.2006.0078 PMID:17518727

Muzzarelli, R. A. (2009). Chitins and chitosans for the repair of wounded skin, nerve, cartilage and bone. *Carbohydrate Polymers*, *76*(2), 167–182. doi:10.1016/j.carbpol.2008.11.002

Muzzarelli, R. A. (2011). Chitosan composites with inorganics, morphogenetic proteins and stem cells, for bone regeneration. *Carbohydrate Polymers*, *83*(4), 1433–1445. doi:10.1016/j.carbpol.2010.10.044

Narayan, R. J. (2010). The next generation of biomaterial development. *Philosophical Transactions of the Royal Society A*, *368*(1917), 1831–1837. doi:10.1098/rsta.2010.0001 PMID:20308105

Nisha, S. A. (2017). Assessment of knowledge and attitude on organ donation among general public at puducherry. *International Journal of Current Research*, *9*(09), 57310–57313.

Nishimura, K., Nishimura, S., Nishi, N., Saiki, I., Tokura, S., & Azuma, I. (1984). Immunological activity of chitin and its derivatives. *Vaccine*, *2*(1), 93–99. doi:10.1016/S0264-410X(98)90039-1 PMID:6397928

Olson, J. L., Atala, A., & Yoo, J. J. (2011). Tissue engineering: Current strategies and future directions. *Chonnam Medical Journal*, *47*(1), 1–13. doi:10.4068/cmj.2011.47.1.1 PMID:22111050

Patel, N., & Poo, M. M. (1982). Orientation of neurite growth by extracellular electric fields. *The Journal of Neuroscience*, *2*(4), 483–496. doi:10.1523/JNEUROSCI.02-04-00483.1982 PMID:6279799

Pei, Q., Zuccarello, G., Ahlskog, M., & Inganäs, O. (1994). Electrochromic and highly stable poly(3, 4-ethylenedioxythiophene) switches between opaque blue-black and transparent sky blue. *Polymer*, *35*(7), 1347–1351. doi:10.1016/0032-3861(94)90332-8

Place, E. S., Evans, N. D., & Stevens, M. M. (2009). Complexity in biomaterials for tissue engineering. *Nature Materials*, *8*(6), 457–470. doi:10.1038/nmat2441 PMID:19458646

Prabhakaran, M. P., Ghasemi-Mobarakeh, L., & Ramakrishna, S. (2011). Electrospun composite nanofibers for tissue regeneration. *Journal of Nanoscience and Nanotechnology*, *11*(4), 3039–3057. doi:10.1166/jnn.2011.3753 PMID:21776670

Qazi, T. H., Rai, R., & Boccaccini, A. R. (2014). Tissue engineering of electrically responsive tissues using polyaniline based polymers: A review. *Biomaterials*, *35*(33), 9068–9086. doi:10.1016/j.biomaterials.2014.07.020 PMID:25112936

Qi, F., Wang, Y., Ma, T., Zhu, S., Zeng, W., Hu, X., ... Luo, Z. (2013). Electrical regulation of olfactory ensheathing cells using conductive polypyrrole/chitosan polymers. *Biomaterials*, *34*(7), 1799–1809. doi:10.1016/j.biomaterials.2012.11.042 PMID:23228424

Rajnicek, A. M., Robinson, K. R., & McCaig, C. D. (1998). The direction of neurite growth in a weak DC electric field depends on the substratum: Contributions of adhesivity and net surface charge. *Developmental Biology*, *203*(2), 412–423. doi:10.1006/dbio.1998.9039 PMID:9808790

Ratner, B. D. (2004). *Biomaterials science: An introduction to materials in medicine*. Amsterdam, Boston: Elsevier Academic Press.

Ratner, B. D., Hoffman, A. S., Schoen, F. J., & Lemons, J. E. (2004). *Introduction-Biomaterials Science: An Evolving, Multidisciplinary Endeavor* (3rd ed.). Academic Press.

Sall, M. L., Diaw, A. K. D., Gningue-Sall, D., Chevillot-Biraud, A., Oturan, N., Oturan, M. A., ... Aaron, J. J. (2018). Removal of lead and cadmium from aqueous solutions by using 4-amino-3-hydroxynaphthalene sulfonic acid-doped polypyrrole films. *Environmental Science and Pollution Research International*. PMID:29318483

Samba, R., Herrmann, T., & Zeck, G. (2015). PEDOT–CNT coated electrodes stimulate retinal neurons at low voltage amplitudes and low charge densities. *Journal of Neural Engineering*, *12*(1), 016014. doi:10.1088/1741-2560/12/1/016014 PMID:25588201

Schmidt, C. E., & Leach, J. B. (2003). Neural tissue engineering: Strategies for repair and regeneration. *Annual Review of Biomedical Engineering*, *5*(1), 293–347. doi:10.1146/annurev.bioeng.5.011303.120731 PMID:14527315

Shamaeli, E., & Alizadeh, N. (2014). Nanostructured biocompatible thermal/electrical stimuli-responsive biopolymer-doped polypyrrole for controlled release of chlorpromazine: Kinetics studies. *International Journal of Pharmaceutics*, *472*(1-2), 327–338. doi:10.1016/j.ijpharm.2014.06.036 PMID:24969668

Shi, C., Zhu, Y., Ran, X., Wang, M., Su, Y., & Cheng, T. (2006). Therapeutic Potential of Chitosan and Its Derivatives in Regenerative Medicine1. *The Journal of Surgical Research*, *133*(2), 185–192. doi:10.1016/j.jss.2005.12.013 PMID:16458923

Shi, G., Zhang, Z., & Rouabhia, M. (2008). The regulation of cell functions electrically using biodegradable polypyrrole-polylactide conductors. *Biomaterials*, *29*(28), 3792–3798. doi:10.1016/j.biomaterials.2008.06.010 PMID:18602689

Shin, M. K., Spinks, G. M., Shin, S. R., Kim, S. I., & Kim, S. J. (2009). Nanocomposite hydrogel with high toughness for Bioactuators. *Advanced Materials*, *21*(17), 1712–1715. doi:10.1002/adma.200802205

Shirakawa, H., Louis, E. J., MacDiarmid, A. G., Chiang, C. K., & Heeger, A. J. (1977). Synthesis of electrically conducting organic polymers: Halogen derivatives of polyacetylene, (CH)x. *Journal of the Chemical Society. Chemical Communications*, *16*(16), 578–580. doi:10.1039/c39770000578

Singla, A. K., & Chawla, M. (2001). Chitosan: Some pharmaceutical and biological aspects-an update. *The Journal of Pharmacy and Pharmacology*, *53*(8), 1047–1067. doi:10.1211/0022357011776441 PMID:11518015

Snook, G. A., Kao, P., & Best, A. S. (2011). Conducting-polymer-based supercapacitor devices and electrodes. *Journal of Power Sources*, *196*(1), 1–12. doi:10.1016/j.jpowsour.2010.06.084

Srivastava, M., Srivastava, S. K., Nirala, N. R., & Prakash, R. (2014). A chitosan-based polyaniline-Au nanocomposite biosensor for determination of cholesterol. *Analytical Methods*, *6*(3), 817–824. doi:10.1039/C3AY41812J

Stavrinidou, E., Winther-Jensen, O., Shekibi, B. S., Armel, V., Rivnay, J., Ismailova, E., ... Winther-Jensen, B. (2014). Engineering hydrophilic conducting composites with enhanced ion mobility. *Physical Chemistry Chemical Physics*, *16*(6), 2275–2279. doi:10.1039/C3CP54061H PMID:24352071

Suh, J. K. F., & Matthew, H. W. (2000). Application of chitosan-based polysaccharide biomaterials in cartilage tissue engineering: A review. *Biomaterials*, *21*(24), 2589–2598. doi:10.1016/S0142-9612(00)00126-5 PMID:11071608

Svirskis, D., Travas-Sejdic, J., Rodgers, A., & Garg, S. (2010). Electrochemically controlled drug delivery based on intrinsically conducting polymers. *Journal of Controlled Release*, *146*(1), 6–15. doi:10.1016/j.jconrel.2010.03.023 PMID:20359512

Tahmasbi Rad, A., Ali, N., Kotturi, H. S. R., Yazdimamaghani, M., Smay, J., Vashaee, D., & Tayebi, L. (2014). Conducting scaffolds for liver tissue engineering. *Journal of Biomedical Materials Research. Part A*, *102*(11), 4169–4181. doi:10.1002/jbm.a.35080 PMID:24443356

Tanigawa, T. (1992). Various biological effects of chitin derivative. *Advances in Chitin and Chitosan,* 206-215.

Teixeira-Dias, B., del Valle, L. J., Aradilla, D., Estrany, F., & Alemán, C. (2012). A conducting polymer/protein composite with bactericidal and electroactive properties. *Macromolecular Materials and Engineering*, *297*(5), 427–436. doi:10.1002/mame.201100172

Teixeira-Dias, B., del Valle, L. J., Estrany, F., Mano, J. F., Reis, R. L., & Alemán, C. (2012). Dextrin and Conducting-Polymer-Containing Biocomposites: Properties and Behavior as Cellular Matrix. *Macromolecular Materials and Engineering*, *297*(4), 359–368. doi:10.1002/mame.201100180

Thrivikraman, G., Madras, G., & Basu, B. (2014). Intermittent electrical stimuli for guidance of human mesenchymal stem cell lineage commitment towards neural-like cells on electroconductive substrates. *Biomaterials*, *35*(24), 6219–6235. doi:10.1016/j.biomaterials.2014.04.018 PMID:24816362

Ulutürk, C., & Alemdar, N. (2018). Electroconductive 3D polymeric network production by using polyaniline/chitosan-based hydrogel. *Carbohydrate Polymers*, *193*, 307–315. doi:10.1016/j.carbpol.2018.03.099 PMID:29773386

Upadhyay, J., Kumar, A., Gupta, K., & Mandal, M. (2015). Investigation of physical and biological properties of polypyrrole nanotubes–chitosan nanocomposites. *Carbohydrate Polymers*, *132*, 481–489. doi:10.1016/j.carbpol.2015.06.028 PMID:26256373

Usami, Y., Okamoto, Y., Minami, S., Matsuhashi, A., Kumazawa, N. H., Tanioka, S. I., & Shigemasa, Y. (1994). Chitin and chitosan induce migration of bovine polymorphonuclear cells. *The Journal of Veterinary Medical Science*, *56*(4), 761–762. doi:10.1292/jvms.56.761 PMID:7999904

Usami, Y., Okamoto, Y., Minami, S., Matsuhashi, A., Kumazawa, N. H., Tanioka, S. I., & Shigemasa, Y. (1994). Migration of canine neutrophils to chitin and chitosan. *The Journal of Veterinary Medical Science*, *56*(6), 1215–1216. doi:10.1292/jvms.56.1215 PMID:7696425

Vacanti, J. P., & Langer, R. (1999). Tissue engineering: The design and fabrication of living replacement devices for surgical reconstruction and transplantation. *Lancet*, *354*, S32–S34. doi:10.1016/S0140-6736(99)90247-7 PMID:10437854

Vellakkat, M., & Hundekal, D. (2017). Electrical conductivity and supercapacitor properties of polyaniline/chitosan/nickel oxide honeycomb nanocomposite. *Journal of Applied Polymer Science*, *134*(9). doi:10.1002/app.44536

Vert, M., Doi, Y., Hellwich, K. H., Hess, M., Hodge, P., Kubisa, P., ... Schué, F. (2012). Terminology for biorelated polymers and applications (IUPAC Recommendations 2012). *Pure and Applied Chemistry*, *84*(2), 377–410. doi:10.1351/PAC-REC-10-12-04

Wallace, G. G., Teasdale, P. R., Spinks, G. M., & Kane-Maguire, L. A. (2002). *Conductive electroactive polymers: Intelligent materials systems* (2nd ed.). CRC press, Taylor & Francis.

Wang, S., Guan, S., Wang, J., Liu, H., Liu, T., Ma, X., & Cui, Z. (2017). Fabrication and characterization of conductive poly(3,4-ethylenedioxythiophene) doped with hyaluronic acid/poly (l-lactic acid) composite film for biomedical application. *Journal of Bioscience and Bioengineering*, *123*(1), 116–125. doi:10.1016/j.jbiosc.2016.07.010 PMID:27498308

Wang, S., Guan, S., Xu, J., Li, W., Ge, D., Sun, C., ... Ma, X. (2017). Neural stem cell proliferation and differentiation in the conductive PEDOT-HA/Cs/Gel scaffold for neural tissue engineering. *Biomaterials Science*, *5*(10), 2024–2034. doi:10.1039/C7BM00633K PMID:28894864

Wang, S., Guan, S., Zhu, Z., Li, W., Liu, T., & Ma, X. (2017). Hyaluronic acid doped-poly (3, 4-ethylenedioxythiophene)/chitosan/gelatin (PEDOT-HA/Cs/Gel) porous conductive scaffold for nerve regeneration. *Materials Science and Engineering C*, *71*, 308–316. doi:10.1016/j.msec.2016.10.029 PMID:27987712

Wang, X., Zhi, L., & Müllen, K. (2008). Transparent, conductive graphene electrodes for dye-sensitized solar cells. *Nano Letters*, *8*(1), 323–327. doi:10.1021/nl072838r PMID:18069877

Wang, Z., Roberge, C., Wan, Y., Dao, L. H., Guidoin, R., & Zhang, Z. (2003). A biodegradable electrical bioconductor made of polypyrrole nanoparticle/poly (D, L-lactide) composite: A preliminary *in vitro* biostability study. *Journal of Biomedical Materials Research. Part A*, *66*(4), 738–746. doi:10.1002/jbm.a.10037 PMID:12926024

Wu, M., He, J., Ren, X., Cai, W. S., Fang, Y. C., & Feng, X. Z. (2014). Development of functional biointerfaces by surface modification of polydimethylsiloxane with bioactive chlorogenic acid. *Colloids and Surfaces. B, Biointerfaces*, *116*, 700–706. doi:10.1016/j.colsurfb.2013.11.010 PMID:24290104

Xia, L., Wei, Z., & Wan, M. (2010). Conducting polymer nanostructures and their application in biosensors. *Journal of Colloid and Interface Science*, *341*(1), 1–11. doi:10.1016/j.jcis.2009.09.029 PMID:19837415

Xia, Y., Lu, X., & Zhu, H. (2013). Natural silk fibroin/polyaniline (core/shell) coaxial fiber: Fabrication and application for cell proliferation. *Composites Science and Technology*, *77*, 37–41. doi:10.1016/j.compscitech.2013.01.008

Xiao, Q., Feng, M., Liu, Y., Lu, S., He, Y., & Huang, S. (2018). The graphene/polypyrrole/chitosan-modified glassy carbon electrode for electrochemical nitrite detection. *Ionics*, *24*(3), 845–859. doi:10.100711581-017-2247-y

Xiao, Y., Li, C. M., Wang, S., Shi, J., & Ooi, C. P. (2010). Incorporation of collagen in poly (3, 4-ethylenedioxythiophene) for a bifunctional film with high bio-and electrochemical activity. *Journal of Biomedical Materials Research. Part A*, *92*(2), 766–772. doi:10.1002/jbm.a.32412 PMID:19274716

Xu, C., Guan, S., Wang, S., Gong, W., Liu, T., Ma, X., & Sun, C. (2018). Biodegradable and electroconductive poly (3, 4-ethylenedioxythiophene)/carboxymethyl chitosan hydrogels for neural tissue engineering. *Materials Science and Engineering C*, *84*, 32–43. doi:10.1016/j.msec.2017.11.032 PMID:29519441

Yan, L., Zhao, B., Liu, X., Li, X., Zeng, C., Shi, H., ... Liu, Y. (2016). Aligned nanofibers from polypyrrole/graphene as electrodes for regeneration of optic nerve via electrical stimulation. *ACS Applied Materials & Interfaces*, *8*(11), 6834–6840. doi:10.1021/acsami.5b12843 PMID:26926578

Yang, S., Leong, K. F., Du, Z., & Chua, C. K. (2001). The design of scaffolds for use in tissue engineering. Part I. *Traditional factors. Tissue Engineering*, *7*(6), 679–689. doi:10.1089/107632701753337645 PMID:11749726

Zhang, J., Neoh, K. G., & Kang, E. T. (2018). Electrical stimulation of adipose-derived mesenchymal stem cells and endothelial cells co-cultured in a conductive scaffold for potential orthopaedic applications. *Journal of Tissue Engineering and Regenerative Medicine*, *12*(4), 878–889. doi:10.1002/term.2441 PMID:28482125

Zhang, R., Xu, X., Fan, X., Yang, R., Wu, T., & Zhang, C. (2018). Application of conducting micelles self-assembled from commercial poly (3, 4-ethylenedioxythiophene): Poly (styrene sulfonate) and chitosan for electrochemical biosensor. *Colloid & Polymer Science*, *296*(3), 495–502. doi:10.100700396-018-4270-6

Zhang, Y., & Zhang, M. (2002). Three-dimensional macroporous calcium phosphate bioceramics with nested Chitosan sponges for load-bearing bone implants. *Journal of Biomedical Materials Research. Part A*, *61*(1), 1–8. doi:10.1002/jbm.10176 PMID:12001239

Zhao, X., Li, P., Guo, B., & Ma, P. X. (2015). Antibacterial and conductive injectable hydrogels based on quaternized chitosan-graft-polyaniline/oxidized dextran for tissue engineering. *Acta Biomaterialia*, *26*, 236–248. doi:10.1016/j.actbio.2015.08.006 PMID:26272777

Zhao, X., Wu, H., Guo, B., Dong, R., Qiu, Y., & Ma, P. X. (2017). Antibacterial anti-oxidant electroactive injectable hydrogel as self-healing wound dressing with hemostasis and adhesiveness for cutaneous wound healing. *Biomaterials*, *122*, 34–47. doi:10.1016/j.biomaterials.2017.01.011 PMID:28107663

Zhong, H., He, A., Lu, J., Sun, M., He, J., & Zhang, L. (2016). Carboxymethyl chitosan/conducting polymer as water-soluble composite binder for LiFePO4 cathode in lithium ion batteries. *Journal of Power Sources*, *336*, 107–114. doi:10.1016/j.jpowsour.2016.10.041

Chapter 7

Polyaniline as Proficient Electrode Material for Supercapacitor Applications:

PANI Nanocomposites for Supercapacitor Applications

Dipanwita Majumdar
Chandernagore College, India

ABSTRACT

Sky-high renewable energy demands urged the revolutionary development of environment benign, portable and miniatured, high power generating energy storage systems in the form of electrochemical capacitors commonly known as "supercapacitors" or "ultracapacitors". Supercapacitor designing requires smart electrode materials for providing higher energy and power densities, mechanical and electrochemical durability, enhanced thermal operating range with minimal production and maintenance cost. Polyaniline, as conducting polymer, particularly in nano-morphology, has been one of the pioneer electroactive materials paving the corridor for commercial development of pseudocapacitors. Considering these points, the chapter initially concentrates on basic concepts of fabrication, designing and performance evaluation procedure of supercapacitors, followed by discussion on the role of PANI as potential supercapacitor electrode material, delineating current achievements and major challenges encountered in the process of progress so as to obtain superior electrochemical energy storage systems in the near future.

DOI: 10.4018/978-1-5225-7838-3.ch007

INTRODUCTION

Aim of Supercapacitors

Ever mounting energy needs, increasing environmental pollution and depleting of conventional energy resources, have urged the progress of renewable, sustainable energy production with environmental benign systems for the last three decades. Great efforts have been committed in developing rechargeable, secondary batteries so as to be able to store energy for autonomy purposes (Yoo, Markevich & Salitra, 2014) (Tarascon & Armand, 2001). However, to cope up with the sky-high demands for portable-miniatured electronics systems and hybrid-electric vehicles, short-pulsed-high power requirements have ideally motivated the current research on electrochemical capacitors (Burke, 2000). The electrochemical capacitors, more famously recognized as supercapacitors or "supercaps" or ultracapacitors, are observed to display longer cycle, faster charging/discharging rate with appreciable energy and power densities along with low maintenance cost. However, energy densities of current supercapacitors are very unsatisfactory compared to secondary batteries (Kotz & Carlen, 2000). So, present research concentrates on improving the limitations of the supercapacitors preserving all its advantages to full extent. Some of the points of superiority of the supercapacitors over secondary batteries are well reflected in the following comparative study as depicted in Table-1: Comparison of Supercapacitors with Secondary Batteries (Chu and Braatz, 2002).

Evolution of Supercapacitors

Electricity has been one of the novel and most amazing discoveries in the last 400 years of human civilization. Since then, efforts were in the hunt to channelize it into public needs. Around that era when the insights of electricity were growing up, "electrochemistry" got discovered through the hands of Luigi Galvani, the Italian physician and anatomist. He proposed "animal electricity", as an unknown vital force contained in the animal tissue, which activated nerves and muscles when touched with metal probes.

Table 1. Comparison of supercapacitors with secondary batteries

Available Performances	Supercapacitors	Secondary Batteries
Cost of production	Lower cost per cycle	Higher cost per cycle
Number of Charging/ Discharging cycles	High	low
Heat build-up during GCD charging/ discharging cycles	Low thermal energy released with maximum GCD efficiency	Serious heating up
Charge protection	No danger of overcharging	Shut-off circuits needed to detect full-charge
Environmental issues	No corrosive chemicals	Chemicals needed to be dispose of safely.
Maximum possible voltage	Low valued. Need to combine in series but that leads to increasing internal resistance.	Maximum voltage obtained with straightforward series combinations.
Energy density	Higher than all trivial capacitors	Superior in energy storage
Power density (rate of energy discharge)	Very high rate of energy discharge	Lower rate of energy discharge
Operating temperature	-40°C to 65°C, more effective in low temperature regions	-20 °C to 100 °C
Operational life	high Longevity	high Longevity

However, Alessandro Giuseppe Antonio Anastasio Volga clarified and showed in the early 1800s that animal tissue was not necessary for the generation of current. He argued that the frog legs in Galvani's experiments acted as an electroscope. The true source of animal stimulation was due to the contact between two different metals present in the probe. He coined the electricity thus produced as "metallic electricity". He also discovered that certain fluids could create continuous current of electrical power on using as a conductor that ultimately led to the invention of the first voltaic cell, the "battery". In 1836, John F. Daniel, an English chemist, designed an improved device to obtain steadier current than earlier ones. In 1859, the French physicist Gaston Planté first proposed the idea of first "rechargeable battery", based on lead acid, that is still employed today. In 1899, Waldmar Jungner from Sweden invented another secondary rechargeable battery nickel-cadmium battery (NiCd), made of nickel and cadmium for positive (cathode) and negative (anode) electrodes respectively. Around 1932 Shlecht and Ackermann improved the durability of Ni-Cd with elevated load currents output. Since then, Ni-Cd cell was the only rechargeable battery for transportable applications. In the early 1950s, General Electric engineers used porous carbon electrodes, in capacitors designing. H. Becker in 1957 developed a "Low voltage electrolytic capacitor" using porous carbon electrodes. In 1966, Standard Oil of Ohio (SOHIO) company developed "electrical energy storage apparatus", while designing experimental fuel cells. Even in 1970, Donald L. Boos patented the electrochemical capacitor with activated carbon electrodes. Around 1980s, B E Conway carried out extensive research works on RuO_2 electrochemical capacitors and distinctively differentiated "Supercapacitor" and "Battery" characteristic behavior in electrochemical energy storage process. The most primitive supercapacitor was fabricated in 1982 by Pinnacle Research Institute (PRI), for military applications and then commercialized under the trade name "PRI Ultracapacitor". Maxwell Laboratories (afterward Maxwell Technologies) further introduced "Boost Caps" in 1992 to highlight their usage in power applications. Around 1994 D A. Evans designed "Electrolytic-Hybrid Electrochemical Capacitor" employing tantalum electrolytic capacitor, to attain high voltage of 200V. In 1999, Conway coined the term "supercapacitor" to enlighten the increased capacitance as a result of faradaic charge transfer between electrodes and ions mediated by electrode-surface redox reactions. Since then, sincere research efforts to improve electrode materials along with development in the electrolytes choices are in the continuous run for seeking better supercapacitor performances. Very recent developments include lithium-ion hybrid capacitors pioneered by FDK since 2007 with superior cell voltage output, and enhanced specific energy. (Wikipedia) (Majumdar, 2018) (Majumdar, 2018)

Basics of Supercapacitors

Supercapacitor Definition

Supercapacitors, considered as the intermediate systems between conventional capacitors and batteries, are electrical devices capable of storing energy more efficiently than commonly obtainable capacitors and capable of delivering ultrafast power supply. The fabrication details of supercapacitor models currently available in the market are displayed in Figure 1.

They are fabricated and outlined on the same fundamental principles and equations as that of conventional capacitors but employ higher surface area electrodes to ensure greater energy densities while thinner dielectrics to attain greater capacitances in addition to higher power densities compared to those of batteries (Burke,2000), (Kotz & Carlen, 2000). This is revealed in the Ragone plot used to compare the energy-storage performances of various devices in Figure 2. (Christen & Carlen, 2000) (Chuizhou, Gall, & Irazoqui, 2013).

Figure 1. Fabrication details of "wound" model and "stacked" model supercapacitors using activated carbon (AC) electrodes.
Created with Universal Document Converter
(Ref: https://en.wikipedia.org/wiki/Supercapacitor)

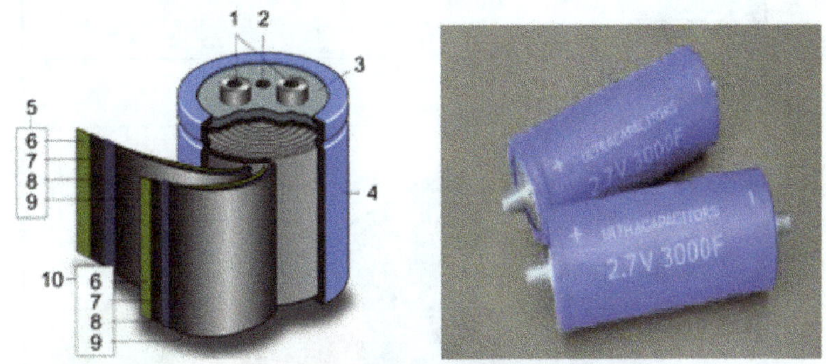

Schematic construction of a wound supercapacitor
1. terminals, 2. safety vent, 3. sealing disc, 4. aluminum can, 5. positive pole, 6. separator, 7. carbon electrode, 8. collector, 9. carbon electrode, 10. negative pole

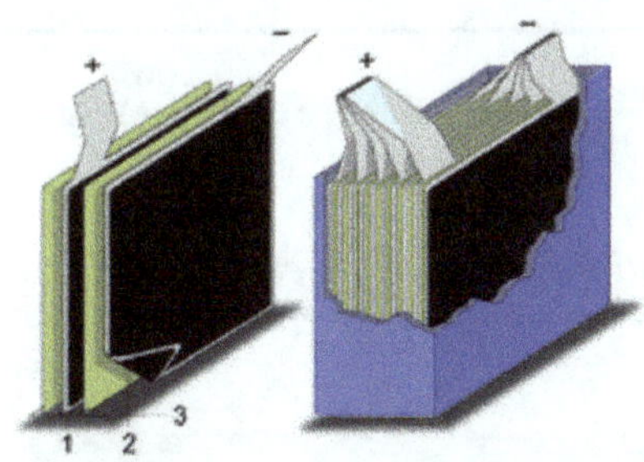

Schematic construction of a supercapacitor with stacked electrodes
1. positive electrode, 2. negative electrode, 3. separator

Supercapacitor-Related Key Terms and Definitions

Typically, supercapacitors have two electrode configuration (displayed in figure 3), where two active electrode materials remain separated by a porous dielectric film in an electrolyte solution. However, electrode materials characterizations of the supercapacitor are conventionally carried out in a three-electrode configuration (shown in figure 3), which comprises a reference electrode, counter electrode and working electrode with other accessories same as that of 2-electrode configuration (Miao, Fan, Chen & Liu, 2013).

1. **Electrode:** An electrical conductor, usually a small metal wire/ rod or other substance coated onto some conductor, employed to contact with a non-metallic portion of a circuit such as a semiconductor, an electrolyte, a vacuum or air, etc. Most of the electrochemical reactions are observed to occur on its active-surface.
2. **Electrolyte:** Refers to an ion-conducting matter, that may be an aqueous or organic based, even dilute acid, base, salt, organic material, ionic liquid, gel and even polymeric substances.
3. **Separator:** Generally, a poor-conducting material placed between the electrodes to prevent electrical short circuit. Typical examples include polymers, papers and glass fiber sheets, etc.
4. **Charging:** It is defined as a course of action of energy storage where a device starts to accumulate energy by means of migration of charged ions to the oppositely poled terminal electrodes respectively as shown in Figure 4.

Figure 2. Ragone plot for comparing the performance comparison of various energy-storing devices.
Created with Universal Document Converter
(Ref: Meng, Chuizhou & Z Gall, Oren & Irazoqui, Pedro, 2013)

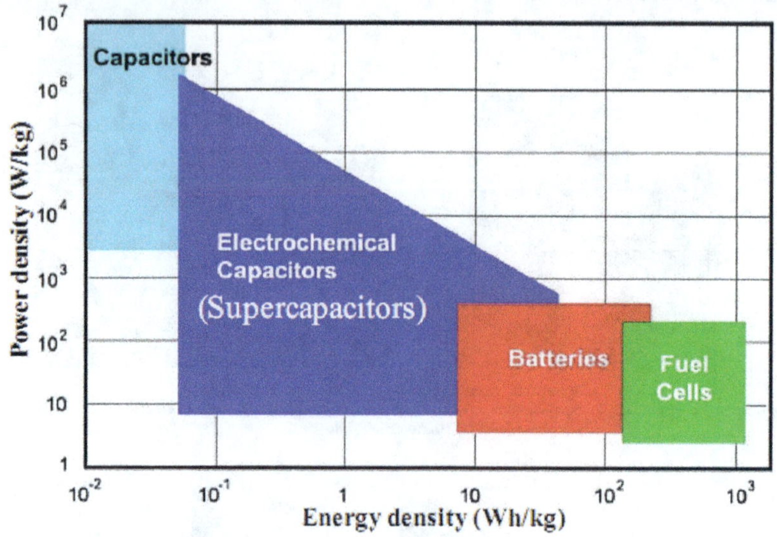

Figure 3. Supercapacitor fabrication: 2-electrode configuration and 3-electrode configuration
Created with Universal Document Converter

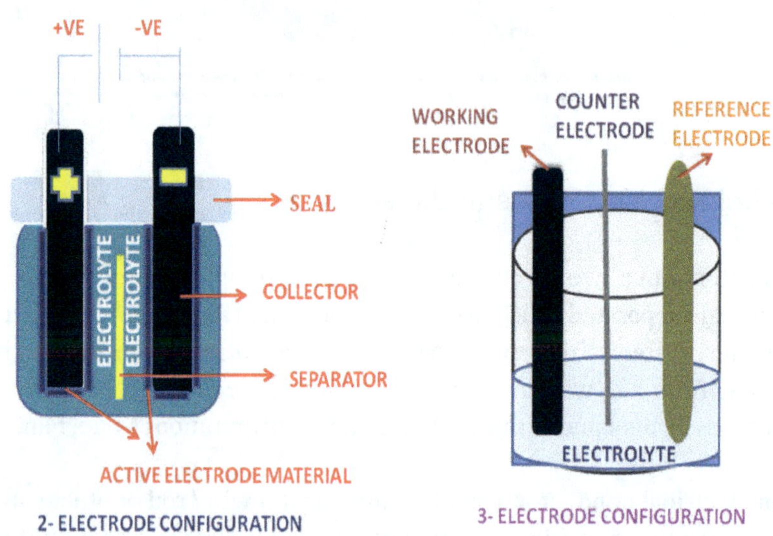

5. **Discharging:** It is energy supplying process in which a device delivers the stored energy to an external system; a state where all the ions are distributed randomly within the cell; as shown in Figure 4(c).
6. **Capacitance (C):** Electrostatic charge storing capacity of a material is associated to the charge (Q) per applied voltage (V), and is mathematically represented as C= (Q/V). It is usually expressed in farads (F). The capacitance of the active electrode materials can be categorized according to its

Figure 4. Charging, fully-charged state and discharging states in supercapacitor
Created with Universal Document Converter

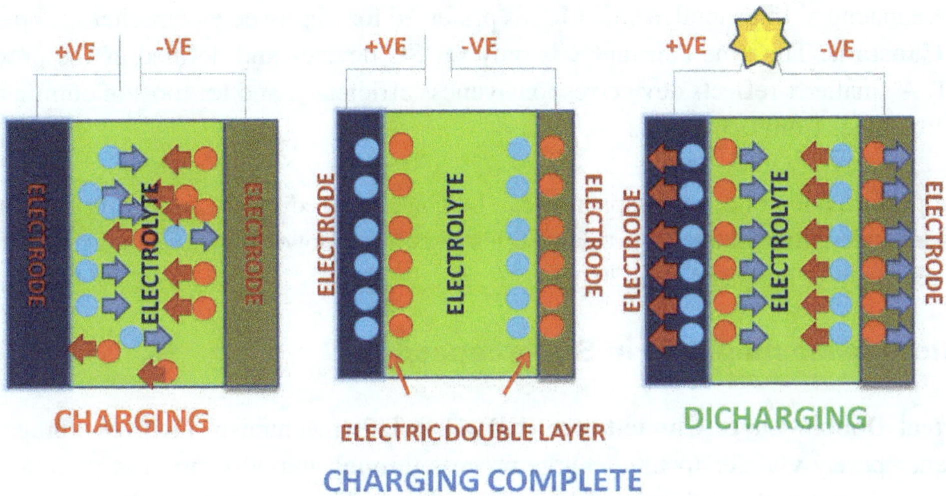

charge storage capability units as follows: (i) gravimetric capacitance, expressed in Fg^{-1}, (ii) volumetric capacitance expressed in Fcm^{-3} and (iii) areal capacitance expressed in Fcm^{-2} respectively.

7. **Energy Density:** Refers to the capacity of an electrode material or a supercapacitor to store energy in watt-hours-per-kilogram (specific energy, $Whkg^{-1}$) or watt-hours-per-litre (energy density, WhL^{-1}). The specific energy in Wh kg^{-1} as well as energy density in WhL^{-1} (E) are given by

$$E = \left(\frac{1}{3600ms}\right)\Delta V\int_{V1}^{Vn} idV \text{ or } E = \left(\frac{1}{3600vs}\right)\Delta V\int_{V1}^{Vn} idV \tag{1}$$

where, *m* and *v* are the mass (kg) and volume (liter, L) of the active materials on both electrodes and *s* is the scan rate, V_n and V_1 are the initial voltages and *i* is the current respectively.

8. **Power Density:** The rate of energy discharged by a supercapacitor. The specific power in Wkg^{-1} and the power density (P) in WL^{-1} are expressed as follows:

$$P = \left(\frac{1}{m}\right)\int_{V1}^{Vn} idV \text{ or } P = \left(\frac{1}{v}\right)\int_{V1}^{Vn} idV \tag{2}$$

where, *m* and *v* are the mass (kg) and volume (liter, L), respectively, of the electro-active materials.

9. **Operating Voltage:** the potential applied to the electrochemical system or the proper potential window within which a given electrochemical cell normally operates.

10. **Equivalent Series Resistance:** The resistance of a resistor is usually called the equivalent series resistance, R_{ES} (ohms), and is essential in reflecting the power performance and energy efficiency of supercapacitors. In general, a small R_{ES} is preferred for improved electrochemical performance.
11. **Time Constant:** The time constant τ is only for SC devices and defined as the product of R_{ES} and CT. A smaller τ reflects device-responsiveness efficiency, and for most of commercial SCs, τ normally ranges from 0.5 to 3.6 s

Thus, the performance of a supercapacitor can be characterized by the following key parameters, - cell capacitance, operating voltage, equivalent series resistance, power density, energy density, and time constant respectively. (Zhang and Pan, 2015)

Charge Storage Mechanisms in Supercapacitors

- **Electrical Double-Layer Capacitance (EDLCs):** This mechanism involves storage of electrical energy purely via electrostatic charge process through ion-adsorption at the interface of the electro-active material and the electrolyte as depicted in Figure 5. Activated carbon, carbon nanofibers, mesoporous carbon, carbon nanotubes and graphene have been consumed extensively in the research and development of EDLCs (Zhang, Ye & Gao, 2016), (An, Kim & Park, 2001), (Du, Yeh & Pan, 2005), (Dai, Chang & Baek, 2012) (Simon & Gogotsi, 2008). Nevertheless, these electrode materials deliver substandard energy density owing to their inadequate interface area exposed to the electrolyte. It was well-established that in most systems, at higher scan rates, charge storage mainly involves double layer capacitance and thus contributes to higher power and energy at these high scan rates.
- **Pseudocapacitance (PC):** Pseudocapacitance mechanism attributes to faster and reversible redox reactions (change in electrochemical state of the substance) at the redox-active materials surface as shown in Figure 5. A typical charge/discharge process involves movement of electrons from the surface of the electrode material to the current collectors or vice versa, which initiates

Figure 5. Electrical double-layer capacitance and pseudocapacitance mechanisms of charge storage in supercapacitors
Created with Universal Document Converter

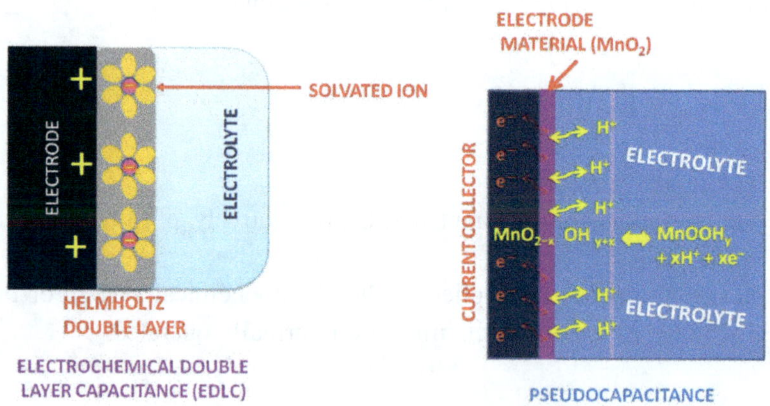

redox reactions of the electrode materials. The materials widely used include electrically conductive polymers (e.g. PANI, polypyrrole, and poly[3,4-ethylenedioxythiophene]) (Frackowiak & Beguin, 2001) (Ramya, Sivasubramanian & Sangaranarayanan, 2013) (Snook, Kao & Best,2011) (Eftekhari, Li & Yang, 2017) and oxides, chalchogenides, pctogenides of transition metals (e.g., MnO_2, NiO, RuO_2, V_2O_5 and TiO_2, CuO, WO_3,MoS_2, VN, etc.) (Peng, Zhou & Chen, 2008), (Kim & Kim, 2001) (Takahashi, Limmer, & Wang, 2004) (Rakhi, Chen & Cha, 2012) (Huang, Xing & Zhuo, 2009) (Duay, Sherrill &Gui, 2013) (Guan, Yu & Chen, 2016).

As per band theory, semiconductors have narrow energy-separated filled valence band and vacant conduction band. On applying suitable potential for electron excitation, valence band electrons get promoted to conduction band and thus, become delocalized. Simultaneously, holes also originate within the valence band that gets mobilized too. Thus, nonstop electron dislocation and production of hole over the wider potential range applied leads to constant capacitive current flow. As the electron displacement in most semiconductor metal oxides occur continuously amongst every actively interacting closely-adjacent redox sites, a "quasi-rectangular" CVs results. This type of charge storage process is the known as "capacitive faradaic process". (Huang, Xing & Zhuo, 2009) (Duay, Sherrill &Gui, 2013) (Guan, Yu & Chen, 2016), (Lokhande, Lokhande & Lokhande, 2016)

Characterization of the Supercapacitors

The electrochemical performances and efficiencies of supercapacitors are commonly characterized by using cyclic voltammetric (CV), chronopotentiometrically by galvanostatic charging/discharging (GCD) and electrochemical impedance spectroscopy (EIS) techniques (Majumdar, Baugh & Bhattacharya, 2017). They help to measure voltage, current, time, internal resistances and other allied parameters, from which specific capacitance, and thus energy density as well as power density values can be evaluated accordingly (Zhang and Pan, 2015) (Majumdar & Bhattacharya, 2017).

Cyclic Voltammetry (CV)

A typical CV graph consists of (i) a linear change of electric potential between two operating potential extremes (potential windows: V_1 and V_n), (ii) the speed/rate of the potential change (scan rates: s, mVs^{-1}). The so-obtained current values during positive and negative sweeps are employed to investigate relevant electrochemical reactions. The curve-shape sometimes indicates the charge storage mechanisms, i.e. rectangular for most EDLCs or quasi-rectangular, with humps or broadened peaks for pseudocapacitors due to redox reactions shown in Figure 6.

For a supercapacitor in two electrode configuration, gravimetric capacitance (C_{wt}, Fg^{-1}), volumetric capacitance (C_{vol}, Fcm^{-3}) and areal capacitance(C_A, Fcm^{-2}) were calculated from its CV on the basis of following equations (Zhang & Pan, 2015), (Yang, Cheng & Wang, 2013)

Figure 6. Cyclic voltammogram of electrode material showing nearly rectangular for most electrochemical double layer charge storage or rectangular-with hump shaped for pseudocapacitors due to redox peaks
Created with Universal Document Converter

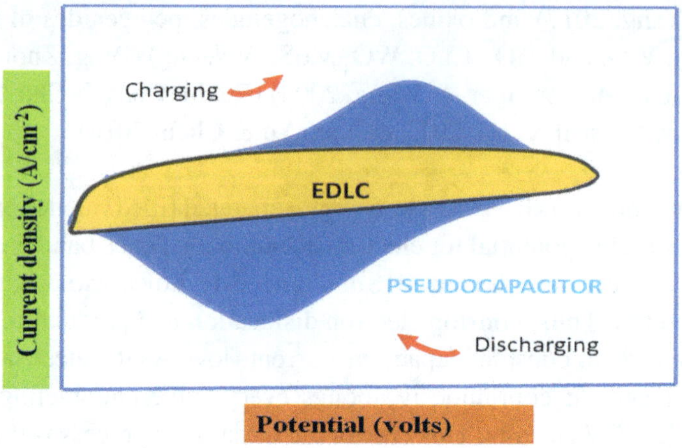

$$C(wt)=\left(\frac{4}{ms\Delta V}\right)\int_{V1}^{Vn} idV$$
$$C(vol)=\left(\frac{4}{ms\Delta V}\right)\int_{V1}^{Vn} idV \quad (3)$$
$$C(A)=\left(\frac{2}{As\Delta V}\right)\int_{V1}^{Vn} idV$$

where ∫ (idV) is the integrated area of the CV curve, ***s*** is the scan rate (V s−1), ΔV is (2 × the voltage window, from V_1 and V_n and then back to V_1, m is the mass of the electrode-active materials (g), ***v*** is the volume of the active materials on both electrodes (cm^3) and **A** is the footprint area of the device (cm^2).] For a supercapacitor in three-electrode configuration, the factor 4 should be replaced by 2 in the above equations, and m and v now represent, respectively, the mass (g) and the volume (in cm^3) of the elcreto-active materials on each electrode.

Chrono-Potentiometry

Chronopotentiometrically galvanostatic charging/discharging cycles are monitored for supercapacitor characterization. In a typical test, charging and discharging at constant current i.e. galvanostatic condition, is carried out for a supercapacitor. During charging process of the supercapacitor as observed in Figure 7, a constant current is maintained for its charging process as the voltage changes from the discharged state (E_1) to the maximum operating voltage (E_2). Again, a negative current of same magnitude is applied to discharge the device to reach to its initial state (E_1). The GCD curve includes two parts: (i) a capacitive part obtained from change in voltage due to the electrochemical energy transformations in the supercapacitor, and (ii) a resistive part representing the voltage drop because of the equivalent series resistance (ESR) of the supercapacitor.

The respectively gravimetric (C_{wt}, Fg^{-1}) or volumetric (C_{vol}, Fcm^{-3}) or areal (C_A, Fcm^{-2}) capacitance values in two-electrode configuration were calculated from GCD curves as per the equations provided, (Zhang & Pan, 2015), (Yang, Cheng & Wang, 2013);

$$C(wt) = \frac{4I}{m\left(\frac{\Delta E}{\Delta t}\right)} \text{ or } C(vol) = \frac{4I}{v\left(\frac{\Delta E}{\Delta t}\right)} \text{ or } C(A) = \frac{4I}{A\left(\frac{\Delta E}{\Delta t}\right)} \quad (4)$$

where, m, V and A are respectively the current (A), the discharge curve slope and the mass (g), volume (cm^3) and footprint area (cm^2) of the two electrodes.

In a three-electrode supercapacitor configuration, the factor 4 is substituted by 2 in the above equations, and *m* and *v* represent the mass (g) and the volume (cm^3) of the active materials on one electrode, respectively. The specific energy (E_{wt}) and the specific power (P_{wt}) are calculated below using the total mass of the electroactive materials on both electrodes as per equation (5) and (6)

$$E(vol)(Whl^{-1}) = \frac{0.125C(vol)(\Delta E)^2}{3.6} \quad (5)$$

$$P(vol)(Wl^{-1}) = \frac{E(vol) \text{ x } 3600}{\Delta t} \quad (6)$$

The energy (E_{vol}) and the power (P_{vol}) density values are calculated below using the total volume of the electroactive materials.

The linear shape of the GCD curves at high current density describes ideal supercapacitor behavior of the cells. Generally, specific capacitance versus number of GCD cycles are plotted to demonstrate the electrochemical stability of supercapacitor systems. However, constant voltage hold tests offer more significant conclusions than respective cycling tests with identical upper voltage limits (Lokhande, Lokhande, & Lokhande, 2016).

Electrochemical Impedance Spectroscopy

Electrochemical impedance spectroscopy (EIS) is also known as the dielectric spectroscopic testing, is one of the most useful tools to investigate as well as evaluate the electrode capacitive activities as well as electrode-electrolyte interfacial behavior of the supercapacitor. The resulting data so-obtained are generally graphically represented by a Nyquist graph (Figure 8) plotted imaginary impedance and real impedance.

A theoretical Nyquist plot of a supercapacitor consists of three frequency- dependent (03) regions. At very high frequency, the supercapacitor acts like a pure resistor. At shorter frequency region, the imaginary part increases sharply and vertically representing pure capacitive-like characteristics. In the middle frequency domain, the electrode porosity influence can be observed. When the frequency is gradually decreased from very high frequency, the ions starts to penetrate deeper inside the pores of the

Figure 7. A typical galvanostatic charging/discharging cycle of a supercapacitor
Created with Universal Document Converter

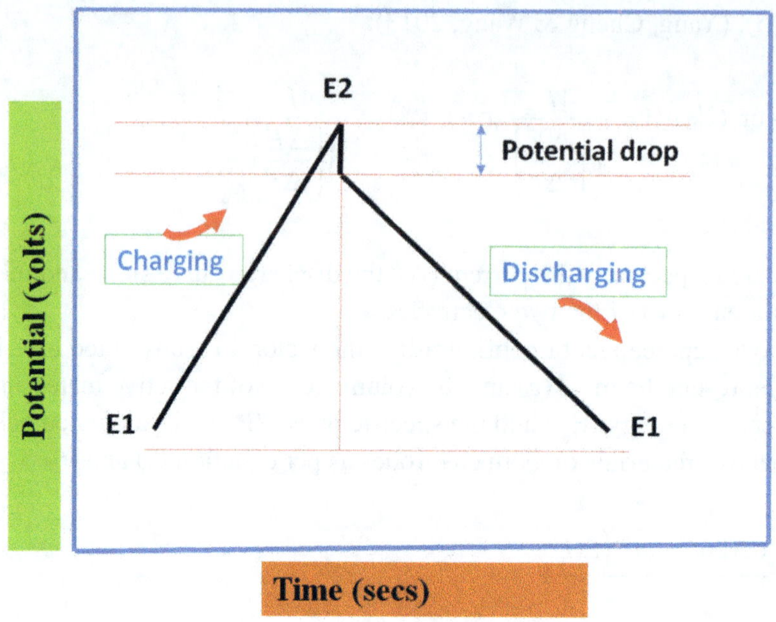

Figure 8. A Nyquist plot (typical) for electrochemical impedance spectroscopy study
Created with Universal Document Converter

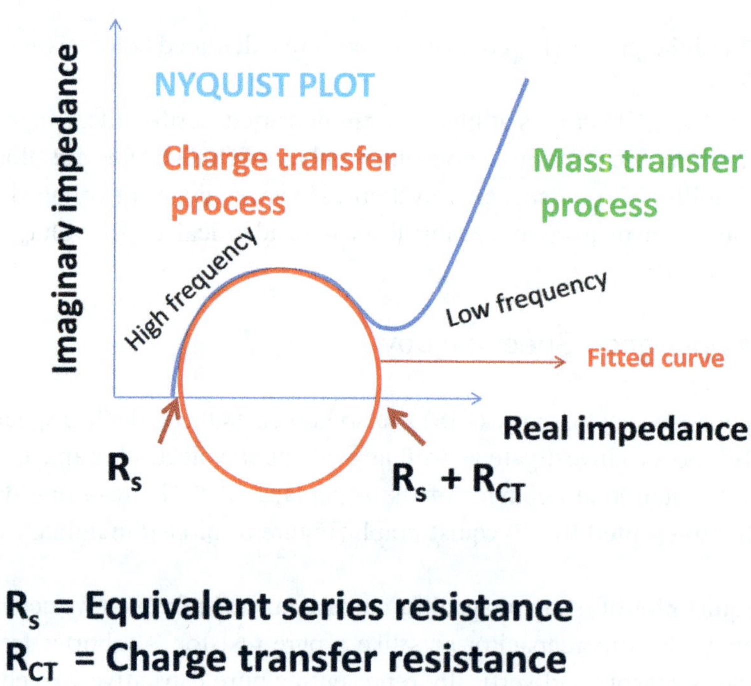

electrode, and thus, more electrode surface becomes accessible for ion adsorption process (Dupont & Donne, 2016) (Wu, Chu & Hu, 2013). Thus, this technique is handful to evaluate the electrode equivalent series resistances (ESR), potential-dependant Faradaic resistances, mechanism of mass transport and in evaluation of electrode capacitance and energy and power densities. Electrochemical Impedance Spectroscopy has several advantages: the frequency response of the supercapacitor is well-known in different conditions of temperature and voltage. This is a more precise method for resistance measurement and complementary to other high current methods (Weingarth, Foelske-Schmitz & Kötz, 2013) (Majumdar, Baugh & Bhattacharya, 2017).

Merits and Applications of Supercapacitors

Ultracapacitor have the following benefits:

- **Safe Power Facility:** Provides reliable short-term power, even on the failure or fluctuation from primary source.
- **Energy Storage Capability:** Enabling support for high power loads.
- **Pulse Power Capacity:** Provides peak power supply to the load while drawing minimum power from the source.
- Minimizes the dimension & weight of the required power source.
- Improves run-time & battery longevity, particularly at low temperature environment.
- Can replace batteries in many situations and also can harvest energy from clean sources.
- Protects against sudden/accidental power loss or fluctuations/interruptions.
- Battery-like replacements are not required (unlimited discharge cycles).
- Environmentally benign & safe.

The various important potential applications of supercapacitors in the present era are outlined and represented in Figure 9.

Supercapacitors in Military Applications

Recent military operations urge heavily on un-interrupted electrical energy supplies. Sensors, various communications systems like radios, radars; viewing and sighting systems, along with innumerable other target equipment are currently supported by heavy-duty batteries. But these existing power supply mechanisms suffer a lot of disadvantages such as - (i) heavy weight and voluminous, hence often immovable; (ii) requires periodic replacement for continuous power demand; (iii) not efficient under extreme temperature conditions; (iv) disposal is a big environmental issue; (v) large installation expenses as well as recurring maintenance costs.

Supercapacitors are blessed with high energy densities and power densities. As far as commercialization is concerned, individual supercapacitors are now collectively employed as "supercapacitor modules" to address maximum stable output voltages supplies. Supercapacitor-based hybrid power systems offer better possibility to deploy powerful yet miniatured, paper-like flexible power generating/store-house equipment with improved weight-burden issues. Moreover, supercapacitors are largely capable of solving current problematic issues related to (i) backing-up power supply for electronics in military vehicles, fire control systems in armored vehicles and tanks, airbag operation; portable and mini- power storage

Figure 9. Various potential applications of supercapacitors
Created with Universal Document Converter

for emergency hand-held radios and radars. Constant, un-interrupted supplies of power to GPS-guided missiles and projectiles will be handful as well. Even with this new technological aid, Unmanned Aerial Vehicles (UAV) and Radars could be installed at the LOCs for constant watch of the enemy movements. Some of the important sectors which "supercaps" are going to contribute significantly in military applications are shown in Figure 10. In the recent future, flexible supercapacitor systems will broadly enforce efficient environmental energy capture schemes to improve the working efficiency of the military devices under extreme environmental conditions as well. Thus, this upcoming technology will certainly improve the blueprints of conventional electronic military equipment as well as modify futuristic weaponries that will surely strengthen the self-defense aspect in the near-coming days.

Limitations of the Existing Commercialized Supercapacitors

- Generally exhibits low per cell voltage. Thus, for fulfilling higher voltage requirements, the cells have to be connected in series which also increases ESR.
- The ultracapacitor fabrication and specifications are solely dependent on intended application and manufacturing concern.
- The electrode materials may also differ significantly between designers or due to specific application requirements.

Figure 10. Potential applications of supercapacitors in the military field
Created with Universal Document Converter

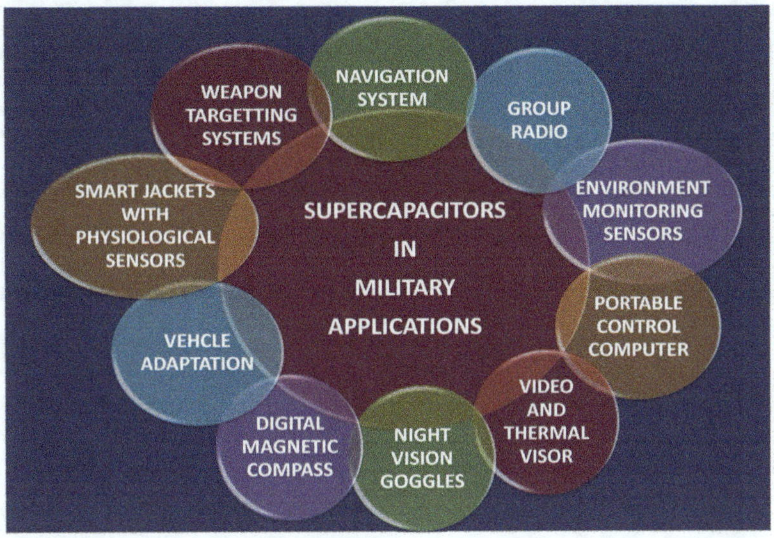

- Cannot be applied in AC and high frequency circuits because their times constant are not appropriate for usage in AC or high frequency circuits.
- Mechanical Stress leads to deformation, lead breakage and ESR increase of the device.
- High temperature efficiency is low as it leads to capacitance loss, ESR increase, bulging possible venting
- Over-voltage leads to capacitance loss, increase of ESR, bulging and possible venting.

POLYANILINE (PANI) AS SUPERCAPACITOR ELECTRODE MATERIAL

Polyaniline (PANI), a p-type conjugated polymer, has well been studied as an electrode material for energy storage devices owing to its high theoretical capacity (assuming full doping, neglecting the mass of the anion: 294 $mAhg^{-1}$), good conductivity, low cost, and ease of synthesis. PANI is found to exhibit variable oxidation states and are accordingly named as fully reduced leuco-emeraldine (LE) ($y = 1$), semi-oxidized emeraldine base (EB) ($y = 0.5$) and completely-oxidized pernigraniline (PE) ($y = 0$) respectively. The conducting nature - along with chemical structure of PANI forms at different oxidation states is shown in Figure 11 (Devillers, Jemei, & Péra, 2014) (Heeger, 2001).

The plausible mechanistic steps involved in aniline oxidative polymerization to form PANI under acidic conditions are shown schematically in Figure 12.

The early researches on PANI were carried out around 1860s on aniline oxidation, isolated only 20 years before then. This was the pioneer work done to develop successfully the commercially route to Aniline Black dye. Then after a long gap of about century, in 1968, report on metal-like electrical conductivity of doped-PANI and its potential electrochemical applications were first reported, followed by revolutionary contribution from MacDiarmid, Heeger and his co-workers, that resulted in the birth

Figure 11. Chemical structure of PANI in different oxidation states: (here: Ins-Insulator, Con- conducting)
Created with Universal Document Converter

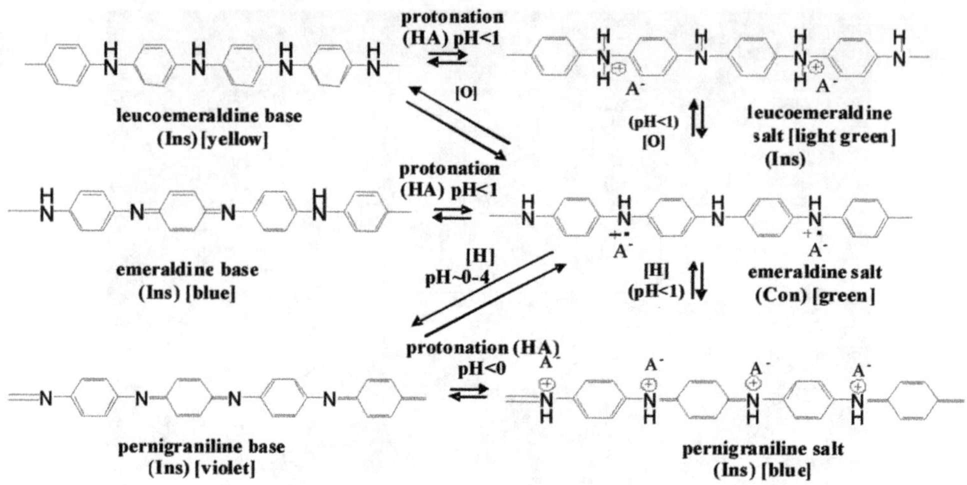

Scheme-1: Chemical structure of PANI in different oxidation states: (here: Ins-Insulator, Con- conducting]

Figure 12. Schematic representation of the steps involved in oxidative polymerization of aniline in acid medium
Created with Universal Document Converter

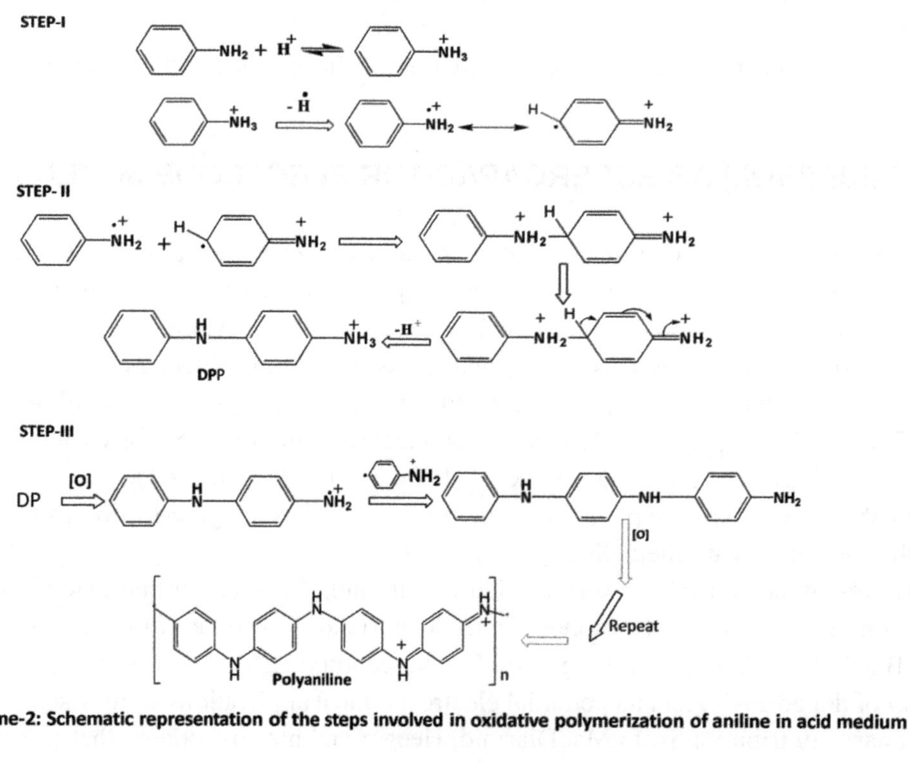

Scheme-2: Schematic representation of the steps involved in oxidative polymerization of aniline in acid medium

of conducting/conjugated polymers that rightly owned the Nobel Prize in Chemistry in the year 2000 (Teasdale, 2003) (Letheby,1862) (MacDiarmid, 2001).

Since then, PANI has been among the more popular and widely employed conducting polymers, especially in the area of electrochemistry (MacDiarmid, 2001) (Chiang, Druy, Gau & Heeger, 1978) (Genies, Tsintavis, 1985) (Genies, Lapkowski, Penneau, 1988). The reversible redox-active behaviors of PANI films formed on various substrate-electrode surfaces through facile electropolymerization that are detected by the modern CV instruments boosted its usage in multiple electrochemical applications (Feldberg, 1984) (Fiordiponti, Pistoia, 1989). In 1984, Feldberg detected the plausible contribution of capacitive currents in the electrochemistry of conjugated polymers through voltammetric and impedance spectroscopic studies that highlighted the capacitive nature of these conducting polymers (Feldberg, 1984). Further intriguing works of Conway research group established PANI as a promising candidate for electrochemical capacitors (ECs) delivering remarkable pseudocapacitive signatures (Conway, 1991).

Kaner and co-researchers also made detailed explorations on PANI as an electroactive material of supercapacitors (Huang, Huang, Ng, & Kaner, 1993). Shortly, Rudge and his coworkers examined the doping effect of PANI systems on its supercapacitor performances (Rudge, Raistrick, Gottesfeld, & Ferraris, 1994). They showed that PANI shows more material characteristics that of natural supercapacitor system rather than battery. Arbizzani and her coworkers explicitly explained that the charge storage behavior of conducting polymers was attributed to redox reactions (Arbizzani, Mastragostino & Meneghello, 1996). Hu and Chu designed an iridium-doped PANI film based supercapacitor that showed strong dependency of specific capacitance of PANI on the dopants nature as well as degree of doping, although this dependency is non-linear (Hu & Chu, 2001). Such variation with nature of dopants has been displayed in Table 2.

Table 2. Variation of electrochemical performances with dopants

PANI Dopants	Discharge Current Density Potential Window and Specific Capacitance	Electrolyte	% Retention (Number of GCD Cycles)	References
PANI-HCl	1.25 mA cm^{-2} 0-1.0 V 70 Fg^{-1}	1M Et_4NBF_4	57% (1000)	(Bian, Yu, 2010)
BF_4-doped	50 mV s^{-1} 0-0.75V 74 Fg^{-1}	4.0 M HBF	74 5% (20,000)	(Bandyopadhyay, Kuila & Balamurugan, 2017)
PANI-$LiPF_6$	3.75 A g^{-1} 0-0.75V 115 Fg^{-1}	1 M Et_4NBF_4	78% (5000)	(Belanger, Ren & Davey, 2000)
Self-doped PANI	1 A g-1 0.6-0.7V 408 Fg^{-1}	0.5 Na_2SO_4	---	(Ryu, Wu, Lee & Chang, 2003)
Sulfonated PANI	1 A g^{-1} 0-1.2V 1107 Fg^{-1}	1 M H_2SO_4	-----	(Ryu, Wu, Lee & Chang, 2003)
De-doped PANI nanofibers	2.5 A g^{-1} 0-0.65V 593 Fg^{-1}	1 M H_2SO_4	70% (5000)	(Rudge, Raistrick, Gottesfeld & Ferraris, 1994)

The possible reason for such non-linearity behavior is that some dopants often gets covalently attached to the PANI chains and thus cannot be electrochemically removed, thereby blocking a part of PANI's capacity. A common difficulty in use of PANI-based supercapacitors arises from the fact that high proton concentration degrades the electroactive material. Thus many authors have promoted self-doped-PANI that contains ionisable negatively-charged functional groups, most frequently sulphonate or carboxylate units, attached to aromatic rings or to the nitrogen atoms through a spacer consisting of a small number of methylene (-CH_2-) units that serve as intramolecular dopant anions neutralizing the positive charge of the protonated nitrogen atoms on the polymer backbone; thus, replacing the need of supporting electrolytes with dopant agents. Such internal-doping property introduced several distinguishing characteristics quite unusual from those of the pristine PANI (Ghenaatian, Mousavi, & Kazemi, 2009). De-doping initiates morphological changes in favor of the initiation of micro-porous structure that promote better accessibility of electrolyte ions to the electroactive material (Bian & Yu, 2010). Thus doping-dedoping mechanism introduces improved electrochemical signature in these systems.

It has been observed that PANI (cationic form as it is p-type semiconductor) stores charge through a pseudocapacitive doping/dedoping mechanism, in which anions transport inside and migrates out of the electrode as PANI is subsequently oxidized and reduced (Bandyopadhyay, Kuila, & Balamurugan, 2017). Accordingly, mass transport of the dopant ion is a potential issue, especially for dense PANI electrodes. In this regard, switchover to nanophase is expected to show promising signatures because of their self-assembling to porous, high surface area morphologies (Belanger, Ren & Davey, 2000) (Ryu, Wu, Lee & Chang, 2003). During chemical polymerization of aniline, various parameters can be controlled to modify the morphology of the resulting PANI. For instance, changing the temperature of polymerization also results in different morphology based-PANI that again have noticeable effects on the specific capacitance and cyclability of the resulting systems (Bian, Luan, Liu & Liu, 2012) (Bláha, Varga & Prokeš 2013) (Gholamian & Contractor, 1988).

Enhancing the electrochemically accessible surface area by designing porous-structured-PANI systems can also significantly improve the supercapacitor performances. Few years back, comparison of electrochemical performances between packed and hollow polyaniline (PANI) nanofibers with controlled wall-thickness fabricated by in-situ polymerization of aniline using electrospun poly(amic acid) fiber membrane as a template was carried out. The hollow porous PANI nanofibers with larger active surface area and optimum wall-thickness showed better capacitive behavior with appreciable capacitance retention. Technique of such nanofiber fabrication and their electrochemical behaviors are displayed in Figure 13 (Miao, Fan, Chen & Liu, 2013).

Sharma *et al.* synthesized a nanoporous PANI with large surface area of 1059 m^2g^{-1} that exhibited specific capacitance ~ 410 Fg^{-1} (Gawli, Banerjee & Dhakras, 2016). It is indeed fascinating that the hyper-cross-linked architecture of PANI promoted an excellent cyclability for retaining ~100% of the initial specific capacitance after 1000 cycles. It also instigates better flexibility and durability of the electrode. Systematic studies of the charge storage efficiencies of various PANI nanostructures like nanotubes, nanowires, nanofibers, nanosheets have been carried out, employing different preparation techniques, some of which has been summarized in Table 3. (Gawli, Banerjee, & Dhakras 2016) (Sharma, Sahoo & Sharma, 2015) (Mu, Ma, Peng, & Li, 2013) (Chen, Rakhi & Alshareef 2013) (Wang, Deng & Duan, 2014) (Miao, Fan& Chen, 2013) (Kuang, Cao, Wang, & Jing, 2013) (Luo, Kong, Jia & Luo, 2013) (Kuila, Nandan, Bo¨hme & Janke, 2009) (Sivakkumar, Kim, Choi & Douglas, 2007) (Jin, Jiang, Zhang, & Li, 2018).

Figure 13. (A) Fabrication technique of Hollow PANI Nanofibers (B) Comparative cyclic voltammograms of packed and hollow PANI Nanofibers at 20mVs^{-1} from -0.2V to +0.8V in 1M H_2SO_4 medium. (C) Comparative specific capacitance plots of packed and hollow PANI nanofibers at different current densities.
Created with Universal Document Converter
[[Reproduced with permission from American Chemical Society, (Miao, Fan, Chen & Liu, 2013)]

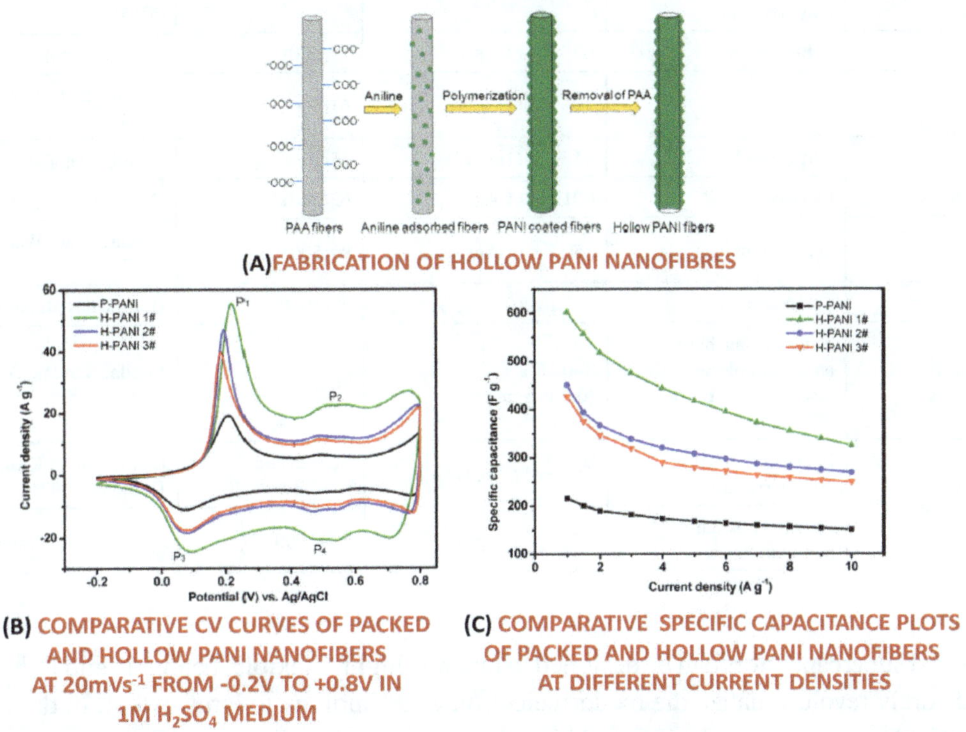

Both theoretical and experimental specific capacitances of PANI systems in sulfuric acid medium were evaluated (Li, Wang & Chu, 2009). The approximate maximum specific capacitance of PANI was calculated to be 2000 Fg^{-1}, while the experimental values evaluated by different methodologies are much smaller than the theoretical value as only a small portion of PANI (mainly the surface in major cases) puts its contribution to the total capacitance and the percentage of effective electroactive phase is again depended on its conductivity as well as diffusion rate of counter-anions which are very much correlated with its morphology (Soni, Kashyap, Nagaraju & Kurungot, 2018). Reverse template technique, recently developed, in which a second component is removed after the synthesis, is another effective method to obtain porous PANI morphology and such porosity introduction nearly doubled the estimated specific capacitance of PANI (Mujawar, Ambade, Battumur & Ambade, 2011) (Cho, Shin & Jang, 2013). It is also feasible to use a base nanofiber as template for polymerizing aniline and subsequently remove it to form porous PANI nanotubes (Xie, Kocaefe, Chen & Kocaefe, 2016).

In the recent past, Chen and co-workers have designed PANI nanostructures with three different morphologies (nanofibers, nanotubes and nanospheres) deploying corresponding MnO_2 templates to tune the morphology accordingly. It was found that the PANI-nanotubes with higher surface area, charge transport features and shorter ion-passage facilities offered better electrochemical behavior in presence of redox- active electrolyte for high performance supercapacitors. Their comparative electrochemical studies are displayed in Figure 14 (Chen, Rakhi & Alshareef, 2013).

Table 3. Variation of electrochemical performances with nano-morphology

Pure PANI Morphology	Method of Synthesis	Specific Capacitance (Discharge Current Density)	% Retention (Number of GCD Cycles)	References
Nanowires	Electrochemical polymerization.	950 Fg^{-1} ($1Ag^{-1}$)		(Sharma, Sahoo & Sharma, 2015)
nanotubes	Self-assembly	625Fg^{-1} ($1Ag^{-1}$)	77(500)	(Mu, Ma, Peng, & Li, 2013)
nanofiber	Template, MnO_2	404 Fg^{-1}($1Ag^{-1}$)	61(5000)	(Chen, Rakhi & Alshareef, 2013)
nanorods	Self-assembly	455 Fg^{-1}($1mVs^{-1}$)	65(1300)	(Wang, Deng & Duan, 2014)
hollow nanofibers	Electrospun	601 Fg^{-1}(1 Ag^{-1})	62(500)	(Miao, Fan& Chen, 2013)
nanorods	Self-assembly	350 Fg^{-1}(1 Ag^{-1})	99(500)	(Kuang, Cao, Wang, & Jing, 2013)
nanoworms	Self assembly	301 Fg^{-1}(0.5 Ag^{-1})	72(1000)	(Luo, Kong, Jia & Luo, 2013)
Ordered vertically aligned nanorods	nanotemplate based on supramolecular assemblies of block copolymer	3407 F g^{-1} (0.5 $mAcm^{-2}$)	500cycles	(Kuila, Nandan, Bo¨hme & Janke, 2009)
nanofibers	Interfacial polymerization	554 Fg^{-1} ($1Ag^{-1}$)	Poor cyclic stability.	(Sivakkumar, Kim, Choi & Douglas, 2007)
nanowhiskers	Chemical oxidative polymerization.	470 Fg^{-1} ($1Ag^{-1}$)	90 (1000)	(Yan, Cheng, Wang& Li, 2011)

PANI based supercapacitors can be designed to become highly flexible, bendable and light-weighted. This would surely revolutionalize the modern electronics technology to large extent. In the recent past, a simple, scalable, paper-based flexible and high-performance all-solid-state SCs based with electrodeposited PANI on Au-coated paper was fabricated with appreciable volume capacitance and high energy and power densities, which will be handful for self-powered portable nanosystems (Yuan, Xiao, Ding & Zhong, 2012). Similar flexible electrodes with high electrochemical performances were reported on carbon-cloth coated PANI nanowires as well (Horng, Lu & Hsu, 2010).

Orientations of the PANI nanotubes have dramatical influence over improvement of its electrochemical efficiency. Single-step, facial template-free strategy for growing vertically aligned arrays of PANI narrow-walled wires on various conducting substrates employing galvanostatic current method. Such alignment facilitated faster and smoother ion diffusion leading to high capacitance of 950 Fg^{-1} with high retention at high current density (780 F g^{-1} @ 40 Ag^{-1}). Further, the capacitances were appreciable both in aqueous as well as in non-aqueous media. Such high capacitance is attributed to the additive effect of the capacitances of the parallel aligned individual PANI nanowires (Wang, Huang & Wei, 2010).

In summary, pristine-PANI applied in supercapacitors has been enormously studied, while its electrochemical signatures especially the cycling stability still cannot meet practical application requirements (Bhowmick, Bain, Maity & Bera, 2012) (Zhang, Li, Zhang & Wang, 2011) (Wu, Pan, Li & Zhao, 2015) (Grover, Goel, Marichi & Sahu, 2016). Inspite of easily designable advantages, light weight and flexibility, they exhibit very poor environmental stabilities during the charging/ discharging process due to swelling and shrinkage of the polymer backbone and also lack long cycle lives. Thus, much further research highlight is demanded for optimizing key issues such as: (i) porosity of the formed-structure, (ii) ordered and well-aligned morphology of the polymer for better charge transport, (iii) optimum doping

Figure 14. Comparative electrochemical performances: (a) cyclic voltammetric curves (b) Galvanostatic charging / Discharging plots (c) specific capacitance plots at different current densities (d) Electrochemical Impedance Spectroscopy (Nyquist plots) of different nano-morphologies of PANI in 1M H_2SO_4-hydroquinone electrolyte.
Created with Universal Document Converter
[Reproduced with permission from ACS publications, (Chen, Rakhi & Alshareef, 2013).

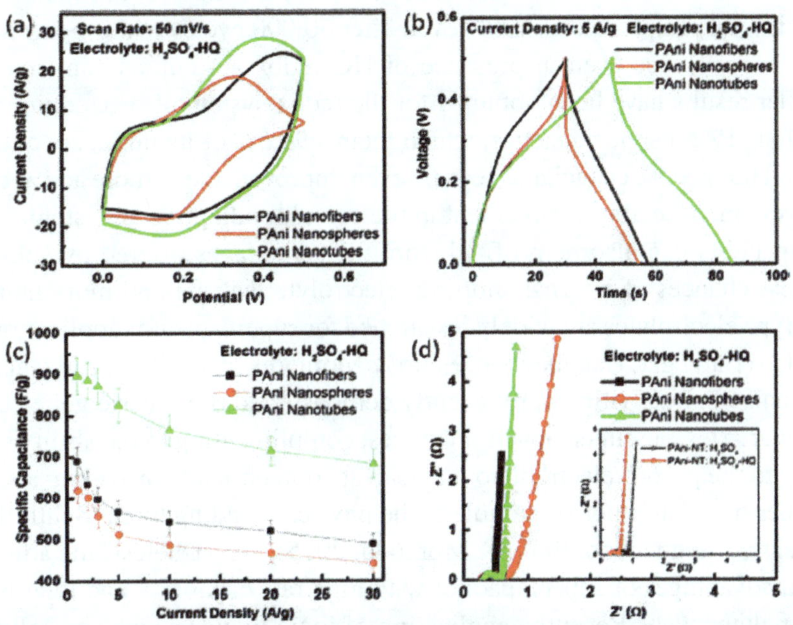

COMPARATIVE ELECTROCHEMICAL PERFORMANCES: (a) CV (b) GALVANOSTATIC CHARGING / DISCHARGING (c) SPECIFIC CAPACITANCE PLOTS AT DIFFERENT CURRENT DENSITIES (d) ELECTROCHEMICAL IMPEDANCE SPECTROSCOPY (NYQUIST PLOTS) OF DIFFERENT NANOMORPHOLOGIES OF PANI IN 1M H_2SO_4-HYDROQUINONE ELECTROLYTE .

level for superior charge transport (iv) stability towards extreme conditions. Hence, researchers have tried to prepare various PANI-based composites through mixing/blending PANI with carbon materials and/or metal oxides and/or their derivatives for the purpose of improving the morphology/electronic structure so as to extract superior electrochemical characteristics for advanced flexible, portable miniatured supercapacitor applications (Tiwari, Tiwari & Kim 2012) (Bhadra & Lee, 2009) (Nguyen & Yoon, 2016).

ROLE OF ELECTROLYTES

Several rigorous studies have accomplished that in order to fabricate smart and improved electrode materials, the device design should also consider proper utilization of electrolytes. In that perspective, an ideal electrolyte should be abundant, less expensive, environmentally benign, electronically insulating, but ionically conductive with appreciable electrochemical stability over wide potential window. Generally, aqueous electrolytes are readily available, cost-effective and show faster ionic conductance. But the organic electrolyte or ionic liquids have achieved more stable overall working potential range.

An innovating idea for escalating the total capacitance of PANI-based supercapacitors is to introduce an additional redox system in the electrolyte. (Chen & Xue, 2015) (Xie, Zhu, Wu & Wu, 2014) (Huang, Humphrey & MacDiarmid, 1986). For instance, addition of hydroquinone (HQ) to the conventional aqueous-H_2SO_4 electrolyte for PANI super-capacitors almost doubled the capacitance owing to Faradaic contribution from HQ (Huang, Humphrey & MacDiarmid, 1986). Although such introduction complicate the electro-chemical processes at the electrode /electrolyte interface, but the GCD cyclability of this supercapacitor is really improved to ~ 90% even after 50,000 cycles without any noticeable change in the charge transfer resistance even in presence of HQ redox system, as confirmed by Impedance Spectroscopy. Similar results have been obtained for the redox system of benzoquinone-hydroquinone (BQ-HQ) for a lifelong PANI supercapacitor, which retains 99.5% of its initial specific capacitance on 13,000 GCD cycles. The specific capacitance can be even improved due to more active electrolytic redox systems that the favors interfacial rearrangement in the corresponding redox system.

Certain supercapacitors can also be profited from the advantages offered by solid or gel polymer electrolytes, with less chances of internal shorting, electrolyte leakage and more handling flexibility. Polymer based gel-type electrolytes are widely researched for supercapacitor application (Senthilkumar, Selvan & Melo, 2013) (Barzegar, Dangbegnon, Bello & Momodu, 2015). Even, advanced polymer electrolytes with high ionic conductivities were recently demonstrated. Polymeric gel electrolyte are often used in designing smart-flexible supercapacitor devices, complimenting to the ability of gel structure to penetrate efficiently to the porous electrode to the nearest contact, while providing a stable structure in maintaining the dimension that is not vulnerable to the physical bending force (Senthilkumar, Selvan & Melo, 2013) (Barzegar, Dangbegnon, Bello & Momodu, 2015). Nevertheless, the admirable flexibility achieved the crucial advantages of supercapacitor with high rate capability and long life cycle must be well maintained simultaneously. Recently, application of ionic liquids as well as solid electrolytes are being highly promoted although these electrolytes have lower ionic conductivities compared to liquid electrolytes, which may hold back significantly the intended power density of supercapacitors (Cheng, Pan, Zhao & Liao, 2018) (Salanne, 2017) (Chuizhou, Gall & Irazoqui, 2013). Thus, considerable explorations are desirable to expand the domain of solid electrolytes with superior properties or to achieve solid-proof encapsulation of liquid electrolytes for leakage free design on next generation –supercapacitors that would be very ideal for robust conditions' usage much appropriate for space and military-based applications (Yang, Zhang, Zhang & Zhang, 2014) (Akinwolemiwa, Wei & Chen, 2017).

SCOPE OF PANI-BASED SUPERCAPACITORS IN MILITARY APPLICATIONS

PANI supercapacitors can be well utilizable for military applications especially taking into account of the following important characteristics:

- PANI based supercapacitors are found to be more flexible and bendable in nature compared to other available inorganic-based supercapacitors (Moussa, El-Kady & Zhao, 2016).
- "Supercap" devices with paper-like design are now being intended that would be light weight as well as readily portable. (Yuan, Xiao, Ding & Zhong, 2012) (Kumar, Kumar, & Awasthi, 2017)
- The installation cost is expected to be minimal as the precursors are readily available.
- Flexible solid-state micro-supercapacitor are now being fabricated which are easy-handled, light, durable, and robust. (Hu, Zhang, Li & Jia, 2014)

- Beside enormous contribution for powered energy hybrid systems, the PANI-based materials can efficiently serve in various multifunctional applications such as for energy absorption, electromagnetic shielding, protective and anti-corrosive coatings, fire retardation agents, chemi-sensors, etc. (Geetha, Kumar, & Rao,2009) (Song & Choi, 2013) (Kurahatti, Surendranathan & Kori 2010) (Sudha, Sivakala & Patel, 2010).

However, issues related to short charging/discharging life cycle and faster environmental decomposition are to be addressed by blending with suitable components to provide "superior" and "smarter" conducting polymer based-composite electrode materials for advanced military applications in the near future.

CONCLUSION AND FUTURE OUTLOOK

The early part of the present chapter focuses on the fundamental concepts and basic principles of fabrication and designing, as well as performance evaluation of supercapacitors. Role of "supercaps" in various applications to the modern society especially towards military and defense sector have been highlighted. Then, we focus on how PANI acts as an efficient electrode material for supercapacitors. As revealed, PANI, as conducting polymer, is one of the pioneer electroactive materials that have paved the corridor for the elementary progress and practical development of pseudocapacitors. More or less all the present discussions carried out can be applied to similar other conducting polymers such as polypyrrole or polythiophene systems, although, PANI still dominates the family of electrode materials for the fabrication of "supercaps". Because of its porous structure compared to other inorganic materials, it is ideally suited for supercapacitor applications as it provides appropriate adsorption /desorption /intercalation sites for electroactive species within its morphology. The chemical flexibility of PANI compared to available inorganic pseudocapacitors towards doping with a variety of species, opens up a wider prospect for tuning device pseudocapacitance characteristics based on both electrode material morphology as well as on opting novel electrolyte choices too. Furthermore, PANI has often been applied as a protective layer, binder or even as additive to enhance the working efficiency of various metal-oxide based supercapacitors. As already discussed, one- and two-dimensional nanostructures of PANI, such as nanofibers, nanowires, nanorods, nanotubes, nanofilms, etc., have attracted an enormous deal of attention across major scientific and engineering disciplines with the prospect that these materials could express superior properties compared to their bulk counterparts in terms of mechanical strength, processibility and charge transport. However, their commercialization is very much restricted by rapid agglomeration during processing, low solubilities in common solvents, poor charging/discharging life cycle and faster environmental decomposition. Again, metal oxides and carbon-based materials like amorphous carbons, carbon nanotubes, graphene systems, owing to their unique redox and versatile properties, such as high electrical conductivity, appreciable chemical stability along with remarkable mechanical behavior, morphology-based surface area, find promising applications in various energy storage devices. But the so-far obtained specific capacitances for the individual components are too low to be applied for realistic applications. Hence, sincere efforts have been made to combine the individual component characteristics so as to obtain high quality materials with superior properties – "nanocomposites" to be applied for future generation energy storage devices, which has been the main objective of the following chapter.

ACKNOWLEDGMENT

DM acknowledges Indian Association for the Cultivation of Science, Jadavpur, Kolkata, WB, India and Department of Chemistry (Physical Chemistry Section), Jadavpur University, Kolkata-700032, WB, India for e-library facilities. DM also acknowledges Chandernagore College, Hooghly, Pin-712136, WB, India for providing permission for carrying out honorary research.

REFERENCES

Akinwolemiwa, B., Wei, C., & Chen, G. Z. (2017). Mechanisms and Designs of Asymmetrical Electrochemical Capacitors. *Electrochimica Acta, 247*, 344–357. doi:10.1016/j.electacta.2017.06.088

An, K. H., Kim, W. S., Park, Y. S., Choi, Y. C., Lee, S. M., Chung, D. C., ... Lee, Y. H. (2001). Supercapacitors using single-walled carbon nanotube electrodes. *Advanced Materials, 13*(7), 497–500. doi:10.1002/1521-4095(200104)13:7<497::AID-ADMA497>3.0.CO;2-H

Arbizzani, C., Mastragostino, M., & Meneghello, L. (1996). Polymer-based redox super-capacitors: A comparative study. *Electrochimica Acta, 41*(1), 21–26. doi:10.1016/0013-4686(95)00289-Q

Bandyopadhyay, P., Kuila, T., Balamurugan, J., Nguyen, T. T., Kim, N. H., & Lee, J. H. (2017). Facile synthesis of novel sulfonated polyaniline functionalized graphene using m-aminobenzene sulfonic acid for asymmetric supercapacitor application. *Chemical Engineering Journal, 308*, 1174–1184. doi:10.1016/j.cej.2016.10.015

Barzegar, F., Dangbegnon, J. K., Bello, A., Momodu, D. Y., & Charlie, A. T. Jr. (2015). Effect of conductive additives to gel electrolytes on activated carbon-based supercapacitors. *AIP Advances, 5*(9), 097171. doi:10.1063/1.4931956

Belanger, D., Ren, X., Davey, J., Uribe, F., & Gottesfeld, S. (2000). Characterization and long-term performance of polyaniline-based electrochemical capacitors. *Journal of the Electrochemical Society, 147*(8), 2923–2929. doi:10.1149/1.1393626

Bhadra, S., & Lee, J. H. (2009). Synthesis of higher soluble nanostructured polyaniline by vapor-phase polymerization and determination of its crystal structure. *Appl. Polymer Science, 114*(1), 331–340.

Bhowmick, B., Bain, M. K., Maity, D., Bera, N. K., Mondal, D., Mollick, M. M. R., ... Chattopadhyay, D. (2012). Synthesis of dendritic polyaniline nanofibers by using soft template of sodium alginate. *Journal of Applied Polymer Science, 123*(3), 1630–1635. doi:10.1002/app.34575

Bian, C., & Yu, A. (2010). De-doped polyaniline nanofibres with micropores for high-rate aqueous electrochemical capacitor. *Synthetic Metals, 160*(13-14), 1579–1583. doi:10.1016/j.synthmet.2010.04.019

Bian, L., Luan, F., Liu, S., & Liu, X. (2012). Self-doped polyaniline on functionalized carbon cloth as electroactive materials for supercapacitor. *Electrochimica Acta, 64*, 17–22. doi:10.1016/j.electacta.2011.12.012

Bláha, M., Varga, M., Prokeš, J., Zhigunov, A., & Vohlídal, J. (2013). Effects of the polymerization temperature on the structure, morphology and conductivity of polyaniline prepared with ammonium peroxodisulfate. *European Polymer Journal, 49*(12), 3904–3911. doi:10.1016/j.eurpolymj.2013.08.018

Burke, A. (2000). Ultracapacitors: Why, how, and where is the technology. *Journal of Power Sources, 91*(1), 37–50. doi:10.1016/S0378-7753(00)00485-7

Chen, K., & Xue, D. (2015). Searching for electrode materials with high electrochemical reactivity. *Journal of Materiomics, 1*(3), 170–187. doi:10.1016/j.jmat.2015.07.001

Chen, W., Rakhi, R. B., & Alshareef, H. N. (2013). Morphology-Dependent Enhancement of the Pseudocapacitance of Template-Guided Tunable Polyaniline Nanostructures. *The Journal of Physical Chemistry C, 117*(29), 15009–15019. doi:10.1021/jp405300p

Cheng, X., Pan, J., Zhao, Y., Liao, M., & Peng, H. (2018). Gel Polymer Electrolytes for Electrochemical Energy Storage. *Advanced Energy Materials, 8*(7), 1702184. doi:10.1002/aenm.201702184

Chiang, C. K., Druy, M. A., Gau, S. C., Heeger, A. J., Louis, E. J., MacDiarmid, A. G., ... Shirakawa, H. (1978). Synthesis of highly conducting films of derivatives of polyacetylene, (CH)x. *Journal of the American Chemical Society, 100*(3), 1013–1015. doi:10.1021/ja00471a081

Chiang, C. K., Fincher, C. R., Park, Y. W., Heeger, A. J., Shirakawa, H., Louis, E. J., ... MacDiarmid, A. G. (1977). Electrical conductivity in doped polyacetylene. *Physical Review Letters, 39*(17), 1098–1101. doi:10.1103/PhysRevLett.39.1098

Cho, S., Shin, K., & Jang, J. (2013). Enhanced electrochemical performance of highly porous supercapacitor electrodes based on solution processed polyaniline thin films. *ACS Applied Materials & Interfaces, 5*(18), 9186–9193. doi:10.1021/am402702y PMID:24032539

Christen, T., & Carlen, M. W. (2000). Theory of Ragone plots. *Journal of Power Sources, 91*(2), 210–216. doi:10.1016/S0378-7753(00)00474-2

Chu, A., & Braatz, P. (2002). Comparison of commercial supercapacitors and high power lithium-ion batteries for power-assist applications in hybrid electric vehicles I. Initial characterization. *Journal of Power Sources, 112*(1), 236–246. doi:10.1016/S0378-7753(02)00364-6

Chuizhou, M., Gall, O. Z., & Irazoqui, P. (2013). A flexible super-capacitive solid-state power supply for miniature implantable medical devices. *Biomedical Microdevices, 15*(6), 973–983. doi:10.100710544-013-9789-1 PMID:23832644

Conway, B. E. (1991). Transition from "supercapacitor" to "battery" behavior in electrochemical energy storage. *Journal of the Electrochemical Society, 138*(6), 1539–1548. doi:10.1149/1.2085829

Dai, L., Chang, D. W., Baek, J. B., & Lu, W. (2012). Carbon nanomaterials for advanced energy conversion and storage. *Small, 8*(8), 1130–1166. doi:10.1002mll.201101594 PMID:22383334

Devillers, N., Jemei, S., Péra, M.-C., Bienaimé, D., & Gustin, F. (2014). Review of characterization methods for supercapacitor modeling. *Journal of Power Sources, 246*, 596–608. doi:10.1016/j.jpowsour.2013.07.116

Diaz, A., & Logan, J. (1980). Electroactive polyaniline films. *Journal of Electroanalytical Chemistry and Interfacial Electrochemistry, 111*(1), 111–114. doi:10.1016/S0022-0728(80)80081-7

Du, C., Yeh, J., & Pan, N. (2005). High power density supercapacitors using locally aligned carbon nanotube electrodes. *Nanotechnology, 16*(4), 350–353. doi:10.1088/0957-4484/16/4/003

Duay, J., Sherrill, S. A., Gui, Z., Gillette, E., & Lee, S. B. (2013). Self-limiting electrodeposition of hierarchical MnO_2 and $M(OH)_2/MnO_2$ nanofibril/nanowires: Mechanism and supercapacitor properties. *ACS Nano, 7*(2), 1200–1214. doi:10.1021/nn3056077 PMID:23327566

Dupont, M. F., & Donne, S. W. (2016). Charge storage mechanisms in electrochemical capacitors: Effects of electrode properties on performance. *Journal of Power Sources, 326*, 613–623. doi:10.1016/j.jpowsour.2016.03.073

Eftekhari, A., Li, L., & Yang, Y. (2017). Polyaniline supercapacitors. *Journal of Power Sources, 347*, 86–107. doi:10.1016/j.jpowsour.2017.02.054

Feldberg, S. W. (1984). Reinterpretation of polypyrrole electrochemistry. Consideration of capacitive currents in redox switching of conducting polymers. *Journal of the American Chemical Society, 106*(17), 4671–4674. doi:10.1021/ja00329a004

Fiordiponti, P., & Pistoia, G. (1989). An impedance study of polyaniline films in aqueous and organic solutions. *Electrochimica Acta, 34*(2), 215–221. doi:10.1016/0013-4686(89)87088-4

Frackowiak, E., & Beguin, F. (2001). Carbon materials for the electrochemical storage of energy in capacitors - review. *Carbon, 39*(6), 937–950. doi:10.1016/S0008-6223(00)00183-4

Frackowiak, E., & Beguin, F. (2001). Carbon materials for the electrochemical storage of energy in capacitors. *Carbon, 39*(6), 937–950. doi:10.1016/S0008-6223(00)00183-4

Gawli, Y., Banerjee, A., Dhakras, D., Deo, M., Bulani, D., Wadgaonkar, P., ... Ogale, S. (2016). 3D polyaniline architecture by concurrent inorganic and organic acid doping for superior and robust high rate supercapacitor performance. *Scientific Reports, 6*(1), 21002. doi:10.1038rep21002 PMID:26867570

Geetha, S., Satheesh Kumar, K. K., Rao, C. R. K., Vijayan, M., & Trivedi, D. C. (2009). EMI shielding: Methods and materials—A review. *Journal of Applied Polymer Science, 112*(4), 2073–2086. doi:10.1002/app.29812

Genies, E., Lapkowski, M., & Penneau, J. (1988). Cyclic voltammetry of polyaniline: Interpretation of the middle peak, J. *Electroanal. Chem. Interf. Electrochem., 249*(1-2), 97–107. doi:10.1016/0022-0728(88)80351-6

Genies, E., & Tsintavis, C. (1985). Redox mechanism and electrochemical behaviour or polyaniline deposits. *Journal of Electroanalytical Chemistry and Interfacial Electrochemistry, 195*(1), 109–128. doi:10.1016/0022-0728(85)80009-7

Ghenaatian, H. R., Mousavi, M. F., Kazemi, S. H., & Shamsipur, M. (2009). Electrochemical investigations of self-doped polyaniline nanofibers as a new electroactive material for high performance redox supercapacitor. *Synthetic Metals, 159*(17-18), 1717–1722. doi:10.1016/j.synthmet.2009.05.014

Gholamian, M., & Contractor, A. (1988). Effect of the temperature of synthesis on the conductivity and electrochemical behaviour of polyaniline. *Journal of Electroanalytical Chemistry and Interfacial Electrochemistry, 252*(2), 291–301. doi:10.1016/0022-0728(88)80217-1

Grover, S., Goel, S., Marichi, R. B., Sahu, V., Singh, G., & Sharma, R. K. (2016). Polyaniline all solid-state pseudocapacitor: Role of morphological variations in performance evolution. *Electrochimica Acta, 196*, 131–139. doi:10.1016/j.electacta.2016.02.157

Guan, L., Yu, L., & Chen, G. Z. (2016). Capacitive and non-capacitive faradaic charge storage. *Electrochimica Acta, 206*, 464–478. doi:10.1016/j.electacta.2016.01.213

Heeger, A. (2001). Nobel Lecture: Semiconducting and metallic polymers: The fourth generation of polymeric materials. *Reviews of Modern Physics, 73*(3), 681–700. doi:10.1103/RevModPhys.73.681

Horng, Y.-Y., Lu, Y.-C., Hsu, Y.-K., Chen, C.-C., Chen, L.-C., & Chen, K.-H. (2010). Flexible supercapacitor based on polyaniline nanowires/carbon cloth with both high gravimetric and area-normalized capacitance. *Journal of Power Sources, 195*(13), 4418–4422. doi:10.1016/j.jpowsour.2010.01.046

Hu, C., & Chu, C. (2001). Electrochemical impedance characterization of polyaniline-coated graphite electrodes for electrochemical capacitors d effects of film coverage/thickness and anions. *Journal of Electroanalytical Chemistry, 503*(1-2), 105–116. doi:10.1016/S0022-0728(01)00385-0

Hu, H., Zhang, K., Li, S., Jia, S., & Ye, C. (2014). Flexible, in-plane, and all-solid-state micro-supercapacitors based on printed interdigital Au/polyaniline network hybrid electrodes on a chip. *Journal of Materials Chemistry. A, Materials for Energy and Sustainability, 2*(48), 20916–20922. doi:10.1039/C4TA05345A

Huang, C.-C., Xing, W., & Zhuo, S.-P. (2009). Capacitive performances of amorphous tungsten oxide prepared by microwave irradiation. *Scripta Materialia, 61*(10), 985–987. doi:10.1016/j.scriptamat.2009.08.009

Huang, S., Huang, S., Ng, H., & Kaner, R. (1993). Polyaniline capacitors. *Synthetic Metals, 57*(1), 4047–4052. doi:10.1016/0379-6779(93)90555-B

Huang, W., Humphrey, B. D., & MacDiarmid, A. G. (1986). Polyaniline, a novel conducting polymer. Morphology and Chemistry of its oxidation and reduction in aqueous electrolytes. *Journal of the Chemical Society, Faraday Transactions, 1*(198), 2385–2400. doi:10.1039/f19868202385

Kim, I. H., & Kim, K. B. (2001). Ruthenium oxide thin film electrodes for supercapacitors. *Electrochemical and Solid-State Letters, 4*(5), A62–A64. doi:10.1149/1.1359956

Kotz, R., & Carlen, M. (2000). Principles and applications of electrochemical capacitors. *Electrochimica Acta, 45*(15-16), 2483–2498. doi:10.1016/S0013-4686(00)00354-6

Kuang, H., Cao, Q., Wang, X., Jing, B., Wang, Q., & Zhou, L. (2013). Influence of the reaction temperature on polyaniline morphology and evaluation of their performance as supercapacitor electrode. *Journal of Applied Polymer Science, 130*(5), 3753–3758. doi:10.1002/app.39650

Kuila, B. K., Nandan, B., Bo¨hme, M., Janke, A., & Stamm, M. (2009). Vertically oriented arrays of polyaniline nanorods and their super electrochemical properties. *Chemical Communications (Cambridge)*, (38): 5749–5751. doi:10.1039/b912513b PMID:19774257

Kumar, A., Kumar, V., & Awasthi, K. (2017). Polyaniline–Carbon Nanotube Composites: Preparation Methods, Properties, and Applications. *Polymer-Plastics Technology and Engineering, 57*(2), 70–97. doi:10.1080/03602559.2017.1300817

Kurahatti, R. V., Surendranathan, A. O., Kori, S. A., Singh, N., Kumar, A. V. R., & Srivastava, S. (2010). Defence Applications of Polymer Nanocomposites. *Defence Science Journal*, *60*(5), 551–563. doi:10.14429/dsj.60.578

Letheby, H. (1862). XXIX.-On the production of a blue substance by the electrolysis of sulphate of aniline. *Journal of the Chemical Society*, *15*(0), 161–163. doi:10.1039/JS8621500161

Li, H., Wang, J., Chu, Q., Wang, Z., Zhang, F., & Wang, S. (2009). Theoretical and experimental specific capacitance of polyaniline in sulfuric acid. *Journal of Power Sources*, *190*(2), 578–586. doi:10.1016/j.jpowsour.2009.01.052

Lokhande, V. C., Lokhande, A. C., Lokhande, C. D., Kim, J. H., & Ji, T. (2016). Supercapacitive composite metal oxide electrodes formed with carbon, metal oxides and conducting polymers. *Journal of Alloys and Compounds*, *682*, 381–403. doi:10.1016/j.jallcom.2016.04.242

Luo, Y., Kong, D., Jia, Y., Luo, J., Lu, Y., Zhang, D., ... Yu, T. (2013). Self-assembled graphene@PANI nanoworm composites with enhanced supercapacitor performance *RSc. Adv.*, *3*(17), 5851–5859.

MacDiarmid, A. G. (2001). "Synthetic Metals": A Novel Role for Organic Polymers (Nobel Lecture). *Angewandte Chemie International Edition*, *40*(14), 2581–2590. doi:10.1002/1521-3773(20010716)40:14<2581::AID-ANIE2581>3.0.CO;2-2

Majumdar, D. (2018). Role of electrochemists as environmental activists: A first-hand idea of how electrochemistry research have evolved over years in search of betterment of human life!! doi:10.13140/RG.2.2.19750.83526

Majumdar, D., Baugh, N., & Bhattacharya, S. K. (2017). Ultrasound assisted formation of reduced graphene oxide-copper (II) oxide nanocomposite for energy storage applications. *Colloids and Surfaces. A, Physicochemical and Engineering Aspects*, *512*, 158–170. doi:10.1016/j.colsurfa.2016.10.010

Majumdar, D., & Bhattacharya, S. K. (2017). Sonochemically synthesized hydroxy-functionalized graphene–MnO2 nanocomposite for supercapacitor applications. *Journal of Applied Electrochemistry*, *47*(7), 789–801. doi:10.100710800-017-1080-3

Miao, Y.-E., Fan, W., Chen, D., & Liu, T. (2013). High-Performance Supercapacitors Based on Hollow Polyaniline Nanofibers by Electrospinning. *ACS Applied Materials & Interfaces*, *5*(10), 4423–4428. doi:10.1021/am4008352 PMID:23586693

Moussa, M., El-Kady, M. F., Zhao, Z., Majewski, P., & Ma, J. (2016). Recent progress and performance evaluation for polyaniline/graphene nanocomposites as supercapacitor electrodes Nanotechnology, 27.

Mu, J., Ma, G., Peng, H., Li, J., Sun, K., & Lei, Z. (2013). Facile fabrication of self-assembled polyaniline nanotubes doped with D-tartaric acid for high-performance supercapacitors. *Journal of Power Sources*, *242*, 797–802. doi:10.1016/j.jpowsour.2013.05.117

Mujawar, S. H., Ambade, S. B., Battumur, T., Ambade, R. B., & Lee, S. (2011). Electro-polymerization of polyaniline on titanium oxide nanotubes for super-capacitor application. *Electrochimica Acta*, *56*(12), 4462–4466. doi:10.1016/j.electacta.2011.02.043

Nguyen, D. N., & Yoon, H. (2016). Recent Advances in Nanostructured Conducting Polymers: From Synthesis to Practical Applications. *Polymers*, *8*(4), 118. doi:10.3390/polym8040118

Peng, C., Zhou, X. H., Chen, G. Z., Moggia, F., Fages, F., Brisset, H., & Roncali, J. (2008). Internally Referenced analysis of charge transfer reactions in a new ferrocenyl bithiophenic conducting polymer through cyclic voltammetry. *Chemical Communications*, (48), 6606–6608. doi:10.1039/b817101g PMID:19057794

Rakhi, R. B., Chen, W., Cha, D., & Alshareef, H. N. (2012). Substrate dependent selforganization of mesoporous cobalt oxide nanowires with remarkable pseudocapacitance. *Nano Letters*, *12*(5), 2559–2567. doi:10.1021/nl300779a PMID:22494065

Ramya, R., Sivasubramanian, R., & Sangaranarayanan, M. V. (2013). Conducting polymers based electrochemical supercapacitors: Progress and prospects. *Electrochimica Acta*, *101*, 109–129. doi:10.1016/j.electacta.2012.09.116

Rudge, A., Raistrick, I., Gottesfeld, S., & Ferraris, J. P. (1994). A study of the electro-chemical properties of conducting polymers for application in electro-chemical capacitors. *Electrochimica Acta*, *39*(2), 273–287. doi:10.1016/0013-4686(94)80063-4

Ryu, K. S., Wu, X., Lee, Y., & Chang, S. H. (2003). Electrochemical capacitor composed of doped polyaniline and polymer electrolyte membrane. *Journal of Applied Polymer Science*, *89*(5), 1300–1304. doi:10.1002/app.12242

Salanne, M. (2017). Ionic Liquids for Supercapacitor Applications. *Top Curr Chem (Z)*, *375*(3), 63. doi:10.100741061-017-0150-7 PMID:28560657

Senthilkumar, S. T., Selvan, R., & Melo, J. (2013). Redox additive/active electrolytes: A novel approach to enhance the performance of supercapacitors. *Journal of Materials Chemistry. A, Materials for Energy and Sustainability*, *1*(40), 12386–12394. doi:10.1039/c3ta11959a

Sharma, V., Sahoo, A., Sharma, Y., & Mohanty, P. (2015). Synthesis of nanoporous hypercrosslinked polyaniline (HCPANI) for gas sorption and electrochemical supercapacitor applications. *RSC Advances*, *5*(57), 45749–45754. doi:10.1039/C5RA03016A

Simon, P., & Gogotsi, Y. (2008). Materials for electrochemical capacitors. *Nature Materials*, *7*(11), 845–854. doi:10.1038/nmat2297 PMID:18956000

Sivakkumar, S. R., Kim, W. J., Choi, J.-A., & Douglas, R. (2007). Electrochemical performance of polyaniline nanofibres and polyaniline/multi-walled carbon nanotube composite as an electrode material for aqueous redox supercapacitors. *Journal of Power Sources*, *171*(2), 1062–1068. doi:10.1016/j.jpowsour.2007.05.103

Snook, G. A., Kao, P., & Best, A. S. (2011). Conducting-polymer-based supercapacitor devices and electrodes. *Journal of Power Sources*, *196*(1), 1–12. doi:10.1016/j.jpowsour.2010.06.084

Song, E., & Choi, J.-W. (2013). Conducting Polyaniline Nanowire and Its Applications in Chemiresistive Sensing. *Nanomaterials (Basel, Switzerland)*, *3*(3), 498–523. doi:10.3390/nano3030498 PMID:28348347

Soni, R., Kashyap, V., Nagaraju, D., & Kurungot, S. (2018). Realizing high capacitance and rate capability in polyaniline by enhancing the electrochemical surface area through induction of superhydrophilicity. *ACS Applied Materials & Interfaces*, *10*(1), 676–686. doi:10.1021/acsami.7b15534 PMID:29243906

Sudha, J. D., Sivakala, S., Patel, K., & Radhakrishnan Nair, P. (2010). Development of electromagnetic shielding materials from the conductive blends of polystyrene 62 polyaniline-clay nanocomposite. *Composites. Part A, Applied Science and Manufacturing*, *41*(11), 1647–1652. doi:10.1016/j.compositesa.2010.07.015

Wikipedia. (n.d.). Supercapacitor. Retrieved from en.wikipedia.org/wiki/Supercapacitor

Takahashi, K., Limmer, S. J., Wang, Y., & Cao, G. (2004). Synthesis and electrochemical properties of single-crystal V_2O_5 nanorod arrays by template-based electrodeposition. *The Journal of Physical Chemistry B*, *108*(28), 9795–9800. doi:10.1021/jp0491820

Tarascon, J.-M., & Armand, M. (2001). Issues and challenges facing rechargeable lithium batteries. *Nature*, *414*(6861), 359–367. doi:10.1038/35104644 PMID:11713543

Teasdale, P. R. (2003). *Properties of Polyanilines. In G. G. Wallace, G. M. Sprinks, L. A. P. Kane-Maguire, P. R. Teasdale (Eds.), Conductive Electroactive Polymers: Intelligent materials Systems* (2nd ed.). Taylor & Francis.

Tiwari, J. N., Tiwari, R. N., & Kim, K. S. (2012). Zero-dimensional, one-dimensional, two-dimensional and three-dimensional nanostructured materials for advanced electrochemical energy devices. *Progress in Materials Science*, *57*(4), 724–803. doi:10.1016/j.pmatsci.2011.08.003

Wang, J., Zhang, S. Q., Guo, Y. Z., Shen, J., Attia, S. M., Zhou, B., ... Gui, Y. S. (2001). Morphological effects on the electrical and electrochemical properties of carbon aerogels. *Journal of the Electrochemical Society*, *148*(6), D75–D77. doi:10.1149/1.1368104

Wang, K., Huang, J., & Wei, Z. (2010). Conducting Polyaniline Nanowire Arrays for High Performance Supercapacitors. *The Journal of Physical Chemistry C*, *114*(17), 8062–8067. doi:10.1021/jp9113255

Wang, X., Deng, J., Duan, X., Liu, D., Guo, J., & Liu, P. (2014). Crosslinked polyaniline nanorods with improved electrochemical performance as electrode material for supercapacitors. *Journal of Materials Chemistry. A, Materials for Energy and Sustainability*, *2*(31), 12323–12329. doi:10.1039/C4TA02231A

Weingarth, D., Foelske-Schmitz, A., & Kötz, R. (2013). Cycle versus voltage hold - Which is the better stability test for electrochemical double layer capacitors? *Journal of Power Sources*, *225*, 84–88. doi:10.1016/j.jpowsour.2012.10.019

Wu, T.-H., Chu, Y.-H., Hu, C.-C., & Hardwick, L. J. (2013). Criteria appointing the highest acceptable cell voltage of asymmetric supercapacitors. *Electrochemistry Communications*, *27*, 81–84. doi:10.1016/j.elecom.2012.10.033

Wu, W., Pan, D., Li, Y., Zhao, G., Jing, L., & Chen, S. (2015). Facile fabrication of polyaniline nanotubes using the self-assembly behavior based on the hydrogen bonding: A mechanistic study and application in high-performance electrochemical supercapacitor electrode. *Electrochimica Acta*, *152*, 126–134. doi:10.1016/j.electacta.2014.11.130

Xie, H., Zhu, Y., Wu, Y., Wu, Z., & Liu, E. (2014). The effect of hydroquinone as an electrolyte additive on electrochemical performance of the polyaniline supercapacitor. *Materials Research Bulletin*, *50*, 303–306. doi:10.1016/j.materresbull.2013.11.032

Xie, Y., Kocaefe, D., Chen, C. & Kocaefe, Y. (2016). Review of Research on Template Methods in Preparation of Nanomaterials. *Journal of Nanomaterials*. doi:. doi:10.1155/2016/2302595

Yan, Y., Cheng, Q., Wang, G., & Li, C. (2011). Growth of polyaniline nanowhiskers on mesoporous carbon for supercapacitor application. *Journal of Power Sources*, *196*(18), 7835–7840. doi:10.1016/j.jpowsour.2011.03.088

Yang, X., Cheng, C., Wang, Y., Qiu, L., & Li, D. (2013). Liquid-mediated dense integration of graphene materials for compact capacitive energy storage. *Science*, *341*(6145), 534–537. doi:10.1126cience.1239089 PMID:23908233

Yang, X., Zhang, L., Zhang, F., Zhang, T., Huang, Y., & Chen, Y. (2014). A high-performance all-solid-state supercapacitor with graphene-doped carbon material electrodes and a graphene oxide-doped ion gel electrolyte. *Carbon*, *72*, 381–386. doi:10.1016/j.carbon.2014.02.029

Yoo, H. D., Markevich, E., Salitra, G., Sharon, D., & Aurbach, D. (2014). On the challenge of developing advanced technologies for electrochemical energy storage and conversion. *Materials Today*, *17*(3), 110–121. doi:10.1016/j.mattod.2014.02.014

Yuan, L., Xiao, X., Ding, T., Zhong, J., Zhang, X., Shen, Y., ... Wang, Z. L. (2012). Paper-Based Supercapacitors for Self-Powered Nanosystems. *Angewandte Chemie International Edition*, *51*(20), 4934–4938. doi:10.1002/anie.201109142 PMID:22473807

Zhang, H., Li, H., Zhang, F., Wang, J., Wang, Z., & Wang, S. (2008). Polyaniline nanofibers prepared by a facile electrochemical approach and their supercapacitor performance. *Journal of Materials Research*, *23*(09), 2326–2332. doi:10.1557/jmr.2008.0304

Zhang, S., & Pan, N. (2015). Supercapacitors Performance Evaluation. *Advanced Energy Materials*, *5*(6), 1401401. doi:10.1002/aenm.201401401

Chapter 8
Polyaniline Nanocomposites:
Innovative Materials for Supercapacitor Applications – PANI Nanocomposites for Supercapacitor Applications

Dipanwita Majumdar
Chandernagore College, India

ABSTRACT

Polyaniline in various forms has been widely explored as an electrode material for supercapacitors due to its high theoretical charge storage capacity, facile-cost-effective synthesis, good mechanical strength and ultrafast charge transport. However, commercialization of such pristine forms is very much restricted by low solubilities, rapid agglomeration during device design accompanied by poor electrochemical life and fast environmental decomposition. The blending with nano-carbon materials, metal oxides and other competent materials, may result in high quality materials– "nanocomposites" with superior features is ideally fit for future generation energy storage devices. The present chapter deals with detailed discussions on designing, the fabrication of such binary and ternary nanocomposites, correlating their morphology with electrochemical behavior, so as to optimize their supercapacitive performances. Such an attempt would help to outline the present status and future aspects of these materials which will be of first-hand assistance especially to the beginners to this field of research.

INTRODUCTION

As reviewed in the previous chapter, Polyaniline (PANI) has high potentiality as supercapacitor electrode material and very recently, it has been able to achieve specific capacitance as high as 950 Fg^{-1} through the involvement of the entire volume in storage of charge, surpassing other conducting polymers that store charge solely on surface (Wang, Huang & Wei, 2010) (Nguyen & Yoon, 2016). However, the theoretically calculated utmost specific capacitance value of PANI is 2000 Fg^{-1}, although the experimental values evaluated by different methodologies are considerably inferior to the predicted value as only a small percentage of PANI make effective contribution to the total capacitance which is again dependent

DOI: 10.4018/978-1-5225-7838-3.ch008

on both – its conductivity state (percentage of emeraldine salt) and extent of counter-anions diffusion through it (Li, Wang & Chu, 2009). Moreover, pseudocapacitive processes of PANI in most cases involve swelling, shrinkage and cracking of the polymer skeleton during doping-dedoping-redoping process, leading to poor GCD cycle efficiency. In addition, at higher applied potentials PANI degradation may result due to over-oxidation, which reduces the electrical conductivity as well as working potential window for pristine PANI electrodes. Furthermore, addressing of the following key issues: (i) ordered and well-aligned morphology; (ii) optimization of doping level for superior charge transport; (iii) introduction and maintenance of porous-structure for large number of stable galvanostatic charging/discharging cycles; is crucial for upgrading the current device-performances. These problems necessitate the development of "nanocomposite" fabrication that combine PANI with other materials such as nano-carbonaceous systems or nano-metal oxides for improving the overall electrochemical and mechanical stability as well as energy storage capacity of the "supercap" electrodes (Yin, Liu & Fan, 2012) (Beidaghia & Gogotsi, 2014) (Shukla, Banerjee & Ravikumar, 2012) (Wu, Chu & Hu, 2013).

The present chapter deals with detailed discussion on fabrication and designing of binary nanocomposites with nano-carbon based systems as well as nano-morphology based-metal oxides and other non-trivial pseudocapacitive substances, correlating the composites' morphology with their electrochemical behavior, so as to bring out the optimum requirements for desired performances. In addition to these two-component nanocomposites, discussion on recently developed ternary-nanocomposites composed of either metal oxide with strong pseudocapacitance and/or a high surface area porous-carbon to act as suitable substrate to anchor the pseudocapacitive materials, and nanophased-PANI as the common material have been carried out in details. These novel systems have become attractive candidates for the developing new-generation supercapacitors. The motivation is that all the three components would contribute synergistically to the overall specific capacitance, making-up their disadvantages so as to achieve superior electrical conductivity, larger active-surface area, and enhanced electrochemically-active redox sites accessibility for improving the device performances.

CARBON-BASED PANI NANOCOMPOSITES

Carbon based nanomaterials (carbon nanotubes, graphene, activated porous carbons and so on) are considered as ideal EDLC-electrode materials due to their controllable porous morphology, high electrical conductivity, appropriate pore distribution and large active-surface area (Salunkhe, Lin, & Malgras, 2015; Tang, Salunkhe & Liu, 2015; Salunkhe, Tang & Kamachi, 2015). Gu et al. pointed out the merits as well as the limits of many nano-carbon EDLCs materials, and gave an overview of the kinetic factors and structure-property correlationship for the electrolyte ion-adsorption and its passageway across such systems (Gu & Yushin, 2014). Largeot group commented that close matching of nano-carbon pore size and the electrolyte ion-size results in the maximum double-layer capacitance (Largeot, Portet & Chmiola, 2008). However, most of the nanocarbons exhibit low capacitance ($< 200\ Fg^{-1}$) that limits their pristine-form usage in this field. But taking the above-said benefits of these materials, the so-obtained nanocomposites with conducting polymers such as PANI considerably modifies the Galvanostatic charging/discharging (GCD) cyclic stability, increases the overall charge transport properties as well as device specific capacitance thereby presenting a better electrochemical behavior and mechanical stability (Chen & Xue, 2015). Such nanocomposites have been discussed in details in the following section.

Porous Carbons Based-PANI Nanocomposites

Early studies on porous carbon systems usually concerned with activated carbons (ACs), ordered mesoporous carbons (OMCs), and other amorphous carbons which were generally blended along with conducting PANI so as to obtain better chemical features and easier film processability. The Electrical double layer capacitance (EDLC) provided by such carbon materials along with the Pseudocapacitance (PC) assistance from PANI would further maximize the total specific capacitance of the electrode material (Mastragostino, Arbizzani & Meneghello, 1996). There are vast varieties of carbonaceous precursors for synthesizing ACs and OMCs for supercapacitor electrode materials nowadays (Liang & Dai, 2006). The nanocomposites are easily fabricated through either by chemical or by electrochemical polymerization techniques (Fan, Hu, & Maier, 2007) (Ryu, Lee & Kim, 2005) Extensive research reveals that dimension-effect of AC particles has great influence on the morphology of composite (Wang, Lin & Shen, 2016). For instance, small sized AC particles prefers to remain wrapped in PANI matrix, while PANI remains coated onto the surface for micro-sized AC carbon particles. Thin and porous coating of PANI onto carbon surface promotes proper utilization of electro-active materials preferring shorter ion-diffusion path, reduced energy/power loss, leading to large specific capacitance. Use of suitable surfactants also becomes effective for controlling/tuning the morphology of the PANI/AC composites for enhanced electrochemical outputs. Some of such works have been delineated in Table 1 (Misoon & Seok, 2012).

Table 1. Electrochemical performances of various PANI/ porous nano-carbons

Material	Discharge Rate	Potential Range/V	Electrolyte	Specific Capacitance F g^{-1}	Cycling/Capacity Retention (Number of Cycles)	Ref.
PANI/OMC	6 Ag^{-1}	2 - 4 vs. Li/ Li^+	1M $LiPF_6$	146	---	(Woo, Dokko & Nakano, 2009)
PANI/ACNF	1 Ag^{-1}	0.0-0.8	1M H_2SO_4	832	92%(1000)	(Yang, Jang & Kim,2013)
PANI/C	2 Ag^{-1}	0.2-0.8	0.5 M H_2SO_4	1486	82% (500)	(Bavio, Acosta & Kessler, 2014)
2D-Mesoporous C/PANI	2 mV s^{-1}	-0.1-1.0	1M H_2SO_4	470	91%(4000)	(Zhou, Mo & Zou, 2011)
PANI/OMC	0.1 Ag^{-1}	0-0.8	1M H_2SO_4	517	91.5% (1000)	(Yan, Cheng & Zhu, 2013)
Bamboo carbon/PANI	0.5 Ag^{-1}	0.2-0.7	1M H_2SO_4	244	92% (1000)	(Zhou, Li & Dong,2011)
Vertically aligned CSA-doped polyaniline nano-whiskers on OMC	1 Ag-1	−0.2–0.8	1M H_2SO_4	470	90.4 (1000)	(Woo, Dokko K., Nakano H., 2009)
Freestanding hierarchically porous carbon framework decorated polyaniline	0.5 Ag^{-1}	-0.6-0.3	1M H_2SO_4	290	78(10000)	(Miao, Shao & Li, 2016)

OMCs have drawn a great deal of technological attention in the past years for its unique mesoporous structure and large surface area. The best electrochemical performance was observed on a PANI-porous carbon monolith composite electrode having specific capacitance of 2200 Fg^{-1} but with much depressed electrochemical stability (Fan, Hu & Maier, 2007) Lately, pore surface of OMCs were coated with thin layer of PANI to obtain specific capacitance as high as 602.5 Fg^{-1} with improved GCD cycle stability (Xing, Zhuo & Cui, 2007). Recently, works based on comparative study on in situ-polymerization versus simple mixing of components revealed that the former process provided thin, intimate and uniform PANI coating on porous carbons, promoting better electrochemical features accordingly (Bleda-Martínez, Morallón & Cazorla-Amorós, 2007). Nanocomposites have been designed using surfactants that can stabilize as well act as in-situ-dopants for PANI coated on the carbon surface (Misoon & Seok, 2012) (Zhou, Li & Wang, 2011). Efforts to coat polyaniline film inside the pores of the activated carbon fiber with a thickness around 0.5 nm yielded much impressive electrochemical stability results (Salinas-Torres, Sieben & Lozano-Castelló, 2013). Such thin PANI film does not hinder the entrance and exit of anions through the sufficiently micro-pored fibers for the composite, thus prevents electrode volume change during the GCD process what is very much advantageous over other less porous PANI/carbon composites. Some researchers also investigated detailed electrochemical properties of PANI/carbon spheres as well as for PANI/carbon particles electrode materials. For example, Shen and co-workers demonstrated the performance of nano-PANI/hollow carbon spheres composite synthesized via in-situ polymerization method that demonstrated improved capacitance of 435 Fg^{-1} at a current density of 0.5 Ag^{-1} with 60% of initial capacitance retention after 2000 GCD cycles (Shen, Ran, Zhang & Liu, 2015). Similar other PANI-nanoporous carbon-based composites show appreciable specific capacitance of due to the combination of both merits of high intrinsic pseudocapacitance of PANI and good stability of OMCs, some of which have been summarized in Table 1.

The above studies revealed that the capacitance retention efficiency in GCD cycles for these composites degrades faster for larger cycle numbers which has been the main commercial constraint (Lee, Kim & Park, 2014) (Shen, Rana & Zhang, 2015) (Zhang, Li & Zhang, 2010). Moreover, they exhibit much lower energy and power densities owing to larger inter-grain resistances thereby presenting poor charge transfer features. Even fabrication methods for such uniform-sized, monodispersed hollow, porous carbons require expensive reagents and equipment designing that add-up to the manufacturing expenditure of supercapacitors (Ning, Zhong & Wana, 2016) (Wang, Li & Xia, 2006) (Fernández-Merino, Guardia & Paredes, 2010) . In addition, redox sites are unified within the mesopores and thus, redox systems often work within narrow potential window, highly undesirable for supercapacitor applications (Zhang, Wang & Li, 2014).

PANI/CNT Nanocomposites

Uncontrolled polymerization during synthesis leads to disordered PANI network initiating sluggish ion diffusion kinetics and poor charge transport within the active material. Therefore, it is crucial to incorporate 1D-nanomaterials that can channelize and promote ordering and uni-directional aligning of PANI chains, for better redox sites accessibility, charge transport and faster ion-diffusion through the network. Carbon Nanotubes (CNTs) assemblies have the advantages of producing freestanding lightweight, high conductive, thin, porous electrodes (Zhao, Sanchez, Dobson & Grant, 2011) (Zhang, Cao & Wang, 2008) (Zhang, Cao & Yang, 2007) (Kim, Javey & Tu, 2005). This eliminates the requirement for additional electrochemically inert binders, minimizing cell assembly procedures to large extent besides preserving the 3D porous arrangement of the nano-composite throughout the lifespan of the capacitor (Kim, Choi, & Shim, 2002) (Kim, Chung &Chu, 2010) (Cinke, Li, Chen & Cassell, 2005) (Dai, Chang & Baek, 2012). Moreover, there are adequate scopes for site-selective chemical interactions between PANI and single-walled carbon nanotubes (SWNTs). The functional groups on SWNT surfaces have certain bonding communications with the active sites (-NH-/=N-) of PANI through various van der Waals –type, π—π, and hydrogen–bonding (H-bond) interactions as shown in Figure 1. (Abdiryim, Ubul, Jamal & Rahman, 2012)

The synergetic combination of these excellent conducting and mechanically strong CNTs and high pseudocapacitance of the PANIs reinforce each other to yield superior electrochemical signatures (Zhu, Wang & Sun, 2011). Accordingly, various methodologies are being adopted to fabricate CNT/PANI composites possessing diverse morphologies. For instances, PANI and single/multi-walled CNTs composites are popularly fabricated using chemical or electrochemical methodologies to form PANI-coated CNTs core-shell structures (Fan, Wang & Zhao, 2012) (Yang, Cheng & Song, 2010). These electrodes illustrate exceptional electrochemical results due to surface-exposed pseudocapacitive PANI and enhanced electrical conductivity as well as mechanical stability of backbone-CNTs. Moreover, interconnected CNTs also improve charge transport process. PANI nanofibers or nanowires are also found to surround the CNTs, giving rise to caterpillar-like hybrids. However, the chemical polymerization route usually requires pre-chemical treatment on CNTs in order to improve the stability of CNT dispersion during the polymerization process (Liu, Sun & Gao, 2010). For instance, Yang *et al.* demonstrated in-situ doping of PANI on SWCNT-grafted poly(4-vinylpyridine) (P4VP) (Yang, Hao & Yuan, 2014). The CNT-modification prevented the self-aggregation without affecting its intrinsic conductivity and morphology, resulting in large-valued specific capacitance of 1065 Fg^{-1} with appreciable GCD stability (~ 92.2% after 1000 GCD cycles). A facile methodology via single step electrospinning process for binder-free PANI-CNT nanofiber electrodes fabrication is reported to display moderate capacitance of ~ 385 Fg^{-1} (Simotwo,

Figure 1. Schematic representation of various chemical connections through hydrogen bonding and $\pi-\pi$ interactions between PANI and single-walled carbon nanotubes (SWNTs)
Created by Universal Document Converter

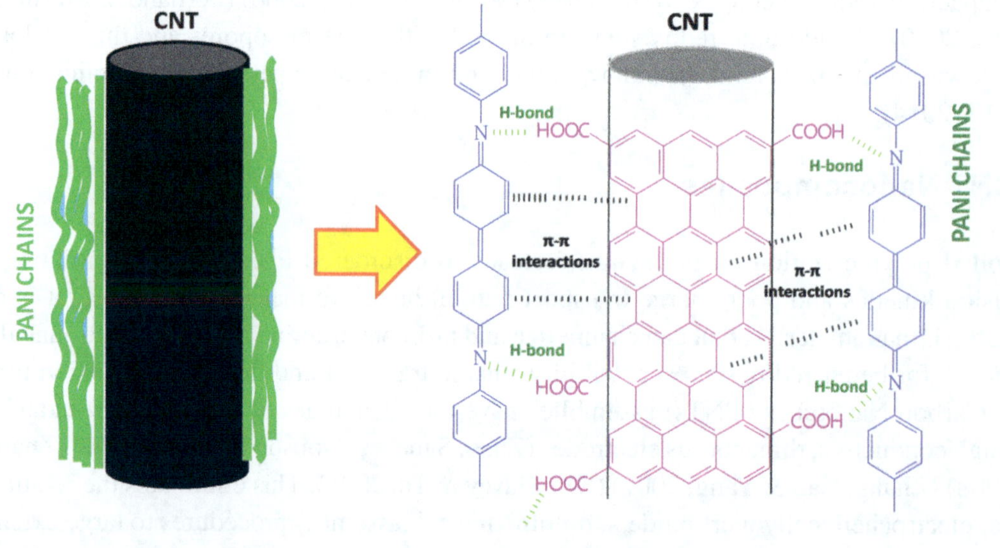

DelRe & Kalra, 2016). Guo *et al.* synthesized PANI/multiwalled-CNTs composite through facile, cost-effective microwave-assisted polymerization method that presented a charge storing capacity of 322 Fg^{-1} with profound specific energy density value approximately 12 times that of pristine-MWCNTs (Guo, Mi & Zhou, 2015). Lately, PANI/MWCNT nanocomposite with tubular structure fabricated by an in-situ low-temperature procedure with only 10%-MWCNT content exhibited appreciable specific capacitance value ~ 552.11 Fg^{-1} measured at 4 $mAcm^{-2}$ current density compared to pure-PANI (411.52 Fg^{-1}) (Imani, & Farzi, 2017).). Here, low-temperature oxidative polymerization process promoted well-oriented-large scale composite with special tubular morphology. Niu and co-workers reported a 'skeleton/skin' strategy for preparation of free-standing, thin and flexible PANI/single-walled-CNTs composite films using in-situ electrochemical polymerization method for "supercap" applications (Niu, Luan & Shao, 2012). Some of the comparable works based on PANI/CNTs composites obtained via various synthetic methodologies have been briefly discussed in Table 2. Needless to mention that these composites achieved extra-ordinary values of specific capacitance with minimum ESRs among all other carbon-based composites but the high cost of electrode-fabrication as well as rigorous synthetic techniques of CNTs manufacture restrict random and commercial usage in energy storage-saving applications.

Carbon nanofibers (CNFs) are an added kind of porous, useful carbon material that could be cost-effective alternatives for CNTs (Liu, Zhou, Chen & Zhang, 2015). Like CNTs, they possess excellent charge transport properties, extraordinary flexible 3D structure, good mechanical stability as well as attractive chemical characteristics. They too often serve as free standing current collectors for chemically and electrochemically polymerized PANI systems (Yan, Tai, Chen& Xue, 2011) (Tran, Singhal & Lawrence, 2015) (Cheng, Tang & M, 2011) Designing of freestanding 3D-supercapacitor electrodes with PANI and porous CNFs were recently reported (Tran, Singhal & Lawrence, 2015). The system exhibited appreciable specific capacitance ~366 Fg^{-1} at 100 mVs^{-1} high scan rate upholding PANI's prominent pseudocapacitive performance. In another work, flexible electrodes obtained by growing uniformly coated PANI nanoparticles onto 3-D interconnected CNFs electrodes displayed impressive electrochemical results (Cheng, Tang & M, 2011). Many similar works on the composites have been explored, some of which are outlined in Table 2. (Dawoud, Altahtamouni & Zagho, 2017).

Table 2. Electrochemical performances of differently designed PANI/CNTs

Material	Discharge Rate	PotentialRange/V	Electrolyte	SpecificCapacitanceF g^{-1}	Cycling/ Capacity Retention (Number of Cycles)	Ref.
PANI on CNF	5 mV s^{-1}	0-0.8	1M H_2SO_4	264		(Yang, Jang, Kim 2013)
CNF/PANI	2 Ag^{-1}	0-0.8	1M H_2SO_4	638	90% (1000)	(Yan, Tai & Chen, 2011)
PANI/CNT	0.5 Ag^{-1}	0.2-0.8	0.5 M H_2SO_4	1065	92% (1000)	(Yang, Hao &Yuan, 2014)
PANI/CNT	5 mV s^{-1}	0.1-0.7	PVA/H_3PO_4	440	93% (1000)	(Otrokhov, Pankratov & Shumakovich,2014)
PANI/MWCNT	5 mA cm^{-2}	0.2-0.8	1M NaNO3	328	94% (1000)	(Dong, He, Xu & Li, 2007)
PANI/SWCNT	5 mA cm^{-2}	0-0.7	1 M H_2SO_4	485	94% (1500)	(Gupta & Miura, 2006)
PANI/MWCNT	1 mV s^{-1}	0-1.0	0.1 M H_2SO_4	560		Zhou, Qin & Li, 2010)

These electrodes discussed above possess several beneficial qualities including large scalable production yields, highly conducting and mechanical robust. However, a recent data sheet on safety for CNFs designates the material to induce eye irritations and the formed dust clouds by these materials are severely toxic to respiratory system organ. Thus, additional safety-care must be adopted while handling CNFs (Pacurari, Lowe, Tchounwou & Kafour, 2016).

The preceding paragraphs demonstrate that PANI-1D-nanocarbon based composites are indeed promising electrode materials for future generation energy storage devices. However, further studies are still desired on novel composite materials obtainable by facile, cost-effective methodologies for fabrication of composites with better contacts and intriguing electrochemical properties for profitable output. It is thus anticipated that more stimulating discoveries will be experienced in this field in the following years.

PANI / Functionalized-Graphene Nanocomposites

The superiority of two-dimensional (2D)-structure over 1D-morphology is due to an increased surface area, geometry and greater exposure of reactive sites that have instigated its extensive study since last decade. Graphene, ideally, a 2D- single layer of sp^2-hybridized carbon, is particularly interesting as an ultra-thin film electrode material for its superfast electron mobility, excellent mechanical potency and large surface area (theoretically up to ~2630 m^2g^{-1}) (Geim & Novoselov, 2007; Guo & Dong, 2011; Matte, Subrahmanyam & Rao, 2011). In energy storage systems, it has been vastly employed as a counter-electrode as working electrode for EDLC type devices as well as negative electrode owing to high hydrogen overpotential in pristine form or as a conductive (σ as high as 200 Sm^{-1}) additive powder on working electrode (Wang & Hu, 2012). Graphene is usually synthesized by cost-effective wet-chemical method from graphene oxide (GO) obtained from modified Hummer's process followed by thermal or chemical or electrochemical reduction (Guo, Wang & Qian, 2009). Even many other graphene synthesis methodologies such as mechanical exfoliation, chemical vapor deposition, epitaxial growth, are available in the literature (Rao, Sood, Subrahmanyam & Govindaraj, 2009; Yi & Shen, 2015; Tetlow, de Boer & Forde, 2014; Zhang, Zhang & Zhou, 2013). But, graphene obtained from wet technique is generally multilayered and largely scalable. Recently, it is established that electrochemical behavior of single layered graphene is no good than multi-layered graphenes (Brownson, Varey & Hussain, 2014; Majumdar & Bhattacharya, 2017). So, it is wise to use less expensive multi-layered graphenes rather than other expensive single-layered graphene systems. Moreover, graphene functionalisation is more effectively carried out using wet-technique. Functionalized graphene stores charge via both EDLC and PC mechanism, latter originating from various redox-active oxygen-containing functionalities like epoxy, hydroxyl and carboxylic acid groups on the surface. However, maximum specific capacitance of graphene today has reached around ~ 200 Fg^{-1} in aqueous (acidic, neutral, alkaline), organic, and even ionic liquid electrolytes although the capacitive retention is impressive for several thousands of GCD cycles. So, graphene-based nanocomposites are opted for the purpose of improving the electrochemical properties in forthcoming supercapacitors (Majumdar, Baugh & Bhattacharya, 2017).

Recently, researchers realized that high aspect ratio and surface defects density of graphene tremendously boosts the interfacial load transfer to host material such as PANI thereby improving the energy storage characteristics (Vaquero, Díaz & Andersona, 2012) (Gao, 2017) (Majumdar, 2016). However, higher amount of either RGO or PANI would result in higher viscosity, leading to low GCD efficiency and poorer gravimetric capacitance (Tyagi, Tripathi& Gupta, 2015). Thus, morphological optimization by varying the preparation techniques as well as syntheses-reaction conditions is vital for acquiring estimated results.

Recently, Hu et al. fabricated a polyaniline nanorod-graphene composite using in-situ electrochemical polymerization technique that exhibited significantly enhanced performance with the maximum specific capacitance of 878.57 Fg^{-1} at appreciable scan rate (Hu, Tu & Jiao, 2012). By growing ordered vertically-aligned PANI nanorods on graphene sheets as shown on Figure 2, specific capacitance as high as 1665 Fg^{-1} with superior electrochemical stability was achieved (Li, Zhao & Yu, 2013).

In another work, graphene paper coated with PANI prepared by in-situ anodic electropolymerization of aniline showed moderate gravimetric capacitance of 233 Fg^{-1} although the electrode mechanical stability got improved by ~ 43% (Wang, Li &Zhao, 2009). Such low capacitance was mainly due to small PC contribution from the surface -polyaniline films that remained coated on the graphene paper. However, the EDLC part from graphene sheets remained almost un-utilized as the underneath pores got blocked by the agglomerated PANI layers on the paper surface.

To avoid such problem, Wang and Liu groups independently fabricated layer-by-layer assembly of reduced graphene oxide nanosheets-polyaniline nanofibers composite by electrodepositing PANI nanofibers onto conductive graphene nanosheets (Gao, Yang & Wang, 2014). This multilayered composite displayed an improved capacitive performance, better rate capability and improved cycle performance, with ~93% charge storage efficiency after 1000 GCD cycles. Parveen et al. reported PANI@GN nanocomposite with highly crystalline regimes obtained via high-temperature-mediated-in-situ-oxidative-polymerization that exhibited gravimetric capacitance of 613 Fg^{-1} measured at 10 mVs^{-1} scan rate; the observed value being

Figure 2. Electrochemical performance of ordered vertically aligned PANI nanorods on graphene sheets

[Reprinted with permission from ACS publication (Li, Zhao & Yu, 2013)]

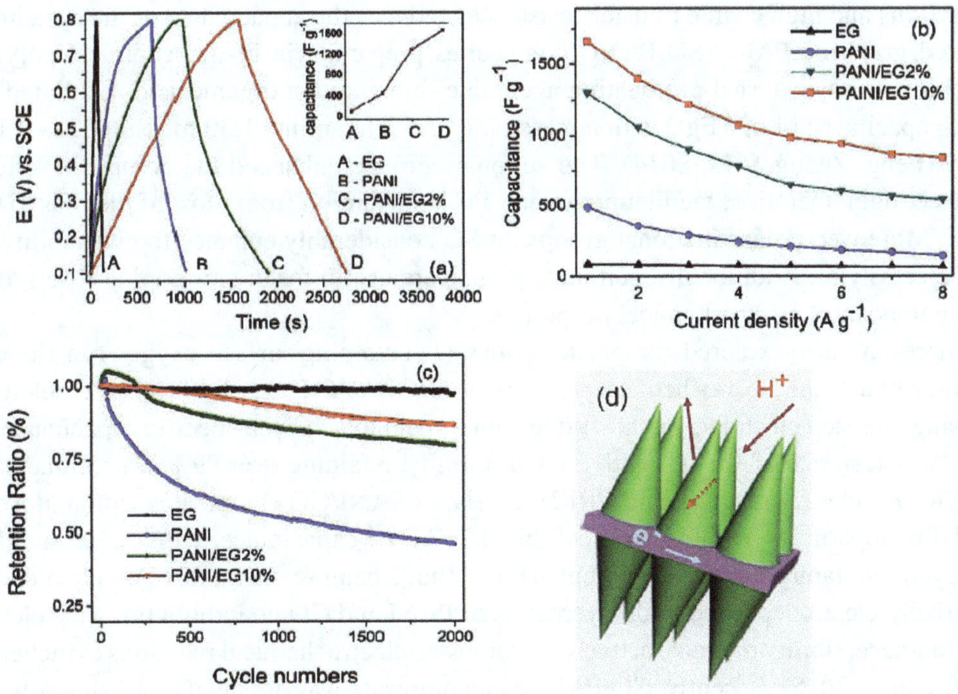

superior to either of the individual components indicating their synergistic contribution (Parveen, Mahato & Ansari, 2016). Graphene/PANI-nanofibers composite was prepared employing a facile, one-pot technique that involved reduction of graphene oxide (GO) by aniline followed by in-situ polymerization process (Jin, Fang & Jia, 2014). Aniline played dual roles in this case: firstly, as a chemical reducing agent to reduce GO to graphene and secondly, as a monomer source for producing PANI-nanofibers. The resultant composite electrode exhibited appreciable gravimetric capacitance of 965 Fg^{-1} at current density of 0.5 Ag^{-1} with 90% specific capacitance retention after 2000 GCD cycles.

Another facile approach was adopted to fabricate hybrid system containing mesoporous PANI film grown on ultra-thin graphene nanosheets via in situ polymerization using graphene-mesoporous silica composite as template (Wang, Yan &, Fan, 2014). Its mesoporous structure and 3D-conductive network promoted ultrafast charge transport as a result of which a maximum specific capacitance of 749 Fg^{-1} at current density ~ 0.5 Ag^{-1} was achieved with excellent GCD efficiency. Similarly, graphene-PANI composite containing graphene sheets (~15 wt %) coated homogeneously with PANI via in-situ-polymerization method exhibited appreciable gravimetric capacitance of 1046 Fg^{-1} measured at 10Ag^{-1} current density; with capacitance retention efficiency of 96% even on raising the current density from 10 to 100 Ag^{-1} (Yan, Wei & Shao, 2010). Lately, pyrolytic graphene-based composites with highly-oriented polyaniline (PANI) nano-thorn arrays prepared by solvothermal in-situ polymerization in reverse micro-emulsion system exhibited high specific capacitances efficiency of 595 Fg^{-1} and 438 Fg^{-1} at 0.5 Ag^{-1} and 2 Ag^{-1} current densities respectively, with good GCD rates (Zhang, Li & Feng, 2013). The nanothorn structure shortened the ion-diffusion pathways resulting in improved electrochemical signature as shown in Figure 2(d).

3-D hybrid systems of hollow graphene balls coated with PANI were fabricated through self-assembling of GO, poly(methyl methacrylate) (PMMA) particles, and PANI followed by subsequent elimination of the PMMA core-particles (Trung, Tam, & Kim, 2014). The so-obtained electrodes showed gravimetric capacitance of 331 Fg^{-1} at a current density of 1Ag^{-1} with average cycling stability. Again, poor inter-grain interactions and high charge transfer resistances reduces the capacitance of the system.

Sulfonated graphene-PANI (SG-PANI) composites prepared via in-situ chemical polymerization without employing any external-protonating agent like- inorganic or organic acids exhibited maximum gravimetric capacitance of 676 Fg^{-1} with a mass ratio of SG to aniline 1:30 measured at ~0.1 Ag^{-1} current density (Feng, Zhang & Li, 2014). The sulfonate groups enhanced the composite solubility and restricted sheet-agglomeration, facilitating greater PC contribution from most of the SGs to the overall capacitance. Moreover, polar functional groups on SG considerably enhance the wettability of carbon phase with electrolytes. Additionally, sulfonate groups are useful for in-situ-doping PANI, resulting in amplifying composite's electrochemical properties.

Researchers have also explored composites with GO containing surface-oxygen functionalities substantially contribute to the electrochemical performances. GO/PANI layer-by-layer assembled composite prepared using one-step electrochemical co-deposition method achieved specific capacitance as high as 1136.4 Fg^{-1} at a scan rate of 1mVs^{-1} with a good stability, retaining over 89% of its initial capacitance after 1000 GCD cycles (Sarker & Hong, 2012). Lately, of PANI/GO composites with uniform distribution of PANI nanoparticles on GO sheets exhibited 797 F g^{-1} capacitance with a good stability during long-term cyclic voltammetric measurements (Luo, Zhu, Zhang & Tang, 2013). Gupta et al. showed that alternatively-electrodeposited multilayer films of PANI and GO on indium tin oxide electrodes enhanced the ion-accessibility of redox-active sites for faster-electrochemical reactions (Mitchell, Candler, De Souza & Gupta, 2015). Recently, GO/PANI nanocomposite was obtained by in-situ-polymerization technique in supercritical carbon dioxide (SC-CO_2) medium. The nanocomposite with 0.1 M aniline concentration displayed appreciable capacitance (425 Fg^{-1}) at current density of 0.2 Ag^{-1} (Xu, Wang & Wei, 2012). Zero surface tension of SC-CO_2 perpetually wets the basal planes of GO with ethanol, helping dissolved-anilinium ions to get easily adsorbed onto GO surface thereby triggering uniform distribution of PANI nanoparticles over it. Bao et al. introduced polyaniline nanowires as "spacers" in-between GO sheets that self-assembled into graphene/PANI hybrid aerogels using hydrothermal process (Yang, Xu & Bao, 2014). The so-obtained morphology facilitated faster ion-diffusion and thus appreciable specific capacitance of 520.3 Fg^{-1} was achieved with reasonable GCD stability. However, topological defects and numerous oxygenated groups on GO, considerably hamper charge transport process, thereby increasing the charge transfer resistance of the system (Yalcin, Galande & Kappera, 2015).

To combat this problem, many recent researchers are replacing graphene oxide or graphene by reduced-graphene-oxide (rGO) to strengthen the adherence of PANI to graphene layers through suitable site-interactions with the oxygen functionalities on rGO surfaces, without compromising with the conductivity. GO reduction could be realized using common chemical reductants such as reduced-ascorbic acid, citric acid, hydrazine, sodium borohydride, or even by electrochemical or thermal process or even enzymatically under inert condition (Amaya, Matsumoto & Flores-Conde, 2016) (Chen, Yan & Bangal, 2010) (Park, An & Potts, 2011).

Chen group constructed 3D-porous electrodes with electrochemically obtained rGO-PANI nanocomposite that exhibited high gravimetric capacitance ~ 716 Fg^{-1} at 0.47 Ag^{-1} current density (Li, Peng, Li & Chen, 2012). Lindfors *et a*l formulated simple electrochemical reduction procedure to fabricate nanostructured PANI-rGO composite films with globular surface morphology (having mean grain size of 50 nm), that exhibited with high areal capacitance of 15 $mFcm^{-2}$ in a symmetric two-electrode cell even after 10,000 GCD cycles between potential window of 0 - 0.4 V (Boeva, Milakin & Pesonen, 2014). Another similar hybrid material, designed by three steps of in-situ polymerization-reduction-dedoping-re-doping process, renders gravimetric capacitance as high as 1126 Fg^{-1} at scan rate 1mVs^{-1} (Wang, Hao, Yang & Lu, 2010). PANI-hollow nano-spheres embedded onto electrochemically-rGO having core shell morphology was fabricated via solution-mediated-co-assembly procedure. Such designing significantly increases specific surface area, providing greater exposure to electroactive sites along with shorter ion-diffusion pathways. Wrapping of rGO sheets on the segregated spheres further introduced conductive channels by bridging them together, thereby enhancing the electrochemical yield of the system (Fan, Zhang & Tjiu, 2013). Apart from the above studies, a large volume of work are available in the literature based on tuning the specific capacitance along with GCD cycle stability of functionalized graphene/PANI nanocomposites, some of which are provided in Table 3.

Although the capacitance value ranges widely, the supercapacitors possess many unique advantages such as easy fabrication, suitable thickness-tuning, flexibility and stable cycling characteristics. Compared with other morphologies, freestanding 3D-structure owns unique advantages and hence it is crucial to synthesize appropriate 3D-structures by simpler techniques (Xia, Hao & Lei, 2012) (Wang, Zhang & Long, 2018) (Mezgebe, Yan & Wei, 2017). Novel approaches for flexible graphene-based PANI composite electrodes are in continuous search for the fabrication of flexible micro-supercapacitors in the near future.

Table 3. Electrochemical performances of differently obtained PANI/graphene composites

Material	Method of Preparation	Discharge Current Density	PotentialRange/V	Electrolyte	SpecificCapacitanceF g^{-1}	Cycling/ Capacity Retention(Number of Cycles)	Ref.
PANI nanorod- – Graphene	electrochemical polymerization in reverse micelle	1 Ag^{-1}	0-0.7	1 M $HClO_4$	878.57	> 80% (1000)	(Tyagi, Tripathi & Gupta, 2015).
polyaniline nanorods/ Graphite nanosheets	in situ polymerization	1 Ag^{-1}	0.1-0.8	1 M H_2SO_4	1665	>97% (2000)	(Hu, Tu & Jiao, 2012)
Polyaniline hollow spheres/ rGO	solution-based co-assembly process	1 Ag^{-1}	0.2-0.8	1M H_2SO_4	614	90% (500)	(Fan, Zhang & Tjiu, 2013)
Polyaniline nanocones / Graphene	electrodeposition	1 Ag^{-1}	0-0.7	1M $HClO_4$	751	93% (1000)	(Yu, Ma & Liu, 2015)
PANI nanowires array on polystyrene microsphere/ reduced graphene film	dilute polymerization	0.5 Ag^{-1}		1M H_2SO_4	740	87% (1000)	(Wang, Ma & Gan, 2015)
PANI nanorod/ rGO	a low temperature oxidation polymerization	0.2 Ag^{-1}	0-0.8	1M H_2SO_4	1045	85.3 (10000)	(Liu, Shang & Zhang, 2014)
Sandwiched-polyaniline nanotube/ Graphene	chemical polymerization	0.1 Ag^{-1}	0.2-0.8	1M H_2SO_4	676	97% (1000)	(Dai, Ma & Gan, 2016)
GO/PANI	interfacial electrochemical polymerization	1 Ag^{-1}	0-0.8	PVA /H_3PO_4	1095		(Li, Li & Feng, 2015)
GO/PANI	Electropolymerization	1 mV s^{-1}	0.1-0.9	1M H_2SO_4	1136	89% (1000)	(Mao, Zhang & Chan, 2012)
Graphene/PANI	Chemical oxidation	0.2 Ag^{-1}	0.2-0.8	2M H_2SO_4	536		(Chen, Rakhi & Alsharee, (2013)
PANI/Graphene	Electropolymerization	1 Ag^{-1}	0-0.8	1M H_2SO_4 +þ-HQ	553	90% (50,000)	(Yu, Zhao, Li &Huang, 2014)
Graphene/PANI	dipping and drying	1 Ag^{-1}	0-0.8	1M H_2SO_4	790	80% (5000)	(Yu, Zhao & Li 2014)
Nanowires-PANI/G	Surfactant assisted-in-situ chemical polymerization	1 Ag^{-1}	0-0.8	1M H_2SO_4	1255	90% (2000)	(Yu, Li & Zhao, 2013)
fish scale-like polyaniline nanosheets/ G	Single-step template-free polymerization.	5 Ag^{-1}	0.2-0.8	1M H_2SO_4	413	81% (1000)	(Ning, Li & Yan, 2013)
Nanofiber-PANI/ Graphite	Microwave-assisted chemical -vapor-induced in-situ polymeri-zation	1 Ag^{-1}	0-0.8	6M H_2SO_4	2136	91% (1000)	(Li, Yang & Lei, 2014)
PANI/GO	nanoscale blending method of layer-by-layer (LBL) assembly	0.5 Ag^{-1}	0.2-0.8	1M H_2SO_4	375	91% (500) at 3 A g^{-1}	(Lee, Yun & Park, 2012)
Nanofiber-PANI/ GO	electrodeposition	0.025 mA cm^{-2}	0-0.7	1M H_2SO_4	662		(Khalid, Tumelero& Zoldan, 2014)
Nanofiber PANI/ GO	in situ polymerization	0.1 Ag^{-1}	0.2-0.8	2M H_2SO_4	480		(Zhang, Zhang & Zhao, 2010)
Graphene/PANI nanorod	steamed water regulation	1Ag^{-1}	0-1.0	1M H_2SO_4	808		(Zhang, Huang & Hu,2017)

continued on following page

Table 3. Continued

Material	Method of Preparation	Discharge Current Density	PotentialRange/V	Electrolyte	SpecificCapacitanceF g^{-1}	Cycling/ Capacity Retention(Number of Cycles)	Ref.
Sandwiched-Graphene/ PANI hollow spheres	Hummers' method+ Sol-gel method+ chemical polymerization	0.5 Ag^{-1}	0-0.8	1M H_2SO_4	529	85% (1000)	(Li, Mengyu & Shiyong, 2016)

METAL OXIDE BASED PANI COMPOSITES

For the last few years, less expensive oxides of transition metals (TMOs), such as MnO_x, V_2O_5, TiO_2, Fe_3O_4, Co_3O_4, NiO, MoO_3, WO_3, etc. have been extensively explored as potential supercapacitor electrode materials (Gao & Fan, 2013) (Augustyn, Simon & Dunn, 2014). However, their usage in pristine form is restricted for commercialization due to poor electronic properties accompanied by slow charge transfer kinetics, average rate capability and low conductivity. Hence to surmount the problems, its coating/blending with stable carbon forms is quite indispensable (Shivhare & Vyas, 2017). Initially, PANI/TMOs were prepared by mixing chemically-polymerized PANI powders with TMO particles obtained from electrodeposition or sol–gel technique (Zhi, Xiang & Li, 2013) (Li, Gan & Zheng, 2012). Finally, the mixture paste was coated on various substrates such as tantalum matrix or polyvinylidene fluoride-co-hexafluoropropylene (Negi & Adhyapak, 2002). However, such procedure is complicated with low adhesion leading to unsatisfactory electrochemical outputs, consequently replaced by binder-free technologies discussed in subsequent sections.

Rigorous research revealed that PANI based metal oxide composites often show improved electrochemical performances due to their synergistic contribution (Cai, Cui & Zu,2017) (Majumdar & Bhattacharya, 2017) (Deshmukh, Patil & Bulakhe, 2015). However, this optimum condition for synergism is very much dependent on morphological architecture of the nanocomposite building the diffusion channels. Covering nanostructured metal oxides with PANI can improve both the electrical charge transport properties with simultaneous increasing the pseudocapacitance contribution as a result of PANI redox systems (Shivhare & Vyas, 2017) (Meng, Cai & Chen, 2017). In addition, PANI matrix is electrochemically permeable, and so it allows faster diffusion of the electrolyte ions to reach the underneath metal oxide so as to contribute maximum capacitance. Moreover, it also increases the electrode stability in acid-electrolytes.

Among various TMOs, MnO_2 has been one of the most interesting system for its electrochemical reversibility, lower price, high surface area, excellent charge storage features and environmental stability (Yang, Xu & Zou, 2016) (Zhang, Zhang & Miao, 2018). Sun et al. prepared PANI/MnO_2 composite with honeycomb-like structure containing PANI coated MnO_2 nano-spheres to avoid electrolytic instability in acidic solution and exhibited high specific density of 565 Fg^{-1} at 0.8 Ag^{-1} discharge current density, with ~ 77% of capacitance retention for 1000 GCD cycles (Sun, Gan & Ma, 2015). PANI/MnO_2 nanofibers composite electrodes by simply mixing of aqueous dispersed solution of PANI nanofibers and $KMnO_4$ aqueous with high current sensitivity were also reported shortly (Prathap, Satpati & Srivastava, 2013). Gemeay group explored the effects of dopants on the degree of crystallinity PANI/MnO_2 composites and concluded that the composite crystallinity follows the dopants order: $HSO_4^- > H_2PO_4^- > Cl^- > NO_3^-$ (Gemeay, Mansour & El-Sharkawy, 2005). Employing surfactants as dopants has become an effective step towards improving the specific capacitance of the composite. Zhang and co-researchers reported that

PANI retained its true conductivity property even after inserting it into noctadecyltrimethylammonium-intercalated MnO_2 layer (Zhang, Ji & Zhang, 2007). The said-composite showed appreciable electrochemical signatures. Chen and co-workers designed PANI nanorod/MnO_2 nanoparticle composite film by employing a silane-based coupling agent to modify-MnO_2 nanoparticles so as to improve the interracial charge process between MnO_2 and PANI (Chen, Sun & Luan, 2010). Apart from MnO_2, researchers have also studied electrochemical charge storing performance of various-morphology based PANI/other TMOs as well as non-transition metal oxide- based composites that have achieved a lot of progress in the recent years as discussed, few of which are discussed in Table 4.

Rigorous studies reveal that various conducting polymers have been employed as electrode substrate for tuning the electrodeposition characteristics of metal oxides too (Prasad & Miura, 2004). Prasad and Miura attempted to electrodeposit nano-phased MnO_2 on PANI-coated electrode and the composite supercapacitor displayed a high capacitance of 715 Fg^{-1}(Prasad & Miura, 2004). Ates et al. intended to make PANI-CuO composite film electrodes electrochemically grown on glassy-carbon electrodes that displayed maximum gravimetric capacitance of 286.35 F g^{-1} at scan rate of 20 mVs^{-1} much higher compared to similarly-fabricated other CuO-based CP composites (Ates, Serin & Ekmen, 2015).

Many experimentalists adopted core-shell structured TMO/PANI nanocomposites usually synthesized via two-step process. Metal oxides are derived through chemical / electrochemical techniques, coupled with annealing process, that was subsequently coated by PANI using chemical or electrochemical pro-

Table 4. Electrochemical performances of differently PANI/metal oxide composites

Composite	Electrochemical Performance				Reference
	Discharge Current Density and PotentialRange/V	Electrolyte	CapacitanceF g^{-1}	SpecificCapacity Retention(Number of Cycles)	
PANI/ MnO_x	1.0 mA cm^{-2} 0-0.65V	1 M $NaNO_3$	588	90% (1000)	(Zhang, Zhang & Miao, 2018)
PANI/TiO_2 nanowire arrays	1 A g^{-1} 0-0.6V	1M HCl	732	86% (2000)	(Xie, Li & Lai, 2011)
PANI/ MnO_2	5 A g^{-1} 0-1V	0.1 M Na_2SO_4	751	96.5% (5000)	(Prasad & Miura, 2004)
PANI/MnO_2	2 A g^{-1} 0-0.7	1 M H_2SO_4	626		(Jaidev, Jafri, & Mishra, 2011)
MnO_2/PANI	2 A g^{-1} 0-0.8	1 M Na_2SO_4	498	95% (5000)	(Sumboj, Foo & Yan, 2012)
PANI/SnO_2	1 A g^{-1} -0.2-+0.8	1 M H_2SO_4	305	95.5% (500)	(Hu, Xie & Wang, 2009)
PANI/TiO_2	2.5Ag^{-1} 0.2-0.7	1 M H_2SO_4	495	49.5% (3000)	(Li, Zhang &Wang, 2010)
PANI/V_2O_5	0.5mAcm^{-2} -0.9- +0.7	5 M LiCl	443	92% (5000)	(Bai, Liu & Luan, 2014)
PANI/MoO_3	300 mAg^{-1} -0.2 - 1.0 V	1 M Na_2SO_4	200	----	(Zheng, Xu & Jin, 2011)
PANI/Fe_3O_4	1 mA cm^{-2} 0-0.75	pH 7.4 buffer	213	-------	(Radhakrishnan, Prakash & Rao, 2009)
Fe_2O_3/PANI nanowire arrays	1 A g^{-1} −0.15 to 0.8 V	1.0 M Na_2SO_4	901	95.77% after 10,000 cycles	(Lu, Chen & Zhou, 2015)

cedure, resulting in the formation of core-shell composites as depicted in Figure 3. Such nanostructures show improved electrochemical signatures compared to the pristine metal oxide or CPs under similar conditions (Ates, Serin & Ekmen, 2015). Similar works on nanocomposites of PANI with Fe_2O_3 or with TiO_2 ensured considerable improvement in capacitive retention efficiency lately (Lu, Chen & Zhou, 2015).

Further progress in improving the interfacial charge transfer process initiated growing of well-ordered aligned PANI chains morphology that may yield smarter composite materials. One such successful instance is the electrochemical-co-deposition of PANI onto MnO_2 designed by Jiang and co-workers by drenching PANI nanofibers in $KMnO_4$ solution to obtain uniform loading of ultrafine MnO_2 floccules on the PANI nanofibers surface (Jiang, Ma & Li, 2012). The so-obtained composite exhibited capacitance of 383 Fg^{-1} with better GCD cycle stability in 1 M Na_2SO_4 aqueous solution Similar works on PANI/MnO_x composites represent good combination conductivity and capacitance, obtaining specific capacitance as high as ~1292 Fg^{-1} in an organic acetonitrile electrolyte (Zou, Wang & He, 2010). Analogous investigations regarding electrochemical-co-deposition of PANI/WO_3, PANI/V_2O_5 and PANI/NiO composites owing to the matching deposition potential window of these TMOs with PANI were lately reported too (Zou, Liang & Liu, 2011) (Bai, Liu & Luan, 2014) (Peng, Liu & Hua, 2010).

Recently, unique PANI-intercalated MOO_3 sandwich nanocomposites with well-oriented PANI chains are confined in the MoO_3 layers, showed 200 Fg^{-1} capacitance at wider potential window of -0.2V to1.0 V (vs. a saturated calomel electrode, SCE) in 1 M Na_2SO_4 electrolyte (Zheng, Xu & Jin, 2011). Lately, RuO_2 embedded PANI electrodes obtained by electrodeposition process imparted porous morphology with larger surface area exhibiting symmetrical, smooth-shaped CV-curves, displaying specific capaci-

Figure 3. Comparative electrochemical performances of α-Fe_2O_3 (black curves) and α-Fe_2O_3@PANI (red curves), tested in three-electrode system with 1.0 M Na_2SO_4 as the electrolyte

Created by Universal Document Converter

[Reprinted with permission from ACS publications, Xue-Feng Lu, Xiao-Yan Chen, Wen Zhou, 2015)]

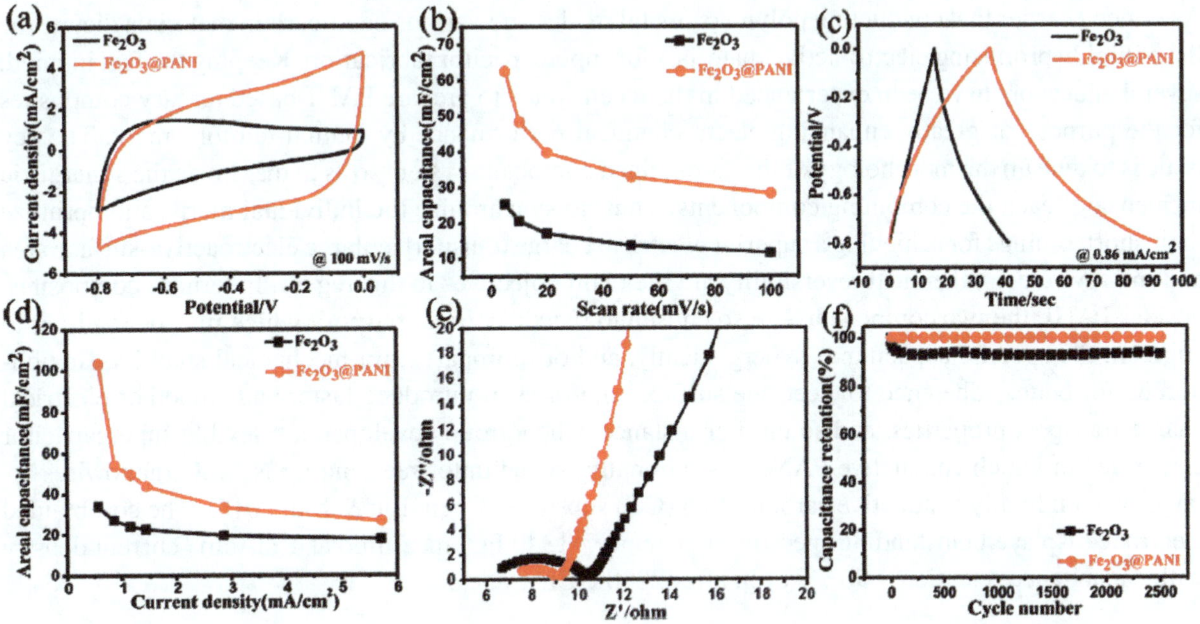

tance ~ 474 Fg^{-1} at 50 mVs^{-1}scan rate in 0.5 M H_2SO_4 electrolyte within wider potential window of -0.6 V to +0.6 V and ~ 88% capacitance efficiency after 1000 CV cycles (Wang, Lin & Shen, 2016).

In recent times, unique core–shell bimetallic oxide based–PANI composites have shown tremendous results. Highly porous $NiCo_2O_4$-PANI nanorod arrays exhibited appreciable capacitance of 901 Fg^{-1} at 1 Ag^{-1} current density in 1 M H_2SO_4 electrolyte with outstanding capacitance retention of ~ 91% even after 3000 GCD cycles measured at high current density of 10 Ag^{-1} (Jabeen, Xia & Yang, 2016). The hetero-structured-morphology of the as-grown $NiCo_2O_4$ provided a conductive core plus support for the electrochemically active nano-PANI layer, as shown in Figure 4, favoring high capacitance while maintaining appreciable cycling stability and rate capability. This work propelled a new direction of utilizing porous matrix as the highly effective support for polymers toward efficient energy storage. Another binary composite of $CoMoO_4$,0.75H_2O/PANI has also been lately synthesized by the in-situ oxidation-polymerization of aniline with virgin mixed-oxide obtained using simple and profitable one-pot hydrothermal process. The composite demonstrated a specific capacitance of 380 F g^{-1}, at a current density of 1 Ag^{-1} with capacitance retention of about 90.4% after 1000 consecutive charge–discharge cycles at a constant current density of 1 A g^{-1}. (Mandal, Ghosh & Giri, 2014). Although spinels have remarkable features such as multiple reversible-redox states, excellent pseudocapacitive properties and superior electrochemical stability but low electrical conductivity and difficulties in structure-to-performance electrode designing limits their frequent usage in supercapacitor applications.

PANI composites based on other metal chalcogenides/ hydroxides/ pnictogenides have also been reported with appreciable performances although relative abundance of the starting materials, chemical stability along with high production cost still questions their spontaneous usage in commercialization devices (Chen, Fan & Zhang, 2015) (Zhang., Shi & Liu, 2015) (Lei, Jiang & Lu, 2015) (Zhu, Sun & Yang, 2015).

PANI BASED TERNARY COMPOSITES

As noticed earlier that conducting polymers, metal oxides and various nano-carbon materials dictate the list of highly promising electro-active materials for supercapacitor fabrication. Keeping the fact in mind, several endeavors have been experienced in the recent years to prepare PANI-based ternary composites for the purpose of greatly enhancing electrochemical performance by mutual reinforcement. The key issue is to tune up the morphology of the so-obtained composite system so as to maximize the synergistic influence of each the combining components. Thus, to synchronize the individual merits and optimize their shortcomings for achieving a superior electrical charge transport, enlarge electroactive-surface area and improve electrochemical reversibility are the main objectives in this regard. In ternary composites, besides PANI, the two components are so chosen to meet a definite purpose which may be to either to augment the specific capacitance synergistically, and/or to impart extra mechanical stability, flexibility and /or better adherence to electrode surface, and/or even introduce faster and smoother electrical charge transport properties, and so on. For instance, Chen group developed a "flexible supercapacitor electrode" in which conductive PANI was rationally coated onto freestanding $NiMoO_4$ nanowires array grown on highly conductive carbon-cloth (CC) substrate (Chen, Liu & Liu, 2015). The so-obtained electrode displayed outstanding specific capacitance ~1340 Fg^{-1} measured at 1 $mAcm^{-2}$ current density with capacitance retention ~ 96.7% even after 2000 GCD cycles.

Figure 4. Formation steps of $NiCo_2O_4$@PANI NRAs and corresponding electrochemical studies of the nanocomposite.
Created by Universal Document Converter
[Reprinted with permission from ACS publications, (Jabeen, Xia & Yang, 2016)]

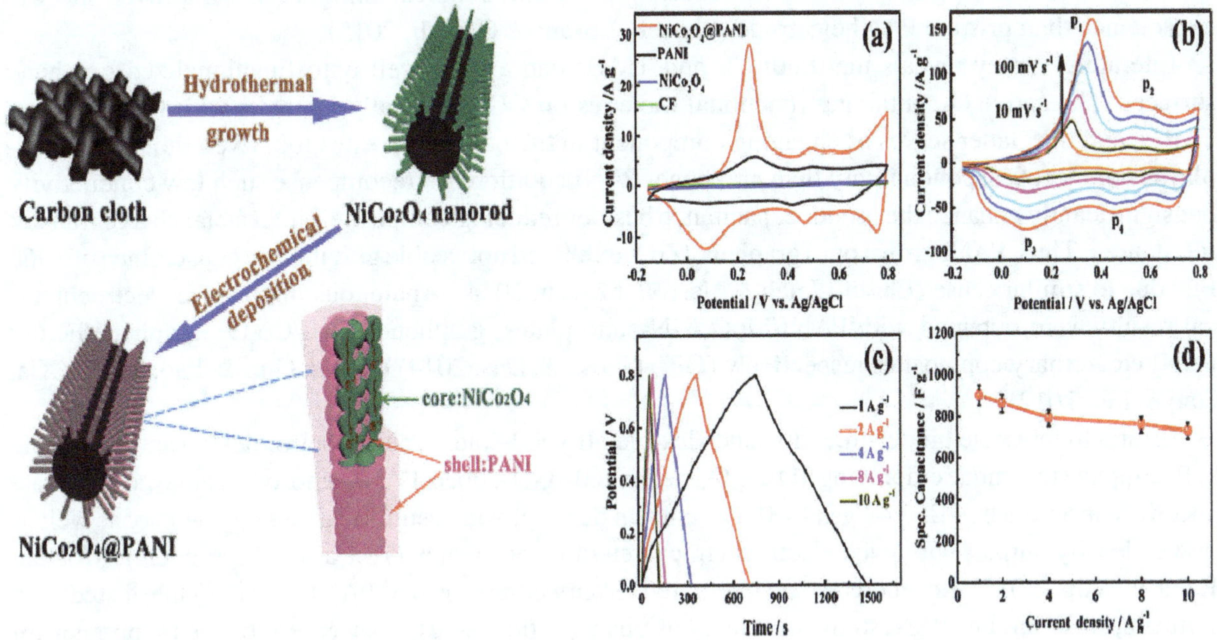

CNT/PANI composites often suffer polarization effects whenever CNT electrodes come in contact with electrolyte lead to low capacitance of <100 Fg^{-1}. Such lower capacitance is partly ascribed to reduced wettable character of CNT electrodes that restricts most of the usable surface for charge accumulation process to form the double-layer. Therefore, to reduce such polarization effects, pseudocapacitive TMOs or conjugated polymers are introduced on CNT surfaces thereby improving the electrochemical performance. Recently a novel MWCNT/MnO_2/PANI nanostructured composite supercapacitor electrode material using in-situ polymerization technique was reported where the components synergistically reinforce appreciable specific capacitance, high energy and power densities respectively (Sk., Yue & Jena, 2015). Lately, thin film electrodes with nanocomposites of WO_3 and PANI electrodeposited on binder-free, mechanically strong SWNT pre-deposited onto glass substrates by vacuum filtration followed by stamping method displayed enhanced the charge storage capacity along with good capacity retention (Yuksel, Durucan & Unalan, 2016). Zhao group attempted to incorporate carbon hollow shells as structural support for PANI/RuO_2 composite electrodes (Zhao, Guo & Gao, 2012). These hollow shells provided mechanical support as well as established faster charge transport via continuous PANI network. Thus, a maximum capacitance of 531 Fg^{-1} was obtained that retained ~87.5% of its initial value even after 1000 GCD cycles.

Recently, nanoparticle-introduction into CP/carbon based systems played an essential role for upgrading electrochemical properties of the existing binary systems. Generally, metal particles (Au, Ag, Bi etc.) are deliberately decorated on CP or nano-carbon surfaces to upgrade their electrical, thermal and mechanical properties (Marjanovic, 2013) (Patil, Shaikh & Dalavi, 2011). Similar evidences were obtained from PANI/GN/Ag nanocomposites obtained via simple, inexpensive in-situ polymerization technique that exhibited considerable specific capacitance value ~591 Fg^{-1} at 5 mVs^{-1} scan rate in 1 M KCl electrolyte with good electrochemical stability and superior energy - power densities respectively (Patil, Shaikh & Dalavi, 2011). Shayeh and co-workers prepared PANI/rGO/Au ternary nanocomposite similarly deposited on glassy carbon electrode by CV method indicated higher and improved specific capacitance than pristine PANI electrode (Shayeh, Ehsani & Ganjali, 2015)

Literature survey reveals that both CP and TMOs can anchor well onto functionalized graphene surfaces. The active O-containing functional moieties on GO chemically interacts with PANI chains/ TMOs, and thus latter serves as engaging component in the nanocomposites too. Even though GO displays poor electrical conductivity than graphene, PANI addition can recompense such low conductivity but significantly enhance the pseudocapacitance besides reducing composite's interfacial charge transfer resistances. Thus, PANI grown onto graphene/ZrO_2 exhibited noticeable gravimetric capacitance of 1360 Fg^{-1} due to similar cause (Ehsani, Jaleh & Nasrollahzadeh, 2014). Analogous impressive electrochemical results were obtained with PANI/MoO_3/GN-nano-plates, graphene/PANI/Co_3O_4, graphene/Fe_2O_3/ PANI etc. ternary composites respectively (Giri, Ghosh & Das, 2014) (Wang, Guo & Yao, 2017) (Xia, Hao & Lei, 2012).

Efforts to fabricate binder-free, thin and flexible all-solid-state supercapacitor device using scalable soft-template technique employing 3D-Fe_3O_4-decorated-rGO-doped-PANI nanorods displayed moderate specific capacitance (~ 283.4 Fg^{-1} at 1.0 Ag^{-1} current density) with a significant energy density as well as power density outputs with good electrochemical retention efficiency (78% after 5000th cycle) (Mondal, Rana & Malik, 2017). Recent report on integrated ternary composites of PANI/TiO_2/GO fabricated with intimate, cross-binding PANI strands, anatase-TiO_2 nanoparticles and GO sheets established synergism for facile electron transport process, that resulted in remarkable charge storing performances. The composite displayed large capacitance ~1020 F g^{-1} measured at 2 mVs^{-1} scan rate with substantial GCD stability (Su, Wang & Zhang, 2012). Hierarchical sulfonated-graphene-nanosheets (sGNS) /carboxylated-MWCNT / PANI nanocomposites were designed through interfacial polymerization technique. Excellent interfacial contact between PANI nanorods and sGNS/cMWCNT substrate through π–π and van der waals interaction shortened ion-diffusion and migration paths for rapid redox reaction, thus leading to high energy density of the device. Unique interpenetrating architecture increased the accessible contact area for rapid charge transport at the electrolyte/electrode interfaces. Moreover, flexible porous structure of sGNS also favors large EDLC initiating superior energy storage capability (Shen, Yang, Li & Wang, 2013). Recently, RGO sheets-wrapped-PANI-nanowire-arrays grown on nitrogen-doped-carbon-fiber-cloth (e-CFC) exhibited large specific capacitance of 1145 Fg^{-1}, substantially higher than that of PANI/e-CFC (1050 Fg^{-1}) and GO/PANI/e-CFC (940 Fg^{-1}) respectively (Yu, Li & Zhao, 2014). Narrow diameters (~ 100 nm) with large surface area encouraged high exposure to electroactive region, short diffusion pathways, and enhanced electrode/electrolyte interfaces. Subsequent wrapping PANI/eCFC composite with RGO sheets layers restricted abrupt volume-change of the PANI nanowires arrays, improving the GCD cycle stability. The formation mechanism of the aforesaid composite is displayed in Figure 5 (Yu, Li & Zhao, 2014). Thus, combination of rGO with PANI loaded on 3D-flexible conducting substrate brings about fascinating electrochemical properties due to synergism.

Figure 5. Schematic representation of Coating RGO Nanosheets on PANI nanowire Arrays Deposited on Nitrogen-Doped eCFC.
Created by Universal Document Converter
[Reprinted with permission from ACS publications, (Yu, Li & Zhao, 2014)]

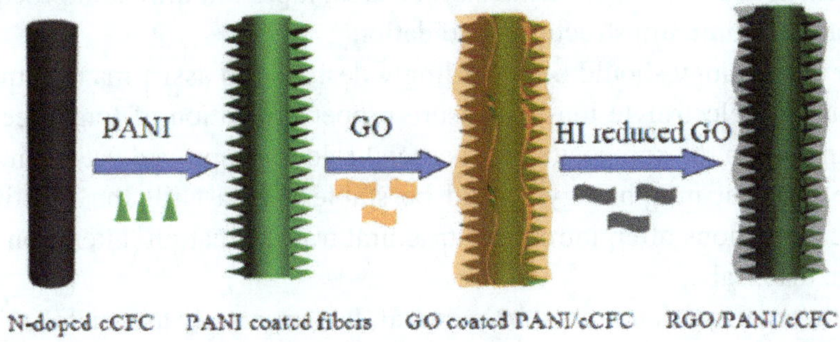

CONCLUSION

This chapter explored the current status of designing, fabrication as well as performance evaluation of PANI-based supercapacitors electrode materials and identified the main challenges and upcoming trends in this field. Studies reveal that binary composites have significantly improved the electrochemical efficiency of the designed-supercapacitors, yet to address issues to flexibility, higher conductivity, mechanical stability, and similar other important issues, addition of third component becomes essential. Till date results signify that ternary composites with better "coordination effect" preferably suited for superior electrochemical device fabrication. These composites show potential electrochemical behaviour for commercial applications to revolutionize the current electronics technology. The obtained high energy and power densities along with light-weighted, paper-like, flexible and highly bendable attracting features of these composites have further promoted its demand for applying in space and military applications. Efforts are made to bring down the production and installation costs which are herein the major obstacles in this regard. Therefore, scientists are now focusing on fabricating electrode materials with desired electrochemical performance in a simple and cost-effective route. Besides, PANI-based supercapacitor designed to well-perform at extreme environmental situations still remain to be ascertained. Recently, solid-state textile supercapacitors are being developed with high flexibility which is obvious to bring new dimensions to military applications. Alongside massive contribution for powered energy hybrid systems, PANI-based nanocomposite materials efficiently offer various multifunctional applications such as energy absorption, electromagnetic shielding, protective and anti-corrosive coatings, fire retardation agents, chemi-sensors, etc. which is obviously an additional advantage especially in defense sectors (Geetha, Kumar, & Rao, 2009; Song & Choi, 2013; Kurahatti, Surendranathan & Kori 2010; Sudha, Sivakala & Patel, 2010). Nonetheless, the advanced properties of the in-hand composite materials have paved an approach to effectively use them in commercial energy storage devices especially in the fields of advanced engineering, heavy electrical, space-aeronautical and military allied purposes, although non-stop searches for more sophisticated higher-energy-system will remain uninterrupted! Fast growing practical applications of micro-supercapacitors will attract more interest to this field and will encourage greater number of researchers to work on addressing the immediate challenges in advancement of these supercapacitor electrode materials in the near-future.

CHALLENGES AND PROSPECTS

- The upcoming studies should be devoted to designing tunable 3D- interconnected porous network structure, for efficient control of internal surface area, rapid ion diffusion process, maximum volume utilization and minimum structure degradation.
- The electrode morphology should be accordingly designed to assist maximum surface reactions, easy accessibility to electrolyte ions, to ensure proper utilization of total electro-active surface. The loading amount in the composite plays crucial role for improved yet optimized results.
- The electrode material morphology should be stable and last till the life time of the device. Electrochemical reactions often introduce structural transformation/ alteration due to collapse or coalesce of the materials.
- The electrode internal resistance must be minimal. It is intimately associated with the capacitance and hence on energy and power densities of the device. Large electrode thickness and poor conductivity contributes to such losses. Moreover, use of binders often increases electrode internal resistances. Thus, manufacture of stable binder-free composite electrodes needs to be worked upon immediately.
- Rapid advance of flexible polymer-based electronics urges development of mechanically well-flexible PANI based- composite systems, highly important for advanced engineering and military applications.
- Development of facile fabrication methods for micro-supercapacitor electrode materials guided by leakage-free electrolytes technologies should be encouraged to integrate these devices with other microelectronic systems on a small chip.
- Widening of operating voltage window is highly desirable for increasing the energy density as well as power performance of electrochemical capacitors. Besides, higher cell potential window would require smaller cell-numbers in series connections for obtaining specified voltage systems.
- Problem of minimizing high "self-discharge rates" owing to internal chemical reactions still remains to be addressed in the skill of electrochemical capacitors.

ACKNOWLEDGMENT

DM acknowledges Indian Association for the Cultivation of Science, Jadavpur, Kolkata, WB, India and Department of Chemistry (Physical Chemistry Section), Jadavpur University, Kolkata-700032, WB, India for e-library facilities. DM also acknowledges Chandernagore College, Hooghly, Pin-712136, WB, India for providing permission to carry out honorary research.

REFERENCES

Abdiryim, T., Ubul, A., Jamal, R., & Rahman, A. (2012). Solid-State Synthesis of Polyaniline/Single-Walled Carbon Nanotubes: A Comparative Study with Polyaniline/Multi-Walled Carbon Nanotubes. *Materials (Basel)*, *5*(7), 1219–1231. doi:10.3390/ma5071219

Amaya, R. O., Matsumoto, Y., Flores-Conde, A., Pérez-Guzmán, M. A., & Ortega-López, M. (2016). In-situ formation of rGO quantum dots during GO reduction via interaction with citric acid in aqueous medium. *Mater. Res. Express*, *3*(10), 105601K. doi:10.1088/2053-1591/3/10/105601

Ates, M., Serin, M. A., Ekmen, I., & Ertas, Y.N. (2015). Supercapacitor behaviors of Polyaniline/CuO, polypyrrole/CuO and PEDOT/CuO nanocomposites. *Polymer Bulletin*, *72*(10), 2573–2589. doi:10.100700289-015-1422-4

Augustyn, V., Simon, P., & Dunn, B. (2014). Pseudocapacitive oxide materials for high-rate electrochemical energy storage. *Energy & Environmental Science*, *7*(5), 1597–1614. doi:10.1039/c3ee44164d

Bai, M.-H., Liu, T.-Y., Luan, F., Li, Y., & Liu, X.-X. (2014). Electrodeposition of vanadium oxide-polyaniline composite nanowire electrodes for high energy density supercapacitors. *Journal of Materials Chemistry. A, Materials for Energy and Sustainability*, *2*(28), 10882–10888. doi:10.1039/C3TA15391F

Bavio, M. A., Acosta, G. G., & Kessler, T. (2014). Polyaniline and polyaniline-carbon black nanostructures as electrochemical capacitor electrode materials. *International Journal of Hydrogen Energy*, *39*(16), 8582–8589. doi:10.1016/j.ijhydene.2014.01.018

Beidaghia, M., & Gogotsi, Y. (2014). Capacitive energy storage in micro-scale devices: Recent advances in design and fabrication of micro-supercapacitors. *Energy & Environmental Science*, *7*(3), 867–884. doi:10.1039/c3ee43526a

Bleda-Martínez, M. J., Morallón, E., & Cazorla-Amorós, D. (2007). Polyaniline/porous carbon electrodes by chemical polymerisation: Effect of carbon surface chemistry. *Electrochimica Acta*, *52*(15), 4962–4968. doi:10.1016/j.electacta.2007.01.073

Boeva, Z. A., Milakin, K. A., Pesonen, M., Ozerin, A. N., Sergeyev, V. G., & Lindfors, T. (2014). Dispersible composites of exfoliated graphite and polyaniline with improved electrochemical behaviour for solid-state chemical sensor applications. *RSC Advances*, *4*(86), 46340–46350. doi:10.1039/C4RA08362H

Brownson, D. A., Varey, S. A., Hussain, F., Haigh, S. J., & Banks, C. E. (2014). Electrochemical properties of CVD grown pristine graphene: Monolayer-vs. quasi-graphene. *Nanoscale*, *6*(3), 1607–1621. doi:10.1039/C3NR05643K PMID:24337073

Cai, X., Cui, X., Zu, L., Zhang, Y., Gao, X., Lian, H., ... Wang, X. (2017). Ultra high electrical performance of nano nickel oxide and polyaniline composite materials. *Polymers*, *9*(7), 288.

Chen, C. N., Fan, W., Zhang, Q., Ma, T., Fu, X. W., & Wang, Z. B. (2015). In situ synthesis of cabbage like polyaniline@hydroquinone nanocomposites and electrochemical capacitance investigations. *Journal of Applied Polymer Science*, *132*(29), 42290. doi:10.1002/app.42290

Chen, K., & Xue, D. (2015). Searching for electrode materials with high electrochemical reactivity. *Journal of Materiomics*, *1*(3), 170–187. doi:10.1016/j.jmat.2015.07.001

Chen, L., Sun, L. J., Luan, F., Liang, Y., Li, Y., & Liu, X. X. (2010). Synthesis and pseudocapacitive studies of composite films of polyaniline and manganese oxide nanoparticles. *Journal of Power Sources*, *195*(11), 3742–3747. doi:10.1016/j.jpowsour.2009.12.036

Chen, W., Rakhi, R. B., & Alshareef, H. N. (2013). Capacitance enhancement of polyaniline coated curved-graphene supercapacitors in a redox-active electrolyte. *Nanoscale*, *5*(10), 4134–4138. doi:10.1039/c3nr00773a PMID:23567992

Chen, W., Yan, L., & Bangal, P. R. (2010). Preparation of graphene by the rapid and mild thermal reduction of graphene oxide induced by microwaves. *Carbon*, *48*(4), 1146–1152. doi:10.1016/j.carbon.2009.11.037

Chen, Y. P., Liu, B. R., Liu, Q., Wang, J., Liu, J. Y., Zhang, H. S., ... Jing, X. Y. (2015). Flexible all-solid-state asymmetric supercapacitor assembled using coaxial NiMoO4 nanowire arrays with chemically integrated conductive coating. *Electrochimica Acta*, *178*, 429–438. doi:10.1016/j.electacta.2015.08.040

Cheng, Q., Tang, J., Ma, J., Zhang, H., Shinya, N., & Qin, L.-C. (2011). Polyaniline-coated electro-etched carbon fiber cloth electrodes for supercapacitors. *The Journal of Physical Chemistry C*, *115*(47), 23584–23590. doi:10.1021/jp203852p

Cinke, M., Li, J., Chen, B., Cassell, A., Delzeit, L., Han, J., & Meyyappan, M. (2002). Pore structure of raw and purified HiPco single-walled carbon nanotubes. *Chemical Physics Letters*, *365*(1-2), 69–74. doi:10.1016/S0009-2614(02)01420-3

Dai, L., Chang, D. W., Baek, J.-B., & Lu, W. (2012). Carbon Nanomaterials for Advanced Energy Conversion and Storage. *Small*, *8*(8), 1130–1166. doi:10.1002mll.201101594 PMID:22383334

Dai, W., Ma, L., Gan, M., Wang, S., Sun, X., Wang, H., ... Zhou, T. (2016). Fabrication of sandwich nanostructure graphene/polyaniline hollow spheres composite and its applications as electrode materials for supercapacitor. *Materials Research Bulletin*, *76*, 344–352. doi:10.1016/j.materresbull.2015.12.045

Dawoud, H.D., Altahtamouni, T.M., Zagho, M.M. & Bensalah, N. (2017). A brief overview of flexible CNT/PANI super capacitors *Mater. Sci. Nanotechnol.*, *1*(2), 23-36.

Deshmukh, P. R., Patil, S. V., Bulakhe, R. N., Pusawale, S. N., Shim, J.-J., & Lokhande, C. D. (2015). Chemical synthesis of PANI–TiO_2 composite thin film for supercapacitor application. *RSC Advances*, *5*(84), 68939–68946. doi:10.1039/C5RA09233G

Dong, B., He, B., Xu, C., & Li, H. (2007). Preparation and electrochemical characterization of polyaniline/multi-walled carbon nanotubes composites for super-capacitor. *Materials Science and Engineering B*, *143*(1-3), 7–13. doi:10.1016/j.mseb.2007.06.017

Ehsani, A., Jaleh, B., & Nasrollahzadeh, M. (2014). Electrochemical properties and electrocatalytic activity of conducting polymer/copper nanoparticles supported on reduced graphene oxide composite. *Journal of Power Sources*, *257*, 300–307. doi:10.1016/j.jpowsour.2014.02.010

Fan, H., Wang, H., Zhao, N., Zhang, X., & Xu, J. (2012). Hierarchical nanocomposite of polyaniline nanorods grown on the surface of carbon nanotubes for high-performance supercapacitor electrode. *Journal of Materials Chemistry*, *22*(6), 2774–2780. doi:10.1039/C1JM14311E

Fan, L. Z., Hu, Y.-S., Maier, J., Adelhelm, P., Smarsly, B., & Antonietti, M. (2007). High Electroactivity of Polyaniline in Supercapacitors by Using a Hierarchically Porous Carbon Monolith as a Support. *Advanced Functional Materials*, *17*(16), 3083–3087. doi:10.1002/adfm.200700518

Fan, W., Zhang, C., Tjiu, W. W., Pramoda, K. P., He, C., & Liu, T. (2013). Graphene-wrapped polyaniline hollow spheres as novel hybrid electrode materials for supercapacitor applications. *ACS Applied Materials & Interfaces*, *5*(8), 3382–3391. doi:10.1021/am4003827 PMID:23517224

Feng, W., Zhang, Q., Li, Y., & Feng, Y. (2014). Preparation of sulfonated graphene/polyaniline composites in neutral solution for high-performance supercapacitors. *Journal of Solid State Electrochemistry*, *18*(4), 1127–1135. doi:10.100710008-013-2369-8

Fernández-Merino, M. J., Guardia, L., Paredes, J. I., Villar-Rodil, S., Solís-Fernández, P., Martínez-Alonso, A., & Tascón, J. M. D. (2010). Vitamin C Is an Ideal Substitute for Hydrazine in the Reduction of Graphene Oxide Suspensions. *J. PhysChemComm*, *114*(14), 6426–6432.

Gao, S & Fan, H. (2013). A review on the metal-oxide-based pseudocapacitors. *Huaxue Tongbao [Chemistry Bulletin]*, *76*, 202-209.

Gao Y, (2017). Graphene and Polymer Composites for Supercapacitor Applications: a Review. *Nanoscale Res Lett.*, *12*, 387.

Gao, Z., Yang, W., Wang, J., Yan, H., Yao, Y., Ma, J., ... Liu, L. (2013). Electrochemical synthesis of layer-by-layer reduced graphene oxide sheets/polyaniline nanofibers composite and its electrochemical performance. *Electrochimica Acta*, *91*, 185–194. doi:10.1016/j.electacta.2012.12.119

Geetha, S., Satheesh Kumar, K. K., Rao, C. R. K., Vijayan, M., & Trivedi, D. C. (2009). EMI shielding: Methods and materials—A review. *Journal of Applied Polymer Science*, *112*(4), 2073–2086. doi:10.1002/app.29812

Geim, A. K., & Novoselov, K. S. (2007). The rise of graphene. *Nature Materials*, *6*(3), 183–191. doi:10.1038/nmat1849 PMID:17330084

Gemeay, A. H., Mansour, I. A., El-Sharkawy, R. G., & Zaki, A. B. (2005). Preparation and characterization of polyaniline/manganese dioxide composites via oxidative polymerization: Effect of acids. *European Polymer Journal*, *41*(11), 2575–2583. doi:10.1016/j.eurpolymj.2005.05.030

Giri, S., Ghosh, D., & Das, C. K. (2014). Growth of vertically aligned tunable polyaniline on graphene/ZrO_2 nanocomposites for supercapacitor energy-storage application. *Advanced Functional Materials*, *24*(9), 1312–1324. doi:10.1002/adfm.201302158

Gu, W. T., & Yushin, G. (2014). Review of nanostructured carbon materials for electrochemical capacitor applications: Advantages and limitations of activated carbon, carbide-derived carbon, zeolite-templated carbon, carbon aerogels, carbon nanotubes, onion-like carbon, and graphene. *WIREs Energy & Environment, 3*, 424–473.

Guo, F., Mi, H., Zhou, J., Zhao, Z., & Qiu, J. (2015). Hybrid pseudocapacitor materials from polyaniline@ multi-walled carbon nanotube with ultrafine nanofiber-assembled network shell. *Carbon*, *95*, 323–329. doi:10.1016/j.carbon.2015.08.052

Guo, H. L., Wang, X. F., Qian, Q. Y., Wang, F. B., & Xia, X. H. (2009). A green approach to the synthesis of graphene nanosheets. *ACS Nano*, *3*(9), 2653–2659. doi:10.1021/nn900227d PMID:19691285

Guo, S., & Dong, S. (2011). Graphene nanosheet: Synthesis, molecular engineering, thin film, hybrids, and energy and analytical applications. *Chemical Society Reviews*, *40*(5), 2644–2672. doi:10.1039/c0cs00079e PMID:21283849

Gupta, V., & Miura, N. (2006). Polyaniline/single-Wall carbon nanotube (PANI/SWCNT) composites for high performance supercapacitors. *Electrochimica Acta*, *52*(4), 1721–1726. doi:10.1016/j.electacta.2006.01.074

He, S., Hu, X., Chen, S., Hu, H., Hanif, M., & Hou, H. (2012). Needle-like polyaniline nanowires on graphite nanofibers: Hierarchical micro/nano-architecture for high performance supercapacitors. *Journal of Materials Chemistry*, *22*(11), 5114–5120. doi:10.1039/c2jm15668g

Hu, L., Tu, J., Jiao, S., Hou, J., Zhu, H., & Fray, D. J. (2012). In situ electrochemical polymerization of a nanorod-PANI–Graphene composite in a reverse micelle electrolyte and its application in a supercapacitor. *Physical Chemistry Chemical Physics*, *14*(45), 15652–15656. doi:10.1039/c2cp42192e PMID:23076399

Hu, Z., Xie, Y., Wang, Y., Mo, L., Yang, Y., & Zhang, Z. (2009). Polyaniline/SnO_2 nanocomposite for supercapacitor applications. *Materials Chemistry and Physics*, *114*(2-3), 990–995. doi:10.1016/j.matchemphys.2008.11.005

Imani, A., & Farzi, G. (2017). Low temperature process of electronic charge transport mechanism in PANI/MWCNT nanocomposites: Tubular morphology. *Journal of Materials Science Materials in Electronics*, *28*(14), 10684–10692. doi:10.100710854-017-6843-4

Jabeen, N., Xia, Q., Yang, M., & Xia, H. (2016). Unique core–shell nanorod arrays with polyaniline deposited into mesoporous $NiCo_2O_4$ support for high-performance supercapacitor electrodes. *ACS Applied Materials & Interfaces*, *8*(9), 6093–6100. doi:10.1021/acsami.6b00207 PMID:26889785

Jaidev, J. R. I., Jafri, R. I., Mishra, A. K., & Ramaprabhu, S. (2011). Polyaniline-MnO_2 nanotube hybrid nanocomposite as supercapacitor electrode material in acidic electrolyte. *Journal of Materials Chemistry*, *21*(44), 17601–17605. doi:10.1039/c1jm13191e

Jiang, H., Ma, J., & Li, C. Z. (2012). Polyaniline–MnO_2 coaxial nanofiber with hierarchical structure for high-performance supercapacitors. *Journal of Materials Chemistry*, *22*(33), 16939–16942. doi:10.1039/c2jm33249c

Jin, Y., Fang, M., & Jia, M. (2014). In situ one-pot synthesis of graphene–polyaniline nanofiber composite for high-performance electrochemical capacitors. *Applied Surface Science*, *308*, 333–340. doi:10.1016/j.apsusc.2014.04.168

Khalid, M., Tumelero, M. A., Zoldan, V. C., Cid, C. C. P., Franceschini, D. F., Timm, R. A., ... Pasa, A. A. (2014). Polyaniline nanofibers -graphene oxide nanoplatelets composite thin film electrodes for electrochemical capacitors. *RSC Advances*, *4*(64), 34168–34178. doi:10.1039/C4RA06145D

Kim, B., Chung, H., Chu, K. S., Yoon, H. G., Lee, C. J., & Kim, W. (2010). Synthesis of vertically-aligned carbon nanotubes on stainless steel by water-assisted chemical vapor deposition and characterization of their electrochemical properties. *Synthetic Metals*, *160*(7-8), 584–587. doi:10.1016/j.synthmet.2009.12.008

Kim, W., Choi, H. C., Shim, M., Li, Y. M., Wang, D. W., & Dai, H. J. (2002). Synthesis of ultralong and high percentage of semiconducting single-walled carbon nanotubes. *Nano Letters*, *2*(7), 703–708. doi:10.1021/nl025602q

Kim, W., Javey, A., Tu, R., Cao, J., Wang, Q. & Dai, H. J. (2005) Electrical contacts to carbon nanotubes down to 1nm in diameter. *Appl. Phys. Lett., 87.*

Kurahatti, R. V., Surendranathan, A. O., Kori, S. A., Singh, N., Kumar, A. V. R., & Srivastava, S. (2010). Defence Applications of Polymer Nanocomposites. *Defence Science Journal*, *60*(5), 551–563. doi:10.14429/dsj.60.578

Largeot, C., Portet, C., Chmiola, J., Taberna, P. L., Gogotsi, Y., & Simon, P. (2008). Relation between the Ion Size and Pore Size for an Electric Double-Layer Capacitor. *Journal of the American Chemical Society*, *130*(9), 2730–2731. doi:10.1021/ja7106178 PMID:18257568

Lee, S., Kim, J., & Park, S. (2014). Activated carbon nanotubes/polyaniline composites as supercapacitor electrodes. *Energy*, *78*, 298–303. doi:10.1016/j.energy.2014.10.011

Lee, T., Yun, T., Park, B., Sharma, B., Song, H., & Kim, B. (2012). Hybrid multilayer thin film supercapacitor of graphene nanosheets with polyaniline: Importance of establishing intimate electronic contact through nanoscale blending. *Journal of Materials Chemistry*, *22*(39), 21092–21099. doi:10.1039/c2jm33111j

Lei, J. Y., Jiang, Z. Q., Lu, X. F., Nie, G. D., & Wang, C. (2015). Synthesis of Few-Layer MoS_2 Nanosheets-Wrapped Polyaniline Hierarchical Nanostructures for Enhanced Electrochemical Capacitance Performance. *Electrochimica Acta*, *176*, 149–155. doi:10.1016/j.electacta.2015.07.028

Li, D., Li, Y., Feng, Y., Hu, W., & Feng, W. (2015). Hierarchical graphene oxide/polyaniline nanocomposites prepared by interfacial electrochemical polymerization for flexible solid-state supercapacitors. *Journal of Materials Chemistry. A, Materials for Energy and Sustainability*, *3*(5), 2135–2143. doi:10.1039/C4TA05643D

Li, H. L., Wang, J. X., Chu, Q. X., Wang, Z., Zhang, F. B., & Wang, S. C. (2009). Theoretical and experimental specific capacitance of polyaniline in sulfuric acid. *Journal of Power Sources*, *190*(2), 578–586. doi:10.1016/j.jpowsour.2009.01.052

Li, W. D., Mengyu, M., Shiyong, G., Xiaowu, W., Huihui, S., Huining, W., & Zhou, W. T. (2016). Fabrication of sandwich nanostructure graphene/polyaniline hollow spheres composite and its applications as electrode materials for supercapacitor materials. *Research Bulletin*, *76*, 344–352. doi:10.1016/j.materresbull.2015.12.045

Li, X., Gan, W. P., Zheng, F., Li, L. L., Zhu, N. N., & Huang, X. Q. (2012). Preparation and electrochemical properties of RuO_2/polyaniline electrodes for supercapacitors. *Synthetic Metals*, *162*(11-12), 953–957. doi:10.1016/j.synthmet.2012.04.002

Li, X., Yang, L., Lei, Y., Gu, L., & Xiao, D. (2014). Microwave-assisted chemical-vapor induced in situ polymerization of polyaniline nanofibers on graphite electrode for high-performance supercapacitor. *ACS Applied Materials & Interfaces*, *6*(22), 19978–19989. doi:10.1021/am505533c PMID:25361469

Li, X., Zhang, H., Wang, G., & Jiang, Z. (2010). A novel electrode material based on a highly homogeneous polyaniline/titanium oxide hybrid for high-rate electrochemical capacitors. *Journal of Materials Chemistry*, *20*(47), 10598–10601. doi:10.1039/c0jm03330h

Li, Y., Peng, H., Li, G., & Chen, K. (2012). Synthesis and electrochemical performance of sandwich-like polyaniline/graphene composite nanosheets. *European Polymer Journal*, *48*(8), 1406–1412. doi:10.1016/j.eurpolymj.2012.05.014

Li, Y., Zhao, X., Yu, P., & Zhang, Q. (2013). Oriented arrays of polyaniline nanorods grown on graphite nanosheets for an electrochemical supercapacitor. *Langmuir*, *29*(1), 493–500. doi:10.1021/la303632d PMID:23205664

Liang, C., & Dai, S. (2006). Synthesis of Mesoporous Carbon Materials via Enhanced Hydrogen-Bonding Interaction. *Journal of the American Chemical Society*, *128*(16), 5316–5317. doi:10.1021/ja060242k PMID:16620083

Liu, J. L., Sun, J., & Gao, L. A. (2010). A promising way to enhance the electrochemical behavior of flexible single-walled carbon nanotube/polyaniline composite films. *The Journal of Physical Chemistry C*, *114*(46), 19614–19620. doi:10.1021/jp1092042

Liu, X., Shang, P., Zhang, Y., Wang, X., Fan, Z., Wang, B., & Zheng, Y. (2014). Three dimensional and stable polyaniline-grafted graphene hybrid materials for supercapacitor electrodes. *Journal of Materials Chemistry. A, Materials for Energy and Sustainability*, *2*(37), 15273–15278. doi:10.1039/C4TA03077J

Liu, Y., Zhou, J., Chen, L., Zhang, P., Fu, W., Zhao, H., ... Xie, E. (2015). Highly Flexible Freestanding Porous Carbon Nanofibers for Electrodes Materials of High-Performance All-Carbon Supercapacitors. *ACS Applied Materials & Interfaces*, *7*(42), 23515–23520. doi:10.1021/acsami.5b06107 PMID:26449440

Lu, X.-F., Chen, X.-Y., Zhou, W., Tong, Y.-X., & Li, G.-R. (2015). α-Fe_2O_3@PANI Core−Shell Nanowire Arrays as Negative Electrodes for Asymmetric Supercapacitors. *ACS Applied Materials & Interfaces*, *7*(27), 14843–14850. doi:10.1021/acsami.5b03126 PMID:26090902

Luo, Z., Zhu, L., Zhang, H., & Tang, H. (2013). Polyaniline uniformly coated on graphene oxide sheets as supercapacitor material with improved capacitive properties. *Materials Chemistry and Physics*, *139*(2-3), 572–579. doi:10.1016/j.matchemphys.2013.01.059

Majumdar, D. (2016). Functionalized-Graphene/Polyaniline Nanocomposites as Proficient Energy Storage Material: An Overview. *Innov Ener Res*, *5*, 145.

Majumdar, D., Baugh, N., & Bhattacharya, S. K. (2017). Ultrasound assisted formation of reduced graphene oxide-copper (II) oxide nanocomposite for energy storage applications. *Colloids and Surfaces. A, Physicochemical and Engineering Aspects*, *512*, 158–170. doi:10.1016/j.colsurfa.2016.10.010

Majumdar, D., & Bhattacharya, S. K. (2017). Sonochemically synthesized hydroxy-functionalized graphene–MnO_2 nanocomposite for supercapacitor applications. *Journal of Applied Electrochemistry*, *47*(7), 789–801. doi:10.100710800-017-1080-3

Mandal, M., Ghosh, D., Giri, S., Shakir, I., & Das, C. K. (2014). Polyaniline-wrapped 1D $CoMoO_4{\cdot}0.75H_2O$ nanorods as electrode materials for supercapacitor energy storage applications. *RSC Advances*, *4*(58), 30832–30839. doi:10.1039/C4RA03399J

Mao, L., Zhang, K., Chan, H. S. O., & Wu, J. (2012). Surfactant-stabilized graphene/ polyaniline nanofiber composites for high performance supercapacitor electrode. *Journal of Materials Chemistry*, *22*(1), 80–85. doi:10.1039/C1JM12869H

Marjanovic, G. C. (2013). Recent advances in polyaniline composites with metals, metalloids and nonmetals. *Synthetic Metals*, *170*, 31–56. doi:10.1016/j.synthmet.2013.02.028

Mastragostino, M., Arbizzani, C., Meneghello, L., & Paraventi, R. (1996). Electronically conducting polymers and activated carbon: Electrode materials in super-capacitor technology. *Advanced Materials*, *8*(4), 331–334. doi:10.1002/adma.19960080409

Meng, Q., Cai, K., Chen, Y., & Chen, L. (2017). Research progress on conducting polymer based supercapacitor electrode materials. *Nano Energy*, *36*, 268–285. doi:10.1016/j.nanoen.2017.04.040

Mezgebe, M. M., Yan, Z., Wei, G., Gong, S., Zhang, F., Guang, S., & Xu, H. (2017). 3D graphene-Fe3O4-polyaniline, a novel ternary composite for supercapacitor electrodes with improved electrochemical properties. *Materials Today Energy*, *5*, 164–172.

Miao, F., Shao, C., Li, X., Wang, K., Lu, N., & Liu, Y. (2016). Freestanding hierarchically porous carbon framework decorated by polyaniline as binder-free electrodes for high performance supercapacitors. *Journal of Power Sources*, *329*, 516–524. doi:10.1016/j.jpowsour.2016.08.111

Misoon, O., & Seok, K. (2012). Effect of dodecyl benzene sulfonic acid on the preparation of polyaniline/activated carbon composites by in situ emulsion polymerization. *Electrochimica Acta*, *59*, 196–201. doi:10.1016/j.electacta.2011.10.058

Mitchell, E., Candler, J., De Souza, F., Gupta, R. K., Gupta, B. K., & Dong, L. F. (2015). High performance supercapacitor based on multilayer of polyaniline and graphene oxide. *Synthetic Metals*, *199*, 214–218. doi:10.1016/j.synthmet.2014.11.028

Mondal, S., Rana, U., & Malik, S. (2017). Reduced Graphene Oxide/Fe_3O_4/Polyaniline Nanostructures as Electrode Materials for an All-Solid-State Hybrid Supercapacitor. *The Journal of Physical Chemistry C*, *121*(14), 7573–7583. doi:10.1021/acs.jpcc.6b10978

Negi, Y. S., & Adhyapak, P. V. (2002). Development in polyaniline conducting polymers. *Journal of Macromolecular Science -. Polymer Reviews (Philadelphia, Pa.)*, *42*, 35–53.

Nguyen, D. N. & Yoon, H. (2016). Recent Advances in Nanostructured Conducting Polymers: from Synthesis to Practical Applications. *Polymers, 8*, 118.

Ning, G., Li, T., Yan, J., Xu, C., Wei, T., & Fan, Z. (2013). Three-dimensional hybrid materials of fish scale-like polyaniline nanosheet arrays on graphene oxide and carbon nanotube for high-performance ultracapacitors. *Carbon*, *54*, 241–248. doi:10.1016/j.carbon.2012.11.035

Ning, X., Zhong, W., & Wana, L. (2016). Ultrahigh specific surface area porous carbon nanospheres and its composite with polyaniline: Preparation and application for supercapacitors. *RSC Advances*, *6*(30), 25519–25524. doi:10.1039/C6RA00596A

Niu, Z., Luan, P., Shao, Q., Dong, H., Li, J., Chen, J., ... Xie, S. (2012). A "skeleton/skin" strategy for preparing ultrathin free-standing single-walled carbon nanotube/polyaniline films for high performance supercapacitor electrodes. *Energy & Environmental Science*, *5*(9), 8726–8733. doi:10.1039/c2ee22042c

Otrokhov, G., Pankratov, D., Shumakovich, G., Khlupova, M., Zeifman, Y., Vasil'eva, I., ... Yaropolov, A. (2014). Enzymatic synthesis of polyaniline/ multi-walled carbon nanotube composite with core shell structure and its electrochemical characterization for supercapacitor application. *Electrochimica Acta*, *123*, 151–157. doi:10.1016/j.electacta.2013.12.089

Pacurari, M., Lowe, K., Tchounwou, P., & Kafoury, R. (2016). A review on the respiratory system toxicity of carbon nanoparticles. *International Journal of Environmental Research and Public Health*, *13*(3), 325.

Park, S., An, J., Potts, J. R., Velamakanni, A., Murali, S., & Ruoff, R. S. (2011). Hydrazine-reduction of graphite- and graphene oxide. *Carbon*, *49*(9), 3019–3023. doi:10.1016/j.carbon.2011.02.071

Parveen, N., Mahato, N., Ansari, M. O., & Cho, M. H. (2016). Enhanced electrochemical behavior and hydrophobicity of crystalline polyaniline@ graphene nanocomposite synthesized at elevated temperature. *Composites. Part B, Engineering*, *87*, 281–290. doi:10.1016/j.compositesb.2015.10.029

Patil, D. S., Shaikh, J. S., Dalavi, D. S., Karanjkar, M. M., Devan, R. S., Ma, Y. R., & Patil, P. S. (2011). An Mn Doped Polyaniline Electrode for Electrochemical Supercapacitor. *Journal of the Electrochemical Society*, *158*(6), A653. doi:10.1149/1.3561428

Peng, X.-Y., Liu, X.-X., Hua, P.-J., Diamond, D., & Lau, K.-T. (2010). Electrochemical codeposition of nickel oxide and polyaniline. *Journal of Solid State Electrochemistry*, *14*(1), 1–7. doi:10.100710008-008-0778-x

Prasad, K. R., & Miura, N. (2004). Polyaniline-MnO_2 composite electrode for high energy density electrochemical capacitor. *Electrochemical and Solid-State Letters*, *7*(11), A425. doi:10.1149/1.1805504

Prasad, K. R., & Miura, N. (2004). Potentiodynamically deposited nanostructured manganese dioxide as electrode material for electrochemical redox supercapacitors. *Journal of Power Sources*, *135*(1-2), 354–360. doi:10.1016/j.jpowsour.2004.04.005

Prathap, M. U. A., Satpati, B., & Srivastava, R. (2013). Facile preparation of polyaniline/MnO_2 nanofibers and its electrochemical application in the simultaneous determination of catechol, hydroquinone, and resorcinol. *Sensors and Actuators. B, Chemical*, *186*, 67–77. doi:10.1016/j.snb.2013.05.076

Radhakrishnan, S., Prakash, S., Rao, C. R. K., & Vijayan, M. (2009). Organically soluble bifunctional Polyaniline Magnetite composites for sensing and supercapacitor applications. *Electrochemical and Solid-State Letters*, *12*(4), A84. doi:10.1149/1.3074315

Ramakrishna Matte, H. S. S., Subrahmanyam, K. S., & Rao, C. N. R. (2011). Synthetic aspects and selected properties of graphene. *Nanomaterials and Nanotechnology*, *1*(1), 3–13.

Rao, C. E. E., Sood, A. E., Subrahmanyam, K. E., & Govindaraj, A. (2009). Graphene: The new two-dimensional nanomaterial. *Angewandte Chemie International Edition*, *48*(42), 7752–7777.

Ryu, K. S., Lee, Y.-G., Kim, K. M., Park, Y. J., Hong, Y.-S., Wu, X., ... Ko, J. M. (2005). Electrochemical capacitor with chemically polymerized conducting polymer based on activated carbon as hybrid electrodes. *Synthetic Metals*, *153*(1-3), 89–92. doi:10.1016/j.synthmet.2005.07.167

Salinas-Torres, D., Sieben, J. M., Lozano-Castelló, D., Cazorla-Amorós, D., & Morallón, E. (2013). Asymmetric hybrid capacitors based on activated carbon and activated carbon fibre–PANI electrodes. *Electrochimica Acta*, *89*, 326–333. doi:10.1016/j.electacta.2012.11.039

Salunkhe, R. R., Lin, J. J., Malgras, V., Dou, S. X., Kim, J. H., & Yamauchi, Y. (2015). Large-scale synthesis of coaxial carbon nanotube/$Ni(OH)_2$ composites for asymmetric supercapacitor application. *Nano Energy*, *11*, 211–218. doi:10.1016/j.nanoen.2014.09.030

Salunkhe, R. R., Tang, J., Kamachi, Y., Nakato, T., Kim, J. H., & Yamauchi, Y. (2015). Asymmetric Supercapacitors Using 3D Nanoporous Carbon and Cobalt Oxide Electrodes Synthesized from a Single Metal–Organic Framework. *ACS Nano*, *9*(6), 6288–6296. doi:10.1021/acsnano.5b01790 PMID:25978143

Sarker, A. K., & Hong, J. D. (2012). Layer-by-layer self-assembled multilayer films composed of graphene/polyaniline bilayers: High-energy electrode materials for supercapacitors. *Langmuir*, *28*(34), 12637–12646. doi:10.1021/la3021589 PMID:22866750

Shayeh, J. S., Ehsani, A., Ganjali, M. R., Norouzi, P., & Jaleh, B. (2015). Conductive polymer/reduced graphene oxide/Au nano particles as efficient composite materials in electrochemical supercapacitors. *Applied Surface Science*, *353*, 594–599. doi:10.1016/j.apsusc.2015.06.066

Shen, J., Yang, C., Li, X., & Wang, G. (2013). High-Performance Asymmetric Supercapacitor Based on Nano architectured Polyaniline/Graphene/Carbon Nanotube and Activated Graphene Electrodes. *ACS Applied Materials & Interfaces*, *5*(17), 8467–8476. doi:10.1021/am4028235 PMID:23931572

Shen, K., Rana, F., Zhang, X., Liu, C., Wang, N., Niu, X., ... Chen, S. (2015). Supercapacitor electrodes based on nano-polyaniline deposited on hollow carbon spheres derived from cross-linked co-polymers. *Synthetic Metals*, *209*, 369–376. doi:10.1016/j.synthmet.2015.08.012

Shivhare, S., & Vyas, S. (2017). Polyaniline (PANI) Metal Oxide Nano Composites as a Conducting Material. *International Journal of Research and Scientific Innovation.*, *IV*, 86–89.

Shukla, A. K., Banerjee, A., Ravikumar, M. K., & Jalajakshi, A. (2012). Electrochemical capacitors: Technical challenges and prognosis for future markets. *Electrochimica Acta*, *84*, 165–173. doi:10.1016/j.electacta.2012.03.059

Simotwo, S. K., DelRe, C., & Kalra, V. (2016). Supercapacitor Electrodes Based on High-Purity Electrospun Polyaniline and Polyaniline–Carbon Nanotube Nanofibers. *ACS Applied Materials & Interfaces*, *8*(33), 21261–21269. doi:10.1021/acsami.6b03463 PMID:27467445

Sk, M. M., Yue, C. Y., & Jena, R. K. (2015). Non-covalent interactions and supercapacitance of pseudocapacitive composite electrode materials (MWCNTCOOH/MnO2/PANI). *Synthetic Metals*, *208*, 2–12. doi:10.1016/j.synthmet.2014.10.026

Soni, R., Kashyap, V., Nagaraju, D., & Kurungot, S. (2018). Realizing High Capacitance and Rate Capability in Polyaniline by Enhancing the Electrochemical Surface Area through Induction of Superhydrophilicity. *ACS Applied Materials & Interfaces*, *10*(1), 676–686. doi:10.1021/acsami.7b15534 PMID:29243906

Su, H., Wang, T., Zhang, S., Song, J., Mao, C., Niu, H., ... Tian, Y. (2012). Facile synthesis of polyaniline/TiO_2/graphene oxide composite for high performance supercapacitors. *Solid State Sciences*, *14*(6), 677–681. doi:10.1016/j.solidstatesciences.2012.03.020

Sudha, J. D., Sivakala, S., Patel, K., & Radhakrishnan Nair, P. (2010). Development of electromagnetic shielding materials from the conductive blends of polystyrene 62 polyaniline-clay nanocomposite. *Composites. Part A, Applied Science and Manufacturing*, *41*(11), 1647–1652. doi:10.1016/j.compositesa.2010.07.015

Sumboja, A., Foo, C. Y., Yan, J., Yan, C., Gupta, R. K., & Lee, P. S. (2012). Significant electrochemical stability of manganese dioxide/polyaniline coaxial nanowires by self-terminated double surfactant polymerization for pseudocapacitor electrode. *Journal of Materials Chemistry*, *22*(45), 23921–23928. doi:10.1039/c2jm32456c

Sun, X. W., Gan, M. Y., Ma, L., Wang, H. H., Zhou, T., Wang, S. Y., ... Wang, H. N. (2015). Fabrication of PANI-coated honeycomb-like MnO_2 nanospheres with enhanced electrochemical performance for energy storage. *Electrochimica Acta*, *180*, 977–982. doi:10.1016/j.electacta.2015.09.056

Tang, J., Salunkhe, R. R., Liu, J., Torad, N. L., Imura, M., Furukawa, S., & Yamauchi, Y. (2015). Thermal Conversion of Core–Shell Metal–Organic Frameworks: A New Method for Selectively Functionalized Nanoporous Hybrid Carbon. *Journal of the American Chemical Society*, *137*(4), 1572–1580. doi:10.1021/ja511539a PMID:25580699

Tetlow, H., de Boer, J. P., Forde, I. J., Vvedensky, D. D., Coraux, J., & Kantorovich, L. (2014). Growth of epitaxial graphene: Theory and experiment. *Physics Reports*, *542*(3), 195–295. doi:10.1016/j.physrep.2014.03.003

Tran, C., Singhal, R., Lawrence, D., & Kalra, V. (2015). Polyaniline-coated freestanding porous carbon nanofibers as efficient hybrid electrodes for supercapacitors. *Journal of Power Sources*, *293*, 373–379. doi:10.1016/j.jpowsour.2015.05.054

Trung, N. B., Tam, T. V., Kim, R. H., Hur, S. H., Kim, E. J., & Choi, W. M. (2014). Three-dimensional hollow balls of graphene–polyaniline hybrids for supercapacitor applications. *Chemical Engineering Journal*, *255*, 89–96. doi:10.1016/j.cej.2014.06.028

Tyagi, A., Tripathi, K. M., & Gupta, R. K. (2015). Recent progress in micro-scale energy storage devices and future aspects. *Journal of Materials Chemistry. A, Materials for Energy and Sustainability*, *3*(45), 22507–22541. doi:10.1039/C5TA05666G

Vaquero, S., Díaz, R., Andersona, M., Palma, J., & Marcilla, R. (2012). Insights into the influence of pore size distribution and surface functionalities in the behaviour of carbon supercapacitors. *Electrochimica Acta*, *86*, 241–247. doi:10.1016/j.electacta.2012.08.006

Wang, D. W., Li, F., Zhao, J., Ren, W., Chen, Z. G., Tan, J., ... Cheng, H.-M. (2009). Fabrication of graphene/polyaniline composite paper via in situ anodic electropolymerization for highperformance flexible electrode. *ACS Nano*, *3*(7), 1745–1752. doi:10.1021/nn900297m PMID:19489559

Wang, H., Guo, Z., Yao, S., Li, Z., & Zhang, W. (2017). Design and Synthesis of Ternary Graphene/Polyaniline/Co_3O_4 Hierarchical Nanocomposites for Supercapacitors. *International Journal of Electrochemical Science*, *12*, 3721–3731. doi:10.20964/2017.05.66

Wang, H., Hao, Q., Yang, X., Lu, L., & Wang, X. (2010). A nanostructured graphene/polyaniline hybrid material for supercapacitors. *Nanoscale*, *2*(10), 2164–2170. doi:10.1039/c0nr00224k PMID:20689894

Wang, H., & Hu, Y. H. (2012). Graphene as a counter electrode material for dye-sensitized solar cells. *Energy & Environmental Science*, *5*(8), 8182–8188. doi:10.1039/c2ee21905k

Wang, H., Lin, J., & Shen, Z. X. (2016). Polyaniline (PANI) based electrode materials for energy storage and conversion. *Journal of Science: Advanced Materials and Devices*, *1*, 225–255.

Wang, K., Huang, J., & Wei, Z. (2010). Conducting Polyaniline Nanowire Arrays for High Performance Supercapacitors. *The Journal of Physical Chemistry C*, *114*(17), 8062–8067. doi:10.1021/jp9113255

Wang, Q., Yan, J., Fan, Z., Wei, T., Zhang, M., & Jing, X. (2014). Mesoporous polyaniline film on ultra-thin graphene sheets for high performance supercapacitors. *Journal of Power Sources*, *247*, 197–203. doi:10.1016/j.jpowsour.2013.08.076

Wang, S., Ma, L., Gan, M., Fu, S., Dai, W., Zhou, T., ... Wang, H. (2015). Free-standing 3D graphene/polyaniline composite film electrodes for high-performance supercapacitors. *Journal of Power Sources*, *299*, 347–355. doi:10.1016/j.jpowsour.2015.09.018

Wang, Y.-G., Li, H.-Q., & Xia, Y.-Y. (2006). Ordered Whisker-like Polyaniline Grown on the Surface of Mesoporous Carbon and Its Electrochemical Capacitance Performance. *Advanced Materials*, *18*(19), 2619–2623. doi:10.1002/adma.200600445

Wang, Z., Zhang, Q., Long, S., Luo, Y., Yu, P., Tan, Z., ... Bai, H. (2018). Three-Dimensional Printing of Polyaniline/Reduced Graphene Oxide Composite for High-Performance Planar Supercapacitor. *ACS Applied Materials & Interfaces*, *10*(12), 10437–10444. doi:10.1021/acsami.7b19635 PMID:29543426

Woo, S., Dokko, K., Nakano, H., & Kanamura, K. (2009). Incorporation of polyaniline into macropores of three-dimensionally ordered macroporous carbon electrode for electrochemical capacitors. *Journal of Power Sources*, *190*(2), 596–600. doi:10.1016/j.jpowsour.2009.01.050

Wu, T.-H., Chu, Y.-H., Hu, C.-C., & Hardwick, L. J. (2013). Criteria appointing the highest acceptable cell voltage of asymmetric supercapacitors. *Electrochemistry Communications*, *27*, 81–84. doi:10.1016/j.elecom.2012.10.033

Xia, X., Hao, Q., Lei, W., Wang, W., Sun, D., & Wang, X. (2012). Nanostructured ternary composites of graphene/Fe2O3/polyaniline for high-performance supercapacitors. *Journal of Materials Chemistry*, *22*(33), 16844–16850. doi:10.1039/c2jm33064d

Xie, K., Li, J., Lai, Y., Zhang, Z., Liu, Y., Zhang, G., & Huang, H. (2011). Polyaniline nanowire array encapsulated in titania nanotubes as a superior electrode for supercapacitors. *Nanoscale*, *3*(5), 2202–2207. doi:10.1039/c0nr00899k PMID:21455534

Xing, W., Zhuo, S., Cui, H., & Yan, Z. (2007). Synthesis of polyaniline-coated ordered mesoporous carbon and its enhanced electrochemical properties. *Materials Letters*, *61*(23-24), 4627–4630. doi:10.1016/j.matlet.2007.02.062

Xu, G., Wang, N., Wei, J., Lv, L., Zhang, J., Chen, Z., & Xu, Q. (2012). Preparation of Graphene Oxide/Polyaniline Nanocomposite with Assistance of Supercritical Carbon Dioxide for Supercapacitor Electrodes. *Industrial & Engineering Chemistry Research*, *51*(44), 14390–14398. doi:10.1021/ie301734f

Yalcin, S. E., Galande, C., Kappera, R., Yamaguchi, H., Martinez, U., Velizhanin, K. A., ... Mohite, A. D. (2015). Direct Imaging of Charge Transport in Progressively Reduced Graphene Oxide Using Electrostatic Force Microscopy. *ACS Nano*, *9*(3), 2981–2988. doi:10.1021/nn507150q PMID:25668323

Yan, J., Wei, T., Shao, B., Fan, Z., Qian, W., Zhang, M., & Wei, F. (2010). Preparation of a graphene nanosheet/polyaniline composite with high specific capacitance. *Carbon*, *48*(2), 487–493. doi:10.1016/j.carbon.2009.09.066

Yan, X., Tai, Z., Chen, J., & Xue, Q. (2011). Fabrication of carbon nanofiber-polyaniline composite flexible paper for supercapacitor. *Nanoscale*, *3*(1), 212–216. doi:10.1039/C0NR00470G PMID:21060936

Yan, Y., Cheng, Q., Zhu, Z., Pavlinek, V., Saha, P., & Li, C. (2013). Controlled synthesis of hierarchical polyaniline nanowires/ordered bimodal mesoporous carbon nanocomposites with high surface area for supercapacitor electrodes. *Journal of Power Sources*, *240*, 544–550. doi:10.1016/j.jpowsour.2013.03.190

Yang, F., Xu, M., Bao, S.-J., Wei, H., & Chai, H. (2014). Self-assembled hierarchical graphene/polyaniline hybrid aerogels for electrochemical capacitive energy storage. *Electrochimica Acta*, *137*, 381–387. doi:10.1016/j.electacta.2014.06.017

Yang, J. E., Jang, I., Kim, M., Baeck, S. H., Hwang, S., & Shim, S. E. (2013). Electrochemically polymerized vine-like nanostructured polyaniline on activated carbon nanofibers for supercapacitor. *Electrochimica Acta*, *111*, 136–143. doi:10.1016/j.electacta.2013.07.183

Yang, M., Cheng, B., Song, H., & Chen, X. (2010). Preparation and electrochemical performance of polyaniline-based carbon nanotubes as electrode material for supercapacitor. *Electrochimica Acta*, *55*(23), 7021–7027. doi:10.1016/j.electacta.2010.06.077

Yang, X., Xu, K., Zou, R., & Hu, J. (2016). A Hybrid Electrode of Co_3O_4@PPy Core/Shell Nanosheet Arrays for High-Performance Supercapacitors. *Nano-Micro Letters*, *8*(2), 143–150. doi:10.100740820-015-0069-x PMID:30460274

Yang, Y., Hao, Y., Yuan, J., Niu, L., & Xia, F. (2014). In situ preparation of caterpillar-likepolyaniline/carbon nanotube hybrids with core shell structure for high performance supercapacitors. *Carbon*, *78*, 279–287. doi:10.1016/j.carbon.2014.07.004

Yi, M., & Shen, Z. (2015). A review on mechanical exfoliation for the scalable production of graphene. *Journal of Materials Chemistry. A, Materials for Energy and Sustainability*, *3*(22), 11700–11715. doi:10.1039/C5TA00252D

Yin, Y., Liu, C., & Fan, S. (2012). Well-Constructed CNT Mesh/PANI Nanoporous Electrode and Its Thickness Effect on the Supercapacitor Properties. *The Journal of Physical Chemistry C*, *116*(50), 26185–26189. doi:10.1021/jp3083387

Yu, M., Ma, Y., Liu, J., & Li, S. (2015). Polyaniline nanocone arrays synthesized on three dimensional graphene network by electrodeposition for supercapacitor electrodes. *Carbon*, *87*, 98–105. doi:10.1016/j.carbon.2015.02.017

Yu, P., Li, Y., Zhao, X., Wu, L., & Zhang, Q. (2013). In situ growth of ordered polyaniline nanowires on surfactant stabilized exfoliated graphene as high-performance supercapacitor electrodes. *Synthetic Metals*, *185*, 89–95. doi:10.1016/j.synthmet.2013.10.010

Yu, P., Li, Y., Zhao, X., Wu, L., & Zhang, Q. (2014). Graphene-wrapped polyaniline nanowire arrays on nitrogen-doped carbon fabric as novel flexible hybrid electrode materials for high-performance supercapacitor. *Langmuir*, *30*(18), 5306–5313. doi:10.1021/la404765z PMID:24761945

Yu, P., Zhao, X., Huang, Z., Li, Y., & Zhang, Q. (2014). Free-standing three-dimensional graphene and polyaniline nanowire arrays hybrid foams for high performance flexible and lightweight supercapacitors. *Journal of Materials Chemistry. A, Materials for Energy and Sustainability*, *2*(35), 14413–14420. doi:10.1039/C4TA02721C

Yuksel, R., Durucan, C., & Unalan, H. E. (2016). Ternary nanocomposite SWNT/WO3/PANI thin film electrodes for supercapacitors. *Journal of Alloys and Compounds*, *658*, 183–189. doi:10.1016/j.jallcom.2015.10.216

Zhang, H., Cao, G., Wang, Z., Yang, Y., Shi, Z., & Gu, Z. (2008). Tube-covering-tube nanostructured polyaniline/carbon nanotube array composite electrode with high capacitance and superior rate performance as well as good cycling stability. *Electrochemistry Communications*, *10*(7), 1056–1059. doi:10.1016/j.elecom.2008.05.007

Zhang, H., Cao, G. P., & Yang, Y. S. (2007). Electrochemical properties of ultra-long, aligned, carbon nanotube array electrode in organic electrolyte. *Journal of Power Sources*, *172*(1), 476–480. doi:10.1016/j.jpowsour.2007.07.060

Zhang, J. L., Shi, L., Liu, H. D., Deng, Z. W., Huang, L. H., Mai, W. J., ... Cai, X. (2015). In situ synthesis of cabbage like polyaniline@hydroquinone nanocomposites and electrochemical capacitance investigations. *Journal of Alloys and Compounds*, *651*, 126–134. doi:10.1016/j.jallcom.2015.08.090

Zhang, K., Zhang, L. L., Zhao, X. S., & Wu, J. (2010). Graphene/polyaniline nanofibers composites as supercapacitor electrodes. *Chemistry of Materials*, *22*(4), 1392–1401. doi:10.1021/cm902876u

Zhang, L., Huang, D., Hu, N., Yang, C., Li, M., Wei, H., ... Zhang, Y. (2017). Three-dimensional structures of graphene/polyaniline hybrid films constructed by steamed water for high-performance supercapacitors. *Journal of Power Sources*, *342*, 1–8. doi:10.1016/j.jpowsour.2016.11.068

Zhang, L. L., Li, S., Zhang, J., Guo, P., Zheng, J., & Zhao, X. S. (2010). Enhancement of Electrochemical Performance of Macroporous Carbon by Surface Coating of Polyaniline. *Chemistry of Materials*, *22*(3), 1195–1202. doi:10.1021/cm902685m

Zhang, Q., Li, Y., Feng, Y., & Feng, W. (2013). Electropolymerization of graphene oxide/polyaniline composite for high-performance supercapacitor. *Electrochimica Acta*, *90*, 95–100. doi:10.1016/j.electacta.2012.11.035

Zhang, Q. Z., Zhang, D., Miao, Z. C., Zhang, X. L., & Chou, S. (2018). Research Progress in MnO_2 -Carbon Based Supercapacitor Electrode Materials. *Small*, *14*(24), 1702883. doi:10.1002mll.201702883 PMID:29707887

Zhang, X., Ji, L. Y., Zhang, S. C., & Yang, W. S. (2007). Synthesis of a novel polyaniline-intercalated layered manganese oxide nanocomposite as electrode material for electrochemical capacitor. *Journal of Power Sources*, *173*(2), 1017–1023. doi:10.1016/j.jpowsour.2007.08.083

Zhang, Y., Zhang, L., & Zhou, C. (2013). Review of Chemical Vapor Deposition of Graphene and Related Applications. *Accounts of Chemical Research*, *46*(10), 2329–2339. doi:10.1021/ar300203n PMID:23480816

Zhang, Z., Wang, G., Li, Y., Zhang, X., Qiao, N., Wang, J., ... Hao, Z. (2014). A new type of ordered mesoporous carbon/polyaniline composites prepared by a two-step nanocasting method for high performance supercapacitor applications. *Journal of Materials Chemistry. A, Materials for Energy and Sustainability*, *2*(39), 16715–16722. doi:10.1039/C4TA03351E

Zhao, D., Guo, X., Gao, Y., & Gao, F. (2012). An Electrochemical Capacitor Electrode Based on Porous Carbon Spheres Hybrided with Polyaniline and Nanoscale Ruthenium Oxide. *ACS Applied Materials & Interfaces*, *4*(10), 5583–5589. doi:10.1021/am301484s PMID:22988980

Zhao, X., Sanchez, B. M., Dobson, P. J., & Grant, P. S. (2011). The role of nanomaterials in redox-based supercapacitors for next generation energy storage devices. *Nanoscale*, *3*(3), 839–855. doi:10.1039/c0nr00594k PMID:21253650

Zheng, L., Xu, Y., Jin, D., & Xie, Y. (2011). Polyaniline-Intercalated Molybdenum Oxide Nanocomposites: Simultaneous Synthesis and their Enhanced Application for Supercapacitor. *Chemistry, an Asian Journal*, *6*(6), 1505–1514. doi:10.1002/asia.201000770 PMID:21557481

Zhi, M., Xiang, C., Li, J., Li, M., & Wu, N. (2013). Nanostructured carbon–metal oxide composite electrodes for supercapacitors: A review. *Nanoscale*, *5*(1), 72–88. doi:10.1039/C2NR32040A PMID:23151936

Zhou, D., Li, Y., Wang, J., Xu, P., & Han, X. (2011). Synthesis of polyaniline nanofibers with high electrical conductivity from CTAB-SDBS mixed surfactants. *Materials Letters*, *65*(23-24), 3601–3604. doi:10.1016/j.matlet.2011.08.021

Zhou, S., Mo, S., Zou, W., Jiang, F., Zhou, T., & Yuan, D. (2011). Preparation of polyaniline/ 2-dimensional hexagonal mesoporous carbon composite for supercapacitor. *Synthetic Metals*, *161*(15-16), 1623–1628. doi:10.1016/j.synthmet.2011.05.028

Zhou, X., Li, L., Dong, S., Chen, X., Han, P., Xu, H., ... Cui, G. (2011). A renewable bamboo carbon/ polyaniline composite for a high-performance supercapacitor electrode material. *Journal of Solid State Electrochemistry*, *16*(3), 877–882. doi:10.100710008-011-1435-3

Zhou, Y., Qin, Z., Li, L., Zhang, Y., Wei, Y., Wang, L., & Zhu, M. (2010). Polyaniline/multi-Walled carbon nanotube composites with Core Shell structures as super-capacitor electrode materials. *Electrochimica Acta*, *55*(12), 3904–3908. doi:10.1016/j.electacta.2010.02.022

Zhu, J. X., Sun, W. P., Yang, D., Zhang, Y., Hoon, H. H., Zhang, H., & Yan, Q. Y. (2015). Multifunctional Architectures Constructing of PANI Nanoneedle Arrays on MoS2 Thin Nanosheets for High-Energy Supercapacitors. *Small*, *11*(33), 4123–4129. doi:10.1002mll.201403744 PMID:26044446

Zhu, Z., Wang, G., Sun, M., Li, X., & Li, C. (2011). Fabrication and electrochemical characterization of polyaniline nanorods modified with sulfonated carbon nanotubes for supercapacitor applications. *Electrochimica Acta*, *56*(3), 1366–1372. doi:10.1016/j.electacta.2010.10.070

Zou, B.-X., Liang, Y., Liu, X.-X., Diamond, D., & Lau, K.-T. (2011). Electrodeposition and pseudocapacitive properties of tungsten oxide/polyaniline composite. *Journal of Power Sources*, *196*(10), 4842–4848. doi:10.1016/j.jpowsour.2011.01.073

Zou, W., Wang, W., He, B., Sun, M., & Yin, Y. (2010). Supercapacitive properties of hybrid films of manganese dioxide and polyaniline based on active carbon in organic electrolyte. *Journal of Power Sources*, *195*(21), 7489–7493. doi:10.1016/j.jpowsour.2010.05.020

Chapter 9
Polymer Nanocomposites Coating for Anticorrosion Application

Sudheer Kumar
Central Institute of Plastics Engineering and Technology (CIPET), India

Sukhila Krishnan
Central Institute of Plastics Engineering and Technology (CIPET), India

Sushanta Kumar Samal
Central Institute of Plastics Engineering and Technology (CIPET), India

Smita Mohanty
Central Institute of Plastics Engineering and Technology (CIPET), India

Sanjay Kumar Nayak
Central Institute of Plastics Engineering and Technology (CIPET), India

ABSTRACT

Corrosion is the foremost reason for the deterioration of metallic components used in harsh environments and hence research on the development of an innovative protective system with admirable performance has become a subject of leading importance. Currently, chromates free organic or the polymer coatings on the metal substrate bestow an efficient barrier amongst the metal and inhibit corrosion and save the environment. Researchers have involved in the development of polymer coating containing nanoparticles because reinforcement of nanoparticles decreases the porosity and provides a zig-zag diffusion path which in turn forms barrier layer on the metal surface to protect the metal from the harmful corrosive species. This present chapter deliberates the application of polymer-based nanocomposite coating to mitigate the corrosion of metals against harsh environment. This chapter covers the utilization of carbon-based nanoparticles, inorganic nanoparticle, conductive polymer, self-heling polymer, etc., and could shows a new insight to for anti-corrosive metal surface protective coating.

DOI: 10.4018/978-1-5225-7838-3.ch009

INTRODUCTION

Currently, corrosion of metals has been becoming a prime problem which extremely spoils the industry with a financial loss and creates an enormous risk to human beings. Because, corrosion cannot be completely prohibited and it's simply obstructed and reduced. Therefore, it is high time to protect the metal from corrosion to minimize economic losses. Steel is extensively utilized materials in various application fields, for example, civil engineering, military, transportation, gas and oil industry owing to its excellent mechanical properties and comparatively lower cost contrasted to advanced alloy materials (El Rayes et al., 2013). Corresponding to the published report of the World Steel Association in October 2017, the world utilization quantity to 145.3 million tonnes (mt) in 2017 with an enhanced of 5.9% evaluated in the year 2016 (World Steel Association et al., 2017). On the other hand, lots of the critical issues of steel arises due to its corrosion, reduction, particularly in marine environments that characteristically have higher salinity and humidity, these two factors influence the steel mechanical performance. An approximated 20% of world steel production is lost every year in the form of rust (Bhandari et al., 2015). Every nation is spending 1-5% gross national product (GNP) on corrosion issues (Koch et al., 2002). In the evaluations of the World Corrosion Organization (WCO) the yearly cost of approximately \$2.4 trillion expenditure worldwide (Natesan et al., 2005). For that reason, to inhibit the mechanical properties of the steels via applying nanocomposites organic coatings to the substrate and improved the product life and reduce the maintenance cost. Then, diffusion of the H_2O molecules through the anticorrosive polymer nanocomposite coatings is recognized as the main contributor to decrease the corrosion, delamination and adhesion in the coatings, which accordingly leads to common coating substitute and notable maintenance costs (Melchers et al., 2008).

Numerous, corrosion approaches are presently accessible like the employ of coatings, corrosion inhibitors, surface treatment, and electrochemical cathodic protection (Chen et al., 2007; Song, 2005; Solmaz et al., 2008; Yun et al., 2007). Amongst all, the utilization of corrosion protection coating is attracting to the overall market interest due to the suitable fabrication and excellent corrosion protection properties (Low et al., 2006; Wang et al., 2009; Gray et al., 2002).

Currently, nanocomposite coatings have attracted a great interest and rapidly developing approaches for scheming and fabrication of durable anticorrosive surface coatings with superior properties. A nanoparticle exhibits high surface area to volume ratio and provides a torturous path for the diffusion of H_2O molecule which leads to better surface protection as compared to bulk materials. The nano materials obtained higher barrier properties to the conventional fillers. Several nanoparticles-based coatings have been reported such as TiO_2 (Shi et al., 2002; Liu et al., 2018), Al_2O_3 (Golru et al., 2014; Dhoke et al., 2012), SiO_2 (Bakhshandeh et al., 2014; Palraj et al., 2015), ZnO (Ramezanzadeh et al., 2011; Sonawane et al., 2010), Fe_2O_3 (Dhoke et al., 2009; Palimi et al., 2015) etc. The fabrication of polymer nanocomposite coatings to protect steel substrates from the corrosion seems to be a fascinating solution. Figure 1 demonstrated the application of polymer nanocomposites coating surfaces in various fields.

The present chapter aimed to deliberate the contribution of nanoparticles to towards the significantly improvement of anticorrosion properties of the polymer based coatings for anticorrosion applications after the modification of the structure of the bulk coating and provides the additional functionality. On the other hand, nanotechnology provides the platform to develop high performance polymeric coatings with excellent corrosion resistance and other important properties. Further, to discussion of polymer nanocomposites coatings influence by the modification of polymer coatings, fabrication techniques and the employ of polymer nanocoatings as a carrier for corrosion inhibitors. This chapter also provides the

Figure 1. Polymer nanocoatings coating application in various fields

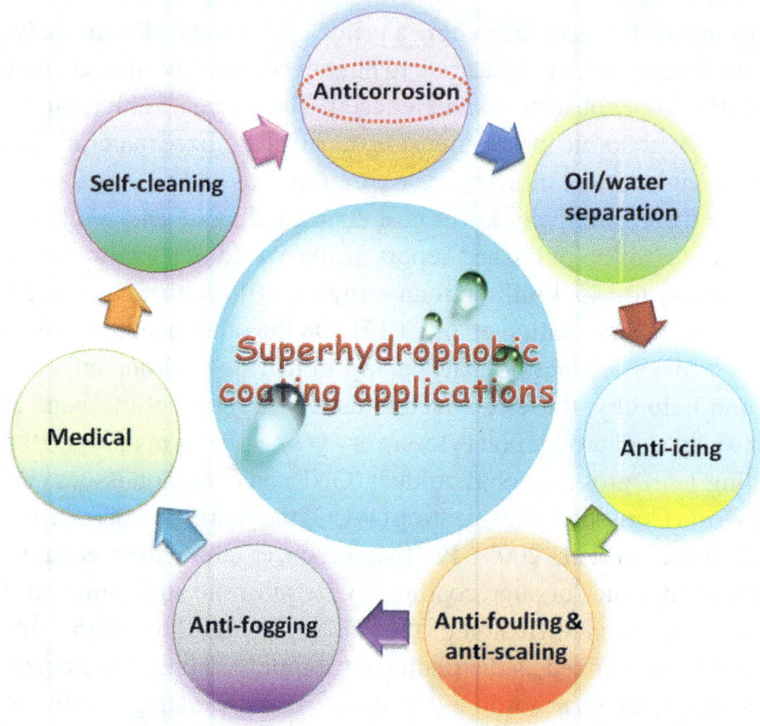

information about the corrosion mechanism and approaches for the prevention of corrosion. The chapter shall extensively cover the use carbon-based nanoparticles and inorganic nanoparticles in polymeric matrix towards the preparation anticorrosion coating and their applications. The chapter shall also cover the conductive polymer-based nanocomposite coating and self-healing coating for corrosion protection. This chapter also focused on the anticorrosion application of polymer nanocomposites coatings in the Military.

NANOMATERIALS IN CORROSION PROTECTION COATING

Recently, the nanomaterials are used as filler to protective barrier to alleviate the corrosion of the metal from various corrosive effects. This phenomenon based on the two main factors: (i) the first aspect is based on the corrosion behaviors of nanostructure materials, whether the material exhibited better corrosion resistance or not (e.g. corrosion/oxidation/cracking resistance of a nanostructured steel substrate/coating) as compared with the microstructured counterpart, (ii) the second aspect is based on how nanosized materials can be successfully used in corrosion protection approaches (e.g. in an military, automobile and aerospace coating), an area where nanomaterials corrosion control advantages are considerable (Saji et al., 2012). Nanomaterials advancements are anticipated to enhance corrosion monitoring and assessment sectors. Figure 2 demonstrates a correlation between nanomaterials and corrosion. Improvement of corrosion resistance by the use of nanostructure materials are also demonstrated higher mechanical and electronics properties. Nanostructure materials provides a stronger, harder and tougher coatings to the metal substrate that are more resistant to corrosion and employed in the useful application in defense armor, aerospace component, medical devices, constriction equipment, sport equipment etc., (Sorensen et al., 2009)

POLYMER NANOCOMPOSITE COATING FOR CORROSION PROTECTION

Currently, lots of approaches employed for the corrosion protection coatings are promoted significantly by nanomaterials. Nanocomposite coating can be specified as having either the thickness of the coating in nanosize range or the second phase particles. These particles are dispersed into the matrix in the nanosclae range or coatings having nanosized grains/phases, etc. Thin film coatings have been accessible for various years. The naonocoating can be divided into nanostructured coatings (coatings with grain sizes in the range of 100 nm) and nanocomposites coatings (Table 1) (Agarwala et al., 2006).

Figure 2. Schematic representation showing the correlation of nanomaterial in corrosion protection

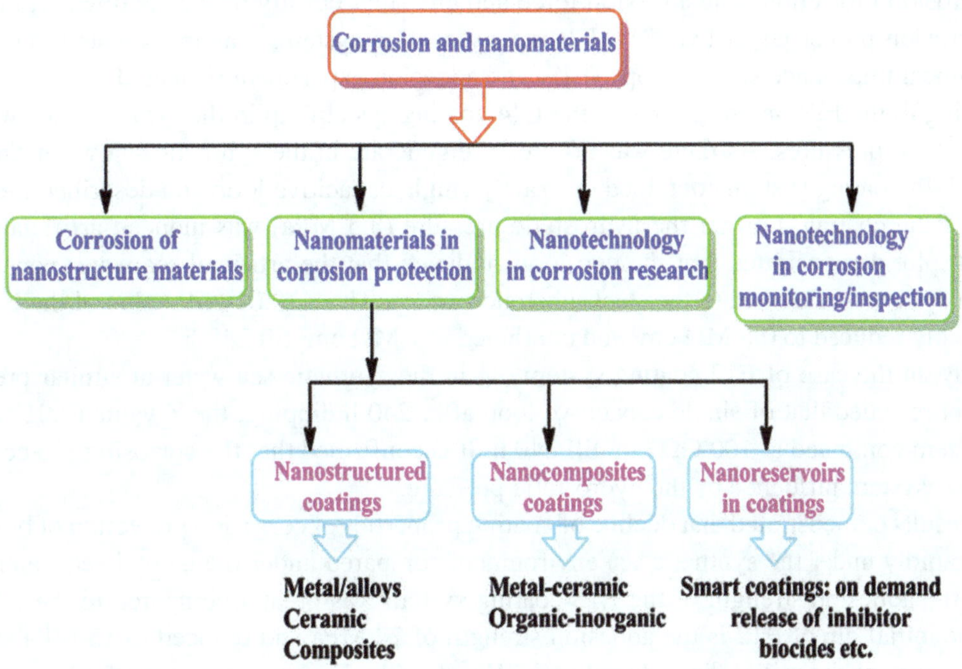

Table 1. Nanocoatings in corrosion protection

Thin Film	Nanograined Coatings	Nanocomposite Coatings	Smart Coatings
Ceramic, metallic, composite (nano/micrograined), self-assembled	Ceramic, metallic, composite (thin/thick films)	Polymer-ceramic, metallic-ceramic, polymer-metallic	Replacement for conversion coatings, smart polymer coatings

Carbon Based Nanoparticles

Recently, the nanoparticles employed as a preparation of anticorrosion nanocomposite coating technology widely applied for the protection of various fields such as marine, automobile, electronics and steel, metal industries etc. Corrosion is an extensively detected phenomenon in metal industries (Jeon et al., 2013). Many studied have been done by the researchers for the corrosion protection of metals but up to now, the perfect corrosion protection technique for inhibiting the corrosion of the metals has not been developed (Aglan et al., 2007). The most frequently employed and suitable fabrication technique for metal protection against corrosion species particularly in polymeric organic coating (Shen et al., 2016). Recently, the carbon nanoparticles, mostly single-walled carbon nanotube (SWCNTs) and multi-walled carbon nanotube (MWCNTs), Graphene oxide (GO) are well-known reinforcement for the corrosion application.

Polymer Nanocomposites Coating With Carbon Nanotube (CNT)

The future direction of organic coatings plays a very important character in the field of ocean engineering for corrosion protection. The adhesion force and crosslink density of the coating system effect on the anticorrosion properties of two E44, E12 epoxy coating systems was investigated via employing electrochemical impedance spectroscopy (EIS) and adhesion experiment (Liu et al., 2013).

According to the EIS investigations of the E44 coating system dip in the synthetic sea water at 3.5 MPa hydrostatic pressures. As there was no pressure generate in the synthetic sea water and Nyquist plot of the E44 coating system contained of exactly single capacitive loop and described the clean capacitance of the coating. Further the hydrostatic pressure (3.5 MPa) was induced artificial sea water the Nyquist plot demonstrated that the two loops indicate that the artificial sea water penetrated into the coating system and arrived to the steel substrate surface. The Z (0.01 HZ) value of the E44 coating system quickly reduced to 0.8 MΩ cm^2 and continued at 1 MΩ cm^2 till 240 h.

Similarly, in the case of E12 coating system dip in the synthetic sea water at similar pressure and Nyquist plot revealed that of single capacitive loop after 240 h dipping, the Z value (0.01 Hz) of E12 coating system continued at 100 GΩ cm^2 till 240 h. It is confirmed that the corrosion protection of the E12 coatings system influence by the hydrostatic pressure.

As the results demonstrated that decline of coating properties of corrosion protection of both systems was more rapidly under the synthetic sea environment compared under the natural sea water pressure. Although, the adhesion strength of the E12 coating system was greater compared to the E44 coating system. The initial dip of E12 is the adhesion strength of 22 MPa and reduced to 15 MPa steadily for E44 coating system the initial dip reduced to 3 MPa steadily. For that reason the E12 coating system showed higher adhesion strength with the substrate and superior anticorrosion properties than E44 coating system. On the other hand, after enhancement of the adhesion strength and crosslink density of the coatings and reduced the diffusion velocity of the water molecules during epoxy coatings. Thus, the potential of corrosion protection can be raised via growing the adhesion strength of the coating systems.

The preparation of epoxy based nanocomposites after incorporation of various wt% MWCNTs via using the innovative dispersion technique at the same time ultrasonic wave and axial flow impeller (Kumar et al., 2017) as shown in Figure 3. The field emission scanning electron microscopy (FESEM) studied of the tensile fracture surface exhibited that MWCNTs well disperse into the epoxy matrix due to the better interfacial interaction. This method able to crack the agglomeration of MWCNTs and provides the homogeneous dispersion in the epoxy matrix. Further, the corrosion investigation of without coating mild steel and coated virgin epoxy and its nanocomposites.

The Tafel plots indicate that the straight line of corrosion current (I_{corr}) and corrosion potential (E_{corr}) along with linear parts from anodic and cathodic curves. The I_{corr} curve shows that intensity decline of cathodic (O_2) and dissolution of anodic of metal ions. The lower value of I_{corr} demonstrated the lower corrosion protection rate. The epoxy/MWCNTs nanocomposites coated mild steel substrate exhibited more positive value of the corrosion potential (E_{corr}) compared to the virgin epoxy coated mild steel substrate and the mild steel substrate. After incorporation of MWCNTs enhancement of the corrosion potential (E_{corr}) become more positive up to 0.75 wt% and after increased the incorporation of MWCNTs wt.% into the epoxy system decreased the corrosion potential positive value. The enhancement of the positive value of the E_{corr} confirms that the enhancement of the corrosion protection of epoxy/MWCNTs nanocomposite coating systems.

The several reaction steps including into the corrosion of mild steel like reduction and oxidation as given in Figure 4 (Chang et al., 2012).

As a result the corrosion route of the mild steel, first the oxidations of Fe takes place with loose the electrons and go after by reduction of O_2. The both reactions also called a half reaction related to the each other and required to enough water (H_2O) and oxygen (O_2) for the corrosion on the substrate. That purposes the coating systems competent to bestow the barrier between the corrosive environment such as H_2O and O_2 with substrate surface owing to the anticorrosive character (Kumar et al., 2016). For that reason this type coating frequently employed on metal surface and corrosive species such as H_2O, O_2,

Figure 3. Schematic preparations of epoxy/MWCNTs nanocomposites via employing concurrently ultrasonic waves and ultrasonic probe
(Kumar et al., 2017)

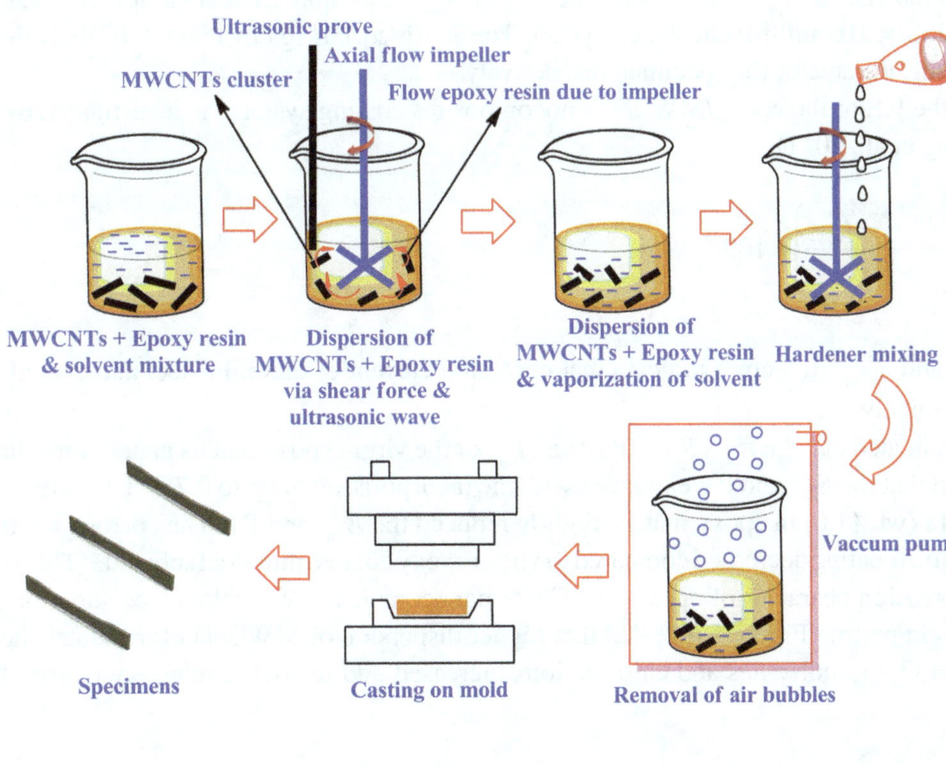

Figure 4.

$$Fe \longrightarrow Fe^{+2} + 2e^{-}$$

$$Fe^{2+} \longrightarrow Fe^{+3} + 1e^{-}$$

$$O_2\,(g) + 2H_2O + 4\,e^{-} \longrightarrow 4OH^{-}$$

$$2Fe^{2+}\,(aq) + O_2\,(g) + 2H_2O \longrightarrow 2FeOOH + 2H^{+}$$

etc., penetrate for the epoxy nanocomposites coating for corrosion. Many researchers have reported that the nano filler well dispersed in the polymer matrix extensively reduced the permeability of O_2 and H_2O in the matrix and demonstrate remarkable enhancement in anticorrosion character (Yu et al., 2014; Wong et al., 2014).

In this investigation, the corrosion rates (R_{corr}) and protection efficiency (P.E) were also determined. The R_{corr} is the total of corrosion reduction in weight and thickness per year, according to as per the ASTM standard (G1, G4, G31). The reduction can be communicated via subsequent Eq. (1) such as Faradays's law (Wessling et al., 1994; Lu et al., 1995) via presuming regular corrosion on the mild steel surface as shown in below.

$$R_{Corr}\left(MPY\right) = \frac{I_{Corr}.K.E_{W}}{d \times A} \tag{1}$$

where corrosion rate is R_{corr} (milli-inches per year), I_{corr} corrosion current (amp), K is the universal constant (1.288 × 105 milli-inches/A-cm-year). The E_w (g/eq.), d (g.cm^3) and A is equivalent weight, density and surface area of the specimen respectively.

Further, the P.E of the epoxy/MWCNTs nocomposites coating systems is investigated by following Eq. (2) (Weng et al., 2011).

$$P.E\left(\%\right) = \left(\frac{I^{0}_{Corr} - I^{C}_{Corr}}{I^{0}_{Corr}}\right) \times 100 \tag{2}$$

where I^{0}_{Corr} and I^{C}_{Corr} (µA cm^{-2}) indicate that corrosion current of the mild steel and coated nanocomposites respectively.

Table 2 indicated that the E_{corr}, R_{corr}, (P.E) and I_{corr} of the virgin epoxy and its nanocomposites. Table 2 demonstrated that the E_{corr} and P.E enhanced with the incorporation of up to 0.75 wt. % further increased the MWCNTs (wt.%1.0) in epoxy matrix slightly reduced the E_{corr} and P.E The corrosion rate of epoxy nanocomposites coating decreased compared to virgin epoxy coated mild steel substrate. This is attributed to the anticorrosion character of epoxy/MWCNTs nanocomposites can take place superior dispersion of MWCNTs in matrix. Figure 5 revealed that higher dispersion of MWCNTs formulates the diffusion pathway of H_2O, O_2 molecules and chloride ions increased and restricted in epoxy matrix. Numerous researchers also reported that the improvement of the anticorrosion character of carbon nanotube (Jeon et al., 2013; Harb et al., 2016; Freguia et al., 2007).

Therefore, the uniform and cluster free homogeneous dispersion of MWCNTs in epoxy matrix reduces the corrosion rate and bestows prevention from corrosion. The degradation in anticorrosion performance of epoxy/MWCNT nanocomposites coating with 1.0 wt.% MWCNTs owing to the development of MWCNTs clusters in epoxy matrix decreased the anticorrosion plus mechanical properties (Meguid et al., 2004).

The corrosion protection of the aluminum substrate was increased by the addition of nanofiller like MWCNTs. The proper dispersion of MWCNTs in the epoxy matrix is the main challenge. For that reason, researcher developed environmentally friendly single-step supercritical processing technique to improve the dispersion of MWCNTs in the matrix and successfully achieve the effective anticorrosion coating (Khanam et al., 2018). The Figure 6 depicts that the preparation of epoxy/MWCNTs nanocomposites coating including supercritical carbon dioxide ($scCO_2$) accommodated dispersion of MWCNTs into the matrix at a particular pressure. In the supercritical stage, CO_2 simply distributes into epoxy matrix owing to its lower viscosity, small molecule size and high diffusivity as shown in Figure 7.

The free energy barrier produces the repulsion and increases the gap between the polymer chain and permits to the easily and quickly MWCNTs intercalate within the epoxy matrix. The homogeneous dispersion of MWCNTs into the epoxy matrix, this is attributed to the dissolving character and rapid diffusion of $scCO_2$ facilitate the fast addition of MWCNTs. At the depressurization time CO_2 rapidly diffuses out and obtain finally remove on vacuum degassing.

Table 2. The values of the E_{corr}, R_{corr}, (P.E) and I_{corr} values of the virgin epoxy and its nanocomposites

Substrate (Mild Steel)	E_{corr} (mV)	I_{corr} (/cm^{-2})	P.E (%)	R_{corr} (MPY)
Uncoated substrate	-689	30.20 µA	-	16.1
Virgin epoxy	-536	1.24 µA	95.9	0.66
Epoxy/MWCNTs-0.25%	-525	458 nA	98.4	0.24
Epoxy/MWCNTs-0.50%	-491	105 nA	99.6	0.05
Epoxy/MWCNTs-0.75%	-484	4.55 nA	99.9	0.025
Epoxy/MWCNTs-1.00%	-580	35 nA	99.8	0.019

Figure 5. Recommended route of O_2, Cl^- ion and water molecules for corrosion protection virgin epoxy/MWCNTs nanocomposites
(Kumar et al., 2017)

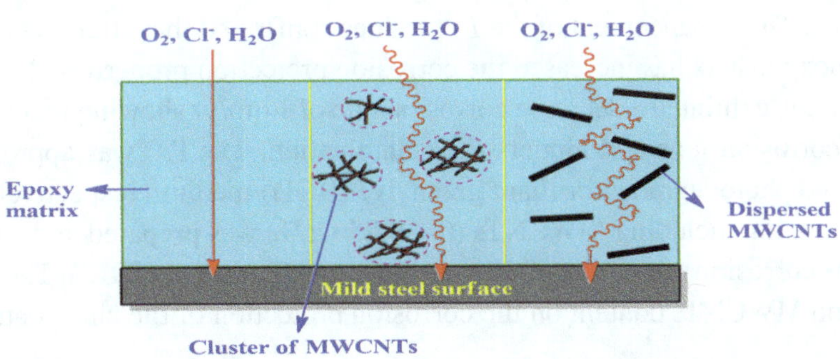

Figure 6. Preparation of $scCO_2$ assist epoxy/MWCNTs nanocomposites coating and mechanism for corrosion protection
(Khanam et al., 2018)

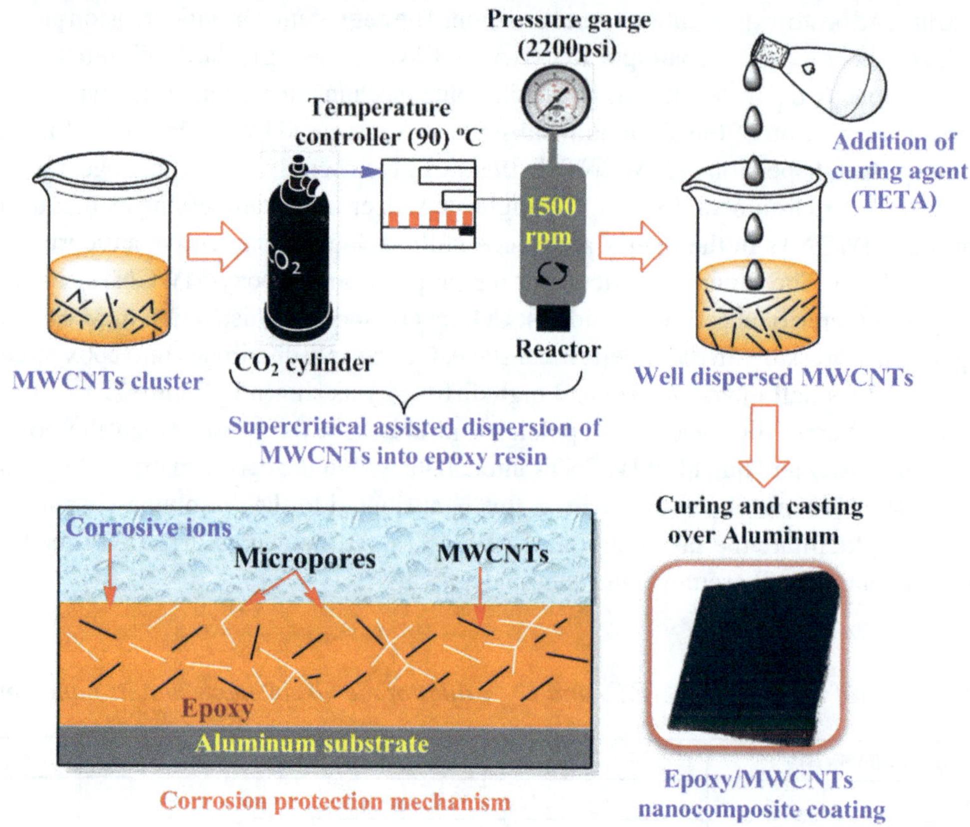

In the potentiodynamic polarization studies of the corrosion protection of uncoated aluminum (Al) substrate, virgin epoxy coating and its nanocomposites coating via Tafel plots. The results revealed that lower and higher values of I_{corr}, E_{corr} respectively, provides to enhancement of corrosion protection properties (Hu et al., 2017; Khun et al., 2014).

At this time, the anodic dissolution of without coated Al substrate was confirmed with higher and lower values of I_{corr} and E_{corr} correspondingly, in the both mediums. The epoxy coated Al substrate exhibit comparatively higher E_{corr} value in sodium chloride solution (NaCl) medium. On the other hand, reduction of the E_{corr} values was detected in the photodegraded medium and its contribute to the delamination of epoxy coating. The enhancement of the E_{corr} values confirmed that after incorporation of 1 wt.% MWCNTs in the epoxy matrix has increased the corrosion protection properties. The epoxy/MWCNTs nanocomposites coating exhibited a decrease corrosion rate of 1 μm/yr showing that MWCNTs bounded the penetration of corrosion ions and hinders the delamination, The P.E was approximated to 99.9%, 91.21 in the NaCl and photodegraded brilliant green dye (BGD) medium respectively.

The epoxy zinc coating including MWCNTs (0.1-0.25 wt.%) was prepared and studied the effect on the MWCNTs on the corrosion protection of the steel substrate (Park et al., 2015). The Figure 8 displayed the effect of zinc and MWCNTs content on the corrosion protection of the zinc coated MWCNTs steel

Figure 7. Schematic diagram uniform dispersion of the MWCNTs into the epoxy matrix under $scCO_2$ (Khanam et al., 2018)

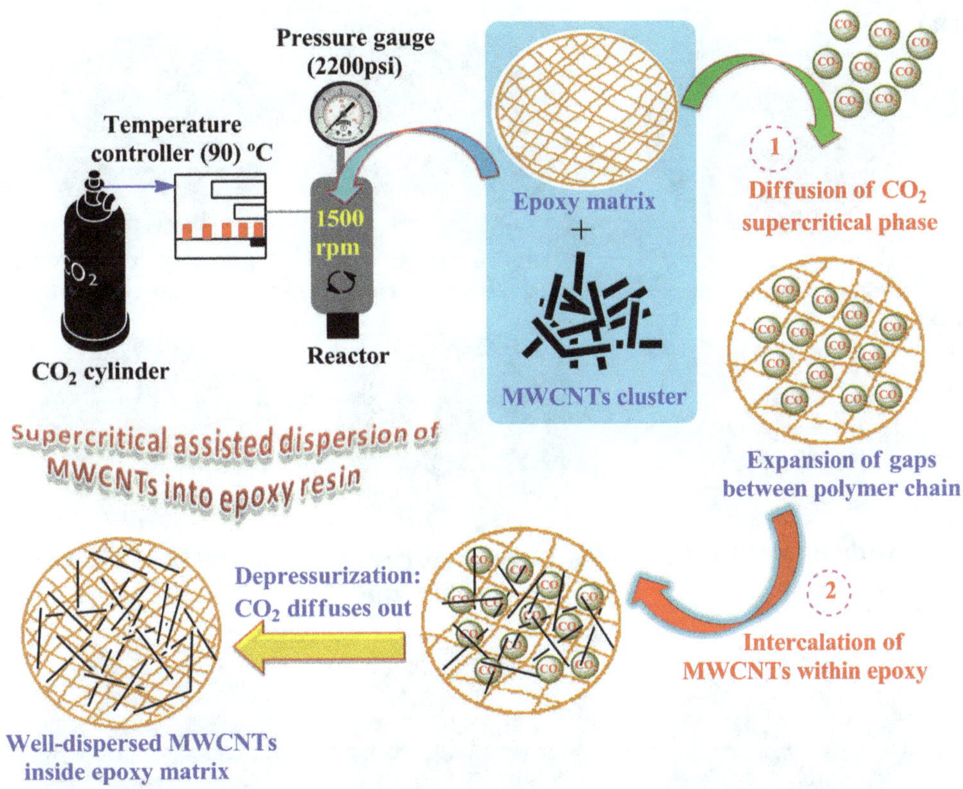

surface. Further, investigation the red rust was examined on the epoxy/zinc coated steel substrate areas without MWCNTs and the area of the corrosion decreased with enhanced the zinc content into the epoxy matrix. The corrosion areas also decreased with enhancing the MWCNTs content into the epoxy/zinc concentration. Finally, the epoxy contains 60 vol.% zinc and 0.25 wt% MWCNTs coated steel substrate was not observed the corrosion on the surface.

Polymer Nanocomposites With Graphene Oxide (GO)

Recently, researchers are more curious in developing graphene oxide (GO) based materials for anticorrosion applications owing to their water repellent properties and structural suitability as depicted in Figure 9 (Xu et al., 2018; Berry et al., 2013; Su et al., 2014). On the other hand, defect free grapheme oxide exhibit very inert surface properties because the non-polar covalent double bonds possess the weak interaction between graphene and polymer matrix molecules (Terrones et al., 2011). Furthermore, GO have strong π–π interactions, van der Waals forces and hydrophobic interactions which results the aggregation of GO in the matrix. (Tang et al., 2015). Therefore, surface modification of grapheme oxide plays a significant character in the synthesis of graphene based polymer nanocomposites with high performance anticorrosion applications (Potts et al., 2011; Kuilla et al., 2010).

Figure 8. Epoxy/zinc/MWCNTs coated steel substrate with different contents of the zinc and MWCNTs (Park et al., 2015)

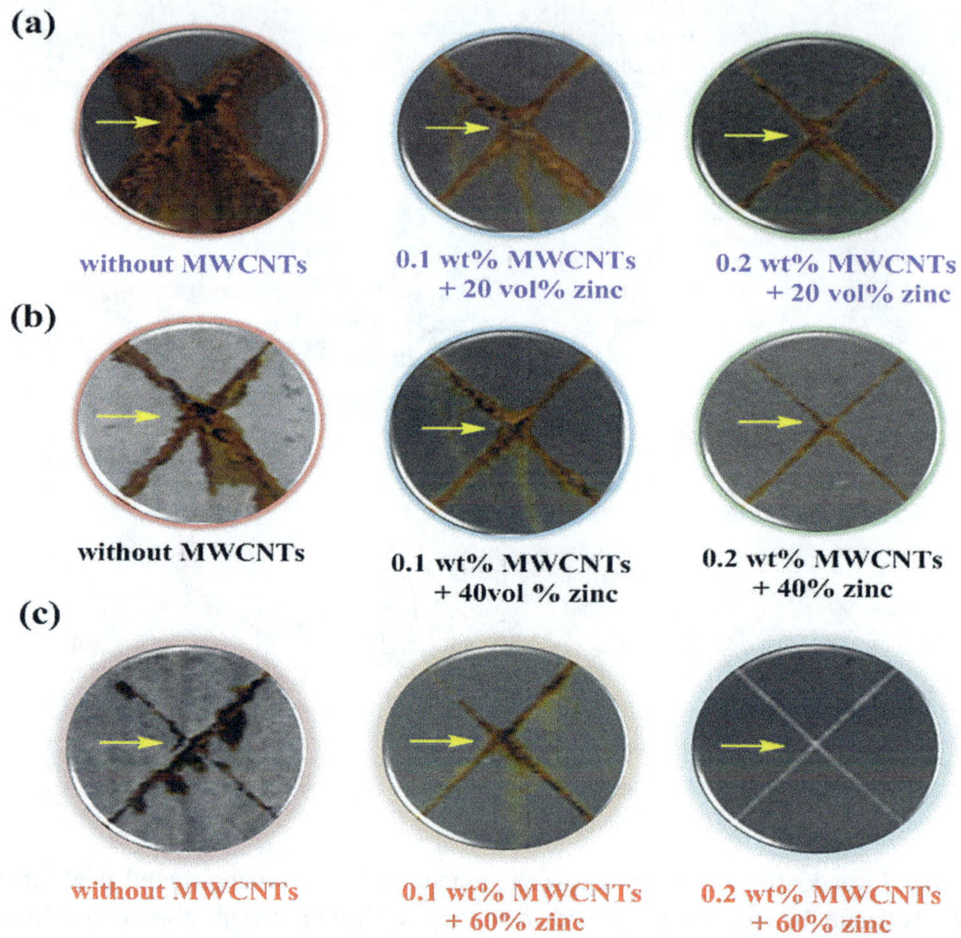

Figure 9. Chemical structure of graphene oxide (GO)

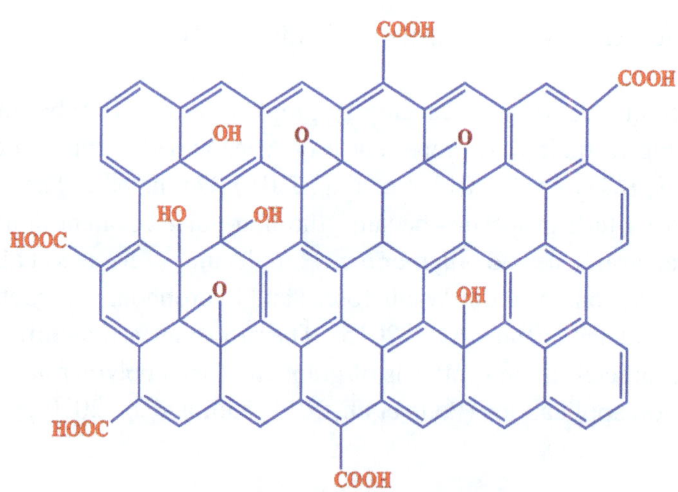

The prepared epoxy coatings after incorporation of poly(styrenesulfonate)-polyaniline/reduced graphene oxide (PSS-PANI/rGO) nanofiller for anticorrosion applications as shown in Figure 10 (Lin et al., 2018.). First the rGO platelets dispersed into the PSS solution by in situ oxidative polymerization of PANI. Further, the PSS-PANI/rGO nanocomposite fillers blend with epoxy resin (DGEBA) at various wt% (0.1, 0.5, 1.0) via tri-roll mill and cured with V009 curing agent.

The results revealed that after incorporation of 0.5 wt% PSS-PANI/rGO filler enhanced the ultimate tensile strength (TS) and tensile modulus (TM) of 64.9 MPa, 1.89 GPa respectively, of epoxy nanocomposites systems compared to virgin epoxy (43.1 MPa, 1.55 GPa) resin and epoxy/rGO systems. This is attributed to the tough interfacial bonding amongst PANI and epoxy resin for that reason the prepared nanocomposites (PSS-PANI/rGO) platelets shows the more compatibility with epoxy matrix compared to rGO.

Furthermore, in potentiodynamic polarization investigation by using Stern-Geary Eqs. (3 & 4): (Bard et al., 1980).

$$R_P = \frac{1}{I_{corr}} \frac{\beta_a \beta_c}{2.303\left(\beta_a + \beta_c\right)} \tag{3}$$

where,

R_P = polarization resistance,
E_{corr} = corrosion potential, density,
$\beta_a \beta_c$ = anodic and cathodic slopes,
$I_{corr} =$ corrosion current.

$$CR = \frac{kMI_{corr}}{\rho_m} \tag{4}$$

where,

CR = corrosion rate,
k = 3268.5 mol/A as a constant
M = M.wt. of iron (56 g/mol)
ρ_m = iron density (7.85 g/cm^3)
$I_{corr^{\prime} b}$ and $I_{corr^{\prime}}$ c = corrosion current density of uncoated coated steel electrodes, respectively.

The higher values of E_{corr}, polarization resistance and lower values of $I_{corr,}$ CR exhibit improved anticorrosion properties of the coatings. Virgin epoxy coated bare carbon steel substrate was revealed the decreased I_{corr} and CR values to 37.1µA/cm^2 and 0.865 mm/year, respectively, demonstrating a better physical insulation of epoxy coating. Further, the addition of *rGO* into the virgin epoxy coating enhanced the anticorrosion performance and after addition of higher 0.5 wt. % of *rGO* again increased the anticorrosion efficiency due to the decreased of I_{corr} and *CR* values of epoxy/*rGO* nanocomposite system from 0.093μA/cm 2 to 2.2μm/year, correspondingly.

Many researchers have been reported that the existence of the graphene (G) platelets can successfully increase the protection efficiency via elongate the zigzag diffusion pathways for corrosive reactant oxygen and Cl^- ions. Additionally, the epoxy/PSS-PANI/rGO nanocomposite coating systems was revealed that superior values of E_{corr} compared to the the virgin epoxy and epoxy/*rGO* nanocomposite coating systems due to the cause of passivation effect of PANI.

While the low % of PANI content into the epoxy/PSS-PANI/rGO nanocomposite system exhibited better dispersion and higher surface area of the platelets and PANI directly attached to the surface of the carbon steel. According to the earlier reported PANI make contact with the iron surface and perform as a catalyst for the oxidation reaction on the iron surface, and formed a heavy passivation layer that was formulated of Fe_2O_3 and Fe_3O_4.

The heavy Fe_3O_4 passivation layers would cause an optimistic change of the E_{corr} as well as prevent the metal below the layer from further corrosion. Recently, lots of researchers developed the 1-3 wt% PANI containing coating shows that remarkable enhancement in corrosion performance (Sakhri et al., 2010; Zhang et al., 2013). During this investigation the epoxy/PSS-PANI/rGO nanocomposite coating systems at a incorporating of 0.5 wt% filler demonstrated the excellent anticorrosion properties amongst all the coating system owing to the dense interface between filler and matrix. Further enhanced the 1.0 wt% of PSS-PANI/rGO filler into the epoxy system reduced the anticorrosion efficiency owing to the poor dispersion of the PSS-PANI/rGO filler. That filler improved the mechanical properties of epoxy nanocomposite materials and provides the promising potential in anticorrosion applications.

Figure 10. Representation of anticorrosion mechanism of epoxy/rGO and epoxy/PSS-PANI/rGO nanocomposites coatings
(Lin et al., 2018)

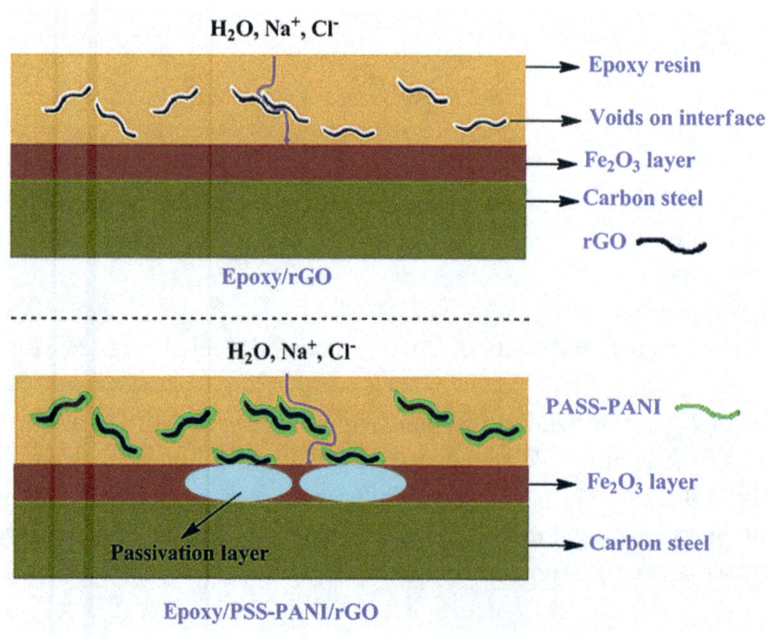

The development of a solvent based epoxy-GO nanocomposites coating with various GO wt.% (0, 0.05, 0.1, 0.3 and 0.5) with improved corrosion protection of steel substrates is described in Figure 11 compared to the virgin epoxy (Pourhashem et al., 2017). The potentiodynamic polarization results demonstrated that the R+H, (R+GO)+H, and R+H, R+(H+GO) coating system in 3.5 sodium chloride solution (NaCl) following stabilization for 2 h.

The E_{corr} and corrosion current density value of the virgin epoxy coatings is -0.543 V vs., 14.6 μA/cm^2 respectively, demonstrated that it is responsible for the corrosion. The nanocomposites coating system revealed moderately dissimilar behavior from virgin epoxy and nanocomposites coatings fabrication has an enormous cause on its corrosion resistance (Chang et al., 2014). The E_{corr} values shift to the more negative value, -0.562 V vs. saturated calomel electrode (SCE) in the case of (R + GO) + H nanocomposite coating system and I_{corr} extensively reduces to 0.88 μA/cm^2. Conversely, R+(H + GO) coated substrate reveals more positive E_{corr} -0.455 V vs. SCE value with significant decline in I_{corr} value of 0.22 μA/cm^2 and the decent shift E_{corr} following to reduce of I_{corr}, imply that the R+(H+GO) nanocomposite coating system performs as a protective layer and enhanced the corrosion resistance of the steel substrate. Moreover, Tafel slopes of anodic and cathodic both have altered via the addition of GO in coating, demonstrating the corrosion prevention properties of coating system has altered by GO

Figure 11. Development of epoxy/GO nanocomposites coating
(Pourhashem et al., 2017)

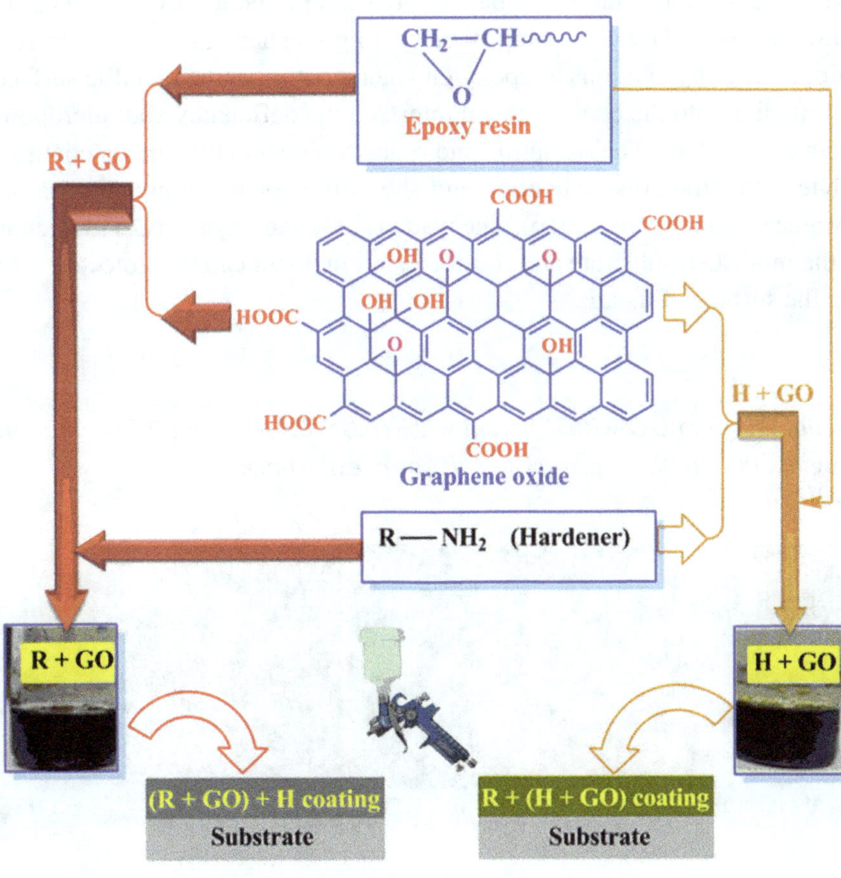

incorporation and based on the fabrication process. For that reason, homogeneous distribution of GO in epoxy matrix improves the corrosion and barrier performance of the coating system

Further, study of potential dynamic polarization of virgin epoxy coated steel substrate and its nanocomposites with the addition of 0.05, 0.1, 0.3 and 0.5 wt% GO. The result shows that corrosion current density (E_{corr}) reduces extensively via the addition of little quantities of GO and all nanocomposites systems demonstrate higher corrosion resistance contrasted to the virgin epoxy coating system. Thus, the 0.1 wt % GO containing nanocomposites system, E_{corr} value shifts toward the positive route compared to the other system and exhibit the smallest value of I_{corr}. As a result, epoxy-0.1 wt.% GO nanocomposaites coating system shows excellent corrosion properties compare to the other system due to the uniform dispersion of the GO in the matrix and enhance the barrier effect next to corrosive electrolyte (Qi et al., 2015).

On the other hand, the salt sprays investigation of 300 h with exposure to the accelerated condition as shown in the Figure 12. The virgin epoxy coating system exhibited rusts and blister appear on the surface and corrosion are enhanced approximately the scratch, representing poor corrosion resistance of epoxy coating system. The nanocomposites coating systems with various GO wt% display different behavior while rust, blister with smaller size are examined on the 0.05 wt.% GO coating system, lower area of corrosion and rust product shows near the scribe of coating with 0.5 wt.% GO.

In the case of 0.1-0.3 wt.% GO nanocomposites system demonstrated no blister and a extremely small number of rust marks on the surface and corrosion approximately the scratch is not extended. So, the decreasing order of the coating system virgin epoxy ˃ 0.05 ˃ 0.5 ˃ 0.3 ˃ 0.1 wt.% GO. The salt spray investigation corroborates that GO influence on corrosion resistance of the nanocomposites coating system and 0.1 wt.% GO system exhibits the better corrosion protection as depicted in the Figure 12.

Figure 13 shows the epoxy/GO nanocomposites coating mechanism on the mild steel substrate for corrosion protection. The delaminations of epoxy/GO nanocoating in the metallic surface as a result the corrosive groups diffusion into the coating-metal interface via deficiency and micropores of nanocoating (Ramezanzadeh et al., 2016). The oxidation and reduction occur at the metal coating interface of by diffusion electrolyte below the organic coating and shows the corrosion phenomena on metal surface (Yu et al., 2014; Ramezanzadeh et al., 2016). The water (H_2O) and oxygen (O_2) molecules to require for the corrosion on the mild steel substrate and the corrosion incident can be protected via retarding H_2O, O_2 from attacking the surface of metal.

Figure 12. The visual image of GO with different wt.% (0.05, 0.1, 0.3 and 0.5) containing epoxy nanocomposites coating at 300 h in the exposure to salt spray experiment
(Pourhashem et al., 2017)

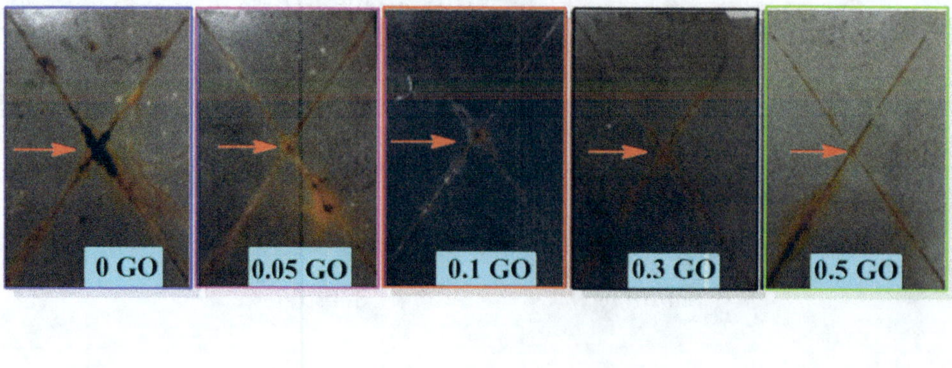

Figure 13. Representation of epoxy-GO nanocomposites coating mechanism on the mild steel substrate for corrosion protection
(Pourhashem et al., 2017)

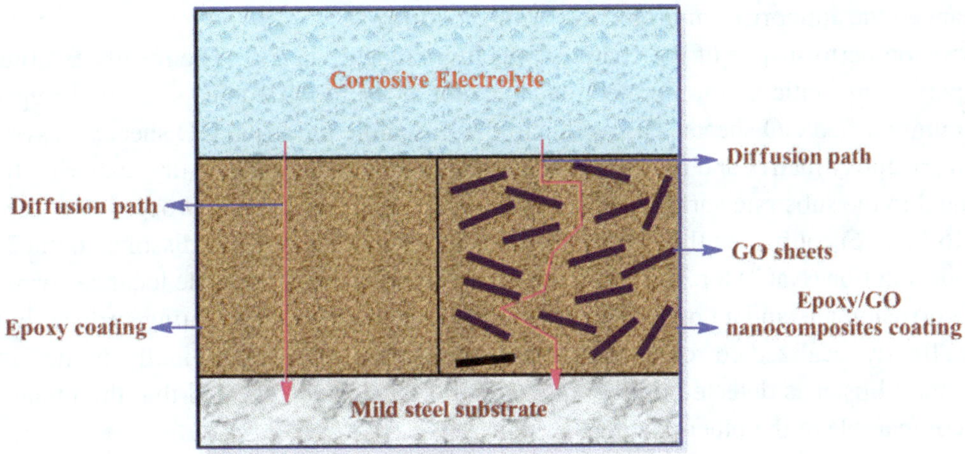

However, GO nanosheet decreases the microscopic deficiencies and pores in the coating, enhances the pathways for electrolyte diffusion to coating-metal interface. As a result, the deficiency of coating decrease due to blocked of local anode and cathode route on metal surface. On the other hand, the GO extensively affects the barrier performance of the epoxy coating because of their sheet like structure create the more zigzag diffusion route of the corrosive groups. The diffusion pathway of the H_2O, O_2 molecules from the coating substrate is increased. This is owing to barrier cause of GO against corrosive groups as depicted in Figure 13.

The development of a novel epoxy/GO nanocomposites coating materials for the corrosion protection of steel substrate in the first time (Rajabi et al., 2015) with various GO wt % incorporation of the epoxy matrix by using mechanical and sonication method. The optical microscopy studied demonstrated that the sonication method is the best for the distribution of GO nanosheets in the epoxy matrix. After increasing the time off the sonication reduced the agglomerates and cluster in the coatings.

The electrochemical observation revealed that, via incorporation of GO nanosheet into the epoxy matrix raise the corrosion resistance of the coating system and achieve the maximum value by employing the 0.25 wt% GO in an epoxy matrix. This is ascribed to the better dispersion of GO nanosheet in epoxy matrix and creates a tortuous pathway for energetic molecules and ions.

On the other hand, GO platelet exhibit high aspect ratio and large surface area, formulate extensive diffusion pathway for penetration of corrosive agent. According to the literature, corrosion resistance of the coating was enhanced via increasing incorporation of GO%. However, the corrosion properties of the coating system containing 0.25 wt% GO is higher than the 0.5 wt% GO coating system this is due to the agglomeration of nanoheet because of high surface energy of the sheet.

The development of a novel method for the enhanced of anticorrosion resistance of the epoxy coatings on the steel substrate with the urea-formaldehyde (UF) resin modified GO by in-situ polycondensation as shown in Figure 14 (Zheng et al., 2017). The UF chemically bonded with a carboxylic group (-COOH) on the GO surface and formation of microspheres. That modification alters the hydrophilic character of GO to lipophilicity and the UF microspheres protected the GO sheets from the huge pressure for the duration of grinding this is the reason reaggregation of GO sheets. Scanning electron microscopy (SEM) analysis and sedimentation test of microstructure in the cross section of grapheme urea-formaldehyde/ epoxy coating (epoxy/GUF) revealed that the GO remained extremely dispersed in epoxy resin as the results enhanced the anticorrosion protection of the coatings.

The corrosion performance of the coatings was investigated via EIS. The results exhibited that the barrier properties and anticorrosion protection character of the modified GO sheet in the epoxy resin is higher than unmodified GO sheets. This is due to the lipophilic modified GO sheet was systematically dispersed in the epoxy matrix and protected the corrosion area from additionally scattering through diffusion channel to the substrate surface. Finally, improved the anticorrosion properties of the coatings.

Figure 15 (a, b, c) confirmed that the epoxy, epoxy/GO and epoxy/GUF dissolve in the 3.5 wt% sodium chloride solution (NaCl) for 3500h. Figure 15 (a) exhibits the observable localize corrosion on EP and lots of corrosion pits and an only some blisters appear on the epoxy coating surface. In the Figure 15 (b) very slightly localized corrosion and blister arise on the epoxy/GO. Finally, in the Figure 15 (c) single one small blister is detected on the epoxy/GUF surface and revealed that the higher corrosion protection comparable to the other coatings systems.

Figure 14. Reaction scheme of the GO modification of the urea-formaldehyde at 60 °C for 3h

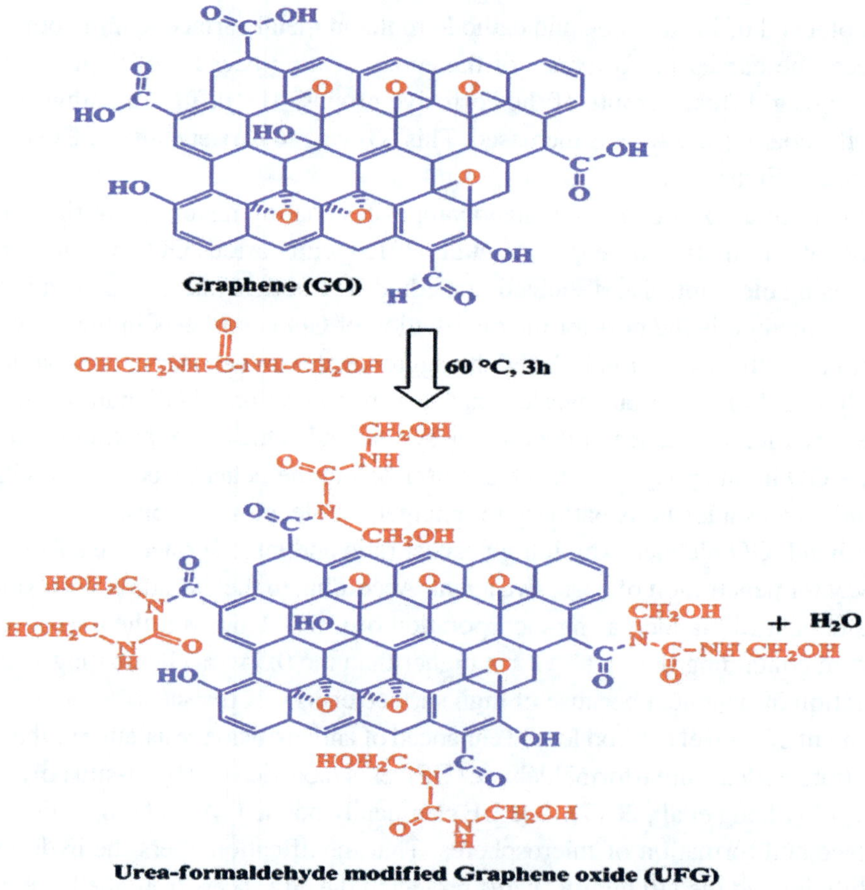

Figure 15. Visual demonstration of the corrosion test of (a) epoxy (b) epoxy/GO and (c) epoxy/GUF immersed in the sodium chloride solution (NaCl)
(Zheng et al., 2017)

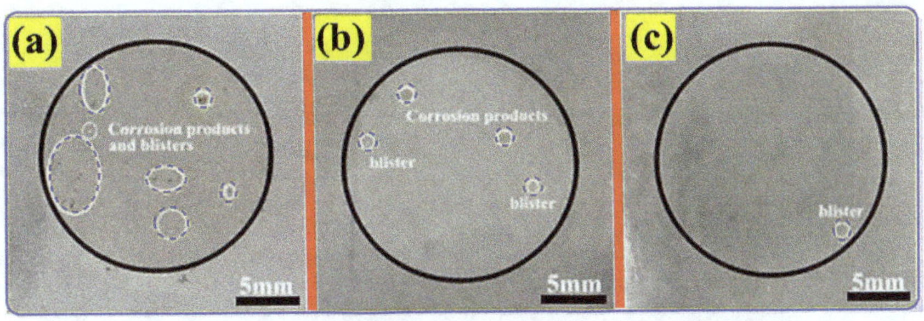

The corrosion protection mechanism of the epoxy/GO and epoxy/GUF coatings system as depicted in Figure 16. After incorporation of unmodified GO into the epoxy resin exhibited poor compatibility and interface bonding between GO and epoxy resin. This is attributed to the interface gap around GO and detectable reuniting. As the corrosive medium increase during the diffusion channel in the coating. The hydrophilicity of the GO effected the corrosive medium to accumulate at the interface between epoxy/GO resins.

In the case of epoxy/GUF coating systems, enhanced the compatibility and the interface bonding between modified GO and epoxy resin. Because of higher dispersion and decreased the nature of re-aggerated after increase corrosion medium throughout the diffusion channels in the coating systems. The modified GO prevents the anticorrosion medium from assembling at the interface among the modified GO and epoxy resin. As a result, the interface was not eliminated and well dispersed modified in the GO, prevented the corrosion medium from further diffusing.

POLYMER NANOCOMPOSITES WITH INORGANIC METAL OXIDE NANOPARTICLES

The preparation of epoxy and its nanocomposites with grapheme oxide (GO), titanium dioxide (TiO_2) and TiO_2/GO hybrid nanoparticles in the 2 wt% by using two step solution technique without employing any additional surfactants (Yu et al., 2015). The structure of TiO_2/GO hybrid is affected naturally via TiO_2 on GO surface and enhances the interlayer spacing. Furthermore, the hybrids not only enhanced the exfoliation and dispersion in epoxy resin but, also raise the corrosion protection of the epoxy coatings in the existence of lower content (2 wt%.) of nanoparticles. According to the electrochemical and water contact angle studied expressed that the 2 wt.% epoxy/TiO_2–GO hybrid coatings exhibited better anticorrosion coating properties compared to the virgin epoxy, epoxy/GO, and epoxy/TiO_2 coatings with similar 2wt.% because of admirable blocking properties for micropores. Figure 17 revealed that the contribution of TiO_2-GO hybrids in the epoxy coatings and for use in the anticorrosion application.

Synthesis of macromolecular coupling agents, including epoxy resin and silane functionalized groups via radical copolymerization of glycidyl methacrylate (GMA) and γ methacryloxypropyltrimethoxysilane (MPS), (F_1–F_3) to form poly(GMA-co-MPS). Further, poly(GMA-co-MPS) were effectively grafted on the titanium dioxide (TiO_2) surface by the direct reaction (TiO_2-F_s) (Pour et al., 2018). The prepared

Figure 16. anticorrosion protection mechanisms of epoxy/GO and epoxy/GUF coating systems
(Zheng et al., 2017)

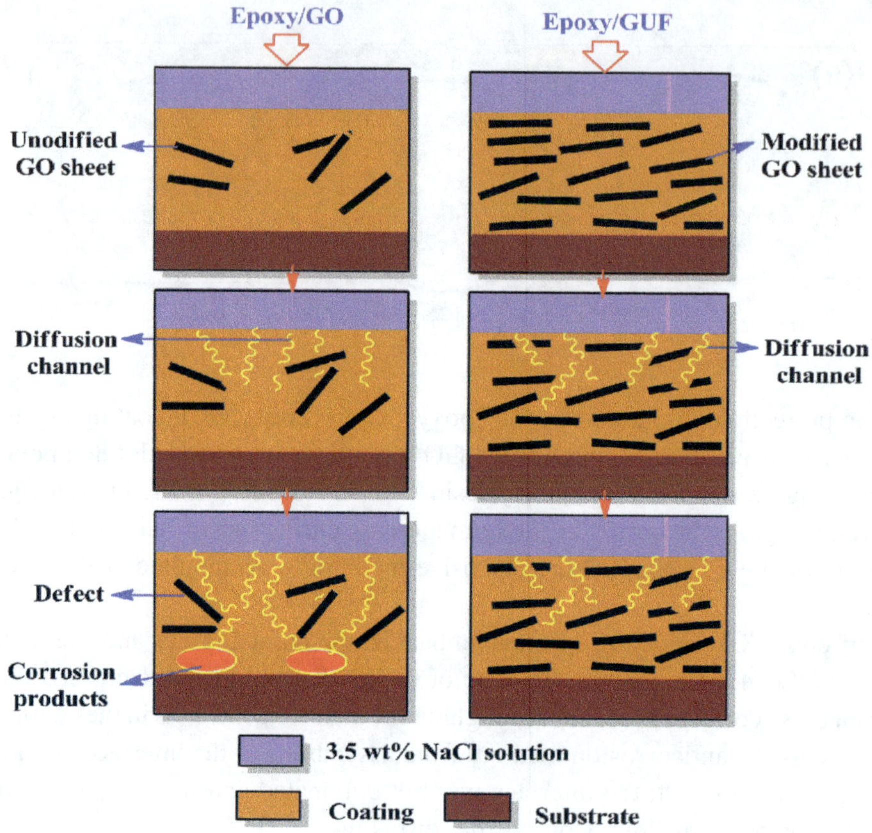

Figure 17. schematic diagrams of epoxy/TiO_2-GO hybrid nanocomposites coating for corrosion protection
(Yu et al., 2015)

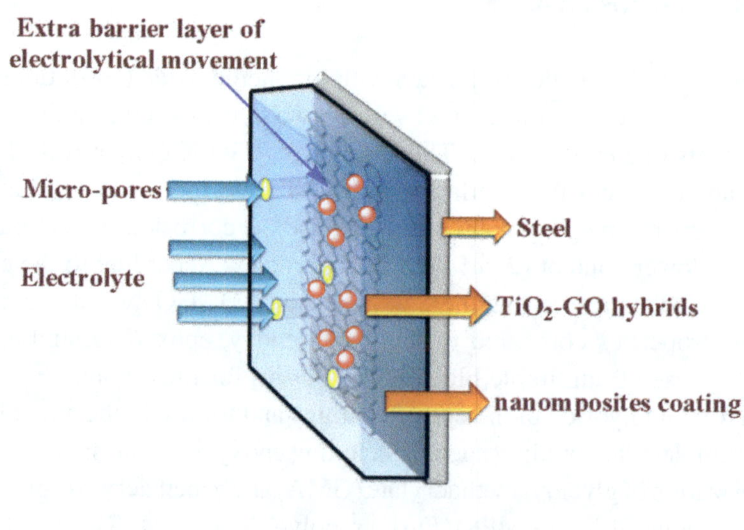

virgin epoxy and TiO_2-F_s homogeneous mixture of nanocomposites was coated on the commercialized mild steel substrate and cured with amine based triethylenetetramine (TETA) curing agent and compared with virgin epoxy, epoxy/TiO_2 coatings cured with the same curing agent. The interfacial binding and uniform mixing amongst TiO_2-F_s nanoparticles and epoxy matrix provides the prolong time anticorrosion properties on the steel substrate via explored EIS prior and after dip in (3.5 wt.%) NaCl solution.

After incorporation of unmodified and modified 1wt% TiO_2 into the epoxy matrix increased the electric resistance (R1) value for 8-146 times following 1day immersion of epoxy/TiO_2 and epoxy/TiO_2-F_3 nanocomposite, compare to virgin epoxy correspondingly. The enhancement of the R_1 value was owing to the decrease of coating porosity and raised barrier properties of corrosion protection.

Similarly, the enrichment in the charge transfer resistant (R_2) in the 2-140 times for unmodified TiO_2 and F_3 modified epoxy/TiO_2-F_3 nanocomposite. This is specified that the higher corrosion protection performance of coating in particular for the TiO_2-F_3 nanocomposites containing coating. The anticorrosion character of the copolymer poly(GMA-co-MPS) modified epoxy/TiO_2 coating owing to the development of Si-O-Si and Si-O-metal bonds after condensation of silanol groups and improve the barrier and coating adhesion properties. Additionally, an epoxy group in GMA segment of the copolymer also created a crosslinked network structure and attached to nanoparticles to the matrix. Thus, the enhancement in the corrosion protection of modified coating due to the formation of very closed packed network. As the results, confirmed that the nanocomposites coating performed approximately as better resistance and barrier properties against penetration of corrosive species like H_2O, O_2 and Cl^- ions into the coating.

Further, the investigation of water absorption of the epoxy resin and its nanocomposites. The diffusion coefficient of the virgin epoxy decreased after incorporation of TiO_2 naoparticles into the coating system as depicted in Table 3. This is attributed to the enhancing the MPS segments (wt.%) in the epoxy/TiO_2-F coating system that is the reason epoxy/TiO_2-F_3 coating system demonstrated higher barrier properties compared to the other coating systems. On the other hand, lower value of the water diffusion coefficient exhibits a better corrosion protection for the substrate. Because of improved the dispersion of the epoxy and copolymer modified TiO_2 nanoparticles improved the stiffness of the coating.

The movement of the H_2O molecules and ions in unmodified epoxy/TiO_2 and copolymer (MPS) modified epoxy/TiO_2 coating as described in Figure 18. The epoxy/TiO_2 coating system had lower barrier properties due to the movement path was short because aggregation of nanoparticles. Whereas TiO_2 nanoparticles grafted copolymer revealed higher dispersion in epoxy matrix and produced a covalent formulation with matrix. For that reason, the coating exhibited resistance against the movement of H_2O molecules and Cl^- ions owing to the development of an extremely fixed honeycomb structure during the epoxy matrix and extended the movement pathway and hindered diffusion.

Table 3. Diffusion coefficient of the virgin epoxy and its nanocomposites

Sample	Diffusion Coefficient(10^{-8} cm^2 s^{-1})
Virgin epoxy	6.79
Epoxy/TiO_2	6.04
Epoxy/TiO_2-F_1	5.50
Epoxy/TiO_2-F_2	5.23
Epoxy/TiO_2-F_3	5.04

Figure 18. Schematic representation of water, oxygen and chlorine ions in the TiO_2 and TiO_2-Fs coating system
(Pour et al., 2018)

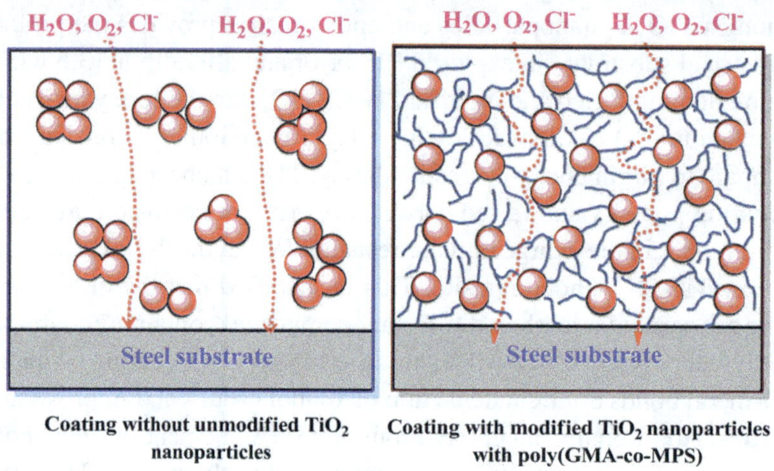

The GO-Al_2O_3 hybrids were effectively fabricated by easy way in the existence of the 3-aminopropyltriethoxysilane (ATPS) followed by prepared epoxy nanocomposites coating with the addition of grapheme oxide (GO), alumina (Al_2O_3) and GO-Al_2O_3 hybrid nanofiller as depicted in Figure 19 (Yu et al., 2015). The Al_2O_3 nanoparticles demonstrated homogeneous dispersion and firmly attached to the GO sheet. The well dispersed exfoliated GO-Al_2O_3 hybrid structure exhibited a remarkable enhancement to increased the corrosion resistance in the epoxy coating at lower content (2 wt.%). After incorporation of 2 wt.% GO, Al_2O_3 and GO/Al_2O_3 of epoxy matrix and examined by EIS. The EIS spectra of the epoxy nocomposites coating after dip of 108, 142, 240 and 60 h, respectively, revealed that organic coating at the final corrosion condition. The results confirmed that the epoxy/GO-Al_2O_3 nanocomposite coating revealed the superior anticorrosion properties compared to the GO and Al_2O_3, this is ascribed to the better dispersion of GO-Al_2O_3 hybrid nanofiller into the epoxy matrix and tremendous inherent plugging micro performance related to the its thin sheet structure.

Figure 20 displayed the corrosion mechanism of GO-Al_2O_3 hybrid. The improvement of the corrosion protection, employing epoxy/GO-Al_2O_3 due to the anticorrosion properties of the Al_2O_3, thin sheet structure bestows an additional barrier layer to mostly hinder micro pores for electrolyte permeation and inhibit the underlying metals from corrosion attack and better dispersion GO-Al_2O_3 thin sheet in the matrix prevent the corrosion. This is due to the comparatively higher aspect ratio than loading o sheet and reinforced the anticorrosion properties of nanocomposites coatings.

According to the another studied the preparation of epoxy/TiO_2 nanocoatings with the various wt.% of TiO_2 such as 5, 10, and 20 (Zhang et al., 2007) was explored by EIS in 3.5% NaCl medium. As a results, the 10 wt.% epoxy/TiO_2 nanocomposite coating exhibited higher ionic resistance during the coating and lower breakpoint frequency, better corrosion resistance with other compositions. Further, the 10 wt.% epoxy/TiO_2 nanocomposite coating revealed that lower diffusion coefficient for the internal movement of energetic approach during the coating in the given system.

Figure 19. Preparation of epoxy/GO-Al_2O_3 nanocomposites coating for anticorrosion application (Yu et al., 2015)

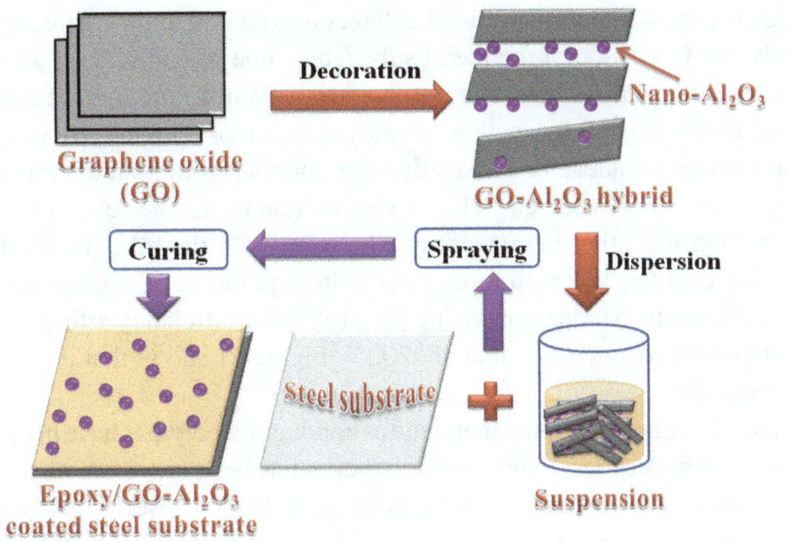

Figure 20. Corrosion resistance mechanisms of epoxy/GO-Al_2O_3 hybrids nanocomposites coating (Yu et al., 2015)

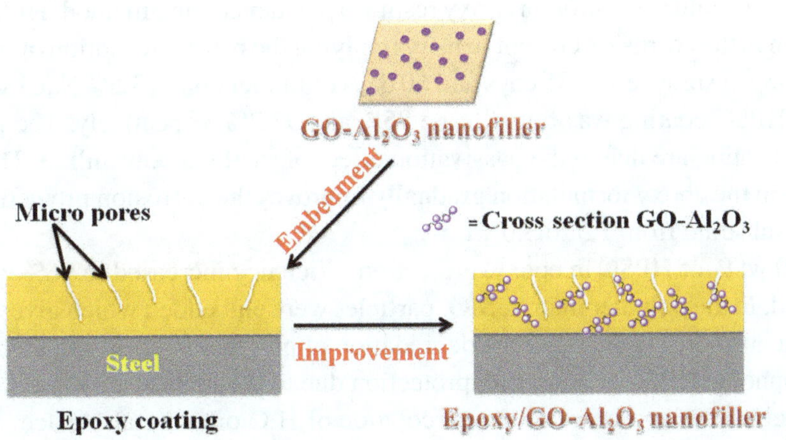

The effect of zirconium oxide (ZrO_2) naoparticels on the corrosion protection after incorporation into the epoxy matrix on the mild steel substrate (Haddadi et al., 2015). The trimethoxymethylsilane (MTMS) was utilized as a surface modifier to enhance the dispersion of the ZrO_2 nanoparticles in the matrix. The dispersion of the modified and unmodified ZrO_2 in epoxy coating was examined by emission-scanning electron microscopy (FE-SEM). The salt spray and EIS were used to analyzed corrosion prevent performance of the epoxy/ZrO_2 nanocomposite coating and the revealed that the enhancement of the corrosion protection in the existence of the surface modifier ZrO_2 nanoparticles.

CONDUCTING POLYMER NANOCOMPOSITE FOR CORROSION PROTECTION

Corrosion protection is a most important topic of enhancing attention to the metallurgical and electronic industries. Currently, the Corrosion of the metal substrate is one of main issues all over of world (Yao et al., 2008). A number of methods have been utilized to prevent to the metal from corrosion. In this time, the conducting polymer coatings method employed as a most extensively in the world (Hosseini et al., 2007). In the last two decades, the conducting polymer deposition on the metal electrode surface has been extremely significance research fields in electrochemistry. The several benefit of conducting polymer and their composites other than coatings like paints. On the other hand, they do not contain any toxic materials harmful for the environment. After incorporation of conducting polymers perform as both physical and electronic barrier, enhancing the protection efficiency afforded via other materials and act simply as physical barrier (Herrasti et al., 2005). Figure 21 shows that the conducting polymers for the corrosion protection.

The applications of the conducting polyacetylene as conducting polymer have been examined (Chiang et al., 1977). In the last few decades, many research works has been reported related to applications of conducting polymer towards corrosion prevention coating. In 1985, DeBerry had also reported the used of conducting polymer in the corrosion protection application, in which the stainless steel coated via PANI was set aside in the passive condition for a prolong time in H_2SO_4 solution (DeBerry et al., 1977)

The polyaniline-SiO_2 hydrophocbic polymer composites (HPSC) have been developed via chemical oxidation polymerization as depict in Figure 22 (Dhawan et al., 2015). Appropriate HPSC composite coating was applied on mild steel through epoxy resin via powder coating method. HPSC coating exhibits abundant reduction in the corrosion current density implying the better prevention of mild steel in marine environment. In the salt spray test of 35 days and 60 days of immersion in 3.5% NaCl solution, protection efficiency of the HPSC coating was found to be 85% and 93.3% respectively. The good properties of HPSC containing coating are due to the passivation effect of steel via polyaniline. The enhancement of the HPSC content in the epoxy formulation gradually improves the corrosion protection effect of HPSC coated mild steel substrate from 1.5 to 3.0 wt. %.

In addition, 6.0 wt % of HPSC in epoxy, protection efficiency increased to 96% when HPSC coated mild steel was used. In the polymer matrix, SiO_2 particles were embedded which gives the reinforcement to PANI chain that attributes to decrease the degradation of polymer in corrosive medium. The metal in presence of hydrophobic HPSC coating gets protection due to the formation of passivating layer which works as a physical barrier and prevents the percolation of H_2O onto metal surface. Moreover, powder coating method also performs a vital role for attaining durable, higher quality, and better anticorrosive

Figure 21. Chemical structures of the conducting polymers

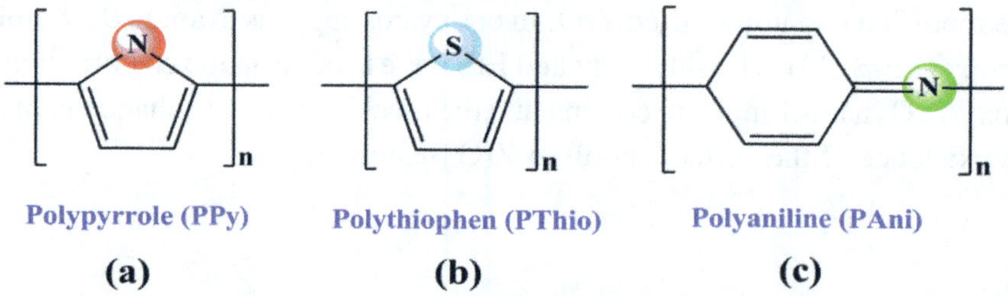

Figure 22. Schematic representations of hydrophobic polyaniline-SiO_2 composites
(Dhawan et al., 2015)

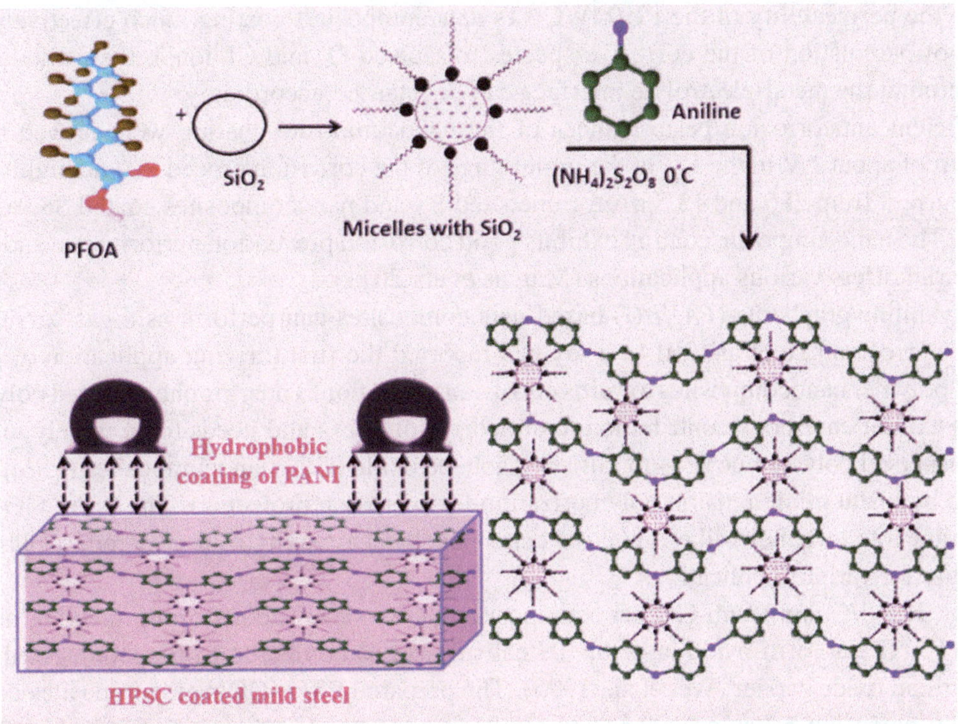

coatings properties. The investigations demonstrate that the polymer has exhibit good corrosion protection character and it can be a potential candidate for used as corrosion protection in a corrosive medium.

Polyurethane (PU) has been extensively employed as coatings, adhesives, elastomers, paints, constructional materials, furnishings, fibers, packing and artificial skins in the medical, industrial and automotive fields, respectively. Now, the CNTs have excellent mechanical properties and suitable reinforcement for the structural application (Gu et al., 2013). For anticorrosion applications, usage of conductive PU/MWCNT nanocomposites has not been earlier reported. By insitu surface-initiated polymerization (SIP) process, the conductive polyurethane (PU) nanocomposites loaded with 5wt% MWCNTs were prepared (Wei et al., 2013). The anticorrosion properties of the PU/MWCNTs nanocomposites coating on the stainless steel (SS) substrate was investigated by 3 wt% NaCl solution medium via observing open circuit potential (E_{ocp}) and following Tafel plots of the coated SS electrode.

The maximum of 97.7% protection efficiency was attained. An equivalent circuit of the coating was suggested to fit the electrochemical impedance spectroscopy (EIS) data, indicating an impressive protection for SS. The results demonstrate that the PU/MWCNTs nanocomposites gives a considerable protection from corrosive ions in the solution for SS while giving a way for its conductivity due to the better dispersion of MWCNTs reinforcement in the PU matrix. EIS illustrates the anticorrosion mechanism (Figure 23) of the conductive PU/MWCNTs nanocomposites coating Organic or inorganic fillers addition to the polymer matrix mostly creates a solid barrier for the passage of gas molecules through the polymer preventing the corrosion process (Weng et al., 2010). With increasing aspect ratio of the filler, the filler-matrix interfacial area also increases, and thus more difficult the passage, resulting to a decreased permeability (Sinha Ray et al., 2003). Both the high aspect ratio (around 1000) (Yakobson et al., 2001) and the excellent homogeneous dispersion of MWCNTs in the polymer matrix attributes to decrease the permeability of the PU/MWCNTs nanocomposites coating much effectively. With the prevention of percolation of the corrosive species (dissolved O_2 and Cl^- ions), the cathodic reactions occurring around the metal-electrolyte interface can be retarded accordingly.

The efficient anticorrosion performances of the nanocomposite coating were proved by a major positive shift of about 1 V in the E_{ocp} at the initial stage of the corrosion procedure and highly decreased corrosion current from 286 and 13.5 nA for uncoated SS and nanocomposites coated SS, respectively, after 312 h. The nanocomposite coating exhibits good corrosion prevention performance and as anticorrosion material offers various applications (Marcus et al., 2011).

The polyaniline/grapheme (*PANI/G*) based nanocomposites can perform as a gas barrier and also prevent the corrosion (Yu et al., 2014). They had reported the first thriving application of grapheme-based vinyl polymer nanocomposites towards corrosion protection. Thus, graphene-based polymer nanocomposites exhibit enormous contribution to anticorrosion areas and needs to be widely investigated. The well-dispersed polystyrene (PS)/modified-graphene oxide (GO) nanocomposites (PS/m-GO) were synthesized by insitu miniemulsion polymerization for corrosion protection. The PS/m-GO nanocomposites exhibit lots of benefit like low cost, better mechanical strength, transparency, plasticity, and processability for various applications.

However, PANI/G and PS/m-GO nanocomposites shows various corrosion protection mechanism; conjugated PANI can form oxide layer on a steel surface, as compared to non-conjugated PS which doesn't forms an oxide barrier (Wei et al., 1995). The prepared PS/m-GO nanocomposites demonstrate excellent anti-corrosion behavior as far as virgin PS is concerned, where with the inclusion of 2 wt% modified GO in the PS matrix, the P.E raises from 37.9% to 99.5%. The results reveal the significant improved properties to almost complete exfoliation of the m-GO dispersion, which highly increases the PS/m-GO gas barrier properties and interfacial bond strength. The alters in E_{corr}, I_{corr}, R_{corr} and P.E (%) bestow indication that PS/m-GO nanocomposites coating enhanced the anticorrosion properties compared to the polyaniline/G materials reported in earlier. This investigation ascribed the higher performance to the homogeneous dispersion of GO with the polymer matrix.

Figure 24 shows that the enhancement mechanism of PS/m-GO nanocomposites properties for corrosion protection on the metal surface. As the corrosion mechanism in a neutral Cl^- solution can be shown with an anodic route of the dissolution of Fe and a reduction in the cathodic route of O_2 and

Figure 23. Schematic representations of PU/MWCNTs nanocomposites anticorrosion coating mechanism (Wei et al., 2013)

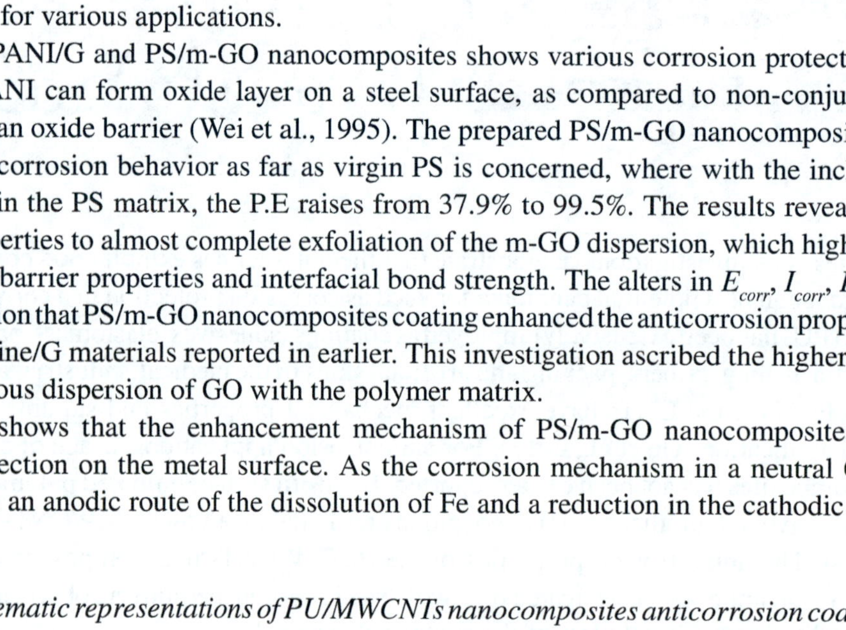

H_2O, OH- can be produced; followed by the formation of Fe_2O_3 through a chemical oxidation series of the process (Kinlen et al., 1997). The diverse materials for corrosion inhibit were produced depending on the prevention of the corrosion reaction, like ceasing O_2 and H_2O diffusion onto the metal surface, and retarding metal oxidation.

The cross-sectional morphologies of the PS and PS/m-GO nanocomposite materials were analyzed using FE-SEM in which the fracture lines are aligned in a homogeneous path in case of PS surface. Whereas, the PS/m-GO nanocomposites morphologies of cross-sectional were comparatively flat and homogeneous, indicating tough interfacial interactions between the uniform dispersion of m-GO and the PS matrix. Consequently, electroactive polymers could be used for corrosion prevention moreover due to the barrier or anodic prevention of metal steels.

SELF-HEALING NANOCOMPOSITE COATING FOR CORROSION PROTECTION

As discussed in earlier section, intrinsically conducting polymers (ICPs) were widely utilized in the field of protective coatings attributed to their character like self-healing and eco-friendly nature, which make them an ideal material to substitute the chromium (VI) based coatings (Tian et al., 2014; Qi et al., 2012; Xing et al., 2014). By adding polypyrrole-carbon black (PPyCB) composite as an inhibiting pigment, organic coating formulations based on self-healing polyvinyl butyral (PVB) were synthesized (Niratiwongkorn et al., 2016). Even in extreme environmental conditions, the redox character and diffusion barrier properties of PPy rendered self-healing to the PVB/PPyCB composite coatings. PPy forms a steady passive layer on the metal surface by the interaction of doped PPy with metal iron oxide which was revealed from SEM images and Raman spectroscopy. In addition to this, reduced PPy prevents the diffusion of H_2O and O_2 into the coating. The incorporation of extra conducting particles such as grapheme promotes the protective character of the PVB/PPyCB composite coatings. These composite

Figure 24. The enhancement of corrosion protection mechanism of nanocomposites on the metal surface (Yu et al., 2014)

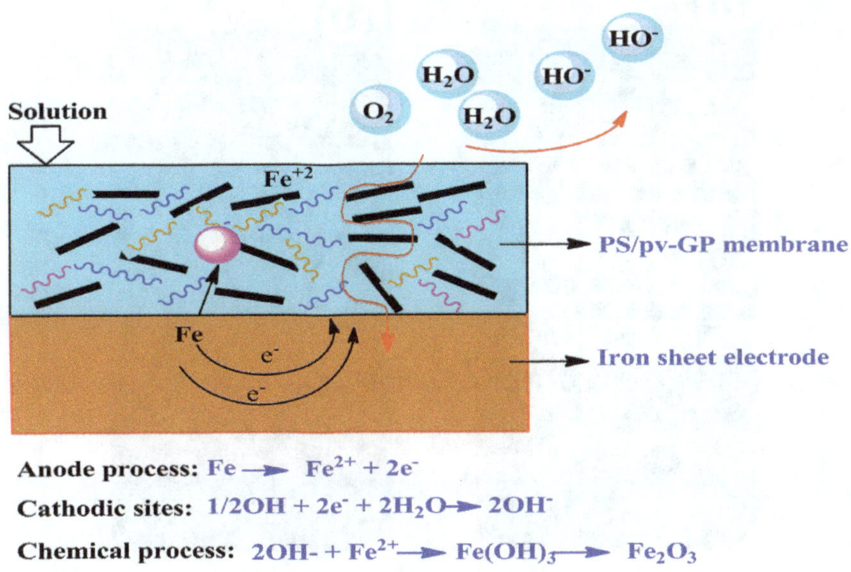

coatings exhibits a potential application of conducting particles in improving the protective character of organic coatings employed commercially for a barrier coating in industries. On the other hand, at elevated temperature a good adhesion of PVB and PPyCB pigments were obtained on the metal surface while fabricating PVB/PPyCB coatings. Figure 25 shows that the cross cut tape test done on PVB/PPyCB20 composite coatings prior and after 1 d dip in 4 wt% NaCl solution.

The Figure 25 (a) indicates that the cut of the ends PVB/PPyCB20 composites were smooth and the rectangles of the networks were not separated, reflecting the strong adhesion of the composite matrix on the metal surface. During the immersion, although the corrosion products were developed mostly at the cross cut areas, main coating regions were not removed during the tape test (Figure 25 b), which indicates the stability of the composites coating, signifying tough binding on the metal surface inhibiting the seepage of H_2O via the corners of the scratches.

Figure 26 displays the corrosion protection mechanism of the PVB/PPyCB composite coatings. In accordance to mechanism, the nature of PPy give an anodic prevention to the steel by forming a passive film at the defect area, consequently increasing the quantity of decreased PPy around the passivated regions. This passive film and hydrophobic character of the decreased PPy acts as a diffusion barrier which lowers the rate of charge transfer reaction. Up to 20 wt% of concentration, PPyCB pigment was effectively added into PVB coating formulation without altering the barrier and adhesion performance of coatings. In addition, of grapheme in the PVB/PPyCB20 composite coatings improved the more number of passive layer deposition on the metal surface, which prevented the electron transfer between metal and polymer along with inhibiting H_2O and gas from reaching the metal surface, thus protecting metal from more corrosion. In summary, by the incorporation of conducting composite particles like PPyCB which acts as a corrosion inhibiting pigments induces a high degree of self-healing inhibit properties to the organic coatings without altering the inherent coating properties.

Figure 25. Representation of cross cut tape analysis on PVC-PPyCB20 nanocomposites coatings (a) before and after (b) immesrsion in 4 wt.% NaCl solution
(Niratiwongkorn et al., 2016)

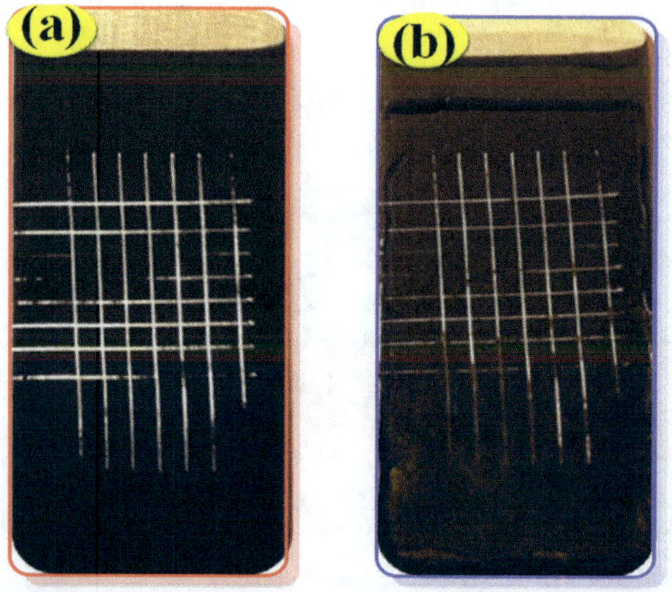

Figure 26. Schematic diagram of corrosion protection impart via PPy dye in PVB-PPyCB composite coatings
(Niratiwongkorn et al., 2016)

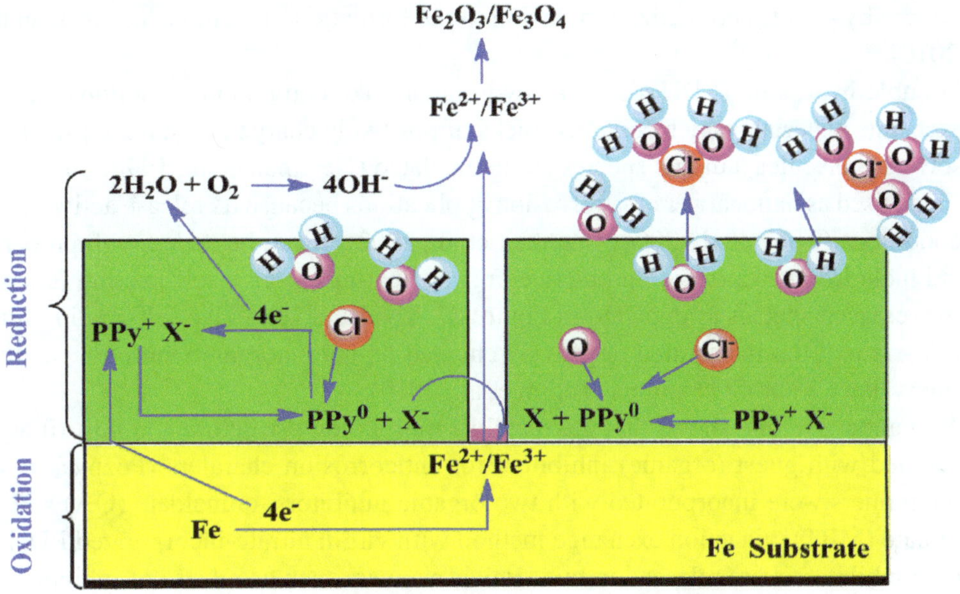

In the development of the self-healing nanocomposites coating with mesoporous silica which release the small molecules accumulate in the nanoporous surface with self-healing surfaces also been developed (Yeganeh et al., 2014). For example, in the addition of molybdate into the particles surface which alternatively transfer to the surface and complete the self-repairing. In the inclusion of 0.05 M MoO_4^{2-} to a 0.1 M H_2SO_4 solution results to a P.E of 71% leading to the development of a Mo-rich protective layer (Wang et al., 2005)

The electrodeposition of novel corrosion resistant zinc-based composite coatings consisting mesoporous silica (MCM-41), with adequate molybdate (Alipour et al., 2017). The molybdate plays a significant role for corrosion inhibitor and "anti-corrosive self-healing" agent to the coating to be liberated while corrosion. In presence of 3-mercopropyltrimethoxysilane (MPTMS) in the coating, enhanced the dispersion and bonding properties of the functionlized MCM-41 particles in the electrodeposition solution.

Galvanostatically Zinc metal base was co-deposited with the MCM-41 particles at different states to obtain the smoothest and highly corrosion resistant films. In a chloride solution, the open circuit potential (OCP) of Zn–MCM-41–silane(SH)–molybdate(Mo) functionalization increases with time due to the release of anodic inhibitor Mo that accomplish the self-healing property to the coating. From the EIS and potentiodynamic polarization experiments, it was revealed that the corrosion resistance raises in the sequence: Zn ˂ Zn–MCM-41 ˂ Zn–MCM-41–Mo ˂ Zn–MCM-41–SH–Mo.

The Zn–MCM-41–Mo nanocoating varies towards more positive potentials with time, attributed to anodic protection of molybdate as a result of anodic dissolution and cathodic prevention of steel base metal. On the other hand, the variations in OCP of Zn–MCM-41 are also positive although the change is very slightly. Mo(VI) is a weak oxidizing agent, which behaves as an anodic inhibitor for the formation of MCM-41–Mo protective layer (Thomas et al., 2013; Treacy et al., 1999). Active agent molybdate controls the process of passivation of Fe (Robertson et al., 1951). Moreover, the amount of anodic polarization reduces by –SH functionalization, owing to the formation of tortuous network on the surface (Liu et al., 2010).

Layered double hydroxide (LDHs) such as hydrotalcite-like compounds or anionic clays are used as a anion exchange materials which contains stacks of positively charged with mixed-metal hydroxide layers and solvent molecules, anionic species are intercalated (Newman et al., 1998; Shao et al., 2013). LDHs have been used as nanocarriers for corrosion applications because its release active species under controlled conditions (Choy et al., 2004; Buchheit et al., 2003). It can be added with proper corrosion inhibitors and included into corrosion protective organic coatings to give self-healing functionality by liberate of the reserved species as depicted in Figure 27. The LDHs can give self-healing functionality by managed liberate of the inactivated negatively charged organic/inorganic inhibitors and trap of the corrosive anions like Cl^- and SO^{-2} anions (Tedim et al., 2010).

The LDHs nanocrystalline shows that the platelike morphology structure and host (inorganic) materials intercalated with guest (organic) inhibitors for anticorrosion charaters (Poznyak et al., 2009). LDHs nanocontainers were incorporated with two organic inhibitors, quinaldate (QA) and 2-mercaptobenzothiazolate (MBT), via anion exchange method with virgin nitrate-incorporated LDH precursor. The organic (-) inhibitors were released in two similar manners such as release of inhibitor and trap of the energetic anions, as long as active response for corrosion prevention control occurs. The AA2024 specimens which were kept in contact with LDH suspension in a 0.05 M NaCl medium for 24 h exhibits a low anticorrosion activity which was due to the particular alkalization of the medium in the existence of hydroxide-rich LDHs particles, that leads to a decline in the protection efficiency of QA and MBT inhibitors. On the other hand, with the MBT-loaded LDHs at longer immersion periods, the corrosion mechanisms are well prevented, owing to the development of a protective film, indicating excellent properties in contrast to QA loaded films.

The synthesis of self-healing water-based epoxy coating after modification with nanocapsules incorporated with 2-benzothiazolethiol (MBT) for corrosion prevention of steel and aluminum alloy were investigated (Plawecka et al., 2014). The nanocapsules containers were synthesized by direct encapsulation of hydrophobic active agent. Used reagents chemical structures are shown in Figure 28. On steel and aluminum specimens, the epoxy coating was applied through dip-coating. The results revealed that the incorporation of nanocapsules to water-based epoxy primers does not alter the barrier character of these coatings. The addition of inhibitor-incorporated nanocapsules enhances the corrosion resistance of steel and aluminum alloys reflecting its utilization in water-based epoxy coatings.

MILITARY APPLICATION OF POLYMER NANOCOMPOSITES COATING

Many researchers have been done to develop and synthesis strategies to fabricate nanocomposites coating surface which can be commercialized for employing in different areas (Kurahatti et al., 2010; Thilagavathi et al., 2008). Therefore, in this part, we will commonly concentrate on anticorrosion coatings for military application.

Figure 27. The liberate of inhibitors (Inh⁻) is prompted via existence of anions in solution (Cl⁻) (b) LDHs perform a dual role, bestowing inhibitors to prevention the metallic substrate and trapping energetic groups from the atmosphere
(Poznyak et al., 2009)

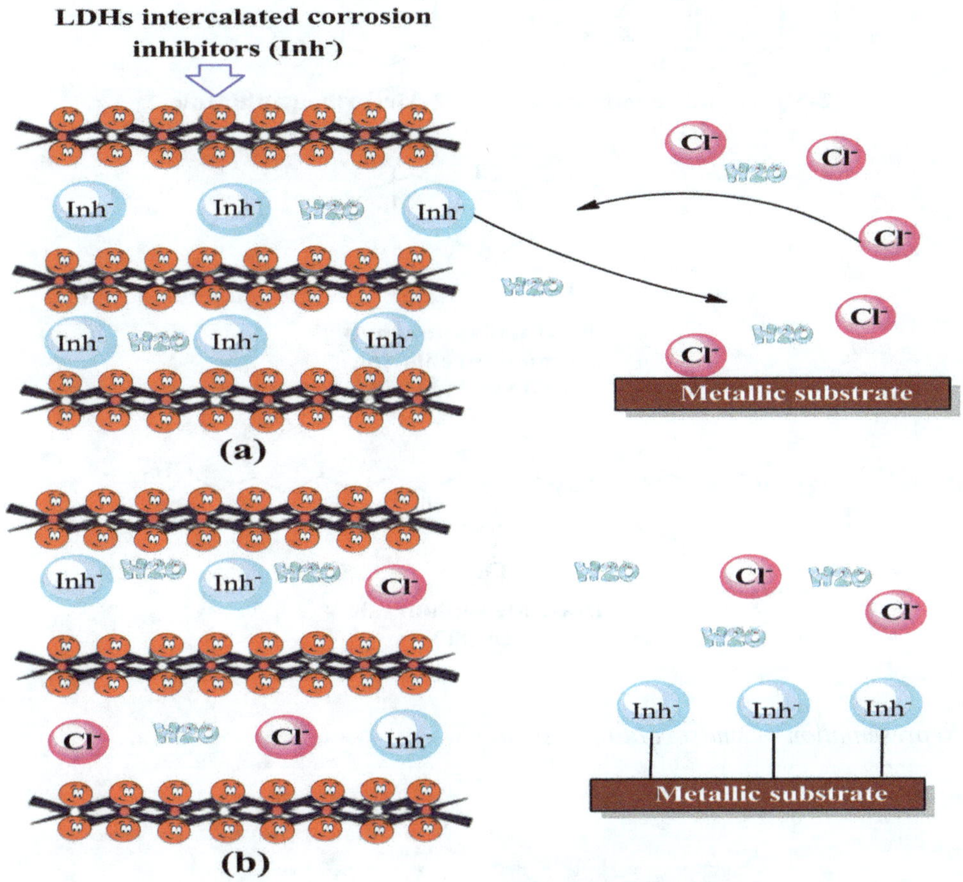

Nanostructured based coatings materials demonstrated a high corrosion resistant after incorporated together with metals, polymer, paint/film, composites etc. This is exhibit one capability to work at the molecular level atom by atom to generate smart structure with basically advanced molecular system and produced innovative materials that will allocate for extensive service life and minor failure rates. These technologies developed to novel coatings materials for the military requirement (Zunino III et al., 2005).

Army corrosion office (ACO) assembled one team for the military and industry representatives. This team is evolving the multilayer active smart coating with several functionalities like visual display, self repair, self management, sensing packing, corrosion inhibitors and artificial intelligence that can modified as requirement as shown in the Figure 29 (Zunino III et al., 2007).

According to the current research with functionalized single carbon nanotubes employed as smart coating materials and inks to initiate self-healing, sensing active switching, color modification and other functionalities as depicted in Figure 30. The corrosion barrier coatings system exhibited two type active p-n (positive and negative doped) junction layer at the bottom and a passive top layer has been

Figure 28. Reagents chemical structure for used in the preparation of inhibitors-incorporated electrolyte capsules
(Plawecka et al., 2014)

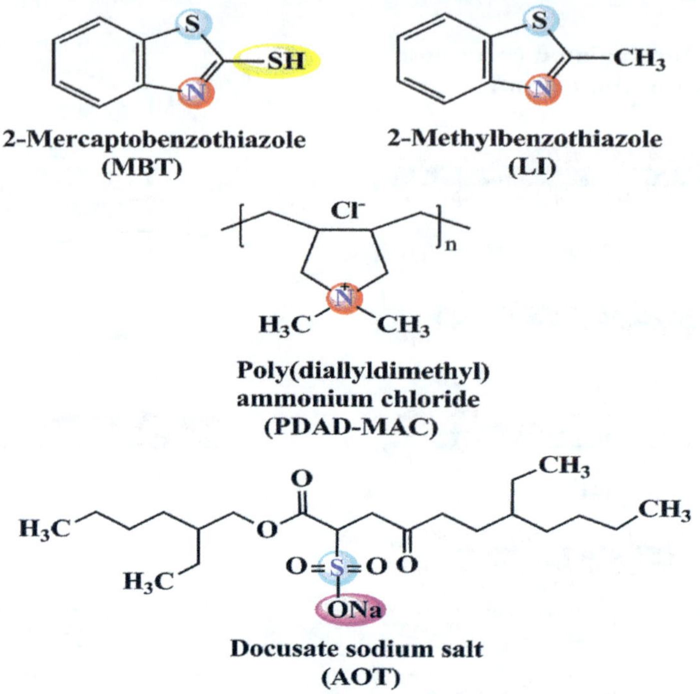

Figure 29. Representation of smart coating system

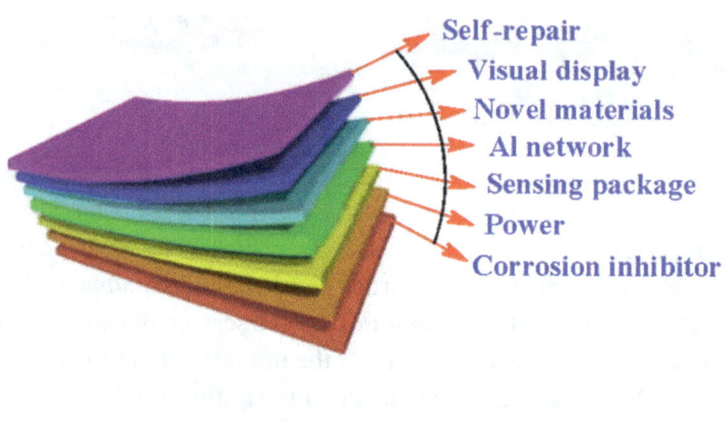

formulated. The junction coating bestows active corrosion protection plus corrosion sensing potential (Zunino III et al., 2007; Zunino III et al., 2008).

These coatings utilized in the current and future weapon systems in military and civilization application. The most important benefit of the coating technology program (ACT) for future development of military grade active sensing to observed damage corrosion substrate and environmental conditions such as radiation temperature, chemical etc. The sensor technology under can be used in the army for the protection of material in the corrosion. The Figure 31 demonstrated that the military application of the polymer nanocomposites coating in various component.

Figure 30. Liquid ink solar cell coating on flexible substrate
(Zunino III et al., 2007)

Figure 31. Military applications of polymer nanocomposites coating

CONCLUSION

This chapter briefed the current development in the preparation of anticorrosion nanocomposites coatings employing different nanofiller such as CNTs, GO, TiO_2, Al_2O_3, ZnO, SiO_2 and ZrO_2, etc. and fabrication methods. It was observed that a nanoparticle plays an imperative character in the enhancement of the corrosion protection of metal substrate. The polymer based nanocomposite anticorrosion coating via hindering the flow of corrosive agents to the metal substrate, prevents the metal surface corrosion for long time. The nanocomposites coating materials demonstrated higher corrosion resistance, minimum discontinuity or porosity in coating, adequate thickness and low diffusion rate of the H_2O, O_2 and Cl^- ions. The polyaniline and metal oxide containing coatings like ZnO, TiO_2 and SiO_2 reportedly improve the corrosion resistance owing to the formation of homogeneous passive layer on the metal surface. Furthermore, the various fabrication methods involved in the preparation of polymer nanocomposites coating. The researchers can enable to understand the technique and practical ideas for achieving superior anticorrosion properties in the future.

REFERENCES

Agarwala, V., Chandra Agarwala, R., & Sunder Daniel, B. S. (2006). Development of nanograined metallic Materials by bulk and coating techniques. *Synthesis and Reactivity in Inorganic, Metal-Organic, and Nano-Metal Chemistry*, *36*(1), 3–16. doi:10.1080/15533170500471128

Aglan, A., Allie, A., Ludwick, A., & Koons, L. (2007). Formulation and evaluation of nano-structured polymeric coatings for corrosion protection. *Surface and Coatings Technology*, *202*(2), 370–378. doi:10.1016/j.surfcoat.2007.05.090

Alipour, K., & Nasirpouri, F. (2017). Smart anti-corrosion self-healing zinc metal-based molybdate functionalized-mesoporous-silica (MCM-41) nanocomposite coatings. *RSC Advances*, *7*(82), 51879–51887. doi:10.1039/C7RA06923E

Bakhshandeh, E., Jannesari, A., Ranjbar, Z., Sobhani, S., & Saeb, M. R. (2014). Anti-corrosion hybrid coatings based on epoxy–silica nano-composites: Toward relationship between the morphology and EIS data. *Progress in Organic Coatings*, *77*(7), 1169–1183. doi:10.1016/j.porgcoat.2014.04.005

Bard, A. J., Faulkner, L. R., Leddy, J., & Zoski, C. G. (1980). *Electrochemical Methods: Fundamentals and Applications* (Vol. 2). New York: Wiley.

Berry, V. (2013). Impermeability of graphene and its applications. *Carbon*, *62*, 1–10. doi:10.1016/j.carbon.2013.05.052

Bhandari, J., Khan, F., Abbassi, R., Garaniya, V., & Ojeda, R. (2015). Modelling of pitting corrosion in marine and offshore steel structures–A technical review. *Journal of Loss Prevention in the Process Industries*, *37*, 39–62. doi:10.1016/j.jlp.2015.06.008

Buchheit, R. G., Guan, H., Mahajanam, S., & Wong, F. (2003). Active corrosion protection and corrosion sensing in chromate-free organic coatings. *Progress in Organic Coatings*, *47*(3-4), 174–182. doi:10.1016/j.porgcoat.2003.08.003

Chang, C. H., Huang, T. C., Peng, C. W., Yeh, T. C., Lu, H. I., Hung, W. I., & Yeh, J. M. (2012). Novel anticorrosion coatings prepared from polyaniline/graphene composites. *Carbon*, *50*(14), 5044–5051. doi:10.1016/j.carbon.2012.06.043

Chang, K. C., Hsu, M. H., Lu, H. I., Lai, M. C., Liu, P. J., Hsu, C. H., & Liu, W. R. (2014). Room-temperature cured hydrophobic epoxy/graphene composites as corrosion inhibitor for cold-rolled steel. *Carbon*, *66*, 144–153. doi:10.1016/j.carbon.2013.08.052

Chen, Y., Wang, X. H., Li, J., Lu, J. L., & Wang, F. S. (2007). Long-term anticorrosion behaviour of polyaniline on mild steel. *Corrosion Science*, *49*(7), 3052–3063. doi:10.1016/j.corsci.2006.11.007

Chiang, C. K., Fincher, C. R. Jr, Park, Y. W., Heeger, A. J., Shirakawa, H., Louis, E. J., & MacDiarmid, A. G. (1977). Electrical conductivity in doped polyacetylene. *Physical Review Letters*, *39*(17), 1098–1101. doi:10.1103/PhysRevLett.39.1098

Choy, J. H., Jung, J. S., Oh, J. M., Park, M., Jeong, J., Kang, Y. K., & Han, O. J. (2004). Layered double hydroxide as an efficient drug reservoir for folate derivatives. *Biomaterials*, *25*(15), 3059–3064. doi:10.1016/j.biomaterials.2003.09.083 PMID:14967539

DeBerry, D. W. (1985). Modification of the electrochemical and corrosion behavior of stainless steels with an electroactive coating. *Journal of the Electrochemical Society*, *132*(5), 1022–1026. doi:10.1149/1.2114008

Dhawan, S. K., Kumar, A., Bhandari, H., Bisht, B. M. S., & Khatoon, F. (2015). Development of Highly Hydrophobic and Anticorrosive Conducting Polymer Composite Coating for Corrosion Protection in Marine Environment. *American Journal of Political Science*, *5*(1A), 7–17.

Dhoke, S. K., & Khanna, A. S. (2009). Effect of nano-Fe_2O_3 particles on the corrosion behavior of alkyd based waterborne coatings. *Corrosion Science*, *51*(1), 6–20. doi:10.1016/j.corsci.2008.09.028

Dhoke, S. K., & Khanna, A. S. (2012). Electrochemical impedance spectroscopy (EIS) study of nano-alumina modified alkyd based waterborne coatings. *Progress in Organic Coatings*, *74*(1), 92–99. doi:10.1016/j.porgcoat.2011.11.020

El Rayes, M. M., Abdo, H. S., & Khalil, K. A. (2013). Erosion-corrosion of cermet coating. *International Journal of Electrochemical Science*, *8*(1), 1117–1137.

Freguia, S., Rabaey, K., Yuan, Z., & Keller, J. (2007). Non-catalyzed cathodic oxygen reduction at graphite granules in microbial fuel cells. *Electrochimica Acta*, *53*(2), 598–603. doi:10.1016/j.electacta.2007.07.037

Golru, S. S., Attar, M. M., & Ramezanzadeh, B. (2014). Studying the influence of nano-Al2O3 particles on the corrosion performance and hydrolytic degradation resistance of an epoxy/polyamide coating on AA-1050. *Progress in Organic Coatings*, *77*(9), 1391–1399. doi:10.1016/j.porgcoat.2014.04.017

Gray, J., & Luan, B. (2002). Protective coatings on magnesium and its alloys—A critical review. *Journal of Alloys and Compounds*, *336*(1-2), 88–113. doi:10.1016/S0925-8388(01)01899-0

Gu, H., Tadakamalla, S., Zhang, X., Huang, Y., Jiang, Y., Colorado, H. A., & Guo, Z. (2013). Epoxy resin nanosuspensions and reinforced nanocomposites from polyaniline stabilized multi-walled carbon nanotubes. *Journal of Materials Chemistry. C, Materials for Optical and Electronic Devices*, *1*(4), 729–743. doi:10.1039/C2TC00379A

Haddadi, S. A., Mahdavian, M., & Karimi, E. (2015). Evaluation of the corrosion protection properties of an epoxy coating containing sol–gel surface modified nano-zirconia on mild steel. *RSC Advances*, *5*(36), 28769–28777. doi:10.1039/C5RA02127H

Harb, S. V., Pulcinelli, S. H., Santilli, C. V., Knowles, K. M., & Hammer, P. (2016). A comparative study on graphene oxide and carbon nanotube reinforcement of PMMA-siloxane-silica anticorrosive coatings. *ACS Applied Materials & Interfaces*, *8*(25), 16339–16350. doi:10.1021/acsami.6b04780 PMID:27266403

Herrasti, P., Recio, F. J., Ocon, P., & Fatas, E. (2005). Effect of the polymer layers and bilayers on the corrosion behaviour of mild steel: Comparison with polymers containing Zn microparticles. *Progress in Organic Coatings*, *54*(4), 285–291. doi:10.1016/j.porgcoat.2005.07.001

Hosseini, M. G., Sabouri, M., & Shahrabi, T. (2007). Corrosion protection of mild steel by polypyrrole phosphate composite coating. *Progress in Organic Coatings*, *60*(3), 178–185. doi:10.1016/j.porgcoat.2007.07.029

Hu, H., He, Y., Long, Z., & Zhan, Y. (2017). Synergistic effect of functional carbon nanotubes and graphene oxide on the anti-corrosion performance of epoxy coating. *Polymers for Advanced Technologies*, *28*(6), 754–762. doi:10.1002/pat.3977

Jeon, H., Park, J., & Shon, M. (2013). Corrosion protection by epoxy coating containing multi-walled carbon nanotubes. *Journal of Industrial and Engineering Chemistry*, *19*(3), 849–853. doi:10.1016/j.jiec.2012.10.030

Jeon, H., Park, J., & Shon, M. (2013). Corrosion protection by epoxy coating containing multi-walled carbon nanotubes. *Journal of Industrial and Engineering Chemistry*, *19*(3), 849–853. doi:10.1016/j.jiec.2012.10.030

Khanam, Z., Singh, V., & Zaidi, M. G. H. (2018). Enhanced corrosion protection performance with MWCNT dispersed epoxy coating prepared under supercritical CO2 assistance. *Polymers for Advanced Technologies*, *29*(9), 2457–2466. doi:10.1002/pat.4357

Khun, N. W., Troconis, B. R., & Frankel, G. S. (2014). Effects of carbon nanotube content on adhesion strength and wear and corrosion resistance of epoxy composite coatings on AA2024-T3. *Progress in Organic Coatings*, *77*(1), 72–80. doi:10.1016/j.porgcoat.2013.08.003

Kinlen, P. J., Silverman, D. C., & Jeffreys, C. R. (1997). Corrosion protection using polyanujne coating formulations. *Synthetic Metals*, *85*(1-3), 1327–1332. doi:10.1016/S0379-6779(97)80257-3

Koch, G. H., Brongers, M. P., Thompson, N. G., Virmani, Y. P., & Payer, J. H. (2002). Corrosion cost and preventive strategies in the United States (No. FHWA-RD-01-156,).

Kuilla, T., Bhadra, S., Yao, D., Kim, N. H., Bose, S., & Lee, J. H. (2010). Recent advances in graphene based polymer composites. *Progress in Polymer Science*, *35*(11), 1350–1375. doi:10.1016/j.progpolymsci.2010.07.005

Kumar, A., Anant, R., Kumar, K., Chauhan, S. S., Kumar, S., & Kumar, R. (2016). Anticorrosive and electromagnetic shielding response of a graphene/TiO_2-epoxy nanocomposite with enhanced mechanical properties. *RSC Advances*, *6*(114), 113405–113414. doi:10.1039/C6RA15273B

Kumar, A., Ghosh, P. K., Yadav, K. L., & Kumar, K. (2017). Thermo-mechanical and anti-corrosive properties of MWCNT/epoxy nanocomposite fabricated by innovative dispersion technique. *Composites. Part B, Engineering*, *113*, 291–299. doi:10.1016/j.compositesb.2017.01.046

Kurahatti, R. V., Surendranathan, A. O., Kori, S. A., Singh, N., Kumar, A. V., & Srivastava, S. (2010). Defence Applications of Polymer Nanocomposites. *Defence Science Journal*, *60*(5), 551–563. doi:10.14429/dsj.60.578

Lin, Y. T., Don, T. M., Wong, C. J., Meng, F. C., Lin, Y. J., Lee, S. Y., & Chiu, W. Y. (2018). Improvement of mechanical properties and anticorrosion performance of epoxy coatings by the introduction of polyaniline/graphene composite. *Surface and Coatings Technology*. doi:10.1016/j.surfcoat.2018.01.050

Liu, B., Fang, Z. G., Wang, H. B., & Wang, T. (2013). Effect of cross linking degree and adhesion force on the anti-corrosion performance of epoxy coatings under simulated deep sea environment. *Progress in Organic Coatings*, *76*(12), 1814–1818. doi:10.1016/j.porgcoat.2013.05.022

Liu, D. L., Yang, Z. G., Wang, Z. Q., & Zhang, C. (2010). Synthesis and evaluation of corrosion resistance of molybdate-based conversion coatings on electroplated zinc. *Surface and Coatings Technology*, *205*(7), 2328–2334. doi:10.1016/j.surfcoat.2010.09.018

Liu, J., Yu, Q., Yu, M., Li, S., Zhao, K., Xue, B., & Zu, H. (2018). Silane modification of titanium dioxide-decorated graphene oxide nanocomposite for enhancing anticorrosion performance of epoxy coatings on AA-2024. *Journal of Alloys and Compounds*, *744*, 728–739. doi:10.1016/j.jallcom.2018.01.267

Low, C. T. J., Wills, R. G. A., & Walsh, F. C. (2006). Electrodeposition of composite coatings containing nanoparticles in a metal deposit. *Surface and Coatings Technology*, *201*(1-2), 371–383. doi:10.1016/j.surfcoat.2005.11.123

Lu, W. K., Elsenbaumer, R. L., & Wessling, B. (1995). Corrosion protection of mild steel by coatings containing polyaniline. *Synthetic Metals*, *71*(1-3), 2163–2166. doi:10.1016/0379-6779(94)03204-J

Marcus, P. (Ed.). (2011). *Corrosion mechanisms in theory and practice*. CRC press. doi:10.1201/b11020

Meguid, S. A., & Sun, Y. (2004). On the tensile and shear strength of nano-reinforced composite interfaces. *Materials & Design*, *25*(4), 289–296. doi:10.1016/j.matdes.2003.10.018

Melchers, R. E., & Jeffrey, R. (2008). Modeling of long-term corrosion loss and pitting for chromium-bearing and stainless steels in seawater. *Corrosion*, *64*(2), 143–154. doi:10.5006/1.3280683

Natesan, M., Venkatachari, G., & Palaniswamy, N. (2005). Corrosivity and durability maps of India. *Corrosion Prevention & Control*, *52*(2), 43–55.

Newman, S. P., & Jones, W. (1998). Synthesis, characterization and applications of layered double hydroxides containing organic guests. *New Journal of Chemistry*, *22*(2), 105–115. doi:10.1039/a708319j

Niratiwongkorn, T., Luckachan, G. E., & Mittal, V. (2016). Self-healing protective coatings of polyvinyl butyral/polypyrrole-carbon black composite on carbon steel. *RSC Advances*, *6*(49), 43237–43249. doi:10.1039/C6RA01619G

Palimi, M. J., Rostami, M., Mahdavian, M., & Ramezanzadeh, B. (2015). A study on the corrosion inhibition properties of silane-modified Fe_2O_3 nanoparticle on mild steel and its effect on the anticorrosion properties of the polyurethane coating. *Journal of Coatings Technology and Research, 12*(2), 277–292. doi:10.100711998-014-9631-6

Palraj, S., Selvaraj, M., Maruthan, K., & Rajagopal, G. (2015). Corrosion and wear resistance behavior of nano-silica epoxy composite coatings. *Progress in Organic Coatings, 81*, 132–139. doi:10.1016/j.porgcoat.2015.01.005

Park, S., & Shon, M. (2015). Effects of multi-walled carbon nano tubes on corrosion protection of zinc rich epoxy resin coating. *Journal of Industrial and Engineering Chemistry, 21*, 1258–1264. doi:10.1016/j.jiec.2014.05.042

Plawecka, M., Snihirova, D., Martins, B., Szczepanowicz, K., Warszynski, P., & Montemor, M. F. (2014). Self healing ability of inhibitor-containing nanocapsules loaded in epoxy coatings applied on aluminium 5083 and galvanneal substrates. *Electrochimica Acta, 140*, 282–293. doi:10.1016/j.electacta.2014.04.035

Potts, J. R., Dreyer, D. R., Bielawski, C. W., & Ruoff, R. S. (2011). Graphene-based polymer nanocomposites. *Polymer, 52*(1), 5–25. doi:10.1016/j.polymer.2010.11.042

Pour, Z. S., Ghaemy, M., Bordbar, S., & Karimi-Maleh, H. (2018). Effects of surface treatment of TiO 2 nanoparticles on the adhesion and anticorrosion properties of the epoxy coating on mild steel using electrochemical technique. *Progress in Organic Coatings, 119*, 99–108. doi:10.1016/j.porgcoat.2018.02.019

Pourhashem, S., Vaezi, M. R., Rashidi, A., & Bagherzadeh, M. R. (2017). Exploring corrosion protection properties of solvent based epoxy-graphene oxide nanocomposite coatings on mild steel. *Corrosion Science, 115*, 78–92. doi:10.1016/j.corsci.2016.11.008

Poznyak, S. K., Tedim, J., Rodrigues, L. M., Salak, A. N., Zheludkevich, M. L., Dick, L. F. P., & Ferreira, M. G. S. (2009). Novel inorganic host layered double hydroxides intercalated with guest organic inhibitors for anticorrosion applications. *ACS Applied Materials & Interfaces, 1*(10), 2353–2362. doi:10.1021/am900495r PMID:20355873

Qi, K., Qiu, Y., Chen, Z., & Guo, X. (2012). Corrosion of conductive polypyrrole: Effects of environmental factors, electrochemical stimulation, and doping anions. *Corrosion Science, 60*, 50–58. doi:10.1016/j.corsci.2012.04.013

Qi, K., Sun, Y., Duan, H., & Guo, X. (2015). A corrosion-protective coating based on a solution-processable polymer-grafted graphene oxide nanocomposite. *Corrosion Science, 98*, 500–506. doi:10.1016/j.corsci.2015.05.056

Rajabi, M., Rashed, G. R., & Zaarei, D. (2015). Assessment of graphene oxide/epoxy nanocomposite as corrosion resistance coating on carbon steel. *Corrosion Engineering, Science and Technology, 50*(7), 509–516. doi:10.1179/1743278214Y.0000000232

Ramezanzadeh, B., Ahmadi, A., & Mahdavian, M. (2016). Enhancement of the corrosion protection performance and cathodic delamination resistance of epoxy coating through treatment of steel substrate by a novel nanometric sol-gel based silane composite film filled with functionalized graphene oxide nanosheets. *Corrosion Science, 109*, 182–205. doi:10.1016/j.corsci.2016.04.004

Ramezanzadeh, B., & Attar, M. M. (2011). Studying the corrosion resistance and hydrolytic degradation of an epoxy coating containing ZnO nanoparticles. *Materials Chemistry and Physics, 130*(3), 1208–1219. doi:10.1016/j.matchemphys.2011.08.065

Ramezanzadeh, B., Niroumandrad, S., Ahmadi, A., Mahdavian, M., & Moghadam, M. M. (2016). Enhancement of barrier and corrosion protection performance of an epoxy coating through wet transfer of amino functionalized graphene oxide. *Corrosion Science, 103*, 283–304. doi:10.1016/j.corsci.2015.11.033

Robertson, W. D. (1951). Molybdate and tungstate as corrosion inhibitors and the mechanism of inhibition. *Journal of the Electrochemical Society, 98*(3), 94–100. doi:10.1149/1.2778118

Saji, V. S., & Cook, R. M. (Eds.). (2012). *Corrosion protection and control using nanomaterials*. Elsevier. doi:10.1533/9780857095800

Sakhri, A., Perrin, F. X., Aragon, E., Lamouric, S., & Benaboura, A. (2010). Chlorinated rubber paints for corrosion prevention of mild steel: A comparison between zinc phosphate and polyaniline pigments. *Corrosion Science, 52*(3), 901–909. doi:10.1016/j.corsci.2009.11.010

Shao, M., Wei, M., Evans, D. G., & Duan, X. (2013). Hierarchical Structures Based on Functionalized Magnetic Cores and Layered Double-Hydroxide Shells: Concept, Controlled Synthesis, and Applications. *Chemistry (Weinheim an der Bergstrasse, Germany), 19*(13), 4100–4108. doi:10.1002/chem.201204205 PMID:23424036

Shen, W., Feng, L., Liu, X., Luo, H., Liu, Z., Tong, P., & Zhang, W. (2016). Multiwall carbon nanotubes-reinforced epoxy hybrid coatings with high electrical conductivity and corrosion resistance prepared via electrostatic spraying. *Progress in Organic Coatings, 90*, 139–146. doi:10.1016/j.porgcoat.2015.10.006

Shi, H., Liu, F., Yang, L., & Han, E. (2008). Characterization of protective performance of epoxy reinforced with nanometer-sized TiO_2 and SiO_2. *Progress in Organic Coatings, 62*(4), 359–368. doi:10.1016/j.porgcoat.2007.11.003

Sinha Ray, S., Yamada, K., Okamoto, M., Ogami, A., & Ueda, K. (2003). New polylactide/layered silicate nanocomposites. 3. High-performance biodegradable materials. *Chemistry of Materials, 15*(7), 1456–1465. doi:10.1021/cm020953r

Solmaz, R., Kardaş, G., Culha, M., & Erbil, M. (2008). Investigation of adsorption and inhibitive effect of 2-mercaptothiazoline on corrosion of mild steel in hydrochloric acid media. *Electrochimica Acta, 53*(20), 5941–5952. doi:10.1016/j.electacta.2008.03.055

Sonawane, S. H., Teo, B. M., Brotchie, A., Grieser, F., & Ashokkumar, M. (2010). Sonochemical synthesis of ZnO encapsulated functional nanolatex and its anticorrosive performance. *Industrial & Engineering Chemistry Research, 49*(5), 2200–2205. doi:10.1021/ie9015039

Song, G. (2005). Recent progress in corrosion and protection of magnesium alloys. *Advanced Engineering Materials, 7*(7), 563–586. doi:10.1002/adem.200500013

Sorensen, P. A., Kiil, S., Dam-Johansen, K., & Weinell, C. E. (2009). Anticorrosive coatings: A review. *Journal of Coatings Technology and Research, 6*(2), 135–176. doi:10.100711998-008-9144-2

Su, Y., Kravets, V. G., Wong, S. L., Waters, J., Geim, A. K., & Nair, R. R. (2014). Impermeable barrier films and protective coatings based on reduced graphene oxide. *Nature Communications*, *5*(1), 4843. doi:10.1038/ncomms5843 PMID:25208890

Tang, M., Wu, T., Na, H., Zhang, S., Li, X., & Pang, X. (2015). Fabrication of graphene oxide aerogels loaded with catalytic AuPd nanoparticles. *Materials Research Bulletin*, *63*, 248–252. doi:10.1016/j.materresbull.2014.12.008

Tedim, J., Poznyak, S. K., Kuznetsova, A., Raps, D., Hack, T., Zheludkevich, M. L., & Ferreira, M. G. S. (2010). Enhancement of active corrosion protection via combination of inhibitor-loaded nanocontainers. *ACS Applied Materials & Interfaces*, *2*(5), 1528–1535. doi:10.1021/am100174t PMID:20455547

Terrones, M., Martín, O., González, M., Pozuelo, J., Serrano, B., Cabanelas, J. C., & Baselga, J. (2011). Interphases in Graphene Polymer-based Nanocomposites: Achievements and Challenges. *Advanced Materials*, *23*(44), 5302–5310. doi:10.1002/adma.201102036 PMID:22299145

Thilagavathi, G., Raja, A. S. M., & Kannaian, T. (2008). Nanotechnology and protective clothing for defence personnel. *Defence Science Journal*, *58*(4), 451–459. doi:10.14429/dsj.58.1667

Thomas, S., Birbilis, N., Venkatraman, M. S., & Cole, I. S. (2013). Self-repairing oxides to protect zinc: Review, discussion and prospects. *Corrosion Science*, *69*, 11–22. doi:10.1016/j.corsci.2013.01.011

Tian, Z., Yu, H., Wang, L., Saleem, M., Ren, F., Ren, P., & Huang, L. (2014). Recent progress in the preparation of polyaniline nanostructures and their applications in anticorrosive coatings. *RSC Advances*, *4*(54), 28195–28208. doi:10.1039/c4ra03146f

Treacy, G. M., Wilcox, G. D., & Richardson, M. O. W. (1999). Behaviour of molybdate-passivated zinc coated steel exposed to corrosive chloride environments. *Journal of Applied Electrochemistry*, *29*(5), 647–654. doi:10.1023/A:1026461924663

Wang, D., & Bierwagen, G. P. (2009). Sol–gel coatings on metals for corrosion protection. *Progress in Organic Coatings*, *64*(4), 327–338. doi:10.1016/j.porgcoat.2008.08.010

Wang, D., Tang, X., Qiu, Y., Gan, F., & Chen, G. Z. (2005). A study of the film formation kinetics on zinc in different acidic corrosion inhibitor solutions by quartz crystal microbalance. *Corrosion Science*, *47*(9), 2157–2172. doi:10.1016/j.corsci.2004.10.003

Wei, H., Ding, D., Wei, S., & Guo, Z. (2013). Anticorrosive conductive polyurethane multiwalled carbon nanotube nanocomposites. *Journal of Materials Chemistry. A, Materials for Energy and Sustainability*, *1*(36), 10805–10813. doi:10.1039/c3ta11966a

Wei, Y., Wang, J., Jia, X., Yeh, J. M., & Spellane, P. (1995). Polyaniline as corrosion protection coatings on cold rolled steel. *Polymer*, *36*(23), 4535–4537. doi:10.1016/0032-3861(95)96866-7

Weng, C. J., Chang, C. H., Peng, C. W., Chen, S. W., Yeh, J. M., Hsu, C. L., & Wei, Y. (2011). Advanced anticorrosive coatings prepared from the mimicked xanthosoma sagittifolium-leaf-like electroactive epoxy with synergistic effects of superhydrophobicity and redox catalytic capability. *Chemistry of Materials*, *23*(8), 2075–2083. doi:10.1021/cm1030377

Weng, C. J., Huang, J. Y., Huang, K. Y., Jhuo, Y. S., Tsai, M. H., & Yeh, J. M. (2010). Advanced anticorrosive coatings prepared from electroactive polyimide-TiO_2 hybrid nanocomposite materials. *Electrochimica Acta*, *55*(28), 8430–8438. doi:10.1016/j.electacta.2010.07.063

Wessling, B. (1994). Passivation of metals by coating with polyaniline: Corrosion potential shift and morphological changes. *Advanced Materials*, *6*(3), 226–228. doi:10.1002/adma.19940060309

Wong, M., Ishige, R., White, K. L., Li, P., Kim, D., Krishnamoorti, R., & Nishimura, R. (2014). Large-scale self-assembled zirconium phosphate smectic layers via a simple spray-coating process. *Nature Communications*, *5*(1), 3589. doi:10.1038/ncomms4589 PMID:24709971

World Steel Association. (2017). October 2017 crude steel production. Retrieved from https://www.worldsteel.org/media-centre/press-releases/2017/october-2017-crude-steel-production.html

Xing, C., Zhang, Z., Yu, L., Zhang, L., & Bowmaker, G. A. (2014). Electrochemical corrosion behavior of carbon steel coated by polyaniline copolymers micro/nanostructures. *RSC Advances*, *4*(62), 32718–32725. doi:10.1039/C4RA05826G

Xu, X., Yi, D., Wang, Z., Yu, J., Zhang, Z., Qiao, R., & Liu, Z. (2018). Greatly enhanced anticorrosion of Cu by commensurate graphene coating. *Advanced Materials*, *30*(6), 1702944. doi:10.1002/adma.201702944 PMID:29266426

Yakobson, B. I., & Avouris, P. (2001). Mechanical properties of carbon nanotubes. In *Carbon Nanotubes* (pp. 287–327). Springer. doi:10.1007/3-540-39947-X_12

Yao, B., Wang, G., Ye, J., & Li, X. (2008). Corrosion inhibition of carbon steel by polyaniline nanofibers. *Materials Letters*, *62*(12-13), 1775–1778. doi:10.1016/j.matlet.2007.10.001

Yeganeh, M., Saremi, M., & Rezaeyan, H. (2014). Corrosion inhibition of steel using mesoporous silica nanocontainers incorporated in the polypyrrole. *Progress in Organic Coatings*, *77*(9), 1428–1435. doi:10.1016/j.porgcoat.2014.05.007

Yu, Y. H., Lin, Y. Y., Lin, C. H., Chan, C. C., & Huang, Y. C. (2014). High-performance polystyrene/graphene-based nanocomposites with excellent anti-corrosion properties. *Polymer Chemistry*, *5*(2), 535–550. doi:10.1039/C3PY00825H

Yu, Y. H., Lin, Y. Y., Lin, C. H., Chan, C. C., & Huang, Y. C. (2014). High-performance polystyrene/graphene-based nanocomposites with excellent anti-corrosion properties. *Polymer Chemistry*, *5*(2), 535–550. doi:10.1039/C3PY00825H

Yu, Z., Di, H., Ma, Y., He, Y., Liang, L., Lv, L., & Luo, Z. (2015). Preparation of graphene oxide modified by titanium dioxide to enhance the anti-corrosion performance of epoxy coatings. *Surface and Coatings Technology*, *276*, 471–478. doi:10.1016/j.surfcoat.2015.06.027

Yu, Z., Di, H., Ma, Y., Lv, L., Pan, Y., Zhang, C., & He, Y. (2015). Fabrication of graphene oxide–alumina hybrids to reinforce the anti-corrosion performance of composite epoxy coatings. *Applied Surface Science*, *351*, 986–996. doi:10.1016/j.apsusc.2015.06.026

Yun, H., Li, J., Chen, H. B., & Lin, C. J. (2007). A study on the N-, S-and Cl-modified nano-TiO2 coatings for corrosion protection of stainless steel. *Electrochimica Acta, 52*(24), 6679–6685. doi:10.1016/j.electacta.2007.04.078

Zhang, H., Wang, J., Liu, X., Wang, Z., & Wang, S. (2013). High performance self-healing epoxy/polyamide protective coating containing epoxy microcapsules and polyaniline nanofibers for mild carbon steel. *Industrial & Engineering Chemistry Research, 52*(30), 10172–10180. doi:10.1021/ie400666a

Zhang, X., Wang, F., & Du, Y. (2007). Effect of nano-sized titanium powder addition on corrosion performance of epoxy coatings. *Surface and Coatings Technology, 201*(16-17), 7241–7245. doi:10.1016/j.surfcoat.2007.01.042

Zheng, H., Shao, Y., Wang, Y., Meng, G., & Liu, B. (2017). Reinforcing the corrosion protection property of epoxy coating by using graphene oxide–poly (urea–formaldehyde) composites. *Corrosion Science, 123*, 267–277. doi:10.1016/j.corsci.2017.04.019

Zunino, J., III. (2005). US Army Development of Active Smart Coatings System for Military Vehicles. NSTI.

Zunino, J. L., III. (2007). Active Coatings Technologies for Customized Military Coating Systems. In *Cleantech 2007: the Cleantech Conference, Venture Forum and Trade Show*, Santa Clara, CA, May 23-24 (p. 234). CRC.

Zunino, J. L. III. (2007). Development of Materials and Sensor for the U.S. Armys Active coatings Technology Program. In *NSTI* (p. 4). Nanotech.

Zunino, J. L. III. (2008). Development of Active Systems for Military Utilization. In *NSTI* (p. 3). Nanotech.

Section 3

Defense Applications of Polymer Nanocomposites

Chapter 10
Scope of Polymer/ Graphene Nanocomposite in Defense Relevance:
Defense Application of Polymer/Graphene

Ayesha Kausar
National University of Science and Technology, Pakistan

ABSTRACT

This chapter outlines important aspects and progression from graphene to polymer/graphene nanocomposite to a relevant defense application. Graphene is unique nanocarbon material having a large surface area, high Young's modulus, thermal conductivity, electrical conductivity, and optical transmittance. Engineering thermoplastic polymers have been employed as matrices for graphene reinforcement. Various routes have been employed for graphene-filled polymeric nanomaterials. Intrinsic physical properties of nanocomposite depend on graphene modification and dispersion techniques. Polymer/graphene nanocomposite may have multifunctional characteristics due to synergistic effect of polymer/graphene. The article mainly discusses nanocomposite with potential uses in soldierly applications including flame resistance, ballistic protection, electromagnetic interference shielding, electrostatic-charge dissipation, sensors, corrosion protection, fuel cell, batteries, etc. The gestalt of defense applications of polymer/graphene nanocomposite may offer future perspective to develop promising materials.

INTRODUCTION

Several potential opportunities have been promised by nanotechnology for defence technologies. In this regard, various advantages have been explored for defence applications. Range of military platforms have been benefited from nanotechnoloy revolution (Kurahatti et al., 2010). In defense technology, power systems, sensing, stealth technology, aviation, marine, aerodynamics, etc., have frequently employed nanomaterials (Domun et al., 2015; Lyu & Choi, 2015; Nambiar & Yeow, 2012). Polymers alone have also been applied in defense applications (Kausar & Siddiq, 2016). However, more success has been

DOI: 10.4018/978-1-5225-7838-3.ch010

achieved with nanomaterials. Design and fabrication of new nanomaterials having high performance multifunctional characteristics may result in amplified recital of defense sector (Singh & Wang, 2015). In polymer nanocomposite, small size of nanofillers may lead to large interfacial area between the matrix and additives. The interfacial interactions may control the dispersion of nanofiller in polymer, and thus control the final properties. Moreover, to implement the properties of nanocomposites, processing methods are important to control the particle size distribution and dispersion. Graphene is a nano allotrope of carbon with single layer of carbon atoms arranged in hexagonal lattice (Ferrari et al., 2006; Zhou et al., 2018). One of the important uses of graphene has been identified for polymer nanocomposite (Kausar, 2016a). Traditional nanofillers such as clay, fibers, and particles have been frequently replaced by this unique nanofiller (Kausar, 2014, 2016b). Several advantages of polymer/graphene nanocomposite have been explored for defense applications (V.C. Verma). Need of this technology in military relevance has been realized over a few decades. This chapter basically presents potential use of polymer/graphene nanocomposite in defense along with significant application areas. Moreover, challenges and opportunities in fabrication and use of polymer/graphene-based engineering materials for defense purposes have been considered. There is hardly any known review article available focusing the essential aspects of polymer/graphene nanocomposite for military applications. There is extensive need of a comprehensive chapter covering all possible aspects and applications of polymer/graphene nanocomposite in military relevance. The chapter presents not only the background related to graphene, polymer/graphene nanocomposite, fabrication, essential properties, and dispersion but also covers the applications ranging from ballistic and fire safety to biomedical applications.

BACKGROUND

Graphene: A Unique Nanocarbon

Graphene is an important type of nanomaterial with atomic thick hexagonally designed sheets of sp^2 hybridized carbon atoms (Mukhopadhyay & Gupta, 2012). It is a one-dimensional nanofiller having large surface area of 2630 m^2g^{-1} (Fig. 1). Its structure has been studied through several theoretical and experimental techniques. Graphene also possesses remarkably high Young's modulus, strength, and thermal conductivity. Graphene has been exploited as an imperative material for solar cell, sensor, and other energy applications (Georgakilas et al., 2012). One of the significant applications of graphene is nanocomposite (Verdejo, Bernal, Romasanta, & Lopez-Manchado, 2011). Polymer nanocomposite may have excellent electrical, mechanical, and structural properties owing to graphene nanofiller. In this regard, various approaches have been developed for graphene synthesis and its incorporation in polymer matrices (Kim, Abdala, & Macosko, 2010). Graphene has been declared as a preferred nanofiller for polymers compared with other nanocarbon materials.

Polymer/Graphene Nanocomposite

Over several previous decades, polymer/graphene nanocomposite has been an exciting zone of research. Owing to range of functional characteristics of graphene, efficient polymeric nanocomposites have been developed (Fig. 2). Polymer/graphene nanosheet have been fabricated using water/ethanol solvent method. The nanocomposite possesses low percolation threshold of 0.070 vol.% due to the formation of conductive

Figure 1. Nanocarbon structure with different dimensionality

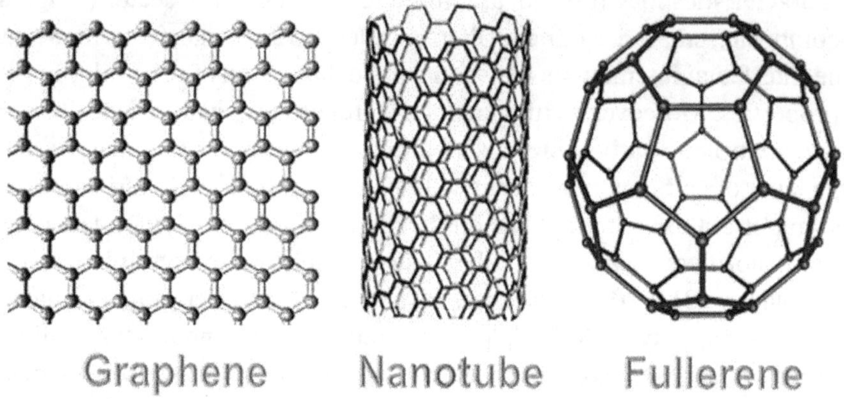

network (Pang, Chen, Zhang, Zeng, & Li, 2010). Similarly, melt technique has also been employed to develop poly(lactic acid)/graphene nanocomposite (Bao, Song, et al., 2012). The nanocomposites have shown fine dispersion and low percolation threshold at 0.08 wt.%. Poly(3,4-ethylenedioxythiophene)/graphene was prepared using *in situ* method (Guo, Wang, Chen, Lou, & Li, 2011). The conjugated polymer/graphene materials were designed for lithium ion battery electrode (Song et al., 2012). Poly(methyl methacrylate) (PMMA) has also been effectively used for graphene-based materials (Pramoda, Hussain, Koh, Tan, & He, 2010). 0.5 wt.% graphene nanofiller enhanced storage modulus from 1.29 to 2 GPa and glass transition from 119 to 131°C. Among all techniques, solution mixing was found as the most efficient and facile method for nanocomposite synthesis (Carpio, Santos, Wei, & Rodrigues, 2012; Ma et al., 2013; Pang, Zhang, Chen, Zeng, & Li, 2010). However, the worth of melt blending has also been analyzed (Istrate et al., 2014). Inclusion of 0.07 wt.% graphene in melt nanocomposite enhanced the elastic modulus by 10% and tensile strength by 40%. Polyethylene, polystyrene and poly(methyl-methacrylate) formed well-ordered layered structure with graphene to enhance the intrinsic nanocomposite properties (Rissanou & Harmandaris, 2014). Thus, graphene acted as a remarkable nanomaterial, which has opened new era of applications for polymer/graphene high performance materials.

PRODUCTION OF POLYMER/GRAPHENE NANOCOMPOSITE

There are three main approaches for the incorporation of graphene in polymeric matrices i.e. *in situ* polymerization, solution mixing, and melt intercalation (Table 1). Selection of desired technique depends on several factors such as interaction between polymer/graphene, hydrophobicity, polarity, reactivity, etc.

In-situ Polymerization

In this polymerization method, monomer and graphene nanofiller are amalgamated in the presence of catalyst. *In-situ* polymerization may need heat or radiation initiation (Zheng, Lu, & Wong, 2004). Common examples for polymer/graphene nanocomposite processed include polystyrene/graphene, polyaniline/graphene, silicon/graphene nanocomposite and many other materials. This technique may offer strong interaction between the matrix and nanofiller leading to higher load transfer and mechanical properties. This technique also enables fine miscibility and homogeneous dispersion of graphene in polymer matrix. However, this method fails for high nanofiller loading in polymer matrices. The nanofiller addition and *in situ* polymerization may drastically increase the viscosity of the system, and so processing becomes difficult. In some cases, solvent removal from the nanocomposite is difficult and needs additional steps for solvent removal.

Solution Mixing

Solution mixing is also a frequently used technique for polymer/graphene nanocomposite. This method has great potential for large scale production of polymer/graphene nanocomposite owing to simplicity (Ramanathan et al., 2007). The polymer is solubilized in a suitable solvent and graphene is dispersed in it. Majority of thermoplastic polymers have been processed using this technique such as poly(methyl methacrylate), polycaprolactone, polystyrene, poly(vinyl alcohol), polyurethane, polyvinyl chloride, etc. In solution mixing, solubility, and dispersion of graphene both are challenging. Functional graphene nanosheets have been used to resolve this problem. However, re-aggregation of nanosheets may occur during solvent evaporation.

Melt Intercalation Technique

In melt intercalation method, polymer and graphene are mixed in molten state (Das & Prusty, 2013). Shear mixing is performed at high temperature. Homogeneous mixing of polymer and graphene is difficult to achieve even at high temperature. Some examples include formation of polyethylene terephthalate/graphene, polystyrene/graphene, and polypropylene/graphene nanocomposite using melt method. Use of high shear force and temperature may cause defects in graphene nanosheets. The main drawbacks associated are dispersion and distribution of nanofiller in polymer matrices.

Table 1. Polymer/graphene nanocomposite

Polymer	Nanofiller	Processing Technique	Reference
Polyethylene	Graphene	Solution mixing	(Pang, Chen, et al., 2010)
Poly(lactic acid)	Graphene	Melt blending	(Bao, Song, et al., 2012)
Poly(3,4-ethylenedioxythiophene)	Graphene	*In situ*	(Guo et al., 2011)
Poly(anthraquinonyl sulfide) and polyimide	Graphene	Solution mixing	(Song et al., 2012)
Poly(methyl methacrylate)	Graphene	*In situ*	(Pramoda et al., 2010)
Polyvinyl-*N*-carbazole	Graphene	Solution mixing	(Carpio et al., 2012)
Epoxy	Graphene	Solution mixing	(Ma et al., 2013)
Ultrahigh molecular weight polyethylene	Graphene nanosheet	Solution mixing	(Pang, Zhang, et al., 2010)
Polyethylene terephthalate	Exfoliated graphene	Melt blending	(Istrate et al., 2014)
Polyethylene, polystyrene and poly(methyl-methacrylate)	Graphene	Solution mixing	(Rissanou & Harmandaris, 2014)

Figure 2. Important properties of polymer/graphene nanocomposite

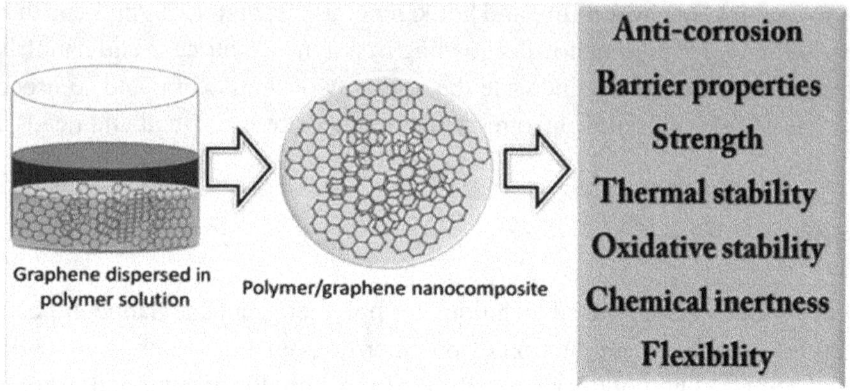

PROPERTIES OF POLYMER/GRAPHENE NANOCOMPOSITE

Important properties of polymer/graphene nanocomposite include high electrical conductivity, capacitance, thermal conductivity, non-flammability, strength, electromagnetic interference, and biomedical features. Poly(vinyl alcohol)/graphene nanocomposites possess improved electrochemical performance (Abdulhakeem et al., 2014). The specific capacitance of 0.62-1.92 fcm^{-2} was observed. Poly(vinyl alcohol)/graphene nanocomposites also revealed maximum energy density of 42 mwh cm^{-2} and power density of 0.5 w cm^{-2} (Bello et al., 2015). Polydimethylsiloxane/graphene nanocomposite possess high thermal conductivity of 0.39 $wm^{-1}k^{-1}$ (Jeong et al., 2015). Polydimethylsiloxane/graphene nanocomposite also own high stretchability and durability. Polyaniline and graphene foam nanocomposite have been synthesized using *in situ* polymerization. The polyaniline/graphene nanocomposite has been reported for stability, high electrical conductivity, pseudocapacitance, and energy densities (Z.-F. Li et al., 2013). Specific capacitance of the nanocomposite was found to be 790 Fg^{-1}. The electromagnetic parameters of polyaniline/graphene nanocomposite have also been studied. The maximum absorption was found as -52.5 dB at 13.8 GHz. Reinforcement of graphene in polystyrene may significantly enhance the tensile strength and young's modulus of the resulting nanocomposite (Fang, Wang, Lu, Yang, & Nutt, 2009). Superior properties of polymer/graphene nanocomposite can be attributed to strong interfacial interaction between the matrix and graphene nanofiller. Here, stacking or aggregation of graphene nanosheet may also affect the intrinsic properties of nanocomposite such as thermal, mechanical, and electrical behavior (Xu, Cao, & Heath, 2009). Different types of morphological states of graphene-based nanocomposites have been mentioned in Fig. 3.

DISPERSION AND SURFACE TREATMENT OF GRAPHENE IN MATRIX

Dispersion of graphene nanofiller in polymers is challenging. During processing, graphene nanosheet may restack to form aggregates. Random stacking of graphene nanosheets may affect the nanocomposite properties and performance. Modified graphene nanofiller has been used to enhance dispersion in the matrix. In this way, homogeneous nanofiller dispersion has been achieved. Surface treatment of graphene has been performed to modify the surface physically or chemically. Treated graphene may form physical or chemical links with the matrix. Thus, interaction in the polymer/graphene system may enhance the nanocomposite dispersion and performance. Interaction between polymer and graphene may be electrostatic, π-π interaction, hydrogen bonding, or covalent linkage. Enhanced connection between the polymer and treated graphene nanosheet can improve the nanofiller dispersion significantly. In this way, polymer/graphene nanocomposite may have high performance to be focused for various technical applications as discussed in the succeeding sections.

MAIN FOCUS

Technical Significance of Polymer/Graphene Nanocomposite

Ballistic Safety

Initial efforts on ballistic protection involve the development of aromatic polyamide (kevlar, nomex)-based body armour and automobile facings (Lee, Wetzel, & Wagner, 2003). Later developments involved the incorporation of nanomaterials such as carbon nanotube and nanofiber in armour materials (Fig. 4). Superior mechanical properties of carbon nanotube was employed for developing armour materials (Micheli et al., 2016). Similarly, graphene-based polymer nanocomposites have been used for defense materials (Y. Wang, Shi, & Yin, 2011; Xiang et al., 2011). However, conventionally, Kevlar-based body armour was prepared using colloidal dispersion of spherical silica nanoparticle (450 nm) in polyethylene glycol (Thilagavathi, Raja, & Kannaian, 2008). The ballistic performance (absorbed energy) of nanocomposite material was much improved compared with the neat Kevlar. Kaolin clay nanoplatelets (500 nm) was also dispersed in armour material (Rosen, Laufer, Kalman, Wetzel, & Wagner, 2007).

Figure 3. Schematic of morphological states of polymer/graphene nanocomposites

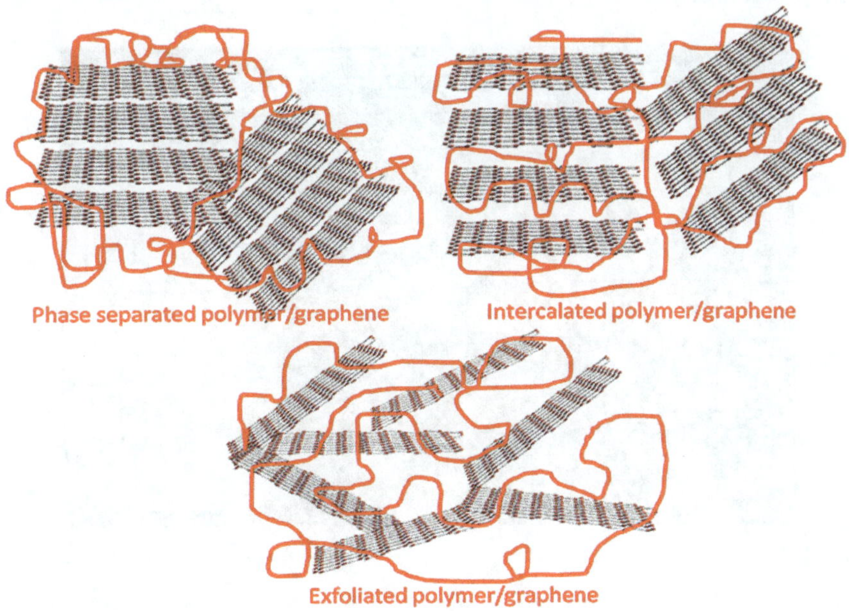

Nevertheless, graphene-based material offered much flexible armour with improved ballistic protection, mechanical strength, and low weight. Graphene reinforcement may provide overall improvement in tensile strength, toughness, thermal stability, gas-barrier properties, and flame retardancy of the nanocomposite (Kashiwagi et al., 2004). The melting/dripping behavior of the graphene-based nanomaterial has been delayed owing to reduced rate of burning. Thus, the use of graphene-based ballistics may enhance the safety of soldiers during impulsive outbreaks. However, commercial polymer/graphene nanomaterials with reduced melting/dripping behavior need further research attempt. Similar to ballistic materials, fire safety of polymer/graphene nanocomposite have been studied.

Fire Retardancy

Today's and future battlefield demand enhanced protection of soldiers from exposure to hazardous fire, chemicals, and other warfare agents (Thilagavathi et al., 2008). Use of graphene-based materials in protective overgarment may provide non-flammable and impermeable barrier materials (Table 2). These nanomaterials also provide protection by acting as physical barriers to chemicals and war agents. The polymer/graphene laminate fabrics have been used in gloves, boots, and clothing purposes. Polymer/graphene nanocomposite have also been coated on nylon fabric to improve the non-flammability and gas barrier property. The coating method also has wide military application (Bao, Guo, Yuan, Hu, & Song, 2012). Fire safety of polystyrene/graphene nanocomposite has been improved by ~60% decrease in peak release rate. In polypropylene nanocomposite, graphene acted as intumescent flame-retardant (Dittrich, Wartig, Hofmann, Mülhaupt, & Schartel, 2013; G. Huang et al., 2014). Graphene reinforcement in poly(vinyl alcohol) affected the flame retardant properties of materials (G. Huang, H. Liang, et al., 2012). Functionalized graphene has also been used as an intumescent flame retardant in epoxy nanocomposite (G. Huang, Chen, Tang, & Gao, 2012). Thus, lightweight and cost effective flame retardant protective overgarment and shelter systems have been developed using polymer/graphene nanocomposite.

Figure 4. Ballistic protection

Similar to ballistic garments demanding heat/pressure protection, non-flammability of materials have been enhanced using polymer/graphene materials. However, ballistic performance of garments demand additional mechanical strength and robustness. Future research is demanded in both the categories.

Electromagnetic Interference Shielding

Electromagnetic interference (EMI) shielding is an important and demanding area of defense. It involves the use of conductive or magnetic materials to reduce electromagnetic field in space. Initially, metal-based sheets and foams have been used for electromagnetic shielding. Conducting polymers have also been frequently employed in electromagnetic interference shielding materials (Kausar & Hussain, 2013). Applications of EMI materials definitely extend from aircraft to electronics industry. Recently, electrical conductivity of these polymers has been enhanced using nanofillers to match the modern device performance (Kausar, 2018). Graphene-based nanofiller has been known to enhance the EMI shielding efficiency of nanocomposite through lowering the percolation values. Epoxy/graphene nanocomposite have shown the EMI efficiency of 20 dB in X-band with 10 wt.% loading and low percolation threshold (Wu et al., 2017). The conducting polymer/graphene nanocomposite offered lightweight materials having fine processability and shielding efficiency (Kausar, Anwar, & Muhammad, 2016). Better graphene dispersion (8 vol.%) and high conductivity of nanocomposite led to enhanced shielding efficiency up to 21 dB at 8.2 GHz (Liang et al., 2009). Functionalized epoxy/graphene had percolation threshold of 0.52 vol.% and EMI shielding effectiveness of 21 dB at 8.2-12.4 GHz (X-band) (X. Huang, Qi, Boey, & Zhang, 2012). Thus, research attraction has been drawn towards graphene carbon nanofiller in EMI shielding nanocomposite (Kuilla et al., 2010). Main aim of the research on EMI shielding must focus the attainment of the effectiveness of above 20 dB for commercial applications. For military relevance, EMI shielding is as important as ballistic protection.

Electrostatic Charge Dissipation

Dissipation of charge on devices and air and space vehicles definitely cause severe defense problems (Baur & Silverman, 2007). The nanomaterials having sufficient electrical conductivity, strength, and environmental stability are demanded in this field. Initially, carbon nanotube and nanofiber have been

Table 2. Non-flammability of polymer/graphene

Nanocomposite	Nanofiller Content (wt.%)	Non-Flammability/Peat Heat Release Rate (PHRR)/kWm^{-2}	Reference
Polystyrene/graphene	0.5	1127	(Bao, Guo, et al., 2012)
Polypropylene/graphene	1	538	(Huang et al., 2014)
Polypropylene/exfoliated graphene	5	1104	(Dittrich et al., 2013)
Poly(vinyl alcohol)/ melamine polyphosphate-graphene	10	144	(Huang, Liang, et al., 2012)
Poly(vinyl alcohol)/graphene	2	209	(Huang, Chen, et al., 2012)
Epoxy/graphene	5	392	(Huang, Chen, et al., 2012)

used in electrostatic charge dissipation (ESD) materials (Kumar, Lively, Sun, Li, & Zhong, 2010). Graphene is also a demanding material in anti-static nanocomposite coatings (Chen, Li, Shao, Du, & Burnell-Gray, 2007; Tong, Bohm, & Song, 2013). Nanocomposite of polyurethane and graphene have been prepared with 0-5.5 vol. % (M. Verma, Verma, Dhawan, & Choudhary, 2015). The nanomaterials have tunable electrical conductivities relying on nanofiller content and processing parameters. The ESD of polyurethane/graphene nanocomposite was studied by decay of voltage as a function of time. The 1.6 vol. % graphene loading showed static decay time of 0.49 sec. Consequently, successful attempts on polyurethane- and epoxy/graphene nanocomposite may lead to useful light weight antistatic coatings for defense applications. Similar to EMI shielding, epoxy/graphene materials have found successful in electrostatic charge dissipation application. However, polymer/graphene is less researched in ESD compared with the EMI sielding materials.

Sensors

Gas sensors have recently employed graphene-based materials. Graphene-based materials may adsorb gas molecules on surface. The graphene-based electronic sensors work on the mechanism of alteration of graphene's electrical conductivity (σ) upon adsorption of molecules. Thus, graphene-based devices own enhanced sensitivity and molecular level detection. Reports have been found on graphene sensing (Schedin et al., 2007). Graphene established good sensing properties for H_2O, NO_2, NH_3 and CO molecules. The graphene sensing properties were regained at 150 °C. Detection of specific gases in mixed gas environment is known as selectivity, which may be enhanced using functional graphene surfaces or nanocomposite designs (Robinson, Perkins, Snow, Wei, & Sheehan, 2008). Thus, gas-based sensors have high specific surface area and surface-to-bulk ratio leading to high sensitivity (Hai-Juan, Shuang, Lian-Zhe, & Guo-Bao, 2009). Sensor selectivity depends on attractive gas adsorption features. In addition to functional graphene, molecularly imprinted polymers have specific molecular recognition capacity (Nicholls et al., 2009). Graphene reinforced molecularly imprinted polymers possess further high sensitivity, high selectivity, and low cost (Zou, Xiang, Sun, & Xu, 2008). Thus, polymer/graphene nanocomposite have been used for sensing, biosensing, and nanoelectronics. Polymer/graphene nanocomposite in sensors forms a unique combination. Mechanism of polymer/graphene nanocomposite functioning in sensors is interesting. Graphene own high optical properties, electrical conductivity, carrier mobility, large surface/volume ratio, and thermal conductivity, valuable for sensor functions. In sensors, large surface area of graphene offers enhanced surface loading of desired gas molecules or biomolecules. The graphene surface also provides small band gap and excellent conductivity favorable for conducting electrons between the electrode surface and molecules.

Corrosion Protection

Corrosion protection of metals and alloys is a hot area of defense research. Polymers coatings have been used to resist chemical attack and mechanical damaged. Polymer nanocomposite have been employed to attain corrosion potential (Savage, 2004; N. B. Singh & Agarwal, 2016). Polymer/graphene nanocomposite possess enhanced abrasion, scratch, and corrosion resistance. Polymer/graphene coating may act as diffusion barrier to corroding molecules, chemicals, water, and other solvents (Fig. 5). In aerospace and electronic devices, corrosion protection >150 °C is greatly appreciated in defense industry. In this regard, epoxy/graphene nanocomposite was developed (Ramezanzadeh, Niroumandrad, Ahmadi, Mahdavian, & Moghadam, 2016). Amine-functional graphene was prepared and dispersed in epoxy through wet transfer method. The nanocomposite was tested on mild steel for barrier and corrosion protection efficiency. Incorporation of 0.1 wt.% nanofiller enhanced the corrosion resistance. Polyaniline/graphene nanocomposites have found to increase the corrosion protection of steel (Chang et al., 2012). Increase in graphene content resulted in improved barrier and anti-corrosion properties. Hence, polymer/graphene has found success in corrosion protection. Anti-corrosion performance of polyaniline/graphene and epoxy/graphene coatings need further investigations before commercialization. New approaches need to be adopted for improving erosion resistance and mechanical-thermal stability of coatings.

Fuel Cell

Polymer electrolyte membrane fuel cells (PEMFC) are found promising materials in energy conversion devices. Polymer-based PEMFC has been engrossed due to rapid start-up time and high efficiency (M. Li & Scott, 2010). Primarily, Nafion (perfluorinated polymer) membranes have been used for high electrical conductivity and chemical resistance. Later, poly(perfluorosulfonic acid) and acid doped membranes have been developed. Design of new membranes having high proton conductivity and low cost are still under research (Nagarale, Shin, & Singh, 2010). The membrane had high strength, flexibility, and ion-transport behavior. Poly(ethylene oxide)/graphene nanocomposites have been used for strengthened and low-temperature fuel cell membrane (Nagarale et al., 2010). The power density of fuel cell was enhanced to >50 mWcm^{-2} with a RH of 100% (Fig. 6). Sulfonated polyimide/graphene-based proton exchange membranes possess thickness of 80 μm. The ionic conductivity of the membrane was 0.1 S cm^{-1} at 100% relative humidity. The limiting current was found as 180 mA. Thus, graphene-based nanomaterials have been effectively used in the polymeric fuel cell membranes. However, very few membranes have been studied so far. Similar to Li-ion battery separators, few polymer electrolyte membranes are known for fuel cells.

Li-ion Battery

Polymer/graphene nanocomposites have been employed in portable electronics such as lithium-ion batteries. Polymer/graphene nanocomposites have been used to construct battery electrodes (cathode and anode), which affect the capacitance, energy density, and overall performance. In usual Li-ion batteries, lithium compounds such as lithium cobalt oxide ($LiCoO_2$) have been used as cathode (Fig. 7). Investiga-

Figure 5. Corrosion inhibition by polymer/graphene layer

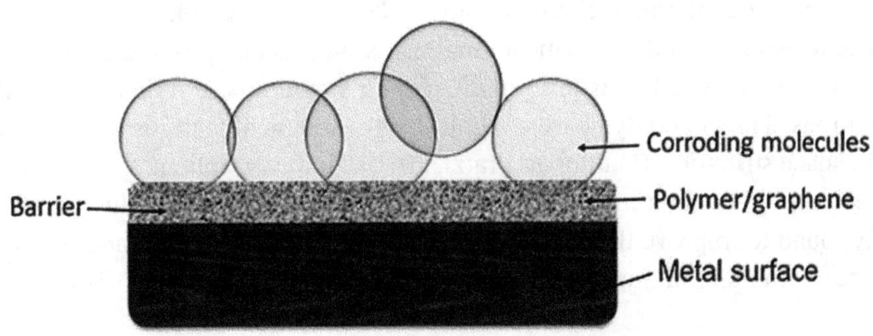

Figure 6. Polymer/graphene in fuel cell

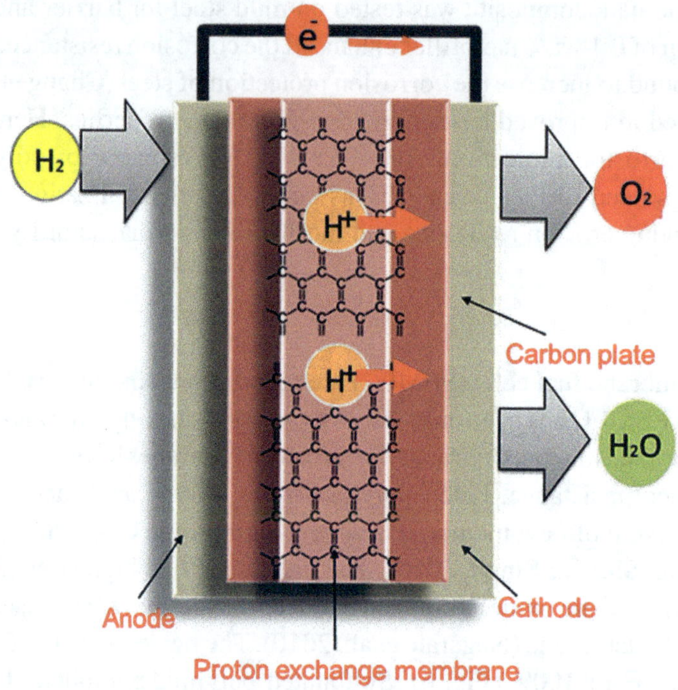

tion on new anode materials have been focused for improved lithium storage. Mechanism involves the enhanced ion transport between the battery electrodes. Anode material is constructed using graphene materials (Whittingham, 2004). More suitable materials are based on graphene reinforced in conducting materials (G. Wang, Shen, Yao, & Park, 2009). The nanosheet enhanced lithium storage capacity as anodes in Li-ion cell. Moreover, polymer/graphene has been used as a separator in Li-ion batteries. Such separators possess high thermal resistance, high dimensional stability, and safety for batteries (H. Liao et al., 2017). Thus, polymer/graphene nanocomposites have been found promising as electrodes and separator membranes in high-power Li-ion batteries. However, not much research is found on polymer/graphene materials for battery electrodes and separators. Further research is needed on new design, electrochemical performance, and rate capability of these nanocomposites.

Aerospace

Owing to unique mechanical, thermal, chemical/physical properties, graphene is an efficient nanostructured nanomaterial to be employed in thermosetting resins such as epoxies and phenolics. The polymer/graphene nanocomposite own high tensile strength, Young's modulus, and thermal stability suitable for aerospace applications. They not only possess significant mechanical features but also own light weight to be employed in space structures (Dunlop & Fratzl, 2010). Epoxy/graphene systems own fine potential to fabricate strong and lightweight materials having enhanced properties. Inclusion of 1-10 vol.% graphene has greatly found to improve the Young's modulus, strength, hardness, and thermal properties of the nanocomposite. Both melt mixing and solution methods have been used to disperse the nanofiller in

Figure 7. Polymer/graphene in Li-ion battery

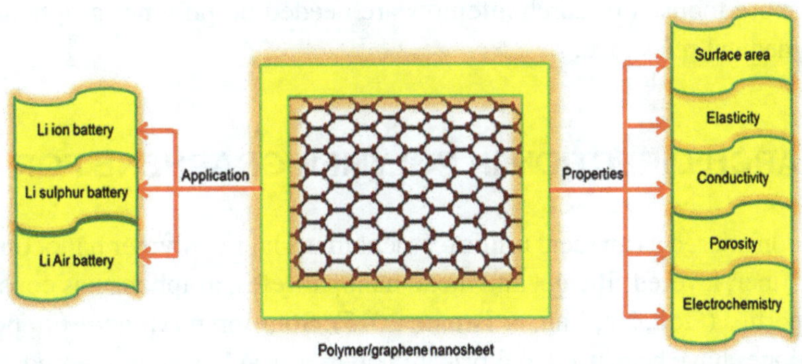

epoxy matrix. Microscopic examinations have been used to see the agglomerates of graphene or dispersion in the matrix (Y.-H. Liao, Marietta-Tondin, Liang, Zhang, & Wang, 2004). Consequently, graphene enhanced materials have been employed to replace the metal components in aerospace. Inclusion of graphene in thermosets has reduced the cure times and enhanced the thermal properties. Robustness and self-healing properties of graphene have also been used to form aerospace coatings. Common challenges in aerospace industry include weight reduction and increased aerodynamic efficiency of spacecraft components. Conventional epoxy/carbon fiber composite have been practices in this industry, however there is still need of polymer/graphene nanocomposite for enhanced space performance. Better compatibility between epoxy matrix and graphene and interfacial adhesion point to better aerospace performance of polymer/graphene compared with the conventional polymer/carbon fiber composite.

Biomedical

Besides other applications, polymer/graphene nanocomposites have also been employed in biomedical fields. Graphene-based nanomaterials have become emerging materials having superior optical, electrical, mechanical, thermal, and chemical performance. Tissue engineering, n wound healing, drug delivery, cancer therapy, imaging, etc. have found scope for polymer/graphene nanocomposite (Shen, Zhang, Liu, & Zhang, 2012). Biomedical field has employed range of polymers with graphene such as poly(methyl methacrylate), poly(caprolactone), polyurethane, polyamide, polystyrene, nylon 6, nylon 6,6, polyimide, etc. (Pei, Travas-Sejdic, & Williams, 2012). The hydrogen bonding and interfacial adhesion between functional graphene and polymers may significantly improve the thermal stability and mechanical properties of matrices for bone tissue engineering (Wu, Ramezani, & Hakkarainen, 2018). The tensile strength and Young's modulus of nanocomposite films were enhanced up to 113 MPa and 4.18 GPa, respectively (Silva, Alves, & Paiva, 2018). Molecular imaging use molecular imaging agents to measure biomarkers at different stages of diseases. Polymer/graphene nanocomposite have been used in several molecular imaging techniques such as magnetic resonance imaging and positron emission tomography. In this regard, PEGylated graphene may exhibit photoluminescence in the visible and infrared range (Zhang, Nayak, Hong, & Cai, 2012). Moreover, sp^2 hybridized carbon area of graphene led to promising drug and gene delivery applications. Owing to π-π stacking interaction between the aromatic rings of polymers and the graphene form efficient drugs delivery tools. For example, polymer/graphene nanocarriers have been used for sequential delivery of siRNA doxorubicin (DOX) for enhanced anticancer efficacy. However, more focused research attempts are needed on polymer/graphene nanomaterials to be employed in biomedical relevance.

FUTURE RESEARCH DIRECTION IN POLYMER/GRAPHENE FOR DEFENSE

Graphene has been considered as an ideal reinforcement material in polymer nanocomposite. Compared with nanotube, nanofiber, layered silicate, and metal nanoparticles, graphene has considerably enhanced nanocomposite properties (Goh, Ng, Lau, & Ismail, 2015). Solution mixing, *in situ* polymerization, and melt blending techniques have been used to fabricate polymer/graphene nanocomposite (Keledi, Hári, & Pukánszky, 2012). Use of suitable technique may considerably enhance mechanical, thermal, electrical, ant-corrosion, and barrier properties. However, there are several challenges to attain polymer/graphene nanocomposite with high degree of reinforcement using suitable fabrication technique. These materials have shown promising consequences for defense nanocomposite. Perfectly aligned nanostructure of graphene reinforced nanomaterial may lead to enhanced features (Gupta, Sakthivel, & Seal, 2015; Tang & Zhou, 2013). 100% corrosion prevention of polymer/graphene has not been attained yet. Similarly, ballistic and flame retardancy need to be enhanced using functional graphene nanofiller. EMI and ESD performance of nanocomposite need further efforts. Production of nanocomposite in large amount is also challenging. Further advances in fabrication strategies may boost future defense application through efficient graphene dispersion. Although there are few polymer/graphene-based materials on the market, polymer/graphene commercialization still need efforts. Production of polymer/graphene has been increased significantly from 2000. Several steps have been taken to warrant the commercialization of polymer/graphene. Economical methods are still needed for the large-scale production of these nanocomposite. A ten-year EU program (Graphene Flagship) of one billion $ has been launched in 2013 (Phiri, Gane, & Maloney, 2017). According to this program, various international research groups were brought together to commercialize their products. The polymer/graphene commercialization program will play an active role in graphene revolution as a material of future.

SUMMARY

This chapter offers various propositions of current state of polymer/graphene nanocomposite in defense applications. Graphene-based nanocomposite possess new future defense domains owing to electrical conductivity, non-flammability, thermal stability, and mechanical robustness. Functionalization of graphene nanosheet has been considered as an indispensable way to enhance the nanocomposite performance. The review mainly discusses polymer/graphene nanocomposite materials with latent applications in soldierly applications such as ballistic protection, flame resistance, corrosion protection, EMI shielding, ESD, sensors, etc. Still extensive endeavors are crucial for attaining new future technical demands in advance defense related research zones.

REFERENCES

Abdulhakeem, B., Farshad, B., Damilola, M., Fatemeh, T., Mopeli, F., Julien, D., & Ncholu, M. (2014). Morphological characterization and impedance spectroscopy study of porous 3D carbons based on graphene foam-PVA/phenol-formaldehyde resin composite as an electrode material for supercapacitors. *RSC Advances*, *4*(73), 39066–39072. doi:10.1039/C4RA05425C

Bao, C., Guo, Y., Yuan, B., Hu, Y., & Song, L. (2012). Functionalized graphene oxide for fire safety applications of polymers: A combination of condensed phase flame retardant strategies. *Journal of Materials Chemistry*, *22*(43), 23057–23063. doi:10.1039/c2jm35001g

Bao, C., Song, L., Xing, W., Yuan, B., Wilkie, C. A., Huang, J., ... Hu, Y. (2012). Preparation of graphene by pressurized oxidation and multiplex reduction and its polymer nanocomposites by masterbatch-based melt blending. *Journal of Materials Chemistry*, *22*(13), 6088–6096. doi:10.1039/c2jm16203b

Baur, J., & Silverman, E. (2007). Challenges and opportunities in multifunctional nanocomposite structures for aerospace applications. *MRS Bulletin*, *32*(4), 328–334. doi:10.1557/mrs2007.231

Bello, A., Barzegar, F., Momodu, D., Dangbegnon, J., Taghizadeh, F., Fabiane, M., & Manyala, N. (2015). Asymmetric supercapacitor based on nanostructured graphene foam/polyvinyl alcohol/formaldehyde and activated carbon electrodes. *Journal of Power Sources*, *273*, 305–311. doi:10.1016/j.jpowsour.2014.09.094

Carpio, I. E. M., Santos, C. M., Wei, X., & Rodrigues, D. F. (2012). Toxicity of a polymer–graphene oxide composite against bacterial planktonic cells, biofilms, and mammalian cells. *Nanoscale*, *4*(15), 4746–4756. doi:10.1039/c2nr30774j PMID:22751735

Chang, C.-H., Huang, T.-C., Peng, C.-W., Yeh, T.-C., Lu, H.-I., Hung, W.-I., ... Yeh, J.-M. (2012). Novel anticorrosion coatings prepared from polyaniline/graphene composites. *Carbon*, *50*(14), 5044–5051. doi:10.1016/j.carbon.2012.06.043

Chen, X., Li, C., Shao, W., Du, H., & Burnell-Gray, J. (2007). The anti-static poly (ethylene terephthalate) nanocomposite fiber by in situ polymerization: The thermo-mechanical and electrical properties. *Journal of Applied Polymer Science*, *105*(3), 1490–1495. doi:10.1002/app.26381

Das, T. K., & Prusty, S. (2013). Graphene-based polymer composites and their applications. *Polymer-Plastics Technology and Engineering*, *52*(4), 319–331. doi:10.1080/03602559.2012.751410

Dittrich, B., Wartig, K.-A., Hofmann, D., Mülhaupt, R., & Schartel, B. (2013). Flame retardancy through carbon nanomaterials: Carbon black, multiwall nanotubes, expanded graphite, multi-layer graphene and graphene in polypropylene. *Polymer Degradation & Stability*, *98*(8), 1495–1505. doi:10.1016/j.polymdegradstab.2013.04.009

Domun, N., Hadavinia, H., Zhang, T., Sainsbury, T., Liaghat, G., & Vahid, S. (2015). Improving the fracture toughness and the strength of epoxy using nanomaterials–a review of the current status. *Nanoscale*, *7*(23), 10294–10329. doi:10.1039/C5NR01354B PMID:26006766

Dunlop, J. W., & Fratzl, P. (2010). Biological composites. *Annual Review of Materials Research*, *40*(1), 1–24. doi:10.1146/annurev-matsci-070909-104421

Fang, M., Wang, K., Lu, H., Yang, Y., & Nutt, S. (2009). Covalent polymer functionalization of graphene nanosheets and mechanical properties of composites. *Journal of Materials Chemistry*, *19*(38), 7098–7105. doi:10.1039/b908220d

Ferrari, A. C., Meyer, J., Scardaci, V., Casiraghi, C., Lazzeri, M., Mauri, F., & ... Geim, A. K. (2006). Raman spectrum of graphene and graphene layers. *Physical Review Letters*, *97*(18), 187401. doi:10.1103/PhysRevLett.97.187401 PMID:17155573

Georgakilas, V., Otyepka, M., Bourlinos, A. B., Chandra, V., Kim, N., Kemp, K. C., ... Kim, K. S. (2012). Functionalization of graphene: Covalent and non-covalent approaches, derivatives and applications. *Chemical Reviews*, *112*(11), 6156–6214. doi:10.1021/cr3000412 PMID:23009634

Goh, P., Ng, B., Lau, W., & Ismail, A. (2015). Inorganic nanomaterials in polymeric ultrafiltration membranes for water treatment. *Separation and Purification Reviews*, *44*(3), 216–249. doi:10.1080/15422119.2014.926274

Guo, C. X., Wang, M., Chen, T., Lou, X. W., & Li, C. M. (2011). A Hierarchically Nanostructured Composite of MnO2/Conjugated Polymer/Graphene for High-Performance Lithium Ion Batteries. *Advanced Energy Materials*, *1*(5), 736–741. doi:10.1002/aenm.201100223

Gupta, A., Sakthivel, T., & Seal, S. (2015). Recent development in 2D materials beyond graphene. *Progress in Materials Science*, *73*, 44–126. doi:10.1016/j.pmatsci.2015.02.002

Hai-Juan, L., Shuang, H., Lian-Zhe, H., & Guo-Bao, X. (2009). Progress in Ru (bpy) 32+ Electrogenerated Chemiluminescence. *Chinese Journal of Analytical Chemistry*, *37*(11), 1557–1565. doi:10.1016/S1872-2040(08)60139-5

Huang, G., Chen, S., Tang, S., & Gao, J. (2012). A novel intumescent flame retardant-functionalized graphene: Nanocomposite synthesis, characterization, and flammability properties. *Materials Chemistry and Physics*, *135*(2-3), 938–947. doi:10.1016/j.matchemphys.2012.05.082

Huang, G., Liang, H., Wang, Y., Wang, X., Gao, J., & Fei, Z. (2012). Combination effect of melamine polyphosphate and graphene on flame retardant properties of poly (vinyl alcohol). *Materials Chemistry and Physics*, *132*(2-3), 520–528. doi:10.1016/j.matchemphys.2011.11.064

Huang, G., Wang, S., Song, P., Wu, C., Chen, S., & Wang, X. (2014). Combination effect of carbon nanotubes with graphene on intumescent flame-retardant polypropylene nanocomposites. *Composites. Part A, Applied Science and Manufacturing*, *59*, 18–25. doi:10.1016/j.compositesa.2013.12.010

Huang, X., Qi, X., Boey, F., & Zhang, H. (2012). Graphene-based composites. *Chemical Society Reviews*, *41*(2), 666–686. doi:10.1039/C1CS15078B PMID:21796314

Istrate, O. M., Paton, K. R., Khan, U., O'Neill, A., Bell, A. P., & Coleman, J. N. (2014). Reinforcement in melt-processed polymer–graphene composites at extremely low graphene loading level. *Carbon*, *78*, 243–249. doi:10.1016/j.carbon.2014.06.077

Jeong, Y. R., Park, H., Jin, S. W., Hong, S. Y., Lee, S. S., & Ha, J. S. (2015). Highly stretchable and sensitive strain sensors using fragmentized graphene foam. *Advanced Functional Materials*, *25*(27), 4228–4236. doi:10.1002/adfm.201501000

Kashiwagi, T., Grulke, E., Hilding, J., Groth, K., Harris, R., Butler, K., ... Douglas, J. (2004). Thermal and flammability properties of polypropylene/carbon nanotube nanocomposites. *Polymer*, *45*(12), 4227–4239. doi:10.1016/j.polymer.2004.03.088

Kausar, A. (2014). Fabrication and properties of polyamide and graphene oxide coated carbon fiber reinforced epoxy composites. *American Journal of Political Science*, *4*(3), 88–93.

Kausar, A. (2016a). Bucky papers of poly (methyl methacrylate-co-methacrylic acid)/polyamide 6 and graphene oxide-montmorillonite. *Journal of Dispersion Science and Technology*, *37*(1), 66–72. doi:10.1080/01932691.2015.1027908

Kausar, A. (2016b). Thermal Conductivity Measurement of Polyvinylpyrrolidone/ Polyethylene/ Graphene Nanocomposite. *Nanoscience and Nanotechnology*, *6*(2), 34–37.

Kausar, A. (2018). Shielding Efficacy of Polymeric Nano-Structure. *Research Journal of Nanoscience and Engineering*, *2*(1), 9–14.

Kausar, A., Anwar, Z., & Muhammad, B. (2016). Recent developments in epoxy/graphite, epoxy/graphene, and epoxy/graphene nanoplatelet composites: A comparative review. *Polymer-Plastics Technology and Engineering*, *55*(11), 1192–1210. doi:10.1080/03602559.2016.1163589

Kausar, A., & Hussain, S. T. (2013). Effect of multi-walled carbon nanotube reinforcement on the physical properties of poly (thiourea-azo-ether)-based nanocomposites. *Journal of Plastic Film & Sheeting*, *29*(4), 365–383. doi:10.1177/8756087913487003

Kausar, A., & Siddiq, M. (2016). Latent of Modified Epoxy Materials in Defence Applications. In A. Méndez-Vilas & A. Solano-Martín (Eds.), Polymer Science: Research Advances, Practical Applications and Educational Aspects (pp. 278-288). Formatex Research Center.

Keledi, G., Hári, J., & Pukánszky, B. (2012). Polymer nanocomposites: Structure, interaction, and functionality. *Nanoscale*, *4*(6), 1919–1938. doi:10.1039/c2nr11442a PMID:22349033

Kim, H., Abdala, A. A., & Macosko, C. W. (2010). Graphene/polymer nanocomposites. *Macromolecules*, *43*(16), 6515–6530. doi:10.1021/ma100572e

Kuilla, T., Bhadra, S., Yao, D., Kim, N. H., Bose, S., & Lee, J. H. (2010). Recent advances in graphene based polymer composites. *Progress in Polymer Science*, *35*(11), 1350–1375. doi:10.1016/j.progpolymsci.2010.07.005

Kumar, S., Lively, B., Sun, L., Li, B., & Zhong, W. (2010). Highly dispersed and electrically conductive polycarbonate/oxidized carbon nanofiber composites for electrostatic dissipation applications. *Carbon*, *48*(13), 3846–3857. doi:10.1016/j.carbon.2010.06.050

Kurahatti, R., Surendranathan, A., Kori, S., Singh, N., Kumar, A. R., & Srivastava, S. (2010). Defence applications of polymer nanocomposites. *Defence Science Journal*, *60*(5), 551–563. doi:10.14429/dsj.60.578

Lee, Y. S., Wetzel, E. D., & Wagner, N. J. (2003). The ballistic impact characteristics of Kevlar® woven fabrics impregnated with a colloidal shear thickening fluid. *Journal of Materials Science*, *38*(13), 2825–2833. doi:10.1023/A:1024424200221

Li, M., & Scott, K. (2010). A polymer electrolyte membrane for high temperature fuel cells to fit vehicle applications. *Electrochimica Acta*, *55*(6), 2123–2128. doi:10.1016/j.electacta.2009.11.044

Li, Z.-F., Zhang, H., Liu, Q., Sun, L., Stanciu, L., & Xie, J. (2013). Fabrication of high-surface-area graphene/polyaniline nanocomposites and their application in supercapacitors. *ACS Applied Materials & Interfaces*, *5*(7), 2685–2691. doi:10.1021/am4001634 PMID:23480549

Liang, J., Wang, Y., Huang, Y., Ma, Y., Liu, Z., Cai, J., ... Chen, Y. (2009). Electromagnetic interference shielding of graphene/epoxy composites. *Carbon*, *47*(3), 922–925. doi:10.1016/j.carbon.2008.12.038

Liao, H., Zhang, H., Qin, G., Li, Z., Li, L., & Hong, H. (2017). A macro-porous graphene oxide-based membrane as a separator with enhanced thermal stability for high-safety lithium-ion batteries. *RSC Advances*, *7*(36), 22112–22120. doi:10.1039/C7RA02950K

Liao, Y.-H., Marietta-Tondin, O., Liang, Z., Zhang, C., & Wang, B. (2004). Investigation of the dispersion process of SWNTs/SC-15 epoxy resin nanocomposites. *Materials Science and Engineering A*, *385*(1-2), 175–181. doi:10.1016/S0921-5093(04)00857-3

Lyu, M.-Y., & Choi, T. G. (2015). Research trends in polymer materials for use in lightweight vehicles. *International Journal of Precision Engineering and Manufacturing*, *16*(1), 213–220. doi:10.100712541-015-0029-x

Ma, J., Meng, Q., Michelmore, A., Kawashima, N., Izzuddin, Z., Bengtsson, C., & Kuan, H.-C. (2013). Covalently bonded interfaces for polymer/graphene composites. *Journal of Materials Chemistry. A, Materials for Energy and Sustainability*, *1*(13), 4255–4264. doi:10.1039/c3ta01277h

Micheli, D., Vricella, A., Pastore, R., Delfini, A., Giusti, A., Albano, M., ... Primiani, V. M. (2016). Ballistic and electromagnetic shielding behaviour of multifunctional Kevlar fiber reinforced epoxy composites modified by carbon nanotubes. *Carbon*, *104*, 141–156. doi:10.1016/j.carbon.2016.03.059

Mukhopadhyay, P., & Gupta, R. K. (2012). *Graphite, Graphene, and their polymer nanocomposites*. CRC Press. doi:10.1201/b13051

Nagarale, R., Shin, W., & Singh, P. K. (2010). Progress in ionic organic-inorganic composite membranes for fuel cell applications. *Polymer Chemistry*, *1*(4), 388–408. doi:10.1039/B9PY00235A

Nambiar, S., & Yeow, J. T. (2012). Polymer-composite materials for radiation protection. *ACS Applied Materials & Interfaces*, *4*(11), 5717–5726. doi:10.1021/am300783d PMID:23009182

Nicholls, I. A., Andersson, H. S., Charlton, C., Henschel, H., Karlsson, B. C., Karlsson, J. G., ... Wikman, S. (2009). Theoretical and computational strategies for rational molecularly imprinted polymer design. *Biosensors & Bioelectronics*, *25*(3), 543–552. doi:10.1016/j.bios.2009.03.038 PMID:19443204

Pang, H., Chen, T., Zhang, G., Zeng, B., & Li, Z.-M. (2010). An electrically conducting polymer/graphene composite with a very low percolation threshold. *Materials Letters*, *64*(20), 2226–2229. doi:10.1016/j.matlet.2010.07.001

Pang, H., Zhang, Y.-C., Chen, T., Zeng, B.-Q., & Li, Z.-M. (2010). Tunable positive temperature coefficient of resistivity in an electrically conducting polymer/graphene composite. *Applied Physics Letters*, *96*(25), 251907. doi:10.1063/1.3457170

Pei, Y., Travas-Sejdic, J., & Williams, D. E. (2012). Reversible electrochemical switching of polymer brushes grafted onto conducting polymer films. *Langmuir*, *28*(21), 8072–8083. doi:10.1021/la301031b PMID:22551237

Phiri, J., Gane, P., & Maloney, T. C. (2017). General overview of graphene: Production, properties and application in polymer composites. *Materials Science and Engineering B*, *215*, 9–28. doi:10.1016/j.mseb.2016.10.004

Pramoda, K., Hussain, H., Koh, H., Tan, H., & He, C. (2010). Covalent bonded polymer–graphene nanocomposites. *Journal of Polymer Science. Part A, Polymer Chemistry*, *48*(19), 4262–4267. doi:10.1002/pola.24212

Ramanathan, T., Stankovich, S., Dikin, D., Liu, H., Shen, H., Nguyen, S., & Brinson, L. (2007). Graphitic nanofillers in PMMA nanocomposites—An investigation of particle size and dispersion and their influence on nanocomposite properties. *Journal of Polymer Science. Part B, Polymer Physics*, *45*(15), 2097–2112. doi:10.1002/polb.21187

Ramezanzadeh, B., Niroumandrad, S., Ahmadi, A., Mahdavian, M., & Moghadam, M. M. (2016). Enhancement of barrier and corrosion protection performance of an epoxy coating through wet transfer of amino functionalized graphene oxide. *Corrosion Science*, *103*, 283–304. doi:10.1016/j.corsci.2015.11.033

Rissanou, A. N., & Harmandaris, V. (2014). Dynamics of various polymer–graphene interfacial systems through atomistic molecular dynamics simulations. *Soft Matter*, *10*(16), 2876–2888. doi:10.1039/c3sm52688g PMID:24667937

Robinson, J. T., Perkins, F. K., Snow, E. S., Wei, Z., & Sheehan, P. E. (2008). Reduced graphene oxide molecular sensors. *Nano Letters*, *8*(10), 3137–3140. doi:10.1021/nl8013007 PMID:18763832

Rosen, B. A., Laufer, C. N., Kalman, D. P., Wetzel, E. D., & Wagner, N. J. (2007). Multi-threat performance of kaolin-based shear thickening fluid (STF)-treated fabrics. *Proceedings of SAMPE*, *3*(7).

Savage, S. (2004). *Defence applications of nanocomposite materials*. Sweden: Swedish Defence Research Agency Kista.

Schedin, F., Geim, A., Morozov, S., Hill, E., Blake, P., Katsnelson, M., & Novoselov, K. (2007). Detection of individual gas molecules adsorbed on graphene. *Nature Materials*, *6*(9), 652–655. doi:10.1038/nmat1967 PMID:17660825

Shen, H., Zhang, L., Liu, M., & Zhang, Z. (2012). Biomedical applications of graphene. *Theranostics*, *2*(3), 283–294. doi:10.7150/thno.3642 PMID:22448195

Silva, M., Alves, N. M., & Paiva, M. C. (2018). Graphene-polymer nanocomposites for biomedical applications. *Polymers for Advanced Technologies*, *29*(2), 687–700. doi:10.1002/pat.4164

Singh, N. B., & Agarwal, S. (2016). Nanocomposites: An overview. *Emerging Materials Research*, *5*(1), 5–43. doi:10.1680/jemmr.15.00025

Singh, V. V., & Wang, J. (2015). Nano/micromotors for security/defense applications. A review. *Nanoscale*, *7*(46), 19377–19389. doi:10.1039/C5NR06254C PMID:26554557

Song, Z., Xu, T., Gordin, M. L., Jiang, Y.-B., Bae, I.-T., Xiao, Q., ... Wang, D. (2012). Polymer–graphene nanocomposites as ultrafast-charge and-discharge cathodes for rechargeable lithium batteries. *Nano Letters*, *12*(5), 2205–2211. doi:10.1021/nl2039666 PMID:22449138

Tang, Q., & Zhou, Z. (2013). Graphene-analogous low-dimensional materials. *Progress in Materials Science*, *58*(8), 1244–1315. doi:10.1016/j.pmatsci.2013.04.003

Thilagavathi, G., Raja, A., & Kannaian, T. (2008). Nanotechnology and protective clothing for defence personnel. *Defence Science Journal*, *58*(4), 451–459. doi:10.14429/dsj.58.1667

Tong, Y., Bohm, S., & Song, M. (2013). Graphene based materials and their composites as coatings. *Austin Journal of Nanomedicine & Nanotechnology*, *1*(1), 1003.

Verdejo, R., Bernal, M. M., Romasanta, L. J., & Lopez-Manchado, M. A. (2011). Graphene filled polymer nanocomposites. *Journal of Materials Chemistry*, *21*(10), 3301–3310. doi:10.1039/C0JM02708A

Verma, M., Verma, P., Dhawan, S., & Choudhary, V. (2015). Tailored graphene based polyurethane composites for efficient electrostatic dissipation and electromagnetic interference shielding applications. *RSC Advances*, *5*(118), 97349–97358. doi:10.1039/C5RA17276D

Verma, V. C. (n.d.). Recent developments in Nanotechnology and their Potential Defense Applications. *Nanotechnology*, *5*.

Wang, G., Shen, X., Yao, J., & Park, J. (2009). Graphene nanosheets for enhanced lithium storage in lithium ion batteries. *Carbon*, *47*(8), 2049–2053. doi:10.1016/j.carbon.2009.03.053

Wang, Y., Shi, Z., & Yin, J. (2011). Kevlar oligomer functionalized graphene for polymer composites. *Polymer*, *52*(16), 3661–3670. doi:10.1016/j.polymer.2011.06.012

Whittingham, M. S. (2004). Lithium batteries and cathode materials. *Chemical Reviews*, *104*(10), 4271–4302. doi:10.1021/cr020731c PMID:15669156

Wu, D., Ramezani, M., & Hakkarainen, M. (2018). *Graphene-Polymer Nanocomposites for Biomedical Applications. In NanoBioMaterials* (pp. 51–66). CRC Press.

Wu, J., Ye, Z., Ge, H., Chen, J., Liu, W., & Liu, Z. (2017). Modified carbon fiber/magnetic graphene/epoxy composites with synergistic effect for electromagnetic interference shielding over broad frequency band. *Journal of Colloid and Interface Science*, *506*, 217–226. doi:10.1016/j.jcis.2017.07.020 PMID:28735195

Xiang, C., Lu, W., Zhu, Y., Sun, Z., Yan, Z., Hwang, C.-C., & Tour, J. M. (2011). Carbon nanotube and graphene nanoribbon-coated conductive kevlar fibers. *ACS Applied Materials & Interfaces*, *4*(1), 131–136. doi:10.1021/am201153b PMID:22117617

Xu, K., Cao, P., & Heath, J. R. (2009). Scanning tunneling microscopy characterization of the electrical properties of wrinkles in exfoliated graphene monolayers. *Nano Letters*, *9*(12), 4446–4451. doi:10.1021/nl902729p PMID:19852488

Zhang, Y., Nayak, T. R., Hong, H., & Cai, W. (2012). Graphene: A versatile nanoplatform for biomedical applications. *Nanoscale*, *4*(13), 3833–3842. doi:10.1039/c2nr31040f PMID:22653227

Zheng, W., Lu, X., & Wong, S. C. (2004). Electrical and mechanical properties of expanded graphite-reinforced high-density polyethylene. *Journal of Applied Polymer Science*, *91*(5), 2781–2788. doi:10.1002/app.13460

Zhou, L., Fox, L., Włodek, M., Flores, L. I., Slastanova, A., Robles, E., ... Cattelan, M. (2018). Surface structure of few layer graphene. *Carbon*, *136*, 255–261. doi:10.1016/j.carbon.2018.04.089

Zou, Y., Xiang, C., Sun, L.-X., & Xu, F. (2008). Glucose biosensor based on electrodeposition of platinum nanoparticles onto carbon nanotubes and immobilizing enzyme with chitosan-SiO2 sol–gel. *Biosensors & Bioelectronics*, *23*(7), 1010–1016. doi:10.1016/j.bios.2007.10.009 PMID:18054479

KEY TERMS AND DEFINITIONS

Electrostatic Charge Dissipation: Flow of electrons between two electrically charged objects in contact.

Electromagnetic Interference Shielding: It is the practice of reducing electromagnetic field in space using certain barriers.

Corrosion: Process of corroding/damaging a metal, stone, or other materials.

Flame Retardancy: Ability of materials to resist flame/fire.

Graphene: It is a nano allotrope of carbon with hexagonally arranged carbon atoms.

Nanocomposite: It is composite material having particle with nanometer size.

Sensor: Device used to detect desired changes according to its type.

Chapter 11
The Use of Polymer Nanocomposites in the Aerospace and the Military/Defence Industries

Emmanuel Rotimi Sadiku
Tshwane University of Technology, South Africa

Oluranti Agboola
Covenant University, Nigeria

Mokgaotsa Jonas Mochane
Tshwane University of Technology, South Africa

Victoria Oluwaseun Fasiku
University of Kwazulu Natal, South Africa

Shesan John Owonubi
University of Zululand, South Africa

Idowu David Ibrahim
Tshwane University of Technology, South Africa

Babul Reddy Abbavaram
Tshwane University of Technology, South Africa

Williams Kehinde Kupolati
Tshwane University of Technology, South Africa

Tippabattini Jayaramudu
Universidad de Talca, Chile

Chukwunonso Aghaegbulam Uwa
Tshwane University of Technology, South Africa

Oluyemi Ojo Daramola
Federal University of Technology Akure, Nigeria & Tshwane University of Technology, South Africa

Nnamdi Chibuike Iheaturu
Federal University of Technology Owerri, Nigeria

Clara Nkuna
Tshwane University of Technology, South Africa

Samuel Eshorame Sanni
Covenant University, Nigeria

Olusesan Frank Biotidara
Yaba College of Technology, Nigeria

Azunna Agwo Eze
Tshwane University of Technology, South Africa

Kokkarachedu Varaprasad
Centro de Investigacion de Polımeros Avanzados (CIPA), Chile

Oladimeji Adetona Adeyeye
Tshwane University of Technology, South Africa

Koena Mantsopa Selatile
Tshwane University of Technology, South Africa

Abongile Sinawo Ndamase
Tshwane University of Technology, South Africa

DOI: 10.4018/978-1-5225-7838-3.ch011

ABSTRACT

Previously, applications of composites were limited to the military aerospace. This is because civilian aircraft with composites inclusions was considered to be too expensive. The use of composite in aircrafts, instead of steel, has resulted in lightweight aircraft structures and has consequently reduced the level of fuel consumption and costs of fuel, thereby reducing CO_2 emissions. Undoubtedly, nanocomposites applications abound in several aspects of human life and the use of nanoparticle in materials dates back to the understanding of the nature of these materials. This chapter will focus on the use of nanopolymers in the aerospace and in the military. Particular attention will be given to nano military weapons, nanocoating for military applications, nanotechnology for military drones, nanotechnology in military suits, gloves, boots and nanotechnology in armored military vehicles, aircraft, and military ships and in military medicine.

INTRODUCTION

In many aspects of human life, nanotechnology plays a crucial role and the enabling advances in defence technologies are simply staggering, even though some of the opportunities may not likely to be realised in the next few decades. However, currently, many advantages are being realized and more explorations are in steady progress, particularly for defence and aerospace applications. An insight into the capabilities offered by nanocomposites in general and polymer nanocomposites in particular in the aerospace and defence industries, which includes: smart materials, light and hard materials, fuel sources and storage and protective devices, will be highlighted in this chapter. It will discuss polymer-based nanocomposite materials, nanoscale fillers and the potential use of nanocomposite materials in the aerospace and defence sectors, with practical examples, where appropriate. In the past, composites were used, mainly in military aerospace applications. Nowadays, composites find applications in civilian aviation and in the military aerospace. This was because, the manufacture of civilian aircrafts with composites was considered, rather too expensive and should composites be a part of their design; they were then used, typically, for non-structural applications. However, when the civilian aircraft industry needed to respond to the rising cost of oil and pressures due to environmental concerns, this had to change for materials of comparative high strength and lighter weight.

There are recent developments in aircraft design, which can obviously be noticeable in the Airbus A380, Bombardier C-series and of course, the Boeing B787 airliners. These developments have led to a phenomenon rise in the use of composites, which has also led to a corresponding increase in adhesive bonding of primary structures in aircraft. The aerospace industry is one of the, undeniably, foremost adopters of advanced composite materials, particularly composites reinforced with carbon fibre. This is so because of the industry demands for low-weight materials with high strength and stiffness. Of course, the ability to mould composite parts and components into curved and complex shapes, is highly important for military applications. Indeed, it can be safely argued that composites constitute at least a quarter (¼) of aircraft by weight, since the introduction of composites in the aerospace industry and polymer composites and polymer nanocomposites, adhesives and sealants, constitute a major part of these materials.

In their review work, Kurahatti et al. (2010) provided an insight into the various capabilities offered by nanocomposites. They discussed some polymer-based nanocomposite materials, nanoscale fillers and provided typical examples of the actual and potential usages of nanocomposite materials in the defence sector, with practical examples.

The use of nanocomposites in aircraft construction induces significant advantages and serves as engineered and advanced solution in aircraft construction. Aircraft construction has greatly tapped into the use of nanocomposites (Armstrong, 2010 and Monti, 2010). There are varied applications of nanocomposites in the aerospace industry. Amongst such applications that are specific in the construction of aircraft, include:

(a) the strengthening element for structures, such as: frames, stringers or as the outer layer for honeycomb-type structures used at fuselage and the wings, (b) when ceramic matrix is combined and integrated with nanoaditions, the nanocomposites can represent a unique solution for the radomes of hypersonic airplanes and (c) when zirconia-based nanocomposites were used as thermal protection for turbo-engines.

The use of nanocomposites in military food packaging has been reported, while carbon nanotube and graphene composites are being vigorously studied for use in ballistic protection and for body armour. Developments are ongoing for military applications of nanocomposites in shock-absorbing materials, solid lubricants, electrostatic charge dissipation, fire retardation, electromagnetic shielding and corrosion protection. (http://www.nanotechmag.com/nanotech-in-the-army/)

Khatavkar and Balasubramanian (2016), in their review work, reported that the encompassing of radar communication in modern avionics, provides the backbone for safe flight and tactical warfare. They systematically and succinctly reviewed composite materials and the existing technologies that are employed for the fabrication of high strength, low dielectric loss sandwich radomes for supersonic aircrafts. There have been considerable and commendable advances in the precision and sophistication, but efficient communication that relies on the effective transmission of electromagnetic waves through materials shielding of the antenna and therefore, exhibiting high transparency for radio/microwaves. Normally, radome structure of aircrafts is often subjected to sudden and drastic acceleration, with substantial level of drag forces, in addition to being subjected to weather elements, such as erosion due to rain and dust. These conditions can change the transmission characteristics due to change in shape/thickness of the structure. Hence, the structure of radomes in supersonic aircraft, is normally dominated by aerodynamic standards. This has resulted in numerous challenges that are often, encountered when selecting the materials that have superior mechanical strength and resilience, while displaying the minimum influence on the transmission profile of radio waves that are pertinent for aircraft construction. They concluded that these challenges are further compounded by the advent of supersonic/hypersonic aircrafts and missiles, where there is the need for high strength preservation and structural integrity at all times, which is highly necessary at temperatures in excess of 350 °C.

It is a common knowledge that carbon-fibre reinforced plastic (CFRP) has always had useful applications in the aerospace industry. Amongst which, is the fuselage and wing structures on the A350 series that are made primarily, of carbon-fibre reinforced plastic (CFRP). The overall weight of the aircraft is, significantly reduced by the use of advanced composite, thereby, offering the operators the advantage of lower fuel burn, increased resistance to corrosion and easier maintenance (Pezzin, et. al., 2010). As a result, of the increase in demand, Airbus has decided to increase production and hence, expecting to their supply chain, to support the increased demand. The aerospace industry is a typical industrial sector that has adapted, very strongly to the use of advanced composite materials, especially composites reinforced with carbon fibre. This is more so, as the sector requires low-weight materials with high strength and stiffness. This is in addition to the ability to mould composite parts and components into curved and complex shapes and this necessity is of utmost importance for military applications. Indeed, the modern aerospace industry uses at least, 25% by weight of composites in a typical aircraft, since the advent of composites. As a result of the recent developments in aircraft design, as exemplified in the Bombardier C-series, Airbus A380 and Boeing B787 airliners, there has been a dramatic rise in the use of composites and consequently, there has also been a corresponding increase in adhesive bonding of the primary structures in aircraft.

Core Polymeric Materials for Aircraft Structures

The core materials for aircraft structures, include, but not limited to:

1. Thermosetting resins utilized in aircraft manufacturing, which employs the core-shell technology to effectively, increase the core strength of fibre-reinforced composite parts or structural adhesives. The technology operates effectively across a wide range of environmental temperatures, from as low as -100 °C, up to the highest usage temperature of the thermosetting resins. The use of this material, can significantly, decrease susceptibility of aircraft structure to fatigue, thereby improving service life. These thermosetting products contain reactive resins, which are distributed with fully cured silicone rubber particles. These rubber particles with a size of between 0.1-3 μm have an organic shell structure that contains the reactive groups.
2. An ever-expanding range of advanced, high quality polymers and composite materials, are needed in order to provide solutions to the critical demands of the aviation and aerospace industries. Rubber-modified epoxy resins are used to strengthen epoxy prepregs for aerospace requirements; this is in addition to improving core toughness and minimizing microcrack development in fibre-reinforced composites and structural adhesives. The use of this epoxy resins can increase fibre wetting and adhesion to aluminium substrates. There are special laminates that have superior flame resistance to those based on epoxy resins; hence, a completely formaldehyde-free aerospace interior components, is a possibility.
3. Bismaleimide resins are functionally, designed for high-performance composites and they are typically and specially suitable for applications in the aerospace industry. These bismaleimide products include bismaleimide monomers, toughening modifiers and formulated bismaleimide resins. Their excellent temperature resistance and high glass transition temperatures are their special characteristics; thereby, providing optimal performance in any environmental condition. These set of resins, have other advantages, which include the preservation of mechanical properties at temperatures of around 250 °C, minimal smoke and toxicant secretions, good resistance to flame and radiation and simple processing.
4. Extruded sheets of polyphenylsulfone (PPSU) are, specially designed to offer infinite interior decoration possibilities for aircraft cabin components. These sheets can be available in opaque and transparent options and they meet critical flame, smoke and toxicity requirements. They also have excellent chemical and impact resistances. PPSU can be manufactured with conventional forming apparatus for a variety of applications, such as: seating, window shades and sidewall dividers for galleys and lavatories. Figure 1, shows the interior of an aircraft cabin where PPSU sheets were used.
5. An acrylic sheet, based on polymethylmethacrylate (PMMA) that is durable, lightweight and with high impact-resistance properties, can be utilized in the military and commercial aircrafts glazing. Such a material can be produced by using equipment, specifically reserved for aviation acrylic products, which should be rigorously inspected, following development. The largest in this

Figure 1. Interior of an aircraft where PPSU was used.
Source: (https://www.azom.com/article.aspx?ArticleID=15795) Retrieved May 20, 2018

industry (2mx3m), which can provide optimal optical quality and thickness tolerance in monolithic production, are the aviation grade sheets.

6. A high-performance polymer, which offers phenomenal physical properties with high temperature and chemical resistances, is polyimide (PI). Polyimide is available in different forms, thus: (a) when it is used as a solvent, it can be employed to make anti-friction coatings or insulating layers for aerospace electronics. This is due to its minimal dielectric constant and high dielectric strength. (b) When used in fibre form, P84® is, commonly employed to produce defensive clothing for pilots or for aerospace insulation and sealing materials. (c) If needed as powder, polyimide can be manufactured into near-net-shape components on the plane or the engine by using the sintering technology. PI offers many interesting and beneficial properties that include: excellent chemical resistance, good temperature stability to 350 °C, good mechanical strength, minimal abrasion and relatively low coefficient of friction.
7. The structural foam made of polymethacrylimide (PMI) has been used in fibre composite technology for more than half-a-century. It is made from extremely robust and rigid composite structures. The foam is compatible with all typical thermoset and thermoplastic polymers. The foam demonstrates many highly beneficial properties, such as: robust and impressive mechanical strength (even at low densities), a wide temperature range and a high temperature resistance of up to 220 °C. In addition, PMI structural foam has a unique compressive creep behaviour and a range of strength and cell sizes that can be specified to meet any specific application needs. Since the foam takes up resin, only in the exposed cut surface cells, there is little or no concern for unnecessary weight accumulation. It is also a suitable candidate to be CNC-milled or thermoformed into intricate geometries.
8. Polyether ether ketone (PEEK) is a high-performance thermoplastic polymer, which is designed for production of durable aerospace components. The material is well suited to withstand elevated temperatures and chemical and conductive environments. Therefore, this material can be utilized for applications in the toughest of conditions. It also offers some interesting mechanical advantages, including high elongation-at-break, good impact strength and low notch sensitivity due to its high ductility and molecular weight. For lifetime durability, this is definitely advantageous. The mate-

rial can be used as a resin matrix for thermoplastic composite unidirectional fibre tapes and fabric prepregs. Under continuous use at temperatures of up to 250 °C, PEEK has excellent durability, just as its good impact resistance and creep behaviour, are well demonstrated. Its lightweight structure in addition to its high mechanical strength make it a good candidate as an effective alternative to traditional thermosets and metals. In addition to all these attributes, PEEK is available as a specially formulated ultra-fine powder and *via* a selective laser sintering technique, it can suitably offer solutions to hot air ducting, clips and brackets in a wide variety of commercial and military applications.

9. In its formulation, polyamide-12 (PA12) often contain a non-migrating flame retardant, free of halogens and phosphorous. It has excellent heat and flammable resistant nature; hence, it is especially suitable for use for aircraft interior parts or for extrusion and injection moulding. PA12 is extremely lightweight and thus, can produce significant weight savings in comparison to the traditional phenolic equivalents. Polyamide 12 (PA 12) is produced in extremely fine powders, via a sintering technology. Therefore, it can provide high quality and reliable processing. It has very advantageous mechanical properties that enable complex geometries for optimum design freedom, part consolidation and simplified logistics, VESTOSINT® ultra-flexible grades are eight times more flexible and five times stronger than their equivalent counterparts (Vacogne, 2010, Apeldorn, et. al., 2010 and Mortara, 2010).

The focus of this chapter will be in the use of nanocomposites, especially polymer nanocomposites, in the aerospace and military/defence industries.

AEROSPACE APPLICATIONS OF NANOCOMPOSITE MATERIALS

Metal Matrix Nanomaterials

The use of nanocomposites in aircraft construction represents an advanced material solution. Nanocomposite material helps (in aircraft construction) as a strengthening element for aircraft structures (stringers, frames) or as skin for the honeycomb type structures used at the wings and fuselage. As base materials for high temperature applications, nano carbon-carbon composites were first developed for the aerospace technology, *viz*: components and missiles, space shuttles, re-entry vehicles, brake lining and as brake disk material for civil and military aircrafts (http://www.romnet.net/nano/2010.02.03_prezentari/14_materiale_tribologie_dlDinca.pdf). These materials with nanoadditivated ceramic matrix, represent a possible and unique solution for the radoms of the hypersonic airplanes. Nanocomposites with zirconia-based complex nanopowders are used as thermal protection for pieces of turboengines. Two types of nanocomposites used in the aircraft, include: polymeric nanocomposites additivated with carbon nanotubes and nanoclays ZrO_2-based nanocoatings as thermal barrier coatings. There are multi-layered structures for thermal barrier coatings (TBC) employed in the coating of turboengines, include structures such as $Me/MeCrA\ell Y/ZrO_2.Y_2O_3$ and multi-layered materials for extreme heat conditions in aeronautics. Other advanced materials include: carbon-carbon composites mesophasic matrix with nanotubes fillers and hybrid fibre/metal composites, polymer-nanocarbon, polymer-nanoclays and polymer nanocomposites-carbon/glass fiber-reinforced structure additivated with carbon or silicon carbide.

The 1-D nanostructures of carbon nanotubes (CN) allow them to have a unique electrical versatility, hence, making them to have the capability of providing remarkable electrical and mechanical properties. By employing the Kryronization process, a new metal alloying technique that uses standard alloys and base metals, carbon nanotube metal matrix composites (CNT-MMC), can be created. This process solves the problems of providing a reliable and cost-effective process for the uniform dispersion of carbon nanotubes in metal. (http://bourqueindustries.com/pages/about_kryron.php)

Metal nanocomposites may contain a metal as the second component and both components are, often very finely, dispersed in each other. Such nanocomposites of these combinations can have improved electrical, optical and magnetic properties, just as the tribological, corrosion resistance and other protective properties can have dramatic improvement. Reinforced metal matrix composites, also known as metal matrix nanocomposites (MMC), can be classified as either continuous or noncontinuous reinforced materials. One of the most crucial nanocomposites as far as their high tensile strength and electrical conductivity are concerned is the carbon nanotube metal matrix composite (CNTMMC). The application of the Kryron technology for the aerospace industry was pioneered by the Kryronization process, which has developed specialized CNTMMC materials from a variety of core substrates. The high strength-to-weight ratio, excellent electrical conductivity, anticorrosive and nonmagnetic qualities and the heat dissipation properties of these materials, allow them to be good and natural candidates that can be fit for the aerospace industry.

In aircraft manufacturing, bulky metals that possess some degree of nanoscale structure enjoy a wide usage within the aircraft industry. Such metals are accepted, since they have significantly better properties than those with larger granular structure.

The primary benefits in the materials used in aircraft construction are: good yield strength, tensile strength, corrosion resistance and low density, hence light weight, resulting from the combined features listed above, which make for strong but lighter aerospace materials. Consequently, with lighter frames for construction, an aircraft consumes less fuel and thus emitting a more compact and reduced carbon footprint. As a result of the excellent strength of nanostructured metals, lightweight does not necessarily come at the expense of durability.

The overriding benefit of using nanotechnology coatings in the aviation industry is their ability to enhance the durability of metal components. The aircraft industry has witnessed an age-old drawback, corrosion of which an aircraft and some metals are unfortunately, prone to. One of the aerospace materials that are benefitting from nanocoating is magnesium alloys, which are substantially lighter than aluminium and steel. Unfortunately, they corrode easily due to the high reactivity of magnesium. Therefore, aerospace nanocoatings protect the normally reactive magnesium alloy from direct contact with its environment. Before to the development of nanocoatings, dipping magnesium alloys into a chromium bath was probably the only remedy. However, chromium is a very highly toxic pollutant; hence, its extensive usage translated into high cost to the environment (Burkit, 2010). Protection against corrosion by using nanocoatings also extends to materials such as: boron oxides, silicon and cobalt-phosphorous nanocrystals. In addition to the structures of aircraft surfaces, nanocoatings are also applied to mechanical components, e.g. the turbine blades. This is so since mechanical component has to endure high temperatures and of course, the friction generated through wear and tear. Tribological coatings can reduce, significantly, the friction coefficient and improves wear resistance, which can lead to significant improvements in the overall engine efficiency (https://dabflow.co/nanocoatings-in-the-aviation-industry/).

Ceramic Matric Composites

Ceramic matrix composite can be lightweight and durable components for use in a jet engine. Amongst their features are: they exhibit composites with low mass and excellent heat resistance when compared to metals. Most often, ceramic matrix composite engines require less cooling air and hence, improving the overall engine efficiency.

The so-called "hot section" of a jet engine, requires durable, lightweight composite components. Ceramic matrix composites (CMC), which are capable of handling the extreme temperatures, find applications in the GE Rolls-Royce Fighter Engine Team's F136 developmental engine for the Joint Strike Fighter (JSF), which contains third-stage, low-pressure turbine vanes. Typically made of silicon carbide ceramic fibers and ceramic resin, CMCs are further enhanced, with proprietary coatings.

In 1995, GE Aviation pioneered the use of composites in the so-called "cold section" of jet engines. This was done by introducing the first composite front fan blade (a carbon/epoxy construction) in its GE90 engine, which powers the Boeing 777. It was then, envisaged that Boeing 787, GE was to use similar composite fan blades and introduce a braided-fiber composite fan blade containment case. As expected, this was to provide huge weight savings. It was reported that the combustor liners that were under consideration for future GEnx production models, was to use CMC (https://www.compositesworld.com/news/ge-aviation-to-use-ceramic-matrix-composites-in-jet-engines). Figure 2 shows an aircraft engine low-pressure turbine vanes, made from ceramic matrix composites (CMC).

Polymer Matrix Nanocomposites

The inclusion of nanoparticles to a polymer matrix enhances the composite material's properties and hence, its performance, which is critical and effective in the production of high-performance composites, when the nanoparticles are finely dispersed. Of course, the nanofiller has to have substantial or rather

Figure 2. Aircraft engine low-pressure turbine vanes made by GE from ceramic matrix composites
Source: (https://www.compositesworld.com/news/ge-aviation-to-use-ceramic-matrix-composites-in-jet-engines) Retrieved May 20, 2018

better properties than the polymer matrix. It has been established that the introduction of nanofillers (with excellent dispersion) in a polymer matrix can induce new physical properties, such as accelerated biodegradability or fire resistance that may be, unintendedly absent in the unfilled matrix, thereby effectively changing the nature of the base matrix. Therefore, the use of nanoparticles in a polymer matrix creates a nanocomposite that can yield an optimal multifunctional material that can be suitable for several applications in the aerospace industry.

POLYMERIC-BASED NANOMATERIALS FOR MILITARY AND AEROSPACE APPLICATIONS

Carbon fibres inclusion in a composite are generally embedded in a matrix of resin or other polymer. They increasingly find usage in aircraft and space vehicles, hence making such structure to be exceptionally strong and characteristically light. However, in the presence of damage by lightning since composites have poor electrical conductivity, exposure to the sun's ultraviolet rays or delamination caused by out-of-plane load, impact or moisture, these composites behaviour are not very well understood. Therefore, a composite wherein nanoparticles are dispersed in a polymer matrix, yields a polymer nanocomposite with improved resistance to fatigue and fracture. Understandably so, the distribution of nanoparticles throughout a polymer matrix can be quite difficult, since strong chemical bonding between the nanoparticles (and nanotubes in particular) and the matrix is essential to the ultimate performance of the nanocomposite material (Goodwin & Stiles, 2010).

It is a matter of concern that during flight, aircraft parts are subjected to varying loads, which can ultimately lead to the development of cracks in high-stress areas. This necessitates that structural parts should be regularly inspected and repaired, otherwise cracks can increase, which can eventually cause structural failure and loss of life, in the course of an accident. Of course, aircraft inspection and repairs are costly to airlines and this is in addition to the fact that high fuel prices and international efforts on climate change have necessitated the unprecedented attention to the need for greater fuel efficiency systems. Therefore, the increasing international competition favours the rapid, low-cost production of reliable, easy-to-maintain and efficient aircrafts that are capable of increased load and range. Hence, the aerospace industry is faced with a challenge to develop advanced materials that are stronger, lighter, safer, fuel-efficient and cost-effective, all at the same time. Therefore, nanotechnology seems to be the possible route to creating near perfect materials that can increase performance and at the same time, passenger safety, while saving cost of production of aircrafts and cost of maintenance when such products are in operation.

Surface-modified silica nanoparticles inclusion in an epoxy resin can be used in many fiber-reinforced composites and structural adhesives for aerospace applications. The resin can significantly enhance the compressive strength and fatigue performance of the fiber-reinforced composites products made with this resin. The nanoparticles are very small and they do not aggregate. This characteristic allows them to infiltrate, very easily, all the fiber structures without comprising the impregnation by increased viscosities. The conventional processing technology, e.g., resin infusion, RTM and VARTM can be used to manufacture the required products and should be appropriate for prepreg manufacturing.

Polymeric-based nanocomposites can be classified into three types, *viz*:

1. Polymer nanocomposites based on the base polymer. These can be sub-classified into:

   a. Polyolefin-based polymer nanocomposites
   b. Nylon-based polymer nanocomposites
   c. Soft polymer nanocomposites and gels
   d. Biodegradable polymer nanocomposites
   e. Poly(vinyl chloride)-based nanocomposites
   f. Clay-containing poly (ethylene terephthalate) (PET)-based polymer nanocomposites
   g. Thermoplastic polyurethane (TPU)-based polymer nanocomposites
2. Polymer nanocomposites based on fillers inclusion. These include
   a. Carbon nanotube-based nanocomposites and its processing
   b. Calcium carbonate nanocomposites
   c. Metal-polymer nanocomposites
   d. Environmental life-cycle assessment (LCA) of polymer nanocomposites
   e. Metal-polymer nanocomposites
   f. Layered double hydroxides (LDHs) as functional fillers in polymer nanocomposites
   g. Cellulose nanoparticles as reinforcement in polymer nanocomposites
3. Polymer nanocomposites based on their applications. These include:
   a. Polymer nanocomposites for aerospace applications
   b. Polymer nanocomposites in fuel cells
   c. Polymer nanocomposite coatings
   d. Flame-retardant polymer nanocomposites
   e. Polymer nanocomposites for optical applications

Some of these classes will be discussed as per their applications in the aerospace and military sectors.

The aerospace and electronics industries have witnessed the rapid usage of polymer composite laminates, since these industries require laminates that are structurally rigid in addition to their high stiffness and good di-electrical properties. Nallayan et al., (2017) reported on the comparison of the effect of curing on the properties of E-glass/cyanate modified epoxy cross-plied laminates. Amongst the requirements for electromagnetic (EM) waves to transmit signal with near zero transmission loss, is that the laminates must be transparent. Hence, the establishment of a potent material combination that has high structural strengths, requires laminates that can adequately respond under different loadings. Such materials are expected to be suitable for applications in areas that deal with signal transmissions (Risi, et. al., 2009 and Oliveira et al. 2009).

Nallayan et al. came up with thin laminates with 4 plies with simple stacking configurations of 0/90/0/90 degrees as applicable to a cross-plied laminates. These laminates were fabricated with cyanate ester modified epoxy resin and 1200GSM E-glass unidirectional fibre. They identified the flexural and impact strengths as properties needed for accessing the structural responses of the laminate at room and oven curing conditions. They employed FESEM images to validate the experimental findings.

Figure 3 shows the different types of nanomaterials applicable in the military and defence sectors. It shows the various sectors of the military and aerospace industries, where nanomaterials can be applied. Nanomaterials-based technology, is playing (and will continue to play) an important and highly strategic role in the effective development of various materials and devices that are play dominating roles in military applications and the society in general.

Figure 3. Different types of nanomaterials applicable in the military and defence sectors
Kharat et al (2006)

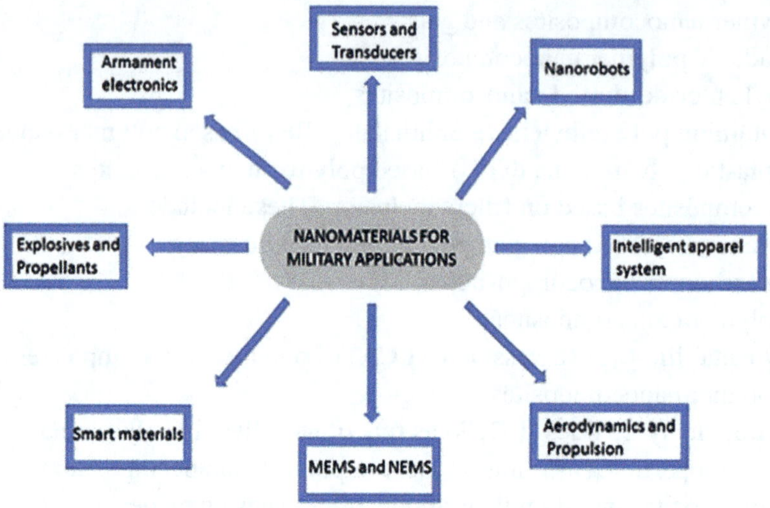

Kausar et al. (2017) presented an overview of polymer and carbon nanotube composites with special emphasis on their applications in technical fields. There are general applications of polymer/carbon nanotube nanocomposite in actuators, sensors, sporting goods etc. They also reported, extensively, on the special features of epoxy and epoxy/carbon nanotube composites. They observed an enhancement in the characteristic features of epoxy matrix with the incorporation of carbon. They inferred that the applications of epoxy/carbon nanotube composites are in different fields, such as: aerospace, automobiles, fuel cells, radar-absorbing material, wind turbine blades and electromagnetic interface shielding.

Aerospace Composites

The aerospace industry, unfortunately, has one of the largest carbon footprints and in order to ameliorate the factors responsible for this, extensive research and development into aerospace materials, especially in the aerospace nanocoatings sector, are driving the industry into the discovery of lighter, more effective materials for the construction of airframes and for enhanced efficiency in the aircraft engines.

There are certain critical considerations for specific aircraft structures that need to be adhered to, very strictly in the design of an aircraft, since each major part of an aircraft involves different considerations. For instance, for the fuselage design, durability and damage tolerance are the primary drivers. In so doing, consideration and rigorous attention must be given to the possible material fatigue, i.e., the crack initiation and growth rate and fracture toughness are the leading materials attributes. In addition, stiffness, strength and corrosion are also key parameters. Certainly, the wing design is influenced by strength, durability and damage-tolerance requirements. Therefore, for polymer matrix composites, fatigue and corrosion resistance are not design drivers. However, materials properties, such as: compressive yield strength, stiffness in compression, fatigue resistance and fracture toughness are key considerations. This is because of the ground-air-ground alternating loads generated during flight. Table 1 shows the comparison of the mechanical properties of aircraft composites.

Table 1. Comparison of the mechanical characteristics of aerospace polymer composites

Type of Material	Code	Tensile Strength (MPa)	Young's Modulus (mPa•10^{13})	Strain (%)
Carbon/Epoxy	IM7/8551-7	2757.903	16.547	1.62
	T300/934	1689.216	13.795	1.0-1.2
	IM6/3501-6	2275.270	15.585	1.5
	AS4/3501-6	689.476	6.895	1.0
	P75/934	930.792	30.337	0.2-0.5
Carbon/PEEK	IM7/APC-2	2888.903	16.547	1.6
Carbon/Phenolic	FM5055	103.424-137.895	1.793-1.931	1.0-1.2
Kevlar/Epoxy	K-49/7934	551.581-586.054	2.748	1.85
Glass/Epoxy	E-glass/934	1034.214-1172.109	4-137-5.516	2.75

Source: (https://www.nanowerk.com/spotlight/spotid=1015.php) "Defence Nanotechnology Research and Development Programs", 2006)

As far as aircraft perform is concerned, the thrust-to-weight ratio is one of the most important matrices. Of course, aerospace and aircraft manufacturers have always aimed to keep this ratio as low as possible. This is achieved by creating lighter aircrafts. However, the metals that were traditionally used in aircraft bodies are heavy. On the other hand, composite materials are considerably lighter than metal and this factor enables manufacturers to create more fuel-efficient aircrafts when mixed with metal.

The history of composites in the aerospace industry and where future innovation is headed, may well be of importance. Previously, composites were used, mainly in military aerospace applications, while the manufacturing of civilian aircrafts with composites was considered too expensive. Indeed, if (and when) composites were parts of the design, they were previously and typically considered for use for non-structural applications.

Materials Considerations for Aeroengines

Traditionally, it was common to have turbine-engine fan blades made from titanium, usually forged Ti-6Aℓ-4V and attached to a forged titanium disk that is also typically Ti-6Aℓ-4V. Undeniably, this alloy has an excellent balance of mechanical properties, including ultimate tensile strength, ductility and fatigue resistance and it can be readily forged, heat-treated and machined. While smaller-diameter engines make use of solid titanium blades, larger-diameter engines use hollow titanium blades that can be made by an SPF/diffusion bonding process. However, large engines, which can have fan blades with lengths >1m, can be made from polymer matrix composites (PMCs), typically carbon fiber in a toughened epoxy matrix.

It is well established that nanoparticulate/fibrous loading confers significant property improvements at very low loading levels of nanoparticles, whereas the traditional micro-particle additives require considerable higher loading levels in order to achieve similar performance. Therefore, the inclusion of nanoparticle can result in significant weight reductions. This is of obvious importance for various military and aerospace applications for similar performance, greater strength for similar structural dimensions and for barrier applications, increased barrier performance for similar material thickness.

However, when the civilian aircraft industry had to respond to the rising cost of oil and pressures due to environmental concerns, it was inevitable for the industry to change and start with the use of composites. It is obvious that the more lightweight aircraft structures that result from substituting composite for metal have lower fuel costs and reduced emissions. By weight, while only 12% of the Boeing 777's weight was composite, the Boeing 787 employed ~50% composite. Of course, composites are, now commonly and routinely used in civil aircraft structures in ever-increasing amounts.

The most significant benefit of composites for the aerospace industry is their lightweight nature. Other benefits of composites in the aerospace industry, include: their good thermal stability, high impact resistance and damage tolerance and these are obvious properties that enable aircrafts constructed with composites to fare better in accidents. In addition, composites are better at resisting fatigue and corrosion than traditional metals, just as they are also easy to assemble. It is expected that as composite technology continues to improve, these stated benefits will become (and indeed are becoming) better pronounced, with the attendant costs of building aircrafts with composites being increasingly reduced.

Further benefits emanating from the development of lighter aircrafts by using composites, allows such aircrafts to become safer and more fuel-efficient.

There are efforts to isolate vibration are integral parts of any engineering project and design. The stoppage (or any significant reduction) of excessive vibration is a necessary must, which knowledgeable and experienced manufacturers should be aware of, otherwise, it can cause unnecessary wastage of energy and damaged equipment. This can lead to broken products, expensive repairs and possibly, reduced profits. It is therefore, necessary for equipment manufacturers to be able to trust the reliability of their anti-vibration mounts that they install on their products. If this is not well ascertained, manufacturers can easily end-up with subpar products and the attendant consequences of unhappy customers.

The comfort of the passengers in the cabin needs continuous attention, since excessive noise is often one of the main sources of cabin discomfort. A noisy ride may be unintendedly introduced by reducing the weight (should weight reduction be prioritized over noise and vibration reduction) of the fuselage, through lighter materials. Effective vibration damping, can however, be achieved by using composites with well-formulated and constructed materials. The incorporation of lightweight insulation and cushioning materials allows composites to provide better vibration damping and airborne noise absorption, thereby, resulting in an aircraft ride that is considerably better enjoyable for the passengers and pilot.

Composites applications on aircraft may include: flight control surfaces, interior components, leading and trailing edge panels on the wing and stabilizer, fairings, floor beams and floor boards, landing gear doors, primary wing and fuselage structure on new generation large aircraft, turbine engine fan blades, vertical and horizontal stabilizer primary structure on large aircraft and in propellers. Figure 4 shows the different sections of a turbofan engine.

Coatings for Aircrafts

Probably, more than 20,000 aircrafts fly our airspaces every day. They are expected to be maintained to the highest standards set by different government agencies around the world. This maintenance includes routine inspection and the maintenance of their protective coatings. As an important contributor to the aerospace economy, the aircraft coatings industry, is probably worth more than $250 million, currently. For the aerospace industry, coating specialists, companies and suppliers are expected to develop and produce products that reduce application times, improve performance and increase durability and many of the new coating innovations are focused on helping airlines become more cost-effective (Ruiz, 2000).

Figure 4. Different sections of a turbofan engine.
Source: (http://aviationfacts.eu/uploads/thema/file_en/585919d470726f2696000000/Composites_in_Commercial_Aircraft_Engines_Fact_sheet.pdf) Retrieved May 15, 2018

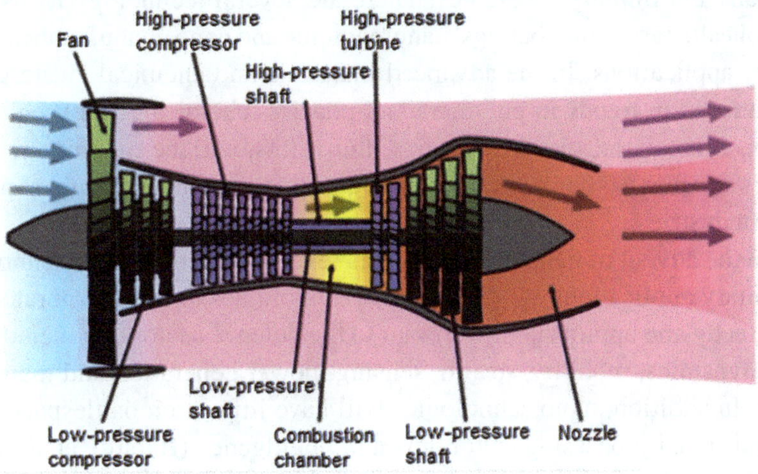

In order to reduce the amount of time it takes for an aircraft to undergo maintenance, often, new coating systems need to be developed. These systems must allow for faster application and drying of exterior topcoat applications. Historically, conventional aerospace coatings have been based on two-component polyurethane coatings that normally require ~72 hours to cure (fully develop their physical properties). To significantly reduce and therefore, accelerate the curing process, many coatings producers have developed coatings formulations that use ultraviolet A (UVA) light. By going through this route, airplanes that undergo a topcoat refurbishing can get back into service, relatively faster, a faster turnaround for aircraft maintenance can significantly enhance the scheduling flexibility and this will ultimately contribute to reducing costs (Huang et al., 2010). With rising fuel costs and a focus on the demand for cost-effectiveness on the running of aircrafts and in ever-increasing costs of aviation fuel, airlines worldwide seek products that can continually contribute to improved aircraft reliability, efficiency and safety. Of course, this applies to commercial aircrafts, military and private aircrafts. Therefore, and clearly so, aliphatic diisocyanates play an enabling and highly complementary roles in the aerospace industry and they are highly regarded in coatings formulations. Surface coatings normally achieve the protection of metals and alloys from corrosion. This coating must resist mechanical damage, emanating from: impact, scratching and abrasion and from chemical attack: acids, bases, salts and solvents. Of course, it should also not be damaged or cracked by having a coefficient of thermal expansion that is significantly and disproportionately different from the metal that needs to be coated (protected).

DEFENCE AND MILITARY APPLICATIONS OF NANOTECHOLOGY

For any functional government, the security (or defence) of its citizens should be the first duty of such government. The safety and protection of a country's citizens and the nation's secrets from foreign interferences are essentially, what a national defence entails. A national military security implies that the country should have the capacity to defend itself in the face of any foreign aggression.

There have been tremendous advances in nanotechnology for military applications, as exemplified in numerous research activities in various fields: in smart textiles, sensors, and lightweight but yet durable materials for weapons and military medicines. There are several technology focus areas that include: nanomaterials, nanohealthcare, nanocoatings, nanomedicine and nano-manufacturing that are vigorously pursued for military applications. In the advanced materials and chemical clusters, are developments and research and innovation trends in polymers and plastics, chemicals and specialty chemicals, bio-chemicals, coatings, composites, metals, coatings, thin films, surface treatments, alloys, oil and gas, fuel additives, fibers and many other related technologies that undoubtedly impact positively on several applications across industries.

It is obvious that the advent of nanotechnology has revolutionised several technological changes for a wide range of military applications and platforms. Technologies to be incorporated within these platforms, which are directly and appropriately relevant to the defence sector that includes: smart structures and materials, aerodynamics, mobility, stealth, sensing, power generation and management, resilience and robustness, etc. In addition, nanotechnologies will have impact on battlespace systems concerned with information and signal processing, autonomy and intelligence (Baratto, et al., 2009).

There are a number of nanomaterials in advanced stages of research and development for aerospace and security applications. They include:

1. Nanocomposite materials in combination with: polymers, metals and ceramics. It is well established that these materials can be multifunctional in behaviour and can be used for special coatings and high-temperature applications.
2. There are nanowires of various metals used as gas sensors, environmental monitoring and nano-actuators.
3. Carbon nanotubes (CNTs) that can be either single walled carbon nanotube (SWNT) or the multi-layer nanotubes and they find applications in space and aircraft manufactures.

The sensitivity of sensors to changes in different environments is used to measure the needed parameters in the required environment. The parameters to be measured can include: chemical activity, thermal conductivity, electrical conductivity, capacitance, etc. These parameters will greatly depend on grain size (microstructure) of the materials used as sensors and a change in the sensor material's chemical, physical or mechanical characteristics is normally exploited for the detection of the parameter to be measured. Smart sensor technologies have been developed for soldiers, for the enhancement of military intelligence gathering in the field. Harmful chemicals and biological weapons can be detected with the use of chemical and biological nanosensors, just as they can also be used as damage detection systems, i.e., physical nanosensors that can detect fractures in military equipment (http://www.nanotechmag.com/nanotech-in-the-army/).

Kurahatti, (2010) succinctly reviewed the defence applications of polymer nanocomposites. The review provided an insight into the possible capabilities of nanocomposites, which include: smart materials, harder/lighter platforms, new fuel sources and storage and novel medical applications. It further discussed the polymer-based nanocomposite materials, nanoscale fillers and provided excellent examples of the actual and potential uses of nanocomposite materials in defence with practical examples. Polymer nanotube mats and/or woven fibres continuous sheets or films of a nanocomposite, easily find applications in the defence sector.

A simple method of where non-woven mat was exposed to a high intensity light source (e.g., from a flash tube), can result in the immediate joining of fibres at cross-over points of contact. By using a mask, fibres can be joined in any desired pattern, it is envisaged that such materials may find wide applications in defence as electrically conductive fabrics, sensors, electromagnetic shielding, microwave absorption, electrical energy storage (capacitors), actuators and materials for micro UAVs. El-Fatatry (2007) reported on the staggering potential opportunities, promised by nanotechnology for enabling advances in defence technologies opined that these opportunities maybe only realised in a few decades to come. However, many advantages are currently obvious, particularly for defence applications. In his lecture, he provided an insight into the capabilities offered by nanotechnology, which will enable new defence capabilities, including: smart materials, new fuel sources, harder/lighter platforms and storage as well as novel medical applications, be realized. Certain polymer nanocomposite materials offer many advantages to the military platforms, ranging from conventional vehicles to more advanced miniaturized platforms (e.g. unmanned autonomous micro-vehicles).

These days, there is virtually no complete materials conference without a nanotechnology theme and a Hamburg conference of 2010, dealt extensively on a cluster of five papers, which included amongst others:

1. Low concentrations of carbon nanofibres in GFRP, serving as a tool for strain and damage sensing for structural health monitoring.
2. Multifunctional composites with carbon nanotubes for space applications.
3. Use of CNFs to improve Z-conductivity (i.e., conductivity in the z-direction in the presence of electric field during curing) in non-woven carbon fibre veil structures in order to develop electromagnetic shielding capabilities.

At the 2nd International Conference on Polymers in Defence and Aerospace Applications, held in Hamburg, Young, (2010) reported that wide-ranging technical programme with sessions including: electronic materials and applications, novel materials and processes, carbon nano fibre-based materials, composites, inorganic nano-materials and coatings.

Lahiff, et al. (2003) developed a controlled method of producing free-standing nanotube-polymer composite films that can be used to form nanosensor, which contains, at least, one conductive channel that comprises an array of aligned CNTs embedded in a matrix (e.g., poly (dimethylsiloxane)). This material can be used to determine a real time physical condition of a material, such as that of an airplane wing or chassis, while the airplane is in flight. Such materials may be used for the detection and identification of toxic gases, such as: chemical warfare agents, flammable gases, solvent vapours, etc., touch sensors for interaction between operators and machines. They reported on polymer/nanotube composites where the density and position of the nanotubes within the composite can be controlled, with the carbon nanotubes growing from organometallic micro-patterns.

Detection of Explosives on Airport Luggage

Considering the threat of terrorists, intent on using explosives, especially at airports and border posts, the need to detect trace amount of explosives cannot be over-emphasized. This is so because of the requirements to provide security to humans and properties. Dutta et al. (2016) reported on the synthesis of carbon and silver nanoparticle impregnated graft polymers of poly(vinylalcohol) and polythiophene *via* a single step free radical polymerization reaction for sensing nitroaromatic compounds *via* fluorescence measurements. The conducting nanoparticles were to enhance the π-electron density of the luminescent polymer. The electron rich nanocomposites are to be used for the highly selective and ultrasensitive detection of electron deficient nitroaromatic explosives, based on a fluorescence quenching method (Colpo, et. al., 2007).

POLYMER NANOCOMPOSITES FOR BALLISTIC, GLOVES, BOOTS AND BODY ARMOUR AND SMART TEXTILES FOR MILITARY AND DEFENCE APPLICATIONS

There has been the adoption of different nanomaterials as filler materials in order to enhance the properties of structural and non-structural polymers. The commonest materials in use for this purpose are: nanofibres, nanoclays, carbon nanotubes and graphene. Significant advantages of carbon nanotubes have been recorded as fillers for many polymers; this is because carbon nanotubes have proven to be exceptionally stiff, tough and they possess outstanding electrical properties. Aside their superior strength-to-weight ratios, they offer enhanced vibration and fire resistance and these characteristics make them ideal for aviation applications (https://dabflow.co/nanocoatings-in-the-aviation-industry/). Polymer nanocomposites have properties that make them highly suitable in aircraft manufacturing. Therefore, they are, increasingly being used as selective replacements for certain metals in the airframes of an aircraft. This is to deliver substantial weight savings, as well as cost savings for manufacturers and in turn, aircraft fleet operators.

There have been the developments, via nanotechnology principles and molecular design that can assist in the alignment of nanofillers in polymeric fibers of super-strong fibers for ballistic armour applications.

The improvements in weaponry that come with similar improvements in armour and nanomaterials can be designed in order to create an exceptionally strong armour. For example, the Kryron Terminator ballistic body and vehicle armour are made from an aluminum alloy that is combined with carbon nanotubes. Such a ballistic armour plate is capable of withstanding multi-impact. It is lightweight and durable and in this regard, nanostructured materials can make lighter, stronger and adaptive armour and nanofiber-based suits that can deliver improved protection against projectiles. In the same vain, the so-called "smart-materials" should be capable to adapt to changes in light, temperature, pressure, stresses or possibly pH. Even though this is still in its infancy, a battle suit in the fields of communications, strength and soldier protection should be able to remain comfortable when worn and lightweight and at the same time be able to: stop bullets, protect the wearer against toxins, monitor vital signs and administer first aid, where possible.

Reports on polymer nanocomposites ballistic testing are rather, scanty. For now, this may be due to the secrecy associated with and the lack of suitable nanocomposite materials for the production of such materials. Therefore, for light protection, e.g. body armour and vehicle liners, woven materials, e.g., Kevlar are commonly used. With recent development in electrospinning, it is likely that electrospun nanofibres, spun carbon nanotubes (CNTs) may be useful in such applications, even though, there are few reports of the promising application of nanocomposites in body armour. Military researches into nanotechnology have no doubt, advanced from improving medical and casualty care for soldiers to producing lightweight, strong and multi-functional materials for use in clothing, for protection and to provide enhanced connectivity. Hence, improved body armour is a major focus for military nanotechnology research and numerous different technologies have been explored; some of which will be operational in just a few years' time or maybe are in operation already. These include, but not limited to having:

1. Si or TiO_2 nanoparticles embedded in epoxy matrix.
2. SiO_2 nanoparticles in a liquid polymer, which hardens on ballistic impact (Shear Thickening Fluid).
3. Iron nanoparticles in inert oil, which hardens when stimulated with an electrical pulse (Magnetorheological Fluid).

Figure 4 shows the futuristic body armour that incorporates nano-enhanced technologies. As far as a decade-and-half ago, scientists at the Massachusetts Institute of Technology, MIT did investigate smart materials that are able to be responsive to the conditions of their environments, including sensors that able to detect chemical or biological warfare agents and lightweight bullet-proof materials. Attempts were also made to incorporate wound detection and treatment systems within uniforms and hence, responsive systems, e.g., the materials hardening that is able to provide an instant splint for a broken bone, were being developed (https://www.azonano.com/article.aspx?ArticleID=1050). Recently, Russia unveiled, during her Stormtrooper 2017 (at the National University of Science and Technology, Moscow), its hi-tech motorised 'Star Wars' armour for soldiers, complete with exo-skeleton, night vision and helmet. The all-black kit included a helmet with a tinted glass visor topped with radio cable, to be worn by model soldier cradling fearsome-looking firearm in a pair of black gloves and with the arms covered with khaki-patterned material and pads, as shown in Figure 5. The Russian advanced high-tech amour item that includes an exo-skeleton or the outer layer, which is designed to boost strength and stamina, is shown in Figure 6.

In addition, fingerprinting has become considerably quicker, cheaper, more effective, and readily available, by using DNA techniques involving nanotechnology and there is the possibility that nano-based sensors will be used as electronic detectors in sniffer dogs, e.g., at the airports or border posts, for improved security and in drug-busting exercises, even in residential places. With increasing advances in nanotechnology, quantum dots, fluorescent nanoparticles, which glow when exposed to ultraviolet light, can be employed as tags and labels in order to prevent theft and counterfeiting of currency, for instance and medically too, to trace the course of drugs within the body.

Promising results were reported by the US Army research laboratory that indicated that the combination of inorganic nanoparticles (of silica) in polyethylene glycol, when used as a shear thickening fluid for impregnation of a conventional Kevlar, resulted in greatly improved ability of the material to absorb energy. The ballistic performance (in terms of absorbed energy) was more than double, such a four layer Kevlar impregnated with the shear thickening fluid could absorb as much energy as would have been absorbed by 10 layers without the shear thickening fluid. Consequently, this will translate into a more flexible armour with reduced weight. It is believed that such materials can easily find applications in body/personal armour, where flexibility of movement is required, aside the protection against blunt weapons (stones, sticks and bars) for the arms and legs.

In their book, "The Science of Armour Materials," Crouch et al. (2017) discussed several aspects of amour materials, including the future of armour materials, ballistic testing methodologies, fibres, textiles and protective apparels, polymers and fibre-reinforced plastics, etc. The different ranges of reinforced and unreinforced polymers that includes a comprehensive review of the many methods of manufacturing armour-quality, composite components were discussed, since each approach usually results in an armour product with very different properties. Spliceless technologies that enable fully formed combat helmet shells to be manufactured in one-step from a flat lay-up of composite materials, especially those based on advanced materials, e.g., ultra-high molecular weight polyethylene (UHMWPE) fibres were discussed. They also highlighted the construction of soft body armour into its constituent elements by

Figure 5. Futuristic body armour incorporating nano-enhanced technologies.
Source: (https://www.azonano.com/article.aspx?ArticleID=3028) (https://www.azonano.com/article.aspx?ArticleID=1466)

comprehensively covering the grades of fibre, styles of fabric, quantities of fabric and the various ways of forming a coherent pack, such as: stitching, quilting and thermoplastically tacking the layers together. The formation of fabrics by the integration of textiles with electronic components in order to provide enhanced functionality to the product, results in smart textiles. Often, smart textiles are used to monitor the location and physiological condition of personnel in order to protect soldiers in conflict zones and in other technological applications. There are nanotechnology-enabled smart fabrics and generally, these fabrics have become the building blocks of advanced smart textiles for various military applications, including military clothing, tents and textiles for military shelter (http://vibrantgujarat.com/writeread-data/images/pdf/project-profiles/Uniform-for-defence-personnel-nanotechnology.pdf).

Polymer/Nanographene Composites for Aerospace and Military Applications

As a result of its unique combination of properties, graphene is particularly suitable for the improvement of the functionality of composites. Because of its 2-D carbon-based structure, the material is in the form of robust and flexible sheets that are a single atom thick. In addition, it is lightweight and at the same time, it has enormous strength and it is an excellent conductor of heat and electricity. Therefore, graphene

Figure 6. Russian advanced high-tech amour item that includes an exo-skeleton or the outer layer, designed to boost strength and stamina. The amour item was pictured as a combat suit on display (left) alongside a Star Wars Stormtrooper (right).
Source: (http://www.dailymail.co.uk/news/article-4653886/Russia-unveils-generation-Star-Wars-combat-uniform.html)

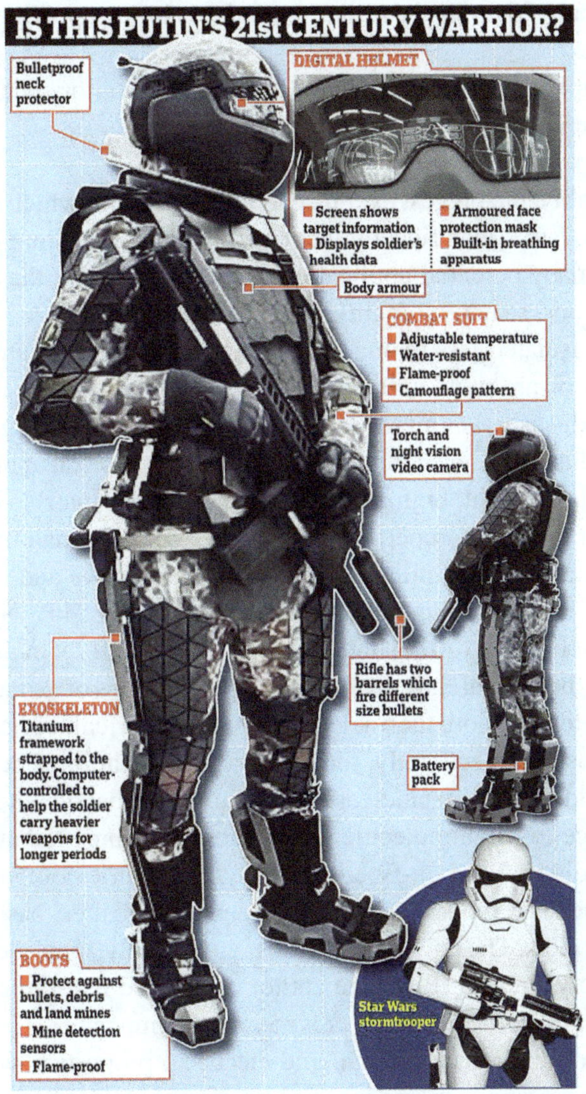

sheets, with a lateral size of several microns, can be produced on a large scale from the exfoliation of graphite. Amongst the desirable properties of graphene are:

1. It is a lightweight material that allows a new generation of high performance of lighter and stronger composites.
2. It is a 2-D material and possesses remarkable optical properties that include a simple transparency for visible light.

3. It is an exceptional good conductor of electricity and heat. This is due to its nanometric thickness that allows electrons and heat to move up to 200x faster.
4. It is an allotrope of carbon that consists of a small number of layers of carbon atoms that are arranged in a hexagonal lattice and each layer is separated from another, only by ~0.33 nanometers.
5. Graphene is 100+ stronger than steel and can be used to reinforce polymers or fiberglass and carbon fibre.
6. It has excellent hydrophobicity capacity that can be used to tailor special coatings in the marine, aerospace, automotive industries.

Graphene attracts, probably, the biggest ever research initiative, on which advanced research into the creation of unique and innovative materials by incorporating graphene into composites, is continually on going, resulting in the many potential application areas that include: the aerospace and automotive industries, energy applications, structural foams, films and coatings.

For a material that was isolated less than a decade ago, graphene has attracted considerable attention and there has been an upward trend in the capabilities and possibilities of graphene. Graphene has potentials in a number of industries, among which, are: aerospace, wind energy, electronics, automotive, medical and biomedical industries. Lynch-Branzoi (2017) reported on an approach that enabled the fabrication of a new class of lightweight, high performance polymeric-matrix composites (PMCs). These PMCs have tuneable, multifunctional properties with simple, versatile manufacturing approach and low materials costs that are suitable for the military, aerospace, automotive and infrastructure applications. Njuguna and Pielichowski, (2003), Njuguna and Pielichowski, (2004a), Njuguna and Pielichowski, (2004b), Njuguna and Pielichowski, (2010) and Njuguna et. al., (2012), in extensive experimental and review works, examined the theoretical and experimental investigations of extracting valuable fundamental parameters that include: field emission, thermal stability and electrical, optical and mechanical properties of polymer nanocomposites, specifically, for aerospace application. They concluded that polymer nanocomposites could provide significant increase in: modulus, gas barrier, thermal performance, atomic oxygen resistance, resistance to small molecule permeation and improved ablative performance when compared to the typical traditional carbon-fiber-reinforced polymeric composites.

Galpaya (2015) developed and reported on polymer nanocomposites, based on graphene and its derivatives, for a range of polymers. The polymers, include: epoxy, poly(vinyl alcohol) (PVA), poly(methyl methacrylate) (PMMA), polyethylene (PE). The author concluded that the superior properties of graphene were reflected in graphene-incorporated polymer nanocomposites. These properties include: the materials exhibiting greater mechanical, thermal and electrical properties, when compared to the neat polymers. These improved properties can lead in the development of: high strength, lightweight structural polymer composites, with possible applications in the fields of: automotive, aerospace, electronics and green energy, (de Farias et al, (2009) Meng and Hu (2008), Ni, et. al. (2007)).

Khan et al (2016) reported, in a comprehensive review work, on the synthesis, structure and properties of graphene oxide nanofillers and their polymeric nanocomposites. They also reported on the hybrid properties and synthesis of: graphene oxide-carbon nanotube, graphene oxide-montmorillonite and graphene oxide-nanodiamond. The hybrids reported on, include: nanobifiller in polymer/graphene oxide-carbon nanotube, polymer/graphene oxide-montmorillonite and polymer/graphene oxide-nanodiamond composites. They are of the view that the aspect of polymer and graphene oxide-based nanobifiller hybrid composites is relatively, less studied/reported. Therefore, they recommended that nanobifiller technology limitations and research challenges should be of concern. This is because; polymer/graphene oxide nanobifiller composites have a wide range of unexplored potentials, especially in the technological fields, such as: aerospace, aerospace, medical energy and industries. Figure 7(a) shows a graphene-coated motorcycle helmet that allows better distribution of heat, hence, making it less susceptible to damage when compared to the traditional helmets, particularly in high temperature conditions. This allows the helmet to improve on the wearer's thermal comfort and safety. Figure 7(b) shows a drone, whose propeller blades were coated with graphene in order to enhance their durability; since graphene can be mixed with thermoplastics, this drone propeller blades was prepared, from graphene-polypropylene. They were produced by melt-mixing graphene nanoplatelets with polypropylene in order to form a master batch, which was subsequently compression-moulded into the desired shape. A series of graphene-ABS filaments, which can be used for 3D printing, has been prepared. It has been proven that the presence of graphene in polymers, improves the stiffness and thermal properties of such composites.

In early 2016, the world's first graphene-skinned aircraft, called Prospero, was flown, successfully, at the Farnborough International Air Show in the UK (https://graphene-flagship.eu/graphene-shows-promise-for-composite-applications).

A collaboration between the Universities of Manchester and Central Lancashire (UK) and a number of SMEs, including Haydale Composite Solutions (UK), envisaged that in the near future, graphene might be used as a potentially disruptive technology in the aerospace sector. The Graphene Flagship stand at Composites Europe 2016 displayed the opportunity to study the details of Prospero on the ground and exhibited, for the first time, a drone with graphene-enhanced propeller blades.

Composites with graphene inclusion can allow the possibility of obtaining the maximum classification in fire resistance criteria in aeronautics, automotive and building industry applications, therefore, graphene-fibre-reinforced composites can enable self-sensing in the aeronautics and automotive industries.

NANOCOATINGS IN AEROSPACE AND MILITARY APPLICATIONS

In addition to being costly, conventional coating in the defence and military fields can be hazardous for the workers in these areas, especially when some of these coatings may need to be touched-up by hand. It is estimated that ~25% of army vehicles are out of service due to coatings damages that often

Figure 7. (a) A motorcycle helmets with a graphene coating that allows better distribution of impact force, making the helmet less susceptible to damage even in high temperature conditions and (b) A drone with graphene-enhanced propeller blades.
Source: (https://graphene-flagship.eu/graphene-shows-promise-for-composite-applications) Retrieved July 01, 2018

(a) (b)

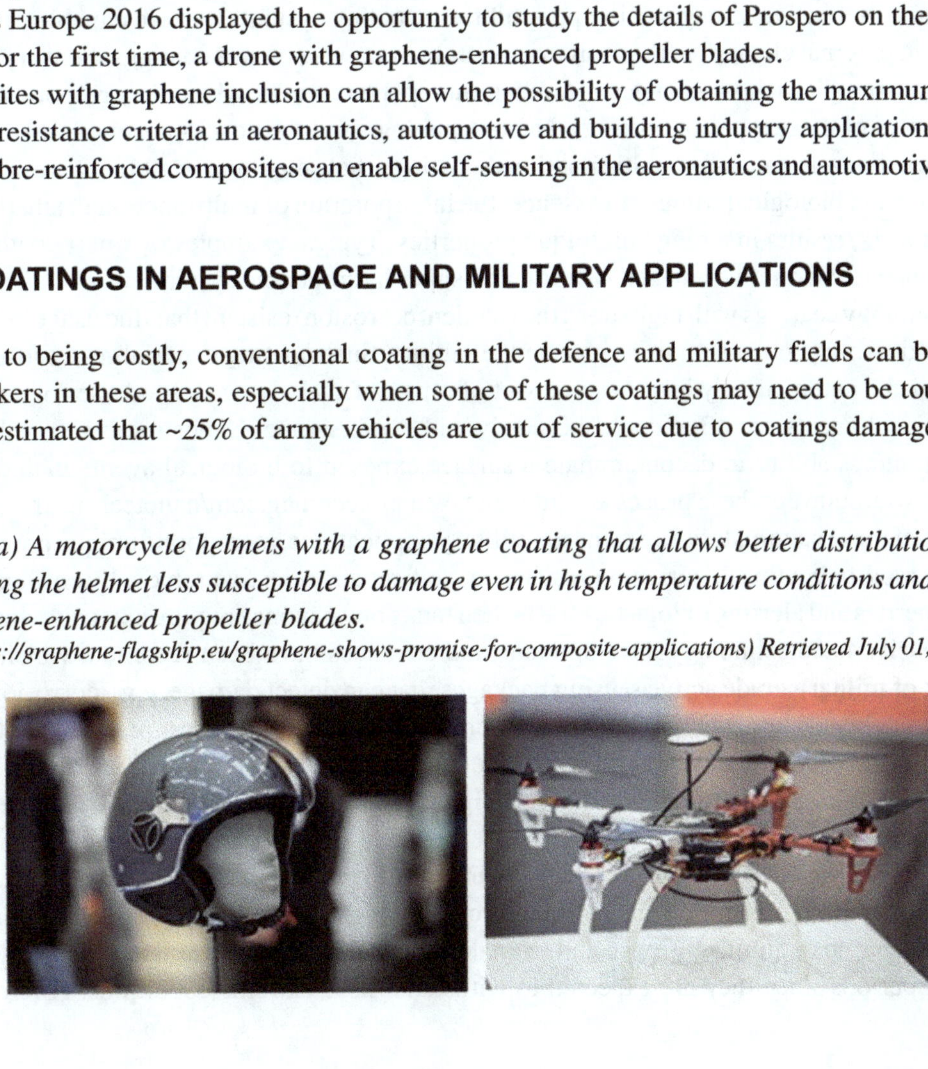

require repainting (https://www.azonano.com/article.aspx?ArticleID=568). The use of smart coatings and nanopaints for military vehicles, when corroded or scratched, can help to detect and mend/heal such defects. In addition, there is also the possibility for vehicles to change colour in the battlefield, thereby, creating an instant camouflage and rendering tanks and other military vehicles, difficult to detect. There are multi-functional nanocoatings for aerospace applications. Such nanocoatings can provide corrosion protection by using environmentally-safe materials. Of crucial importance, is there lightweight, high-strength, thermally stable nanomaterials that are highly desirable in aircraft engines. They can also sense corrosion and mechanical damage of aircraft skin. They can initiate responses to sensed damage (chemical and physical). In addition, they can achieve optimal adhesion by using environmentally-compliant materials and they can improve fatigue resistance and display colour on demand. The use of nanomaterials will make it possible for military vehicles, when corroded or scratched, to detect and heal such damaged surfaces. Overall, nanomaterials employed for coatings, allow for considerable cost savings in addition to improved quality coating.

The use of nanocoating finds application in military and commercial steel ship hulls and steel, aluminium or fiberglass boat hulls. When these structures are nanocoated, they become less- or non-susceptible to permanent adhesion of marine fouling organisms, such as: barnacles, sea-grasses etc. When organisms, such as: bacteria, barnacles and algae stick to the surfaces of ships, the coating renders such organisms ineffective to any fouling possibilities. Coating systems that are capable of responding dynamically to external changes in their environments are known as smart coatings. Sensory response to environmental stimuli, such as: changes in temperature, pH or current, where they respond accordingly, are the attributes of these types of coatings. Nanotechnology and the development of advanced materials have allowed new avenues for multi-functional coatings that sense and react to: corrosion, radiation stress, pressure and biological stimuli. In essence, the incorporation of multifunctional, smart nanofillers to polymer matrix, results in a range of unique properties. Typical examples of smart coatings include stimuli responsive, conductive and self-healing and super hydrophobic systems.

Nanotechnology coatings with high strength, excellent corrosion resistant that find usage in the military, can improve: durability, corrosion resistance and reliability. Such coating materials can sense damage or corrosion and can automatically initiate some sort of repair of such damage. Processes of decontamination of biological agents are often complex, costly and cumbersome. Nanostructured coating systems with the continuous ability to decontaminate a surface exposed to biological agents including spores, are available to circumvent these processes (http://www.nanotechmag.com/nanotech-in-the-army/). The development of these nanomaterial systems for military applications, incorporate unique properties, such as: ability to modify coatings', corrosion resistance, self-repair, selective removal, sensing, colouring, physical properties and alerting of logistics staff when tanks or weaponry require extensive and expensive repairs. Extensive nanotechnology developments and advances have been recorded, which enable the development of military grade active sensing packages that can detect damage, e.g., corrosion, substrate integrity, etc. and environmental conditions e.g., chemicals, radiation, gases, pH, temperature, strain, etc. With all the foregoing, nanotechnology, will no doubt, result in a significant impact on battlespace systems that are concerned with information and signal processing, autonomy and intelligence (http://www.nanotechmag.com/nanotech-in-the-army/).

Highly demanding and excruciating conditions are possible environments that military personnel and equipment, sometimes encounter. Nanostructured coating technology in military aircraft and turbine-powered vehicles and equipment is expected to enable these facilities to operate uninterrupted for relatively longer periods since they are expected to withstand these extreme conditions. Aircraft cockpits have witnessed the application of scratch resistant coatings. Generally, coatings are commonly used for the protection of the structures and surfaces of the aircraft from harsh environments. These environments can vary from wide range of temperature conditions, high pressure and providing drag reduction. Unfortunately, conventional aerospace coatings are grossly deficient in terms of high mechanical integrity, constant environmental constraints and financial drawbacks. Therefore, these constraints give rise to new opportunities that are required to tackle and have excellent thermal barrier properties, icephobic and protective nanocoatings.

For the reduction of noise, shock and thermal energy transmitted in the cabin by using firewall, hood/tunnel sidewall insulation, engine isolation treatments, trim treatments and floor mats made of hydrophobic melamine foam, HMF (see Figure 8a-c). This is in an attempt to improve on energy management. The materials have good flame resistance, reduced smoke generation, highly durable and with low toxicity emissions. A flexible acoustic grade, open-cell, ether-based urethane foam (Figure 8a) by POLYDAMP®, is designed to render maximum sound absorption of a given foam thickness. Ether foams are known to offer excellent resistance to moisture, heat and chemicals and at the same time, hydrolytically stable. A melamine-based foam (PMF) by POLYDAMP, is an extremely lightweight insulation material (Figure 8b). It boasts some exceptional resistance to heat, low flame propagation and smoke, just as it possesses excellent thermal and acoustical insulation properties. It meets the aircraft, transportation vehicles flame, toxicity and smoke standards, while it is resilient to compression set and vibration. Typically, applications include: environmental control ducting systems (ECS), heating, ventilation and air conditioning (HVAC) and wall/fuselage insulation in aircrafts. An aluminium foil-faced fiberglass fabric and polysulfonamide (PSA) composite (Figure 8c), by POLYTECH® Heat Shield, was designed to reflect radiant heat and at the same time, to provide excellent degree of protection to high heat exposure. This product has applications for fuel tanks, any other areas exposed to extreme heat from exhaust pipes, turbos and it is, primarily employed in the aerospace and shipboard industries.

Figure 9 shows a High Mobility Multi-purpose Wheeled Vehicle (HMMWV), where polymeric-based materials were employed for acoustic and vibration damping purposes.

Wound dressings that are coated with nanomaterials can allow for the targeted and controlled release of drugs and proteins for a specified time to a specific target. This is possible because of the unique property of nanoparticle aggregates suitable for wound dressing, which offers and allows the exciting opportunity of controlling the release of growth factors and other actives in order to accelerate wound healing.

There seems to be a great potential in the use energetic nano-sized particles in propulsion applications. This is achievable through the unique combustion properties of nanoparticles. These properties include: the very rapid ignition and short combustion times that make them to be exceptionally valuable for propulsion systems; since they can be included in solid propellants, solid fuels or even as energetic gellant in liquid rocket propellant systems and there is the feasibility of their applications in rocket motors. It has been reported that the US Air Force uses NanoCeram® alumina filter and Seldon filters for water filtration in portable water purifiers and for air filters inside aircraft, in addition to other applications that include biological warfare filters for the military, hospitals and other critical buildings (http://www.nanotechmag.com/nanotech-in-the-army/). As mentioned earlier, the potential for some special nanocoatings to change colour when required is a strong possibility and this can include the adaptive camouflage for tanks moving from the jungle to open fields or into urban areas.

Figure 8. (a and b): A variety of products that provide thermal and acoustic insulation for military engine compartments and firewalls. These include POLYDAMP® acousticfoams, POLYDAMP® acoustical foam barrier composites and urethane or vinyl-based materials where motor vehicle safety standards are highly essential and required
Source: (https://www.polytechinc.com/products/acoustic-foam#PHM) Retrieved July 5, 2018 and (c) POLYTECH® Heat Shield. Source: (https://www.polytechinc.com/products/thermal-insulation) Retrieved July 6, 2018

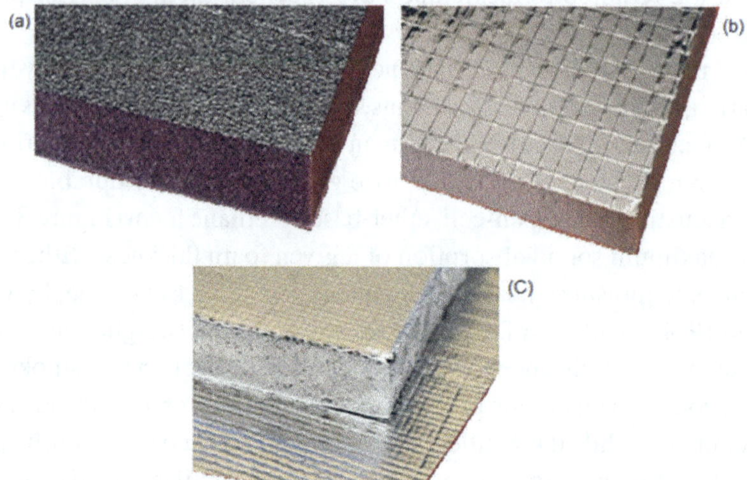

Figure 9. High Mobility Multi-purpose Wheeled Vehicle (HMMWV) where polymeric-based materials were employed for acoustic and vibration purposes.
Source: (https://www.polytechinc.com/industries/military-vehicles) Retrieved July 6, 2018

Advantages of Nanocoatings in the Aviation Industry

The aerospace industry relies, very highly, on high performance and low resistance (friction, drag etc.). In this respect, nanocoating has the potential to transform and significantly reduce fuel consumption and maintenance costs (Hu, 2006). The harsh high altitude environments that aircrafts operate in, demands good protection to surfaces, which is highly critical to ensuring predictable airflow over the aircraft outer skin. Low friction will be generated by the passage of air over the aircraft and hence, there will be a corresponding low drag coefficient, if the external surfaces of the aircraft are smooth (https://dabflow.co/nanocoatings-in-the-aviation-industry/). Consequently, this will contribute and translate to a considerable level of fuel economy and reduce the frequency of the aircraft going for routine maintenance on the airframe paintwork.

NANOTECHNOLOGY IN MILITARY AIRCRAFTS

Globally, research and development activities in nanotechnology keep increasing at such a phenomenon rate. In the military and in military aircrafts, especially, nanotechnology has potential applications for defence, especially in the field of: sensors, memory storage, nanorobotics, transducers, nanoelectronics, propellants and explosives in order to enhance the performance of devices and weapon systems. The nanomaterials needed for aircraft electro-communication components, i.e. carbon nanotubes (CNTs) require a unique set of properties that include ballistic electron transport and a huge current-carrying capacity. These properties make them to be of great interest for nanoelectronics. Polymer nanocomposites in the form of magnetic nanoparticles-iron oxide nanoparticles, i.e., Fe_2O_3 and Fe_3O_4-are incorporated in polymer films and composites that can be used in various data storage media/devices. On the other hand, ceramic nanoparticles, such as: barium titanate, barium strontium titanateare are used for making super capacitors. MEMS (Micro Electro Mechanical Systems) and NEMS (Nano Electro Mechanical Systems) offer the possibility of developing a standard fuel management unit, which controls the fuel control for aero-engines.

Research efforts are in progress towards the manipulation of the flow of energy within and between the molecules; this area of nanoscience is called "nano-energetics". It enables the building of extremely lethal weapons, e.g. the "cave-buster bombs". These bombs are envisaged, to have several times the detonation force of conventional bombs, such as the daisy bomb or the MOAB (mother of all bombs).

NANOTECHNOLOGY IN MILITARY UNMANNED AERIAL VEHICLES (UAVS)

Unmanned Aircraft Systems (UAS) has a few other names, which include: unmanned aerial vehicles (UAVs), drones and remotely piloted aircrafts (RPAs) and these nomenclatures may be used interchangeably. They have a long background or history in the defence and military applications and the coming of nanotechnology, has further advanced the usefulness and importance of UAVs in the military. There are different sizes of UASs, while the military and defence bodies have their own standard, civilians do have theirs. Classifications and sub-classes may be by: size, endurance or range and use a tier system that is employed by the military. The classifications of UAVs can be represented, adequately in Table 2. They can also be classified in terms of range of travel possible and their endurance in flight.

With the multiple asymmetrical security threats, facing many societies, UVSs may be useful tools to gather intelligence before terrorists strike. With such necessities, drones are (and will continue to be) effective assets in the anti-terror warfare. The Turkish military, recently invested into two types of drones: multicopter type strike drones and nano drones. Table 2 shows the different classes of UAVs.

UAVs with dimensions and with sizes similar to a large insect have lengths of between 30-50 cm are and classified as small UAVs, with the insect-like types that have flapping or rotary wings, being the more popular micro design. They are very light and very small in size and can be employed in spying and for biological warfare. On the other hand, the Small UAV class, sometimes called mini-UAV, applies to UAVs that have at least, one dimension greater than 50 cm, but not larger than 2 m. This category boasts of fixed-wing design and are often launched with hand, i.e. throwing them in the air. The deadly, insect-sized drones of the future is unobtrusive, invasive and lethal. It is believed that the US Air Force is reportedly developing winged drones that can sneak up on a suspected enemy as stealthily as a mosquito.

Even though smaller than aircraft, the medium class UAVs are rather too heavy to be carried by an individual. Usually, with a wingspan of between 5-10 m, they can carry a payload of between 100-200 kg. Examples include the: medium fixed-wing UAVs are the Israeli-US Hunter and the British Watchkeeper, US Boeing Eagle Eye, the RQ-2 Pioneer, the BAE systems Skyeye R4E and the RQ-5A Hunter. With a wingspan of ~10.2 m and length of ~6.9 m, it weighs ~885 kg at take-off. A typical crossover UAV that spans the specifications of a small and medium sized-UAV, is the RS-20 by the American Aerospace.

Large UAVs are used mainly for combat operations by the military. Typical examples of large UAVs are the US General Atomics Predator A and B and the US Northrop Grumman Global Hawk. Undoubtedly, the most interesting thing about the UAVs is that they can eliminate the need for an on-board pilot. However, many of those used in the military have the same combat and surveillance capabilities as a human piloted aircraft. Nanodrones, such as Black Hornet Nano and Hummingbird, are commonly used by the military to survey and spy on enemy territories and danger areas that are ordinarily not seen. Nanodrones can provide considerable tactical support and advantage and can therefore, assist in preventing or minimising causalities. Probably, in many countries, the most advanced nanodrones in existence have been designed for military use. Nanodrones designed for recreational use, however, are similar to the designs of their larger counterparts and they operate similarly, just as many of them share all of the same features in the average-sized drone. Affordability of these devices is the most important advantage to these nanodrones; they are relatively less prone to damages and they are not subject to FAA regulations as long as they weigh < 255.15 g. The increasing efforts to produce smaller and lighter subsystems, has no doubt, driven the development of nanotechnology and nanoelectromechanical systems. However, the implementation of nano-scale technologies in UASs, in particular, presents unique challenges and over-

Table 2. Classification of UAVs

Class	Size	Take-off Weight (kg)	Operating Altitude (m)	Air Speed (knots)
1	Very Small and Small	<9	<365.8#	<100
2	Medium	9.5-25	<1066.8	<250
3	Large	<600	<5486.4##	<250
4	X-Large	>600	>5486.4	Any air speed

#AGL = Above Ground Level,##MSL = Mean Sea Level.

Source: (https://www.e-education.psu.edu/geog892/node/5) Accessed June 20, 2018

Figure 10. (a) The Nano Hummingbird UAS by AeroVironment: weighs >½ kg and a wingspan of ~16 cm, a typical replica feature of a real hummingbird. Its two flapping wings.
Source: (https://www.defenceiq.com/defence-technology/articles/nano-drone-tech-is-advancing) Retrieved June 29, 2018 and (b) the Dragonfly insect, by the DragonflEye group that enables its propulsion and control.
Source: (https://www.defenceiq.com/defence-technology/articles/nano-drone-tech-is-advancing) Retrieved June 29, 2018

coming these difficulties have resulted in attempts to replicate the aerodynamic systems found in insects and birds. Figure 10 (a) shows the Nano Hummingbird UAS by AeroVironment and (b) Dragonfly insect.

FUTURE TREND/PERSPECTIVE, *VIS-À-VIS* THE FUTURE OF COMPOSITES AND NANOCOMPOSITES IN AIRCRAFTS CONSTRUCTION

It is common knowledge that composites were first introduced to the military aerospace industry. In future, the industry will keep continually demand for lighter and more efficient aircrafts, which may eventually allow aircrafts to be constructed of purely composite material instead of metal. However, for now, aircrafts are constructed mainly, in parts, with metals and composites. It is envisaged that materials manufacturers will have to upgrade the current composite systems so that they will be able to provide better and effective vibrational damping along with their lightweight properties. It is therefore, pertinent that aerospace manufacturers will have the need to continue working with material scientists and materials manufacturers in order to further develop and come up with innovations that will facilitate, invigorate and enhance the production of better and more efficient, safer and quieter aircrafts for military and civilian usages.

Nanotechnology that includes the development of military grade active sensing packages to detect damage (corrosion, substrate integrity, etc.) and environmental conditions, i.e. radiation, chemicals, temperature, gases, strain, etc., will be a common feature. Without a doubt, nanotechnology already has an impact on battlespace systems that are concerned with information and signal processing, autonomy and intelligence. The US Navy announced, in the year 2013, the deployment of a laser weapon on the USS Ponce (http://www.theglobaljournal.net/article/view/1132/), which may very well be expected to be employed in future warfare. It is expected that these laser weapons will be more widely deployed, with the aim to strengthen defence capabilities against missiles, artillery and mortars.

CHALLENGES

It is envisaged that nanomaterials will play a disruptive role in the next-generation thermal management for high power electronics in aerospace platforms, according to Barako et al. (2018). The wise usage of small receptor-enhancers that are designed to increase human alertness and reduce the reaction times can cause addiction and/or subsequent chronic fatigue syndrome. This, consequently, can lead to weakness and possibly neural damage and even maybe death. Hence, there deployment should be carefully monitored. When large quantities of smart weapons, especially miniaturized, robotic weapons and intelligent target-seeking ammunitions are deplored without reliable remote off-switches, this may lead to unexpected injuries to combatants and possibly civilians. This can lead to the destruction to infrastructure and the attendant environmental pollution. Technically, the performance of an airplane depends, mainly on number of variables, such as: structural efficiency, propulsion efficiency, aerodynamic efficiency and the speed of the aircraft. If these parameters can be estimated, the prediction of fuel economy/consumption may be relatively easy.

CONCLUSION

The more than 2 decades of basic nanoscience research and more than fifteen years of tailored research and developments, have led to various applications of nanotechnology that are delivering at an alarming rate, multiple results in areas envisaged and in unexpected manners that nanotechnology dreams have benefited the society, not with its unintended consequences, though.

In the not-too-distant a future, it is envisaged that there will be the development of nanomaterial systems for military applications by incorporating unique properties, such as: self-repair, selective removal, corrosion resistance, sensing, ability to modify coatings' physical properties, colorizing and alerting logistics staff when tanks or weaponry require more extensive repair. Significant proportion of the ~60kg load of a 21st century soldier emanates from the many electronic devices that are needed for communications, etc. An important goal of the military researches into nanomaterials is to aim to reduce such load to ~20kg, by following the route in the development of lighter and higher gravimetric energy density (Wh/kg) batteries. Since nanocoatings are applied onto the surface to be protected, in the event of a fire, a ceramic layer forms within seconds, which provides a protective lining against heat and hence reducing, very radically, the amount of smoke generated from a fire.

Nanocoatings in the aviation industry are find expanding and extensive applications in areas with a need to: increase an aircraft's resistance to extreme temperatures, increase its exposure corrosion in addition to an increase in abrasion and wear resistance and tear on the engine parts. Resulting from the reliable high-performance, there are increasing demands for nanocoatings because of their enduring intrinsic characteristics that offer many substantial advantages over and above the conventional coatings. The advent of nanocoating and its robust adoption by aerospace companies and their maintenance agencies, have added to the aviation industry, superior and distinct facial characteristics to aircraft interiors, structural frames and hence, extending them to their use to engine parts and component surfaces. Filters in airplane cabin and many other types of air filters are nanotechnology-based that allow "mechanical filtration". In this process, the fibre material creates nanoscale pores that trap particles larger than the size of the pores. Such filters may also contain charcoal layers that can remove odours. More than ever before, in the security and safety spheres, nanotechnology-enabled sensors and solutions allow high level of sensitivity, which can detect and identify chemical or biological agents in the air, water and soil (for agricultural purposes). In general, nanotechnology can offer the much-anticipated promise of developing multifunctional materials that can contribute to the building and maintaining of: lighter aircraft, spacecraft and ships, safer, smarter and more efficient and faster air, land and see vehicles, with improved structural integrity.

REFERENCES

Apeldorn, T., Fabris, F. W., & Altstädt, V., Lühnen, O., Hörhold, A. & Ranocha, K. (2010). Development of a Thermoplastic Printed Circuit Board for High Performance Applications in the Aviation Industry. *Polymers in Defence & Aerospace Applications*, *2*, 1-12.

Armstrong, G. (2010). Engineered Coatings for Composites and Polymers used in Defence and Aerospace–now and the future. In 2nd *International Conference on: Polymers in Defence & Aerospace Applications*, Hamburg, Germany, February 10-11 (pp. 162-171).

AzoNano. (2005). From nautalius to nanobo(a)ts: the visual construction of nanoscience. *AZO nano*. Retrieved from https://www.azonano.com/article.aspx?ArticleID=1466

AzoNano. (2004). Nanotechnology coatings in the defence and aerospace industry. *AZO nano*. Retrieved from https://www.azonano.com/article.aspx?ArticleID=568

Barako, M. T., Gambin, V., & Tice, J. (2018). Integrated nanomaterials for extreme thermal management: A perspective for aerospace applications. *Nanotechnology*, *29*(15).

Baratto, C., Todros, S., Faglia, G., Comini, E., Sberveglieri, G., Lettieri, S., ... Maddalena, P. (2009). Luminescence response of ZnO nanowires to gas adsorption. *Sensors and Actuators. B, Chemical*, *140*(2), 461–466. doi:10.1016/j.snb.2009.05.018

Borque Industries. (n.d.). Kryron CNT-MMC Alloys. Retrieved from http://bourqueindustries.com/pages/about_kryron.php

Burkitt, B., Malave, V., & Thomaier, R. (2012, March). Silicone based coatings for aircraft applications. In *Abstracts of papers of the American Chemical Society* (Vol. 243). Washington, DC: Amer. Chemical Soc.

Colpo, F., Humbert, L., & Botsis, J. (2007). Characterisation of residual stresses in a single fibre composite with FBG sensor. *Composites Science and Technology*, *67*(9), 1830–184. doi:10.1016/j.compscitech.2006.10.024

Composites World. (2009). GE aviation to use ceramic matrix composites in jet engines. Retrieved from https://www.compositesworld.com/news/ge-aviation-to-use-ceramic-matrix-composites-in-jet-engines

Crouch, I. G., Sandlin, J., & Thomas, S. (2017). Polymers and fibre-reinforced plastics. In I.G. Crouch (Ed.), The Science of Armour Materials (Ch. 5, pp. 203-268) Woodhead Publishing.

Dabflow. (n.d.). Nanocoateings in the aviation industry. Retrieved from https://dabflow.co/nanocoatings-in-the-aviation-industry/

de Farias, M. A., Coelho, L. A., & Pezzin, S. H. (2009). Synthesis of Epoxy-copolysilsesquioxanes Organicinorganic Hybrid Materials: Evaluation of Catalyst and Reaction Time in the Sol-Gel Process. In *Proceedings of the 10th Brazilian Polymer Congress*, Foz do Iguaçu.

Dinca, I., Manoliu, V., Stefan, A., Stan, A. & Llina, S. (2010). Nanocomposites and nanotechnologies in aerospace research. *National Institute for aerospace research and development*. Retrieved from http://www.romnet.net/nano/2010.02.03_prezentari/14_materiale_tribologie_dlDinca.pdf

Dutta, P., Chakravarty, S., & Sarma, N. S. (2016). Detection of nitroaromatic explosives using π-electron rich luminescent polymeric nanocomposites. *RSC Advances*, *6*(5), 3680–3689. doi:10.1039/C5RA20347C

El-Fatatry, A. (2007). Defence Applications. In Nanotechnology Aerospace Applications.

Evonik Industries. (2018). Material solutions for aviation and aerospace markets. *AZO Materials*. Retrieved from https://www.azom.com/article.aspx?ArticleID=15795

Galpaya, D. (2015). *Synthesis, Characterization and Applications of Graphene Oxide-Polymer Nanocomposites* [PhD Thesis]. Queensland University of Technology.

Goodwin, B. W., & Stiles, R. E. (2010). Phosphazene Elastomer use in Defence and Aerospace. In 2nd *International Conference on: Polymers in Defence & Aerospace Applications*, Hamburg, Germany, February 10-11 (pp. 14-21).

Government of Gujarat. (2017). Manufacturing of Uniform for Defence Personnel (Nanotechnology). Retrieved from http://vibrantgujarat.com/writereaddata/images/pdf/project-profiles/Uniform-for-defence-personnel-nanotechnology.pdf

Graphene flagship. (n.d.). Graphene shows promise for composite applications. Retrieved from https://graphene-flagship.eu/graphene-shows-promise-for-composite-applications

Hu, L. (2016). Introduction to Active Coatings for Smart Textiles. In *Active Coatings for Smart Textiles* (pp. 1–7). Woodhead Publishing Series in Textiles. doi:10.1016/B978-0-08-100263-6.00001-0

Huang, W. M., Fu, Y. Q., & Zhao, Y. (2010). High Performance Polyurethane Shape-Memory Polymer and Composites. In 2nd *International Conference on: Polymers in Defence & Aerospace Applications*, Hamburg, Germany, February 10-11 (pp. 136-144).

Jackson, R. (2017). Small is beautiful: nano drone tech is advancing. *DefenceIQ*. Retrieved from https://www.defenceiq.com/defence-technology/articles/nano-drone-tech-is-advancing

Kausar, A., Rafique, I., & Muhammad, B. (2017). Aerospace Application of Polymer Nanocomposite with Carbon Nanotube, Graphite, Graphene Oxide and Nanoclay. *Polymer-Plastics Technology and Engineering*, *56*(13), 1438–1456.

Khan, D. M., Kausar, A., & Salman, S. M. (2016). Exploitation of Nanobifiller in Polymer/Graphene Oxide-Carbon Nanotube, Polymer/Graphene Oxide-Nanodiamond, and Polymer/Graphene Oxide-Montmorillonite Composite: A Review. *Polymer-Plastics Technology and Engineering*, *55*(7), 744–768.

Kharat, D. K., Muthurajan, H., & Praveenkumar, B. (2006). Present and Futuristic Military Applications of Nanodevices, Synthesis and Reactivity in Inorganic. *Metal-Organic and Nano-Metal Chemistry*, *36*(2), 231–235. doi:10.1080/15533170500524801

Khatavkar, N., & Balasubramanian, K. (2016). Composite materials for supersonic aircraft radomes with ameliorated radio frequency transmission-a review. *RSC Advances*, *6*(8), 6709–6718. doi:10.1039/C5RA18712E

Kurahatti, R. V., Surendranathan, A. O., Kori, S. A., Singh, N., Kumar, A., & Srivastava, S. (2010). Defence Applications of Polymer Nanocomposites. *Defence Science Journal*, *60*(5), 551–563. doi:10.14429/dsj.60.578

Lahiff, E., Ryu, C. Y., Curran, S. A., Minett, A., Blau, W. J., & Ajayan, P. M. (2003). Selective Positioning and Density Control of Nanotubes within a Polymer Thin Film. *Nano Letters*, *3*(10), 1333–1337. doi:10.1021/nl034273e

Luchtvaarfeiten. (n.d.). Composites in commercial aircraft engines fact sheet. Retrieved from http://aviationfacts.eu/uploads/thema/file_en/585919d470726f2696000000/Composites_in_Commercial_Aircraft_Engines_Fact_sheet.pdf

Lynch-Branzoi, J. (2017). Modulus enhancement of polymer nanocomposites reinforced with graphene nanoflakes. In *19th International Conference on Nanotechnology and Expo*, Atlanta, GA, Rutgers University, November 13-14. 10.4172/2157-7439-C1-061

Meng, Q., & Hu, J. (2008). Self-organizing alignment of carbon nanotube in shape memory segmented fiber prepared by in situ polymerization and melt spinning. *Part A*, *39*(2), 314–321. doi:10.1016/j.compositesa.2007.10.007

Monti, M. (2010). Development of Multifunctional Advanced Composites with Polymer Nanocomposite Matrices for Aerospace Applications. In 2nd *International Conference on: Polymers in Defence & Aerospace Applications*, Hamburg, Germany, February 10-11 (pp. 86-91).

Mortara, S. (2010). Use of Fluoropolymers in Aerospace and Defence: New Applications and Advantages. In 2nd *International Conference on: Polymers in Defence & Aerospace Applications*, Hamburg, Germany, February 10-11 (pp. 60-67).

Nallayan, W. A., & Vijayakumar, K. R. (2017). Comparison of the Effect of Curing on the Properties of E-Glass/Cyanate modified Epoxy Cross Plied Laminates. *Materials Science and Engineering*, *197*(1).

Nano work. (2006). Military nanotechnology- how worried should we be? Retrieved from https://www.nanowerk.com/spotlight/spotid=1015.php

Nanotech. (2014). Nanotechnology in military applications. Retrieved from http://www.nanotechmag.com/nanotech-in-the-army/

Nau, H. (2013). The future of nanotechnology in warfare. *Global*. Retrieved from http://www.theglobaljournal.net/article/view/1132/

Ni, Q. Q., Zhang, C., Fu, Y., Dai, G., & Kimura, T. (2007). Shape memory effect and mechanical properties of carbon nanotube/shape memory polymer nanocomposites. *Composite Structures*, *81*(2), 176–184. doi:10.1016/j.compstruct.2006.08.017

Njuguna, J., & Pielichowsk, K. (2010). Ageing and Performance Predictions of Polymer Nanocomposites for Exterior Aerospace and Defence Applications. In *2nd International Conference on: Polymers in Defence & Aerospace Applications*, Hamburg, Germany, February 10-11 (pp. 146-157).

Njuguna, J., & Pielichowski, K. (2003). Polymer Nanocomposites for Aerospace Applications: Properties. *Advanced Engineering Materials*, *5*(11), 769–778. doi:10.1002/adem.200310101

Njuguna, J., & Pielichowski, K. (2004a). Polymer Nanocomposites for Aerospace Applications: Fabrication. *Adv. Mater. Sci.*, *6*(4), 193-203.

Njuguna, J., & Pielichowski, K. (2004b). Polymer Nanocomposites for Aerospace Applications: Characterization. *Adv. Mater. Sci.*, *6*(4), 204-210.

Njuguna, J., Pielichowski, K., & Fan, J. (2012). Polymer nanocomposites for aerospace applications. In Adv. Polymer Nanocomposites, Types and Applications, Science and Engineering (Ch. 15, pp. 472-539). Woodhead Publishing.

Oliveira, A., Amico, S. C., Risi, C. L. S., Pezzin, S. H., & Coelho, L. A. F. (2009). Production of Tricomponent Composites (Carbon fiber/Epoxy/Carbon nanotubes) by Resin Transfer Molding. In *Proceedings of the 11th International Conference on Advanced Materials*, Rio de Janeiro.

Penn State. (n.d.). Classification of unmanned aerial systems. Retrieved from https://www.e-education.psu.edu/geog892/node/5

Pezzin, S. H., Coelho, L. A. F. & Amico, S.C. (2010). New Generation of Multi-functional Composites with Carbon Nanotubes for Space Applications. In *2nd International Conference on: Polymers in Defence & Aerospace Applications*, Hamburg, Germany, February 10-11 (pp. 102-108).

Polytechinc. (n.d.). Acoustic foam. Retrieved from https://www.polytechinc.com/products/acoustic-foam#PHM

Polytechinc. (n.d.). Thermal insulation. Retrieved from https://www.polytechinc.com/products/thermal-insulation

Risi, C. L. S., Hattenhauer, I., Coelho, L. A. F., Ramos, A., & Pezzin, S. H. (2009). *The Role of Curing rate on the Growth of Percolated Networks in Epoxy/MWNT Nanocomposites under an Electrical Field.* In Proceedings of the 11th International Conference on Advanced Materials, Rio de Janeiro.

Ruiz, F.A. (2000). *Proceedings of Polymers Laminations and Coatings Conference.*

Soutter, W. (2012). Nanotechnology in the military. *AZO nano*. Retrieved from https://www.azonano.com/article.aspx?ArticleID=3028

Tingle, R. (2017). Stormtrooper 2017: Russia unveils its hi-tech motorised 'Star Wars' armour for soldiers complete with exoskeleton, night vision and helmet. *Daily Mail*. Retrieved from http://www.dailymail.co.uk/news/article-4653886/Russia-unveils-generation-Star-Wars-combat-uniform.html

Vacogne, C. (2010). Rapid Manufacture of Structural Thermoplastic Composite Components for Aerospace and Defence Applications. In *2nd International Conference on: Polymers in Defence & Aerospace Applications*, Hamburg, Germany, February 10-11.

Young, R. (2010). Polymers in Defence and Aerospace Applications. In *Hamburg Conference*, February 10-11.

Compilation of References

Abbas, M., Iftikhar, H., Malik, M. H., & Nazir, A. (2018). Surface coatings of TiO_2 nanoparticles onto the designed fabrics for enhanced self-cleaning properties. *Coatings*, *8*(35), 1–9.

Abd El-Lateef, H. M., & Khalaf, M. M. (2015). Corrosion resistance of ZrO_2-TiO_2 nanocomposite multilayer thin films coated on carbon steel in hydrochloric acid solution. *Materials Characterization*, *108*, 29–41. doi:10.1016/j.matchar.2015.08.010

Abdin, Z., Alim, M. A., Saidur, R., Islam, M. R., Rashmi, W., Mekhilef, S., & Wadi, A. (2013). Solar energy harvesting with the application of nanotechnology. *Renewable & Sustainable Energy Reviews*, *26*, 837–852. doi:10.1016/j.rser.2013.06.023

Abdiryim, T., Ubul, A., Jamal, R., & Rahman, A. (2012). Solid-State Synthesis of Polyaniline/Single-Walled Carbon Nanotubes: A Comparative Study with Polyaniline/Multi-Walled Carbon Nanotubes. *Materials (Basel)*, *5*(7), 1219–1231. doi:10.3390/ma5071219

Abdollahnejad, Z., Pacheco-torgal, F., Félix, T., Tahri, W., & Aguiar, J. B. (2015). Mix design, properties and cost analysis of fly ash-based geopolymer foam. *Construction & Building Materials*, *80*, 18–30. doi:10.1016/j.conbuildmat.2015.01.063

Abdulhakeem, B., Farshad, B., Damilola, M., Fatemeh, T., Mopeli, F., Julien, D., & Ncholu, M. (2014). Morphological characterization and impedance spectroscopy study of porous 3D carbons based on graphene foam-PVA/phenol-formaldehyde resin composite as an electrode material for supercapacitors. *RSC Advances*, *4*(73), 39066–39072. doi:10.1039/C4RA05425C

Abraham, T. N., & George, K. E. (2009). Studies on recyclable nylon-reinforced PP composites: Effect of fiber diameter. *Journal of Thermoplastic Composite Materials*, *22*(1), 5–20. doi:10.1177/0892705708091603

Abu-Thabit, N. Y., & Makhlouf, A. S. H. (2015). Recent advances in nanocomposite coatings for corrosion protection application. In *Handbook of nanoceramics and nanocomposite coatings and materials* (pp. 515–549). Amsterdam: Elsevier. doi:10.1016/B978-0-12-799947-0.00024-9

Agarwala, V., Chandra Agarwala, R., & Sunder Daniel, B. S. (2006). Development of nanograined metallic Materials by bulk and coating techniques. *Synthesis and Reactivity in Inorganic, Metal-Organic, and Nano-Metal Chemistry*, *36*(1), 3–16. doi:10.1080/15533170500471128

Aglan, A., Allie, A., Ludwick, A., & Koons, L. (2007). Formulation and evaluation of nano-structured polymeric coatings for corrosion protection. *Surface and Coatings Technology*, *202*(2), 370–378. doi:10.1016/j.surfcoat.2007.05.090

Ahluwalia, R. K., & Wang, X. (2008). Fuel cell systems for transportation: Status and trends. *Journal of Power Sources*, *177*(1), 167–176. doi:10.1016/j.jpowsour.2007.10.026

Ahuja, A. (2017, November 26). Lack of organ donation in India: Here is why half a million people die annually in India due to unavailability of organs. *NDTV*. Retrieved from http://sites.ndtv.com/moretogive/lack-organ-donation-india-half-million-people-die-annually-india-due-unavailability-organs-2107/

Ajayan, P. M., Schadler, L. S., & Braun, P. V. (2003). Nanocomposite Science and Technology. Wiley. doi:10.1002/3527602127

Akafuah, N. K., Poozesh, S., Salaimeh, A., Patrick, C., Lawler, K., & Saito, K. (2016). Evolution of the Automotive Body Coating Process-A Review. *Coatings*, *6*(24), 1–22.

Akinwolemiwa, B., Wei, C., & Chen, G. Z. (2017). Mechanisms and Designs of Asymmetrical Electrochemical Capacitors. *Electrochimica Acta*, *247*, 344–357. doi:10.1016/j.electacta.2017.06.088

Alcock, B., Cabrera, N. O., Barkoula, N.-M., Loos, J., & Peijs, T. (2006). The mechanical properties of unidirectional all-polypropylene composites. *Composites. Part A, Applied Science and Manufacturing*, *37*(5), 716–726. doi:10.1016/j.compositesa.2005.07.002

Alex, R. (2010). Nanofilters in rubber and rubber blends. In Rubber nanocomposites-Preparation, properties, and applications (pp. 231-233). Singapore: John Wiley & Sons (Asia) Pte Ltd.

Alexandre, M., & Dubois, P. (2000). Polymer-layered silicate nanocomposites: Preparation, properties and uses of a new class of materials. *Materials Science and Engineering*, *28*(1-2), 1–63. doi:10.1016/S0927-796X(00)00012-7

Al-Harahsheh, M., Al-Zboon, K., Al-Makhadmeh, L., Muhannad, H., & Mehaysen, M. (2015). Fly ash based geopolymer for heavy metal removal: A case study on copper removal. *Journal of Environmental Chemical Engineering*, *3*(3), 1669–1677. doi:10.1016/j.jece.2015.06.005

Alipanah, M., & Li, X. (2016). Numerical studies of lithium-ion battery thermal management systems using phase change materials and metal foams. *International Journal of Heat and Mass Transfer*, *102*, 1159–1168. doi:10.1016/j.ijheatmasstransfer.2016.07.010

Alipour, K., & Nasirpouri, F. (2017). Smart anti-corrosion self-healing zinc metal-based molybdate functionalized-mesoporous-silica (MCM-41) nanocomposite coatings. *RSC Advances*, *7*(82), 51879–51887. doi:10.1039/C7RA06923E

Al-Majed, A. A., Neumann, C. M., Brushart, T. M., & Gordon, T. (2000). Brief electrical stimulation promotes the speed and accuracy of motor axonal regeneration. *The Journal of Neuroscience*, *20*(7), 2602–2608. doi:10.1523/JNEUROSCI.20-07-02602.2000 PMID:10729340

AL-Oqla, F. M., & Sapuan, S. M. (2014). Natural fiber reinforced polymer composites in industrial applications: Feasibility of date palm fibers for sustainable automotive industry. *Journal of Cleaner Production*, *66*, 347–354. doi:10.1016/j.jclepro.2013.10.050

Al-Zboon, K., Al-Harahsheh, M. S., & Falah, B. H. (2011). Fly ash-based geopolymer for Pb removal from aqueous solution. *Journal of Hazardous Materials*, *188*(1-3), 414–421. doi:10.1016/j.jhazmat.2011.01.133 PMID:21349635

Amaya, R. O., Matsumoto, Y., Flores-Conde, A., Pérez-Guzmán, M. A., & Ortega-López, M. (2016). In-situ formation of rGO quantum dots during GO reduction via interaction with citric acid in aqueous medium. *Mater. Res. Express*, *3*(10), 105601K. doi:10.1088/2053-1591/3/10/105601

Amerio, E., Fabbri, P., Malucelli, G., Messori, M., Sangermano, M., & Taurino, R. (2008). Scratch resistance of nano-silica reinforced acrylic coatings. *Progress in Organic Coatings*, *62*(2), 129–133. doi:10.1016/j.porgcoat.2007.10.003

Ammala, A., Hill, A. J., Meakin, P., Pas, S. A., & Turney, T. W. (2002). Degradation studies of polyolefins incorporating transparent nanoparticulate zinc oxide UV stabilizers. *Journal of Nanoparticle Research*, *4*(1-2), 167–174. doi:10.1023/A:1020121700825

Anandhan, S., & Bandyopadhyay, S. (2011). Polymer Nanocomposites: From Synthesis to Applications, Nanocomposites and Polymers with Analytical Methods. InTech.

Andrejkovicová, S., Sudagar, A., Rocha, J., Patinha, C., Hajjaji, W., Ferreira da Silva, E., ... Rocha, F. (2016). The effect of natural zeolite on microstructure, mechanical and heavy metals adsorption properties of metakaolin based geopolymers. *Applied Clay Science*, *126*, 141–152. doi:10.1016/j.clay.2016.03.009

Andrews, R., & Weisenberger, M. C. (2004). Carbon nanotube polymer composites. *Current Opinion in Solid State and Materials Science*, *8*(1), 31–37. doi:10.1016/j.cossms.2003.10.006

An, K. H., Kim, W. S., Park, Y. S., Choi, Y. C., Lee, S. M., Chung, D. C., ... Lee, Y. H. (2001). Supercapacitors using single-walled carbon nanotube electrodes. *Advanced Materials*, *13*(7), 497–500. doi:10.1002/1521-4095(200104)13:7<497::AID-ADMA497>3.0.CO;2-H

Ansari, R., & Keivani, M. B. (2006). Polyaniline conducting electroactive polymers thermal and environmental stability studies. *Journal of Chemistry*, *3*(4), 202–217.

Anya, C. C. (1999). Microstructural nature of strengthening and toughening in Al2O3-SiC(P) nanocomposites. *Journal of Materials Science*, *34*(22), 5557–5567. doi:10.1023/A:1004729015686

Apeldorn, T., Fabris, F. W., & Altstädt, V., Lühnen, O., Hörhold, A. & Ranocha, K. (2010). Development of a Thermoplastic Printed Circuit Board for High Performance Applications in the Aviation Industry. *Polymers in Defence & Aerospace Applications*, *2*, 1-12.

Arbizzani, C., Mastragostino, M., & Meneghello, L. (1996). Polymer-based redox super-capacitors: A comparative study. *Electrochimica Acta*, *41*(1), 21–26. doi:10.1016/0013-4686(95)00289-Q

Armstrong, G. (2010). Engineered Coatings for Composites and Polymers used in Defence and Aerospace–now and the future. In 2nd *International Conference on: Polymers in Defence & Aerospace Applications*, Hamburg, Germany, February 10-11 (pp. 162-171).

Ashby, M. F., Ferreira, P. J., & Schodek, D. L. (2009). *Nanomaterials, Nanotechnologies, and design: An introduction for engineers and architects.* Elsevier.

Ashley, M., Magiera, C., Ramidi, P., Blackburn, G., Scott, T. G., Gupta, R., ... Biswas, A. (2012). Nanomaterials and processes for carbon capture and conversion into useful by-products for a sustainable energy future. *Greenhouse Gases: Science and Technology*, *2*(6), 419–444. doi:10.1002/ghg.1317

Asmatulu, R., Diouf, D., Moniruddin, M., & Nuraje, N. (2016). Enhanced anti-weathering of nanocomposite coatings with silanized graphene nanomaterials. *International Journal of Engineering Research and Application*, *6*(6), 23–36.

Ateh, D. D., Navsaria, H. A., & Vadgama, P. (2006). Polypyrrole-based conducting polymers and interactions with biological tissues. *Journal of the Royal Society, Interface*, *3*(11), 741–752. doi:10.1098/rsif.2006.0141 PMID:17015302

Ates, M., Serin, M. A., Ekmen, I., & Ertas, Y.N. (2015). Supercapacitor behaviors of Polyaniline/CuO, polypyrrole/CuO and PEDOT/CuO nanocomposites. *Polymer Bulletin*, *72*(10), 2573–2589. doi:10.100700289-015-1422-4

Augustyn, V., Simon, P., & Dunn, B. (2014). Pseudocapacitive oxide materials for high-rate electrochemical energy storage. *Energy & Environmental Science*, *7*(5), 1597–1614. doi:10.1039/c3ee44164d

Ayad, M. M., Amer, W. A., & Kotp, M. G. (2017). Magnetic polyaniline-chitosan nanocomposite decorated with palladium nanoparticles for enhanced catalytic reduction of 4-nitrophenol. *Molecular Catalysis*, *439*, 72–80. doi:10.1016/j.mcat.2017.06.023

Ayad, M. M., Amer, W. A., Kotp, M. G., Minisy, I. M., Rehab, A. F., Kopecký, D., & Fitl, P. (2017). Synthesis of silver-anchored polyaniline–chitosan magnetic nanocomposite: A smart system for catalysis. *RSC Advances*, *7*(30), 18553–18560. doi:10.1039/C7RA02575K

Azam, M., & Samad, M. (2018). A novel organoclay reinforced UHMWPE nanocomposite coating for tribological applications. *Progress in Organic Coatings*, *118*, 97–107. doi:10.1016/j.porgcoat.2018.01.028

AzoNano. (2004). Nanotechnology coatings in the defence and aerospace industry. *AZO nano*. Retrieved from https://www.azonano.com/article.aspx?ArticleID=568

AzoNano. (2005). From nautalius to nanobo(a)ts: the visual construction of nanoscience. *AZO nano*. Retrieved from https://www.azonano.com/article.aspx?ArticleID=1466

Azo-Nano. (2015). Properties and Applications of Graphene Oxide and Functionalized Graphene. Retrieved from https://www.azonano.com/article.aspx?ArticleID=4044

Babaei, A., & Arefazar, A. (2015). Structural, rheological, and mechanical properties of PA6/SAN/SEBS ternary blend/organoclay nanocomposites. *Journal of Applied Polymer Science*, *132*(20). doi:10.1002/app.41969

Bae, S. H., Che, J. H., Seo, J. M., Jeong, J., Kim, E. T., Lee, S. W., ... Chung, H. (2012). *In vitro* biocompatibility of various polymer-based microelectrode arrays for retinal prosthesis. *Investigative Ophthalmology & Visual Science*, *53*(6), 2653–2657. doi:10.1167/iovs.11-9341 PMID:22427592

Bahr, J. L., & Tour, J. M. (2002). Covalent chemistry of single-wall carbon nanotubes. *Journal of Materials Chemistry*, *12*(7), 1952–1958. doi:10.1039/b201013p

Bai, C., & Colombo, P. (2016). High-porosity geopolymer membrane supports by peroxide route with the addition of egg white as surfactant. *Ceramics International*, 1–7. doi:10.1016/j.ceramint.2016.10.205

Bai, C., & Colombo, P. (2018). Processing, properties and applications of highly porous geopolymers: A review. *Ceramics International*, *44*(14), 16103–16118. doi:10.1016/j.ceramint.2018.05.219

Bai, M.-H., Liu, T.-Y., Luan, F., Li, Y., & Liu, X.-X. (2014). Electrodeposition of vanadium oxide-polyaniline composite nanowire electrodes for high energy density supercapacitors. *Journal of Materials Chemistry. A, Materials for Energy and Sustainability*, *2*(28), 10882–10888. doi:10.1039/C3TA15391F

Bakhshandeh, E., Jannesari, A., Ranjbar, Z., Sobhani, S., & Saeb, M. R. (2014). Anti-corrosion hybrid coatings based on epoxy–silica nano-composites: Toward relationship between the morphology and EIS data. *Progress in Organic Coatings*, *77*(7), 1169–1183. doi:10.1016/j.porgcoat.2014.04.005

Balandin, A. A., Ghosh, S., Bao, W., Calizo, I., Teweldebrhan, D., Miao, F., & Lau, C. N. (2008). Superior thermal conductivity of single-layer graphene. *Nano Letters*, *8*(3), 902–907. doi:10.1021/nl0731872 PMID:18284217

Balint, R., Cassidy, N. J., & Cartmell, S. H. (2014). Conductive polymers: Towards a smart biomaterial for tissue engineering. *Acta Biomaterialia*, *10*(6), 2341–2353. doi:10.1016/j.actbio.2014.02.015 PMID:24556448

Bandurin, D. A., Kleshch, V. I., Smolnikova, E. A., Obronov, I. V., Nasibulin, A. G., Kauppinen, E. I., & Obraztsov, A. (2013). Scanning anode field emission microscopy of nanocarbons. *Journal of Nanoelectronics and Optoelectronics*, *8*(1), 114–118. doi:10.1166/jno.2013.1431

Bandyopadhyay, P., Kuila, T., Balamurugan, J., Nguyen, T. T., Kim, N. H., & Lee, J. H. (2017). Facile synthesis of novel sulfonated polyaniline functionalized graphene using m-aminobenzene sulfonic acid for asymmetric supercapacitor application. *Chemical Engineering Journal*, *308*, 1174–1184. doi:10.1016/j.cej.2016.10.015

Baneyx, F. (2011). Review of biomimetic and bioinspired nanomaterials. *Journal of the American Chemical Society*, *133*(13), 5161–5161. doi:10.1021/ja2018338

Baniasadi, H., SA, A. R. & Mashayekhan, S. (2015). Fabrication and characterization of conductive chitosan/gelatin-based scaffolds for nerve tissue engineering. *International Journal of Biological Macromolecules*, *74*, 360–366. doi:10.1016/j.ijbiomac.2014.12.014 PMID:25553968

Bao, C., Guo, Y., Yuan, B., Hu, Y., & Song, L. (2012). Functionalized graphene oxide for fire safety applications of polymers: A combination of condensed phase flame retardant strategies. *Journal of Materials Chemistry*, *22*(43), 23057–23063. doi:10.1039/c2jm35001g

Bao, C., Song, L., Xing, W., Yuan, B., Wilkie, C. A., Huang, J., ... Hu, Y. (2012). Preparation of graphene by pressurized oxidation and multiplex reduction and its polymer nanocomposites by masterbatch-based melt blending. *Journal of Materials Chemistry*, *22*(13), 6088–6096. doi:10.1039/c2jm16203b

Bao, Z., Dodabalapur, A., & Lovinger, A. J. (1996). Soluble and processable regioregular poly (3-hexylthiophene) for thin film field-effect transistor applications with high mobility. *Applied Physics Letters*, *69*(26), 4108–4110. doi:10.1063/1.117834

Barako, M. T., Gambin, V., & Tice, J. (2018). Integrated nanomaterials for extreme thermal management: A perspective for aerospace applications. *Nanotechnology*, *29*(15).

Baratto, C., Todros, S., Faglia, G., Comini, E., Sberveglieri, G., Lettieri, S., ... Maddalena, P. (2009). Luminescence response of ZnO nanowires to gas adsorption. *Sensors and Actuators. B, Chemical*, *140*(2), 461–466. doi:10.1016/j.snb.2009.05.018

Barber, P., Balasubramanian, S., Anguchamy, Y., Gong, S., Wibowo, A., Gao, H., ... zur Loye, H.-C. (2009). Polymer composite and nanocomposite dielectric materials for pulse power energy storage. *Materials (Basel)*, *2*(4), 1697–1733. doi:10.3390/ma2041697

Bard, A. J., Faulkner, L. R., Leddy, J., & Zoski, C. G. (1980). *Electrochemical Methods: Fundamentals and Applications* (Vol. 2). New York: Wiley.

Barltrop, J. A., Owen, T. C., Cory, A. H., & Cory, J. G. (1991). 5-(3-carboxymethoxyphenyl)-2-(4, 5-dimethylthiazolyl)-3-(4-sulfophenyl) tetrazolium, inner salt (MTS) and related analogs of 3-(4, 5-dimethylthiazolyl)-2, 5-diphenyltetrazolium bromide (MTT) reducing to purple water-soluble formazans as cell-viability indicators. *Bioorganic & Medicinal Chemistry Letters*, *1*(11), 611–614. doi:10.1016/S0960-894X(01)81162-8

Barzegar, F., Dangbegnon, J. K., Bello, A., Momodu, D. Y., & Charlie, A. T. Jr. (2015). Effect of conductive additives to gel electrolytes on activated carbon-based supercapacitors. *AIP Advances*, *5*(9), 097171. doi:10.1063/1.4931956

Baur, J., & Silverman, E. (2007). Challenges and opportunities in multifunctional nanocomposite structures for aerospace applications. *MRS Bulletin*, *32*(4), 328–334. doi:10.1557/mrs2007.231

Bautista, Y., Gonzalez, J., Gilabert, J., Ibanez, M. J., & Sanz, V. (2011). Correlation between the wear resistance, and the scratch resistance, for nanocomposite coatings. *Progress in Organic Coatings*, *70*(4), 178–185. doi:10.1016/j.porgcoat.2010.09.022

Bavio, M. A., Acosta, G. G., & Kessler, T. (2014). Polyaniline and polyaniline-carbon black nanostructures as electrochemical capacitor electrode materials. *International Journal of Hydrogen Energy*, *39*(16), 8582–8589. doi:10.1016/j.ijhydene.2014.01.018

Beidaghia, M., & Gogotsi, Y. (2014). Capacitive energy storage in micro-scale devices: Recent advances in design and fabrication of micro-supercapacitors. *Energy & Environmental Science*, *7*(3), 867–884. doi:10.1039/c3ee43526a

Belanger, D., Ren, X., Davey, J., Uribe, F., & Gottesfeld, S. (2000). Characterization and long-term performance of polyaniline-based electrochemical capacitors. *Journal of the Electrochemical Society*, *147*(8), 2923–2929. doi:10.1149/1.1393626

Bello, A., Barzegar, F., Momodu, D., Dangbegnon, J., Taghizadeh, F., Fabiane, M., & Manyala, N. (2015). Asymmetric supercapacitor based on nanostructured graphene foam/polyvinyl alcohol/formaldehyde and activated carbon electrodes. *Journal of Power Sources*, *273*, 305–311. doi:10.1016/j.jpowsour.2014.09.094

Bendorf, A., Kelly, P. J., Kerridge, I. H., McCaughan, G. W., Myerson, B., Stewart, C., & Pussell, B. A. (2013). An international comparison of the effect of policy shifts to organ donation following cardiocirculatory death (DCD) on donation rates after brain death (DBD) and transplantation rates. *PLoS One*, *8*(5), e62010. doi:10.1371/journal.pone.0062010 PMID:23667452

Berry, V. (2013). Impermeability of graphene and its applications. *Carbon*, *62*, 1–10. doi:10.1016/j.carbon.2013.05.052

Beyer, G. (2002). Nanocomposites: A new class of flame retardants for polymers. *Plastics Additives & Compounding*, *4*(10), 22–28. doi:10.1016/S1464-391X(02)80151-9

Bhadra, S., & Lee, J. H. (2009). Synthesis of higher soluble nanostructured polyaniline by vapor-phase polymerization and determination of its crystal structure. *Appl. Polymer Science*, *114*(1), 331–340.

Bhandari, J., Khan, F., Abbassi, R., Garaniya, V., & Ojeda, R. (2015). Modelling of pitting corrosion in marine and offshore steel structures–A technical review. *Journal of Loss Prevention in the Process Industries*, *37*, 39–62. doi:10.1016/j.jlp.2015.06.008

Bhang, S. H., Jeong, S. I., Lee, T. J., Jun, I., Lee, Y. B., Kim, B. S., & Shin, H. (2012). Electroactive Electrospun Polyaniline/Poly [(L-lactide)-co-(ε-caprolactone)] Fibers for Control of Neural Cell Function. *Macromolecular Bioscience*, *12*(3), 402–411. doi:10.1002/mabi.201100333 PMID:22213547

Bhattacharya, S., Saha, I., Mukhopadhyay, A., Chattopadhyay, D., Ghosh, U.C. & Chatterjee, D. (2013). Role of nanotechnology in water treatment and purification: Potential applications and implications. *International Journal of Chemical Science and Technology.*

Bhattacharya, M. (2016). Polymer nanocomposites-A comparison between carbon nanotubes, graphene, and clay as nanofillers. *Materials (Basel)*, *9*(4), 262. doi:10.3390/ma9040262 PMID:28773388

Bhowmick, B., Bain, M. K., Maity, D., Bera, N. K., Mondal, D., Mollick, M. M. R., ... Chattopadhyay, D. (2012). Synthesis of dendritic polyaniline nanofibers by using soft template of sodium alginate. *Journal of Applied Polymer Science*, *123*(3), 1630–1635. doi:10.1002/app.34575

Bian, C., & Yu, A. (2010). De-doped polyaniline nanofibres with micropores for high-rate aqueous electrochemical capacitor. *Synthetic Metals*, *160*(13-14), 1579–1583. doi:10.1016/j.synthmet.2010.04.019

Bian, L., Luan, F., Liu, S., & Liu, X. (2012). Self-doped polyaniline on functionalized carbon cloth as electroactive materials for supercapacitor. *Electrochimica Acta*, *64*, 17–22. doi:10.1016/j.electacta.2011.12.012

Bidez, P. R., Li, S., MacDiarmid, A. G., Venancio, E. C., Wei, Y., & Lelkes, P. I. (2006). Polyaniline, an electroactive polymer, supports adhesion and proliferation of cardiac myoblasts. *Journal of Biomaterials Science. Polymer Edition*, *17*(1-2), 199–212. doi:10.1163/156856206774879180 PMID:16411609

Bigg, D. M. (1987). Mechanical properties of particulate filled polymers. *Polymer Composites*, *8*(2), 115–122. doi:10.1002/pc.750080208

Bi, M., Hao, Y., Zhang, J., Lei, M., & Bi, K. (2017). Particle size effect of $BaTiO_3$ nanofillers on the energy storage performance of polymer nanocomposites. *Nanoscale*, *9*(42), 16386–16395. doi:10.1039/C7NR05212J PMID:29053167

Bláha, M., Varga, M., Prokeš, J., Zhigunov, A., & Vohlídal, J. (2013). Effects of the polymerization temperature on the structure, morphology and conductivity of polyaniline prepared with ammonium peroxodisulfate. *European Polymer Journal*, *49*(12), 3904–3911. doi:10.1016/j.eurpolymj.2013.08.018

Bleda-Martínez, M. J., Morallón, E., & Cazorla-Amorós, D. (2007). Polyaniline/porous carbon electrodes by chemical polymerisation: Effect of carbon surface chemistry. *Electrochimica Acta*, *52*(15), 4962–4968. doi:10.1016/j.electacta.2007.01.073

Bober, P., Humpolíček, P., Pacherník, J., Stejskal, J., & Lindfors, T. (2015). Conducting polyaniline based cell culture substrate for embryonic stem cells and embryoid bodies. *RSC Advances*, *5*(62), 50328–50335. doi:10.1039/C5RA07504A

Boddu, V. M., Abburi, K., Talbott, J. L., & Smith, E. D. (2003). Removal of hexavalent chromium from wastewater using a new composite chitosan biosorbent. *Environmental Science & Technology*, *37*(19), 4449–4456. doi:10.1021/es021013a PMID:14572099

Boeva, Z. A., Milakin, K. A., Pesonen, M., Ozerin, A. N., Sergeyev, V. G., & Lindfors, T. (2014). Dispersible composites of exfoliated graphite and polyaniline with improved electrochemical behaviour for solid-state chemical sensor applications. *RSC Advances*, *4*(86), 46340–46350. doi:10.1039/C4RA08362H

Bogdanovich, S., Kuznetsova, U., Usova, V., & Pesetskii, S. (2018). Influence of technology of melt compounding of PA6 / organoclay nanocomposites on their structure, mechanical and rheological properties. *Polymer Materials and Technologies*, *4*(2), 50–60. doi:10.32864/polymmattech-2018-4-2-50-60

Böke, N., Birch, G. D., Nyale, S. M., & Petrik, L. F. (2015). New synthesis method for the production of coal fly ash-based foamed geopolymers. *Construction & Building Materials*, *75*, 189–199. doi:10.1016/j.conbuildmat.2014.07.041

Borah, R. (2018). Synthesis and surface functionalization of conducting polymer nanostructures for biomedical applications [Ph.D. Thesis]. Tezpur University.

Borque Industries. (n.d.). Kryron CNT-MMC Alloys. Retrieved from http://bourqueindustries.com/pages/about_kryron.php

Boysen, E. (n.d.). Fuel and nanotechnology. *Understanding Nano*. Retrieved from http://www.understandingnano.com/fuel.html

Brassard, J.-D., Sarkar, D.K. & Perron, J. (2012). Surface modification and functionalization of oxide nanoparticles for superhydrophobic applications. *Advanced Materials Research*, *409*, 469-473. doi:10.4028/www.scientific.net/AMR.409.469

Brownson, D. A., Varey, S. A., Hussain, F., Haigh, S. J., & Banks, C. E. (2014). Electrochemical properties of CVD grown pristine graphene: Monolayer-vs. quasi-graphene. *Nanoscale*, *6*(3), 1607–1621. doi:10.1039/C3NR05643K PMID:24337073

Buchheit, R. G., Guan, H., Mahajanam, S., & Wong, F. (2003). Active corrosion protection and corrosion sensing in chromate-free organic coatings. *Progress in Organic Coatings*, *47*(3-4), 174–182. doi:10.1016/j.porgcoat.2003.08.003

Buggy, M., Bradley, G., & Sullivan, A. (2005). Polymer–filler interactions in kaolin/nylon 6,6 composites containing a silane coupling agent. *Composite Part A*, *36*(4), 437–442. doi:10.1016/j.compositesa.2004.10.002

Bumanis, G., Vitola, L., Bajare, D., Dembovska, L., & Pundiene, I. (2017). Impact of reactive SiO2/Al2O3 ratio in precursor on durability of porous alkali activated materials. *Ceramics International*, *43*(7), 5471–5477.

Burchardt, T., Gouérec, P., Sanchez-Cortezon, E., Karichev, Z., & Miners, J. H. (2002). Alkaline fuel cells: Contemporary advancement and limitations. *Fuel*, *81*(17), 2151–2155. doi:10.1016/S0016-2361(02)00163-1

Burke, A. (2000). Ultracapacitors: Why, how, and where is the technology. *Journal of Power Sources*, *91*(1), 37–50. doi:10.1016/S0378-7753(00)00485-7

Burke, A. F. (2007). Batteries and ultracapacitors for electric, hybrid, and fuel cell vehicles. *Proceedings of the IEEE*, *95*(4), 806–820. doi:10.1109/JPROC.2007.892490

Burkitt, B., Malave, V., & Thomaier, R. (2012, March). Silicone based coatings for aircraft applications. In *Abstracts of papers of the American Chemical Society* (Vol. 243). Washington, DC: Amer. Chemical Soc.

Burris, D., Zhao, S., Duncan, R., Lowitz, J., Perry, S., Schadler, L., & Sawyer, W. (2009). A route to wear resistant PTFE via trace loadings of functionalized nanofillers. *Wear*, *267*(1-4), 653–660. doi:10.1016/j.wear.2008.12.116

Bu, Y., Xu, H. X., Li, X., Xu, W. J., Yin, Y. X., Dai, H. L., ... Xu, P. H. (2018). A conductive sodium alginate and carboxymethyl chitosan hydrogel doped with polypyrrole for peripheral nerve regeneration. *RSC Advances*, *8*(20), 10806–10817. doi:10.1039/C8RA01059E

Cai, X., Cui, X., Zu, L., Zhang, Y., Gao, X., Lian, H., ... Wang, X. (2017). Ultra high electrical performance of nano nickel oxide and polyaniline composite materials. *Polymers*, *9*(7), 288.

Camargo, P. H. C., Satyanarayana, K. G., & Wypych, F. (2009). Nanocomposites: Synthesis, structure, properties and new application opportunities. *Materials Research*, *12*(1), 1–39. doi:10.1590/S1516-14392009000100002

Carbajal, G., Martínez-Villafañe, A., Gonzalez-Rodriguez, J. G., & Castaño, V. M. (2001). Corrosion-resistant coatings: A nanotechnology approach. *Anti-Corrosion Methods and Materials*, *48*(4), 241–244. doi:10.1108/00035590110398899

Carpio, I. E. M., Santos, C. M., Wei, X., & Rodrigues, D. F. (2012). Toxicity of a polymer–graphene oxide composite against bacterial planktonic cells, biofilms, and mammalian cells. *Nanoscale*, *4*(15), 4746–4756. doi:10.1039/c2nr30774j PMID:22751735

Chai, C., & Leong, K. W. (2007). Biomaterials approach to expand and direct differentiation of stem cells. *Molecular Therapy*, *15*(3), 467–480. doi:10.1038j.mt.6300084 PMID:17264853

Chandra, A. K. (2017). Nanomaterial adding to efficiency. *Rubber Asia*. Retrieved from https://www.rubberasia.com/2017/02/13/nano-materials-add-efficiency/

Chandra, A. K., & Kumar, N. R. (2017). Polymer nanocomposites for automobile engineering applications. In *Properties and applications of polymer nanocomposites*. Germany: Springer. doi:10.1007/978-3-662-53517-2_7

Chandy, T., & Sharma, C. P. (1990). Chitosan-as a biomaterial. *Biomaterials, Artificial Cells, and Artificial Organs*, *18*(1), 1–24. doi:10.3109/10731199009117286 PMID:2185854

Chang, C.-H., Huang, T.-C., Peng, C.-W., Yeh, T.-C., Lu, H.-I., Hung, W.-I., ... Yeh, J.-M. (2012). Novel anticorrosion coatings prepared from polyaniline/graphene composites. *Carbon*, *50*(14), 5044–5051. doi:10.1016/j.carbon.2012.06.043

Chang, K. C., Hsu, M. H., Lu, H. I., Lai, M. C., Liu, P. J., Hsu, C. H., & Liu, W. R. (2014). Room-temperature cured hydrophobic epoxy/graphene composites as corrosion inhibitor for cold-rolled steel. *Carbon*, *66*, 144–153. doi:10.1016/j.carbon.2013.08.052

Chan, S., Bantang, J. P., & Camacho, D. (2015). Influence of nanomaterial fillers in biopolymer electrolyte system for squaraine-based dye-sensitized solar cells. *International Journal of Electrochemical Science*, *10*, 7696–7706.

Chaudhary, D., Prasad, R., Gupta, R., & Bhattacharya, S. (2005). Clay intercalation and influence on crystallinity of EVA-based clay nanocomposites. *Thermochimica Acta*, *433*(1-2), 187–195. doi:10.1016/j.tca.2005.02.031

Cheever, L., & Koshland, D. E. (1992). Retention of habituation in PC12 cells. *Proceedings of the National Academy of Sciences of the United States of America*, *89*(21), 10084–10088. doi:10.1073/pnas.89.21.10084 PMID:1438197

Chen, C. N., Fan, W., Zhang, Q., Ma, T., Fu, X. W., & Wang, Z. B. (2015). In situ synthesis of cabbage like polyaniline@ hydroquinone nanocomposites and electrochemical capacitance investigations. *Journal of Applied Polymer Science*, *132*(29), 42290. doi:10.1002/app.42290

Chen, D., Feng, H., & Li, J. (2012). Graphene oxide: Preparation, functionalization, and electrochemical applications. *Chemical Reviews*, *112*(11), 6027–6053. doi:10.1021/cr300115g PMID:22889102

Chen, G., Ushida, T., & Tateishi, T. (2002). Scaffold design for tissue engineering. *Macromolecular Bioscience*, *2*(2), 67–77. doi:10.1002/1616-5195(20020201)2:2<67::AID-MABI67>3.0.CO;2-F

Cheng, Q., Tang, J., Ma, J., Zhang, H., Shinya, N., & Qin, L.-C. (2011). Polyaniline-coated electro-etched carbon fiber cloth electrodes for supercapacitors. *The Journal of Physical Chemistry C*, *115*(47), 23584–23590. doi:10.1021/jp203852p

Cheng, T. W., Lee, M. L., Ko, M. S., Ueng, T. H., & Yang, S. F. (2012). The heavy metal adsorption characteristics on metakaolin-based geopolymer. *Applied Clay Science*, *56*, 90–96. doi:10.1016/j.clay.2011.11.027

Cheng, X., Pan, J., Zhao, Y., Liao, M., & Peng, H. (2018). Gel Polymer Electrolytes for Electrochemical Energy Storage. *Advanced Energy Materials*, *8*(7), 1702184. doi:10.1002/aenm.201702184

Chen, J. M., & Hao, O. J. (1998). Microbial chromium (VI) reduction. *Critical Reviews in Environmental Science and Technology*, *28*(3), 219–251. doi:10.1080/10643389891254214

Chen, K., & Xue, D. (2015). Searching for electrode materials with high electrochemical reactivity. *Journal of Materiomics*, *1*(3), 170–187. doi:10.1016/j.jmat.2015.07.001

Chen, L., Sun, L. J., Luan, F., Liang, Y., Li, Y., & Liu, X. X. (2010). Synthesis and pseudocapacitive studies of composite films of polyaniline and manganese oxide nanoparticles. *Journal of Power Sources*, *195*(11), 3742–3747. doi:10.1016/j.jpowsour.2009.12.036

Chen, L., Wang, K., Toh, M., & He, C. (2005). Poly(propylene)/clay nanocomposites prepared by reactive compounding with an epoxy based masterbatch. *Macromolecular Materials and Engineering*, *290*(10), 1029–1036. doi:10.1002/mame.200500176

Chen, W., Rakhi, R. B., & Alshareef, H. N. (2013). Capacitance enhancement of polyaniline coated curved-graphene supercapacitors in a redox-active electrolyte. *Nanoscale*, *5*(10), 4134–4138. doi:10.1039/c3nr00773a PMID:23567992

Chen, W., Rakhi, R. B., & Alshareef, H. N. (2013). Morphology-Dependent Enhancement of the Pseudocapacitance of Template-Guided Tunable Polyaniline Nanostructures. *The Journal of Physical Chemistry C*, *117*(29), 15009–15019. doi:10.1021/jp405300p

Chen, W., Yan, L., & Bangal, P. R. (2010). Preparation of graphene by the rapid and mild thermal reduction of graphene oxide induced by microwaves. *Carbon*, *48*(4), 1146–1152. doi:10.1016/j.carbon.2009.11.037

Chen, X., Li, C., Shao, W., Du, H., & Burnell-Gray, J. (2007). The anti-static poly (ethylene terephthalate) nanocomposite fiber by in situ polymerization: The thermo-mechanical and electrical properties. *Journal of Applied Polymer Science*, *105*(3), 1490–1495. doi:10.1002/app.26381

Chen, X., Wang, J., Lin, M., Zhong, W., Feng, T., Chen, X., ... Xue, F. (2008). Mechanical and thermal properties of epoxy nanocomposites reinforced with amino-functionalized multi-walled carbon nanotubes. *Materials Science and Engineering A*, *492*(1-2), 236–242. doi:10.1016/j.msea.2008.04.044

Chen, Y. P., Liu, B. R., Liu, Q., Wang, J., Liu, J. Y., Zhang, H. S., ... Jing, X. Y. (2015). Flexible all-solid-state asymmetric supercapacitor assembled using coaxial NiMoO4 nanowire arrays with chemically integrated conductive coating. *Electrochimica Acta*, *178*, 429–438. doi:10.1016/j.electacta.2015.08.040

Chen, Y., Wang, X. H., Li, J., Lu, J. L., & Wang, F. S. (2007). Long-term anticorrosion behaviour of polyaniline on mild steel. *Corrosion Science*, *49*(7), 3052–3063. doi:10.1016/j.corsci.2006.11.007

Chen, Z., Mo, X., & Qing, F. (2007). Electrospinning of collagen–chitosan complex. *Materials Letters*, *61*(16), 3490–3494. doi:10.1016/j.matlet.2006.11.104

Cherusseri, J., Sharma, R., & Kar, K. K. (2015). Nanotechnology advancement on carbon nanotube/polypyrrole composite electrodes for supercapacitors. In *Handbook of polymer nanocomposites. Processing, performance and application* (p. 483). London: Springer. doi:10.1007/978-3-642-45229-1_22

Chiang, C. K., Druy, M. A., Gau, S. C., Heeger, A. J., Louis, E. J., MacDiarmid, A. G., ... Shirakawa, H. (1978). Synthesis of highly conducting films of derivatives of polyacetylene, (CH)x. *Journal of the American Chemical Society*, *100*(3), 1013–1015. doi:10.1021/ja00471a081

Chiang, C. K., Fincher, C. R., Park, Y. W., Heeger, A. J., Shirakawa, H., Louis, E. J., ... MacDiarmid, A. G. (1977). Electrical conductivity in doped polyacetylene. *Physical Review Letters*, *39*(17), 1098–1101. doi:10.1103/PhysRevLett.39.1098

Chinese Academy of Science. (2014, July 10). Wind turbine blade carbon nanotubes / polymer nano composites developed. JEC Group. Retrieved from http://www.jeccomposites.com/knowledge/international-composites-news/wind-turbine-blade-carbon-nanotubes-polymer-nano-composites

Chiou, N. R. (2006). Aligned and oriented polyaniline nanofibers: fabrication and applications [PhD thesis]. The Ohio State University.

Chiou, N. R., & Epstein, A. J. (2005). Polyaniline nanofibers prepared by dilute polymerization. *Advanced Materials*, *17*(13), 1679–1683. doi:10.1002/adma.200401000

Cho, S.-J., Uddin, M.-J., & Alaboina, P. (2017). Review of nanotechnology for cathode materials in batteries. In Emerging nanotechnologies in rechargeable energy storage systems (pp. 83–129).

Choa, Y. H., Yang, J. K., Kim, B. H., Jeong, Y. K., Lee, J. S., Nakayama, T., ... Niihara, K. (2003). Preparation and characterization of metal: Ceramic nanoporous nanocomposite powders. *Journal of Magnetism and Magnetic Materials*, *266*(1-2), 12–19. doi:10.1016/S0304-8853(03)00450-5

Cho, J., & Paul, D. (2001). Nylon 6 nanocomposites by melt compounding. *Polymer*, *42*(3), 1083–1094. doi:10.1016/S0032-3861(00)00380-3

Cho, S., Shin, K., & Jang, J. (2013). Enhanced electrochemical performance of highly porous supercapacitor electrodes based on solution processed polyaniline thin films. *ACS Applied Materials & Interfaces*, *5*(18), 9186–9193. doi:10.1021/am402702y PMID:24032539

Choudalakis, G., & Gotsis, A. D. (2009). Permeability of polymer/clay nanocomposites: A review. *European Polymer Journal*, *45*(4), 967–984. doi:10.1016/j.eurpolymj.2009.01.027

Chow, W., & Mohd Ishak, Z. (2015). Polyamide blend-based nanocomposites: A review. *Express Polymer Letters*, *9*(3), 211–232. doi:10.3144/expresspolymlett.2015.22

Choy, J. H., Jung, J. S., Oh, J. M., Park, M., Jeong, J., Kang, Y. K., & Han, O. J. (2004). Layered double hydroxide as an efficient drug reservoir for folate derivatives. *Biomaterials*, *25*(15), 3059–3064. doi:10.1016/j.biomaterials.2003.09.083 PMID:14967539

Christen, T., & Carlen, M. W. (2000). Theory of Ragone plots. *Journal of Power Sources, 91*(2), 210–216. doi:10.1016/S0378-7753(00)00474-2

Chronakis, I. S. (2005). Novel nanocomposites and nanoceramics based on polymer nanofibers using electrospinning process-A review. *Journal of Materials Processing Technology, 167*(2-3), 283–293. doi:10.1016/j.jmatprotec.2005.06.053

Chu, A., & Braatz, P. (2002). Comparison of commercial supercapacitors and high power lithium-ion batteries for power-assist applications in hybrid electric vehicles I. Initial characterization. *Journal of Power Sources, 112*(1), 236–246. doi:10.1016/S0378-7753(02)00364-6

Chua, T., Mariatti, M., Azizan, A., & Rashid, A. (2010). Effects of surface-functionalized multi-walled carbon nanotubes on the properties of poly(dimethyl siloxane) nanocomposites. *Composites Science and Technology, 70*(4), 671–677. doi:10.1016/j.compscitech.2009.12.023

Chuizhou, M., Gall, O. Z., & Irazoqui, P. (2013). A flexible super-capacitive solid-state power supply for miniature implantable medical devices. *Biomedical Microdevices, 15*(6), 973–983. doi:10.100710544-013-9789-1 PMID:23832644

Chung, T. W., Lai, D. M., Chen, S. D., & Lin, Y. I. (2014). Poly (ε-caprolactone) scaffolds functionalized by grafting NGF and GRGD promote growth and differentiation of PC12 cells. *Journal of Biomedical Materials Research. Part A, 102*(2), 315–323. doi:10.1002/jbm.a.34693 PMID:23468336

Cilla, M. S., de Mello Innocentini, M. D., Morelli, M. R., & Colombo, P. (2017). Geopolymer foams obtained by the saponification/peroxide/gelcasting combined route using different soap foam precursors. *Journal of the American Ceramic Society, 100*(8), 3440–3450.

Cinke, M., Li, J., Chen, B., Cassell, A., Delzeit, L., Han, J., & Meyyappan, M. (2002). Pore structure of raw and purified HiPco single-walled carbon nanotubes. *Chemical Physics Letters, 365*(1-2), 69–74. doi:10.1016/S0009-2614(02)01420-3

Coelho, M. C., Torrão, G., Emami, N., & Grácio, J. (2012). Nanotechnology in automotive Industry: Research strategy and trends for the future-Small Objects, big impacts. *Journal of Nanoscience and Nanotechnology, 12*(8), 1–10. doi:10.1166/jnn.2012.4573 PMID:22962798

Coiai, S., Passaglia, E., Pucci, A., & Ruggeri, G. (2015). Nanocomposites based on thermoplastic polymers and functional nanofiller for sensor applications. *Materials (Basel), 8*(6), 3377–3427. doi:10.3390/ma8063377

Colpo, F., Humbert, L., & Botsis, J. (2007). Characterisation of residual stresses in a single fibre composite with FBG sensor. *Composites Science and Technology, 67*(9), 1830–184. doi:10.1016/j.compscitech.2006.10.024

Composites World. (2009). GE aviation to use ceramic matrix composites in jet engines. Retrieved from https://www.compositesworld.com/news/ge-aviation-to-use-ceramic-matrix-composites-in-jet-engines

Contractor, A. Q., Sureshkumar, T. N., Narayanan, R., Sukeerthi, S., Lal, R., & Srinivasa, R. S. (1994). Conducting polymer-based biosensors. *Electrochimica Acta, 39*(8-9), 1321–1324. doi:10.1016/0013-4686(94)E0054-4

Conway, B. E. (1991). Transition from "supercapacitor" to "battery" behavior in electrochemical energy storage. *Journal of the Electrochemical Society, 138*(6), 1539–1548. doi:10.1149/1.2085829

Corté, L., & Leibler, L. (2005). Analysis of polymer blend morphologies from transmission electron micrographs. *Polymer, 46*(17), 6360–6368. doi:10.1016/j.polymer.2005.05.109

CountryFocus. (2017). *Healthcare*. GlobalData, France: Regulatory and Reimbursement Landscape-France.

Courter, J. L. (1997). Mar resistance of automotive clearcoats: I. Relationship to coating mechanical properties. *Journal of Coatings Technology, 69*, 57–63.

Creton, C., Kramer, E. J., Brown, H. R., & Hui, C. Y. (2001). Adhesion and fracture of interfaces between immiscible polymers: From the molecular to the continuum scale. *Advances in Polymer Science, 156*, 53–136. doi:10.1007/3-540-45141-2_2

Crouch, I. G., Sandlin, J., & Thomas, S. (2017). Polymers and fibre-reinforced plastics. In I.G. Crouch (Ed.), The Science of Armour Materials (Ch. 5, pp. 203-268) Woodhead Publishing.

Cui, X., Lee, V. A., Raphael, Y., Wiler, J. A., Hetke, J. F., Anderson, D. J., & Martin, D. C. (2001). Surface modification of neural recording electrodes with conducting polymer/biomolecule blends. *Journal of Biomedical Materials Research. Part A, 56*(2), 261–272. doi:10.1002/1097-4636(200108)56:2<261::AID-JBM1094>3.0.CO;2-I PMID:11340598

Cui, Z., Ni, N. C., Wu, J., Du, G. Q., He, S., Yau, T. M., ... Li, R. K. (2018). Polypyrrole-chitosan conductive biomaterial synchronizes cardiomyocyte contraction and improves myocardial electrical impulse propagation. *Theranostics, 8*(10), 2752–2764. doi:10.7150/thno.22599 PMID:29774073

Daberkow, A., Ehlert, M., & Kaise, D. (2013). Electric car operation and flywheel energy storage. In M. Lienkamp (Ed.), *Conference on Future Automotive Technology* (pp. 19–30). 10.1007/978-3-658-01141-3_2

Dabflow. (n.d.). Nanocoateings in the aviation industry. Retrieved from https://dabflow.co/nanocoatings-in-the-aviation-industry/

Daemen, J., & Serruys, P. W. (2007). Drug-eluting stent update 2007: part II: unsettled issues. *Circulation, 116*(8), 961–968. doi:10.1161/CIRCULATIONAHA.107.691451 PMID:17709651

Dahotre, N. B., & Nayak, S. (2005). Nanocoatings for engine application. *Surface and Coatings Technology, 194*(1), 58–67. doi:10.1016/j.surfcoat.2004.05.006

Dai, G., & Mishnaevsky, L. M. Jr. (2015). Carbon nanotube reinforced hybrid composites: Computational modeling of environmental fatigue and usability for wind blades. *Composites. Part B, Engineering, 78*, 349–360. doi:10.1016/j.compositesb.2015.03.073

Dai, L., Chang, D. W., Baek, J. B., & Lu, W. (2012). Carbon nanomaterials for advanced energy conversion and storage. *Small, 8*(8), 1130–1166. doi:10.1002mll.201101594 PMID:22383334

Dai, W., Ma, L., Gan, M., Wang, S., Sun, X., Wang, H., ... Zhou, T. (2016). Fabrication of sandwich nanostructure graphene/polyaniline hollow spheres composite and its applications as electrode materials for supercapacitor. *Materials Research Bulletin, 76*, 344–352. doi:10.1016/j.materresbull.2015.12.045

Dang, Z. M., Yuan, J. K., Yao, S. H., & Liao, R. J. (2013). Flexible nanodielectric materials with high permittivity for power energy storage. *Advanced Materials, 25*(44), 6334–6365. doi:10.1002/adma.201301752 PMID:24038139

Dang, Z. M., Yuan, J. K., Zha, J. W., Zhou, T., Li, S. T., & Hu, G. H. (2012). Fundamentals, processes and applications of high-permittivity polymer–matrix composites. *Progress in Materials Science, 57*(4), 660–723. doi:10.1016/j.pmatsci.2011.08.001

Dante Rene Guerra, B. S. (2009). Influence of nanoparticles on the physical properties of fiber reinforced polymer composites [Master Thesis]. The Graduate School of The Ohio State University.

Das, A., Stockelhuber, K. W., Jurk, R., & Heinrich, G. (2008). Modified and unmodified multiwalled carbon nanotubes in high performance solution-styrene-butadiene and butadiene rubber blends. *Polymer, 49*(24), 5276–5283. doi:10.1016/j.polymer.2008.09.031

Das, T. K., & Prusty, S. (2013). Graphene-based polymer composites and their applications. *Polymer-Plastics Technology and Engineering, 52*(4), 319–331. doi:10.1080/03602559.2012.751410

Davidovits, J. (1994, October). Properties of geopolymer cements. In *First international conference on alkaline cements and concretes* (Vol. 1, pp. 131-149). Kiev State Technical University, Ukraine: Scientific Research Institute on Binders and Materials.

Davidovits, J. (2017). Geopolymers: Ceramic-like inorganic polymers. *J. Ceram. Sci. Technol.*, *8*, 335–350. doi:10.4416/JCST2017-00038

Davis, C., Mathias, L., Gilman, J., Schiraldi, D., Shields, J., Trulove, P., ... Delong, H. C. (2002). Effects of melt-processing conditions on the quality of poly(ethylene terephthalate) montmorillonite clay nanocomposites. *Journal of Polymer Science. Part B, Polymer Physics*, *40*(23), 2661–2666. doi:10.1002/polb.10331

Davis, J. L., Guzan, K., Mills, K., & Han, L. (2010). *Advanced Nanocomposite Materials Tailored to Solid-State Lighting Applications.*

Davoodi, M. M., Sapuan, S. M., Ahmad, D., Aidy, A., Khalina, A., & Jonoobi, M. (2011). Concept selection of car bumper beam with developed hybrid bio-composite material. *Materials & Design*, *32*(10), 4857–4865. doi:10.1016/j.matdes.2011.06.011

Dawoud, H.D., Altahtamouni, T.M., Zagho, M.M. & Bensalah, N. (2017). A brief overview of flexible CNT/PANI super capacitors *Mater. Sci. Nanotechnol.*, *1*(2), 23-36.

de Farias, M. A., Coelho, L. A., & Pezzin, S. H. (2009). Synthesis of Epoxy-copolysilsesquioxanes Organicinorganic Hybrid Materials: Evaluation of Catalyst and Reaction Time in the Sol-Gel Process. In *Proceedings of the 10th Brazilian Polymer Congress*, Foz do Iguaçu.

DeBerry, D. W. (1985). Modification of the electrochemical and corrosion behavior of stainless steels with an electroactive coating. *Journal of the Electrochemical Society*, *132*(5), 1022–1026. doi:10.1149/1.2114008

Deepak, K., Reddy, N. S., & Naidu, T. V. S. (2015). Thermosetting polymer and nano clay based natural fiber biocomposites. *Procedia Materials Science*, *10*, 626–631. doi:10.1016/j.mspro.2015.06.095

Demirbas, A. (2008). Direct use of methanol in fuel cells. *Energy Sources. Part A, Recovery, Utilization, and Environmental Effects*, *30*(6), 529–535. doi:10.1080/15567030600817159

Dennis, H., Hunter, D., Chang, D., Kim, S., White, J., Cho, J., & Paul, D. (2001). Effect of melt processing conditions on the extent of exfoliation in organoclay-based nanocomposites. *Polymer*, *42*(23), 9513–9522. doi:10.1016/S0032-3861(01)00473-6

Derkacs, D., Lim, S. H., Matheu, P., Mar, W., & Yu, E. T. (2006). Improved performance of amorphous silicon solar cells via scattering from surface plasmon polaritons in nearby metallic nanoparticles. *Applied Physics Letters*, *89*(9), 0931031–0931033. doi:10.1063/1.2336629

Deshmukh, P. R., Patil, S. V., Bulakhe, R. N., Pusawale, S. N., Shim, J.-J., & Lokhande, C. D. (2015). Chemical synthesis of PANI–TiO_2 composite thin film for supercapacitor application. *RSC Advances*, *5*(84), 68939–68946. doi:10.1039/C5RA09233G

Devillers, N., Jemei, S., Péra, M.-C., Bienaimé, D., & Gustin, F. (2014). Review of characterization methods for supercapacitor modeling. *Journal of Power Sources*, *246*, 596–608. doi:10.1016/j.jpowsour.2013.07.116

Dhawan, S. K., Kumar, A., Bhandari, H., Bisht, B. M. S., & Khatoon, F. (2015). Development of Highly Hydrophobic and Anticorrosive Conducting Polymer Composite Coating for Corrosion Protection in Marine Environment. *American Journal of Political Science*, *5*(1A), 7–17.

Dhoke, S. K., & Khanna, A. S. (2009). Effect of nano-Fe_2O_3 particles on the corrosion behavior of alkyd based waterborne coatings. *Corrosion Science*, *51*(1), 6–20. doi:10.1016/j.corsci.2008.09.028

Dhoke, S. K., & Khanna, A. S. (2012). Electrochemical impedance spectroscopy (EIS) study of nano-alumina modified alkyd based waterborne coatings. *Progress in Organic Coatings*, *74*(1), 92–99. doi:10.1016/j.porgcoat.2011.11.020

Dhoke, S. K., Sinha, T. J. M., & Khanna, A. S. (2009). Effect of nano-Al_2O_3 particles on the corrosion behavior of alkyd based waterborne coatings. *Journal of Coatings Technology and Research*, *6*(3), 353–368. doi:10.100711998-008-9127-3

Di Martino, A., Sittinger, M., & Risbud, M. V. (2005). Chitosan: A versatile biopolymer for orthopaedic tissue-engineering. *Biomaterials*, *26*(30), 5983–5990. doi:10.1016/j.biomaterials.2005.03.016 PMID:15894370

Diaz, A., & Logan, J. (1980). Electroactive polyaniline films. *Journal of Electroanalytical Chemistry and Interfacial Electrochemistry*, *111*(1), 111–114. doi:10.1016/S0022-0728(80)80081-7

Dinca, I., Manoliu, V., Stefan, A., Stan, A. & Llina, S. (2010). Nanocomposites and nanotechnologies in aerospace research. *National Institute for aerospace research and development*. Retrieved from http://www.romnet.net/nano/2010.02.03_prezentari/14_materiale_tribologie_dlDinca.pdf

Ding, Y., Fan, F., Tian, Z., & Wang, Z. L. (2010). Atomic structure of Au-Pd bimetallic alloyed nanoparticles. *Journal of the American Chemical Society*, *132*(35), 12480–12486. doi:10.1021/ja105614q PMID:20712315

Dittrich, B., Wartig, K.-A., Hofmann, D., Mülhaupt, R., & Schartel, B. (2013). Flame retardancy through carbon nanomaterials: Carbon black, multiwall nanotubes, expanded graphite, multi-layer graphene and graphene in polypropylene. *Polymer Degradation & Stability*, *98*(8), 1495–1505. doi:10.1016/j.polymdegradstab.2013.04.009

Divya G.S., & Suresha, B. (2016). Recent developments of natural fiber reinforced thermoset polymer composites and their mechanical properties. *Indian Journal of Advances in Chemical Science, S1*, 267-274.

Di, Y., Iannace, S., Di Maio, E., & Nicolais, L. (2003). Nanocomposites by melt intercalation based on polycaprolactone and organoclay. *Journal of Polymer Science. Part B, Polymer Physics*, *41*(7), 670–678. doi:10.1002/polb.10420

Djobo, J. N. Y., Elimbi, A., Tchakouté, H. K., & Kumar, S. (2016). Reactivity of volcanic ash in alkaline medium, microstructural and strength characteristics of resulting geopolymers under different synthesis conditions. *Journal of Materials Science*, *51*(22), 10301–10317. doi:10.100710853-016-0257-1

Domenech, T., Peuvrel-Disdier, E., & Vergnes, B. (2011). Influence of twin-screw processing conditions on structure and properties of polypropylene-organoclay nanocomposites. In *27th World Congress of the Polymer Processing Society* (pp. 1-10). Marrakech: Polymer Processing Society.

Domun, N., Hadavinia, H., Zhang, T., Sainsbury, T., Liaghat, G., & Vahid, S. (2015). Improving the fracture toughness and the strength of epoxy using nanomaterials–a review of the current status. *Nanoscale*, *7*(23), 10294–10329. doi:10.1039/C5NR01354B PMID:26006766

Dong, B., He, B., Xu, C., & Li, H. (2007). Preparation and electrochemical characterization of polyaniline/multi-walled carbon nanotubes composites for super-capacitor. *Materials Science and Engineering B*, *143*(1-3), 7–13. doi:10.1016/j.mseb.2007.06.017

Doula, M., Ioannou, A., & Dimirkou, A. (2002). Copper adsorption and Si, Al, Ca, Mg, and Na release from clinoptilolite. *Journal of Colloid and Interface Science*, *245*(2), 237–250. doi:10.1006/jcis.2001.7961 PMID:16290357

Duay, J., Sherrill, S. A., Gui, Z., Gillette, E., & Lee, S. B. (2013). Self-limiting electrodeposition of hierarchical MnO_2 and $M(OH)_2/MnO_2$ nanofibril/nanowires: Mechanism and supercapacitor properties. *ACS Nano*, *7*(2), 1200–1214. doi:10.1021/nn3056077 PMID:23327566

Du, C., Yeh, J., & Pan, N. (2005). High power density supercapacitors using locally aligned carbon nanotube electrodes. *Nanotechnology*, *16*(4), 350–353. doi:10.1088/0957-4484/16/4/003

Dunlop, J. W., & Fratzl, P. (2010). Biological composites. *Annual Review of Materials Research*, *40*(1), 1–24. doi:10.1146/annurev-matsci-070909-104421

Dupont, M. F., & Donne, S. W. (2016). Charge storage mechanisms in electrochemical capacitors: Effects of electrode properties on performance. *Journal of Power Sources*, *326*, 613–623. doi:10.1016/j.jpowsour.2016.03.073

Dutta, P., Chakravarty, S., & Sarma, N. S. (2016). Detection of nitroaromatic explosives using π-electron rich luminescent polymeric nanocomposites. *RSC Advances*, *6*(5), 3680–3689. doi:10.1039/C5RA20347C

Duxson, P., Fernández-Jiménez, A., Provis, J. L., Lukey, G. C., Palomo, A., & Van Deventer, J. S. J. (2007). Geopolymer technology: The current state of the art. *Journal of Materials Science*, *42*(9), 2917–2933. doi:10.100710853-006-0637-z

Duygu, D., Baykal, T., Açikgöz, I., & Yildiz, K. (2009). Fourier transform infrared (FT-IR) spectroscopy for biological studies. *G.U. Journal of Science*, *22*(3), 117–121.

Eatemadi, A., Daraee, H., Karimkhanloo, H., Kouhi, M., Zarghami, N., Akbarzadeh, A., ... Joo, S. W. (2014). Carbon nanotubes: Properties, synthesis, purification, and medical applications. *Nanoscale Research Letters*, *9*(393), 1–13. PMID:25170330

Eftekhari, A., Li, L., & Yang, Y. (2017). Polyaniline supercapacitors. *Journal of Power Sources*, *347*, 86–107. doi:10.1016/j.jpowsour.2017.02.054

Egerton, R. F. (2013). Control of radiation damage in the TEM. *Ultramicroscopy*, *127*, 100–108. doi:10.1016/j.ultramic.2012.07.006 PMID:22910614

Ehsani, A., Jaleh, B., & Nasrollahzadeh, M. (2014). Electrochemical properties and electrocatalytic activity of conducting polymer/copper nanoparticles supported on reduced graphene oxide composite. *Journal of Power Sources*, *257*, 300–307. doi:10.1016/j.jpowsour.2014.02.010

El Rayes, M. M., Abdo, H. S., & Khalil, K. A. (2013). Erosion-corrosion of cermet coating. *International Journal of Electrochemical Science*, *8*(1), 1117–1137.

Elbert, D. L., & Hubbell, J. A. (1996). Surface treatments of polymers for biocompatibility. *Annual Review of Materials Science*, *26*(1), 365–394. doi:10.1146/annurev.ms.26.080196.002053

El-Eswed, B., Alshaaer, M., Yousef, R. I., Hamadneh, I., & Khalili, F. (2012). Adsorption of Cu(II), Ni(II), Zn(II), Cd(II) and Pb(II) onto kaolin/zeolite based-geopolymers. *Adv. Mater. Phys. Chem.*, *2*(04), 119–125. doi:10.4236/ampc.2012.24B032

El-Fatatry, A. (2007). Defence Applications. In Nanotechnology Aerospace Applications.

Eslami-Farsani, R., Reza Khalili, S., Hedayatnasab, Z., & Soleimani, N. (2014). Influence of thermal conditions on the tensile properties of basalt fiber reinforced polypropylene–clay nanocomposites. *Materials & Design*, *53*, 540–549. doi:10.1016/j.matdes.2013.07.012

Esrafilzadeh, D., Jalili, R., Liu, X., Gilmore, K. J., Razal, J. M., Moulton, S. E., & Wallace, G. G. (2016). A novel and facile approach to fabricate a conductive and biomimetic fibrous platform with sub-micron and micron features. *Journal of Materials Chemistry. B, Materials for Biology and Medicine*, *4*(6), 1056–1063. doi:10.1039/C5TB02237A

Evonik Industries. (2018). Material solutions for aviation and aerospace markets. *AZO Materials*. Retrieved from https://www.azom.com/article.aspx?ArticleID=15795

Fan, D., Zhu, X., Xu, M., & Yan, J. (2006). Adsorption properties of chromium (VI) by chitosan coated montmorillonite. *The Journal of Biological Sciences*, *6*(5), 941–945. doi:10.3923/jbs.2006.941.945

Fang, M., Wang, K., Lu, H., Yang, Y., & Nutt, S. (2009). Covalent polymer functionalization of graphene nanosheets and mechanical properties of composites. *Journal of Materials Chemistry*, *19*(38), 7098–7105. doi:10.1039/b908220d

Fan, H., Wang, H., Zhao, N., Zhang, X., & Xu, J. (2012). Hierarchical nanocomposite of polyaniline nanorods grown on the surface of carbon nanotubes for high-performance supercapacitor electrode. *Journal of Materials Chemistry*, *22*(6), 2774–2780. doi:10.1039/C1JM14311E

Fan, L. Z., Hu, Y.-S., Maier, J., Adelhelm, P., Smarsly, B., & Antonietti, M. (2007). High Electroactivity of Polyaniline in Supercapacitors by Using a Hierarchically Porous Carbon Monolith as a Support. *Advanced Functional Materials*, *17*(16), 3083–3087. doi:10.1002/adfm.200700518

Fan, W., Zhang, C., Tjiu, W. W., Pramoda, K. P., He, C., & Liu, T. (2013). Graphene-wrapped polyaniline hollow spheres as novel hybrid electrode materials for supercapacitor applications. *ACS Applied Materials & Interfaces*, *5*(8), 3382–3391. doi:10.1021/am4003827 PMID:23517224

Farid, M. M., Khudhair, A. M., Razack, S. A., & Al-Hallaj, S. (2004). A review on phase change energy storage: Materials and applications. *Energy Conversion and Management*, *45*(9-10), 1597–1615. doi:10.1016/j.enconman.2003.09.015

Farshi Azhar, F., Olad, A., & Salehi, R. (2014). Fabrication and characterization of chitosan–gelatin/nanohydroxyapatite–polyaniline composite with potential application in tissue engineering scaffolds. *Designed Monomers and Polymers*, *17*(7), 654–667. doi:10.1080/15685551.2014.907621

Fayette, M., & Robinson, R. D. (2014). Chemical transformations of nanomaterials for energy applications. *Journal of Materials Chemistry. A, Materials for Energy and Sustainability*, *2*(17), 5965–5978. doi:10.1039/C3TA13982D

Feldberg, S. W. (1984). Reinterpretation of polypyrrole electrochemistry. Consideration of capacitive currents in redox switching of conducting polymers. *Journal of the American Chemical Society*, *106*(17), 4671–4674. doi:10.1021/ja00329a004

Feng, W., Zhang, Q., Li, Y., & Feng, Y. (2014). Preparation of sulfonated graphene/polyaniline composites in neutral solution for high-performance supercapacitors. *Journal of Solid State Electrochemistry*, *18*(4), 1127–1135. doi:10.100710008-013-2369-8

Fernández-Merino, M. J., Guardia, L., Paredes, J. I., Villar-Rodil, S., Solís-Fernández, P., Martínez-Alonso, A., & Tascón, J. M. D. (2010). Vitamin C Is an Ideal Substitute for Hydrazine in the Reduction of Graphene Oxide Suspensions. *J. PhysChemComm*, *114*(14), 6426–6432.

Fernando, W., & Satyanarayana, K. G. (2005). Functionalization of single layers and nanofibers: A new strategy to produce polymer nanaocomposites with optimized properties. *Journal of Colloid and Interface Science*, *285*(1), 532–543. PMID:15837469

Ferrari, A. C., Meyer, J., Scardaci, V., Casiraghi, C., Lazzeri, M., Mauri, F., & ... Geim, A. K. (2006). Raman spectrum of graphene and graphene layers. *Physical Review Letters*, *97*(18), 187401. doi:10.1103/PhysRevLett.97.187401 PMID:17155573

Fiedler, B., Gojny, F. H., Wichmann, M. H., Nolte, M. C., & Schulte, K. (2006). Fundamental aspects of nano-reinforced composites. *Composites Science and Technology*, *66*(16), 3115–3125. doi:10.1016/j.compscitech.2005.01.014

Finnigan, B., Martin, D., Halley, P., Truss, R., & Campbell, K. (2004). Morphology and properties of thermoplastic polyurethane nanocomposites incorporating hydrophilic layered silicates. *Polymer*, *45*(7), 2249–2260. doi:10.1016/j.polymer.2004.01.049

Fiordiponti, P., & Pistoia, G. (1989). An impedance study of polyaniline films in aqueous and organic solutions. *Electrochimica Acta*, *34*(2), 215–221. doi:10.1016/0013-4686(89)87088-4

Fischer, H. (2003). Polymer nanocomposites: From fundamental research to specific applications. *Materials Science and Engineering C*, *23*(6-8), 763–772. doi:10.1016/j.msec.2003.09.148

Floody, M. C., Theng, B. K. G., Reyes, P., & Mora, M. L. (2009). Natural nanoclays: Applications and future trends - a Chilean perspective. *Clay Minerals*, *44*(02), 161–176. doi:10.1180/claymin.2009.044.2.161

Fogorasi, M. S., & Barbu, I. (2017). The potential of natural fibres for automotive sector – Review. IOP Conf. Series. *Materials Science and Engineering*, *252*(012044), 1–11.

Fornes, T., Yoon, P., Keskkula, H., & Paul, D. (2001). Nylon 6 nanocomposites: The effect of matrix molecular weight. *Polymer*, *42*(25), 09929–09940. doi:10.1016/S0032-3861(01)00552-3

Forrest, S. R. (2005). The Limits to organic photovoltaic cell efficiency. *MRS Bulletin*, *30*(1), 28–32. doi:10.1557/mrs2005.5

Frackowiak, E., & Beguin, F. (2001). Carbon materials for the electrochemical storage of energy in capacitors - review. *Carbon*, *39*(6), 937–950. doi:10.1016/S0008-6223(00)00183-4

Frackowiak, E., Khomenko, V., Jurewicz, K., Lota, K., & Béguin, F. (2006). Supercapacitors based on conducting polymers/nanotubes composites. *Journal of Power Sources*, *153*(2), 413–418. doi:10.1016/j.jpowsour.2005.05.030

Freed, L. E., Marquis, J. C., Nohria, A., Emmanual, J., Mikos, A. G., & Langer, R. (1993). Neocartilage formation in vitro and in vivo using cells cultured on synthetic biodegradable polymers. *Journal of Biomedical Materials Research. Part A*, *27*(1), 11–23. doi:10.1002/jbm.820270104 PMID:8380593

Freguia, S., Rabaey, K., Yuan, Z., & Keller, J. (2007). Non-catalyzed cathodic oxygen reduction at graphite granules in microbial fuel cells. *Electrochimica Acta*, *53*(2), 598–603. doi:10.1016/j.electacta.2007.07.037

Frontini, P., & Pouzara, A. (2015). Trends in the multifunctional performance of polyolefin/clay nanocomposite injection moldings. In K. Friedrich & U. Breuer (Eds.), *Multifunctionality of polymer composites. Challenges and new solutions* (pp. 213–245). Oxford: Elsevier. doi:10.1016/B978-0-323-26434-1.00006-4

Fukushima, Y., & Inagaki, S. (1987). Synthesis of an intercalated compound of montmorillonite and 6-Polyamide. *Journal of Inclusion Phenomena*, *5*(4), 473–482. doi:10.1007/BF00664105

Fu, S.-Y., Feng, X.-Q., Lauke, B., & Mai, Y.-W. (2008). Effects of particle size, particle/matrix interface adhesion and particle loading on mechanical properties of particulate–polymer composites. *Composites. Part B, Engineering*, *39*(6), 933–961. doi:10.1016/j.compositesb.2008.01.002

Fu, X. A., & Qutubuddin, S. (2001). Polymer–clay nanocomposites: Exfoliation of organophilic montmorillonite nanolayers in polystyrene. *Polymer*, *42*(2), 807–813. doi:10.1016/S0032-3861(00)00385-2

Fu, X. A., & Qutubuddin, S. (2004). Synthesis of unsaturated polyester-clay nanocomposites using reactive organoclays. *Polymer Engineering and Science*, *44*(2), 345–351. doi:10.1002/pen.20031

Gadd, G. M. (2010). Metals, minerals and microbes: Geomicrobiology and bioremediation. *Microbiology*, *156*(3), 609–643. doi:10.1099/mic.0.037143-0 PMID:20019082

Galimberti, M. (Ed.). (2011). *Rubber-Clay Nanocomposites: Science, Technology, Applications.* New Jersey: Wiley & Sons. doi:10.1002/9781118092866

Galimberti, M., Cipolletti, V. R., & Coomb, M. (2013). *Applications of clay-polymer nanocomposites. In Handbook of clay science.* (*Vol. 5B*, p. 557). The Netherlands: Elsevier. doi:10.1016/B978-0-08-098259-5.00020-2

Galpaya, D. (2015). *Synthesis, Characterization and Applications of Graphene Oxide-Polymer Nanocomposites* [PhD Thesis]. Queensland University of Technology.

Gandhi, M. R., Viswanathanb, N., & Meenakshi, S. (2010). Preparation and application of alumina/chitosan biocomposite. *International Journal of Biological Macromolecules*, *47*(2), 146–154. doi:10.1016/j.ijbiomac.2010.05.008 PMID:20471415

Gangopadhyay, R., & Amitabha, D. (2000). Conducting polymer nanocomposites. A brief overview. *Chemistry of Materials*, *12*(7), 608–622. doi:10.1021/cm990537f

Gao Y, (2017). Graphene and Polymer Composites for Supercapacitor Applications: a Review. *Nanoscale Res Lett.*, *12*, 387.

Gao, S & Fan, H. (2013). A review on the metal-oxide-based pseudocapacitors. *Huaxue Tongbao [Chemistry Bulletin], 76*, 202-209.

Gao, Z., Yang, W., Wang, J., Yan, H., Yao, Y., Ma, J., ... Liu, L. (2013). Electrochemical synthesis of layer-by-layer reduced graphene oxide sheets/polyaniline nanofibers composite and its electrochemical performance. *Electrochimica Acta*, *91*, 185–194. doi:10.1016/j.electacta.2012.12.119

Garces, J. M., Moll, D. J., Bicerano, J., Fibiger, R., & McLeod, D. G. (2000). Polymeric nanocomposites for automotive applications. *Advanced Materials*, *12*(23), 1835–1839. doi:10.1002/1521-4095(200012)12:23<1835::AID-ADMA1835>3.0.CO;2-T

García-López, D. (2003). Polypropylene–clay nanocomposites: Effect of compatibilizing agents on clay dispersion. *European Polymer Journal*, *39*(5), 945–950. doi:10.1016/S0014-3057(02)00333-6

Garcia, M., Netto, A., & Pontes, A. (2017). Experimental study of shrinkage and ejection forces of reinforced polypropylene based on nanoclays and short glass fibers. *Polymer Engineering and Science*, *58*(1), 55–62. doi:10.1002/pen.24530

Gary, A. (2003). Nanotubes of advanced polymer products. *Plastics Additives & Compounding*, *5*(4), 12. doi:10.1016/S1464-391X(03)00414-8

Gawli, Y., Banerjee, A., Dhakras, D., Deo, M., Bulani, D., Wadgaonkar, P., ... Ogale, S. (2016). 3D polyaniline architecture by concurrent inorganic and organic acid doping for superior and robust high rate supercapacitor performance. *Scientific Reports*, *6*(1), 21002. doi:10.1038rep21002 PMID:26867570

Geetha, S., Satheesh Kumar, K. K., Rao, C. R. K., Vijayan, M., & Trivedi, D. C. (2009). EMI shielding: Methods and materials—A review. *Journal of Applied Polymer Science*, *112*(4), 2073–2086. doi:10.1002/app.29812

Geim, A. K., & Novoselov, K. S. (2007). The rise of graphene. *Nature Materials*, *6*(3), 183–191. doi:10.1038/nmat1849 PMID:17330084

Gemeay, A. H., Mansour, I. A., El-Sharkawy, R. G., & Zaki, A. B. (2005). Preparation and characterization of polyaniline/manganese dioxide composites via oxidative polymerization: Effect of acids. *European Polymer Journal*, *41*(11), 2575–2583. doi:10.1016/j.eurpolymj.2005.05.030

Genies, E., Lapkowski, M., & Penneau, J. (1988). Cyclic voltammetry of polyaniline: Interpretation of the middle peak, J. *Electroanal. Chem. Interf. Electrochem.*, *249*(1-2), 97–107. doi:10.1016/0022-0728(88)80351-6

Genies, E., & Tsintavis, C. (1985). Redox mechanism and electrochemical behaviour or polyaniline deposits. *Journal of Electroanalytical Chemistry and Interfacial Electrochemistry*, *195*(1), 109–128. doi:10.1016/0022-0728(85)80009-7

Georgakilas, V., Otyepka, M., Bourlinos, A. B., Chandra, V., Kim, N., Kemp, K. C., ... Kim, K. S. (2012). Functionalization of graphene: Covalent and non-covalent approaches, derivatives and applications. *Chemical Reviews*, *112*(11), 6156–6214. doi:10.1021/cr3000412 PMID:23009634

Georgakilas, V., Tiwari, J. N., Kemp, K. C., Perman, J. A., Bourlinos, A. B., Kim, K. S., & Zboril, R. (2016). Noncovalent Functionalization of Graphene and Graphene Oxide for Energy Materials, Biosensing, Catalytic, and Biomedical Applications. *Chemical Reviews*, *116*(9), 5464–5519. doi:10.1021/acs.chemrev.5b00620 PMID:27033639

Gerard, M., Chaubey, A., & Malhotra, B. D. (2002). Application of conducting polymers to biosensors. *Biosensors & Bioelectronics*, *17*(5), 345–359. doi:10.1016/S0956-5663(01)00312-8 PMID:11888724

Gerasin, V., Antipov, E., Karbushev, V., Kulichikhin, V., Karpacheva, G., Talroze, R., & Kudryavtsev, Y. (2013). New approaches to the development of hybrid nanocomposites: From structural materials to high-tech applications. *Russian Chemical Reviews*, *82*(4), 303–332. doi:10.1070/RC2013v082n04ABEH004322

Gerasin, V., Zubova, T., Bakhov, F., Barannikov, A., Merekalova, N., Korolev, Y., & Antipov, E. (2007). Structure of polymer/Na+-montmorillonite nanocomposites, produced by blending in melt. *Nanotechnologies in Russia*, *2*(1-2), 90–105.

Ge, Y., Yuan, Y., Wang, K., He, Y., & Cui, X. (2015). Preparation of geopolymer-based inorganic membrane for removing Ni2+ from wastewater. *Journal of Hazardous Materials*, *299*, 711–718.

Ghasemi-Kahrizsangi, A., Shariatpanahi, H., Neshati, J., & Akbarinezhad, E. (2015). Degradation of modified carbon black/epoxy nanocomposite coatings under ultraviolet exposure. *Applied Surface Science*, *353*, 530–539. doi:10.1016/j.apsusc.2015.06.029

Ghasemi-Mobarakeh, L., Prabhakaran, M. P., Morshed, M., Nasr-Esfahani, M. H., Baharvand, H., Kiani, S., ... Ramakrishna, S. (2011). Application of conductive polymers, scaffolds and electrical stimulation for nerve tissue engineering. *Journal of Tissue Engineering and Regenerative Medicine*, *5*(4), e17–e35. doi:10.1002/term.383 PMID:21413155

Ghasemi-Mobarakeh, L., Prabhakaran, M. P., Morshed, M., Nasr-Esfahani, M. H., & Ramakrishna, S. (2009). Electrical stimulation of nerve cells using conductive nanofibrous scaffolds for nerve tissue engineering. *Tissue Engineering. Part A*, *15*(11), 3605–3619. doi:10.1089/ten.tea.2008.0689 PMID:19496678

Ghenaatian, H. R., Mousavi, M. F., Kazemi, S. H., & Shamsipur, M. (2009). Electrochemical investigations of self-doped polyaniline nanofibers as a new electroactive material for high performance redox supercapacitor. *Synthetic Metals*, *159*(17-18), 1717–1722. doi:10.1016/j.synthmet.2009.05.014

Gholamian, M., & Contractor, A. (1988). Effect of the temperature of synthesis on the conductivity and electrochemical behaviour of polyaniline. *Journal of Electroanalytical Chemistry and Interfacial Electrochemistry*, *252*(2), 291–301. doi:10.1016/0022-0728(88)80217-1

Ghosh, S., Bao, W., Nika, D. L., Subrina, S., Pokatilov, E. P., Lau, C. N., & Balandin, A. A. (2010). Dimensional crossover of thermal transport in few-layer graphene. *Nature Materials*, *9*(7), 555–558. doi:10.1038/nmat2753 PMID:20453845

Giannelis, E. P. (1996). Polymer layered silicate nanocomposites. *Advanced Materials*, *8*(1), 29–35. doi:10.1002/adma.19960080104

Gillis, J. (2018). Fuel economy standards: there is no tradeoff with safety, cost, and fleet turnover. Consumer Federation of America. Retrieved from https://consumerfed.org/wp-content/uploads/2018/05/fuel-efficiency-vs-safety-cost-and-fleet-turnover-1.pdf

Gilmore, K. J., Kita, M., Han, Y., Gelmi, A., Higgins, M. J., Moulton, S. E., ... Wallace, G. G. (2009). Skeletal muscle cell proliferation and differentiation on polypyrrole substrates doped with extracellular matrix components. *Biomaterials, 30*(29), 5292–5304. doi:10.1016/j.biomaterials.2009.06.059 PMID:19643473

Giri, S., Ghosh, D., & Das, C. K. (2014). Growth of vertically aligned tunable polyaniline on graphene/ZrO_2 nanocomposites for supercapacitor energy-storage application. *Advanced Functional Materials, 24*(9), 1312–1324. doi:10.1002/adfm.201302158

Gladwin, K. M., Whitby, R. L., Mikhalovsky, S. V., Tomlins, P., & Adu, J. (2013). *In vitro* biocompatibility of multiwalled carbon nanotubes with sensory neurons. *Advanced Healthcare Materials, 2*(5), 728–735. doi:10.1002/adhm.201200233 PMID:23184463

Gleiter, H. (1992). Materials with ultrafine microstructures: Retrospectives and perspectives. *Nanostructured Materials, 1*(1), 1–19. doi:10.1016/0965-9773(92)90045-Y

Godt, J., Scheidig, F., Grosse-Siestrup, C., Esche, V., Brandenburg, P., Reich, A., ... Stournaras, C. J. (2007). Copper and cadmium adsorption on pellets made from fired coal fly ash. *Journal of Hazardous Materials, 148*(3), 538–547. doi:10.1016/j.jhazmat.2007.03.020 PMID:17416461

Goh, P., Ng, B., Lau, W., & Ismail, A. (2015). Inorganic nanomaterials in polymeric ultrafiltration membranes for water treatment. *Separation and Purification Reviews, 44*(3), 216–249. doi:10.1080/15422119.2014.926274

Goikolea, E., & Mysyk, R. (2017). *Nanotechnology in Electrochemical Capacitors*. In *Emerging Nanotechnologies in Rechargeable Energy Storage Systems* (pp. 131–169).

Goklany, T. (2017, August 22). In Numbers: The Status of Organ Donation in India. NDTV. from http://sites.ndtv.com/moretogive/numbers-status-organ-donation-india-221/

Goli, P., Legedza, S., Dhar, A., Salgado, R., Renteria, J., & Balandin, A. A. (2014). Graphene-enhanced hybrid phase change materials for thermal management of li-ion batteries. *Journal of Power Sources, 248*, 37–43. doi:10.1016/j.jpowsour.2013.08.135

Golru, S. S., Attar, M. M., & Ramezanzadeh, B. (2014). Studying the influence of nano-Al2O3 particles on the corrosion performance and hydrolytic degradation resistance of an epoxy/polyamide coating on AA-1050. *Progress in Organic Coatings, 77*(9), 1391–1399. doi:10.1016/j.porgcoat.2014.04.017

Goodwin, B. W., & Stiles, R. E. (2010). Phosphazene Elastomer use in Defence and Aerospace. In *2nd International Conference on: Polymers in Defence & Aerospace Applications*, Hamburg, Germany, February 10-11 (pp. 14-21).

Gopakumar, T., Lee, J., Kontopoulou, M., & Parent, J. (2002). Influence of clay exfoliation on the physical properties of montmorillonite/polyethylene composites. *Polymer, 43*(20), 5483–5491. doi:10.1016/S0032-3861(02)00403-2

Government of Gujarat. (2017). Manufacturing of Uniform for Defence Personnel (Nanotechnology). Retrieved from http://vibrantgujarat.com/writereaddata/images/pdf/project-profiles/Uniform-for-defence-personnel-nanotechnology.pdf

Graphene flagship. (n.d.). Graphene shows promise for composite applications. Retrieved from https://graphene-flagship.eu/graphene-shows-promise-for-composite-applications

Graupner, N., Herrmann, A. S., & Mussig, J. (2009). Natural and man-made cellulose fibre-reinforced poly (lactic acid) (PLA) composites: An overview about mechanical characteristics and application areas. *Composites. Part A, Applied Science and Manufacturing, 40*(6), 810–821. doi:10.1016/j.compositesa.2009.04.003

Gray, J., & Luan, B. (2002). Protective coatings on magnesium and its alloys—A critical review. *Journal of Alloys and Compounds, 336*(1-2), 88–113. doi:10.1016/S0925-8388(01)01899-0

Grdadolnik, J. (2002). ATR-FTIR spectroscopy: Its advantages and limitations. *Acta Chimica Slovenica*, *49*(3), 631–642.

Green, R. A., Lovell, N. H., & Poole-Warren, L. A. (2009). Cell attachment functionality of bioactive conducting polymers for neural interfaces. *Biomaterials*, *30*(22), 3637–3644. doi:10.1016/j.biomaterials.2009.03.043 PMID:19375160

Green, R. A., Lovell, N. H., Wallace, G. G., & Poole-Warren, L. A. (2008). Conducting polymers for neural interfaces: Challenges in developing an effective long-term implant. *Biomaterials*, *29*(24-25), 3393–3399. doi:10.1016/j.biomaterials.2008.04.047 PMID:18501423

Grossiord, N., & Hermant, M. C. (2016). *Latex technology for conductive polymer nanocomposites*. John Wiley & Sons, Inc. doi:10.1002/0471440264.pst506.pub2

Grover, S., Goel, S., Marichi, R. B., Sahu, V., Singh, G., & Sharma, R. K. (2016). Polyaniline all solid-state pseudocapacitor: Role of morphological variations in performance evolution. *Electrochimica Acta*, *196*, 131–139. doi:10.1016/j.electacta.2016.02.157

Gualtieri, M. L., Cavallini, A., & Romagnoli, M. (2016). Interactive powder mixture concept for the preparation of geopolymers with fine porosity. *Journal of the European Ceramic Society*, *2*, 2–7. doi:10.1016/j.jeurceramsoc.2016.03.030

Guan, L., Yu, L., & Chen, G. Z. (2016). Capacitive and non-capacitive faradaic charge storage. *Electrochimica Acta*, *206*, 464–478. doi:10.1016/j.electacta.2016.01.213

Guarino, V., Zuppolini, S., Borriello, A., & Ambrosio, L. (2016). Electro-active polymers (EAPs): A promising route to design bio-organic/bioinspired platforms with on demand functionalities. *Polymers*, *8*(5), 185. doi:10.3390/polym8050185

Gu, B. K., Park, S. J., & Kim, C. H. (2018). Beneficial effect of aligned nanofiber scaffolds with electrical conductivity for the directional guide of cells. *Journal of Biomaterials Science. Polymer Edition*, *29*(7-9), 1053–1065. doi:10.1080/09205063.2017.1364097 PMID:28771114

Gu, H., Tadakamalla, S., Zhang, X., Huang, Y., Jiang, Y., Colorado, H. A., & Guo, Z. (2013). Epoxy resin nanosuspensions and reinforced nanocomposites from polyaniline stabilized multi-walled carbon nanotubes. *Journal of Materials Chemistry. C, Materials for Optical and Electronic Devices*, *1*(4), 729–743. doi:10.1039/C2TC00379A

Guo, C. X., Wang, M., Chen, T., Lou, X. W., & Li, C. M. (2011). A Hierarchically Nanostructured Composite of MnO2/Conjugated Polymer/Graphene for High-Performance Lithium Ion Batteries. *Advanced Energy Materials*, *1*(5), 736–741. doi:10.1002/aenm.201100223

Guo, F., Mi, H., Zhou, J., Zhao, Z., & Qiu, J. (2015). Hybrid pseudocapacitor materials from polyaniline@multi-walled carbon nanotube with ultrafine nanofiber-assembled network shell. *Carbon*, *95*, 323–329. doi:10.1016/j.carbon.2015.08.052

Guo, H. L., Wang, X. F., Qian, Q. Y., Wang, F. B., & Xia, X. H. (2009). A green approach to the synthesis of graphene nanosheets. *ACS Nano*, *3*(9), 2653–2659. doi:10.1021/nn900227d PMID:19691285

Guo, S., & Dong, S. (2011). Graphene nanosheet: Synthesis, molecular engineering, thin film, hybrids, and energy and analytical applications. *Chemical Society Reviews*, *40*(5), 2644–2672. doi:10.1039/c0cs00079e PMID:21283849

Guo, Z., Lei, K., Li, Y., Ng, H. W., Prikhodko, S., & Hahn, H. T. (2008). Fabrication and characterization of iron oxide nanoparticles reinforced vinyl-ester resin nanocomposites. *Composites Science and Technology*, *68*(6), 1513–1520. doi:10.1016/j.compscitech.2007.10.018

Guo, Z., Wei, S., Shedd, B., Scaffaro, R., Pereira, T., & Hahn, H. T. (2007). Particle surface engineering effect on the mechanical, optical and photoluminescent properties of ZnO/vinyl-ester resin nanocomposites. *Journal of Materials Chemistry*, *17*(8), 806–813. doi:10.1039/B613286C

Gupta, A., Sakthivel, T., & Seal, S. (2015). Recent development in 2D materials beyond graphene. *Progress in Materials Science, 73*, 44–126. doi:10.1016/j.pmatsci.2015.02.002

Gupta, T. K., & Kumar, S. (2018). 4 - Fabrication of Carbon Nanotube/Polymer Nanocomposites. In *Carbon nanotubes-reinforced polymers* (pp. 61–81). doi:10.1016/B978-0-323-48221-9.00004-2

Gupta, V. K., & Pangannaya, N. B. (2000). Carbon nanotubes: Bibliometric analysis of patents. *World Patent Information, 22*(3), 185–189. doi:10.1016/S0172-2190(00)00040-5

Gupta, V., & Miura, N. (2006). Polyaniline/single-Wall carbon nanotube (PANI/SWCNT) composites for high performance supercapacitors. *Electrochimica Acta, 52*(4), 1721–1726. doi:10.1016/j.electacta.2006.01.074

Gustafsson, G., Cao, Y., Treacy, G. M., Klavetter, F., Colaneri, N., & Heeger, A. J. (1992). Flexible light-emitting diodes made from soluble conducting polymers. *Nature, 357*(6378), 477–479. doi:10.1038/357477a0

Gu, W. T., & Yushin, G. (2014). Review of nanostructured carbon materials for electrochemical capacitor applications: Advantages and limitations of activated carbon, carbide-derived carbon, zeolite-templated carbon, carbon aerogels, carbon nanotubes, onion-like carbon, and graphene. *WIREs Energy & Environment, 3*, 424–473.

Guyomar, D., Lebrun, L., Putson, C., Cottinet, P. J., Guiffard, B., & Muensit, S. (2009). Electrostrictive energy conversion in polyurethane nanocomposites. *Journal of Applied Physics, 106*(1), 014910–014919. doi:10.1063/1.3159900

Haastert-Talini, K., & Grothe, C. (2013). Electrical stimulation for promoting peripheral nerve regeneration. *International Review of Neurobiology, 109*, 111–124. doi:10.1016/B978-0-12-420045-6.00005-5 PMID:24093609

Habibi, Y. (2014). Key advances in the chemical modification of nanocelluloses. *Chemical Society Reviews, 43*(5), 1519–1542. Retrieved from http://xlink.rsc.org/?DOI=C3CS60204D doi:10.1039/C3CS60204D PMID:24316693

Haddadi, S. A., Mahdavian, M., & Karimi, E. (2015). Evaluation of the corrosion protection properties of an epoxy coating containing sol–gel surface modified nano-zirconia on mild steel. *RSC Advances, 5*(36), 28769–28777. doi:10.1039/C5RA02127H

Hai-Juan, L., Shuang, H., Lian-Zhe, H., & Guo-Bao, X. (2009). Progress in Ru (bpy) 32+ Electrogenerated Chemiluminescence. *Chinese Journal of Analytical Chemistry, 37*(11), 1557–1565. doi:10.1016/S1872-2040(08)60139-5

Haipeng, G., Yinghui, Z., Jianchun, L., Yandao, G., Nanming, Z., & Xiufang, Z. (2000). Studies on nerve cell affinity of chitosan-derived materials. *Journal of Biomedical Materials Research, 52*(2), 285–295. doi:10.1002/1097-4636(200011)52:2<285::AID-JBM7>3.0.CO;2-G PMID:10951367

Ha, J. U., & Xanthos, M. (2009). Functionalization of nanoclays with ionic liquids for polypropylene composites. *Polymer Composites, 30*(5), 534–542. doi:10.1002/pc.20583

Hamidi, R. M., Man, Z., & Azizli, K. A. (2016). Concentration of NaOH and the Effect on the Properties of Fly Ash Based Geopolymer. *Procedia Engineering, 148*, 189–193. doi:10.1016/j.proeng.2016.06.568

Hao, Y. N., Wang, X. H., O'Brien, S., Lombardi, J., & Li, L. T. (2015). Flexible BaTiO3/PVDF gradated multilayer nanocomposite film with enhanced dielectric strength and high energy density. *Journal of Materials Chemistry. C, Materials for Optical and Electronic Devices*, 37.

Hao, Y., Wang, X., Bi, K., Zhang, J., Huang, Y., Wu, L., ... Li, L. (2017). Significantly enhanced energy storage performance promoted by ultimate sized ferroelectric BaTiO3 fillers in nanocomposite films. *Nano Energy, 31*, 49–56. doi:10.1016/j.nanoen.2016.11.008

Harb, S. V., Pulcinelli, S. H., Santilli, C. V., Knowles, K. M., & Hammer, P. (2016). A comparative study on graphene oxide and carbon nanotube reinforcement of PMMA-siloxane-silica anticorrosive coatings. *ACS Applied Materials & Interfaces*, *8*(25), 16339–16350. doi:10.1021/acsami.6b04780 PMID:27266403

Hardy, J. G., Lee, J. Y., & Schmidt, C. E. (2013). Biomimetic conducting polymer-based tissue scaffolds. *Current Opinion in Biotechnology*, *24*(5), 847–854. doi:10.1016/j.copbio.2013.03.011 PMID:23578463

Hardy, J. G., Mouser, D. J., Arroyo-Currás, N., Geissler, S., Chow, J. K., Nguy, L., ... Schmidt, C. E. (2014). Biodegradable electroactive polymers for electrochemically-triggered drug delivery. *Journal of Materials Chemistry. B, Materials for Biology and Medicine*, *2*(39), 6809–6822. doi:10.1039/C4TB00355A

Harish, K. K., Venkatesh, N., Bhowmik, H., & Kuila, A. (2018). Metallic Nanoparticle: A Review. *Biomedical Journal of Scientific and Technical Research*, *4*(2), 1–11. PMID:30370423

Hasan, S., Krishnaiah, A., Ghosh, T. K., Viswanath, D. S., Boddu, V. M., & Smith, E. D. (2003). Adsorption of chromium (VI) on chitosan-coated perlite. *Separation Science and Technology*, *38*(15), 3775–3793. doi:10.1081/SS-120024229

Haupert, F., & Wetzel, B. (2005). Reinforcement of thermosetting polymers by the incorporation of micro- and nanoparticles. In *Polymer composites: from nano- to macroscale* (pp. 45–50). USA: Springer. doi:10.1007/0-387-26213-X_3

Hausner, J., Ziadeh, M., Fischer, B., Kalo, H., Schmid, J., Kunz, R., ... Breu, J. (2015). Transfer batch blending, an innovative solvent/solid assisted method for melt compounding to achieve good dispersion quality for polymerclay-nanocomposites. *Composites Science and Technology*, *114*, 34–41. doi:10.1016/j.compscitech.2015.03.019

Heeger, A. (2001). Nobel Lecture: Semiconducting and metallic polymers: The fourth generation of polymeric materials. *Reviews of Modern Physics*, *73*(3), 681–700. doi:10.1103/RevModPhys.73.681

He, K., Lin, F., Zhu, Y., Yu, X., Li, J., Lin, R., ... Su, D. (2015). Sodiation kinetics of metal oxide conversion electrodes: A comparative study with lithiation. *Nano Letters*, *15*(9), 5755–5763. doi:10.1021/acs.nanolett.5b01709 PMID:26288360

Hench, L. L., & Polak, J. M. (2002). Third-generation biomedical materials. *Science*, *295*(5557), 1014–1017. doi:10.1126cience.1067404 PMID:11834817

Hench, L. L., & Thompson, I. (2010). Twenty-first century challenges for biomaterials. *Journal of the Royal Society, Interface*, *7*(Suppl. 4), S379–S391. doi:10.1098/rsif.2010.0151.focus PMID:20484227

Hernandez-Alejandro, R., Caumartin, Y., Marotta, P. J., Ghent, C., Levstik, M. A., Quan, D., ... Luke, P. P. (2010). Kidney and liver transplants from donors after cardiac death: Initial experience at the London Health Sciences Centre. *Canadian Journal of Surgery*, *53*(2), 93. PMID:20334741

Hernandez, Y., Nicolosi, V., Lotya, M., Blighe, F. M., Sun, Z., McGovern, I. T., ... Byrne, M. (2008). High-yield production of graphene by liquid-phase exfoliation of graphite. *Nature Nanotechnology*, *3*(9), 563–568. doi:10.1038/nnano.2008.215 PMID:18772919

Herrasti, P., Recio, F. J., Ocon, P., & Fatas, E. (2005). Effect of the polymer layers and bilayers on the corrosion behaviour of mild steel: Comparison with polymers containing Zn microparticles. *Progress in Organic Coatings*, *54*(4), 285–291. doi:10.1016/j.porgcoat.2005.07.001

He, S., Hu, X., Chen, S., Hu, H., Hanif, M., & Hou, H. (2012). Needle-like polyaniline nanowires on graphite nanofibers: Hierarchical micro/nano-architecture for high performance supercapacitors. *Journal of Materials Chemistry*, *22*(11), 5114–5120. doi:10.1039/c2jm15668g

Hintze-Bruening, H., & Leroux, F. (2012). Nanocomposite based multifunctional coatings. In *New advances in vehicular technology and automotive engineering* (pp. 56–70). Croatia: Intech Publisher. doi:10.5772/48567

Hirsch, A., & Vostrowsky, O. (2005). Functionalization of carbon nanotubes. *Topics in Current Chemistry*, *245*, 193–237. doi:10.1007/b98169

Hirsch, P., Kässens, M., Püttmann, M., & Reimer, L. (1994). Contamination in a scanning electron microscope and the influence of specimen cooling. *Scanning*, *16*(2), 101–110. doi:10.1002ca.4950160207

Holbery, J., & Houston, D. (2006). Natural fiber reinforced polymer composites in automotive applications. *JOM*, *58*(11), 80–86. doi:10.100711837-006-0234-2

Holbrey, J. D., & Seddon, K. R. (1999). The phase behaviour of 1-alkyl-3-methylimidazolium tetrafluoroborates; ionic liquids and ionic liquid crystals. *Journal of the Chemical Society, Dalton Transactions: Inorganic Chemistry*, (13): 2133–2140. doi:10.1039/a902818h

Hong, C. Y., You, Y. Z., & Pan, C. Y. (2006). A new approach to functionalize multi walled carbon nanotubes by the use of functional polymers. *Polymer*, *47*(2), 4300–4309. doi:10.1016/j.polymer.2006.04.006

Horng, Y.-Y., Lu, Y.-C., Hsu, Y.-K., Chen, C.-C., Chen, L.-C., & Chen, K.-H. (2010). Flexible supercapacitor based on polyaniline nanowires/carbon cloth with both high gravimetric and area-normalized capacitance. *Journal of Power Sources*, *195*(13), 4418–4422. doi:10.1016/j.jpowsour.2010.01.046

Hossain, M. Y. (2014). Natural Textile fibers (Cotton, Jute, Silk, Wool). Retrieved from https://www.slideshare.net/YousufTE/natural-textile-fibers-cotton-jute-silk-wool

Hosseini, M. G., Sabouri, M., & Shahrabi, T. (2007). Corrosion protection of mild steel by polypyrrole phosphate composite coating. *Progress in Organic Coatings*, *60*(3), 178–185. doi:10.1016/j.porgcoat.2007.07.029

Hostler, S., Kaul, P., Qu, V., Cullen, C., & Abramson, A. (2006). Thermal and electrical characterization of nanocomposites for thermoelectrics. In *Proceedings of The Tenth Intersociety Conference on Thermal and Thermomechanical Phenomena in Electronics Systems, ITHERM'06*, San Diego, CA (pp. 1400–1405). 10.1109/ITHERM.2006.1645508

Howard, D., Buttery, L. D., Shakesheff, K. M., & Roberts, S. J. (2008). Tissue engineering: Strategies, stem cells and scaffolds. *Journal of Anatomy*, *213*(1), 66–72. doi:10.1111/j.1469-7580.2008.00878.x PMID:18422523

Huang, W. M., Fu, Y. Q., & Zhao, Y. (2010). High Performance Polyurethane Shape-Memory Polymer and Composites. In 2nd *International Conference on: Polymers in Defence & Aerospace Applications*, Hamburg, Germany, February 10-11 (pp. 136-144).

Huang, C.-C., Xing, W., & Zhuo, S.-P. (2009). Capacitive performances of amorphous tungsten oxide prepared by microwave irradiation. *Scripta Materialia*, *61*(10), 985–987. doi:10.1016/j.scriptamat.2009.08.009

Huang, C., Liou, M. R., & Liu, C. B. (1994). In C. P. Huang (Ed.), *Hazardous Industrial Wastes* (p. 275). Lancaster: Technomic Publishing Company.

Huang, G., Chen, S., Tang, S., & Gao, J. (2012). A novel intumescent flame retardant-functionalized graphene: Nanocomposite synthesis, characterization, and flammability properties. *Materials Chemistry and Physics*, *135*(2-3), 938–947. doi:10.1016/j.matchemphys.2012.05.082

Huang, G., Liang, H., Wang, Y., Wang, X., Gao, J., & Fei, Z. (2012). Combination effect of melamine polyphosphate and graphene on flame retardant properties of poly (vinyl alcohol). *Materials Chemistry and Physics*, *132*(2-3), 520–528. doi:10.1016/j.matchemphys.2011.11.064

Huang, G., Wang, S., Song, P., Wu, C., Chen, S., & Wang, X. (2014). Combination effect of carbon nanotubes with graphene on intumescent flame-retardant polypropylene nanocomposites. *Composites. Part A, Applied Science and Manufacturing*, *59*, 18–25. doi:10.1016/j.compositesa.2013.12.010

Huang, L., Jia, Z., Kymissis, I., & O'Brien, S. (2010). High K capacitors and OFET gate dielectrics from self-assembled BaTiO3 and (Ba, Sr) TiO3 nanocrystals in the superparaelectric limit. *Advanced Functional Materials*, *20*(4), 554–560.

Huang, S., Huang, S., Ng, H., & Kaner, R. (1993). Polyaniline capacitors. *Synthetic Metals*, *57*(1), 4047–4052. doi:10.1016/0379-6779(93)90555-B

Huang, W. S., Humphrey, B. D., & MacDiarmid, A. G. (1986). Polyaniline, a novel conducting polymer. Morphology and chemistry of its oxidation and reduction in aqueous electrolytes. *Journal of the Chemical Society, Faraday Transactions 1. Physical Chemistry in Condensed Phases*, *82*(8), 2385–2400.

Huang, W., Humphrey, B. D., & MacDiarmid, A. G. (1986). Polyaniline, a novel conducting polymer. Morphology and Chemistry of its oxidation and reduction in aqueous electrolytes. *Journal of the Chemical Society, Faraday Transactions*, *1*(198), 2385–2400. doi:10.1039/f19868202385

Huang, X., Qi, X., Boey, F., & Zhang, H. (2012). Graphene-based composites. *Chemical Society Reviews*, *41*(2), 666–686. doi:10.1039/C1CS15078B PMID:21796314

Hu, C., & Chu, C. (2001). Electrochemical impedance characterization of polyaniline-coated graphite electrodes for electrochemical capacitors d effects of film coverage/thickness and anions. *Journal of Electroanalytical Chemistry*, *503*(1-2), 105–116. doi:10.1016/S0022-0728(01)00385-0

Hu, H., He, Y., Long, Z., & Zhan, Y. (2017). Synergistic effect of functional carbon nanotubes and graphene oxide on the anti-corrosion performance of epoxy coating. *Polymers for Advanced Technologies*, *28*(6), 754–762. doi:10.1002/pat.3977

Hu, H., Zhang, K., Li, S., Jia, S., & Ye, C. (2014). Flexible, in-plane, and all-solid-state micro-supercapacitors based on printed interdigital Au/polyaniline network hybrid electrodes on a chip. *Journal of Materials Chemistry. A, Materials for Energy and Sustainability*, *2*(48), 20916–20922. doi:10.1039/C4TA05345A

Hu, J., Huang, L., Zhuang, X., Zhang, P., Lang, L., Chen, X., ... Jing, X. (2008). Electroactive aniline pentamer cross-linking chitosan for stimulation growth of electrically sensitive cells. *Biomacromolecules*, *9*(10), 2637–2644. doi:10.1021/bm800705t PMID:18698845

Hu, L. (2016). Introduction to Active Coatings for Smart Textiles. In *Active Coatings for Smart Textiles* (pp. 1–7). Woodhead Publishing Series in Textiles. doi:10.1016/B978-0-08-100263-6.00001-0

Hu, L., Tu, J., Jiao, S., Hou, J., Zhu, H., & Fray, D. J. (2012). In situ electrochemical polymerization of a nanorod-PANI–Graphene composite in a reverse micelle electrolyte and its application in a supercapacitor. *Physical Chemistry Chemical Physics*, *14*(45), 15652–15656. doi:10.1039/c2cp42192e PMID:23076399

Hu, Y., Lu, H., & Song, L. (2008). Polyethylene/layered silicate nanocomposites: preparation and properties. In D. Nwabunma & T. Kyu (Eds.), *Polyolefin Composites* (pp. 228–250). Hoboken, N.J.: Wiley-Interscience.

Huynh, W. U., Dittmer, J. J., & Alivisatos, A. P. (2002). Hybrid nanorod-polymer solar cells. *Science*, *295*(5564), 2425–2427. doi:10.1126cience.1069156 PMID:11923531

Hu, Z., Xie, Y., Wang, Y., Mo, L., Yang, Y., & Zhang, Z. (2009). Polyaniline/SnO_2 nanocomposite for supercapacitor applications. *Materials Chemistry and Physics*, *114*(2-3), 990–995. doi:10.1016/j.matchemphys.2008.11.005

Hÿtch, M. J., Snoeck, E., & Kilaas, R. (1998). Quantitative measurement of displacement and strain fields from HREM micrographs. *Ultramicroscopy*, *74*(3), 131–146. doi:10.1016/S0304-3991(98)00035-7

Iijima, S. (1991). Helical microtubes of graphitic carbon. *Science*, *50*(8), 962–1079.

Imai, T., Komiya, G., Murayama, K., Ozaki, T., Sawa, F., Shimizu, T., ... Tanaka, T. (2008) Improving Epoxy-based Insulating Materials with Nano-fillers toward Practical Application. In *IEEE International Symposium on Electrical Insulation*. 10.1109/ELINSL.2008.4570310

Imani, A., & Farzi, G. (2017). Low temperature process of electronic charge transport mechanism in PANI/MWCNT nanocomposites: Tubular morphology. *Journal of Materials Science Materials in Electronics*, *28*(14), 10684–10692. doi:10.100710854-017-6843-4

Imtiaz, S. (2014). Glass Fibre. Retrieved from https://www.academia.edu/7463236/GLASS_Fibres_manufacturing_properties_and_applications

International Energy Agency. (2017). *World Energy Outlook 2017*. Retrieved from https://www.iea.org/weo2017/

ISO. (2015). ISO/TS80004-2 Nanotechnologies—Vocabulary—Part 2: Nano-objects.

Istrate, O. M., Paton, K. R., Khan, U., O'Neill, A., Bell, A. P., & Coleman, J. N. (2014). Reinforcement in melt-processed polymer–graphene composites at extremely low graphene loading level. *Carbon*, *78*, 243–249. doi:10.1016/j.carbon.2014.06.077

Jabeen, N., Xia, Q., Yang, M., & Xia, H. (2016). Unique core–shell nanorod arrays with polyaniline deposited into mesoporous $NiCo_2O_4$ support for high-performance supercapacitor electrodes. *ACS Applied Materials & Interfaces*, *8*(9), 6093–6100. doi:10.1021/acsami.6b00207 PMID:26889785

Jackson, R. (2017). Small is beautiful: nano drone tech is advancing. *DefenceIQ*. Retrieved from https://www.defenceiq.com/defence-technology/articles/nano-drone-tech-is-advancing

Jadhav, M. V., Todkar, A. S., Gambhire, V. R., & Sawant, S. Y. (2011). Nanotechnology for powerful solar energy. *International Journal of Advanced Biotechnology and Research*, *2*(1), 208–212.

Jaidev, J. R. I., Jafri, R. I., Mishra, A. K., & Ramaprabhu, S. (2011). Polyaniline-MnO_2 nanotube hybrid nanocomposite as supercapacitor electrode material in acidic electrolyte. *Journal of Materials Chemistry*, *21*(44), 17601–17605. doi:10.1039/c1jm13191e

Jakubovics, J. P. (2008). Lorentz Microscopy. In *Handbook of Microscopy Set* (pp. 505–514). Wiley.

Janaki, V., Oh, B. T., Shanthi, K., Lee, K. J., Ramasamy, A. K., & Kamala-Kannan, S. (2012). Polyaniline/chitosan composite: An eco-friendly polymer for enhanced removal of dyes from aqueous solution. *Synthetic Metals*, *162*(11-12), 974–980. doi:10.1016/j.synthmet.2012.04.015

Jancar, J., Douglas, J. F., Starr, F. W., Kumar, S. K., Cassagnau, P., Lesser, A. J., ... Buehler, M. J. (2010). Current issues in research on structure-property relationships in polymer nanocomposites. *Polymer*, *51*(15), 3321–3343. doi:10.1016/j.polymer.2010.04.074

Jarosovaa, R., & Swain, G. M. (2015). Rapid preparation of room temperature ionic liquids with low water content as characterized with a ta-C:N Electrode. *Journal of the Electrochemical Society*, *162*(8), H507–H511. doi:10.1149/2.0191508jes

Jawaid, M., & Abdul Khalil, H. P. S. (2011). Cellulosic/synthetic fibre reinforced polymer hybrid composites: A review. *Carbohydrate Polymers*, *86*(1), 1–18. doi:10.1016/j.carbpol.2011.04.043

Jeong, S. I., Jun, I. D., Choi, M. J., Nho, Y. C., Lee, Y. M., & Shin, H. (2008). Development of Electroactive and Elastic Nanofibers that contain Polyaniline and Poly (L-lactide-co-ε-caprolactone) for the Control of Cell Adhesion. *Macromolecular Bioscience*, *8*(7), 627–637. doi:10.1002/mabi.200800005 PMID:18401867

Jeong, Y. R., Park, H., Jin, S. W., Hong, S. Y., Lee, S. S., & Ha, J. S. (2015). Highly stretchable and sensitive strain sensors using fragmentized graphene foam. *Advanced Functional Materials*, *25*(27), 4228–4236. doi:10.1002/adfm.201501000

Jeon, H., Park, J., & Shon, M. (2013). Corrosion protection by epoxy coating containing multi-walled carbon nanotubes. *Journal of Industrial and Engineering Chemistry*, *19*(3), 849–853. doi:10.1016/j.jiec.2012.10.030

Jeon, I.-Y., Chang, D. W., Kumar, A. S., & Beak, J.-B. (2011). *Functionalization of Carbon Nanotubes* (p. 96). London: Intech.

Jeon, Y., Noh, H. K., & Song, H.-K. (2017). A lithium-ion battery using partially lithiated graphite anode and amphi-redox $LiMn_2O_4$ cathode. *Scientific Reports*, *7*(1), 1–9. doi:10.103841598-017-14741-x PMID:29093574

Jeyanthi, S., & Rani, J. J. (2012). Influence of natural long fiber in mechanical, thermal and recycling properties of thermoplastic composites in automotive components. *International Journal of Physical Sciences*, *7*(43), 5765–5771.

Jiang, H., Ma, J., & Li, C. Z. (2012). Polyaniline–MnO_2 coaxial nanofiber with hierarchical structure for high-performance supercapacitors. *Journal of Materials Chemistry*, *22*(33), 16939–16942. doi:10.1039/c2jm33249c

Jiang, N. (2013). Damage mechanisms in electron microscopy of insulating materials. *Journal of Physics. D, Applied Physics*, *46*(30), 305502. doi:10.1088/0022-3727/46/30/305502

Jiang, N., Zhou, S., Su, D., & Qiu, J. (2012). Do Eu dopants prefer the precipitated LaF_3 nanocrystals in glass ceramics? *Physica Status Solidi. Rapid Research Letters*, *6*(12), 487–489. doi:10.1002/pssr.201206476

Ji, J., Xiong, H., Zhu, Z., Li, L., Huang, Y., & Yu, X. (2018). Fabrication of Polypyrrole/Chitosan Nanocomposite Aerogel Monolith for Removal of Cr (VI). *Journal of Polymers and the Environment*, *26*(5), 1979–1985. doi:10.100710924-017-1095-1

Jimenez, G., Ogata, N., Kawai, H., & Ogihara, T. (1997). Structure and thermal/mechanical properties of poly (ε-caprolactone) - clay blend. *Journal of Applied Polymer Science*, *64*(11), 2211–2220. doi:10.1002/(SICI)1097-4628(19970613)64:11<2211::AID-APP17>3.0.CO;2-6

Jinschek, J. R., Batenburg, K. J., Calderon, H. A., Kilaas, R., Radmilovic, V., & Kisielowski, C. (2008). 3-D reconstruction of the atomic positions in a simulated gold nanocrystal based on discrete tomography: Prospects of atomic resolution electron tomography. *Ultramicroscopy*, *108*(6), 589–604. doi:10.1016/j.ultramic.2007.10.002 PMID:18082327

Jin, Y., Fang, M., & Jia, M. (2014). In situ one-pot synthesis of graphene–polyaniline nanofiber composite for high-performance electrochemical capacitors. *Applied Surface Science*, *308*, 333–340. doi:10.1016/j.apsusc.2014.04.168

Ji, X., Xu, Y., Zhang, W., Cui, L., & Liu, J. (2016). Review of functionalization, structure and properties of graphene/polymer composite fibers. *Composite: Part A*, *87*, 29–45. doi:10.1016/j.compositesa.2016.04.011

Jlassi, K., Chehimi, M., & Thomas, S. (Eds.). (2017). *Clay-polymer nanocomposites*. Philadelphia, PA: Elsevier Science.

Jo, B., Park, S., & Kim, D. (2008). Mechanical properties of nano-MMT reinforced polymer composite and polymer concrete. *Construction & Building Materials*, *22*(1), 14–20. doi:10.1016/j.conbuildmat.2007.02.009

Johnson, C. L., Snoeck, E., Ezcurdia, M., Rodríguez-González, B., Pastoriza-Santos, I., Liz-Marzán, L. M., & Hÿtch, M. J. (2008). Effects of elastic anisotropy on strain distributions in decahedral gold nanoparticles. *Nature Materials*, *7*(2), 120–124. doi:10.1038/nmat2083 PMID:18084296

Johnson, P. D., Watson, M. A., Brown, J., & Jefcoat, I. A. (2002). Peanut hull pellets as a single use sorbent for the capture of Cu(II) from wastewater. *Waste Management (New York, N.Y.)*, *22*(5), 471–480. doi:10.1016/S0956-053X(01)00036-8 PMID:12092755

Kaiser, A., & Park, Y. (2005). Current-voltage characteristics of conducting polymers and carbon nanotubes. *Synthetic Metals*, *152*(1-3), 181–184. doi:10.1016/j.synthmet.2005.07.245

Kakac, S., & Pramuanjaroenkij, A. (2009). Review of convective heat transfer enhancement with nanofluids. *International Journal of Heat and Mass Transfer*, *52*(13-14), 3187–3196. doi:10.1016/j.ijheatmasstransfer.2009.02.006

Kalaiselvam, S., & Parameshwaran, R. (2014). Nanotechnology in thermal energy storage. In Thermal Energy Storage Technologies for Sustainability (pp. 163-202). doi:10.1016/B978-0-12-417291-3.00008-6

Kannusamy, P., & Sivalingam, T. (2013). Synthesis of porous chitosan–polyaniline/ZnO hybrid composite and application for removal of reactive orange 16 dye. *Colloids and Surfaces. B, Biointerfaces*, *108*, 229–238. doi:10.1016/j.colsurfb.2013.03.015 PMID:23563288

Kara, I., Yilmazer, D., & Akar, S. T. (2017). Metakaolin based geopolymer as an effective adsorbent for adsorption of zinc(II) and nickel(II) ions from aqueous solutions. *Applied Clay Science*, *139*, 54–63. doi:10.1016/j.clay.2017.01.008

Kargarzadeh, H., Mariano, M., Huang, J., Lin, N., Ahmad, I., Dufresne, A., & Thomas, S. (2017). Recent developments on nanocellulose reinforced polymer nanocomposites: A review. *Polymer*, *132*, 368–393. doi:10.1016/j.polymer.2017.09.043

Kashiwagi, T. (2005). Flammability of nanocomposites: Effect of the shape of nanoparticles. In Fire retardancy polymer: New applications of mineral fillers (p. 81). The royal society of chemistry, UK

Kashiwagi, T., Grulke, E., Hilding, J., Groth, K., Harris, R., Butler, K., ... Douglas, J. (2004). Thermal and flammability properties of polypropylene/carbon nanotube nanocomposites. *Polymer*, *45*(12), 4227–4239. doi:10.1016/j.polymer.2004.03.088

Kathi, J., Rhee, K. Y., & Lee, J. H. (2009). Effect of chemical functionalization of multi-walled carbon nanotubes with 3-aminopropyltriethoxysilane on mechanical and morphological properties of epoxy nanocomposites. *Composites. Part A, Applied Science and Manufacturing*, *40*(6-7), 800–809. doi:10.1016/j.compositesa.2009.04.001

Kato, M., Matsushita, M., & Fukumori, K. (2004). Development of a new production method for a polypropylene-clay nanocomposite. *Polymer Engineering and Science*, *44*(7), 1205–1211. doi:10.1002/pen.20115

Kaur, G., Pandey, O. P., Singh, K., Homa, D., Scott, B., & Pickrell, G. (2014). A review of bioactive glasses: Their structure, properties, fabrication and apatite formation. *Journal of Biomedical Materials Research. Part A*, *102*(1), 254–274. doi:10.1002/jbm.a.34690 PMID:23468256

Kausar, A., & Siddiq, M. (2016). Latent of Modified Epoxy Materials in Defence Applications. In A. Méndez-Vilas & A. Solano-Martín (Eds.), Polymer Science: Research Advances, Practical Applications and Educational Aspects (pp. 278-288). Formatex Research Center.

Kausar, A. (2014). Fabrication and properties of polyamide and graphene oxide coated carbon fiber reinforced epoxy composites. *American Journal of Political Science*, *4*(3), 88–93.

Kausar, A. (2016a). Bucky papers of poly (methyl methacrylate-co-methacrylic acid)/polyamide 6 and graphene oxide-montmorillonite. *Journal of Dispersion Science and Technology*, *37*(1), 66–72. doi:10.1080/01932691.2015.1027908

Kausar, A. (2016b). Thermal Conductivity Measurement of Polyvinylpyrrolidone/ Polyethylene/ Graphene Nanocomposite. *Nanoscience and Nanotechnology*, *6*(2), 34–37.

Kausar, A. (2018). Shielding Efficacy of Polymeric Nano-Structure. *Research Journal of Nanoscience and Engineering*, *2*(1), 9–14.

Kausar, A., Anwar, Z., & Muhammad, B. (2016). Recent developments in epoxy/graphite, epoxy/graphene, and epoxy/graphene nanoplatelet composites: A comparative review. *Polymer-Plastics Technology and Engineering*, *55*(11), 1192–1210. doi:10.1080/03602559.2016.1163589

Kausar, A., & Hussain, S. T. (2013). Effect of multi-walled carbon nanotube reinforcement on the physical properties of poly (thiourea-azo-ether)-based nanocomposites. *Journal of Plastic Film & Sheeting*, *29*(4), 365–383. doi:10.1177/8756087913487003

Kausar, A., Rafique, I., & Muhammad, B. (2017). Aerospace Application of Polymer Nanocomposite with Carbon Nanotube, Graphite, Graphene Oxide and Nanoclay. *Polymer-Plastics Technology and Engineering*, *56*(13), 1438–1456.

Kaushal, A., & Singh, S. K. (2017). Removal of heavy metals by nanoadsorbents. A review. *Journal of Environmental Biotechnology Research*, *6*(1), 96–104.

Kawasumi, M., Hasegawa, N., Kato, M., Usuki, A., & Okada, A. (1997). Preparation and mechanical properties of polypropylene-clay hybrids. *Macromolecules*, *30*(20), 6333–6338. doi:10.1021/ma961786h

Keast, V. J., Scott, A. J., Brydson, R., Williams, D. B., & Bruley, J. (2001). Electron energy-loss near-edge structure - A tool for the investigation of electronic structure on the nanometre scale. *Journal of Microscopy*, *203*(2), 135–175. doi:10.1046/j.1365-2818.2001.00898.x PMID:11489072

Keledi, G., Hári, J., & Pukánszky, B. (2012). Polymer nanocomposites: Structure, interaction, and functionality. *Nanoscale*, *4*(6), 1919–1938. doi:10.1039/c2nr11442a PMID:22349033

Kenawy, E., Ghfar, A., Wabaidur, S., Khan, M., Siddiqui, M., Alothman, Z., ... Hamid, M. (2018). Cetyltrimethylammonium bromide intercalated and branched polyhydroxystyrene functionalized montmorillonite clay to sequester cationic dyes. *Journal of Environmental Management*, *219*, 285–293. doi:10.1016/j.jenvman.2018.04.121 PMID:29751259

Ke, Z., & Yongping, B. (2005). Improve the gas barrier property of PET film with montmorillonite by in situ interlayer polymerization. *Materials Letters*, *59*(27), 3348–3351. doi:10.1016/j.matlet.2005.05.070

Khalid, M., Tumelero, M. A., Zoldan, V. C., Cid, C. C. P., Franceschini, D. F., Timm, R. A., ... Pasa, A. A. (2014). Polyaniline nanofibers -graphene oxide nanoplatelets composite thin film electrodes for electrochemical capacitors. *RSC Advances*, *4*(64), 34168–34178. doi:10.1039/C4RA06145D

Khan, W. S., Hamadneh, N. N., & Khan, W. A. (2016). Polymer nanocomposites–synthesis techniques, classification and properties. One Central Press.

Khanam, Z., Singh, V., & Zaidi, M. G. H. (2018). Enhanced corrosion protection performance with MWCNT dispersed epoxy coating prepared under supercritical CO2 assistance. *Polymers for Advanced Technologies*, *29*(9), 2457–2466. doi:10.1002/pat.4357

Khan, D. M., Kausar, A., & Salman, S. M. (2016). Exploitation of Nanobifiller in Polymer/Graphene Oxide-Carbon Nanotube, Polymer/Graphene Oxide-Nanodiamond, and Polymer/Graphene Oxide-Montmorillonite Composite: A Review. *Polymer-Plastics Technology and Engineering*, *55*(7), 744–768.

Khan, M. I., Min, T. K., Azizli, K., Sufian, S., Ullah, H., & Man, Z. (2015). Effective removal of methylene blue from water using phosphoric acid based geopolymers: Synthesis, characterizations and adsorption studies. *RSC Advances*, *5*(75), 61410–61420. doi:10.1039/C5RA08255B

Khan, S., & Narula, A. K. (2016). Bio-hybrid blended transparent and conductive films PEDOT:PSS:Chitosan exhibiting electro-active and antibacterial properties. *European Polymer Journal*, *81*, 161–172. doi:10.1016/j.eurpolymj.2016.06.005

Khan, W. S., Hamadneh, N. N., & Khan, W. A. (2017). Polymer nanocomposites - synthesis techniques, classification and properties. In *Science and applications of tailored nanostructures* (p. 63). UK: One Central Press.

Kharat, D. K., Muthurajan, H., & Praveenkumar, B. (2006). Present and Futuristic Military Applications of Nanodevices, Synthesis and Reactivity in Inorganic. *Metal-Organic and Nano-Metal Chemistry*, *36*(2), 231–235. doi:10.1080/15533170500524801

Khatavkar, N., & Balasubramanian, K. (2016). Composite materials for supersonic aircraft radomes with ameliorated radio frequency transmission-a review. *RSC Advances*, *6*(8), 6709–6718. doi:10.1039/C5RA18712E

Khun, N. W., Troconis, B. R., & Frankel, G. S. (2014). Effects of carbon nanotube content on adhesion strength and wear and corrosion resistance of epoxy composite coatings on AA2024-T3. *Progress in Organic Coatings*, *77*(1), 72–80. doi:10.1016/j.porgcoat.2013.08.003

Kim, W., Javey, A., Tu, R., Cao, J., Wang, Q. & Dai, H. J. (2005) Electrical contacts to carbon nanotubes down to 1nm in diameter. *Appl. Phys. Lett., 87.*

Kim, B., Chung, H., Chu, K. S., Yoon, H. G., Lee, C. J., & Kim, W. (2010). Synthesis of vertically-aligned carbon nanotubes on stainless steel by water-assisted chemical vapor deposition and characterization of their electrochemical properties. *Synthetic Metals*, *160*(7-8), 584–587. doi:10.1016/j.synthmet.2009.12.008

Kim, D. H., Richardson-Burns, S. M., Hendricks, J. L., Sequera, C., & Martin, D. C. (2007). Effect of immobilized nerve growth factor on conductive polymers: Electrical properties and cellular response. *Advanced Functional Materials*, *17*(1), 79–86. doi:10.1002/adfm.200500594

Kim, H., Abdala, A. A., & Macosko, C. W. (2010). Graphene/polymer nanocomposites. *Macromolecules*, *43*(16), 6515–6530. doi:10.1021/ma100572e

Kim, I. H., & Kim, K. B. (2001). Ruthenium oxide thin film electrodes for supercapacitors. *Electrochemical and Solid-State Letters*, *4*(5), A62–A64. doi:10.1149/1.1359956

Kim, I. Y., Seo, S. J., Moon, H. S., Yoo, M. K., Park, I. Y., Kim, B. C., & Cho, C. S. (2008). Chitosan and its derivatives for tissue engineering applications. *Biotechnology Advances*, *26*(1), 1–21. doi:10.1016/j.biotechadv.2007.07.009 PMID:17884325

Kim, P., Jones, S., Hotchkiss, P., Haddock, J., Kippelen, B., Marder, S., & Perry, J. (2007). Phosphonic acid-modified barium titanate polymer nanocomposites with high permittivity and dielectric strength. *Advanced Materials*, *19*(7), 1001–1005. doi:10.1002/adma.200602422

Kim, T., Kim, J. H., Kang, T. E., Lee, C., Kang, H., Shin, M., ... Kim, B. J. (2015). Flexible, highly efficient all-polymer solar cells. *Nature Communications*, 6. PMID:26449658

Kim, W., Choi, H. C., Shim, M., Li, Y. M., Wang, D. W., & Dai, H. J. (2002). Synthesis of ultralong and high percentage of semiconducting single-walled carbon nanotubes. *Nano Letters*, *2*(7), 703–708. doi:10.1021/nl025602q

Kim, W., Nair, N., Lee, C. Y., & Strano, M. S. (2008). Covalent functionalization of single-walled carbon nanotubes alters their densities allowing electronic and other types of separation. *The Journal of Physical Chemistry C*, *112*(19), 7326–7331. doi:10.1021/jp710919b

Kinlen, P. J., Silverman, D. C., & Jeffreys, C. R. (1997). Corrosion protection using polyanujne coating formulations. *Synthetic Metals*, *85*(1-3), 1327–1332. doi:10.1016/S0379-6779(97)80257-3

Koba, Y., & Ishizaki, A. (2014). Chemical Composition of Palm Fiber and Its Feasibility as Cellulosic Raw Material for Sugar Production. *Agricultural and Biological Chemistry*, *54*(5), 1183–1187.

Koch, G. H., Brongers, M. P., Thompson, N. G., Virmani, Y. P., & Payer, J. H. (2002). Corrosion cost and preventive strategies in the United States (No. FHWA-RD-01-156,).

Kolasińska, E., & Kolasiński, P. (2016). A review on electroactive polymers for waste heat recovery. *Materials (Basel)*, *9*(6), 485. doi:10.3390/ma9060485 PMID:28773605

Kole, M., & Dey, T. K. (2010). Thermal conductivity and viscosity of Al_2O_3 nanofluid based on car engine coolant. *Journal of Physics D*, *43*(31), 315501. doi:10.1088/0022-3727/43/31/315501

Kong, D. L. Y., Sanjayan, J. G., & Sagoe-Crentsil, K. (2007). Comparative performance of geopolymers made with metakaolin and fly ash after exposure to elevated temperatures. *Cement and Concrete Research*, *37*(12), 1583–1589. doi:10.1016/j.cemconres.2007.08.021

Koniuszewska, A. G., & Kaczmar, J. W. (2016). Application of polymer based composite materials in transportation. *Progress in Rubber, Plastics and Recycling Technology*, *32*(1), 1–24. doi:10.1177/147776061603200101

Koopmans, G., Hasse, B., & Sinis, N. (2009). The role of collagen in peripheral nerve repair. *International Review of Neurobiology*, *87*, 363–379. doi:10.1016/S0074-7742(09)87019-0 PMID:19682648

Kornmann, X., Berglund, L. A., Sterte, J., & Giannelis, E. P. (1998). Nanocomposites based on montmorillonite and unsaturated polyester. *Polymer Engineering and Science*, *38*(8), 1351–1358. doi:10.1002/pen.10305

Koronis, G., Silva, A., & Fontul, M. (2013). Green composites: A review of adequate materials for automotive applications. *Composites. Part B, Engineering*, *44*(1), 120–127. doi:10.1016/j.compositesb.2012.07.004

Kotal, M., & Bhowmick, A. (2015). Polymer nanocomposites from modified clays: Recent advances and challenges. *Progress in Polymer Science*, *51*, 127–187. doi:10.1016/j.progpolymsci.2015.10.001

Kotek, J., Kelnar, I., Studenovský, M., & Baldrian, J. (2005). Chlorosulfonated polypropylene: Preparation and its application as a coupling agent in polypropylene–clay nanocomposites. *Polymer*, *46*(13), 4876–4881. doi:10.1016/j.polymer.2005.02.119

Kotsilkova, R. (2007). *Thermoset nanocomposites for engineering applications*. UK: Smithers Rapra Technology Limited Publisher.

Kotz, R., & Carlen, M. (2000). Principles and applications of electrochemical capacitors. *Electrochimica Acta*, *45*(15-16), 2483–2498. doi:10.1016/S0013-4686(00)00354-6

Kovářík, T., Křenek, T., Rieger, D., Pola, M., Říha, J., Svoboda, M., ... Kadlec, J. (2017). Synthesis of open-cell ceramic foam derived from geopolymer precursor via replica technique. *Materials Letters*, *209*, 497–500. doi:10.1016/j.matlet.2017.08.081

Ko, Y. H., Kim, J.-H., Jin, L.-H., Ko, S.-M., Kwon, B.-J., Kim, J., ... Cho, Y.-H. (2011). Electrically driven quantum dot/wire/well hybrid light-emitting diodes. *Advanced Materials*, *23*(45), 5364–5369. doi:10.1002/adma.201102534 PMID:22002261

Kozłowski, R., & Władyka-przybylak, M. (2008). Review flammability and fire resistance of composites reinforced by natural fibers. *Polymers for Advanced Technologies*, *19*(6), 446–453. doi:10.1002/pat.1135

Krajewska, B. (2005). Membrane-based processes performed with use of chitin/chitosan materials. *Separation and Purification Technology*, *41*(3), 305–312. doi:10.1016/j.seppur.2004.03.019

Kuang, H., Cao, Q., Wang, X., Jing, B., Wang, Q., & Zhou, L. (2013). Influence of the reaction temperature on polyaniline morphology and evaluation of their performance as supercapacitor electrode. *Journal of Applied Polymer Science*, *130*(5), 3753–3758. doi:10.1002/app.39650

Kubát, J., Rigdahl, M., & Welander, M. (1990). Characterization of interfacial interactions in high density polyethylene filled with glass spheres using dynamic-mechanical analysis. *Journal of Applied Polymer Science*, *39*(7), 1527–1539. doi:10.1002/app.1990.070390711

Kudin, K. N., Ozbas, B., Schniepp, H. C., Prud'homme, R. K., Aksay, I. A., & Car, R. (2008). Raman spectra of graphite oxide and functionalized graphene sheets. *Nano Letters*, *8*(1), 36–41. doi:10.1021/nl071822y PMID:18154315

Kuila, B. K., Nandan, B., Bo¨hme, M., Janke, A., & Stamm, M. (2009). Vertically oriented arrays of polyaniline nanorods and their super electrochemical properties. *Chemical Communications (Cambridge)*, (38): 5749–5751. doi:10.1039/b912513b PMID:19774257

Kuilla, T., Bhadra, S., Yao, D., Kim, N. H., Bose, S., & Lee, J. H. (2010). Recent advances in graphene based polymer composites. *Progress in Polymer Science*, *35*(11), 1350–1375. doi:10.1016/j.progpolymsci.2010.07.005

Kumar, A. M., Suresh, B., Das, S., Obot, I. B., Adesina, A. Y., & Ramakrishna, S. (2017). Promising bio-composites of polypyrrole and chitosan: Surface protective and *in vitro* biocompatibility performance on 316L SS implants. *Carbohydrate Polymers*, *173*, 121–130. doi:10.1016/j.carbpol.2017.05.083 PMID:28732850

Kumar, A., Anant, R., Kumar, K., Chauhan, S. S., Kumar, S., & Kumar, R. (2016). Anticorrosive and electromagnetic shielding response of a graphene/TiO_2-epoxy nanocomposite with enhanced mechanical properties. *RSC Advances*, *6*(114), 113405–113414. doi:10.1039/C6RA15273B

Kumar, A., & Borah, R. (2016). Effect of plasma irradiation on biocompatibility and cell adhesion of polyaniline/chitosan nanocomposites towards Hep G2 and PBMC cells. *Advanced Materials & Processes*, *1*(2), 146–155. doi:10.5185/amp.2016/206

Kumar, A., Ghosh, P. K., Yadav, K. L., & Kumar, K. (2017). Thermo-mechanical and anti-corrosive properties of MWCNT/epoxy nanocomposite fabricated by innovative dispersion technique. *Composites. Part B, Engineering*, *113*, 291–299. doi:10.1016/j.compositesb.2017.01.046

Kumar, A., Kumar, V., & Awasthi, K. (2017). Polyaniline–Carbon Nanotube Composites: Preparation Methods, Properties, and Applications. *Polymer-Plastics Technology and Engineering*, *57*(2), 70–97. doi:10.1080/03602559.2017.1300817

Kumar, G. S. (1992). *Azo functional polymers: Functional group approach in macromolecular design*. USA: Technomic Publishing.

Kumar, H. V., Woltornist, S. J., & Adamson, D. H. (2015). Fractionation and characterization of graphene oxide by oxidation extent through emulsion stabilization. *Carbon*, *98*, 491–495. doi:10.1016/j.carbon.2015.10.083

Kumar, S. R., Juan, C.-H., Liao, G.-M., Lin, J.-S., Yang, C.-C., Ma, W.-T., ... Lue, S. J. (2016). Fumed silica nanoparticles incorporated in quaternized poly(vinyl alcohol) nanocomposite membrane for enhanced power densities in direct alcohol alkaline fuel cells. *Energies*, *9*(15), 1–19. doi:10.3390/en9010015

Kumar, S. R., Ma, W.-T., Lu, H.-C., Teng, L.-W., Hsu, H.-C., Shih, C.-M., ... Lue, S. J. (2017). Surfactant-assisted perovskite nanofillers incorporated in quaternized poly (vinyl alcohol) composite membrane as an effective hydroxide-conducting electrolyte. *Energies*, *10*(615), 1–22. doi:10.3390/en10050615

Kumar, S., & Kumar, R. (2011). Mechanical activation of fly ash: Effect on reaction, structure and properties of resulting geopolymer. *Ceramics International*, *37*(2), 533–541. doi:10.1016/j.ceramint.2010.09.038

Kumar, S., Lively, B., Sun, L., Li, B., & Zhong, W. (2010). Highly dispersed and electrically conductive polycarbonate/oxidized carbon nanofiber composites for electrostatic dissipation applications. *Carbon, 48*(13), 3846–3857. doi:10.1016/j.carbon.2010.06.050

Kurahatti, R. V., Surendranathan, A. O., Kori, S. A., Singh, N., Kumar, A. V. R., & Srivastava, S. (2010). Defence applications of polymer nanocomposites. *Defence Science Journal, 60*(5), 551–563. doi:10.14429/dsj.60.578

Kuralay, F., Demirci, S., Kiristi, M., Oksuz, L., & Oksuz, A. U. (2014). Poly(3,4-ethylenedioxythiophene) coated chitosan modified disposable electrodes for DNA and DNA-drug interaction sensing. *Colloids and Surfaces. B, Biointerfaces, 123*, 825–830. doi:10.1016/j.colsurfb.2014.10.021 PMID:25454660

Kurauchi, T., Okada, A., Nomura, T., Nishio, T., Saegusa, S., & Deguchi, R. (1991). Nylon 6-clay hybrid-Synthesis, properties and application to automotive timing belt cover. *SAE.*

Lahiff, E., Ryu, C. Y., Curran, S. A., Minett, A., Blau, W. J., & Ajayan, P. M. (2003). Selective Positioning and Density Control of Nanotubes within a Polymer Thin Film. *Nano Letters, 3*(10), 1333–1337. doi:10.1021/nl034273e

Laine, L., & Rozite, L. (2010). State of the art, eco-efficient composite materials. Retrieved from https://people.arcada.fi/~rene/biocomposites/lectures/STATE%20OF%20THE%20ART_material_examples.pdf

Largeot, C., Portet, C., Chmiola, J., Taberna, P. L., Gogotsi, Y., & Simon, P. (2008). Relation between the Ion Size and Pore Size for an Electric Double-Layer Capacitor. *Journal of the American Chemical Society, 130*(9), 2730–2731. doi:10.1021/ja7106178 PMID:18257568

Lari, A., Sun, T., & Sultana, N. (2016). PEDOT: PSS-Containing Nanohydroxyapatite/Chitosan Conductive Bionanocomposite Scaffold: Fabrication and Evaluation. *Journal of Nanomaterials.*

Larraza, I., López-Gónzalez, M., Corrales, T., & Marcelo, G. (2012). Hybrid materials: Magnetite-polyethylenimine-montmorillonite, as magnetic adsorbents for Cr(VI) water treatment. *Journal of Colloid and Interface Science, 385*(1), 24–33. doi:10.1016/j.jcis.2012.06.050 PMID:22841705

Layek, R. K., & Nandi, A. K. (2013). A review on synthesis and properties of polymer functionalized graphene. *Polymer, 54*(19), 5087–5103. doi:10.1016/j.polymer.2013.06.027

Lee, B. H., & Kim, H. J. (2006). Influence of isocyanate type of acrylated urethane oligomer and of additives on weathering of UV-cured films. *Polymer Degradation & Stability, 91*(5), 1025–1035. doi:10.1016/j.polymdegradstab.2005.08.002

Lee, H., Fasulo, P., Rodgers, W., & Paul, D. (2005). TPO based nanocomposites. Part 1. Morphology and mechanical properties. *Polymer, 46*(25), 11673–11689. doi:10.1016/j.polymer.2005.09.068

Lee, J. Y., Bashur, C. A., Goldstein, A. S., & Schmidt, C. E. (2009). Polypyrrole-coated electrospun PLGA nanofibers for neural tissue applications. *Biomaterials, 30*(26), 4325–4335. doi:10.1016/j.biomaterials.2009.04.042 PMID:19501901

Lee, J.-H., Rhee, K. Y., & Park, S. J. (2011). Silane modification of carbon nanotubes and its effects on the material properties of carbon/CNT/epoxy three-phase composites. *Composites. Part A, Applied Science and Manufacturing, 42*(5), 478–483. doi:10.1016/j.compositesa.2011.01.004

Lee, S. T., Mi, F.-L., Shen, Y.-J., & Shyu, S.-S. (2001). Equilibrium and kinetic studies of copper(II) ion uptake by chitosan-tripolyphosphate chelating resin. *Polymer, 42*(5), 1879–1892. doi:10.1016/S0032-3861(00)00402-X

Lee, S., Kim, J., & Park, S. (2014). Activated carbon nanotubes/polyaniline composites as supercapacitor electrodes. *Energy, 78*, 298–303. doi:10.1016/j.energy.2014.10.011

Lee, T., Yun, T., Park, B., Sharma, B., Song, H., & Kim, B. (2012). Hybrid multilayer thin film supercapacitor of graphene nanosheets with polyaniline: Importance of establishing intimate electronic contact through nanoscale blending. *Journal of Materials Chemistry*, *22*(39), 21092–21099. doi:10.1039/c2jm33111j

Lee, Y. S., Wetzel, E. D., & Wagner, N. J. (2003). The ballistic impact characteristics of Kevlar® woven fabrics impregnated with a colloidal shear thickening fluid. *Journal of Materials Science*, *38*(13), 2825–2833. doi:10.1023/A:1024424200221

Lei, J. Y., Jiang, Z. Q., Lu, X. F., Nie, G. D., & Wang, C. (2015). Synthesis of Few-Layer MoS_2 Nanosheets-Wrapped Polyaniline Hierarchical Nanostructures for Enhanced Electrochemical Capacitance Performance. *Electrochimica Acta*, *176*, 149–155. doi:10.1016/j.electacta.2015.07.028

Lertwimolnun, W., & Vergnes, B. (2005). Influence of compatibilizer and processing conditions on the dispersion of nanoclay in a polypropylene matrix. *Polymer*, *46*(10), 3462–3471. doi:10.1016/j.polymer.2005.02.018

Le, T. H., Kim, Y., & Yoon, H. (2017). Electrical and electrochemical properties of conducting polymers. *Polymers*, *9*(4), 150. doi:10.3390/polym9040150

Letheby, H. (1862). XXIX.-On the production of a blue substance by the electrolysis of sulphate of aniline. *Journal of the Chemical Society*, *15*(0), 161–163. doi:10.1039/JS8621500161

Le, V. T., Ngo, C. L., Le, Q. T., Ngo, T. T., Nguyen, D. N., & Vu, M. T. (2013). Surface modification and functionalization of carbon nanotube with some organic compounds. *Advances in Natural Sciences: Nanoscience and Nanotechnology*, *4*, 1–5.

Leyden, D. E., & Murthy, R. S. S. (1988). Diffuse reflectance fourier transform IR spectroscopy. *Trends in Analytical Chemistry*, *7*(5), 164–169. doi:10.1016/0165-9936(88)85044-1

Liang, C., & Dai, S. (2006). Synthesis of Mesoporous Carbon Materials via Enhanced Hydrogen-Bonding Interaction. *Journal of the American Chemical Society*, *128*(16), 5316–5317. doi:10.1021/ja060242k PMID:16620083

Liang, J., Wang, Y., Huang, Y., Ma, Y., Liu, Z., Cai, J., ... Chen, Y. (2009). Electromagnetic interference shielding of graphene/epoxy composites. *Carbon*, *47*(3), 922–925. doi:10.1016/j.carbon.2008.12.038

Liao, G.-M., Yang, C.-C., Hu, C.-C., Pai, Y.-L., & Lue, S. J. (2015). Novel quaternized polyvinyl alcohol/quaternized chitosan nano-composite as an effective hydroxide-conducting electrolyte. *Journal of Membrane Science*, *485*, 17–29. doi:10.1016/j.memsci.2015.02.043

Liao, H., Zhang, H., Qin, G., Li, Z., Li, L., & Hong, H. (2017). A macro-porous graphene oxide-based membrane as a separator with enhanced thermal stability for high-safety lithium-ion batteries. *RSC Advances*, *7*(36), 22112–22120. doi:10.1039/C7RA02950K

Liao, Y.-H., Marietta-Tondin, O., Liang, Z., Zhang, C., & Wang, B. (2004). Investigation of the dispersion process of SWNTs/SC-15 epoxy resin nanocomposites. *Materials Science and Engineering A*, *385*(1-2), 175–181. doi:10.1016/S0921-5093(04)00857-3

Liaw, W., Huang, P., Chen, C., Lo, C., & Chang, J. (2008). PPgMA/APTS compound coupling compatabilizer in PP/clay hybrid nanocomposite. *Journal of Applied Polymer Science*, *109*(3), 1871–1880. doi:10.1002/app.28294

Li, D., Li, Y., Feng, Y., Hu, W., & Feng, W. (2015). Hierarchical graphene oxide/polyaniline nanocomposites prepared by interfacial electrochemical polymerization for flexible solid-state supercapacitors. *Journal of Materials Chemistry. A, Materials for Energy and Sustainability*, *3*(5), 2135–2143. doi:10.1039/C4TA05643D

Li, D., Muller, M. B., Gilje, S., Kaner, R. B., & Wallace, G. G. (2008). Processable aqueous dispersions of graphene nanosheets. *Nature Nanotechnology*, *3*(2), 101–105. doi:10.1038/nnano.2007.451 PMID:18654470

Li, D., Wang, Y. L., & Xia, Y. N. (2004). Electrospinning nanofibers as uniaxially aligned arrays and layer-by-layer stacked films. *Advanced Materials*, *16*(4), 361–366. doi:10.1002/adma.200306226

Li, H., Wang, J., Chu, Q., Wang, Z., Zhang, F., & Wang, S. (2009). Theoretical and experimental specific capacitance of polyaniline in sulfuric acid. *Journal of Power Sources*, *190*(2), 578–586. doi:10.1016/j.jpowsour.2009.01.052

Li, J., Claude, J., Norena-Franco, L. E., Seok, S. I., & Wang, Q. (2008). Electrical energy storage in ferroelectric polymer nanocomposites containing surface-functionalized $BaTiO_3$ nanoparticles. *Chemistry of Materials*, *20*(20), 6304–6306. doi:10.1021/cm8021648

Li, J., Haung, X., Cui, L., Chen, N., & Qu, L. (2016). Preparation and supercapacitor performance of assembled graphene fiber and foam [Review]. *Progress in Natural Science: Materials International*, *26*(3), 212–220. doi:10.1016/j.pnsc.2016.05.006

Li, J., Pan, J., Zhang, L., Guo, X., & Yu, Y. (2003). Culture of primary rat hepatocytes within porous chitosan scaffolds. *Journal of Biomedical Materials Research. Part A*, *67*(3), 938–943. doi:10.1002/jbm.a.10076 PMID:14613242

Li, L. L., Li, M., & Chen, D. X. (2006). Solvothermal synthesis and characterization of Sb-doped SnO2 nanoparticles used as transparent conductive films. *Materials Research Bulletin*, *41*(3), 541–546. doi:10.1016/j.materresbull.2005.09.011

Li, L., Liu, J., Su, Y., Li, G., Chen, X., Qiu, X., & Yan, T. (2009). Surface doping for photocatalytic purposes: Relations between particle size, surface modifications, and photoactivity of SnO_2:Zn^{2+}nanocrystals. *Nanotechnology*, *20*(15), 155706. doi:10.1088/0957-4484/20/15/155706

Li, L., Wang, S., & Zhu, Z. (2006). Geopolymeric adsorbents from fly ash for dye removal from aqueous solution. *Journal of Colloid and Interface Science*, *300*(1), 52–59. doi:10.1016/j.jcis.2006.03.062 PMID:16626729

Li, M., & Scott, K. (2010). A polymer electrolyte membrane for high temperature fuel cells to fit vehicle applications. *Electrochimica Acta*, *55*(6), 2123–2128. doi:10.1016/j.electacta.2009.11.044

Lincoln, D., Vaia, R., & Krishnamoorti, R. (2004). Isothermal Crystallization of Nylon-6/Montmorillonite Nanocomposites. *Macromolecules*, *37*(12), 4554–4561. doi:10.1021/ma049768k

Lingayat, A. B., & Suple, Y. R. (2013). Review on phase change material as thermal energy storage medium: Materials, application. *International Journal of Engineering Research and Applications*, *3*(4), 916–921.

Lin, Y. T., Don, T. M., Wong, C. J., Meng, F. C., Lin, Y. J., Lee, S. Y., & Chiu, W. Y. (2018). Improvement of mechanical properties and anticorrosion performance of epoxy coatings by the introduction of polyaniline/graphene composite. *Surface and Coatings Technology*. doi:10.1016/j.surfcoat.2018.01.050

Liu, B., Fang, Z. G., Wang, H. B., & Wang, T. (2013). Effect of cross linking degree and adhesion force on the anticorrosion performance of epoxy coatings under simulated deep sea environment. *Progress in Organic Coatings*, *76*(12), 1814–1818. doi:10.1016/j.porgcoat.2013.05.022

Liu, C., Sungchul Lee, S., Su, D., Zhang, Z., Pfefferle, L., & Haller, G. L. (2012). Synthesis and characterization of nanocomposites with strong interfacial interaction: Sulfated ZrO_2 nanoparticles supported on multiwalled carbon nanotubes. *The Journal of Physical Chemistry C*, *116*(41), 21742–21752. doi:10.1021/jp302377p

Liu, D. L., Yang, Z. G., Wang, Z. Q., & Zhang, C. (2010). Synthesis and evaluation of corrosion resistance of molybdate-based conversion coatings on electroplated zinc. *Surface and Coatings Technology*, *205*(7), 2328–2334. doi:10.1016/j.surfcoat.2010.09.018

Liu, J. L., Sun, J., & Gao, L. A. (2010). A promising way to enhance the electrochemical behavior of flexible single-walled carbon nanotube/polyaniline composite films. *The Journal of Physical Chemistry C, 114*(46), 19614–19620. doi:10.1021/jp1092042

Liu, J., Yu, Q., Yu, M., Li, S., Zhao, K., Xue, B., & Zu, H. (2018). Silane modification of titanium dioxide-decorated graphene oxide nanocomposite for enhancing anticorrosion performance of epoxy coatings on AA-2024. *Journal of Alloys and Compounds, 744*, 728–739. doi:10.1016/j.jallcom.2018.01.267

Liu, P. (2005). Modifications of carbon nanotubes with polymers. *European Polymer Journal, 41*(11), 2693–2703. doi:10.1016/j.eurpolymj.2005.05.017

Liu, X., Shang, P., Zhang, Y., Wang, X., Fan, Z., Wang, B., & Zheng, Y. (2014). Three dimensional and stable polyaniline-grafted graphene hybrid materials for supercapacitor electrodes. *Journal of Materials Chemistry. A, Materials for Energy and Sustainability, 2*(37), 15273–15278. doi:10.1039/C4TA03077J

Liu, X., & Wu, Q. (2001). PP/clay nanocomposites prepared by grafting-melt intercalation. *Polymer, 42*(25), 10013–10019. doi:10.1016/S0032-3861(01)00561-4

Liu, X., & Wu, Q. (2002). Polyamide 66/Clay Nanocomposites via Melt Intercalation. *Macromolecular Materials and Engineering, 287*(3), 180. doi:10.1002/1439-2054(20020301)287:3<180::AID-MAME180>3.0.CO;2-T

Liu, X., Yue, Z., Higgins, M. J., & Wallace, G. G. (2011). Conducting polymers with immobilised fibrillar collagen for enhanced neural interfacing. *Biomaterials, 32*(30), 7309–7317. doi:10.1016/j.biomaterials.2011.06.047 PMID:21745688

Liu, Y., Zhou, J., Chen, L., Zhang, P., Fu, W., Zhao, H., ... Xie, E. (2015). Highly Flexible Freestanding Porous Carbon Nanofibers for Electrodes Materials of High-Performance All-Carbon Supercapacitors. *ACS Applied Materials & Interfaces, 7*(42), 23515–23520. doi:10.1021/acsami.5b06107 PMID:26449440

Li, X., & Coleman, M. R. (2008). Functionalization of carbon nanofibers with diamine and polyimide oligmer. *Carbon, 46*(8), 1115–1125. doi:10.1016/j.carbon.2008.01.035

Li, X., Gan, W. P., Zheng, F., Li, L. L., Zhu, N. N., & Huang, X. Q. (2012). Preparation and electrochemical properties of RuO_2/polyaniline electrodes for supercapacitors. *Synthetic Metals, 162*(11-12), 953–957. doi:10.1016/j.synthmet.2012.04.002

Li, X., Liu, P. F., Zhang, L., Zu, M. Y., Yang, Y. X., & Yang, H. G. (2016). Enhancing alkaline hydrogen evolution reaction activity through Ni–Mn_3O_4 nanocomposites. *Chemical Communications.* PMID:27500290

Li, X., Yang, L., Lei, Y., Gu, L., & Xiao, D. (2014). Microwave-assisted chemical-vapor induced in situ polymerization of polyaniline nanofibers on graphite electrode for high-performance supercapacitor. *ACS Applied Materials & Interfaces, 6*(22), 19978–19989. doi:10.1021/am505533c PMID:25361469

Li, X., Zhang, H., Wang, G., & Jiang, Z. (2010). A novel electrode material based on a highly homogeneous polyaniline/titanium oxide hybrid for high-rate electrochemical capacitors. *Journal of Materials Chemistry, 20*(47), 10598–10601. doi:10.1039/c0jm03330h

Li, Y., Peng, H., Li, G., & Chen, K. (2012). Synthesis and electrochemical performance of sandwich-like polyaniline/graphene composite nanosheets. *European Polymer Journal, 48*(8), 1406–1412. doi:10.1016/j.eurpolymj.2012.05.014

Li, Y., Zhao, X., Yu, P., & Zhang, Q. (2013). Oriented arrays of polyaniline nanorods grown on graphite nanosheets for an electrochemical supercapacitor. *Langmuir, 29*(1), 493–500. doi:10.1021/la303632d PMID:23205664

Li, Z.-F., Zhang, H., Liu, Q., Sun, L., Stanciu, L., & Xie, J. (2013). Fabrication of high-surface-area graphene/polyaniline nanocomposites and their application in supercapacitors. *ACS Applied Materials & Interfaces*, *5*(7), 2685–2691. doi:10.1021/am4001634 PMID:23480549

Lloyd, S. M., & Lave, L. B. (2003). Life cycle economic and environmental implications of using nanocomposites in automobiles. *Environmental Science & Technology*, *37*(15), 3458–3466. doi:10.1021/es026023q PMID:12966996

Lo, C.-F., Wu, J.-F., Li, H.-Y., Hung, W.-S., Shih, C.-M., Hu, C.-C., ... Lue, S. J. (2013). Novel polyvinyl alcohol nanocomposites containing carbon nano-tubes with Fe_3O_4 pendants for alkaline fuel cell applications. *Journal of Membrane Science*, *444*, 41–49. doi:10.1016/j.memsci.2013.05.001

Lokhande, V. C., Lokhande, A. C., Lokhande, C. D., Kim, J. H., & Ji, T. (2016). Supercapacitive composite metal oxide electrodes formed with carbon, metal oxides and conducting polymers. *Journal of Alloys and Compounds*, *682*, 381–403. doi:10.1016/j.jallcom.2016.04.242

Lonkar, S. P., Deshmukh, Y. S., & Abdala, A. A. (2014). Recent Advances in chemical modifications of graphene. *Nano Research.*

Loos, M. R., & Schulte, K. (2011). Is it worth the effort to reinforce polymers with carbon nanotubes? *Macromolecular Theory and Simulations*, *20*(5), 350–362. doi:10.1002/mats.201100007

Louda, P. (2007). Application of thin coating in automotive industry. *Journal of Achievements in Materials and Manufacturing Engineering*, *24*(1), 50–56.

Low, C. T. J., Wills, R. G. A., & Walsh, F. C. (2006). Electrodeposition of composite coatings containing nanoparticles in a metal deposit. *Surface and Coatings Technology*, *201*(1-2), 371–383. doi:10.1016/j.surfcoat.2005.11.123

Lu, A. H., Salabas, E. L., & Schüth, F. (2007). Magnetic nanoparticles: Synthesis, protection, functionalization, and application. *Angewandte Chemie International Edition*, *46*(8), 1222–1244. doi:10.1002/anie.200602866 PMID:17278160

Luchtvaarfeiten. (n.d.). Composites in commercial aircraft engines fact sheet. Retrieved from http://aviationfacts.eu/uploads/thema/file_en/585919d470726f2696000000/Composites_in_Commercial_Aircraft_Engines_Fact_sheet.pdf

Luo, B., Wang, X., Wang, Y., & Li, L. (2014). Fabrication, characterization, properties and theoretical analysis of ceramic/PVDF composite flexible films with high dielectric constant and low dielectric loss. *Journal of Materials Chemistry. A, Materials for Energy and Sustainability*, *2*(2), 510–519. doi:10.1039/C3TA14107A

Luo, Y., Kong, D., Jia, Y., Luo, J., Lu, Y., Zhang, D., ... Yu, T. (2013). Self-assembled graphene@PANI nanoworm composites with enhanced supercapacitor performance *RSc. Adv.*, *3*(17), 5851–5859.

Luo, Z., Zhu, L., Zhang, H., & Tang, H. (2013). Polyaniline uniformly coated on graphene oxide sheets as supercapacitor material with improved capacitive properties. *Materials Chemistry and Physics*, *139*(2-3), 572–579. doi:10.1016/j.matchemphys.2013.01.059

Lu, P., Leong, Y. W., Pallathadka, P. K., & He, C. B. (2013). Effective moduli of nanoparticle reinforced composites considering interphase effect by extended double-inclusion model – Theory and explicit expressions. *International Journal of Engineering Science*, *73*, 33–55. doi:10.1016/j.ijengsci.2013.08.003

Lu, W. K., Elsenbaumer, R. L., & Wessling, B. (1995). Corrosion protection of mild steel by coatings containing polyaniline. *Synthetic Metals*, *71*(1-3), 2163–2166. doi:10.1016/0379-6779(94)03204-J

Lu, X.-F., Chen, X.-Y., Zhou, W., Tong, Y.-X., & Li, G.-R. (2015). α-Fe_2O_3@PANI Core–Shell Nanowire Arrays as Negative Electrodes for Asymmetric Supercapacitors. *ACS Applied Materials & Interfaces*, *7*(27), 14843–14850. doi:10.1021/acsami.5b03126 PMID:26090902

Lu, X., Wang, K., & Chua, Y. (2008). Polypropylene/Clay Nanocomposites. In D. Nwabunma & T. Kyu (Eds.), *Polyolefin composites* (pp. 251–282). Hoboken, NJ: Wiley-Interscience.

Lv, R., Sun, Y., Yu, F., & Zhang, H. (2012). Fabrication of poly (3, 4-ethylenedioxythiophene)-polysaccharide composites. *Journal of Applied Polymer Science*, *124*(1), 855–863. doi:10.1002/app.35117

Lynch-Branzoi, J. (2017). Modulus enhancement of polymer nanocomposites reinforced with graphene nanoflakes. In *19th International Conference on Nanotechnology and Expo*, Atlanta, GA, Rutgers University, November 13-14. 10.4172/2157-7439-C1-061

Lynch, P. A., Stevenson, A. W., Liang, D., Parry, D., Wilkins, S., & Tamura, N. (2007). A laboratory based system for Laue micro x-ray diffraction. *The Review of Scientific Instruments*, *78*(2), 023904. doi:10.1063/1.2437777 PMID:17578120

Lyu, M.-Y., & Choi, T. G. (2015). Research trends in polymer materials for use in lightweight vehicles. *International Journal of Precision Engineering and Manufacturing*, *16*(1), 213–220. doi:10.100712541-015-0029-x

MacDiarmid, A. G. (2001). "Synthetic Metals": A Novel Role for Organic Polymers (Nobel Lecture). *Angewandte Chemie International Edition*, *40*(14), 2581–2590. doi:10.1002/1521-3773(20010716)40:14<2581::AID-ANIE2581>3.0.CO;2-2

MacDiarmid, A. G. (2001). Nobel Lecture: "Synthetic metals": A novel role for organic polymers. *Reviews of Modern Physics*, *73*(3), 701–712. doi:10.1103/RevModPhys.73.701

Madihally, S. V., Flake, A. W., & Matthew, H. W. (1999). Maintenance of CD34 expression during proliferation of CD34+ cord blood cells on glycosaminoglycan surfaces. *Stem Cells (Dayton, Ohio)*, *17*(5), 295–305. doi:10.1002tem.170295 PMID:10527464

Madihally, S. V., & Matthew, H. W. (1999). Porous chitosan scaffolds for tissue engineering. *Biomaterials*, *20*(12), 1133–1142. doi:10.1016/S0142-9612(99)00011-3 PMID:10382829

Maingi, F. M., Mbuvi, H. M., Ng'ang'a, M. M., & Mwangi, H. (2017). Adsorption Kinetics and Isotherms of Methylene Blue by Geopolymers Derived from Common Clay and Rice Husk Ash. *Phys. Chem.*, *7*(4), 87–97. doi:10.5923/j.pc.20170704.02

Maingi, F. M., Mbuvi, H. M., Ng'ang'a, M. M., & Mwangi, H. (2018). Adsorption of Cadmium ions on Geopolymers derived from ordinary clay and rice husk. *Int. J. Mater. Chem.*, *8*(1), 1–9. doi:10.5923/j.ijmc.20180801.01

Ma, J., Meng, Q., Michelmore, A., Kawashima, N., Izzuddin, Z., Bengtsson, C., & Kuan, H.-C. (2013). Covalently bonded interfaces for polymer/graphene composites. *Journal of Materials Chemistry. A, Materials for Energy and Sustainability*, *1*(13), 4255–4264. doi:10.1039/c3ta01277h

Majumdar, D. (2018). Role of electrochemists as environmental activists: A first-hand idea of how electrochemistry research have evolved over years in search of betterment of human life!! doi:10.13140/RG.2.2.19750.83526

Majumdar, D. (2016). Functionalized-Graphene/Polyaniline Nanocomposites as Proficient Energy Storage Material: An Overview. *Innov Ener Res*, *5*, 145.

Majumdar, D., Baugh, N., & Bhattacharya, S. K. (2017). Ultrasound assisted formation of reduced graphene oxide-copper (II) oxide nanocomposite for energy storage applications. *Colloids and Surfaces. A, Physicochemical and Engineering Aspects*, *512*, 158–170. doi:10.1016/j.colsurfa.2016.10.010

Majumdar, D., & Bhattacharya, S. K. (2017). Sonochemically synthesized hydroxy-functionalized graphene–MnO2 nanocomposite for supercapacitor applications. *Journal of Applied Electrochemistry*, *47*(7), 789–801. doi:10.100710800-017-1080-3

Majumder, D. D., & Karan, S. (2013). Magnetic properties of ceramic nanocomposites. In *Ceramic Nanocomposites* (pp. 51–91). Elsevier.

Maleki, A., Hamesadeghi, U., Daraei, H., Hayati, B., Najafi, F., McKay, G., & Rezaee, R. (2017). Amine functionalized multi-walled carbon nanotubes: Single and binary systems for high capacity dye removal. *Chemical Engineering Journal, 313*, 826–835. doi:10.1016/j.cej.2016.10.058

Mandal, M., Ghosh, D., Giri, S., Shakir, I., & Das, C. K. (2014). Polyaniline-wrapped 1D $CoMoO_4 \cdot 0.75H_2O$ nanorods as electrode materials for supercapacitor energy storage applications. *RSC Advances, 4*(58), 30832–30839. doi:10.1039/C4RA03399J

Mao, L., Zhang, K., Chan, H. S. O., & Wu, J. (2012). Surfactant-stabilized graphene/ polyaniline nanofiber composites for high performance supercapacitor electrode. *Journal of Materials Chemistry, 22*(1), 80–85. doi:10.1039/C1JM12869H

Ma, P. C., Kim, J.-K., & Tang, B. Z. (2007). Effects of silane functionalization on the properties of carbon nanotube/ epoxy nanocomposites. *Composites Science and Technology, 67*(14), 2965–2972. doi:10.1016/j.compscitech.2007.05.006

Ma, P. C., Siddiqui, N. A., Marom, G., & Kim, J.-K. (2010). Dispersion and functionalization of carbon nanotubes for polymer-based nanocomposites: A review. *Composites. Part A, Applied Science and Manufacturing, 41*(10), 1345–1367. doi:10.1016/j.compositesa.2010.07.003

Ma, P.-C., & Kim, J.-K. (2011). *Carbon nanotubes for polymer reinforcement.* Taylor and Francis Group, LLC.

Ma, P.-C., & Zhang, Y. (2014). Perspectives of carbon nanotubes/polymer nanocomposites for wind blade materials. *Renewable & Sustainable Energy Reviews, 30*, 651–660. doi:10.1016/j.rser.2013.11.008

Marcus, P. (Ed.). (2011). *Corrosion mechanisms in theory and practice.* CRC press. doi:10.1201/b11020

Maria, C. (2012). Application of FTIR spectroscopy in environmental studies. In Advanced Aspects of Spectroscopy. doi:10.5772/48331

Marjanovic, G. C. (2013). Recent advances in polyaniline composites with metals, metalloids and nonmetals. *Synthetic Metals, 170*, 31–56. doi:10.1016/j.synthmet.2013.02.028

Marquis, D. M., Guillaume, E., & Chivas-Joly, C. (2011). Properties of nanofillers in polymer. In J. Cuppoletti (Ed.), *Nanocomposites and Polymers with Analytical Methods.* IntechOpen; doi:10.5772/21694

Martino, S., D'Angelo, F., Armentano, I., Kenny, J. M., & Orlacchio, A. (2012). Stem cell-biomaterial interactions for regenerative medicine. *Biotechnology Advances, 30*(1), 338–351. doi:10.1016/j.biotechadv.2011.06.015 PMID:21740963

Mastragostino, M., Arbizzani, C., Meneghello, L., & Paraventi, R. (1996). Electronically conducting polymers and activated carbon: Electrode materials in super-capacitor technology. *Advanced Materials, 8*(4), 331–334. doi:10.1002/adma.19960080409

Matabola, K. P., Vries, A. R., Moolman, F. S., & Luyt, A. S. (2009). Single polymer composites: A review. *Journal of Materials Science, 44*(23), 6213–6222. doi:10.100710853-009-3792-1

Mattioli-Belmonte, M., Gigante, A., Muzzarelli, R. A. A., Politano, R., De Benedittis, A., Specchia, N., ... Greco, F. (1999). N, N-dicarboxymethyl chitosan as delivery agent for bone morphogenetic protein in the repair of articular cartilage. *Medical & Biological Engineering & Computing, 37*(1), 130–134. doi:10.1007/BF02513279 PMID:10396855

Mc Carthy, B., Coleman, J. N., Czerw, R., Dalton, A. B., Carroll, D. L., & Blau, W. J. (2001). Microscopy studies of nanotube-conjugated polymer interactions. *Synthetic Metals, 121*(1-3), 1225–1226. doi:10.1016/S0379-6779(00)00906-1

McCaig, C. D., & Zhao, M. (1997). Physiological electrical fields modify cell behaviour. *BioEssays*, *19*(9), 819–826. doi:10.1002/bies.950190912 PMID:9297973

McNally, T., Raymond Murphy, W., Lew, C., Turner, R., & Brennan, G. (2003). Polyamide-12 layered silicate nanocomposites by melt blending. *Polymer*, *44*(9), 2761–2772. doi:10.1016/S0032-3861(03)00170-8

Medpelli, D., Sandoval, R., Sherrill, L., Hristovski, K., & Seo, D.-K. (2015). Iron oxide-modified nanoporous geopolymers for arsenic removal from ground water. *Resour. Efficient Technol.*, *1*(1), 19–27. doi:10.1016/j.reffit.2015.06.007

Medpelli, D., Seo, J. M., & Seo, D. K. (2014). Geopolymer with hierarchically meso-/microporous structures from reactive emulsion templating. *Journal of the American Ceramic Society*, *97*(1), 70–73. doi:10.1111/jace.12724

Medri, V., Papa, E., Dedecek, J., Jirglova, H., Benito, P., Vaccari, A., & Landi, E. (2013). Effect of metallic Si addition on polymerization degree of in situ foamed alkali-aluminosilicates. *Ceramics International*, *39*(7), 7657–7668. doi:10.1016/j.ceramint.2013.02.104

Meguid, S. A., & Sun, Y. (2004). On the tensile and shear strength of nano-reinforced composite interfaces. *Materials & Design*, *25*(4), 289–296. doi:10.1016/j.matdes.2003.10.018

Melchers, R. E., & Jeffrey, R. (2008). Modeling of long-term corrosion loss and pitting for chromium-bearing and stainless steels in seawater. *Corrosion*, *64*(2), 143–154. doi:10.5006/1.3280683

Meng, L., Fu, C., & Lu, Q. (2009). Advanced technology for functionalization of carbon nanotubes: A Review. *Progress in Natural Science*, *19*(7), 801–810. doi:10.1016/j.pnsc.2008.08.011

Meng, Q., Cai, K., Chen, Y., & Chen, L. (2017). Research progress on conducting polymer based supercapacitor electrode materials. *Nano Energy*, *36*, 268–285. doi:10.1016/j.nanoen.2017.04.040

Meng, Q., & Hu, J. (2008). Self-organizing alignment of carbon nanotube in shape memory segmented fiber prepared by in situ polymerization and melt spinning. *Part A*, *39*(2), 314–321. doi:10.1016/j.compositesa.2007.10.007

Merlatti, C., Perrin, F. X., Aragon, E., & Margaillan, A. (2008). Natural and artificial weathering characteristics of stabilized acrylicurethane paints. *Polymer Degradation & Stability*, *93*(5), 896–903. doi:10.1016/j.polymdegradstab.2008.02.008

Messori, M., Toselli, M., Pilati, F., Fabbri, E., Fabbri, P., & Busoli, S. (2003). Poly (caprolactone)/silica organic–inorganic hybrids as protective coatings for poly (methyl methacrylate) substrates. *Surface Coatings International Part B-Coatings Transactions*, *86*(3), 181–186. doi:10.1007/BF02699651

Mezgebe, M. M., Yan, Z., Wei, G., Gong, S., Zhang, F., Guang, S., & Xu, H. (2017). 3D graphene-Fe3O4-polyaniline, a novel ternary composite for supercapacitor electrodes with improved electrochemical properties. *Materials Today Energy*, *5*, 164–172.

Miao, F., Shao, C., Li, X., Wang, K., Lu, N., & Liu, Y. (2016). Freestanding hierarchically porous carbon framework decorated by polyaniline as binder-free electrodes for high performance supercapacitors. *Journal of Power Sources*, *329*, 516–524. doi:10.1016/j.jpowsour.2016.08.111

Miao, Y.-E., Fan, W., Chen, D., & Liu, T. (2013). High-Performance Supercapacitors Based on Hollow Polyaniline Nanofibers by Electrospinning. *ACS Applied Materials & Interfaces*, *5*(10), 4423–4428. doi:10.1021/am4008352 PMID:23586693

Micheli, D., Vricella, A., Pastore, R., Delfini, A., Giusti, A., Albano, M., ... Primiani, V. M. (2016). Ballistic and electromagnetic shielding behaviour of multifunctional Kevlar fiber reinforced epoxy composites modified by carbon nanotubes. *Carbon*, *104*, 141–156. doi:10.1016/j.carbon.2016.03.059

Mihic, A., Cui, Z., Wu, J., Vlacic, G., Miyagi, Y., Li, S. H., ... Li, R. K. (2015). A conductive polymer hydrogel supports cell electrical signaling and improves cardiac function after implantation into myocardial infarct. *Circulation, 132*(8), 772–784. doi:10.1161/CIRCULATIONAHA.114.014937 PMID:26304669

Mirabedini, S. M., Mohseni, M., PazokiFard, S., & Esfandeh, M. (2008). Effect of TiO_2 on the mechanical and adhesion properties of RTV silicone elastomer coatings. *Colloids and Surfaces. A, Physicochemical and Engineering Aspects, 317*(1-3), 80–86. doi:10.1016/j.colsurfa.2007.09.044

Mirjalili, F., Chuah, L., & Salahi, E. (2014). Mechanical and Morphological Properties of Polypropylene/Nano α-$Al2_03$ Composites. T*he Scientific World Journal, 2014,* 1–12. doi:10.1155/2014/718765 PMID:24688421

Mishra, R., Jamshaid, H., & Militky, K. (2017). Basalt nanoparticle reinforced hybrid woven composites: Mechanical and thermo-mechanical performance. *Fibers and Polymers, 18*(12), 2433–2442. doi:10.100712221-017-7545-5

Misoon, O., & Seok, K. (2012). Effect of dodecyl benzene sulfonic acid on the preparation of polyaniline/activated carbon composites by in situ emulsion polymerization. *Electrochimica Acta, 59*, 196–201. doi:10.1016/j.electacta.2011.10.058

Mitchell, E., Candler, J., De Souza, F., Gupta, R. K., Gupta, B. K., & Dong, L. F. (2015). High performance supercapacitor based on multilayer of polyaniline and graphene oxide. *Synthetic Metals, 199*, 214–218. doi:10.1016/j.synthmet.2014.11.028

Mittal, V. (2012). In-situ synthesis of polymer nanocomposites (1st ed.). Wiley-VCH Verlag GmbH & Co.

Mizuno, K., Yamamura, K., Yano, K., Osada, T., Saeki, S., Takimoto, N., ... Nimura, Y. (2003). Effect of chitosan film containing basic fibroblast growth factor on wound healing in genetically diabetic mice. *Journal of Biomedical Materials Research. Part A, 64*(1), 177–181. doi:10.1002/jbm.a.10396 PMID:12483711

Mo, B. H., Zhu, H., Cui, X. M., He, Y., & Gong, S. Y. (2014). Effect of curing temperature on geopolymerization of metakaolin-based geopolymers. *Applied Clay Science, 99*, 144–148. doi:10.1016/j.clay.2014.06.024

Mocuta, C., Stangl, J., Mundboth, K., Metzger, T. H., Bauer, G., Vartanyants, I. A., ... Boeck, T. (2008). Beyond the ensemble average: X-ray microdiffraction analysis of single SiGe islands. *Physical Review B: Condensed Matter and Materials Physics, 77*(24), 245425. doi:10.1103/PhysRevB.77.245425

Modesti, M., Lorenzetti, A., Bon, D., & Besco, S. (2005). Effect of processing conditions on morphology and mechanical properties of compatibilized polypropylene nanocomposites. *Polymer, 46*(23), 10237–10245. doi:10.1016/j.polymer.2005.08.035

Mogilevsky, G., Chen, Q., Kleinhammes, A., & Wu, Y. (2008). The structure of multilayered titania nanotubes based on delaminated anatase. *Chemical Physics Letters, 460*(4-6), 517–520. doi:10.1016/j.cplett.2008.06.063

Mohamadpour, S., Pourabbas, B., & Fabbri, P. (2011). Anti-scratch and adhesion properties of photo-curable polymer/clay nanocomposite coatings based on methacrylate monomers. *Scientia Iranica F, 18*(3), 765–771. doi:10.1016/j.scient.2011.06.001

Mohammed, L., Ansari, M. N. M., Pau, G., Jawaid, M., & Saiful Islam, M. (2015). A Review on natural fiber reinforced polymer composite and Its applications. *International Journal of Polymer Science.*

Mohan, A., Khanna, N., Hwu, J., & Joy, D. C. (2006). Secondary electron imaging in the variable pressure scanning electron microscope. *Scanning, 20*(6), 436–441. doi:10.1002ca.1998.4950200603

Mohseni, M., Ramezanzadeh, B., Yari, H., & Gudarzi, M. M. (2012). The role of nanotechnology in automotive industries. In J. Carmo (Ed.), *New Advances in Vehicular Technology and Automotive Engineering*. IntechOpen. doi:10.5772/49939

Molapo, K. M., Ndangili, P. M., Ajayi, R. F., Mbambisa, G., Mailu, S. M., Njomo, N., ... Iwuoha, E. I. (2012). Electronics of conjugated polymers (I): Polyaniline. *International Journal of Electrochemical Science*, *7*(12), 11859–11875.

Moncada, E., Quijada, R., Lieberwirth, I., & Yazdani-Pedram, M. (2006). Use of PP grafted with itaconic acid as a new compatibilizer for PP/Clay nanocomposites. *Macromolecular Chemistry and Physics*, *207*(15), 1376–1386. doi:10.1002/macp.200600150

Mondal, S., Rana, U., & Malik, S. (2017). Reduced Graphene Oxide/Fe_3O_4/Polyaniline Nanostructures as Electrode Materials for an All-Solid-State Hybrid Supercapacitor. *The Journal of Physical Chemistry C*, *121*(14), 7573–7583. doi:10.1021/acs.jpcc.6b10978

Monti, M. (2010). Development of Multifunctional Advanced Composites with Polymer Nanocomposite Matrices for Aerospace Applications. In 2nd *International Conference on: Polymers in Defence & Aerospace Applications*, Hamburg, Germany, February 10-11 (pp. 86-91).

Morgan, A., & Gilman, J. (2003). Characterization of polymer-layered silicate (clay) nanocomposites by transmission electron microscopy and X-ray diffraction: A comparative study. *Journal of Applied Polymer Science*, *87*(8), 1329–1338. doi:10.1002/app.11884

Mori, T., Okumura, M., Matsuura, M., Ueno, K., Tokura, S., Okamoto, Y., ... Fujinaga, T. (1997). Effects of chitin and its derivatives on the proliferation and cytokine production of fibroblasts in vitro. *Biomaterials*, *18*(13), 947–951. doi:10.1016/S0142-9612(97)00017-3 PMID:9199765

Moroni, L., De Wijn, J. R., & Van Blitterswijk, C. A. (2008). Integrating novel technologies to fabricate smart scaffolds. *Journal of Biomaterials Science. Polymer Edition*, *19*(5), 543–572. doi:10.1163/156856208784089571 PMID:18419938

Mortara, S. (2010). Use of Fluoropolymers in Aerospace and Defence: New Applications and Advantages. In 2nd *International Conference on: Polymers in Defence & Aerospace Applications*, Hamburg, Germany, February 10-11 (pp. 60-67).

Mortimer, R. J., Dyer, A. L., & Reynolds, J. R. (2006). Electrochromic organic and polymeric materials for display applications. *Displays*, *27*(1), 2–18. doi:10.1016/j.displa.2005.03.003

Moussa, M., El-Kady, M. F., Zhao, Z., Majewski, P., & Ma, J. (2016). Recent progress and performance evaluation for polyaniline/graphene nanocomposites as supercapacitor electrodes Nanotechnology, 27.

Moutsatsou, P., Coopman, K., & Georgiadou, S. (2017). Biocompatibility assessment of conducting PANI/Chitosan nanofibers for wound healing applications. *Polymers*, *9*(12), 687. doi:10.3390/polym9120687

Mubarak, N. M., Abdullah, E. C., Jayakumar, N. S., & Sahu, J. N. (2014). An overview on methods for the production of carbon nanotubes. *Journal of Industrial and Engineering Chemistry*, *20*(4), 1186–1197. doi:10.1016/j.jiec.2013.09.001

Mu, J., Ma, G., Peng, H., Li, J., Sun, K., & Lei, Z. (2013). Facile fabrication of self-assembled polyaniline nanotubes doped with D-tartaric acid for high-performance supercapacitors. *Journal of Power Sources*, *242*, 797–802. doi:10.1016/j.jpowsour.2013.05.117

Mujawar, S. H., Ambade, S. B., Battumur, T., Ambade, R. B., & Lee, S. (2011). Electro-polymerization of polyaniline on titanium oxide nanotubes for super-capacitor application. *Electrochimica Acta*, *56*(12), 4462–4466. doi:10.1016/j.electacta.2011.02.043

Mukhopadhyay, P., & Gupta, R. K. (2012). *Graphite, Graphene, and their polymer nanocomposites*. CRC Press. doi:10.1201/b13051

Müller, K., Bugnicourt, E., Latorre, M., Jorda, M., Sanz, Y. E., Lagaron, J. M., ... Schmid, M. (2017). Review on the processing and properties of polymer nanocomposites and nanocoatings and their applications in the packaging, automotive and solar energy fields [Review]. *Nanomaterials (Basel, Switzerland)*, *7*(74), 1–47. PMID:28362331

Mullins, L. (1969). Softening of rubber by deformation. *Rubber Chemistry and Technology*, *42*(1), 339–362. doi:10.5254/1.3539210

Mu, Q. X., Liu, W., Xing, Y. H., Zhou, H., Li, Z., Zang, Y., ... Yan, B. (2008). Protein binding by functionalized multiwalled carbon nanotubes is governed by the surface chemistry of both parties and the nanotube diameter. *The Journal of Physical Chemistry C*, *112*(9), 3300–3307. doi:10.1021/jp710541j

Murugan, R., & Ramakrishna, S. (2007). Design strategies of tissue engineering scaffolds with controlled fiber orientation. *Tissue Engineering*, *13*(8), 1845–1866. doi:10.1089/ten.2006.0078 PMID:17518727

Mustafa Al Bakria, A. M., Kamarudin, H., Bin Hussain, M., Khairul Nizar, I., Zarina, Y., & Rafiza, A. R. (2011). The effect of curing temperature on physical and chemical properties of geopolymers. *Physics Procedia*, *22*, 286–291. doi:10.1016/j.phpro.2011.11.045

Muzzarelli, R. A. (2009). Chitins and chitosans for the repair of wounded skin, nerve, cartilage and bone. *Carbohydrate Polymers*, *76*(2), 167–182. doi:10.1016/j.carbpol.2008.11.002

Muzzarelli, R. A. (2011). Chitosan composites with inorganics, morphogenetic proteins and stem cells, for bone regeneration. *Carbohydrate Polymers*, *83*(4), 1433–1445. doi:10.1016/j.carbpol.2010.10.044

Nagarale, R., Shin, W., & Singh, P. K. (2010). Progress in ionic organic-inorganic composite membranes for fuel cell applications. *Polymer Chemistry*, *1*(4), 388–408. doi:10.1039/B9PY00235A

Naghsh, M., & Shams, K. (2017). Synthesis of a kaolin-based geopolymer using a novel fusion method and its application in effective water softening. *Applied Clay Science*, *146*, 238–245. doi:10.1016/j.clay.2017.06.008

Nah, C., & Abdul Kader, M. (2010). Barrier properties of rubber nanocomposites. In Rubber nanocomposites-Preparation, properties, and applications (pp. 500-521). Singapore: John Wiley & Sons (Asia) Pte Ltd.

Naik, R. B., Jagtap, S. B., & Ratna, D. (2015). Effect of carbon nano - fillers on anticorrosive and physico-mechanical properties of hyperbranched urethane alkyd coatings. *Progress in Organic Coatings*, *87*, 28–35. doi:10.1016/j.porgcoat.2015.05.001

Nallayan, W. A., & Vijayakumar, K. R. (2017). Comparison of the Effect of Curing on the Properties of E-Glass/Cyanate modified Epoxy Cross Plied Laminates. *Materials Science and Engineering*, *197*(1).

Nambiar, S., & Yeow, J. T. (2012). Polymer-composite materials for radiation protection. *ACS Applied Materials & Interfaces*, *4*(11), 5717–5726. doi:10.1021/am300783d PMID:23009182

nano werk. (2012). Polymer nanocomposites drive opportunities in the automotive sector. Retrieved from https://www.nanowerk.com/spotlight/spotid=23934.php

Nano work. (2006). Military nanotechnology- how worried should we be? Retrieved from https://www.nanowerk.com/spotlight/spotid=1015.php

Nanotech. (2014). Nanotechnology in military applications. Retrieved from http://www.nanotechmag.com/nanotech-in-the-army/

Narayan, R. J. (2010). The next generation of biomaterial development. *Philosophical Transactions of the Royal Society A*, *368*(1917), 1831–1837. doi:10.1098/rsta.2010.0001 PMID:20308105

Naskar, A. K., Keum, J. K., & Boeman, R. G. (2016). Polymer matrix nanocomposites for automotive structural components. *Nature Nanotechnology*, *11*(12), 1026–1030. doi:10.1038/nnano.2016.262 PMID:27920443

Natesan, M., Venkatachari, G., & Palaniswamy, N. (2005). Corrosivity and durability maps of India. *Corrosion Prevention & Control*, *52*(2), 43–55.

Nau, H. (2013). The future of nanotechnology in warfare. *Global*. Retrieved from http://www.theglobaljournal.net/article/view/1132/

Negi, Y. S., & Adhyapak, P. V. (2002). Development in polyaniline conducting polymers. *Journal of Macromolecular Science -. Polymer Reviews (Philadelphia, Pa.)*, *42*, 35–53.

Neves, N. M., Campos, R., Pedro, A., Cunha, J., Macedo, F., & Reis, R. L. (2007). Patterning of polymer nanofiber meshes by electrospinning for biomedical applications. *International Journal of Nanomedicine*, *2*(3), 433–448. PMID:18019842

Newman, S. P., & Jones, W. (1998). Synthesis, characterization and applications of layered double hydroxides containing organic guests. *New Journal of Chemistry*, *22*(2), 105–115. doi:10.1039/a708319j

Nguyen, D. N. & Yoon, H. (2016). Recent Advances in Nanostructured Conducting Polymers: from Synthesis to Practical Applications. *Polymers, 8*, 118.

Nguyen, D. N., & Yoon, H. (2016). Recent Advances in Nanostructured Conducting Polymers: From Synthesis to Practical Applications. *Polymers*, *8*(4), 118. doi:10.3390/polym8040118

Nicholls, I. A., Andersson, H. S., Charlton, C., Henschel, H., Karlsson, B. C., Karlsson, J. G., ... Wikman, S. (2009). Theoretical and computational strategies for rational molecularly imprinted polymer design. *Biosensors & Bioelectronics*, *25*(3), 543–552. doi:10.1016/j.bios.2009.03.038 PMID:19443204

Ning, G., Li, T., Yan, J., Xu, C., Wei, T., & Fan, Z. (2013). Three-dimensional hybrid materials of fish scale-like polyaniline nanosheet arrays on graphene oxide and carbon nanotube for high-performance ultracapacitors. *Carbon*, *54*, 241–248. doi:10.1016/j.carbon.2012.11.035

Ning, X., Zhong, W., & Wana, L. (2016). Ultrahigh specific surface area porous carbon nanospheres and its composite with polyaniline: Preparation and application for supercapacitors. *RSC Advances*, *6*(30), 25519–25524. doi:10.1039/C6RA00596A

Ni, Q. Q., Zhang, C., Fu, Y., Dai, G., & Kimura, T. (2007). Shape memory effect and mechanical properties of carbon nanotube/shape memory polymer nanocomposites. *Composite Structures*, *81*(2), 176–184. doi:10.1016/j.compstruct.2006.08.017

Niratiwongkorn, T., Luckachan, G. E., & Mittal, V. (2016). Self-healing protective coatings of polyvinyl butyral/polypyrrole-carbon black composite on carbon steel. *RSC Advances*, *6*(49), 43237–43249. doi:10.1039/C6RA01619G

Nisha, S. A. (2017). Assessment of knowledge and attitude on organ donation among general public at puducherry. *International Journal of Current Research*, *9*(09), 57310–57313.

Nishimura, K., Nishimura, S., Nishi, N., Saiki, I., Tokura, S., & Azuma, I. (1984). Immunological activity of chitin and its derivatives. *Vaccine*, *2*(1), 93–99. doi:10.1016/S0264-410X(98)90039-1 PMID:6397928

Niu, Z., Luan, P., Shao, Q., Dong, H., Li, J., Chen, J., ... Xie, S. (2012). A "skeleton/skin" strategy for preparing ultrathin free-standing single-walled carbon nanotube/polyaniline films for high performance supercapacitor electrodes. *Energy & Environmental Science*, *5*(9), 8726–8733. doi:10.1039/c2ee22042c

Njuguna, J., & Pielichowsk, K. (2010). Ageing and Performance Predictions of Polymer Nanocomposites for Exterior Aerospace and Defence Applications. In *2nd International Conference on: Polymers in Defence & Aerospace Applications*, Hamburg, Germany, February 10-11 (pp. 146-157).

Njuguna, J., Pielichowski, K., & Fan, J. (2012). Polymer nanocomposites for aerospace applications. In Adv. Polymer Nanocomposites, Types and Applications, Science and Engineering (Ch. 15, pp. 472-539). Woodhead Publishing.

Njuguna, J., & Pielichowski, K. (2003). Polymer Nanocomposites for Aerospace Applications: Properties. *Advanced Engineering Materials*, *5*(11), 769–778. doi:10.1002/adem.200310101

Njuguna, J., & Pielichowski, K. (2004a). Polymer Nanocomposites for Aerospace Applications: Fabrication. *Adv. Mater. Sci.*, *6*(4), 193-203.

Nmiri, A., Hamdi, N., Yazoghli-Marzouk, O., Duc, M., & Srasra, E. (2017). Synthesis and characterization of kaolinite-based geopolymer: Alkaline activation effect on calcined kaolinitic clay at different temperatures. *J. Mater. Environ. Sci.*, *8*, 276–290.

Noh, M. W., & Lee, D. C. (1999). Synthesis and characterization of ps-clay nanocomposite by emulsion polymerization. *Polymer Bulletin*, *42*(5), 619–626. doi:10.1007002890050510

Nordin, T. (1980). Development of energy-saving vehicles. *ASEA Journal*, *53*(4-5).

Novais, R. M., Buruberri, L. H., Seabra, M. P., & Labrincha, J. A. (2016). Novel porous fly-ash containing geopolymer monoliths for lead adsorption from wastewaters. *Journal of Hazardous Materials*, *318*, 631–640.

Nuraje, N., Khan, S. I., Misak, H., & Asmatulu, R. (2013). The addition of graphene to polymer coatings for improved weathering. *ISRN Polymer Science*.

Nuzzo, A., Bilotti, E., Peijs, T., Acierno, D., & Filippone, G. (2014). Nanoparticle-induced co-continuity in immiscible polymer blends – A comparative study on bio-based PLA-PA11 blends filled with organoclay, sepiolite, and carbon nanotubes. *Polymer*, *55*(19), 4908–4919. doi:10.1016/j.polymer.2014.07.036

Ogasawara, T., Ishida, Y., Ishikawa, T., & Yokota, R. (2004). Characterization of multi-walled carbon nanotube/phenylethynyl terminated polyimide composites. *Composites part A - Applied Science*, *35*(1), 67–74. doi:10.1016/j.compositesa.2003.09.003

Okada, A., Fukushima, Y., Kawasumi, M., Inagaki, S., Usuki, A., Sugiyama, S., . . . Kamigaito, O. (1988). U.S. Patent No. 4,739,007. Washington, DC: U.S. Patent and Trademark Office.

Okada, A., & Kawasumi, M. (1987). Synthesis and characterization of a nylon 6-clay hybrid. *Polymeric Preprints*, *28*(2), 447–448.

Okada, A., & Usuki, A. (2006). Twenty years of polymer-clay nanocomposites. *Macromolecular Materials and Engineering*, *291*(12), 1449–1476. doi:10.1002/mame.200600260

Oliveira, A., Amico, S. C., Risi, C. L. S., Pezzin, S. H., & Coelho, L. A. F. (2009). Production of Tri-component Composites (Carbon fiber/Epoxy/Carbon nanotubes) by Resin Transfer Molding. In *Proceedings of the 11th International Conference on Advanced Materials*, Rio de Janeiro.

Oliveira, G. E., Clarindo, J. E. S., Santo, K. S. E., & Souza, F. G. Jr. (2013). Chemical modification of cobalt ferrite nanoparticles with possible application as asphaltene flocculant agent. *Materials Research*, *16*(3), 668–671. Retrieved from http://www.scielo.br/scielo.php?script=sci_arttext&pid=S1516-14392013000300020&nrm=iso doi:10.1590/S1516-14392013005000048

Olson, J. L., Atala, A., & Yoo, J. J. (2011). Tissue engineering: Current strategies and future directions. *Chonnam Medical Journal*, *47*(1), 1–13. doi:10.4068/cmj.2011.47.1.1 PMID:22111050

Otrokhov, G., Pankratov, D., Shumakovich, G., Khlupova, M., Zeifman, Y., Vasil'eva, I., ... Yaropolov, A. (2014). Enzymatic synthesis of polyaniline/ multi-walled carbon nanotube composite with core shell structure and its electrochemical characterization for supercapacitor application. *Electrochimica Acta*, *123*, 151–157. doi:10.1016/j.electacta.2013.12.089

Öztürk, A. (2007). Removal of nickel from aqueous solution by the bacterium Bacillus thuringiensis. *Journal of Hazardous Materials*, *147*(1-2), 518–523.

Packham, D. E. (1996). Work of adhesion: Contact angles and contact mechanics. *Int. J. Adhes*, *16*(2), 121–128. doi:10.1016/0143-7496(95)00034-8

Pacurari, M., Lowe, K., Tchounwou, P., & Kafoury, R. (2016). A review on the respiratory system toxicity of carbon nanoparticles. *International Journal of Environmental Research and Public Health*, *13*(3), 325.

Paduchowicz, M., Górski, A., Czmochowski, J., & Rusiński, E. (2012). Numerical and experimental identification of vibration convection chamber of fluid power bowler. *Journal of Vibroengineering*, *14*, 151–157.

Paiva, J. M., Fox-Rabinovich, G., Locks Junior, E., Stolf, P., Ahmed, Y. S., Martins, M. M., ... Veldhuis, S. (2018). Tribological and wear performance of nanocomposite PVD hard coatings deposited on aluminum die casting tool. *Materials (Basel)*, *11*(358), 1–15. PMID:29495620

Pal, S. K., & Pal, K. (2009). Status of abrasion resistant tyres for mining industries: A review. In Tyres in mining and allied sectors: Status and outlook. Allied Publishers, Pvt Ltd.

Palimi, M. J., Rostami, M., Mahdavian, M., & Ramezanzadeh, B. (2015). A study on the corrosion inhibition properties of silane-modified Fe_2O_3 nanoparticle on mild steel and its effect on the anticorrosion properties of the polyurethane coating. *Journal of Coatings Technology and Research*, *12*(2), 277–292. doi:10.100711998-014-9631-6

Palraj, S., Selvaraj, M., Maruthan, K., & Rajagopal, G. (2015). Corrosion and wear resistance behavior of nano-silica epoxy composite coatings. *Progress in Organic Coatings*, *81*, 132–139. doi:10.1016/j.porgcoat.2015.01.005

Pandey, J. K., Kumar, A. P., Misra, M., Mohanty, A. K., Drzal, L. T., & Singh, R. P. (2005). Recent advances in biodegradable nanocomposites. *Journal of Nanoscience and Nanotechnology*, *5*(4), 497–526. doi:10.1166/jnn.2005.111 PMID:16004113

Pandey, S., & Mishra, S. B. (2011). Organic-inorganic hybrid of chitosan/organoclaybionanocomposites for hexavalent chromium uptake. *Journal of Colloid and Interface Science*, *361*(2), 509–520. doi:10.1016/j.jcis.2011.05.031 PMID:21679960

Pang, H., Chen, T., Zhang, G., Zeng, B., & Li, Z.-M. (2010). An electrically conducting polymer/graphene composite with a very low percolation threshold. *Materials Letters*, *64*(20), 2226–2229. doi:10.1016/j.matlet.2010.07.001

Pang, H., Zhang, Y.-C., Chen, T., Zeng, B.-Q., & Li, Z.-M. (2010). Tunable positive temperature coefficient of resistivity in an electrically conducting polymer/graphene composite. *Applied Physics Letters*, *96*(25), 251907. doi:10.1063/1.3457170

Pan, S., & Gupta, A. (2012). Surface-enhanced solar energy conversion systems using gold and silver nanoparticles. *Material Matters*, *7*(4), 64–66.

Papageorgiou, D. G., Kinloch, I. A., & Young, R. J. (2017). Mechanical properties of graphene and graphene-based nanocomposites. *Progress in Materials Science*, *90*, 75–127. doi:10.1016/j.pmatsci.2017.07.004

Paramsothy, M., Gupta, M. M., & Srikanth, N. (2008). Processing, microstructure, and properties of a Mg/Al bimetal macrocomposite. *Journal of Composite Materials*, *42*(24), 2567–2584. doi:10.1177/0021998308098369

Park, S., An, J., Potts, J. R., Velamakanni, A., Murali, S., & Ruoff, R. S. (2011). Hydrazine-reduction of graphite- and graphene oxide. *Carbon*, *49*(9), 3019–3023. doi:10.1016/j.carbon.2011.02.071

Park, S.-J., & Seo, M.-K. (2011). Types of composites. *Interface Science and Technology*, *18*, 501–629. doi:10.1016/B978-0-12-375049-5.00007-4

Park, S., & Shon, M. (2015). Effects of multi-walled carbon nano tubes on corrosion protection of zinc rich epoxy resin coating. *Journal of Industrial and Engineering Chemistry*, *21*, 1258–1264. doi:10.1016/j.jiec.2014.05.042

Parveen, N., Mahato, N., Ansari, M. O., & Cho, M. H. (2016). Enhanced electrochemical behavior and hydrophobicity of crystalline polyaniline@ graphene nanocomposite synthesized at elevated temperature. *Composites. Part B, Engineering*, *87*, 281–290. doi:10.1016/j.compositesb.2015.10.029

Patel, N., & Poo, M. M. (1982). Orientation of neurite growth by extracellular electric fields. *The Journal of Neuroscience*, *2*(4), 483–496. doi:10.1523/JNEUROSCI.02-04-00483.1982 PMID:6279799

Patil, D. S., Shaikh, J. S., Dalavi, D. S., Karanjkar, M. M., Devan, R. S., Ma, Y. R., & Patil, P. S. (2011). An Mn Doped Polyaniline Electrode for Electrochemical Supercapacitor. *Journal of the Electrochemical Society*, *158*(6), A653. doi:10.1149/1.3561428

Patnukao, P., Kongsuwan, A., & Pavasant, P. (2008). Batch studies of adsorption of copper and lead on activated carbon from Eucalyptus camaldulensisDehn.bark. *Journal of Environmental Sciences (China)*, *20*(9), 1028–1034. doi:10.1016/S1001-0742(08)62145-2 PMID:19143307

Paul, D., & Robeson, L. (2008). Polymer nanotechnology: Nanocomposites. *Polymer*, *49*(15), 3187–3204. doi:10.1016/j.polymer.2008.04.017

Pavlidou, S., & Papaspyrides, C. (2008). A review on polymer–layered silicate nanocomposites. *Progress in Polymer Science*, *33*(12), 1119–1198. doi:10.1016/j.progpolymsci.2008.07.008

Pavlovska, I., Malnieks, K., Mezinskis, G., Bidermanis, L., & Karpe, M. (2014). Hard TiO_2-SiO_2 sol-gel coatings for enamel against chemical corrosion. *Surface and Coatings Technology*, *258*, 206–210. doi:10.1016/j.surfcoat.2014.09.027

Pazourková, L., Martynková, G.S., Plachá, D. (2015). Preparation and mechanical properties of polymeric nanocomposites with hydroxyapatite and hydroxyapatite/clay mineral fillers – review. *HSOA Journal of Nanotechnology: Nanomedicine & Nanobiotechnology.*

Pedrazzoli, D., & Pegoretti, A. (2013). Silica nanoparticles as coupling agents for polypropylene/glass composites. *Composites Science and Technology*, *76*, 77–83. doi:10.1016/j.compscitech.2012.12.016

Pegoretti, A., Kolarik, J., Peroni, C., & Migliaresi, C. (2004). Recycled poly(ethylene terephthalate)/layered silicate nanocomposites: Morphology and tensile mechanical properties. *Polymer*, *45*(8), 2751–2759. doi:10.1016/j.polymer.2004.02.015

Peighambardoust, S., Rowshanzamir, S., & Amjadi, M. (2010). Review of the proton exchange membranes for fuel cell applications. *International Journal of Hydrogen Energy*, *35*(17), 9349–9384. doi:10.1016/j.ijhydene.2010.05.017

Pei, Q., Zuccarello, G., Ahlskog, M., & Inganäs, O. (1994). Electrochromic and highly stable poly(3, 4-ethylenedioxythiophene) switches between opaque blue-black and transparent sky blue. *Polymer*, *35*(7), 1347–1351. doi:10.1016/0032-3861(94)90332-8

Pei, S., & Cheng, H.-M. (2012). The reduction in graphene oxide. *Carbon*, *50*(9), 3210–3228. doi:10.1016/j.carbon.2011.11.010

Pei, Y., Travas-Sejdic, J., & Williams, D. E. (2012). Reversible electrochemical switching of polymer brushes grafted onto conducting polymer films. *Langmuir*, *28*(21), 8072–8083. doi:10.1021/la301031b PMID:22551237

Peng, C., Zhou, X. H., Chen, G. Z., Moggia, F., Fages, F., Brisset, H., & Roncali, J. (2008). Internally Referenced analysis of charge transfer reactions in a new ferrocenyl bithiophenic conducting polymer through cyclic voltammetry. *Chemical Communications*, (48), 6606–6608. doi:10.1039/b817101g PMID:19057794

Peng, W., & Riedl, B. (1995). Thermosetting resins. *Journal of Chemical Education*, *72*(7), 587–592. doi:10.1021/ed072p587

Peng, X.-Y., Liu, X.-X., Hua, P.-J., Diamond, D., & Lau, K.-T. (2010). Electrochemical codeposition of nickel oxide and polyaniline. *Journal of Solid State Electrochemistry*, *14*(1), 1–7. doi:10.100710008-008-0778-x

Penn State. (n.d.). Classification of unmanned aerial systems. Retrieved from https://www.e-education.psu.edu/geog892/node/5

Pesetskii, S., Bogdanovich, S., Aderikha, V., & Koval, V. (2016). Low-temperature extrusion in technology of polyethylene terephtalate/clay nanocomposites. *Doklady of the National Academy of Sciences of Belarus*, *60*(1), 102–108.

Pesetskii, S., Bogdanovich, S., Dubrovskii, V., Sodyleva, T., Aderikha, V., & Usova, V. (2016). Morphology and properties of PA6 hybrid composites filled with short carbon fibers and organoclay. *Polymer Materials and Technologies*, *2*(3), 45–57. doi:10.32864/polymmattech-2016-2-3-47-57

Pesetskii, S., Bogdanovich, S., & Myshkin, N. (2013). Tribological behavior of polymer nanocomposites produced by dispersion of nanofillers in molten thermoplastics. In K. Friedrich & A. Schlarb (Eds.), *Tribology of polymeric nanocomposites* (2nd ed., pp. 119–162). Oxford: Elsevier. doi:10.1016/B978-0-444-59455-6.00005-2

Pesetskii, S., Bogdanovich, S., & Sodyleva, T. (2017). Polyamide 6 reinforcement by hybrid short basalt fiber and organoclay filling. *Doklady of the National Academy of Sciences of Belarus*, *61*(2), 74–83.

Pesetskii, S., Shevchenko, V., & Dubrovsky, V. (2015). Reactive compatibilization in technology of poly(alkylene terephthalate)-based composites: polyester blends, short fiber-filled materials, and nanomaterials. In K. Friedrich & U. Breuer (Eds.), *Multifunctionality of Polymer Composites. Challenges and New Solutions* (1st ed., pp. 302–337). Oxford: Elsevier. doi:10.1016/B978-0-323-26434-1.00009-X

Pezzin, S. H., Coelho, L. A. F. & Amico, S.C. (2010). New Generation of Multi-functional Composites with Carbon Nanotubes for Space Applications. In 2nd *International Conference on: Polymers in Defence & Aerospace Applications*, Hamburg, Germany, February 10-11 (pp. 102-108).

Phiri, J., Gane, P., & Maloney, T. C. (2017). General overview of graphene: Production, properties and application in polymer composites. *Materials Science and Engineering B*, *215*, 9–28. doi:10.1016/j.mseb.2016.10.004

Pike, M. G., & Oskay, C. (2014). Modeling random short nanofiber- and microfiber-reinforced composites using the extended finite-element method. *Journal of Nanomechanics & Micromechanics*, *5*(1), 1–11.

Pilla, S., & Lu, Y. C. (2015). The use of nano composites in automotive applications. SAE International Publisher. doi:10.4271/PT-165

Place, E. S., Evans, N. D., & Stevens, M. M. (2009). Complexity in biomaterials for tissue engineering. *Nature Materials*, *8*(6), 457–470. doi:10.1038/nmat2441 PMID:19458646

Plastics Make It Possible. (2015). Make it possible. Automakers turn to plastics for lighter vehicles, increased fuel efficiency. Retrieved from https://www.plasticsmakeitpossible.com/whats-new-cool/automotive/fuel-efficiency/automakers-turn-to-plastics-for-lighter-vehicles-increased-fuel-efficiency/

Plastics Make It Possible. (2016). Make it possible. Automotive plastics play key role as automakers seek to build fuel-efficient vehicles. Retrieved from https://www.plasticsmakeitpossible.com/whats-new-cool/automotive/plastics-play-key-role-automakers-seek-build-fuel-efficient-vehicles/

Plawecka, M., Snihirova, D., Martins, B., Szczepanowicz, K., Warszynski, P., & Montemor, M. F. (2014). Self healing ability of inhibitor-containing nanocapsules loaded in epoxy coatings applied on aluminium 5083 and galvanneal substrates. *Electrochimica Acta*, *140*, 282–293. doi:10.1016/j.electacta.2014.04.035

Pleşa, I., Noţingher, P. V., Schlögl, S., Sumereder, C., & Muhr, M. (2016). Properties of polymer composites used in high-voltage applications. *Polymers*.

Pluta, M. (2004). Morphology and properties of polylactide modified by thermal treatment, filling with layered silicates and plasticization. *Polymer*, *45*(24), 8239–8251. doi:10.1016/j.polymer.2004.09.057

Polytechinc. (n.d.). Acoustic foam. Retrieved from https://www.polytechinc.com/products/acoustic-foam#PHM

Polytechinc. (n.d.). Thermal insulation. Retrieved from https://www.polytechinc.com/products/thermal-insulation

Pop, E., Varshney, V., & Roy, A. K. (2012). Thermal properties of graphene: Fundamentals and applications. *MRS Bulletin*, *37*(12), 1273–1281. doi:10.1557/mrs.2012.203

Postek, M. T. (2006). An approach to the reduction of hydrocarbon contamination in the scanning electron microscope. *Scanning*, *18*(4), 269–274. doi:10.1002ca.1996.4950180402

Pothan, L. A., George, C. N., Jacob, M., & Thomas, S. (2007). Effect of chemical modification on the mechanical and electrical properties of banana fiber polyester composites. *Journal of Composite Materials*, *41*(19), 2371–2386. doi:10.1177/0021998307075456

Potts, J. R., Dreyer, D. R., Bielawski, C. W., & Ruoff, R. S. (2011). Graphene-based polymer nanocomposites. *Polymer*, *52*(1), 5–25. doi:10.1016/j.polymer.2010.11.042

Pourhashem, S., Vaezi, M. R., Rashidi, A., & Bagherzadeh, M. R. (2017). Exploring corrosion protection properties of solvent based epoxy-graphene oxide nanocomposite coatings on mild steel. *Corrosion Science*, *115*, 78–92. doi:10.1016/j.corsci.2016.11.008

Pour, Z. S., Ghaemy, M., Bordbar, S., & Karimi-Maleh, H. (2018). Effects of surface treatment of TiO 2 nanoparticles on the adhesion and anticorrosion properties of the epoxy coating on mild steel using electrochemical technique. *Progress in Organic Coatings*, *119*, 99–108. doi:10.1016/j.porgcoat.2018.02.019

Poznyak, S. K., Tedim, J., Rodrigues, L. M., Salak, A. N., Zheludkevich, M. L., Dick, L. F. P., & Ferreira, M. G. S. (2009). Novel inorganic host layered double hydroxides intercalated with guest organic inhibitors for anticorrosion applications. *ACS Applied Materials & Interfaces*, *1*(10), 2353–2362. doi:10.1021/am900495r PMID:20355873

Prabaharan, S. R. S., & Michael, M. S. (Eds.). (2013). *Nanotechnology in advanced electrochemical power sources*. Taylor & Francis Group, LLC.

Prabhakaran, M. P., Ghasemi-Mobarakeh, L., & Ramakrishna, S. (2011). Electrospun composite nanofibers for tissue regeneration. *Journal of Nanoscience and Nanotechnology*, *11*(4), 3039–3057. doi:10.1166/jnn.2011.3753 PMID:21776670

Pramoda, K., Hussain, H., Koh, H., Tan, H., & He, C. (2010). Covalent bonded polymer–graphene nanocomposites. *Journal of Polymer Science. Part A, Polymer Chemistry*, *48*(19), 4262–4267. doi:10.1002/pola.24212

Prasad, K. R., & Miura, N. (2004). Polyaniline-MnO_2 composite electrode for high energy density electrochemical capacitor. *Electrochemical and Solid-State Letters*, *7*(11), A425. doi:10.1149/1.1805504

Prasad, K. R., & Miura, N. (2004). Potentiodynamically deposited nanostructured manganese dioxide as electrode material for electrochemical redox supercapacitors. *Journal of Power Sources*, *135*(1-2), 354–360. doi:10.1016/j.jpowsour.2004.04.005

Prathap, M. U. A., Satpati, B., & Srivastava, R. (2013). Facile preparation of polyaniline/MnO_2 nanofibers and its electrochemical application in the simultaneous determination of catechol, hydroquinone, and resorcinol. *Sensors and Actuators. B, Chemical*, *186*, 67–77. doi:10.1016/j.snb.2013.05.076

Provis, J. L., Lukey, G. C., & van Deventer, J. S. J. (2005). Do Geopolymers Actually Contain Nanocrystalline Zeolites? A Reexamination of Existing Results. *Chemistry of Materials*, *17*(12), 3075–3085. doi:10.1021/cm050230i

Provis, J. L., Yong, C. Z., Duxson, P., & van Deventer, J. S. J. (2009). Correlating mechanical and thermal properties of sodium silicate-fly ash geopolymers. *Colloids Surfaces A Physicochem. Eng. Asp.*, *336*(1-3), 57–63. doi:10.1016/j.colsurfa.2008.11.019

Prud'homme, E., Joussein, E., & Rossignol, S. (2015). Use of silicon carbide sludge to form porous alkali-activated materials for insulating application. *The European Physical Journal. Special Topics*, *224*(9), 1725–1735.

Prud, E., Michaud, P., Joussein, E., Peyratout, C., Smith, A., Arrii-clacens, S., ... Rossignol, S. (2010). Silica fume as porogent agent in geo-materials at low temperature. *Journal of the European Ceramic Society*, *30*(7), 1641–1648. doi:10.1016/j.jeurceramsoc.2010.01.014

Puglia, D., Biagiotti, J., & Kenny, J. M. (2004). A review on natural fibre-based Composites-Part II: Application of natural reinforcements in composite materials for automotive industry. *Journal of Natural Fibers*, *1*(3), 23–64. doi:10.1300/J395v01n03_03

Qazi, T. H., Rai, R., & Boccaccini, A. R. (2014). Tissue engineering of electrically responsive tissues using polyaniline based polymers: A review. *Biomaterials*, *35*(33), 9068–9086. doi:10.1016/j.biomaterials.2014.07.020 PMID:25112936

Qi, F., Wang, Y., Ma, T., Zhu, S., Zeng, W., Hu, X., ... Luo, Z. (2013). Electrical regulation of olfactory ensheathing cells using conductive polypyrrole/chitosan polymers. *Biomaterials*, *34*(7), 1799–1809. doi:10.1016/j.biomaterials.2012.11.042 PMID:23228424

Qi, K., Qiu, Y., Chen, Z., & Guo, X. (2012). Corrosion of conductive polypyrrole: Effects of environmental factors, electrochemical stimulation, and doping anions. *Corrosion Science*, *60*, 50–58. doi:10.1016/j.corsci.2012.04.013

Qi, K., Sun, Y., Duan, H., & Guo, X. (2015). A corrosion-protective coating based on a solution-processable polymer-grafted graphene oxide nanocomposite. *Corrosion Science*, *98*, 500–506. doi:10.1016/j.corsci.2015.05.056

Quintanilha, R. C., Rocha, I., Vichessi, R. B., Lucht, E., Naidek, K., Winnischofer, H., & Vidotti, M. (2014). Electrochromism: Basis and application of nanomaterials in development of high performance electrodes. *Quimica Nova*, *37*(4), 677–688. doi:10.5935/0100-4042.20140112

Radhakrishnan, S., Prakash, S., Rao, C. R. K., & Vijayan, M. (2009). Organically soluble bifunctional Polyaniline Magnetite composites for sensing and supercapacitor applications. *Electrochemical and Solid-State Letters*, *12*(4), A84. doi:10.1149/1.3074315

Raffaelle, R. P., Castro, S. L., Hepp, A. F., & Bailey, S. G. (2002). Quantum dot solar cells. *Progress in Photovoltaics: Research and Applications*, *10*(6), 433–439. doi:10.1002/pip.452

Rahmiati, T., Azizli, K. A., Man, Z., Ismail, L., & Nuruddin, M. F. (2014, July). Effect of Solid/liquid ratio during curing time fly ash based geopolymer on mechanical property. In *Materials Science Forum* (Vol. 803).

Raj, R., & Shah, S. R. (2012). U.S. Patent No. 8,232,234. Washington, DC: U.S. Patent and Trademark Office.

Rajabi, M., Rashed, G. R., & Zaarei, D. (2015). Assessment of graphene oxide/epoxy nanocomposite as corrosion resistance coating on carbon steel. *Corrosion Engineering, Science and Technology*, *50*(7), 509–516. doi:10.1179/1743278214Y.0000000232

Rajan, T. P. D., Pillai, R. M., & Pai, B. C. (1998). Reinforcement coatings and interfaces in aluminium metal matrix composites [Review]. *Journal of Materials Science*, *33*(14), 3491–3503. doi:10.1023/A:1004674822751

Rajnicek, A. M., Robinson, K. R., & McCaig, C. D. (1998). The direction of neurite growth in a weak DC electric field depends on the substratum: Contributions of adhesivity and net surface charge. *Developmental Biology*, *203*(2), 412–423. doi:10.1006/dbio.1998.9039 PMID:9808790

Rakhi, R. B., Chen, W., Cha, D., & Alshareef, H. N. (2012). Substrate dependent selforganization of mesoporous cobalt oxide nanowires with remarkable pseudocapacitance. *Nano Letters*, *12*(5), 2559–2567. doi:10.1021/nl300779a PMID:22494065

Ramakrishna Matte, H. S. S., Subrahmanyam, K. S., & Rao, C. N. R. (2011). Synthetic aspects and selected properties of graphene. *Nanomaterials and Nanotechnology*, *1*(1), 3–13.

Ramanathan, T., Abdala, A. A., Stankovich, S. D., Dikin, A., Herrera-Alonso, M., Piner, R. D., ... Aksay, I. A. (2008). Functionalized graphene sheets for polymer nanocomposites. *Nature Nanotechnology*, *3*(6), 327–331. doi:10.1038/nnano.2008.96 PMID:18654541

Ramanathan, T., Stankovich, S., Dikin, D., Liu, H., Shen, H., Nguyen, S., & Brinson, L. (2007). Graphitic nanofillers in PMMA nanocomposites—An investigation of particle size and dispersion and their influence on nanocomposite properties. *Journal of Polymer Science. Part B, Polymer Physics*, *45*(15), 2097–2112. doi:10.1002/polb.21187

Ramezanzadeh, B., Ahmadi, A., & Mahdavian, M. (2016). Enhancement of the corrosion protection performance and cathodic delamination resistance of epoxy coating through treatment of steel substrate by a novel nanometric sol-gel based silane composite film filled with functionalized graphene oxide nanosheets. *Corrosion Science*, *109*, 182–205. doi:10.1016/j.corsci.2016.04.004

Ramezanzadeh, B., & Attar, M. M. (2011). Studying the corrosion resistance and hydrolytic degradation of an epoxy coating containing ZnO nanoparticles. *Materials Chemistry and Physics*, *130*(3), 1208–1219. doi:10.1016/j.matchemphys.2011.08.065

Ramezanzadeh, B., Niroumandrad, S., Ahmadi, A., Mahdavian, M., & Moghadam, M. M. (2016). Enhancement of barrier and corrosion protection performance of an epoxy coating through wet transfer of amino functionalized graphene oxide. *Corrosion Science*, *103*, 283–304. doi:10.1016/j.corsci.2015.11.033

Ramya, R., Sivasubramanian, R., & Sangaranarayanan, M. V. (2013). Conducting polymers based electrochemical supercapacitors: Progress and prospects. *Electrochimica Acta*, *101*, 109–129. doi:10.1016/j.electacta.2012.09.116

Rao, C. E. E., Sood, A. E., Subrahmanyam, K. E., & Govindaraj, A. (2009). Graphene: The new two-dimensional nanomaterial. *Angewandte Chemie International Edition*, *48*(42), 7752–7777.

Rashvand, M., Ranjbar, Z., & Rastegar, S. (2012). Preserving anticorrosion properties of epoxy based coatings simultaneously exposed to humidity and UV-radiation using nano zinc oxide. *Journal of the Electrochemical Society*, *159*(3), 129–132. doi:10.1149/2.093203jes

Rathod, V. T., Kumar, J. S., & Jain, A. (2017). Polymer and ceramic nanocomposites for aerospace applications. *Applied Nanoscience*, *7*(8), 519–548. doi:10.100713204-017-0592-9

Ratner, B. D. (2004). *Biomaterials science: An introduction to materials in medicine*. Amsterdam, Boston: Elsevier Academic Press.

Ratner, B. D., Hoffman, A. S., Schoen, F. J., & Lemons, J. E. (2004). *Introduction-Biomaterials Science: An Evolving, Multidisciplinary Endeavor* (3rd ed.). Academic Press.

Ray, S. (2015). Multifunctional nanobiocomposites of biodegradable polylactide and nanoclay. In K. Friedrich & U. Breuer (Eds.), *Multifunctionality of polymer composites. Challenges and new solutions* (pp. 144–212). Oxford: Elsevier. doi:10.1016/B978-0-323-26434-1.00005-2

Ray, S. S., & Bousmina, M. (2005). *Biodegradable polymers and their layered silicate nanocomposites: in greening the 21st century materials world*. Progress in Matererials.

Ray, S. S., & Okamoto, M. (2003). Polymer/layered silicate nanocomposites: A review from preparation to processing. *Progress in Polymer Science*, *28*(11), 1539–1641. doi:10.1016/j.progpolymsci.2003.08.002

Reichert, P., Nitz, H., Klinke, S., Brandsch, R., Thomann, R., & Mülhaupt, R. (2000). Poly(propylene)/organoclay nanocomposite formation: Influence of compatibilizer functionality and organoclay modification. *Macromolecular Materials and Engineering*, *275*(1), 8–17. doi:10.1002/(SICI)1439-2054(20000201)275:1<8::AID-MAME8>3.0.CO;2-6

Renteria, J. D., Nika, D. L., & Balandin, A. A. (2014). Graphene thermal properties: Applications in Thermal management and energy storage. *Applied Sciences*, *4*(4), 525–547. doi:10.3390/app4040525

Reza, F., Yamamuro, J. A., & Batson, G. B. (2004). Electrical resistance change in compact tension specimens of carbon fiber cement composites. *Cement and Concrete Composites*, *26*(7), 873–881. doi:10.1016/j.cemconcomp.2003.06.002

Richard, C. (2018, March 22). Renewables fastest growing energy source in 2017. *Wind Power Monthly*. Retrieved from https://www.windpowermonthly.com/article/1460160/renewables-fastest-growing-energy-source-2017

Risi, C. L. S., Hattenhauer, I., Coelho, L. A. F., Ramos, A., & Pezzin, S. H. (2009). *The Role of Curing rate on the Growth of Percolated Networks in Epoxy/MWNT Nanocomposites under an Electrical Field*. In Proceedings of the 11th International Conference on Advanced Materials, Rio de Janeiro.

Rissanou, A. N., & Harmandaris, V. (2014). Dynamics of various polymer–graphene interfacial systems through atomistic molecular dynamics simulations. *Soft Matter*, *10*(16), 2876–2888. doi:10.1039/c3sm52688g PMID:24667937

Robertson, F. N. (1989). Arsenic in ground water under oxidizing conditions, southwest United States. *Environmental Geochemistry and Health*, *11*(3-4), 171–176. doi:10.1007/BF01758668 PMID:24202428

Robertson, W. D. (1951). Molybdate and tungstate as corrosion inhibitors and the mechanism of inhibition. *Journal of the Electrochemical Society*, *98*(3), 94–100. doi:10.1149/1.2778118

Robinson, J. T., Perkins, F. K., Snow, E. S., Wei, Z., & Sheehan, P. E. (2008). Reduced graphene oxide molecular sensors. *Nano Letters*, *8*(10), 3137–3140. doi:10.1021/nl8013007 PMID:18763832

Rocha, F. S., Mota, H. O., & Vasconcelos, F. H. (2017). The Monte Carlo method to uncertainty calculation of the displacement measurement interferometry in FTIR spectrometry systems. In *I2MTC 2017 - 2017 IEEE International Instrumentation and Measurement Technology Conference, Proceedings*. 10.1109/I2MTC.2017.7969829

Rogers, S. A., & Kaiser, A. B. (2004). Thermopower and resistivity of carbon nanotube networks and organic conducting polymers. *Current Applied Physics*, *4*(2-4), 407–410. doi:10.1016/j.cap.2003.11.060

Romero, M. J., & Al-Jassim, M. M. (2003). Advantages of using piezoelectric quantum structures for photovoltaics. *Journal of Applied Physics*, *93*(1), 626–631. doi:10.1063/1.1524705

Rong, M. Z., Zhang, M. Q., Pan, S. L., Lehmann, B., & Friedrich, K. (2004). Analysis of the interfacial interactions in polypropylene/silica nanocomposites. *Polymer International*, *53*(2), 176–183. doi:10.1002/pi.1307

Rongzhi, X., Yandao, G., Xiufang, Z., & Nanming, Z. (1998). Fibrinogen adsorption on biomaterials as characterized by transmission FTIR spectroscopy. *Tsinghua Science and Technology*, *3*(3), 1115–1120.

Rosen, B. A., Laufer, C. N., Kalman, D. P., Wetzel, E. D., & Wagner, N. J. (2007). Multi-threat performance of kaolin-based shear thickening fluid (STF)-treated fabrics. *Proceedings of SAMPE*, *3*(7).

Roslaniec, Z., Broza, G., & Schulte, K. (2003). Nanocomposites based on multiblock polyester elastomers (PEE) and carbon nanotubes (CNT). *Composite Interfaces*, *10*(1), 95–102. doi:10.1163/156855403763586819

Roy, R., Roy, R. A., & Roy, D. M. (1986). Alternative perspectives on "quasi-crystallinity": Non- uniformity and nano-composites. *Materials Letters*, *4*(8-9), 323–328. doi:10.1016/0167-577X(86)90063-7

Rudge, A., Raistrick, I., Gottesfeld, S., & Ferraris, J. P. (1994). A study of the electro-chemical properties of conducting polymers for application in electro-chemical capacitors. *Electrochimica Acta*, *39*(2), 273–287. doi:10.1016/0013-4686(94)80063-4

Ruiz, F.A. (2000). *Proceedings of Polymers Laminations and Coatings Conference.*

Rutherford, K. L., Trezona, R. I., Ramamurthy, A. C., & Hutchings, I. M. (1997). The abrasive and erosive wear of polymeric paint films. *Wear*, *203-204*, 325–334. doi:10.1016/S0043-1648(96)07369-3

Rytter, E., & Einarsrud, M. A. (1988). Band distortions in emission and specular reflection FTIR spectroscopy. *Mikrochimica Acta*, *95*(1-6), 307–309. doi:10.1007/BF01349775

Ryu, K. S., Lee, Y.-G., Kim, K. M., Park, Y. J., Hong, Y.-S., Wu, X., ... Ko, J. M. (2005). Electrochemical capacitor with chemically polymerized conducting polymer based on activated carbon as hybrid electrodes. *Synthetic Metals*, *153*(1-3), 89–92. doi:10.1016/j.synthmet.2005.07.167

Ryu, K. S., Wu, X., Lee, Y., & Chang, S. H. (2003). Electrochemical capacitor composed of doped polyaniline and polymer electrolyte membrane. *Journal of Applied Polymer Science*, *89*(5), 1300–1304. doi:10.1002/app.12242

Saba, N., Tahir, P. M., & Jawaid, M. (2014). A Review on Potentiality of Nano Filler/Natural Fiber Filled Polymer Hybrid Composites. *Polymers*, *6*(8), 2247–2273. doi:10.3390/polym6082247

Safarova, K., Dvorak, A., Kubinek, R., Vujtek, M., & Rek, A. (2007). Usage of AFM, SEM and TEM for the research of carbon nanotubes. *Modern Research and Educational Topics in Microscopy*, *2*, 513–519.

Saha, N., Roy, I., Sarkar, G., Bhattacharyya, A., Das, R., Rana, D., ... Chattopadhyay, D. (2018). Development of active packaging material based on cellulose acetate butyrate/polyethylene glycol/aryl ammonium cation modified clay. *Carbohydrate Polymers*, *187*, 8–18. doi:10.1016/j.carbpol.2018.01.065 PMID:29486847

Sahaym, U., & Norton, M. (2008). Advances in the application of nanotechnology in enabling a 'hydrogen economy.' *Journal of Materials Science*, *43*(16), 5395–5429. doi:10.100710853-008-2749-0

Saisirirat, P., & Joommanee, B. (2017). Study on the performance of the micro direct ethanol fuel cell (Micro-DEFC) for applying with the portable electronic devices. *Energy Procedia*, *138*, 187–192. doi:10.1016/j.egypro.2017.10.148

Saji, V. S., & Cook, R. M. (Eds.). (2012). *Corrosion protection and control using nanomaterials*. Elsevier. doi:10.1533/9780857095800

Sakhri, A., Perrin, F. X., Aragon, E., Lamouric, S., & Benaboura, A. (2010). Chlorinated rubber paints for corrosion prevention of mild steel: A comparison between zinc phosphate and polyaniline pigments. *Corrosion Science*, *52*(3), 901–909. doi:10.1016/j.corsci.2009.11.010

Salanne, M. (2017). Ionic Liquids for Supercapacitor Applications. *Top Curr Chem (Z)*, *375*(3), 63. doi:10.100741061-017-0150-7 PMID:28560657

Salinas-Torres, D., Sieben, J. M., Lozano-Castelló, D., Cazorla-Amorós, D., & Morallón, E. (2013). Asymmetric hybrid capacitors based on activated carbon and activated carbon fibre–PANI electrodes. *Electrochimica Acta*, *89*, 326–333. doi:10.1016/j.electacta.2012.11.039

Sall, M. L., Diaw, A. K. D., Gningue-Sall, D., Chevillot-Biraud, A., Oturan, N., Oturan, M. A., ... Aaron, J. J. (2018). Removal of lead and cadmium from aqueous solutions by using 4-amino-3-hydroxynaphthalene sulfonic acid-doped polypyrrole films. *Environmental Science and Pollution Research International*. PMID:29318483

Salunkhe, R. R., Lee, Y. H., Chang, K. H., Li, J. M., Simon, P., Tang, J., ... Yamauchi, Y. (2014). Nanoarchitectured graphene-based supercapacitors for next-generation energy-storage applications. *Chemistry (Weinheim an der Bergstrasse, Germany)*, *20*(43), 13838–13852. doi:10.1002/chem.201403649 PMID:25251360

Salunkhe, R. R., Lin, J. J., Malgras, V., Dou, S. X., Kim, J. H., & Yamauchi, Y. (2015). Large-scale synthesis of coaxial carbon nanotube/$Ni(OH)_2$ composites for asymmetric supercapacitor application. *Nano Energy*, *11*, 211–218. doi:10.1016/j.nanoen.2014.09.030

Salunkhe, R. R., Tang, J., Kamachi, Y., Nakato, T., Kim, J. H., & Yamauchi, Y. (2015). Asymmetric Supercapacitors Using 3D Nanoporous Carbon and Cobalt Oxide Electrodes Synthesized from a Single Metal–Organic Framework. *ACS Nano*, *9*(6), 6288–6296. doi:10.1021/acsnano.5b01790 PMID:25978143

Samakande, A. (2005). Synthesis and Characterization of Surfmers for the Synthesis of Polystrene-Clay Nanocomposites [Masters' Thesis]. University of Stellenbosch.

Samba, R., Herrmann, T., & Zeck, G. (2015). PEDOT–CNT coated electrodes stimulate retinal neurons at low voltage amplitudes and low charge densities. *Journal of Neural Engineering*, *12*(1), 016014. doi:10.1088/1741-2560/12/1/016014 PMID:25588201

Samimi, F., Babapoor, A., Azizi, M., & Karimi, G. (2016). Thermal management analysis of a Li-ion battery cell using phase change material loaded with carbon fibers. *Energy*, *96*, 355–371. doi:10.1016/j.energy.2015.12.064

Sanchez-Solis, A., Garcia-Rejon, A., & Manero, O. (2003). Production of nanocomposites of PET-montmorillonite clay by an extrusion process. *Macromolecular Symposia*, *192*(1), 281–292. doi:10.1002/masy.200390038

Sanjay, M. R., Arpitha, G. R., Naik, L. L., Gopalakrishna, K., & Yogesha, B. (2016). Applications of natural fibers and its composites: An overview. *Natural Resources*, *7*(03), 108–114. doi:10.4236/nr.2016.73011

Sarker, A. K., & Hong, J. D. (2012). Layer-by-layer self-assembled multilayer films composed of graphene/polyaniline bilayers: High-energy electrode materials for supercapacitors. *Langmuir*, *28*(34), 12637–12646. doi:10.1021/la3021589 PMID:22866750

Sathishkumar, T., Naveen, J., & Satheeshkumar, S. (2014). Hybrid fiber reinforced polymer composites-A review. *Journal of Reinforced Plastics and Composites*, *33*(5), 454–471. doi:10.1177/0731684413516393

Savage, S. (2004). *Defence applications of nanocomposite materials*. Sweden: Swedish Defence Research Agency Kista.

Schedin, F., Geim, A., Morozov, S., Hill, E., Blake, P., Katsnelson, M., & Novoselov, K. (2007). Detection of individual gas molecules adsorbed on graphene. *Nature Materials*, *6*(9), 652–655. doi:10.1038/nmat1967 PMID:17660825

Schmid, M., Sängerlaub, S., Miesbauer, O., Jost, V., Werthan, J., Stinga, C., ... Müller, K. (2014). Water repellence and oxygen and water vapor barrier of pvoh-coated substrates before and after surface esterification. *Polymers*, *6*(11), 2764–2783. doi:10.3390/polym6112764

Schmidt, C. E., & Leach, J. B. (2003). Neural tissue engineering: Strategies for repair and regeneration. *Annual Review of Biomedical Engineering*, *5*(1), 293–347. doi:10.1146/annurev.bioeng.5.011303.120731 PMID:14527315

Schmidt, D., Shah, D., & Giannelis, E. P. (2002). New advances in polymer/layered silicate nanocomposites. *Current Opinion in Solid State and Materials Science*, *6*(3), 205–212. doi:10.1016/S1359-0286(02)00049-9

Schmidt, G., Nakatani, A., Butler, P., & Han, C. (2002). Small-angle neutron scattering from viscoelastic polymer–clay solutions. *Macromolecules*, *35*(12), 4725–4732. doi:10.1021/ma0115141

Schubel, P. J., Johnson, M. S., Warrior, N. A., & Rudd, C. D. (2006). Characterisation of thermoset laminates for cosmetic automotive applications: Part III - shrinkage control via nanoscale reinforcement. *Composites. Part A, Applied Science and Manufacturing*, *37*(10), 1757–1772. doi:10.1016/j.compositesa.2005.09.014

Senthilkumar, S. T., Selvan, R., & Melo, J. (2013). Redox additive/active electrolytes: A novel approach to enhance the performance of supercapacitors. *Journal of Materials Chemistry. A, Materials for Energy and Sustainability*, *1*(40), 12386–12394. doi:10.1039/c3ta11959a

Serrano, E., Rus, G., & García-Martínez, J. (2009). Nanotechnology for sustainable energy. *Renewable & Sustainable Energy Reviews*, *13*(9), 2373–2384. doi:10.1016/j.rser.2009.06.003

Sethi, V. K., Pandey, M., & Shukla, P. (2011). Use of nanotechnology in solar PV cell. *International Journal of Chemical Engineering and Applications*, *2*(2).

Seyhan, A., Gojny, F., Tanoglu, M., & Schulte, K. (2007b). Rheological and dynamic-mechanical behavior of carbon nanotube/vinyl ester-polyester suspensions and their nanocomposites. *European Polymer Journal*, *43*(7), 2836–2847. doi:10.1016/j.eurpolymj.2007.04.022

Shadlou, S., Ahmadi-Moghadam, B., & Taheri, F. (2014). Nanoenhanced adhesives: A critical review. *Review of Adhesion and Adhesives*, *2*(3), 371–412. doi:10.7569/RAA.2014.097307

Shah, R., & Paul, D. (2006). Organoclay degradation in melt processed polyethylene nanocomposites. *Polymer*, *47*(11), 4075–4084. doi:10.1016/j.polymer.2006.02.031

Shamaeli, E., & Alizadeh, N. (2014). Nanostructured biocompatible thermal/electrical stimuli-responsive biopolymer-doped polypyrrole for controlled release of chlorpromazine: Kinetics studies. *International Journal of Pharmaceutics*, *472*(1-2), 327–338. doi:10.1016/j.ijpharm.2014.06.036 PMID:24969668

Sham, M. L., & Kim, J. K. (2006). Surface functionalities of multi-wall carbon nanotubes after UV/Ozone and TETA treatments. *Carbon, 44*(4), 768–777. doi:10.1016/j.carbon.2005.09.013

Shao, M., Wei, M., Evans, D. G., & Duan, X. (2013). Hierarchical Structures Based on Functionalized Magnetic Cores and Layered Double-Hydroxide Shells: Concept, Controlled Synthesis, and Applications. *Chemistry (Weinheim an der Bergstrasse, Germany), 19*(13), 4100–4108. doi:10.1002/chem.201204205 PMID:23424036

Sharma, R. (2001). Design and applications of environmental cell transmission electron microscope for in situ observations of gas-solid reactions. *Microscopy and Microanalysis, 7*(6), 494–506. PMID:12597794

Sharma, S., & Sagara, K. (2005). Latent heat storage materials and systems: A review. *International Journal of Green Energy, 2*(1), 1–56. doi:10.1081/GE-200051299

Sharma, V., Sahoo, A., Sharma, Y., & Mohanty, P. (2015). Synthesis of nanoporous hypercrosslinked polyaniline (HCPANI) for gas sorption and electrochemical supercapacitor applications. *RSC Advances, 5*(57), 45749–45754. doi:10.1039/C5RA03016A

Sharma, Y. C. (1995). Economic treatment of Cadmium (II)-rich hazardous waste by indigenous materials. *Journal of Colloid and Interface Science, 173*(1), 66–70. doi:10.1006/jcis.1995.1297

Shayeh, J. S., Ehsani, A., Ganjali, M. R., Norouzi, P., & Jaleh, B. (2015). Conductive polymer/reduced graphene oxide/Au nano particles as efficient composite materials in electrochemical supercapacitors. *Applied Surface Science, 353*, 594–599. doi:10.1016/j.apsusc.2015.06.066

Shen, H., Zhang, L., Liu, M., & Zhang, Z. (2012). Biomedical applications of graphene. *Theranostics, 2*(3), 283–294. doi:10.7150/thno.3642 PMID:22448195

Shen, J., Huang, W., Wu, L., Hu, Y., & Ye, M. (2007). The reinforcement role of different amino-functionalized multi-walled carbon nanotubes in epoxy nanocomposites. *Composites Science and Technology, 67*(15-16), 3041–3050. doi:10.1016/j.compscitech.2007.04.025

Shen, J., Yang, C., Li, X., & Wang, G. (2013). High-Performance Asymmetric Supercapacitor Based on Nano architectured Polyaniline/Graphene/Carbon Nanotube and Activated Graphene Electrodes. *ACS Applied Materials & Interfaces, 5*(17), 8467–8476. doi:10.1021/am4028235 PMID:23931572

Shen, K., Rana, F., Zhang, X., Liu, C., Wang, N., Niu, X., ... Chen, S. (2015). Supercapacitor electrodes based on nano-polyaniline deposited on hollow carbon spheres derived from cross-linked co-polymers. *Synthetic Metals, 209*, 369–376. doi:10.1016/j.synthmet.2015.08.012

Shen, W., Feng, L., Liu, X., Luo, H., Liu, Z., Tong, P., & Zhang, W. (2016). Multiwall carbon nanotubes-reinforced epoxy hybrid coatings with high electrical conductivity and corrosion resistance prepared via electrostatic spraying. *Progress in Organic Coatings, 90*, 139–146. doi:10.1016/j.porgcoat.2015.10.006

Shi, C., Zhu, Y., Ran, X., Wang, M., Su, Y., & Cheng, T. (2006). Therapeutic Potential of Chitosan and Its Derivatives in Regenerative Medicine1. *The Journal of Surgical Research, 133*(2), 185–192. doi:10.1016/j.jss.2005.12.013 PMID:16458923

Shi, G., Zhang, Z., & Rouabhia, M. (2008). The regulation of cell functions electrically using biodegradable polypyrrole-polylactide conductors. *Biomaterials, 29*(28), 3792–3798. doi:10.1016/j.biomaterials.2008.06.010 PMID:18602689

Shi, H., Liu, F., Yang, L., & Han, E. (2008). Characterization of protective performance of epoxy reinforced with nanometer-sized TiO_2 and SiO_2. *Progress in Organic Coatings, 62*(4), 359–368. doi:10.1016/j.porgcoat.2007.11.003

Shikuku, V. O., Achieng', G. O., Ng'eno, E., Okowa, G. M., Masitsa, G. A., & Owuor, J. J. (2017). Distribution and Removal Efficiency of Heavy Metals by a Conventional Activated Sludge at a Municipal Wastewater Treatment Plant in Kisumu City-Kenya. *Res. J. Chem. Sci.*, *7*(8), 19–25.

Shin, M. K., Spinks, G. M., Shin, S. R., Kim, S. I., & Kim, S. J. (2009). Nanocomposite hydrogel with high toughness for Bioactuators. *Advanced Materials*, *21*(17), 1712–1715. doi:10.1002/adma.200802205

Shirakawa, H., Louis, E. J., MacDiarmid, A. G., Chiang, C. K., & Heeger, A. J. (1977). Synthesis of electrically conducting organic polymers: Halogen derivatives of polyacetylene, (CH)x. *Journal of the Chemical Society. Chemical Communications*, *16*(16), 578–580. doi:10.1039/c39770000578

Shirazi, A., Mohebbi, F., Kakavand, M. A., He, B., & Rabczuk, T. (2016). Paraffin nanocomposites for heat management of lithium-ion batteries: A computational investigation. *Journal of Nanomaterials*, *2016*, 1–12. doi:10.1155/2016/2131946

Shivhare, S., & Vyas, S. (2017). Polyaniline (PANI) Metal Oxide Nano Composites as a Conducting Material. *International Journal of Research and Scientific Innovation.*, *IV*, 86–89.

Shi, Y., Peng, L., Ding, Y., Zhao, Y., & Yu, G. (2015). Nanostructured conductive polymers for advanced energy storage. *Chem. Soc. Rev. Chem. Soc. Rev*, *6684*(44), 6684–6696. doi:10.1039/C5CS00362H PMID:26119242

Shukla, A. K., Banerjee, A., Ravikumar, M. K., & Jalajakshi, A. (2012). Electrochemical capacitors: Technical challenges and prognosis for future markets. *Electrochimica Acta*, *84*, 165–173. doi:10.1016/j.electacta.2012.03.059

Silva, M., Alves, N. M., & Paiva, M. C. (2018). Graphene-polymer nanocomposites for biomedical applications. *Polymers for Advanced Technologies*, *29*(2), 687–700. doi:10.1002/pat.4164

Silvano, J., Rodrigues, S., Marini, J., Bretas, R., Canevarolo, S., Carvalho, B., & Pinheiro, L. (2013). Effect of reprocessing and clay concentration on the degradation of polypropylene/montmorillonite nanocomposites during twin screw extrusion. *Polymer Degradation & Stability*, *98*(3), 801–808. doi:10.1016/j.polymdegradstab.2012.12.009

Simon, P., & Gogotsi, Y. (2008). Materials for electrochemical capacitors. *Nature Materials*, *7*(11), 845–854. doi:10.1038/nmat2297 PMID:18956000

Simotwo, S. K., DelRe, C., & Kalra, V. (2016). Supercapacitor Electrodes Based on High-Purity Electrospun Polyaniline and Polyaniline–Carbon Nanotube Nanofibers. *ACS Applied Materials & Interfaces*, *8*(33), 21261–21269. doi:10.1021/acsami.6b03463 PMID:27467445

Singh, N. B., & Agarwal, S. (2016). Nanocomposites: An overview. *Emerging Materials Research*, *5*(1), 5–43. doi:10.1680/jemmr.15.00025

Singh, V. V., & Wang, J. (2015). Nano/micromotors for security/defense applications. A review. *Nanoscale*, *7*(46), 19377–19389. doi:10.1039/C5NR06254C PMID:26554557

Singla, A. K., & Chawla, M. (2001). Chitosan: Some pharmaceutical and biological aspects-an update. *The Journal of Pharmacy and Pharmacology*, *53*(8), 1047–1067. doi:10.1211/0022357011776441 PMID:11518015

Sinha Ray, S., Maiti, P., Okamoto, M., Yamada, K., & Ueda, K. (2002). New polylactide/layered silicate nanocomposites. 1. Preparation, characterization, and properties. *Macromolecules*, *35*(8), 3104–3110. doi:10.1021/ma011613e

Sinha Ray, S., Yamada, K., Okamoto, M., Ogami, A., & Ueda, K. (2003). New polylactide/layered silicate nanocomposites. 3. High-performance biodegradable materials. *Chemistry of Materials*, *15*(7), 1456–1465. doi:10.1021/cm020953r

Sivakkumar, S. R., Kim, W. J., Choi, J.-A., & Douglas, R. (2007). Electrochemical performance of polyaniline nanofibres and polyaniline/multi-walled carbon nanotube composite as an electrode material for aqueous redox supercapacitors. *Journal of Power Sources*, *171*(2), 1062–1068. doi:10.1016/j.jpowsour.2007.05.103

Sk, M. M., Yue, C. Y., & Jena, R. K. (2015). Non-covalent interactions and supercapacitance of pseudo-capacitive composite electrode materials (MWCNTCOOH/MnO2/PANI). *Synthetic Metals*, *208*, 2–12. doi:10.1016/j.synthmet.2014.10.026

Smart and BASF. (2018). Retrieved from http://www.smartforvision.basf.com/#polymer_composite_frame

Snook, G. A., Kao, P., & Best, A. S. (2011). Conducting-polymer-based supercapacitor devices and electrodes. *Journal of Power Sources*, *196*(1), 1–12. doi:10.1016/j.jpowsour.2010.06.084

Sodyleva, T., Kuznetsova, Y., Bogdanovich, S., Okhlopkova, A., & Pesetskii, S. (2016). On effect of excess surfactant in the organoclay on properties of nanocomposites based on PA6, obtained by reactive extrusion. *Polymer Materials and Technologies*, *2*(1), 36–47. doi:10.32864/polymmattech-2016-2-1-36-47

Soliman, H. M. A., Kashyout, A. B., El Nouby, M. S., & Abosehly, A. M. (2010). Preparation and characterizations of tungsten oxide electrochromic nanomaterials. *Journal of Materials Science Materials in Electronics*, *21*(12), 1313–1321. doi:10.100710854-010-0068-0

Solmaz, R., Kardaş, G., Culha, M., & Erbil, M. (2008). Investigation of adsorption and inhibitive effect of 2-mercapto-thiazoline on corrosion of mild steel in hydrochloric acid media. *Electrochimica Acta*, *53*(20), 5941–5952. doi:10.1016/j.electacta.2008.03.055

Soltannia, B., Haji Gholami, I., Masajedian, S., Mertiny, P., Sameoto, D., & Taheri, F. (2016). Parametric study of strain rate effects on nanoparticle-reinforced polymer composites. *Journal of Nanomaterials*, *2016*, 1–9. doi:10.1155/2016/9841972

Sonawane, S. H., Teo, B. M., Brotchie, A., Grieser, F., & Ashokkumar, M. (2010). Sonochemical synthesis of ZnO encapsulated functional nanolatex and its anticorrosive performance. *Industrial & Engineering Chemistry Research*, *49*(5), 2200–2205. doi:10.1021/ie9015039

Song, E., & Choi, J.-W. (2013). Conducting Polyaniline Nanowire and Its Applications in Chemiresistive Sensing. *Nanomaterials (Basel, Switzerland)*, *3*(3), 498–523. doi:10.3390/nano3030498 PMID:28348347

Song, G. (2005). Recent progress in corrosion and protection of magnesium alloys. *Advanced Engineering Materials*, *7*(7), 563–586. doi:10.1002/adem.200500013

Song, Z., Xu, T., Gordin, M. L., Jiang, Y.-B., Bae, I.-T., Xiao, Q., ... Wang, D. (2012). Polymer–graphene nanocomposites as ultrafast-charge and-discharge cathodes for rechargeable lithium batteries. *Nano Letters*, *12*(5), 2205–2211. doi:10.1021/nl2039666 PMID:22449138

Soni, R., Kashyap, V., Nagaraju, D., & Kurungot, S. (2018). Realizing high capacitance and rate capability in polyaniline by enhancing the electrochemical surface area through induction of superhydrophilicity. *ACS Applied Materials & Interfaces*, *10*(1), 676–686. doi:10.1021/acsami.7b15534 PMID:29243906

Sorensen, P. A., Kiil, S., Dam-Johansen, K., & Weinell, C. E. (2009). Anticorrosive coatings: A review. *Journal of Coatings Technology and Research*, *6*(2), 135–176. doi:10.100711998-008-9144-2

Soultanidis, N., Zhou, W., Psarras, A. C., Gonzalez, A. J., Iliopoulou, E. F., Kiely, C. J., ... Wong, M. S. (2010). Relating n-pentane isomerization activity to the tungsten surface density of WOx/ZrO_2. *Journal of the American Chemical Society*, *132*(38), 13462–13471. doi:10.1021/ja105519y PMID:20815386

Soutter, W. (2012). Nanotechnology in the automotive industry. *AZO NANO*. Retrieved from https://www.azonano.com/article.aspx?ArticleID=3031

Soutter, W. (2012). Nanotechnology in the military. *AZO nano*. Retrieved from https://www.azonano.com/article.aspx?ArticleID=3028

Soutter, W. (2013). Nanotechnology in electric vehicle batteries. *AZO NANO*. Retrieved from https://www.azonano.com/article.aspx?ArticleID=3157

Spitalsky, Z., Tasis, D., Papagelis, K., & Galiotis, C. (2010). Carbon nanotube-polymer composites: Chemistry, processing, mechanical and electrical properties. *Progress in Polymer Science*, *35*(3), 357–401. doi:10.1016/j.progpolymsci.2009.09.003

Srivastava, M., Srivastava, S. K., Nirala, N. R., & Prakash, R. (2014). A chitosan-based polyaniline-Au nanocomposite biosensor for determination of cholesterol. *Analytical Methods*, *6*(3), 817–824. doi:10.1039/C3AY41812J

Srivastava, S., Pramanik, M., & Acharya, H. (2005). Ethylene/vinyl acetate copolymer/clay nanocomposites. *Journal of Polymer Science. Part B, Polymer Physics*, *44*(3), 471–480. doi:10.1002/polb.20702

Stachewicz, U., Modaresifar, F., Bailey, R. J., Peijs, T., & Barber, A. H. (2012). Manufacture of void-free electrospun polymer nanofiber composites with optimized mechanical properties. *ACS Applied Materials & Interfaces*, *4*(5), 2577–2582. doi:10.1021/am300235r PMID:22524442

Standridge, S. D., Schatz, G. C., & Hupp, J. T. (2009). Distance dependence of plasmon-enhanced photocurrent in dye-sensitized solar cells. *Journal of the American Chemical Society*, *131*(24), 8407–8409. doi:10.1021/ja9022072 PMID:19473006

Stavrinidou, E., Winther-Jensen, O., Shekibi, B. S., Armel, V., Rivnay, J., Ismailova, E., ... Winther-Jensen, B. (2014). Engineering hydrophilic conducting composites with enhanced ion mobility. *Physical Chemistry Chemical Physics*, *16*(6), 2275–2279. doi:10.1039/C3CP54061H PMID:24352071

Stephen, R., & Thomas, S. (2010). Nanocomposites: State of the art, new challenges and opportunities. In Rubber nanocomposites-Preparation, properties, and applications (pp 8-13). Singapore: John Wiley & Sons (Asia) Pte Ltd.

Stieberova, B., Zillka, M., Ticha, M., Freiberg, F., Caramazana-Gonzalez, P., Mckechnie, J., & Lester, E. (2017). Application of ZnO Nanoparticles in a self-cleaning coating on a metal panel: An assessment of environmental benefits. *ACS Sustainable Chemistry & Engineering*, *5*(3), 2493–2500. doi:10.1021/acssuschemeng.6b02848

Strozi, M., Raymundo, M., & Colombo, P. (2014). Open cell geopolymer foams by a novel saponification / peroxide / gelcasting combined route. *Journal of the European Ceramic Society*, *1–5*. doi:10.1016/j.jeurceramsoc.2014.04.001

Su, D. (2017). Advanced electron microscopy characterization of nanomaterials for catalysis. *Green Energy & Environment*, *2*(2), 70–83. doi:10.1016/j.gee.2017.02.001

Su, D., Wang, F., Ma, C., & Jiang, N. (2013). Engineering nano-composite $Li_4Ti_5O_{12}$ anodes via scanning electron-probe fabrication. *Nano Energy*, *2*(3), 343–350. doi:10.1016/j.nanoen.2012.10.009

Sudha, J. D., Sivakala, S., Patel, K., & Radhakrishnan Nair, P. (2010). Development of electromagnetic shielding materials from the conductive blends of polystyrene 62 polyaniline-clay nanocomposite. *Composites. Part A, Applied Science and Manufacturing*, *41*(11), 1647–1652. doi:10.1016/j.compositesa.2010.07.015

Su, H., Wang, T., Zhang, S., Song, J., Mao, C., Niu, H., ... Tian, Y. (2012). Facile synthesis of polyaniline/TiO_2/graphene oxide composite for high performance supercapacitors. *Solid State Sciences*, *14*(6), 677–681. doi:10.1016/j.solidstatesciences.2012.03.020

Suh, J. K. F., & Matthew, H. W. (2000). Application of chitosan-based polysaccharide biomaterials in cartilage tissue engineering: A review. *Biomaterials*, *21*(24), 2589–2598. doi:10.1016/S0142-9612(00)00126-5 PMID:11071608

Sumboja, A., Foo, C. Y., Yan, J., Yan, C., Gupta, R. K., & Lee, P. S. (2012). Significant electrochemical stability of manganese dioxide/polyaniline coaxial nanowires by self-terminated double surfactant polymerization for pseudocapacitor electrode. *Journal of Materials Chemistry*, *22*(45), 23921–23928. doi:10.1039/c2jm32456c

Sun, X. W., Gan, M. Y., Ma, L., Wang, H. H., Zhou, T., Wang, S. Y., ... Wang, H. N. (2015). Fabrication of PANI-coated honeycomb-like MnO_2 nanospheres with enhanced electrochemical performance for energy storage. *Electrochimica Acta*, *180*, 977–982. doi:10.1016/j.electacta.2015.09.056

Su, Y., Kravets, V. G., Wong, S. L., Waters, J., Geim, A. K., & Nair, R. R. (2014). Impermeable barrier films and protective coatings based on reduced graphene oxide. *Nature Communications*, *5*(1), 4843. doi:10.1038/ncomms5843 PMID:25208890

Svirskis, D., Travas-Sejdic, J., Rodgers, A., & Garg, S. (2010). Electrochemically controlled drug delivery based on intrinsically conducting polymers. *Journal of Controlled Release*, *146*(1), 6–15. doi:10.1016/j.jconrel.2010.03.023 PMID:20359512

Szabo, T., Berkesi, O., Forgo, P., Josepovits, K., Sanakis, Y., Petridis, D., & Dekany, I. (2006). Evolution of Surface Functional Groups in a Series of Progressively Oxidized Graphite Oxides. *Chemistry of Materials*, *18*(11), 2740–2749. doi:10.1021/cm060258+

Tahmasbi Rad, A., Ali, N., Kotturi, H. S. R., Yazdimamaghani, M., Smay, J., Vashaee, D., & Tayebi, L. (2014). Conducting scaffolds for liver tissue engineering. *Journal of Biomedical Materials Research. Part A*, *102*(11), 4169–4181. doi:10.1002/jbm.a.35080 PMID:24443356

Takahashi, K., Limmer, S. J., Wang, Y., & Cao, G. (2004). Synthesis and electrochemical properties of single-crystal V_2O_5 nanorod arrays by template-based electrodeposition. *The Journal of Physical Chemistry B*, *108*(28), 9795–9800. doi:10.1021/jp0491820

Tana, F., Messori, M., Contini, D., Cigada, A., Valente, T., Variola, F., ... Bondioli, F. (2017). Synthesis and characterization of scratch-resistant hybrid coatings based on non-hydrolytic sol-gel ZrO_2 nanoparticles. *Progress in Organic Coatings*, *103*, 60–68. doi:10.1016/j.porgcoat.2016.11.022

Tanahashi, M. (2010). Development of fabrication methods of filler/polymer nanocomposites: With focus on simple melt-compounding based approach without surface modification of nanofillers. *Materials (Basel)*, *3*(3), 1593–1619. doi:10.3390/ma3031593

Tang, J., Salunkhe, R. R., Liu, J., Torad, N. L., Imura, M., Furukawa, S., & Yamauchi, Y. (2015). Thermal Conversion of Core–Shell Metal–Organic Frameworks: A New Method for Selectively Functionalized Nanoporous Hybrid Carbon. *Journal of the American Chemical Society*, *137*(4), 1572–1580. doi:10.1021/ja511539a PMID:25580699

Tanglumlert, W., Prasassarakich, P., Supaphol, P., & Wongkasemjit, S. (2006). Hard-coating materials for poly (methyl methacrylate) from glycidoxypropyltrimethoxysilane-modified silatrane via a sol-gel process. *Surface and Coatings Technology*, *200*(8), 2784–2790. doi:10.1016/j.surfcoat.2004.11.018

Tang, M., Wu, T., Na, H., Zhang, S., Li, X., & Pang, X. (2015). Fabrication of graphene oxide aerogels loaded with catalytic AuPd nanoparticles. *Materials Research Bulletin*, *63*, 248–252. doi:10.1016/j.materresbull.2014.12.008

Tang, Q., & Zhou, Z. (2013). Graphene-analogous low-dimensional materials. *Progress in Materials Science*, *58*(8), 1244–1315. doi:10.1016/j.pmatsci.2013.04.003

Tanigawa, T. (1992). Various biological effects of chitin derivative. *Advances in Chitin and Chitosan*, 206-215.

Tarascon, J.-M., & Armand, M. (2001). Issues and challenges facing rechargeable lithium batteries. *Nature*, *414*(6861), 359–367. doi:10.1038/35104644 PMID:11713543

Tchakoute, H. K., Elimbi, A., Diffo Kenne, B. B., Mbey, J. A., & Njopwouo, D. (2013). Synthesis of geopolymers from volcanic ash via the alkaline fusion method: Effect of Al_2O_3/Na_2O molar ratio of soda–volcanic ash. *Ceramics International*, *39*(1), 269–276. doi:10.1016/j.ceramint.2012.06.021

Teasdale, P. R. (2003). *Properties of Polyanilines. In G. G. Wallace, G. M. Sprinks, L. A. P. Kane-Maguire, P. R. Teasdale (Eds.), Conductive Electroactive Polymers: Intelligent materials Systems* (2nd ed.). Taylor & Francis.

Tedim, J., Poznyak, S. K., Kuznetsova, A., Raps, D., Hack, T., Zheludkevich, M. L., & Ferreira, M. G. S. (2010). Enhancement of active corrosion protection via combination of inhibitor-loaded nanocontainers. *ACS Applied Materials & Interfaces*, *2*(5), 1528–1535. doi:10.1021/am100174t PMID:20455547

Teixeira-Dias, B., del Valle, L. J., Aradilla, D., Estrany, F., & Alemán, C. (2012). A conducting polymer/protein composite with bactericidal and electroactive properties. *Macromolecular Materials and Engineering*, *297*(5), 427–436. doi:10.1002/mame.201100172

Teixeira-Dias, B., del Valle, L. J., Estrany, F., Mano, J. F., Reis, R. L., & Alemán, C. (2012). Dextrin and Conducting-Polymer-Containing Biocomposites: Properties and Behavior as Cellular Matrix. *Macromolecular Materials and Engineering*, *297*(4), 359–368. doi:10.1002/mame.201100180

Terrones, M., Martín, O., González, M., Pozuelo, J., Serrano, B., Cabanelas, J. C., & Baselga, J. (2011). Interphases in Graphene Polymer-based Nanocomposites: Achievements and Challenges. *Advanced Materials*, *23*(44), 5302–5310. doi:10.1002/adma.201102036 PMID:22299145

Tetlow, H., de Boer, J. P., Forde, I. J., Vvedensky, D. D., Coraux, J., & Kantorovich, L. (2014). Growth of epitaxial graphene: Theory and experiment. *Physics Reports*, *542*(3), 195–295. doi:10.1016/j.physrep.2014.03.003

Thilagavathi, G., Raja, A. S. M., & Kannaian, T. (2008). Nanotechnology and protective clothing for defence personnel. *Defence Science Journal*, *58*(4), 451–459. doi:10.14429/dsj.58.1667

Thio, Y. S., Argon, A. S., & Cohen, R. E. (2004). Role of interfacial adhesion strength on toughening polypropylene with rigid particles. *Polymer*, *45*(10), 3139–3147. doi:10.1016/j.polymer.2004.02.064

Thomann, I., Pinaud, B. A., Chen, Z., Clemens, B. M., Jaramillo, T. F., & Brongersma, M. L. (2011). Plasmon enhanced solar-to-fuel energy conversion. *Nano Letters*, *11*(8), 3440–3446. doi:10.1021/nl201908s PMID:21749077

Thomas, S., Birbilis, N., Venkatraman, M. S., & Cole, I. S. (2013). Self-repairing oxides to protect zinc: Review, discussion and prospects. *Corrosion Science*, *69*, 11–22. doi:10.1016/j.corsci.2013.01.011

Thrivikraman, G., Madras, G., & Basu, B. (2014). Intermittent electrical stimuli for guidance of human mesenchymal stem cell lineage commitment towards neural-like cells on electroconductive substrates. *Biomaterials*, *35*(24), 6219–6235. doi:10.1016/j.biomaterials.2014.04.018 PMID:24816362

Tian, Z., Hu, H., & Sun, Y. (2013). A molecular dynamics study of effective thermal conductivity in nanocomposites. *International Journal of Heat and Mass Transfer*, *61*, 577–582. doi:10.1016/j.ijheatmasstransfer.2013.02.023

Tian, Z., Yu, H., Wang, L., Saleem, M., Ren, F., Ren, P., & Huang, L. (2014). Recent progress in the preparation of polyaniline nanostructures and their applications in anticorrosive coatings. *RSC Advances*, *4*(54), 28195–28208. doi:10.1039/c4ra03146f

Tingle, R. (2017). Stormtrooper 2017: Russia unveils its hi-tech motorised 'Star Wars' armour for soldiers complete with exoskeleton, night vision and helmet. *Daily Mail*. Retrieved from http://www.dailymail.co.uk/news/article-4653886/Russia-unveils-generation-Star-Wars-combat-uniform.html

Tiwari, J. N., Tiwari, R. N., & Kim, K. S. (2012). Zero-dimensional, one-dimensional, two-dimensional and three-dimensional nanostructured materials for advanced electrochemical energy devices. *Progress in Materials Science*, *57*(4), 724–803. doi:10.1016/j.pmatsci.2011.08.003

Tkalya, E. E., Ghislandi, M., de With, G., & Koning, C. E. (2012). The use of surfactants for dispersing carbon nanotubes and graphene to make conductive nanocomposites. *Current Opinion in Colloid & Interface Science*, *17*(4), 225–232. doi:10.1016/j.cocis.2012.03.001

Tomer, V., Polizos, G., Randall, C. A., & Manias, E. (2011). Polyethylene nanocomposite dielectrics: Implications of nanofiller orientation on high field properties and energy storage. *Journal of Applied Physics*, *109*(7), 074113. doi:10.1063/1.3569696

Tome, S., Etoh, M., Etame, J., & Sanjay, K. (2018). Characterization and leachability behaviour of geopolymer cement synthesized from municipal solid waste incinerator fly ash and volcanic ash blends. *Recycling*, *3*(4), 50. doi:10.3390/recycling3040050

Tong, Y., Bohm, S., & Song, M. (2013). Graphene based materials and their composites as coatings. *Austin Journal of Nanomedicine & Nanotechnology*, *1*(1), 1003.

Tran, C., Singhal, R., Lawrence, D., & Kalra, V. (2015). Polyaniline-coated freestanding porous carbon nanofibers as efficient hybrid electrodes for supercapacitors. *Journal of Power Sources*, *293*, 373–379. doi:10.1016/j.jpowsour.2015.05.054

Treacy, G. M., Wilcox, G. D., & Richardson, M. O. W. (1999). Behaviour of molybdate-passivated zinc coated steel exposed to corrosive chloride environments. *Journal of Applied Electrochemistry*, *29*(5), 647–654. doi:10.1023/A:1026461924663

Trung, N. B., Tam, T. V., Kim, R. H., Hur, S. H., Kim, E. J., & Choi, W. M. (2014). Three-dimensional hollow balls of graphene–polyaniline hybrids for supercapacitor applications. *Chemical Engineering Journal*, *255*, 89–96. doi:10.1016/j.cej.2014.06.028

Tsakalakos, L. (2010). *Nanotechnology for photovoltaics*. CRC Press. doi:10.1201/9781420076752

Twigg, M. V., & Richardson, J. T. (2007). Fundamentals and applications of structured ceramic foamcatalysts. *Industrial & Engineering Chemistry Research*, *46*(12), 4166–4177. doi:10.1021/ie061122o

Tyagi, A., Tripathi, K. M., & Gupta, R. K. (2015). Recent progress in micro-scale energy storage devices and future aspects. *Journal of Materials Chemistry. A, Materials for Energy and Sustainability*, *3*(45), 22507–22541. doi:10.1039/C5TA05666G

Ulutürk, C., & Alemdar, N. (2018). Electroconductive 3D polymeric network production by using polyaniline/chitosan-based hydrogel. *Carbohydrate Polymers*, *193*, 307–315. doi:10.1016/j.carbpol.2018.03.099 PMID:29773386

Understanding Nano. (2010, September). *Rice U. formula could make fuel manufacturing better, greener*. Retrieved from http://www.understandingnano.com/nanocatalyst-tunsten-oxide.html

Upadhyay, J., Kumar, A., Gupta, K., & Mandal, M. (2015). Investigation of physical and biological properties of polypyrrole nanotubes–chitosan nanocomposites. *Carbohydrate Polymers*, *132*, 481–489. doi:10.1016/j.carbpol.2015.06.028 PMID:26256373

Usami, Y., Okamoto, Y., Minami, S., Matsuhashi, A., Kumazawa, N. H., Tanioka, S. I., & Shigemasa, Y. (1994). Chitin and chitosan induce migration of bovine polymorphonuclear cells. *The Journal of Veterinary Medical Science*, *56*(4), 761–762. doi:10.1292/jvms.56.761 PMID:7999904

Usami, Y., Okamoto, Y., Minami, S., Matsuhashi, A., Kumazawa, N. H., Tanioka, S. I., & Shigemasa, Y. (1994). Migration of canine neutrophils to chitin and chitosan. *The Journal of Veterinary Medical Science*, *56*(6), 1215–1216. doi:10.1292/jvms.56.1215 PMID:7696425

Vacanti, J. P., & Langer, R. (1999). Tissue engineering: The design and fabrication of living replacement devices for surgical reconstruction and transplantation. *Lancet*, *354*, S32–S34. doi:10.1016/S0140-6736(99)90247-7 PMID:10437854

Vacogne, C. (2010). Rapid Manufacture of Structural Thermoplastic Composite Components for Aerospace and Defence Applications. In 2nd *International Conference on: Polymers in Defence & Aerospace Applications*, Hamburg, Germany, February 10-11.

Vaia, R., Ishii, H., & Giannelis, E. (1993). Synthesis and properties of two-dimensional nanostructures by direct intercalation of polymer melts in layered silicates. *Chemistry of Materials*, *5*(12), 1694–1696. doi:10.1021/cm00036a004

Vaia, R., Liu, W., & Koerner, H. (2003). Analysis of small-angle scattering of suspensions of organically modified montmorillonite: Implications to phase behavior of polymer nanocomposites. *Journal of Polymer Science. Part B, Polymer Physics*, *41*(24), 3214–3236. doi:10.1002/polb.10698

Vaia, R., Teukolsky, R., & Giannelis, E. (1994). Interlayer structure and molecular environment of alkylammonium layered silicates. *Chemistry of Materials*, *6*(7), 1017–1022. doi:10.1021/cm00043a025

Vaisman, L., Wagner, H. D., & Marom, G. (2006). The role of surfactants in dispersion of carbon nanotubes. *Advances in Colloid and Interface Science*, *128-130*, 37–46. doi:10.1016/j.cis.2006.11.007 PMID:17222381

Van Aert, S., Batenburg, K. J., Rossell, M. D., Erni, R., & Van Tendeloo, G. (2011). Three-dimensional atomic imaging of crystalline nanoparticles. *Nature*, *470*(7334), 374–377. doi:10.1038/nature09741 PMID:21289625

van Deventer, J. S. J., Provis, J. L., Duxson, P., & Lukey, G. C. (2007). Reaction mechanisms in the geopolymeric conversion of inorganic waste to useful products. *Journal of Hazardous Materials*, *139*(3), 506–513. doi:10.1016/j.jhazmat.2006.02.044 PMID:16600483

VanderHart, D., Asano, A., & Gilman, J. (2001). Solid-State NMR Investigation of Paramagnetic Nylon-6 Clay Nanocomposites. 1. Crystallinity, Morphology, and the Direct Influence of Fe3+on Nuclear Spins. *Chemistry of Materials*, *13*(10), 3781–3795. doi:10.1021/cm0110775

Vannini, M., Marchese, P., Celli, A., Marega, C., & Lorenzetti, C. (2018). Strategy to improve PA6 performances by melt compounding. *Polymer Testing*, *67*, 84–91. doi:10.1016/j.polymertesting.2018.02.021

Vaquero, S., Díaz, R., Andersona, M., Palma, J., & Marcilla, R. (2012). Insights into the influence of pore size distribution and surface functionalities in the behaviour of carbon supercapacitors. *Electrochimica Acta*, *86*, 241–247. doi:10.1016/j.electacta.2012.08.006

Vatanpour, V., Madaeni, S. S., Rajabi, L., Zinadini, S., & Derakhshan, A. A. (2012). Boehmite nanoparticles as a new nanofiller for preparation of antifouling mixed matrix membranes. *Journal of Membrane Science*, *401*, 132–143. doi:10.1016/j.memsci.2012.01.040

Vega, F. L. A., Olaya, J. J., & Ruiz, J. B. (2018). Synthesis and corrosion resistance of SiO_2-TiO_2-ZrO_2-Bi_2O_3 coatings spin-coated on Ti6Al4V alloy. *Ceramics International*, *44*(2), 2123–2131. doi:10.1016/j.ceramint.2017.10.161

Vellakkat, M., & Hundekal, D. (2017). Electrical conductivity and supercapacitor properties of polyaniline/chitosan/nickel oxide honeycomb nanocomposite. *Journal of Applied Polymer Science*, *134*(9). doi:10.1002/app.44536

Verdejo, R., Bernal, M. M., Romasanta, L. J., & Lopez-Manchado, M. A. (2011). Graphene filled polymer nanocomposites. *Journal of Materials Chemistry*, *21*(10), 3301–3310. doi:10.1039/C0JM02708A

Verma, M., Verma, P., Dhawan, S., & Choudhary, V. (2015). Tailored graphene based polyurethane composites for efficient electrostatic dissipation and electromagnetic interference shielding applications. *RSC Advances*, *5*(118), 97349–97358. doi:10.1039/C5RA17276D

Verma, V. C. (n.d.). Recent developments in Nanotechnology and their Potential Defense Applications. *Nanotechnology*, *5*.

Vert, M., Doi, Y., Hellwich, K. H., Hess, M., Hodge, P., Kubisa, P., ... Schué, F. (2012). Terminology for biorelated polymers and applications (IUPAC Recommendations 2012). *Pure and Applied Chemistry*, *84*(2), 377–410. doi:10.1351/PAC-REC-10-12-04

Vijayan, P. P., El-Gawady, Y. M. H., & Al-Maadeed, M. A. S. A. (2017). A comparative study on long term stability of self-healing epoxy coating with different inorganic nanotubes as healing agent reservoirs. *Express Polymer Letters*, *11*(11), 863–872. doi:10.3144/expresspolymlett.2017.83

Villmow, T., Pötschke, P., Pegel, S., Häussler, L., & Kretzschmar, B. (2008). Influence of twin-screw extrusion conditions on the dispersion of multi-walled carbon nanotubes in a poly(lactic acid) matrix. *Polymer*, *49*(16), 3500–3509. doi:10.1016/j.polymer.2008.06.010

Vladár, A. E., Radi, Z., Postek, M. T., & Joy, D. C. (2006). Nanotip electron gun for the scanning electron microscope. *Scanning*, *28*(3), 133–141. doi:10.1002ca.4950280301 PMID:16878784

Wagner, J. B., Cavalca, F., Damsgaard, C. D., Duchstein, L. D. L., & Hansen, T. W. (2012). Exploring the environmental transmission electron microscope. *Micron (Oxford, England)*, *43*(11), 1169–1175. doi:10.1016/j.micron.2012.02.008 PMID:22459250

Wallace, G. G., Teasdale, P. R., Spinks, G. M., & Kane-Maguire, L. A. (2002). *Conductive electroactive polymers: Intelligent materials systems* (2nd ed.). CRC press, Taylor & Francis.

Wang, D. W., Li, F., Zhao, J., Ren, W., Chen, Z. G., Tan, J., ... Cheng, H.-M. (2009). Fabrication of graphene/polyaniline composite paper via in situ anodic electropolymerization for highperformance flexible electrode. *ACS Nano*, *3*(7), 1745–1752. doi:10.1021/nn900297m PMID:19489559

Wang, D., & Bierwagen, G. P. (2009). Sol–gel coatings on metals for corrosion protection. *Progress in Organic Coatings*, *64*(4), 327–338. doi:10.1016/j.porgcoat.2008.08.010

Wang, D., Tang, X., Qiu, Y., Gan, F., & Chen, G. Z. (2005). A study of the film formation kinetics on zinc in different acidic corrosion inhibitor solutions by quartz crystal microbalance. *Corrosion Science*, *47*(9), 2157–2172. doi:10.1016/j.corsci.2004.10.003

Wang, D., & Wilkie, C. (2003). In-situ reactive blending to prepare polystyrene–clay and polypropylene–clay nanocomposites. *Polymer Degradation & Stability*, *80*(1), 171–182. doi:10.1016/S0141-3910(02)00399-3

Wang, G., Shen, X., Yao, J., & Park, J. (2009). Graphene nanosheets for enhanced lithium storage in lithium ion batteries. *Carbon*, *47*(8), 2049–2053. doi:10.1016/j.carbon.2009.03.053

Wang, H. (2009). Dispersing carbon nanotubes using surfactants. *Current Opinion in Colloid & Interface Science*, *14*(5), 364–371. doi:10.1016/j.cocis.2009.06.004

Wang, H., Guo, Z., Yao, S., Li, Z., & Zhang, W. (2017). Design and Synthesis of Ternary Graphene/Polyaniline/Co_3O_4 Hierarchical Nanocomposites for Supercapacitors. *International Journal of Electrochemical Science*, *12*, 3721–3731. doi:10.20964/2017.05.66

Wang, H., Hao, Q., Yang, X., Lu, L., & Wang, X. (2010). A nanostructured graphene/polyaniline hybrid material for supercapacitors. *Nanoscale*, *2*(10), 2164–2170. doi:10.1039/c0nr00224k PMID:20689894

Wang, H., & Hu, Y. H. (2012). Graphene as a counter electrode material for dye-sensitized solar cells. *Energy & Environmental Science*, *5*(8), 8182–8188. doi:10.1039/c2ee21905k

Wang, H., Lin, J., & Shen, Z. X. (2016). Polyaniline (PANI) based electrode materials for energy storage and conversion. *Journal of Science: Advanced Materials and Devices*, *1*, 225–255.

Wang, J., Ma, F., Liang, W., & Sun, M. (2017). Electrical properties and applications of graphene, hexagonal boron nitride (h-BN), and graphene/h-BN heterostructures. *Materials Today Physics*, *2*, 6–34. doi:10.1016/j.mtphys.2017.07.001

Wang, J., Zhang, S. Q., Guo, Y. Z., Shen, J., Attia, S. M., Zhou, B., ... Gui, Y. S. (2001). Morphological effects on the electrical and electrochemical properties of carbon aerogels. *Journal of the Electrochemical Society*, *148*(6), D75–D77. doi:10.1149/1.1368104

Wang, K., Choi, M., Koo, C., Choi, Y., & Chung, I. (2001). Synthesis and characterization of maleated polyethylene/clay nanocomposites. *Polymer*, *42*(24), 9819–9826. doi:10.1016/S0032-3861(01)00509-2

Wang, K., Huang, J., & Wei, Z. (2010). Conducting Polyaniline Nanowire Arrays for High Performance Supercapacitors. *The Journal of Physical Chemistry C*, *114*(17), 8062–8067. doi:10.1021/jp9113255

Wang, L., Lu, X., Lei, S., & Song, Y. (2014). Graphene-based polyaniline nanocomposites: Preparation, properties and applications. *Journal of Materials Chemistry. A, Materials for Energy and Sustainability*, *2*(13), 4491–4509. doi:10.1039/C3TA13462H

Wang, Q., Yan, J., Fan, Z., Wei, T., Zhang, M., & Jing, X. (2014). Mesoporous polyaniline film on ultra-thin graphene sheets for high performance supercapacitors. *Journal of Power Sources*, *247*, 197–203. doi:10.1016/j.jpowsour.2013.08.076

Wang, S., Guan, S., Wang, J., Liu, H., Liu, T., Ma, X., & Cui, Z. (2017). Fabrication and characterization of conductive poly(3,4-ethylenedioxythiophene) doped with hyaluronic acid/poly (l-lactic acid) composite film for biomedical application. *Journal of Bioscience and Bioengineering*, *123*(1), 116–125. doi:10.1016/j.jbiosc.2016.07.010 PMID:27498308

Wang, S., Guan, S., Xu, J., Li, W., Ge, D., Sun, C., ... Ma, X. (2017). Neural stem cell proliferation and differentiation in the conductive PEDOT-HA/Cs/Gel scaffold for neural tissue engineering. *Biomaterials Science*, *5*(10), 2024–2034. doi:10.1039/C7BM00633K PMID:28894864

Wang, S., Guan, S., Zhu, Z., Li, W., Liu, T., & Ma, X. (2017). Hyaluronic acid doped-poly (3, 4-ethylenedioxythiophene)/chitosan/gelatin (PEDOT-HA/Cs/Gel) porous conductive scaffold for nerve regeneration. *Materials Science and Engineering C*, *71*, 308–316. doi:10.1016/j.msec.2016.10.029 PMID:27987712

Wang, S., Li, L., & Zhu, Z. H. (2007). Solid-state conversion of fly ash to effective adsorbents for Cu removal from wastewater. *Journal of Hazardous Materials*, *139*(2), 254–259. doi:10.1016/j.jhazmat.2006.06.018 PMID:16839666

Wang, S., Ma, L., Gan, M., Fu, S., Dai, W., Zhou, T., ... Wang, H. (2015). Free-standing 3D graphene/polyaniline composite film electrodes for high-performance supercapacitors. *Journal of Power Sources*, *299*, 347–355. doi:10.1016/j.jpowsour.2015.09.018

Wang, X. Q., & Mujumdar, A. S. (2008). A review on nanofluids—part I: Theoretical and numerical investigations. *Brazilian Journal of Chemical Engineering*, *25*(4), 613–630. doi:10.1590/S0104-66322008000400001

Wang, X., Deng, J., Duan, X., Liu, D., Guo, J., & Liu, P. (2014). Crosslinked polyaniline nanorods with improved electrochemical performance as electrode material for supercapacitors. *Journal of Materials Chemistry. A, Materials for Energy and Sustainability*, *2*(31), 12323–12329. doi:10.1039/C4TA02231A

Wang, X., Du, Y., Yang, J., Tang, Y., & Luo, J. (2008). Preparation, characterization, and antimicrobial activity of quaternized chitosan/organic montmorillonite nanocomposites. *Journal of Biomedical Materials Research*, *84A*(2), 384–390. doi:10.1002/jbm.a.31326 PMID:17618481

Wang, X., Zhi, L., & Müllen, K. (2008). Transparent, conductive graphene electrodes for dye-sensitized solar cells. *Nano Letters*, *8*(1), 323–327. doi:10.1021/nl072838r PMID:18069877

Wang, Y., Chen, F., & Wu, K. (2004). Twin-screw extrusion compounding of polypropylene/organoclay nanocomposites modified by maleated polypropylenes. *Journal of Applied Polymer Science*, *93*(1), 100–112. doi:10.1002/app.20407

Wang, Y.-G., Li, H.-Q., & Xia, Y.-Y. (2006). Ordered Whisker-like Polyaniline Grown on the Surface of Mesoporous Carbon and Its Electrochemical Capacitance Performance. *Advanced Materials*, *18*(19), 2619–2623. doi:10.1002/adma.200600445

Wang, Y.-P., Gao, X.-H., Wang, R.-M., Liu, H.-G., Yang, C., & Xiong, Y.-B. (2008). Effect of functionalized montmorillonite addition on the thermal properties and ionic conductivity of PVDF–PEG polymer electrolyte. *Reactive & Functional Polymers*, *68*(7), 1170–1177. doi:10.1016/j.reactfunctpolym.2008.04.002

Wang, Y., Shi, Z., & Yin, J. (2011). Kevlar oligomer functionalized graphene for polymer composites. *Polymer*, *52*(16), 3661–3670. doi:10.1016/j.polymer.2011.06.012

Wang, Y., Xing, C., Guan, J., & Li, Y. (2017). Towards flexible dielectric materials with high dielectric constant and low loss: PVDF nanocomposites with both homogenously dispersed CNTs and ionic liquids nanodomains. *Polymers*, *9*(562). doi:10.3390/polym9110562

Wang, Z., Nakajima, H., Manias, E., & Chung, T. (2003). Exfoliated PP/clay nanocomposites using ammonium-terminated PP as the organic modification for montmorillonite. *Macromolecules*, *36*(24), 8919–8922. doi:10.1021/ma0352911

Wang, Z., Roberge, C., Wan, Y., Dao, L. H., Guidoin, R., & Zhang, Z. (2003). A biodegradable electrical bioconductor made of polypyrrole nanoparticle/poly (D, L-lactide) composite: A preliminary *in vitro* biostability study. *Journal of Biomedical Materials Research. Part A*, *66*(4), 738–746. doi:10.1002/jbm.a.10037 PMID:12926024

Wang, Z., Zhang, Q., Long, S., Luo, Y., Yu, P., Tan, Z., ... Bai, H. (2018). Three-Dimensional Printing of Polyaniline/Reduced Graphene Oxide Composite for High-Performance Planar Supercapacitor. *ACS Applied Materials & Interfaces*, *10*(12), 10437–10444. doi:10.1021/acsami.7b19635 PMID:29543426

Ward, I. M., & Sweeney, J. (2004). *An introduction to the mechanical properties of solid polymers* (2nd ed.). Wiley.

Wastiels, J., Wu, X., Faignet, S., & Patfoort, G. (1994). Mineral polymer based on fly ash. *J. Resour. Manag. Technol.*, *22*, 135–141.

Wei, H., Ding, D., Wei, S., & Guo, Z. (2013). Anticorrosive conductive polyurethane multiwalled carbon nanotube nanocomposites. *Journal of Materials Chemistry. A, Materials for Energy and Sustainability*, *1*(36), 10805–10813. doi:10.1039/c3ta11966a

Weingarth, D., Foelske-Schmitz, A., & Kötz, R. (2013). Cycle versus voltage hold - Which is the better stability test for electrochemical double layer capacitors? *Journal of Power Sources*, *225*, 84–88. doi:10.1016/j.jpowsour.2012.10.019

Weisenberger, M. C., Grulke, E. A., Jacques, D., Ramtell, T., & Andrews, R. (2003). Enhanced mechanical properties of polyacrylonitrile: Multiwall carbon nanotube composite fibers. *Journal of Nanoscience and Nanotechnology*, *3*(6), 535–539. doi:10.1166/jnn.2003.239 PMID:15002136

Wei, Y., Wang, J., Jia, X., Yeh, J. M., & Spellane, P. (1995). Polyaniline as corrosion protection coatings on cold rolled steel. *Polymer*, *36*(23), 4535–4537. doi:10.1016/0032-3861(95)96866-7

Weng, C. J., Chang, C. H., Peng, C. W., Chen, S. W., Yeh, J. M., Hsu, C. L., & Wei, Y. (2011). Advanced anticorrosive coatings prepared from the mimicked xanthosoma sagittifolium-leaf-like electroactive epoxy with synergistic effects of superhydrophobicity and redox catalytic capability. *Chemistry of Materials*, *23*(8), 2075–2083. doi:10.1021/cm1030377

Weng, C. J., Huang, J. Y., Huang, K. Y., Jhuo, Y. S., Tsai, M. H., & Yeh, J. M. (2010). Advanced anticorrosive coatings prepared from electroactive polyimide-TiO_2 hybrid nanocomposite materials. *Electrochimica Acta*, *55*(28), 8430–8438. doi:10.1016/j.electacta.2010.07.063

Wessling, B. (1994). Passivation of metals by coating with polyaniline: Corrosion potential shift and morphological changes. *Advanced Materials*, *6*(3), 226–228. doi:10.1002/adma.19940060309

Whittingham, M. S. (2004). Lithium batteries and cathode materials. *Chemical Reviews*, *104*(10), 4271–4302. doi:10.1021/cr020731c PMID:15669156

WHO. (2001). Environmental Health Criteria 221: Zinc. Geneva. *Health*, *11*, 171–176.

WHO. (2008). *Guidelines for Drinking-Water Quality*. Geneva, Switzerland: World Health Organization.

Wikipedia. (n.d.). Supercapacitor. Retrieved from en.wikipedia.org/wiki/Supercapacitor

Will Soutter. (2012). Nanotechnology in fuel cells. Retrieved from https://www.azonano.com/article.aspx?ArticleID=3032

Wong, M., Ishige, R., White, K. L., Li, P., Kim, D., Krishnamoorti, R., & Nishimura, R. (2014). Large-scale self-assembled zirconium phosphate smectic layers via a simple spray-coating process. *Nature Communications*, *5*(1), 3589. doi:10.1038/ncomms4589 PMID:24709971

Woo, S., Dokko, K., Nakano, H., & Kanamura, K. (2009). Incorporation of polyaniline into macropores of three-dimensionally ordered macroporous carbon electrode for electrochemical capacitors. *Journal of Power Sources*, *190*(2), 596–600. doi:10.1016/j.jpowsour.2009.01.050

World Steel Association. (2017). October 2017 crude steel production. Retrieved from https://www.worldsteel.org/media-centre/press-releases/2017/october-2017-crude-steel-production.html

Wu, C. L., Zhang, M. Q., Rong, M. Z., & Friedrich, K. (2005). Silica nanoparticles filled polypropylene: Effects of particle surface treatment, matrix ductility and particle species on mechanical performance of the composites. *Composites Science and Technology*, *65*(3-4), 635–645. doi:10.1016/j.compscitech.2004.09.004

Wu, D., Ramezani, M., & Hakkarainen, M. (2018). *Graphene-Polymer Nanocomposites for Biomedical Applications. In NanoBioMaterials* (pp. 51–66). CRC Press.

Wu, D., Sun, Y., & Zhang, M. (2009). Kinetics study on melt compounding of carbon nanotube/polypropylene nanocomposites. *Journal of Polymer Science. Part B, Polymer Physics*, *47*(6), 608–618. doi:10.1002/polb.21668

Wu, J., Ye, Z., Ge, H., Chen, J., Liu, W., & Liu, Z. (2017). Modified carbon fiber/magnetic graphene/epoxy composites with synergistic effect for electromagnetic interference shielding over broad frequency band. *Journal of Colloid and Interface Science*, *506*, 217–226. doi:10.1016/j.jcis.2017.07.020 PMID:28735195

Wu, M., He, J., Ren, X., Cai, W. S., Fang, Y. C., & Feng, X. Z. (2014). Development of functional biointerfaces by surface modification of polydimethylsiloxane with bioactive chlorogenic acid. *Colloids and Surfaces. B, Biointerfaces, 116*, 700–706. doi:10.1016/j.colsurfb.2013.11.010 PMID:24290104

Wu, T.-H., Chu, Y.-H., Hu, C.-C., & Hardwick, L. J. (2013). Criteria appointing the highest acceptable cell voltage of asymmetric supercapacitors. *Electrochemistry Communications, 27*, 81–84. doi:10.1016/j.elecom.2012.10.033

Wu, W., Pan, D., Li, Y., Zhao, G., Jing, L., & Chen, S. (2015). Facile fabrication of polyaniline nanotubes using the self-assembly behavior based on the hydrogen bonding: A mechanistic study and application in high-performance electrochemical supercapacitor electrode. *Electrochimica Acta, 152*, 126–134. doi:10.1016/j.electacta.2014.11.130

Xia, L., Wei, Z., & Wan, M. (2010). Conducting polymer nanostructures and their application in biosensors. *Journal of Colloid and Interface Science, 341*(1), 1–11. doi:10.1016/j.jcis.2009.09.029 PMID:19837415

Xiang, C., Lu, W., Zhu, Y., Sun, Z., Yan, Z., Hwang, C.-C., & Tour, J. M. (2011). Carbon nanotube and graphene nanoribbon-coated conductive kevlar fibers. *ACS Applied Materials & Interfaces, 4*(1), 131–136. doi:10.1021/am201153b PMID:22117617

Xiao, J., Hu, Y., Wang, Z., Tang, Y., Chen, Z., & Fan, W. (2005). Preparation and characterization of poly(butylene terephthalate) nanocomposites from thermally stable organic-modified montmorillonite. *European Polymer Journal, 41*(5), 1030–1035. doi:10.1016/j.eurpolymj.2004.11.025

Xiao, Q., Feng, M., Liu, Y., Lu, S., He, Y., & Huang, S. (2018). The graphene/polypyrrole/chitosan-modified glassy carbon electrode for electrochemical nitrite detection. *Ionics, 24*(3), 845–859. doi:10.100711581-017-2247-y

Xiao, Y., Li, C. M., Wang, S., Shi, J., & Ooi, C. P. (2010). Incorporation of collagen in poly (3, 4-ethylenedioxythiophene) for a bifunctional film with high bio-and electrochemical activity. *Journal of Biomedical Materials Research. Part A, 92*(2), 766–772. doi:10.1002/jbm.a.32412 PMID:19274716

Xia, X., Hao, Q., Lei, W., Wang, W., Sun, D., & Wang, X. (2012). Nanostructured ternary composites of graphene/Fe2O3/polyaniline for high-performance supercapacitors. *Journal of Materials Chemistry, 22*(33), 16844–16850. doi:10.1039/c2jm33064d

Xia, Y., Lu, X., & Zhu, H. (2013). Natural silk fibroin/polyaniline (core/shell) coaxial fiber: Fabrication and application for cell proliferation. *Composites Science and Technology, 77*, 37–41. doi:10.1016/j.compscitech.2013.01.008

Xia, Y., Xiong, Y., Lim, B., & Skrabalak, S. E. (2008). Shape-controlled synthesis of metal nanocrystals: Simple chemistry meets complex physics? *Angewandte Chemie International Edition, 47*, 2–46. PMID:19053095

Xie, Y., Kocaefe, D., Chen, C. & Kocaefe, Y. (2016). Review of Research on Template Methods in Preparation of Nanomaterials. *Journal of Nanomaterials*. doi:. doi:10.1155/2016/2302595

Xie, H., Zhu, Y., Wu, Y., Wu, Z., & Liu, E. (2014). The effect of hydroquinone as an electrolyte additive on electrochemical performance of the polyaniline supercapacitor. *Materials Research Bulletin, 50*, 303–306. doi:10.1016/j.materresbull.2013.11.032

Xie, K., Li, J., Lai, Y., Zhang, Z., Liu, Y., Zhang, G., & Huang, H. (2011). Polyaniline nanowire array encapsulated in titania nanotubes as a superior electrode for supercapacitors. *Nanoscale, 3*(5), 2202–2207. doi:10.1039/c0nr00899k PMID:21455534

Xie, X. L., Mai, Y. W., & Zhou, X. P. (2005). Dispersion and alignment of carbon nanotubes in polymer matrix: A review. *Materials Science and Engineering R Reports, 49*(4), 89–112. doi:10.1016/j.mser.2005.04.002

Xing, C., Zhang, Z., Yu, L., Zhang, L., & Bowmaker, G. A. (2014). Electrochemical corrosion behavior of carbon steel coated by polyaniline copolymers micro/nanostructures. *RSC Advances*, *4*(62), 32718–32725. doi:10.1039/C4RA05826G

Xing, J., Deng, B., & Liu, Q. (2015). Effect of benzimidazolium salt on dispersion and properties of polyphenylene sulfide/organic clay nanocomposites via melt intercalation. *Fibers and Polymers*, *16*(6), 1220–1229. doi:10.100712221-015-1220-5

Xing, W., Wilson, A. M., Eguchi, K., Zank, G., & Dahn, J. R. (1997). Pyrolyzed polysiloxanes for use as anode materials in lithium-ion batteries. *Journal of the Electrochemical Society*, *144*(7), 2410–2416. doi:10.1149/1.1837828

Xing, W., Zhuo, S., Cui, H., & Yan, Z. (2007). Synthesis of polyaniline-coated ordered mesoporous carbon and its enhanced electrochemical properties. *Materials Letters*, *61*(23-24), 4627–4630. doi:10.1016/j.matlet.2007.02.062

Xu, H. (2002). *Geopolymerisation of Aluminosilicate Minerals* [PhD Thesis]. The University of Melbourne.

Xu, C., Guan, S., Wang, S., Gong, W., Liu, T., Ma, X., & Sun, C. (2018). Biodegradable and electroconductive poly (3, 4-ethylenedioxythiophene)/carboxymethyl chitosan hydrogels for neural tissue engineering. *Materials Science and Engineering C*, *84*, 32–43. doi:10.1016/j.msec.2017.11.032 PMID:29519441

Xu, G., Nie, P., Dou, H., Ding, B., Li, L., & Zhang, X. (2017). Exploring metal organic frameworks for energy storage in batteries and supercapacitors. *Materials Today*, *20*(4), 191–209. doi:10.1016/j.mattod.2016.10.003

Xu, G., Wang, N., Wei, J., Lv, L., Zhang, J., Chen, Z., & Xu, Q. (2012). Preparation of Graphene Oxide/Polyaniline Nanocomposite with Assistance of Supercritical Carbon Dioxide for Supercapacitor Electrodes. *Industrial & Engineering Chemistry Research*, *51*(44), 14390–14398. doi:10.1021/ie301734f

Xu, K., Cao, P., & Heath, J. R. (2009). Scanning tunneling microscopy characterization of the electrical properties of wrinkles in exfoliated graphene monolayers. *Nano Letters*, *9*(12), 4446–4451. doi:10.1021/nl902729p PMID:19852488

Xu, W., Yu, M., Shi, S., Shi, H., & Nie, M. (2018). Simultaneous durability and strength enhancement of β-polypropylene through montmorillonite and melt-soluble β-nucleating agent addition. *Polymer Testing*, *65*, 150–155. doi:10.1016/j.polymertesting.2017.11.024

Xu, X., Yi, D., Wang, Z., Yu, J., Zhang, Z., Qiao, R., & Liu, Z. (2018). Greatly enhanced anticorrosion of Cu by commensurate graphene coating. *Advanced Materials*, *30*(6), 1702944. doi:10.1002/adma.201702944 PMID:29266426

Xu, Y., Li, Z., & Duan, W. (2014). Thermal and thermoelectric properties of graphene: A review. *Small*, *10*(11), 2182–2199. doi:10.1002mll.201303701 PMID:24610791

Yakimov, A. & Forrest, S.R. (2002). High photovoltage multiple-heterojunction organic solar cells incorporating interfacial metallic nanoclusters. *Applied Physics Letters, 80*(1667). doi:10.1063/1.1457531

Yakobson, B. I., & Avouris, P. (2001). Mechanical properties of carbon nanotubes. In *Carbon Nanotubes* (pp. 287–327). Springer. doi:10.1007/3-540-39947-X_12

Yalcin, S. E., Galande, C., Kappera, R., Yamaguchi, H., Martinez, U., Velizhanin, K. A., ... Mohite, A. D. (2015). Direct Imaging of Charge Transport in Progressively Reduced Graphene Oxide Using Electrostatic Force Microscopy. *ACS Nano*, *9*(3), 2981–2988. doi:10.1021/nn507150q PMID:25668323

Yang, C., Wei, H., Guan, L., Guo, J., Wang, Y., Yan, X., Zhang, X., Wei, S. & Guo, Z. (2015). Polymer nanocomposite for energy storage, energy saving, and anticorrosion. *Journal of material Science Chemistry A, 3*(29), 14929-14941.

Yang, C.-C., & Lin, C.-T. (2008). Preparation and characterization of the PVA/TiO_2 composite polymer membrane and its application on alkaline DMFC. *ECS Transactions*, *13*(23). doi:10.1149/1.3010874

Yang, F., Xu, M., Bao, S.-J., Wei, H., & Chai, H. (2014). Self-assembled hierarchical graphene/polyaniline hybrid aerogels for electrochemical capacitive energy storage. *Electrochimica Acta*, *137*, 381–387. doi:10.1016/j.electacta.2014.06.017

Yang, J. E., Jang, I., Kim, M., Baeck, S. H., Hwang, S., & Shim, S. E. (2013). Electrochemically polymerized vine-like nanostructured polyaniline on activated carbon nanofibers for supercapacitor. *Electrochimica Acta*, *111*, 136–143. doi:10.1016/j.electacta.2013.07.183

Yang, J., Lee, S., Choi, W. J., Seo, H., Kim, P., Kim, G.-J., ... Shin, J. (2015). Thermoset elastomers derived from Carvomenthide. *Biomacromolecules*, *16*(1), 246–256. doi:10.1021/bm501450c PMID:25514368

Yang, M., Cheng, B., Song, H., & Chen, X. (2010). Preparation and electrochemical performance of polyaniline-based carbon nanotubes as electrode material for supercapacitor. *Electrochimica Acta*, *55*(23), 7021–7027. doi:10.1016/j.electacta.2010.06.077

Yang, S., Leong, K. F., Du, Z., & Chua, C. K. (2001). The design of scaffolds for use in tissue engineering. Part I. *Traditional factors. Tissue Engineering*, *7*(6), 679–689. doi:10.1089/107632701753337645 PMID:11749726

Yang, X., Cheng, C., Wang, Y., Qiu, L., & Li, D. (2013). Liquid-mediated dense integration of graphene materials for compact capacitive energy storage. *Science*, *341*(6145), 534–537. doi:10.1126cience.1239089 PMID:23908233

Yang, X., Xu, K., Zou, R., & Hu, J. (2016). A Hybrid Electrode of Co_3O_4@PPy Core/Shell Nanosheet Arrays for High-Performance Supercapacitors. *Nano-Micro Letters*, *8*(2), 143–150. doi:10.100740820-015-0069-x PMID:30460274

Yang, X., Zhang, L., Zhang, F., Zhang, T., Huang, Y., & Chen, Y. (2014). A high-performance all-solid-state supercapacitor with graphene-doped carbon material electrodes and a graphene oxide-doped ion gel electrolyte. *Carbon*, *72*, 381–386. doi:10.1016/j.carbon.2014.02.029

Yang, Y., Hao, Y., Yuan, J., Niu, L., & Xia, F. (2014). In situ preparation of caterpillar-likepolyaniline/carbon nanotube hybrids with core shell structure for high performance supercapacitors. *Carbon*, *78*, 279–287. doi:10.1016/j.carbon.2014.07.004

Yan, J., Wei, T., Shao, B., Fan, Z., Qian, W., Zhang, M., & Wei, F. (2010). Preparation of a graphene nanosheet/polyaniline composite with high specific capacitance. *Carbon*, *48*(2), 487–493. doi:10.1016/j.carbon.2009.09.066

Yankwa Djobo, J. N., Elimbi, A., Tchakouté, H. K., & Kumar, S. (2016). Mechanical activation of volcanic ash for geopolymer synthesis: Effect on reaction kinetics, gel characteristics, physical and mechanical properties. *RSC Advances*, *6*(45), 39106–39117. doi:10.1039/C6RA03667H

Yan, L., Zhao, B., Liu, X., Li, X., Zeng, C., Shi, H., ... Liu, Y. (2016). Aligned nanofibers from polypyrrole/graphene as electrodes for regeneration of optic nerve via electrical stimulation. *ACS Applied Materials & Interfaces*, *8*(11), 6834–6840. doi:10.1021/acsami.5b12843 PMID:26926578

Yan, X., Tai, Z., Chen, J., & Xue, Q. (2011). Fabrication of carbon nanofiber-polyaniline composite flexible paper for supercapacitor. *Nanoscale*, *3*(1), 212–216. doi:10.1039/C0NR00470G PMID:21060936

Yan, Y., Cheng, Q., Wang, G., & Li, C. (2011). Growth of polyaniline nanowhiskers on mesoporous carbon for supercapacitor application. *Journal of Power Sources*, *196*(18), 7835–7840. doi:10.1016/j.jpowsour.2011.03.088

Yan, Y., Cheng, Q., Zhu, Z., Pavlinek, V., Saha, P., & Li, C. (2013). Controlled synthesis of hierarchical polyaniline nanowires/ordered bimodal mesoporous carbon nanocomposites with high surface area for supercapacitor electrodes. *Journal of Power Sources*, *240*, 544–550. doi:10.1016/j.jpowsour.2013.03.190

Yao, H., & Kimura, K. (2007). Field emission scanning electron microscopy for structural characterization of 3D gold nanoparticle superlattices. In Modern Research and Educational Topics in Microscopy (pp. 568-575).

Yao, B., Wang, G., Ye, J., & Li, X. (2008). Corrosion inhibition of carbon steel by polyaniline nanofibers. *Materials Letters*, *62*(12-13), 1775–1778. doi:10.1016/j.matlet.2007.10.001

Yao, J., Guang, L., Bastiaansen, C. W. M., & Peijs, T. (2015). High performance co-polyimide nanofiber reinforced composites. *Polymer*, *76*, 46–51. doi:10.1016/j.polymer.2015.08.054

Yao, X., Zhang, Z., Zhu, H., & Chen, Y. (2009). Geopolymerization process of alkali-metakaolinite characterized by isothermal calorimetry. *Thermochimica Acta*, *493*(1-2), 49–54. doi:10.1016/j.tca.2009.04.002

Yariv, S., & Cross, H. (Eds.). (2002). Organo-clay complexes and interactions. New York: Marcel Dekker.

Yavari, F., Rafiee, M. A., Rafiee, J., Yu, Z. Z., & Koratkar, N. (2010). Dramatic increase in fatigue life in hierarchical graphene composites. *ACS Applied Materials & Interfaces*, *2*(10), 2738–2743. doi:10.1021/am100728r PMID:20863061

Yavuz, Ö., Altunkaynak, Y., & Güzel, F. (2003). Removal of copper, nickel, cobalt and Manganese from aqueous solution by kaolinite. *Water Research*, *37*(4), 948–952. doi:10.1016/S0043-1354(02)00409-8 PMID:12531278

Yeganeh, M., Saremi, M., & Rezaeyan, H. (2014). Corrosion inhibition of steel using mesoporous silica nanocontainers incorporated in the polypyrrole. *Progress in Organic Coatings*, *77*(9), 1428–1435. doi:10.1016/j.porgcoat.2014.05.007

Yeum, J. H., Park, S. M., Kwon, I. I. J., Kim, J. W., Kim, H. W., Rabbani, M. M., … Oh, W. (2012). Carbon Nanotube Embedded Multi-Functional Polymer Nanocomposites. In Nanocomposites. London: Intech.

Yi, M., & Shen, Z. (2015). A review on mechanical exfoliation for the scalable production of graphene. *Journal of Materials Chemistry. A, Materials for Energy and Sustainability*, *3*(22), 11700–11715. doi:10.1039/C5TA00252D

Yinghong, X., Xin, W., Xujie, Y., & Lude, L. (2003). Nanometre-sized TiO_2 as applied to the modification of unsatured polyester resin. *Materials Chemistry and Physics*, *77*(2), 609–611. doi:10.1016/S0254-0584(02)00114-1

Yin, Y., Liu, C., & Fan, S. (2012). Well-Constructed CNT Mesh/PANI Nanoporous Electrode and Its Thickness Effect on the Supercapacitor Properties. *The Journal of Physical Chemistry C*, *116*(50), 26185–26189. doi:10.1021/jp3083387

Yi, W., Malkovskiy, A., Xu, Y., Wang, X.-Q., Sokolov, A. P., Lebron-Colon, M., ... Pang, Y. (2010). Polymer conformation-assisted wrapping of single-walled carbon nanotube: The impact of cis-vinylene linkage. *Polymer*, *51*(2), 475–481. doi:10.1016/j.polymer.2009.11.052

Yoo, H. D., Markevich, E., Salitra, G., Sharon, D., & Aurbach, D. (2014). On the challenge of developing advanced technologies for electrochemical energy storage and conversion. *Materials Today*, *17*(3), 110–121. doi:10.1016/j.mattod.2014.02.014

Yoon, P., Hunter, D., & Paul, D. (2003). Polycarbonate nanocomposites: Part 2. Degradation and color formation. *Polymer*, *44*(18), 5341–5354. doi:10.1016/S0032-3861(03)00523-8

Yoo, Y., Tiwari, R., Yoo, Y., & Paul, D. (2010). Effect of organoclay structure and mixing protocol on the toughening of amorphous polyamide/elastomer blends. *Polymer*, *51*(21), 4907–4915. doi:10.1016/j.polymer.2010.08.036

Young, R. (2010). Polymers in Defence and Aerospace Applications. In *Hamburg Conference*, February 10-11.

Yuan, L., Xiao, X., Ding, T., Zhong, J., Zhang, X., Shen, Y., ... Wang, Z. L. (2012). Paper-Based Supercapacitors for Self-Powered Nanosystems. *Angewandte Chemie International Edition*, *51*(20), 4934–4938. doi:10.1002/anie.201109142 PMID:22473807

Yu, J. G., Huang, K. L., Liu, S. Q., & Tang, J. C. (2008). Preparation and characterization of polycarbonate modified multiple-walled carbon nanotubes. *Chinese Journal of Chemistry*, *26*(3), 560–563. doi:10.1002/cjoc.200890105

Yuksel, R., Durucan, C., & Unalan, H. E. (2016). Ternary nanocomposite SWNT/WO3/PANI thin film electrodes for supercapacitors. *Journal of Alloys and Compounds*, *658*, 183–189. doi:10.1016/j.jallcom.2015.10.216

Yu, M., Ma, Y., Liu, J., & Li, S. (2015). Polyaniline nanocone arrays synthesized on three dimensional graphene network by electrodeposition for supercapacitor electrodes. *Carbon*, *87*, 98–105. doi:10.1016/j.carbon.2015.02.017

Yun, H., Li, J., Chen, H. B., & Lin, C. J. (2007). A study on the N-, S-and Cl-modified nano-TiO2 coatings for corrosion protection of stainless steel. *Electrochimica Acta*, *52*(24), 6679–6685. doi:10.1016/j.electacta.2007.04.078

Yu, P., Li, Y., Zhao, X., Wu, L., & Zhang, Q. (2013). In situ growth of ordered polyaniline nanowires on surfactant stabilized exfoliated graphene as high-performance supercapacitor electrodes. *Synthetic Metals*, *185*, 89–95. doi:10.1016/j.synthmet.2013.10.010

Yu, P., Li, Y., Zhao, X., Wu, L., & Zhang, Q. (2014). Graphene-wrapped polyaniline nanowire arrays on nitrogen-doped carbon fabric as novel flexible hybrid electrode materials for high-performance supercapacitor. *Langmuir*, *30*(18), 5306–5313. doi:10.1021/la404765z PMID:24761945

Yu, P., Zhao, X., Huang, Z., Li, Y., & Zhang, Q. (2014). Free-standing three-dimensional graphene and polyaniline nanowire arrays hybrid foams for high performance flexible and lightweight supercapacitors. *Journal of Materials Chemistry. A, Materials for Energy and Sustainability*, *2*(35), 14413–14420. doi:10.1039/C4TA02721C

Yu, W., & Xie, H. (2011). A review on nanofluids: Preparation, stability mechanisms, and applications. *Journal of Nanomaterials*. doi:10.1155/2012/435873

Yu, Y. H., Lin, Y. Y., Lin, C. H., Chan, C. C., & Huang, Y. C. (2014). High-performance polystyrene/graphene-based nanocomposites with excellent anti-corrosion properties. *Polymer Chemistry*, *5*(2), 535–550. doi:10.1039/C3PY00825H

Yu, Z., Di, H., Ma, Y., He, Y., Liang, L., Lv, L., & Luo, Z. (2015). Preparation of graphene oxide modified by titanium dioxide to enhance the anti-corrosion performance of epoxy coatings. *Surface and Coatings Technology*, *276*, 471–478. doi:10.1016/j.surfcoat.2015.06.027

Yu, Z., Di, H., Ma, Y., Lv, L., Pan, Y., Zhang, C., & He, Y. (2015). Fabrication of graphene oxide–alumina hybrids to reinforce the anti-corrosion performance of composite epoxy coatings. *Applied Surface Science*, *351*, 986–996. doi:10.1016/j.apsusc.2015.06.026

Yu, Z., Yang, M., Zhang, Q., Zhao, C., & Mai, Y. (2003). Dispersion and distribution of organically modified montmorillonite in nylon-66 matrix. *Journal of Polymer Science. Part B, Polymer Physics*, *41*(11), 1234–1243. doi:10.1002/polb.10480

Zafeiropoulos, N. E., Williams, D. R., Baillie, C. A., & Matthews, F. L. (2002). Engineering and characterisation of the interface in flax fibre/polypropylene composite materials. Part I. Development and investigation of surface treatments. *Composites. Part A, Applied Science and Manufacturing*, *33*(8), 1083–1093. doi:10.1016/S1359-835X(02)00082-9

Zakaria, Z., Kamarudin, S., & Timmiati, S. (2016). Membranes for direct ethanol fuel cells: An overview. *Applied Energy*, *163*, 334–342. doi:10.1016/j.apenergy.2015.10.124

Zanetti, M., Camino, G., Canavese, D., Morgan, A., Lamelas, F., & Wilkie, C. (2002). Fire retardant halogen–antimony–clay synergism in polypropylene layered silicate nanocomposites. *Chemistry of Materials*, *14*(1), 189–193. doi:10.1021/cm011124t

Zang, Z., Breidt, C., Chang, L., Haupert, F., & Friedrich, K. (2004). *Enhancement of the wear resistance of epoxy: short carbon fibre, graphite, ptfe and nano-TiO_2. In Composites Part A Applied Science and Manufacturing*. Elsevier. doi:10.1016/j.compositesa.2004.05.005

Zerda, A., & Lesser, A. (2001). Intercalated clay nanocomposites: Morphology, mechanics, and fracture behavior. *Journal of Polymer Science. Part B, Polymer Physics*, *39*(11), 1137–1146. doi:10.1002/polb.1090

Zhang, G., Wu, T., Lin, W., Tan, Y., Chen, R., Huang, Z., ... Qu, J. (2017). Preparation of polymer/clay nanocomposites via melt intercalation under continuous elongation flow. *Composites Science and Technology*, *145*, 157–164. doi:10.1016/j.compscitech.2017.04.005

Zhang, H., Cao, G. P., & Yang, Y. S. (2007). Electrochemical properties of ultra-long, aligned, carbon nanotube array electrode in organic electrolyte. *Journal of Power Sources*, *172*(1), 476–480. doi:10.1016/j.jpowsour.2007.07.060

Zhang, H., Cao, G., Wang, Z., Yang, Y., Shi, Z., & Gu, Z. (2008). Tube-covering-tube nanostructured polyaniline/carbon nanotube array composite electrode with high capacitance and superior rate performance as well as good cycling stability. *Electrochemistry Communications*, *10*(7), 1056–1059. doi:10.1016/j.elecom.2008.05.007

Zhang, H., Li, H., Zhang, F., Wang, J., Wang, Z., & Wang, S. (2008). Polyaniline nanofibers prepared by a facile electrochemical approach and their supercapacitor performance. *Journal of Materials Research*, *23*(09), 2326–2332. doi:10.1557/jmr.2008.0304

Zhang, H., Wang, J., Liu, X., Wang, Z., & Wang, S. (2013). High performance self-healing epoxy/polyamide protective coating containing epoxy microcapsules and polyaniline nanofibers for mild carbon steel. *Industrial & Engineering Chemistry Research*, *52*(30), 10172–10180. doi:10.1021/ie400666a

Zhang, J. L., Shi, L., Liu, H. D., Deng, Z. W., Huang, L. H., Mai, W. J., ... Cai, X. (2015). In situ synthesis of cabbage like polyaniline@hydroquinone nanocomposites and electrochemical capacitance investigations. *Journal of Alloys and Compounds*, *651*, 126–134. doi:10.1016/j.jallcom.2015.08.090

Zhang, J., Neoh, K. G., & Kang, E. T. (2018). Electrical stimulation of adipose-derived mesenchymal stem cells and endothelial cells co-cultured in a conductive scaffold for potential orthopaedic applications. *Journal of Tissue Engineering and Regenerative Medicine*, *12*(4), 878–889. doi:10.1002/term.2441 PMID:28482125

Zhang, K., Heo, N., Shi, X., & Park, J. H. (2013). Chemically modified graphene oxide-wrapped quasi-micro Ag decorated silver trimolybdate nanowires for photocatalytic applications. *The Journal of Physical Chemistry C*, *117*(45), 24023–24032. doi:10.1021/jp407038c

Zhang, K., Zhang, L. L., Zhao, X. S., & Wu, J. (2010). Graphene/polyaniline nanofibers composites as supercapacitor electrodes. *Chemistry of Materials*, *22*(4), 1392–1401. doi:10.1021/cm902876u

Zhang, L. L., Li, S., Zhang, J., Guo, P., Zheng, J., & Zhao, X. S. (2010). Enhancement of Electrochemical Performance of Macroporous Carbon by Surface Coating of Polyaniline. *Chemistry of Materials*, *22*(3), 1195–1202. doi:10.1021/cm902685m

Zhang, L., Huang, D., Hu, N., Yang, C., Li, M., Wei, H., ... Zhang, Y. (2017). Three-dimensional structures of graphene/polyaniline hybrid films constructed by steamed water for high-performance supercapacitors. *Journal of Power Sources*, *342*, 1–8. doi:10.1016/j.jpowsour.2016.11.068

Zhang, Q. Z., Zhang, D., Miao, Z. C., Zhang, X. L., & Chou, S. (2018). Research Progress in MnO_2 -Carbon Based Supercapacitor Electrode Materials. *Small*, *14*(24), 1702883. doi:10.1002mll.201702883 PMID:29707887

Zhang, Q., Dandeneau, C. S., Zhou, X., & Cao, G. (2009). ZnO nanostructures for dye-sensitized solar cells. *Advanced Materials*, *21*(41), 4087–4108. doi:10.1002/adma.200803827

Zhang, Q., Li, Y., Feng, Y., & Feng, W. (2013). Electropolymerization of graphene oxide/polyaniline composite for high-performance supercapacitor. *Electrochimica Acta*, *90*, 95–100. doi:10.1016/j.electacta.2012.11.035

Zhang, R., Xu, X., Fan, X., Yang, R., Wu, T., & Zhang, C. (2018). Application of conducting micelles self-assembled from commercial poly (3, 4-ethylenedioxythiophene): Poly (styrene sulfonate) and chitosan for electrochemical biosensor. *Colloid & Polymer Science*, *296*(3), 495–502. doi:10.100700396-018-4270-6

Zhang, S., & Pan, N. (2015). Supercapacitors Performance Evaluation. *Advanced Energy Materials*, *5*(6), 1401401. doi:10.1002/aenm.201401401

Zhang, X., Ji, L. Y., Zhang, S. C., & Yang, W. S. (2007). Synthesis of a novel polyaniline-intercalated layered manganese oxide nanocomposite as electrode material for electrochemical capacitor. *Journal of Power Sources*, *173*(2), 1017–1023. doi:10.1016/j.jpowsour.2007.08.083

Zhang, X., & Samori, P. (2017). Graphene/Polymer nanocomposites for supercapacitors. *ChemNanoMat*, *3*(6), 362–372. doi:10.1002/cnma.201700055

Zhang, X., Wang, F., & Du, Y. (2007). Effect of nano-sized titanium powder addition on corrosion performance of epoxy coatings. *Surface and Coatings Technology*, *201*(16-17), 7241–7245. doi:10.1016/j.surfcoat.2007.01.042

Zhang, Y., Chi, Q., Liu, L., Zhang, C., Chen, C., Wang, X., & Lei, Q. (2017). Enhanced electric polarization and breakdown strength in the all-organic sandwich-structured poly(vinylidene fluoride)-based dielectric film for high energy density capacitor. *APL Materials*, *5*(7), 076109. doi:10.1063/1.4995653

Zhang, Y., & Liu, L. (2013). Fly ash-based geopolymer as a novel photocatalyst for degradation of dye from wastewater. *Particuology*, *11*(3), 353–358.

Zhang, Y., Nayak, T. R., Hong, H., & Cai, W. (2012). Graphene: A versatile nanoplatform for biomedical applications. *Nanoscale*, *4*(13), 3833–3842. doi:10.1039/c2nr31040f PMID:22653227

Zhang, Y., Zhang, L., & Zhou, C. (2013). Review of Chemical Vapor Deposition of Graphene and Related Applications. *Accounts of Chemical Research*, *46*(10), 2329–2339. doi:10.1021/ar300203n PMID:23480816

Zhang, Y., & Zhang, M. (2002). Three-dimensional macroporous calcium phosphate bioceramics with nested Chitosan sponges for load-bearing bone implants. *Journal of Biomedical Materials Research. Part A*, *61*(1), 1–8. doi:10.1002/jbm.10176 PMID:12001239

Zhang, Z., Wang, G., Li, Y., Zhang, X., Qiao, N., Wang, J., ... Hao, Z. (2014). A new type of ordered mesoporous carbon/polyaniline composites prepared by a two-step nanocasting method for high performance supercapacitor applications. *Journal of Materials Chemistry. A, Materials for Energy and Sustainability*, *2*(39), 16715–16722. doi:10.1039/C4TA03351E

Zhao, D., Guo, X., Gao, Y., & Gao, F. (2012). An Electrochemical Capacitor Electrode Based on Porous Carbon Spheres Hybrided with Polyaniline and Nanoscale Ruthenium Oxide. *ACS Applied Materials & Interfaces*, *4*(10), 5583–5589. doi:10.1021/am301484s PMID:22988980

Zhao, H., & Burke, A. (2010). *Fuel cell powered vehicles using supercapacitors: Device characteristics, control strategies, and simulation results. Research Report – UCD-ITS-RR-10-01*. Davis: Institute of Transportation Studies, University of California. Retrieved from www.its.ucdavis.edu

Zhao, L., Li, J., Guo, S., & Du, Q. (2006). Ultrasonic oscillations induced morphology and property development of polypropylene/montmorillonite nanocomposites. *Polymer*, *47*(7), 2460–2469. doi:10.1016/j.polymer.2006.02.011

Zhao, X. G., Jin, E. M., & Gu, H.-B. (2013). Increased charge transfer of PVDF-HFP based electrolyte by addition of graphite nanofiber and its application in dye-sensitized solar cells. *Applied Surface Science*, *287*, 8–12. doi:10.1016/j.apsusc.2013.09.017

Zhao, X., Li, P., Guo, B., & Ma, P. X. (2015). Antibacterial and conductive injectable hydrogels based on quaternized chitosan-graft-polyaniline/oxidized dextran for tissue engineering. *Acta Biomaterialia*, *26*, 236–248. doi:10.1016/j.actbio.2015.08.006 PMID:26272777

Zhao, X., Sanchez, B. M., Dobson, P. J., & Grant, P. S. (2011). The role of nanomaterials in redox-based supercapacitors for next generation energy storage devices. *Nanoscale*, *3*(3), 839–855. doi:10.1039/c0nr00594k PMID:21253650

Zhao, X., Wu, H., Guo, B., Dong, R., Qiu, Y., & Ma, P. X. (2017). Antibacterial anti-oxidant electroactive injectable hydrogel as self-healing wound dressing with hemostasis and adhesiveness for cutaneous wound healing. *Biomaterials*, *122*, 34–47. doi:10.1016/j.biomaterials.2017.01.011 PMID:28107663

Zheng, H., Meng, Y. S., & Zhu, Y. (2015). Frontiers of in situ electron microscopy. *MRS Bulletin*, *40*(1), 12–18. doi:10.1557/mrs.2014.305

Zheng, H., Shao, Y., Wang, Y., Meng, G., & Liu, B. (2017). Reinforcing the corrosion protection property of epoxy coating by using graphene oxide–poly (urea–formaldehyde) composites. *Corrosion Science*, *123*, 267–277. doi:10.1016/j.corsci.2017.04.019

Zheng, L., Xu, Y., Jin, D., & Xie, Y. (2011). Polyaniline-Intercalated Molybdenum Oxide Nanocomposites: Simultaneous Synthesis and their Enhanced Application for Supercapacitor. *Chemistry, an Asian Journal*, *6*(6), 1505–1514. doi:10.1002/asia.201000770 PMID:21557481

Zheng, W., Lu, X., & Wong, S. C. (2004). Electrical and mechanical properties of expanded graphite-reinforced high-density polyethylene. *Journal of Applied Polymer Science*, *91*(5), 2781–2788. doi:10.1002/app.13460

Zhi, C., Bando, Y., Tang, C., Xie, R., Sekiguchi, T., & Golberg, D. (2005). Perfectly dissolved boron nitride nanotubes due to polymer wrapping. *Journal of the American Chemical Society*, *127*(46), 15996–15997. doi:10.1021/ja053917c PMID:16287265

Zhi, M., Xiang, C., Li, J., Li, M., & Wu, N. (2013). Nanostructured carbon–metal oxide composite electrodes for supercapacitors: A review. *Nanoscale*, *5*(1), 72–88. doi:10.1039/C2NR32040A PMID:23151936

Zhong, H., He, A., Lu, J., Sun, M., He, J., & Zhang, L. (2016). Carboxymethyl chitosan/conducting polymer as water-soluble composite binder for LiFePO4 cathode in lithium ion batteries. *Journal of Power Sources*, *336*, 107–114. doi:10.1016/j.jpowsour.2016.10.041

Zhou, D., Li, Y., Wang, J., Xu, P., & Han, X. (2011). Synthesis of polyaniline nanofibers with high electrical conductivity from CTAB-SDBS mixed surfactants. *Materials Letters*, *65*(23-24), 3601–3604. doi:10.1016/j.matlet.2011.08.021

Zhou, L., Fox, L., Włodek, M., Flores, L. I., Slastanova, A., Robles, E., ... Cattelan, M. (2018). Surface structure of few layer graphene. *Carbon*, *136*, 255–261. doi:10.1016/j.carbon.2018.04.089

Zhou, S., Mo, S., Zou, W., Jiang, F., Zhou, T., & Yuan, D. (2011). Preparation of polyaniline/ 2-dimensional hexagonal mesoporous carbon composite for supercapacitor. *Synthetic Metals*, *161*(15-16), 1623–1628. doi:10.1016/j.synthmet.2011.05.028

Zhou, W., Zhou, Z., Song, S., Li, W., Sun, G., Tsiakaras, P., & Xin, Q. (2003). Pt based anode catalysts for direct ethanol fuel cells. *Applied Catalysis B: Environmental*, *46*(2), 273–285. doi:10.1016/S0926-3373(03)00218-2

Zhou, X., Li, L., Dong, S., Chen, X., Han, P., Xu, H., ... Cui, G. (2011). A renewable bamboo carbon/polyaniline composite for a high-performance supercapacitor electrode material. *Journal of Solid State Electrochemistry*, *16*(3), 877–882. doi:10.100710008-011-1435-3

Zhou, X., Zhu, Y., Liang, J., & Yu, S. (2010). New fabrication and mechanical properties of styrene-butadiene rubber/carbon nanotubes nanocomposite. *Journal of Materials Science and Technology*, *26*(12), 1127–1132. doi:10.1016/S1005-0302(11)60012-1

Zhou, Y., Qin, Z., Li, L., Zhang, Y., Wei, Y., Wang, L., & Zhu, M. (2010). Polyaniline/multi-Walled carbon nanotube composites with Core Shell structures as super-capacitor electrode materials. *Electrochimica Acta*, *55*(12), 3904–3908. doi:10.1016/j.electacta.2010.02.022

Zhu, J. X., Sun, W. P., Yang, D., Zhang, Y., Hoon, H. H., Zhang, H., & Yan, Q. Y. (2015). Multifunctional Architectures Constructing of PANI Nanoneedle Arrays on MoS2 Thin Nanosheets for High-Energy Supercapacitors. *Small*, *11*(33), 4123–4129. doi:10.1002mll.201403744 PMID:26044446

Zhu, Y., Murali, S., Cai, W., Li, X., Suk, J. W., Potts, J. R., & Ruoff, R. S. (2010). Graphene and graphene oxide: Synthesis, properties, and applications. *Advanced Materials*, *22*(35), 3906–3924. doi:10.1002/adma.201001068 PMID:20706983

Zhu, Z., Wang, G., Sun, M., Li, X., & Li, C. (2011). Fabrication and electrochemical characterization of polyaniline nanorods modified with sulfonated carbon nanotubes for supercapacitor applications. *Electrochimica Acta*, *56*(3), 1366–1372. doi:10.1016/j.electacta.2010.10.070

Zindani, D., & Kumar, K. (2018). Material application of polymer composite material. In *Composites and advanced materials for industrial applications* (p. 8). Hershey, PA: IGI Global. doi:10.4018/978-1-5225-5216-1.ch001

Zocca, A., Colombo, P., Gomes, C. M., & Jens, G. (2015). Additive Manufacturing of Ceramics. *Issues, Potentialities, and Opportunities*, *2001*. doi:10.1111/jace.13700

Zou, B.-X., Liang, Y., Liu, X.-X., Diamond, D., & Lau, K.-T. (2011). Electrodeposition and pseudocapacitive properties of tungsten oxide/polyaniline composite. *Journal of Power Sources*, *196*(10), 4842–4848. doi:10.1016/j.jpowsour.2011.01.073

Zou, W., Du, Z., Liu, Y., Yang, X., Li, H., & Zhang, C. (2008). Functionalization of MWNTs using polyacryloyl chloride and the properties of CNT–epoxy matrix nanocomposites. *Composites Science and Technology*, *68*(15-16), 3259–3264. doi:10.1016/j.compscitech.2008.08.011

Zou, W., Wang, W., He, B., Sun, M., & Yin, Y. (2010). Supercapacitive properties of hybrid films of manganese dioxide and polyaniline based on active carbon in organic electrolyte. *Journal of Power Sources*, *195*(21), 7489–7493. doi:10.1016/j.jpowsour.2010.05.020

Zou, Y., Xiang, C., Sun, L.-X., & Xu, F. (2008). Glucose biosensor based on electrodeposition of platinum nanoparticles onto carbon nanotubes and immobilizing enzyme with chitosan-SiO2 sol–gel. *Biosensors & Bioelectronics*, *23*(7), 1010–1016. doi:10.1016/j.bios.2007.10.009 PMID:18054479

Zuhua, Z., Xiao, Y., Huajun, Z., & Yue, C. (2009). Role of water in the synthesis of calcined kaolin-based geopolymer. *Applied Clay Science*, *43*(2), 218–223. doi:10.1016/j.clay.2008.09.003

Zunino, J. L., III. (2007). Active Coatings Technologies for Customized Military Coating Systems. In *Cleantech 2007: the Cleantech Conference, Venture Forum and Trade Show*, Santa Clara, CA, May 23-24 (p. 234). CRC.

Zunino, J., III. (2005). US Army Development of Active Smart Coatings System for Military Vehicles. NSTI.

Zunino, J. L. III. (2007). Development of Materials and Sensor for the U.S. Armys Active coatings Technology Program. In *NSTI* (p. 4). Nanotech.

Zunino, J. L. III. (2008). Development of Active Systems for Military Utilization. In *NSTI* (p. 3). Nanotech.

About the Contributors

Noureddine Ramdani, Doctor of Philosophy (PhD) in Materials Science and Engineering, 2016, Harbin Engineering University, China. Research focuses on preparation and properties of Particle- and Fiber-reinforced Polybenzoxazine Composites. He has an engineering degree (Eng.) in chemical engineering.

* * *

Olubunmi G. Abatan is a chemical engineer with B.Eng. from Igbinedion University in 2007 and a master's degree in chemical engineering from Covenant University in 2013. She is currently running her Ph.D. program in chemical engineering (kinetics and catalysis). She started working in Covenant University in March 2014 as an assistant lecturer and became a lecturer II in February 2015. She has published a few works in some journals. She is a registered engineer in Nigeria (COREN). She is also a member of the Nigerian Society of Engineers and Nigerian Society of Chemical Engineers. She is married with children. She is a Christian and enjoys reading and traveling.

Vladimir Aderikha was born in 1950 in Belarus, of the former Soviet Union. Vladimir N. Aderikha graduated from the People's Friendship University in 1972 (MSci, Chemistry). He received a Candidate of Science degree (Polymer Chemistry) from Nesmeyanov Institute of Organoelement Compounds of the Academy of Sciences of the USSR (Moscow) in 1982. Since then, he has worked at the Metal Polymer Research Institute of the National Academy of Sciences of Belarus. He has co-authored over 50 articles and several patents. His research interests include tribology of polymer-based composites, and technology of engineering composites, including hybrid nanomaterials.

Oladimeji Adeyeye is an MTech student in Polymer Technology in the Department of Chemical, Metallurgical & Materials Engineering (Polymer Division).

Oluranti Agboola, Dr., is a researcher and a senior lecturer in the Department of Chemical, Covenant University of Technology, Ota, Nigeria. She obtained her first academic degree (B. Eng.) in Chemical Engineering from the department of Chemical Engineering of the Federal University of Technology, Minna, Niger State. Nigeria. She obtained her M. Tech and D. Tech. Chemical Engineering, from Tshwane University of Technology, South Africa. Her main research interests are in modeling and simulation; membrane technology, nanotechnology and activated carbon for desalination. She has over 171 Google Scholar citations, an h-index of 6 and an i-10 index of 5. She has authored over 43 publications with 30 indexed in Scopus, Thomson Reuters and other databases.

Gbemisola Akinsipe is a graduate of chemical engineering with interests in energy, power and sustainability solutions.

Augustine Ayeni, B.Sc., M.Sc., and PhD degrees (Chemical Engineering). He is currently a lecturer and a researcher at Covenant University, Ota, Ogun State, Nigeria.

Damilola Elizabeth Babatunde holds a first and a second degree in chemical engineering. She is currently a lecturer and researcher in the Department of Chemical Engineering, Covenant University, Canaanland, Ota, Ogun State. Her research interests include studies on energy and environment

O. M. Babatunde is a Lecturer at the Department Electrical/Electronic Engineering University of Lagos. He has experience as a mechanical and electrical (M&E) consultant for building services. He specializes in power systems, hybrid energy systems and energy-efficient designs for buildings.

Olusesan Biotidara, Dr., received a PhD degree in Chemistry at the Federal University of Agriculture, Abeokuta, Ogun State, Nigeria. He is a Chief Lecturer at the Yaba College of Technology, Yaba, Lagos, Nigeria.

Sergei Bogdanovich, born in 1979, graduated with honors and a bachelor's degree (Techn.) from the Mechanical Department of the State University of Transport, Belarus (BelSUT) in 2001. In 2002, he completed the Magistracy course at BelSUT, presented Master's Dissertation and was awarded the academic degree of Master (Techn). From 2002 to 2005 he was taking a post-graduate course at BelSUT in speciality "Materials Science in Mechanical Engineering." On October 12, 2006 he defended his PhD Dissertation "Triboengineering composite materials based on compatibilized blends of aliphatic polyamides and polyolefins." The results of the dissertation work were, in part, published in international scientific journals (European Polymer Journal, Journal of Applied Polymer Science, Journal of Friction and Wear, etc.) and presented at international conferences. At the present time, S. Bogdanovich has been engaged in research dealing with creating new composite materials based on thermoplastics as well as their blends filled with anti-frictional, electro- and thermal conductive and reinforcing modifiers, among them nano-dimensional additives. In 2013, he got the rank of associate professor. S. Bogdanovich took an active part in the development of creative relations with Poznan University of Technology, Poland and Institute of Polymer Materials, Azerbaijan: to carry out joint research and prepare cooperative publications.

Oluyemi Ojo Daramola, Dr., received a PhD degree in Metallurgical and Materials Engineering from the Federal University of Technology, Akure, Ondo State, Nigeria. He is interested in Composite Materials (PMC and MMC), Nanotechnology, Physical Metallurgy, Materials Design and Characterization, Mechanical Behaviour of Materials, Materials Degradation and Damage Analysis, Corrosion, and Tribology. He is currently a Postdoctoral Research Fellow at the Tshwane University of Technology, Pretoria, South Africa.

Iheanacho Henry Denwigwe holds a B.Sc. (Hons) degree in Electrical and Electronics Engineering from the prestigious University of Lagos, Nigeria. His area of research specialization is Power systems analysis and energy management."

Francis Elehinafe, B.Sc., M.Sc., PhD, earned from Obafemi Awolowo University, Ile-Ife, Osun State, Nigeria. Currently a lecturer and researcher at Covenant University, Ota, Ogun State, Nigeria.

Azunna Agwo Eze is currently doing his doctoral degree program in Mechanical Engineering in Tswane University of Technology Pretoria, South Africa. His research areas are: advanced engineering materials, nanotechnology, sustainable and renewable energy. AkanuIbiam Federal Polytechnics Unwana Afikpo North Local government Area, Ebonyi State Nigeria (2003-2008, obtained National Diploma and Higher national Diploma in Mechanical Engineering), Tshwane University of Technology Pretoria, South Africa (2014-2016, obtained MTech. in Mechanical Engineering), Tshwane University of Technology Pretoria South Africa (2016-), to obtain a DEng. in Mechanical Engineering. Spouse: Chioma A. Eze (Married 2012). Awards: Institutional award (Best researcher of the year 2016) for patent work.

Victoria Fasiku has a Bsc in Biochemistry in 2013 from the Federal University of Technology, Minna, Nigeria and a Msc. in Biology from North West University, Mafikeng, South Africa. She is currently undertaking her PhD in the Department of Pharmaceutical Sciences, Kwazulu Natal University, South Africa.

Sunday Fayomi is a senior researcher in the Department of Mechanical Engineering at Covenant University and a research and innovation associate with the Tshwane University of Technology, Pretoria, South Africa. His research interests are centred on surface structural integrity, corrosion engineering, mechanical metallurgy, environmental science and engineering, materials development and processing, nanotechnology and triboxidation processing. He has published over 170 research paper in reputable journals.

Idowu Ibrahim received a BSc (Hons) degree in Mechanical Engineering from the Federal University of Technology, Minna, Nigeria and an MSc degree in Mechanical Engineering from the Tshawne University of Technology, Pretoria, South Africa. He is in the final stages of his PhD research in mechanical engineering from the Tshawne University of Technology.

Nnamdi Iheaturu was born on the 13th day of January 1975, has a Bachelor of Engineering (B.Eng.) degree in Polymer & Textile Engineering, 1999, M.Sc. Polymer Science & Engineering, 2006, from Federal University of Technology, Owerri, (FUTO), Imo, Nigeria. He was a Research Attaché, for the Shell Petroleum Development Company (SPDC) of Nigeria, Port Harcourt, in 2007, where he worked at the Well Engineering department before traveling to Europe on Erasmus Mundus Scholarship administered by the European Commission. He was European Erasmus Mundus Scholar at 3 European Innovative Universities – UA, Portugal, AAU, Denmark and TUHH, Germany, 2008 – 2010, where he was jointly awarded the European Masters in Materials Science (EMMS) degree. Having been admitted into the PhD programme of Postgraduate School, FUTO, he carried out his PhD research work on "Synthesis, Preparation and Characterization of Nanoporous Core-Shell Clay Epoxy Composites", at AAU, Denmark, as a Doctoral Research Fellow on study leave in the Mechanical and Manufacturing Engineering Department, before returning to Nigeria in 2012. On his return to Nigeria, he vehemently pursued the concluding part of his Ph.D. programme at FUTO and defended his project on the 2nd of October 2014. He was subsequently awarded a Ph.D. in Polymer Science by the Federal University of Technology, Owerri (FUTO), Imo, Nigeria. While studying in Europe, he was elected Vice President,

Europe and became Chair of the European Hub of the FUTO Alumni Association in Diaspora (FAAD), 2009 – 2012. While a resident in Aalborg Denmark, in 2011, he participated in the Aalborg University Match Making and the Solution Camp organized in collaboration with Storvring Business Council. In 2013, he was the only Nigerian selected to attend the 1st Advanced School on Materials for Photonic Applications: glasses, optic fibers, and sol-gel materials tagged "1st SAMPA" held at UNESP, Araraquara, Sao Paolo, Brazil. He is a tenured academic staff of the Department of Polymer & Textile Engineering & School of Engineering Boards of Studies, FUTO. As a registered engineer in Nigeria, he specializes in advanced materials and polymer composites. His current research challenge is "Substituting for Glass/Carbon fibre in Wind Turbine Blades and Composite Load Bearing Systems" geared towards driving down the cost of energy. He is also amongst the research team at the Centre for Energy and Power Systems Research (CEPSR), FUTO is involved in minimizing losses in power transformers through the Nigerian Content Initiative to enhance quality and efficiency. He is a member of so many professional bodies including Polymer Institute of Nigeria (PIN), Council for the Regulation of Engineering in Nigeria (COREN) and the Nigerian Society of Engineers (NSE). In 2015, he was elected into the National Executive Council (NEC) of the Polymer Institute of Nigeria (PIN) as the Assistant Secretary-General, a post he holds till now.

Tippabattini Jayaramudu, Dr., received his Doctor of philosophy from the Sri Krishnadevaraya University, India in 2014; he joined as a Postdoctoral Research Fellow at Tshwane University of Technology, Pretoria, South Africa in June 2014-May 2015. He worked as a Post-Doctoral Research Fellow in Inha University June 2015-2016 and is currently with the Laboratory of Material Sciences, Instituto de Quimica de Recursos Naturales, Universidad de Talca, 747, Talca, Chile.

Ayesha Kausar is currently employed at the National University of Sciences and Technology, Islamabad, Pakistan. She has previously worked for the National Centre for Physics and Quaid-i-Azam University, Islamabad, Pakistan. Her research interests include: synthesis, characterization and structure-property relationship of new polymeric materials; synthesis of nanomaterials including organic–inorganic nanocomposites/hybrid materials; polymeric blends via incorporating nanoparticles into thin polymer films; exploration of practical and potential prospects of the novel synthesized materials (new polymers and nanomaterials) counting morphological, mechanical, thermal, electrical, conducting, etc.; flame retardant materials; proton conducting fuel cell membranes; nanocomposites for polymer Li-ion battery electrodes; composites based on electrospun nanofibers; production of various polymeric nanoparticles and their composites for solar cells; polymer/carbon nanotube/nanoparticle composites for water treatment; potential of polymer/graphite nanocomposites; polymer/graphene hybrids; polymer/fullerene nanomaterials; fabrication of epoxy-based nanocomposite for various applications; radar absorbing materials; aerospace relevance, and so on.

Sukhila Krishnan, Dr., obtained Phd degree in Polymer Science and Technology from Utkal University Bhubaneswar, after finishing M. Tech degree in Polymer Science from Cochin University. Currently she is working as a research fellow in Laboratory for Advanced Research in Polymeric Materials (LARPM), Central Institute of Plastic Engineering and Technology (CIPET), Bhubaneswar. Her main research interests focus on biopolymers from natural resources and renewable resource-based epoxy resins.

Ashok Kumar, Prof., obtained his M. Tech and Ph.D. degree from IIT, Kanpur in Material Science. He had done a master's in physics with electronics as specialization from Meerut University. His research interest involves materials science/condensed matter physics: polymer electrolytes and conducting polymers. He has been a faculty member of the department of physics since 1999.

Sudheer Kumar, Dr., obtained Ph.D degree in Polymer Science and Technology from Utkal University Bhubaneswar, after finishing M. Tech degree in Polymer Nanotechnology from Biju Patnaik University of Technology Odisha. Currently he is working as a research fellow in Laboratory for Advanced Research in Polymeric Materials (LARPM), Central Institute of Plastic Engineering and Technology (CIPET), Bhubaneswar. His main research interests are synthesis of epoxy resin from renewable resources and its blend, nanocomposites, synthesis of end-cap oligomer and polymer coatings.

Williams Kehinde Kupolati, Prof., holds a B.Sc. (Hons) and a Ph.D. in Civil Engineering. He also holds a M. Sc. in Construction Management. He is an Associate Professor at the Department of Civil Engineering, Tshwane University of Technology, Pretoria, South Africa. He is involved with research on the sustainable and appropriate use of polymers, composites, nanomaterials, waste and construction materials for civil engineering infrastructure. In addition, he participates in research dealing with transportation and structural systems, modelling and technopreneurship; and also engages in exploring the bottom top approach to sustainability and community participatory approach to project identification and successful execution from a cultural perspective.

Mokgaotsa Mochane, Dr., received a PhD in Polymer Science from the University of the Free State, South Africa. He is currently a Postdoctoral Research Fellow at the Tshwane University of Technology, Pretoria, South Africa.

Smita Mohanty, Dr., is working as a senior scientist at Laboratory for Advanced Research in Polymeric Materials (LARPM), Central Institute Plastic Engineering and Technology, Bhubaneswar and has more than 12 years of research experience. She has also initiated advanced research at LARPM including E-waste recycling, biopolymers from natural resources, and polymer electrolytes.

Sanjay K. Nayak, Prof., Dr., holds the position of Professor and Chair of R&D wings of the Central Institute Plastic Engineering and Technology with 30 years of experience in teaching and research. He is an active researcher in the field of polymer science and technology, with several publications in peer-reviewed international journals and patented technologies in the area of advanced polymeric materials and characterization.

Abongile Ndamase received a MTech degree in Polymer Technology from the Department of Chemical, Metallurgical & Materials Engineering (Polymer Division) of the Tshwane University of Technology, Pretoria, South Africa.

Clara Nkuna is an MEng degree student in Polymer Technology at the Tshwane University of Technology, where she obtained a BTech degree in Polymer Technology.

Shesan Owonubi obtained the MTech degree in Polymer Technology from the Tshwane University of Technology and a PhD degree in Biology at the North West University, Mafikeng, South Africa. He is currently a Postdoctoral Research fellow at the University of Zululand, KwaDlangezwa, South Africa.

Patricia Popoola, Prof., is an astute researcher in the field of advanced engineering materials. She is currently a Research Professor at the Tshwane University of Technology, Pretoria, South Africa. She has over 2034 Google Scholar citations, an h-index of 23 and an i-10 index of 68. Her first degree is in Metallurgical and Materials Engineering (BSc Hons -Obafemi Awolowo University, Ile Ife, Nigeria); Masters and Doctoral degrees in Metallurgical Engineering -Tshwane University of Technology, South Africa. She has authored over 375 publications with 264 indexed in Scopus, Thomson Reuters and other data bases. Prof Patricia Popoola has graduated 8 doctoral and 15 masters' degrees candidates.

Rotimi Sadiku, Prof., has been a renowned researcher in the field of polymer science and engineering. He is currently a research professor in Tshwane University of Technology, Pretoria South Africa. He has over 3518 Google Scholar citations, h-index of 27 and i-10 index of 103. His first and last degrees are in Polymer Science (B. Eng.) and PhD, respectively, from the University of Strathclyde. He has authored over 627 publications with 170 indexed in Scopus, Thomson Reuters and other databases.

Sushanta Kumar Samal, Dr., obtained a PhD in Polymer Science in 2009. He is now working as a scientist in Laboratory for Advanced Research in Polymeric Materials, Central Institute of Plastics Engineering and Technology, Bhubaneswar. His research interests include development of new advanced materials, biopolymer, composite and nanocomposites, coating, foam, and water purification membranes.

Samuel Eshorame Sanni is a Nigerian who began his career as a lecturer in 2013 and presently works as an academic staff and researcher in the Department of Chemical Engineering of Covenant University. His areas of interest include chemical reaction engineering and kinetics, process modeling and simulation, multiphase transport, petroleum processing, corrosion technology, energy and waste valorization. He has worked as PTDF / NNPC attaché in Kaduna Refining and Petrochemical Company (2011), as the C.E.O of Mattesh Clothier from 2012-2014 where he doubled as Director and producer of Mattesh insecticide, Mattesh Air Fresheners, Mattesh facial cleansers, Kaduna, Nigeria. He has B.Tech, M.Sc and Ph.D. in Chemical Engineering from Ladoke Akintola University of Technology (LAUTECH, Nigeria), Ahmadu Bello University and Covenant University, respectively. He has authored 30 publications with 18 indexed in databases such as Scopus and Thomson Reuters. He is the author of the book titled "Modelling and Simulation of Sand Entrainment and Deposit in Horizontal Oil Transport" and is currently a reviewer for the Bulletin of Chemical Reaction Engineering and Catalysis (Jointly Published by Diponegoro University & Masyarakat Katalis Indonesia - Indonesian Catalyst Society (MKICS), Progress in Industrial Ecology (Publisher- Inderscience), Carbon Letters (Publisher- Korean Carbon Society), the International Journal of Sustainable Engineering (Publisher- Springer), Anticorrosion Methods and Materials (Publisher- Emerald Publishing Co.).

Koena Selatile received an MTech degree in Polymer Technology in the Department of Chemical, Metallurgical & Materials Engineering (Polymer Division) at the Tshwane University of Technology. She is currently undergoing a DEng degree in the same department of the same University.

Victor Shikuku is a lecturer in the department of Chemistry, Kaimosi Friends University College. He received his PhD degree in Chemistry from Maseno University, Kenya. His research interests are in physical chemistry, materials science and environmental chemistry.

Tome Sylvain is an early-career scientist in material chemistry with special interests in geopolymer chemistry CSIR-TWAS 2016 and a PhD Fellow.

Chukwunonso Uwa received a MEng degree in Mechanical Engineering at the Tshwane University of Technology. He is about to commence a DEng degree in the same field at the same University.

Kokkarachedu Varaprasad received his Doctor of philosophy from the Sri Krishnadevaraya University, India. He is a Postdoctoral Research Fellow at the Tshwane University of Technology, Pretoria, South Africa, in June 2013 and another Postdoctoral Research Fellowship at the Universitat Concepcion, Chile in 2013 and 2014. He is currently a research scientist at the Centro de Investigacion de Polımeros Avanzados (CIPA), Avenida Collao, Edificio de Laboratorio, Concepcion, Chile.

Index

H

I

L

M

N

P

R

S

T

U

W

www.ingramcontent.com/pod-product-compliance
Lightning Source LLC
LaVergne TN
LVHW060259011026
849161LV00011B/2738